Lecture Notes in Computer Science   9870

Commenced Publication in 1973
Founding and Former Series Editors:
Gerhard Goos, Juris Hartmanis, and Jan van Leeuwen

More information about this series at http://www.springer.com/series/7411

Olga Galinina · Sergey Balandin
Yevgeni Koucheryavy (Eds.)

Internet of Things, Smart Spaces, and Next Generation Networks and Systems

16th International Conference, NEW2AN 2016
and 9th Conference, ruSMART 2016
St. Petersburg, Russia, September 26–28, 2016
Proceedings

 Springer

Editors
Olga Galinina
Tampere University of Technology
Tampere
Finland

Yevgeni Koucheryavy
Tampere University of Technology
Tampere
Finland

Sergey Balandin
FRUCT Oy
Helsinki
Finland

ISSN 0302-9743   ISSN 1611-3349   (electronic)
Lecture Notes in Computer Science
ISBN 978-3-319-46300-1   ISBN 978-3-319-46301-8   (eBook)
DOI 10.1007/978-3-319-46301-8

Library of Congress Control Number: 2016950882

LNCS Sublibrary: SL5 – Computer Communication Networks and Telecommunications

Printed on acid-free paper

This Springer imprint is published by Springer Nature
The registered company is Springer International Publishing AG
The registered company address is: Gewerbestrasse 11, 6330 Cham, Switzerland

Preface

We welcome you to the joint proceedings of the 16th NEW2AN (Next Generation Teletraffic and Wired/Wireless Advanced Networks and Systems) and the 9th conference on Internet of Things and Smart Spaces ruSMART (Are You Smart?), held in St. Petersburg, Russia, on September 26–28, 2016.

Originally, the NEW2AN conference was launched by ITC (International Teletraffic Congress) in St. Petersburg in June 1993 as an ITC-Sponsored Regional International Teletraffic Seminar. The first event was entitled "Traffic Management and Routing in SDH Networks" and held by R&D LONIIS. In 2002, the event received its current name, the NEW2AN. In 2008, NEW2AN acquired a new companion in Smart Spaces, ruSMART, hence boosting interaction between researchers, practitioners, and engineers across different areas of ICT. From 2012, the scope of the ruSMART conference has been extended to cover the Internet of Things and related aspects.

Presently, NEW2AN and ruSMART are well-established conferences with a unique cross-disciplinary mixture of telecommunications-related research and science. NEW2AN/ruSMART is accompanied by outstanding keynotes from universities and companies across Europe, USA, and Russia.

The 16th NEW2AN technical program addresses various aspects of next-generation data networks. This year, special attention is given to advanced wireless networking and applications as well as to lower-layer communication enablers. In particular, the authors have demonstrated novel and innovative approaches to performance and efficiency analysis of ad hoc and machine-type systems, employed game-theoretical formulations, Markov chain models, and advanced queuing theory. It is also worth mentioning the rich coverage of graphene and other emerging materials, photonics and optics, generation and processing of signals, as well as business aspects.

The 9th conference on Internet of Things and Smart Spaces, ruSMART 2016, provides a forum for academic and industrial researchers to discuss new ideas and trends in the emerging areas of the Internet of Things and Smart Spaces that create new opportunities for fully-customized applications and services. The conference brought together leading experts from top affiliations around the world. This year, ruSMART enjoyed active participation from representatives of various players in the field, including academic teams and industrial world-leader companies, particularly representatives of Russian R&D centers, which have a good reputation for high-quality research and business in innovative service creation and applications development.

We would like to thank the Technical Program Committee members of both conferences, as well as the associated reviewers, for their hard work and important contribution to the conference. This year, the conference program met the highest quality criteria with an acceptance ratio of around 35 %.

The conferences were organized in cooperation with the Open Innovations Association FRUCT, IEEE Communications Society Russia NorthWest Chapter, Tampere University of Technology, St. Petersburg State Polytechnical University, Peoples'

Friendship University of Russia, St. Petersburg State University of Telecommunications, and the Popov Society. This year the conference was held in conjunction with the 40th Interdisciplinary Conference and School Information Technology and Systems 2016. The support of these organizations is gratefully acknowledged.

We also wish to thank all those who contributed to the organization of the conferences. In particular, we are grateful to Aleksandr Ometov for his substantial work on supporting the conference website and his excellent job on the compilation of camera-ready papers and interaction with Springer.

We believe that the 16th NEW2AN and 9th ruSMART conferences delivered an informative, high-quality, and up-to-date scientific program. We also hope that participants enjoyed both technical and social conference components, the Russian hospitality, and the beautiful city of St. Petersburg.

September 2016

Olga Galinina
Sergey Balandin
Yevgeni Koucheryavy

oring Institutions

СПб|ГУТ)))

Organization

Ninosl
Nataraj
Pedro I
Edmun
Antonii
Athana
Edison
Nichol:
Simon
Ales S
Takesh

Jouni 1
Denis
Katarzy
Wei W

Spon:

NEW2AN International Advisory Committee

Igor Faynberg	Alcatel Lucent, USA
Jarmo Harju	Tampere University of Techno
Villy B. Iversen	Technical University of Denmᵢ
Andrey Koucheryavy	State University of Telecommᵤ
Kyu Ouk Lee	ETRI, South Korea
Sergey Makarov	St. Petersburg State Polytechni
Mohammad S. Obaidat	Monmouth University, USA
Andrey I. Rudskoy	St. Petersburg State Polytechni
Konstantin Samouylov	Peoples' Friendship University
Michael Smirnov	Fraunhofer FOKUS, Germany
Manfred Sneps-Sneppe	Ventspils University College, I
Sergey Stepanov	MTUCI, Russia

NEW2AN and ruSMART Technical Program C

Ozgur Akan	Koc University, Turkey
Hassen Alsafi	IIUM, Malaysia
Angelos Antonopoulos	Telecommunications Technolog
	of Catalonia (CTTC), Spain
Konstantin Avrachenkov	Inria Sophia Antipolis, France
Francisco Barcelo-Arroyo	Universitat Politecnica de Catal
Boris Bellalta	Universitat Pompeu Fabra, Spaᵢ
Torsten Braun	University of Bern, Switzerland
Raffaele Bruno	IIT-CNR, Italy
Paulo Carvalho	Centro Algoritmi, Universidade
Wei Koong Chai	University College London, UK
Chrysostomos	Frederick University, Cyprus
Chrysostomou	
Roman Dunaytsev	The Bonch-Bruevich Saint-Peter
	of Telecommunications, Rus:
Dieter Fiems	Ghent University, Belgium
Ivan Ganchev	University of Limerick, Ireland
Giovanni Giambene	University of Siena, Italy
Jarmo Harju	Tampere University of Technol(
Andreas J. Kassler	Karlstad University, Sweden
Andrey Krendzel	Huawei Technologies, Finland
Tatiana Madsen	Aalborg University, Denmark

Contents

ruSMART: Smart Services in Automotive Industry

NEW2AN: Cooperative Communications

NEW2AN: Wireless Networks

NEW2AN: Wireless Sensor Networks

NEW2AN: Security Issues

NEW2AN: IoT and Industrial IoT

NEW2AN: NoC and Positioning

ruSMART: New Generation of Smart Services

Forecasting Youth Unemployment in Korea with Web Search Queries

Chi-Myung Kwon[1] and Jae Un Jung[2(✉)]

[1] Department of MIS, Dong-A University, Busan, Korea
cmkwon@dau.ac.kr
[2] BK21Plus Groups, Dong-A University, Busan, Korea
imhere@dau.ac.kr

Abstract. Governments increase budget expenditure for youth job creation, but youth job markets tightened by prolonged recession are not improved as expected. To ease the problem of youth unemployment, developing relevant policies is important but more accurate and rapid prediction is also critical. This research develops a prediction model additionally utilizing web query information in classical statistical prediction model. Often ARIMA model is applied to estimate unemployment rate. For identified ARIMA model for Korean youth unemployment rate, we apply web query information to improve the accuracy of prediction. Our suggested model shows better performance than ARIMA model with respect to mean squared errors of estimate and prediction. We hope this research will be useful in developing a more improved model to estimate variable of interest.

Keywords: Youth unemployment · Predictive analytics · ARIMA model · Time series analysis · Web search query

1 Introduction

The youth unemployment rate is the number of unemployed 15–24 year-olds expressed as a percentage of the youth labour force. Unemployed people are those who report that they are without work, that they are available for work and that they have taken active steps to find work in the last four weeks [1]. According to the World Bank [2], the world youth unemployment persistently increased to 14.0 % in 2014 from 12.5 % in 2007 and for the same period, the total unemployment in the world grew to 5.9 % in 2014 from 5.5 % in 2007. The overall unemployment shows the decreasing trend from 2009 (6.3 %) unlike the youth indicator on the increase.

Generally, youth unemployment shows about twice higher than total unemployment but in this context, notable is the facts that the youth unemployment is the upward trend unlike the total unemployment with the downward trend, and that the gap between youth and total unemployment is continuously widened. Such phenomena may cause social and economic conflicts between generations [3].

For dealing with the youth unemployment issue, job creation and training for youth are important. If the unemployment prediction for youth is more accurate and rapid, it would help governments to ensure the more positive effects through the more agile and anticipative actions.

© Springer International Publishing AG 2016
O. Galinina et al. (Eds.): NEW2AN/ruSMART 2016, LNCS 9870, pp. 3–14, 2016.
DOI: 10.1007/978-3-319-46301-8_1

With regard to this issue, recent researches [4–6] suggest applying web search trend information for improving the predictability of unemployment prediction models. They used the Google search engine, which has about 90 % worldwide market share [7], to collect the web search query information. Fondeur and Karame estimated French youth unemployment using a modified version of the Kalman filter with Google query information [4]. Choi and Varian [5], and Anvik and Gjelstad [6], respectively, predicted the US initial jobless claims and Norwegian unemployment by use of Google query information in the ARIMA model. The Google search engine has the highest market share in the most countries like EU, US, etc. but not in China and Korea. Naver (Korean representative portal site) specific in Korean has about 75 % market share for PC in Korea. Using the web query information collected from Naver, Kwon et al. [8] developed a prediction model of the Korea's total unemployment.

With a similar direction to the recent studies, this research aims to develop a prediction model for estimating the Korea's youth unemployment and identify whether applying web query information is effective in improving the prediction model. For this end, we use the classical autoregressive integrated moving average (ARIMA) model to estimate the Korea's youth unemployment rate, and then we develop a model additionally utilizing web query information in selected ARIMA model. Finally we intend to compare the performances of developed models with respect to measure of predictability.

2 Korea's Youth Unemployment Data

According to the youth unemployment data of OECD (Organisation for Economic Co-operation and Development), as of 2015, Japan (5.6 %) and Greece (49.8 %) show the lowest and highest rates, respectively among the 34 OCED members. The rate of the Korea's youth unemployment presents 10.5 %, which is below the OECD average (13.9 %) [1].

Table 1. Korea's youth unemployment rate (%)

	2010	2011	2012	2013	2014	2015
January		8.5	8.0	7.5	8.7	9.2
February		8.5	8.3	9.1	10.9	11.1
March		9.5	8.3	8.6	9.9	10.7
April		8.7	8.5	8.4	10.0	10.2
May	6.4	7.3	8.0	7.4	8.7	
June	8.3	7.6	7.7	7.9	9.5	
July	8.5	7.6	7.3	8.3	8.9	
August	7.0	6.3	6.4	7.6	8.4	
September	7.2	6.3	6.7	7.7	8.5	
October	7.0	6.7	6.9	7.8	8.0	
November	6.4	6.8	6.7	7.5	7.9	
December	8.0	7.7	7.5	8.5	9.0	

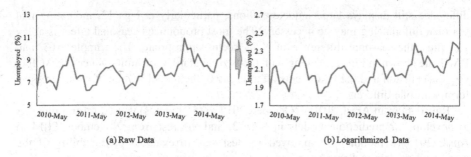

(a) Raw Data   (b) Logarithmized Data

Fig. 1. Trends of Korea's youth unemployment rate

In the definition of youth unemployment, global organizations like ILO (International Labour Organization) and OECD consider the youth as 15–24 year-olds but the Korean government regards the Korean youth as 15–29 year-olds. Table 1 presents Korean youth unemployment percentages from May, 2010 to April, 2015 [9]. Figure 1 (a) shows the change of the Korea's youth unemployment rate. From examining Fig. 1 (a), we note that the variability of the time series increase as its general level increases throughout the period 2010–2015. This suggests that some transformation such as logarithms of raw data should be analyzed, rather than the raw data. Also seasonal variation is observed in the series.

3 Identification of ARIMA Model

3.1 Stationarity of Youth Unemployment Data

We employed the autoregressive integrated moving average (ARIMA) process to describe the change of the youth unemployment rate. The ARIMA(p, d, q) process represents the dth differences of original series as a process containing p autoregressive and q moving average parameters [10–12].

To identify an appropriate ARIMA model for youth unemployment rates, we first transform the original data given Table 1 by taking logarithm. The logarithm data of the youth employment rate, $Y_t = \ln(X_t)$, display less spatial variability than the raw series, where X_t is the value at time t in original time series (see Fig. 1(b)). An increase in the level of the log series indicates that at least one difference will be required to achieve stationarity. Through the logarithm transformation and differencing the 1° of logarithm series, we adjusted the time series given in Fig. 1(a).

For developing a time series model for this process, we compute the sample autocorrelation function (ACF) and partial autocorrelation function (PACF) shown in Fig. 2. For assistance in interpreting these functions, two-standard–error limits are plotted on the graph as dashed lines. We see from Fig. 2(a) that the sample ACF tails off with a sinusoidal decay, while the sample PACF cuts off after lag 0 (except at 12 lag). The increase in the level of the log series indicates at least one difference is required to achieve stationarity. Figure 2(b) presents the sample ACF and PACF after one difference of log transformed series (Y_t). The sample ACF for the series of first difference still displays large autocorrelations, particularly at lags 12, indicating that seasonal differencing may be necessary. The most pronounced seasonal effect is at lag 12. Thus, the seasonal difference of Y_{t-12} seems appropriate. The sample ACF and PACF are shown in Fig. 2(c) along with the two standard error limits. Since the ACF in Fig. 2(c) tails off and the PACF cuts off after lag 0, the model will be of autoregressive form with one difference.

Raw data from April, 2010 to October, 2014 to (54 months) is used for training sets to develop our prediction models in Sect. 3, and the rest from November, 2014 to April, 2015 (6 months) is employed for test sets to evaluate predictability of the developed models in Sect. 4.

To ensure the stationarity of the Korea's youth unemployment data in Fig. 1(a), we perform the logarithm transformation, differencing and seasonal adjustment (remove trend and seasonality) consecutively. Finally, we could get the transformed series shown in Fig. 3.

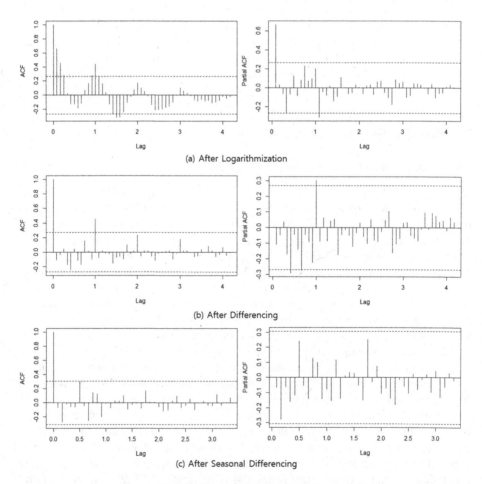

(a) After Logarithmization

(b) After Differencing

(c) After Seasonal Differencing

Fig. 2. Sample ACFs and PACFs

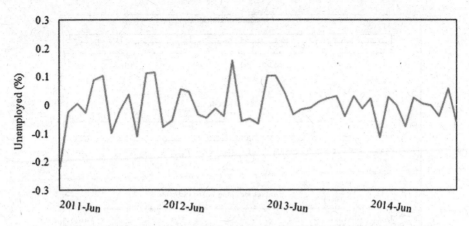

Fig. 3. Stationary process of Korea's youth unemployment rate

3.2 Estimating Parameters of ARIMA Model

We apply the ARIMA program of auto.arima() function in R [13] to identify an appropriate model that fits the youth unemployment adjusted rate. Using youth unemployment rates for period from May 2010 to October 2014 (54 months), we identify the ARIMA model. The remained data of 6 months will be used for estimating the model predictability.

Table 2 shows the results of estimated ARIMA model for Korea's youth unemployment rate. The suggested model is given by $ARIMA(1,0,0)(1,1,0)_{12}$. This result agrees with our discussions about appropriate model derivation. We call this baseline model as Model 1 to distinguish an extended model with web-query information in next Section (Model 2(a) and Model 2(b)).

Table 2. Estimated ARIMA model

	Model 1
Model	$ARIMA(1,0,0)(0,1,0)_{12}$
Coefficient	AR1: 0.7374
AIC	−101.45

After estimating Model 1, we diagnosed the model 1 in terms of residual, ACF and p-value. Residuals show the regularity of standardized residuals, and ACF values are inside the white noise area. All of the p-values are outside the significant range. These results indicate that any problem was not found to confirm the estimated Model 1.

The estimated Model 1, $ARIMA(1,0,0)(0,1,0)_{12}$ can be represented as a linear regression equation:

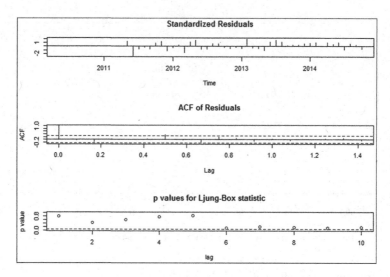

Fig. 4. Diagnostic of Model 1

$$Model1 : \ln(X_t) = \emptyset_0 + \emptyset_1 \ln(X_{t-1}) + \emptyset_2 \ln(X_{t-12}) + e_t. \qquad (1)$$

Here Φ_k (k = 0, 1, 2) is a coefficient of variable and the e_t is an error term at time t. We summarize the results of regression analysis of Eq. (1).

Table 3. Estimated coefficients of Model 1

Predictors	Unstandardized β (Std. Error)	Standardized β	t	Sig.
Constant	−0.098(0.227)		−0.431	0.669
$\ln(X_{t-1})$	0.558(0.096)	0.562	5.841	0.000
$\ln(X_{t-12})$	0.498(0.112)	0.427	4.435	0.000
R^2 (Adjusted)		0.709 (0.694)		
Residual mean square		0.005		

Substitution of model parameters gives Model 1 as

$$Model1 : \ln(X_t) = -0.098 + 0.558 \ln(X_{t-1}) + 0.498 \ln(X_{t-12}) + e_t. \qquad (2)$$

Model 1 shows 70.9 % of the explanatory power (R-squared: 0.709) for the real data. Figure 5 presents the fitted graph of unemployment rates of ARIMA model and their observations in Table 1. Fitted curve shows similar pattern to observed rate of unemployment.

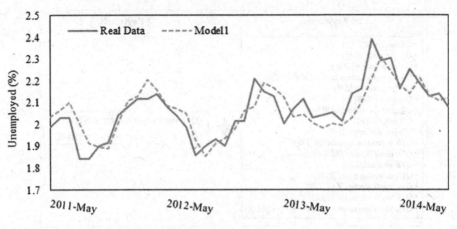

Fig. 5. Fitting of Model 1

4 Improving Baseline Model with Web Query Information

4.1 Collecting and Preprocessing Web Query Information

We try to apply the web query information to improve the predictability of the baseline model (Model 1). We retrieved keywords for web queries associated with the Korea's youth unemployment rate through the analysis of Korean SNS (social network service) and blog data generated from April 8 to May 8 (one month) in 2015. We find 16 Korean words simultaneously mentioned with 'youth unemployment rate' among 577 online documents or messages (twitter: 279, blog: 298). Collected query includes company, economy, employment, enter-graduate-school, government, get-a-job, graduate school, job, join-the-army, permanent position, rental house, support, unemployment, youth startup, youth unemployment, and youth unemployment rate (translated to English from Korean and alphabetized).

In addition to the 16 keywords, we added another two keywords; 'unemployment benefits', which showed a high correlation with 'unemployment rate' in previous our study [8], and 'youth-get-a-job', which is an antonym of 'youth unemployment'. The formation of total keywords' set is depicted in Fig. 6.

We collected web search trends of the 18 queries (keywords) from May 2010 to April 2015 through the Naver query engine. Web query information provides the value scaled to a range from 0 to 100 as a relative frequency like the Google Trends. The web query information is weekly data but youth unemployment data provided by Statistics Korea are generated monthly. For additional use of the web query information in developed ARIMA model, we convert the time interval unit of the web query information from 'week' to 'month'.

For instance, collected weekly data of web query information in Table 4 are converted as follows:

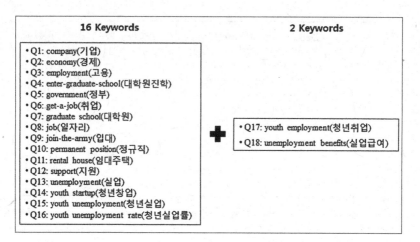

Fig. 6. Keywords associated with Korea's youth unemployment rate

Table 4. Converting time unit of web query information

Month (Days)	Period (yyyymmdd)	Days	Weighted rate(a)	Keyword value(b)	(a)X(b)	Converted value (Total)
2011 September (30)	20100830~20100905	5	5/30	97	16.2	82.4
	20100906~20100912	7	7/30	92	21.5	
	20100913~20100919	7	7/30	81	18.9	
	20100920~20100926	7	7/30	55	12.8	
	20100927~ 20101003	4	4/30	98	13.1	

- We convert the weekly data to monthly data by taking a weighted average dependent upon how many days there are in a specific month. A specific month may not include days of the current month. To align web query data with current month's unemployment rate, we use the weighted average of each week's keyword value.
- Suppose we want to construct data for September 2011 for our category "Q6: get-a-job." The month will include data from 2010.08.30 until 2010.10.03. The first week (20100830–20100905) consists of 2 days of August and 5 days of September. September has 30 days. So weighted rate of the first week is 5/30. In the same way, we compute the weighted rates of all weeks in September. That is, the weighted average of all weeks' keyword values is used for monthly value of keyword.

After converting the values of 18 queries information, we selected significant five queries which have correlations greater than 0.5 with youth unemployment rate. Table 5 presents selected five keywords to be used in model construction of Sect. 4.2; Q1 (company), Q3 (employment), Q14 (youth startup), Q17 (youth employment), and Q18 (unemployment benefits).

Table 5. Selected web queries

Query	Correlation
Q1	0.633^{**}
Q3	0.740^{**}
Q14	0.634^{**}
Q17	0.748^{**}
Q18	0.822^{**}

**Correlation is significant at the 0.01 level (2-tailed).

4.2 ARIMA Model with Web Query Information

For a better prediction of unemployment rate, we consider the way utilizing web query information correlated with unemployment rate. To this end, we combine the information of web query values and autoregressive function of time series model of (1), thus build a linear model (Model 2) as follows:

$$Model 2 : \ln(X_t) = \emptyset_0 + \emptyset_1 \ln(X_{t-1}) + \emptyset_3 \ln(X_{t-12}) + \sum_{k=1}^{n} \beta_k \ln(q_t^k) + e_t. \quad (3)$$

Here $\Phi_k (k = 0, 1, 2)$ are same as those in (1), β_k ($k = 1, .. , n$) are coefficients of query values, q_t^k ($k = 1, .. , n$) are the values of search volume index and the e_t is an error term at time t. Stepwise regression [14] on linear model of (3) provides the estimated model as follows:

We note that in estimated model of (4), the variable of $\ln(X_{t-1})$ is not selected. When we include the variable of $\ln(X_{t-1})$ in the estimated model, the best model is given as

Table 6. Estimated coefficients of Model 2

Model 2(a)	Predictors	Unstandardized β (Std. error)	Standardized β	t	Sig.
	Constant	$-0.333(0.211)$		-1.575	0.123
	$\ln(X_{t-12})$	0.418(0.094)	0.358	4.443	0.000
	$\ln(q_t^{17})$	0.124(0.042)	0.308	0.308	0.006
	$\ln(q_t^{18})$	0.280(0.076)	0.416	0.416	0.001
	R^2 (Adjusted)		0.813(0.799)		
	Residual mean square		0.003		
Model 2(b)	Predictors	Unstandardized β (Std. error)	Standardized β	t	Sig.
	Constant	$-0.491(0.210)$		-2.336	0.025
	$\ln(X_{t-1})$	0.263(0.105)	0.265	2.494	0.017
	$\ln(X_{t-12})$	0.382(0.097)	0.327	3.922	0.000
	$\ln(q_t^{18})$	0.315(0.074)	0.468	4.272	0.000
	R^2 (Adjusted)		0.803(0.788)		
	Residual mean square		0.003		

$$Model2(b) : \ln(X_t) = -0.491 + 0.263 \ln(X_{t-1}) + 0.3820.2 \ln(X_{t-12})$$
$$+ 0.315 \ln(q_t^{18}) + e_t. \tag{4}$$

Table 6 summarizes the regression results of Model 2(a) and Model 2(b). Model 2(a) and Model 2(b), respectively, show 81.3 % and 77.1 % of the explanatory power. Figure 7 presents the fitted graph of unemployment rates from Model 1 and Model 2(a) together with trend of real observations. Fitted curve shows similar pattern of observed rates of unemployment.

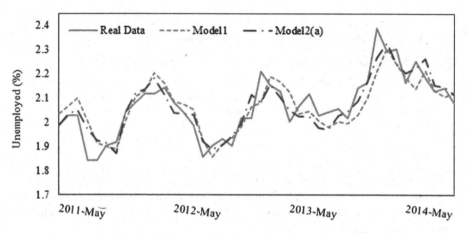

Fig. 7. Fitted curves of Model 1 and Model 2(a)

4.3 Performance Comparison of Model 1 and Model 2

To compare the performances of three developed models (Model 1, Model 2(a) and Model 2(b)), we simply consider the measures of the coefficient of determination, standard error of estimates and mean squared prediction error of models. In computation of mean squared prediction errors, we use the 6 month-dataset from November 2014 to April 2015 after training period of 54 months. Table 7 summarizes the considered measures for three models. Among three models, Model 2(a) shows a little better performance than those of Model 1 and Model 2(b).

Specifically, Model 2(a) and Model 2(b) better explain the youth unemployment rate by 10 % than Model 1. Standard error of estimates and mean squared prediction error of Model 2(a) decrease by about 20 % and 40 %, respectively compared to those

Table 7. Model performance comparison

	Model 1	Model 2(a)	Model 2(b)
Coefficient of determination	0.709	0.813	0.803
Standard error of estimates	0.06890	0.05589	0.05737
Mean squared prediction error	0.005	0.003	0.003

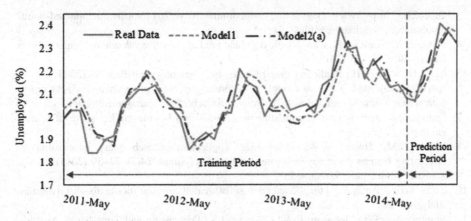

Fig. 8. Predictability of Model 1, Model 2(a) and Model 2(b)

of Model 1. Figure 8 shows the fitted graphs of unemployment rates for training period of 54 months and for prediction period of 6 month together with real observations. Three fitted curves of Model 1, Model 2(a) and Model 2(b) show similar patterns to observed rate of unemployment.

5 Discussion

Our research focuses on development of models for predicting the youth unemployment rate in Korea. Classical method of ARIMA is often applied to estimate unemployment rates. Another direction to forecast such rates suggests utilizing web query information together with statistical models such as ARIMA model. We develop a combined model of ARIMA process and web query information. Estimated model utilizing query information shows better performance than that of classical ARIMA model in prediction of unemployment rate. We consider that appropriate query information associated with response of interest can be usefully applied in developing estimation model. Often accurate and rapid prediction of certain variable is required for early decision of relevant policies. Recently, obtaining appropriate information from web query tends to be easier, thus we expect that effectively utilizing of collected web query information would be helpful for more accurate estimation and prediction of interested variables.

Acknowledgments. This work was supported by the Dong-A University research fund.

References

1. Organization for Economic Co-operation and Development. https://data.oecd.org/unemp/youth-unemployment-rate.htm
2. The Word Bank. http://data.worldbank.org

3. Guardian. https://www.theguardian.com/commentisfree/2011/apr/04/unemployed-youth-revolution-generational-conflict
4. Fondeur, Y., Karame, F.: Can Google data help predict French youth unemployment? Econ. Model. **30**, 117–125 (2013)
5. Choi, H., Varian, H.: Predicting Initial Claims for Unemployment Benefits (2009)
6. Anvik, C., Gjelstad, K.: "Just Google it" - forecasting Norwegian unemployment figures with web queries. Master's thesis, BI Norwegian School of Management (2010)
7. Statista. http://www.statista.com/statistics/216573/worldwide-market-share-of-search-engines/
8. Kwon, C.M., Hwang, S.W., Jung, J.U.: Application of web query information for forecasting Korean unemployment rate. J. Korea Soc. Simul. **24**(2), 31–39 (2015)
9. Statistics Korea. http://kostat.go.kr
10. Time Series Analysis. https://documents.software.dell.com/statistics/textbook/time-series-analysis
11. Montgomery, D.C., Johnson, L.A., Gardiner, J.S.: Forecasting and Time Series Analysis. McGraw-Hill, Texas (1990)
12. Box, G.E.P., Jenkins, G.M., Reinsel, G.C.: Time Series Analysis. Wiley, London (2013)
13. The R Project for Statistical Computing. https://www.r-project.org/
14. Myers, R.H.: Classical and Modern Regression with Applications. Duxbury Press, North Scituate (2000)

Competency Management System for Technopark Residents: Smart Space-Based Approach

Alexander Smirnov[1,2], Alexey Kashevnik[1,2(✉)], Segey Balandin[3],
Olesya Baraniuc[2], and Vladimir Parfenov[2]

[1] SPIIRAS, St. Petersburg, Russia
{smir,alexey}@iias.spb.su
[2] ITMO University, St. Petersburg, Russia
ob@itc.vuztc.ru, parfenov@mail.ifmo.ru
[3] FRUCT Oy, Helsinki, Finland
Sergey.Balandin@fruct.org

Abstract. Technopark is a union of companies called residents that implement innovation activities. Development of the competence management system for technoparks is currently a relevant task. Such system allows automating the process of resident's search that can satisfy potential customer tasks. The paper presents a smart space-based approach for competence management system development and implementation. Every resident is described by a profile that is shared with the smart space and becomes accessible for other competence management system users. The profile consists of several competencies and evidences with skills levels characterized their degree of possession.

Keywords: Competence management · Skills · Smart space · Technopark residents

1 Introduction

Technopark is a union of companies called residents that implement innovation activities. For joint collaboration of residents and for finding of potential customers for a resident or for a group of residents it is useful to acquire resident competencies and use it to represent them. Last years, competency management of companies is a popular research and development topic (see [1, 2]). At the moment, advantages in a global competition is determined by learning and deployment speed of new knowledge into modern technologies and production. In accordance with [3], the main aspects involved in competence management are related to: development of concepts, skills and attitudes (formation); work practices, ability to mobilize resources, which distinguishes it from others; combination of resources; search for better performances; permanent questioning; individual learning process in which the higher responsibility should be attributed to the individual him/herself; relationship to other people. Such system allows automating the process of residents searching that satisfy potential customer tasks. Competence management of employee and companies at all is a popular research

© Springer International Publishing AG 2016
O. Galinina et al. (Eds.): NEW2AN/ruSMART 2016, LNCS 9870, pp. 15–24, 2016.
DOI: 10.1007/978-3-319-46301-8_2

direction in the last years. The paper presents a relate work of modern competence management systems. Main requirements for such systems are formulated based on this related work. An approach for competence management system and its implementation were developed and described based on smart space technology that provides possibilities to organize semantic based information exchange among technopark residents and other competence management system customers.

At the moment, the system is developed and deployed for ITMO University Technopark. This technopart is a member of international association of technoparks and presents residents for potential customers. Thereby, development of the competence management system for ITMO University Technopark is the actual at the moment task.

The rest of the paper is structured as follows. Related work is presented in Sect. 2. Section 3 describes a reference model of the competence management system. Section 4 presents the approach evaluation. The results are summarized in Conclusion.

2 Related Work

The article [4] presents the outcomes of studies, which resulted the creation of the information system for the storage and evaluation of competencies of university students. The system is based on fine-grained representation of the skills and competencies through ontologies. The system supports students in planning their courses, fills a gap in the analysis of competencies and creates profiles for application forms when applying for a job. Presentation profiles based on XML HR for data exchange. For example, if students give the access to their profiles to recruitment companies, it allows recruiters to find the desired employee faster. Due to the high accuracy of stored data, encrypted XML data store is used.

As part of a university course, it was selected about 60 students with relevant competences in the field of computers to work with economic tasks/information systems. Each student has three courses for himself/herself, which he/she later takes. For each course there is a number of competencies, which are prerequisites for the taking of a course and a number of competencies (postconditions), which were obtained during the course of implementation.

The proposed system can only be regarded as the first prototype, as it considered only certain important competences. For example, the competence "programming" was developed in parallel courses and was not represented in the system. Profile was considering only the post-conditions, without taking into account the student's competence in programming.

The article [5] presented the results of studies, conducted over 10 years, which led to the creation of applications for ontology learning, based on competencies and knowledge management. Based on this ontology, the structure of the software for e-learning systems managed by ontology is presented. The researchers concluded, that in order to meet the challenges of the information society, it is necessary to maintain process of competence development in the context of lifelong learning. More flexible, adaptive learning systems are required. The article tells about the experience, of using MISA method (Instructional Engineering Method) - training engineering development method. It was first used in 1992 with the aim of integrating the knowledge of modeling and competence within the framework of this method.

Also, researchers present TELOS system (TelELearning Operating System), which is a teaching operating system with an ontologically-driven architecture. The system integrates computer and human agents using two basic processes: semantic representation of resources and resource aggregation. Acquisition of new competencies is the important task of competence management. This process should be integrated into the software system as an educational engineering tool, allowing to inform the operational tools of the new acquired qualities and their position in relation to the achievement of the objective of acquired new competencies.

The authors [6] address the problem of knowledge modeling in multi-agent systems that allow agents and users, to perceive alike and accept the concept of a domain. Ontologies are offered as a solution that allows to develop a coherent rules for specific domains. The researchers presented a multi-agent system that allows to manage, search and map existing competences of the user with represented ones, on the basis of relevant ontologies identified by specific domain. The authors examined examples for using the competencies in the relationship between universities and prospective students, between companies and future employees. The model considered in the framework of the article allowed universities, students and employees of companies to build and maintain their own competence to assess their knowledge to comply with their mandates and to search for the desired competencies in the respective areas. The direction that the researchers plan to expand within the next scientific work - refinement of quantitative ontologies of the component. It is important for requests and offers matching. Long-term studies will be focused on the model's ability to match two different ontologies of the domain.

Authors of the article [7] pay attention to the lack at the moment of successful indicators in the field of competence management, which could provide promising tools for a more efficient allocation of resources, knowledge management, support for training and human resources development in general, especially in the individual entrepreneurs level. Pilot applications, such as detection of an expert often fail in the long run, because of the incomplete or outdated databases. In order to overcome this problem, scientists have proposed an approach of joint management competencies. In this approach, they have joined in the Web 2.0 technology processes, running from the bottom up with the organizational processes that run from top to bottom. They solved this problem as the task of constructing a joint ontology, which is the basis for the model of ontology aging process. In order to implement model of ontology aging process for competence management, the researchers have built a semantically-social application SOBOLEO that offers competence ontology aging and easy to use interface. Thus, in the article the researchers show how ontology competencies can be developed to cover less formal tag topics. It was proved that it guarantees value and timeliness during application. Easy to use in everyday activities SOBOLEO motivates employees to fill the data into the system.

The article [8] presents a common framework for intelligent competence management system based on ontologies for an information technology company. In the first phase, it was tested in small enterprises working in the field of information technology, and then it applied for other organizations of the same type. Competence management system according to the authors has to achieve the following key objectives: (a) maintain complete and systematic acquisition of knowledge about the competencies of employees of the enterprise; (b) ensure knowledge about the competencies and their owners; (c) apply existing knowledge to achieve the goal.

The core competencies of information management system is an ontology, which plays the role of declarative knowledge base repository, containing the basic concepts (such as company work, competence, domain, group, person, etc.) and their relationship to other concepts, examples and properties. Protégé framework was used to create such ontology. Ontology structure is conceived in such a way that the logic description can be used to represent the concept of determining the subject area in a structured and widespread form. Acquiring knowledge in this approach is performed by enriching ontology, in accordance with IT-company requirements. According to the authors, the advantage of using the system on the basis of ontology, is the ability to identify new relationships among concepts, based on logical conclusions, starting from existing knowledge. The user may choose for request examples of one type concept. The article also provides some examples of using such system.

Authors of the article [9] have focused on the analysis of dynamic competence management system. The system takes into account the changes of competences with the time, caused by diffuse processes in project groups. Authors emphasize that the management and control of knowledge and skills, and, more recently, the companies competence, have become an essential factor of the production process in terms of the strategic human capital management purposes. Knowledge and competence management is becoming increasingly important subject of research for educational institutions. It is necessary to focus on a detailed description of the achievements of the student in the form of their competences, as well as the analysis of competences of companies employees where intellectual capital is equal to the investment to the competence, that allows the employer to make decision regarding trainings, attracting to new projects and recruitment.

Considered article describes the concept of dynamic competence management system, which contributes to a better dynamic nature competences guidance. The authors cite several arguments in favor of use of this system in the organization: (1) the system provides the identification of skills, knowledge, behaviors and capabilities needed to meet current and future staffing needs, (2) it can focus on the individual and group development plans.

Offering the formal approach in building a competence management system, the author of the paper [10] addresses the problem of competence profiles management. Competence management in recent years has become very topical, because it contributes to the achievement of organizational goals and solves problems such as improving the information flow or the competences generation. In the paper were proposed a lot of competence modeling approaches and the use of competency models.

It was revealed that there was no examination of the structures and the use of competence profiles in competence management system. The author has represented the ontological realization of the abstract model, including software architecture of competence profile management system. The main contribution of this work is that the authors consider the formalization of competency profiles operations and ontological implementation of these operations.

The authors of the paper [11] focuses on the role of user behavior modeling and semantically enhanced submissions for personalization of its interaction with the system. The work represents the general ontological foundations of user modeling OntobUMf (Ontology-based User Modeling framework) its components and processes, associated with the user behavior modeling. The authors present them as wireframes, shaping user behavior and classifying people according to their behavior. The basis of OntobUMf is user ontology, which was developed in accordance with the information system management, information package IMS LIP (Information Management System Learning Information Package). Custom ontology includes behavioral concept, extends to IMS LIP specification, defines the users characteristics, interacting with the system. The paper gives examples of OntobUMf in the context of a knowledge management system. Also, in the scientific work, the background of the ontological modeling creation, user behavior for semantically enhanced knowledge management systems are discussed. According to the authors of this article, the results of presented research, may contribute to the development of other frameworks of user behavior, other semantically enhanced user modeling systems or other semantically enhanced information systems.

According to the authors of the paper [12] learning management system Moodle (Modular Object - Oriented Dynamic Learning Environment) is currently the most popular software solution that provides a variety of modules for various educational purposes. However, there are some aspects related to competence management, which are missing in Moodle. Article [10] offers an application that is designed as an extension of Moodle to support the development and evaluation of competences within the course. The article provides detailed information about the competence ontology, adopted for course structure development, based on competence, as well as competence management features built into Moodle. The authors show how these functions, embedded in the learning management system, allow the controlling of the target competencies together with associated elements and evaluate the level of skills, achieved by students within each of the target competencies. In addition, it becomes possible to generate different types of competency reports, depending on the target role (teacher, student or administrator). The application, offered by authors, satisfies the need for practical and convenient way to manage and evaluate the competencies, associated with learning management system Moodle.

In the paper [13] authors analyze the various approaches, presented in the literature, related to the competence modeling and offers a competence ontology as a formal description of the competence characteristics, agents, and educational resources in the educational networks. The proposed by authors ontology also seeks to simulate aspects, related to competence management and tracking to support the development of competences in educational networks throughout life. The authors believe that with the introduction of a paradigm of continuous education and dissemination of the terms "knowledge society", "civil mobility", "globalization", competence based learning and training, interest in technologies, improving the quality of education is growing, as it provides an important advantage for individuals and organizations, supporting the transformation of learning outcomes into permanent and valuable asset - knowledge.

In this context, in order to facilitate the acquisition and ongoing development of new competencies, educational networks have revealed the need to provide a variety of learning opportunities throughout life.

Based on the presented related work the following basic requirements for the competence management system of the technopark residents were highlighted:

- competence management system has to be based on smart space that contains residents competencies modelled in terms of ontology;
- storing residents profiles, information about customer tasks and the ability to handle them in competencies management system;
- separation of user rights to the following key roles: user, administrator and resident;
- web interface support;
- comparison between the residents competence profiles;
- comparison of a task with a profile.

3 Reference Model

Proposed competence management system is based on smart space technology that provides possibilities to share the semantic information among technopark residents and potential customers. Residents share with the smart space their competence profiles that represents main information about a resident and list of references to the resident competencies represented by ontologies (see Fig. 1). A competence profile of a resident is the set of skills with associated levels. A resident is used the competence editor to transfer their competencies into ontological representation. This ontology is shared with the smart space and describes the model of the resident in the smart space.

When a customer would like to collaborate with the technopark he/she opens technopark web portal and search for resident models in smart space who have required

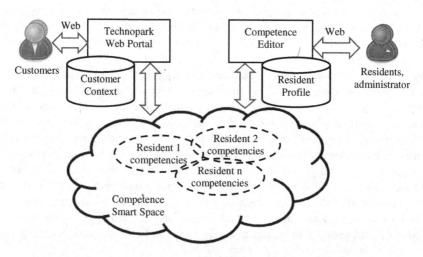

Fig. 1. Reference model of the proposed competence management system

competence or describe his/her context to implement this searching automatically. Context is a set of customer competencies that are required at the moment to solve for him/her task.

Competence editor module supports the following main operations:

- skill lexicon management;
- residents management;
- competence profile management.

Skill lexicon management provides possibilities for a resident/administrator to add new, remove, update items in the technopark skill tree. These skills can be added by a resident to the competence profile. Every skill is characterized by levels that represent degree of possession of the skill. A resident/administrator can specify count of levels and title for every level. Resident management operation provides possibilities for administrator to add, remove and edit a resident profile. The resident profile includes: company name, web site, address, short description, and contact e-mail. Competence management operation provides possibilities for residents to add and remove skills from a resident profile and determine degree of possession by choosing the skill level.

Technopark web portal supports the following main operations:

- determine a customer task;
- compare the customer task with resident profiles;
- compare profiles of different residents;
- range resident profiles based on similarity to the customer task.

A customer can determine a task and specify requirements that a company has to have for implement the task. Compare the customer task with resident competence profiles is used to determine if the resident can perform the selected task or not. For each requirement, the competency that implements the same skill is extracted from the competence profiles. If this skill level is less than the according skill level for the task requirement or profile does not have any competence with required skill, then selected profile cannot perform selected task, otherwise it can.

4   Implementation

Application has been implemented using Java programming language and Spring Framework technology stack for Technopark of ITMO University. Customers, residents, and administrator are working with competence management system through the web interface. Skills tree is shown in Fig. 2. It covers all ITMO University Technopark resident skills. Example of a resident profile is shown in Fig. 3. It includes the resident name, web site, address, resident description and set of competence linked to the resident profile. Implementation of the competence management system in details is described in [14].

Fig. 2. Skills tree for ITMO Univerisy Technopark residents

The following main scenarios are supported by application:

- User knows what competencies he/she needs. In this scenario user uses the search a resident by needed competencies.
- User knows the resident but should know it possibilities. In this scenario user can aggregate all tasks the resident can implement.
- User knows of two residents and he/she would like to compare their profiles. Which company can better implement the needed task.
- User knows the resident and and task and he/she would like to compare the resident and the task to understand if this resident can implement this task.

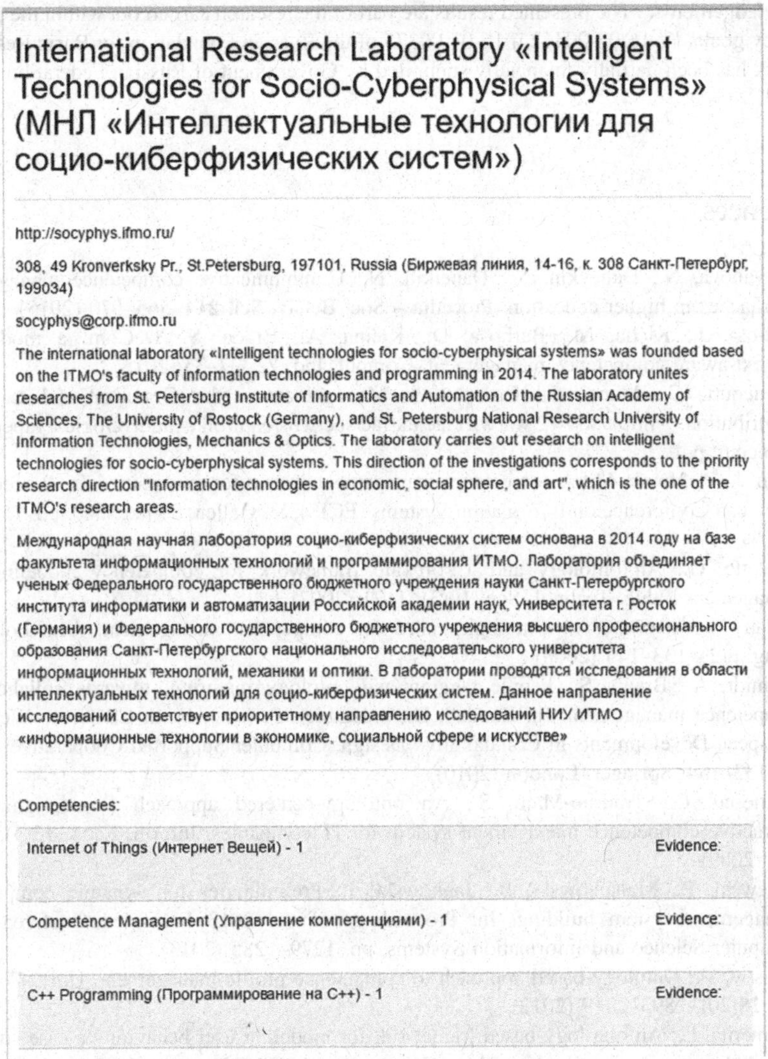

International Research Laboratory «Intelligent Technologies for Socio-Cyberphysical Systems» (МНЛ «Интеллектуальные технологии для социо-киберфизических систем»)

http://socyphys.ifmo.ru/

308, 49 Kronverksky Pr., St.Petersburg, 197101, Russia (Биржевая линия, 14-16, к. 308 Санкт-Петербург, 199034)

socyphys@corp.ifmo.ru

The international laboratory «Intelligent technologies for socio-cyberphysical systems» was founded based on the ITMO's faculty of information technologies and programming in 2014. The laboratory unites researches from St. Petersburg Institute of Informatics and Automation of the Russian Academy of Sciences, The University of Rostock (Germany), and St. Petersburg National Research University of Information Technologies, Mechanics & Optics. The laboratory carries out research on intelligent technologies for socio-cyberphysical systems. This direction of the investigations corresponds to the priority research direction "Information technologies in economic, social sphere, and art", which is the one of the ITMO's research areas.

Международная научная лаборатория социо-киберфизических систем основана в 2014 году на базе факультета информационных технологий и программирования ИТМО. Лаборатория объединяет ученых Федерального государственного бюджетного учреждения науки Санкт-Петербургского института информатики и автоматизации Российской академии наук, Университета г. Росток (Германия) и Федерального государственного бюджетного учреждения высшего профессионального образования Санкт-Петербургского национального исследовательского университета информационных технологий, механики и оптики. В лаборатории проводятся исследования в области интеллектуальных технологий для социо-киберфизических систем. Данное направление исследований соответствует приоритетному направлению исследований НИУ ИТМО «информационные технологии в экономике, социальной сфере и искусстве»

Competencies:

Internet of Things (Интернет Вещей) - 1	Evidence:
Competence Management (Управление компетенциями) - 1	Evidence:
C++ Programming (Программирование на C++) - 1	Evidence:

Fig. 3. Example of a resident profile

5 Conclusion

The paper presents an approach and implementation for competency management system for technopark residents. The system has been implemented for Technopark of ITMO University and accessible by the following link: http://77.234.220.70:8080/. At the moment, the system is being filled by the information about Technopark residents and then this information will be used to generate information pages for Technopark residents in web portal: http://technopark.ifmo.ru/en/.

Acknowledgements. The presented results are part of the research carried out within the project funded by grants ## 16-07-00462 и 16-07-00375 of the Russian Foundation for Basic Research. The work has been partially financially supported by Government of Russian Federation, Grant 074-U01.

References

1. Lukyanova, N., Daneykin, Y., Daneikin, N.: Communicative competence management approaches in higher education. Procedia – Soc. Behav. Sci. **214**, 565–570 (2015)
2. Barbosa, J., Kicha, M., Barbosa, D., Kleina, A., Rigoa, S.: DeCom: a model for context-aware competence management. Comput. Ind. **72**, 27–35 (2015)
3. Bitencourt, C.: Managerial Competence Management – the Organizational Learning Contribution. http://www2.warwick.ac.uk/fac/soc/wbs/conf/olkc/archive/oklc5/papers/i-3_bitencourt.pdf
4. Dorn, J., Pichlmair, M.: A competence management system for universities. In: Proceedings European Conference on Information Systems (ECIS), St. Gallen, Switzerland, pp. 759–770 (2007)
5. Paquette, G.: An ontology and a software framework for competency modeling and management. Educ. Technol. Soc. **10**(3), 1–21 (2007)
6. Iordan, V., Cicortas, A.: Ontologies used for competence management. Acta Polytech. Hung. **5**(2), 133–144 (2008)
7. Schmidt, A., Braun, S.: People tagging and ontology maturing: towards collaborative competence management. In: Randall, D., Salembier, P. (eds.) From CSCW to Web 2.0: European Developments in Collaborative Design. Computer Supported Cooperative Work, pp. 133–154. Springer, London (2010)
8. Niculescu, C., Trausan-Matu, S.: An ontology-centered approach for designing an interactive competence management system for IT companies. Inform. Econ. **13**(4), 159–167 (2009)
9. Różewski, P., Małachowski, B., Jankowski, J.: Preliminaries for dynamic competence management system building. In: Proceedings of the 2013 Federated Conference on Computer Science and Information Systems, pp. 1279–1285 (2013)
10. Tarasov, V.: Ontology-based approach to competence profile management. Univ. Comput. Sci. **18**(20), 2893–2919 (2012)
11. Razmerita, L.: An ontology-based framework for modeling user behavior - a case study in knowledge management systems. Man Cybern. Part A: IEEE Trans. Syst. Hum. **41**(4), 772–783 (2011)
12. Rezgui, K., Mhiri, H., Ghédira, K.: Extending moodle functionalities with ontology-based competency management. Proc. Comput. Sci. **35**, 570–579 (2014)
13. Rezgui, K., Mhiri, H., Ghédira, K.: An ontology-based approach to competency modeling and management in learning networks. In: Jezic, G., Kusek, M., Lovrek, I., Howlett, R.J., Jain, L.C. (eds.) Agent and Multi-Agent Systems: Technologies and Applications. AISC, vol. 296, pp. 257–266. Springer, Heidelberg (2014)
14. Gordeev, B., Baraniuc, O., Kashevnik, A.: Web-based competency management system for Technopark of ITMO University. In: 18th FRUCT & ISPIT Conference, 18–22 April 2016, Technopark of ITMO University, Saint-Petersburg, Russia, pp. 463–466

Data Mining for the Internet of Things with Fog Nodes

Ivan Kholod[(✉)], Ilya Petuhov, and Maria Efimova

Saint Petersburg Electrotechnical University "LETI", Saint Petersburg, Russia
iiholod@mail.ru, ioprst@gmail.com,
maria.efimova@hotmail.com

Abstract. The paper describes an approach of applying an actor model that executes Data Mining algorithms to analyze data in IoT systems with a distributed architecture (with Fog Computing). The approach allows to move computational load closer to the data, thus increasing performance of the analysis and decreasing network traffic. Execution of the 1R algorithm in an IoT system with a distributed architecture and the results of the comparison of distributed and centralized architectures are shown in the paper.

Keywords: Internet of Things · Fog Computing · Data Mining · Distributed data mining · Actor model

1 Introduction

Currently there is a rapid growth of stored information volumes obtained from different devices: sensors, cameras, mobile phones and others. These devices, connected by the Internet, are called Internet of Things (IoT). Cisco analysts consider the period of 2008–2009 to be the birth of the Internet of Things because during this period the number of devices connected to the Internet exceeded the population of the Earth [1], thus making the 'Internet of People' the 'Internet of Things'. According to Gartner, Inc. (a technology research and advisory corporation), there will be nearly 26 billion devices in the Internet of Things by 2020 [2]. Therefore the amount of information coming from those devices will increase over time.

Today this kind of information is referred to as Big data. It is characterized by large volumes of data, a variety of types and rapid generation. Such data is collected from sensors in IoT systems. Data analysis is an important task in such systems.

Scalable data processing systems are used to perform analysis (including intellectual analysis). Examples of such systems are Apache Hadoop and Apache Spark. They are used to process huge amounts of data like those in the systems by Google, Yandex and other popular social networks. However they do not require a centralized storage of the processed data and do not allow to relocate computational load closer towards the data sources, which would reduce traffic and therefore increase speed of the analysis.

Lately, IoT systems with fog nodes have become more popular. They are an alternative to the IoT systems with a centralized architecture. The systems use fog nodes to preprocess data. This paper describes an approach that allows to distribute analysis between nodes and move it closer to the data.

© Springer International Publishing AG 2016
O. Galinina et al. (Eds.): NEW2AN/ruSMART 2016, LNCS 9870, pp. 25–36, 2016.
DOI: 10.1007/978-3-319-46301-8_3

The paper is organized as follows. Section 2 is a review of the approaches to creating data mining systems for the IoT. The third section contains the description of a general approach that allows to map the decomposed algorithm onto blocks of the actors model. The fourth section describes the proposed approach to implementing the data mining system for IoT with a distributed architecture. The last section discusses the experiments and compares the approach with similar solutions.

2 Related Work

Most of the data mining systems for the IoT have a multilayer architecture (Fig. 1) of four levels [3, 4]:

1. The devices layer is the bottom layer. It can be viewed as a hardware or physical layer which performs data collection.
2. The data gathering layer is responsible for connecting the devices layer and the application layer enabling data transfer between them. It also performs cross platform communication, if required.
3. The data processing layer is responsible for critical functions such as device and information management and also takes care of such issues as data filtering, data aggregation, semantic analysis, access control and information discovery.
4. The layer of data analysis services provides services or applications that integrate or analyze the data received from the other two layers.

The last level provides services to execute different analytical tasks. The majority of existing IoT systems have a centralized architecture. The data there is collected in a single storage and is processed by the analytical services, which are also executed on a single computing cluster. There are two approaches to building a centralized analytical service:

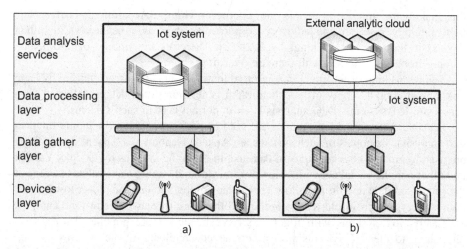

Fig. 1. Data mining for IoT systems with centralized architecture: (a) using internal data mining system, (b) using an external data mining cloud

- integration within existing cloud storages and analytic services [5];
- implementation of new services based on existing scalable analysis systems [4].

Cloud analysis services provided by established companies can be used to implement the first approach.

Azure Machine Learning (Azure ML) [6] is a SaaS cloud-based predictive analytics service from Microsoft Inc. It has been launched in February 2015. Azure ML provides paid services, which allow users to execute the full cycle of Data Mining: data collection, preprocessing, features definitions, choice and application of algorithms, model evaluation and publication. The service is for experienced users with knowledge in machine learning algorithms.

Azure ML can import data from local files, online sources and other cloud-projects (experiments). The reader module allows to load data from external sources, the Internet or other file storages.

In April 2015 Amazon has launched their **Amazon Machine Learning** service that allows users to train predictive models in the cloud [7]. This service provides all stages required for data analysis: data preparation, construction of a machine learning model, its settings, and eventually the prediction. The user can build and fine-tune predictive models using large amounts of data.

It allows users to analyze data stored in other Amazon services (Amazon Simple Storage Service, Amazon Redshift, or in Amazon Relational Database Service). To scale computations, the service uses Apache Hadoop.

Google made its **Cloud Machine Learning** platform [8], which is used by Google Photos, Translate, and Inbox, available to developers in March 2016. It is a managed platform that empowers users to build machine learning models. The platform provides pretrained models and helps to generate customized models. It allows users to apply neural network based machine learning methods, which are used by other Google-services including Photos (image search), the Google app (voice search), Translate, and Inbox (Smart Reply).

All of these services are provided by REST API for client applications. Users can only analyze data stored in Google storage and cannot add new machine learning algorithms.

Scalable data analysis systems can be used to implement the second approach.

Apache Spark Machine Learning Library (MLlib) [9] is a scalable machine learning library for the Apache Spark platform. It consists of common learning algorithms: classification, regression, clustering, collaborative filtering and other. It has an own implementation of MapReduce, which uses memory for data storage (versus Apache Hadoop that uses disk storage). It allows to increase the efficiency of the algorithm performance.

Apache Mahout [10] is also a data mining library concerning the MadReduce paradigm. It can be executed on Apache Hadoop or Spark based platforms. It contains only a few data mining algorithms for distributed execution: collaborative filtering, classification, clustering and dimensionality reduction. Users can extend the library by adding new data mining algorithms. The core libraries are highly optimized and also show good performance for non-distributed execution.

The disadvantage of the IoT systems with centralized approach is that it is necessary to send all the data from the sources to the place where it will be analyzed. This increases the network traffic and the time that the analysis takes in whole. This becomes a significant restriction when analyzing big data in real-time.

Fog Computing that became popular recently is an alternative to the Cloud Computing [11]. Fog Computing enables a new breed of applications and services, and that there is a fruitful interplay between the Cloud and the Fog, particularly when it comes to data management and analytics. The IoT systems that use Fog Computing have intermediate fog nodes at the level of intermediate levels of the IoT systems (Fig. 2) where the data analysis is performed without the data being sent to the centralized storage.

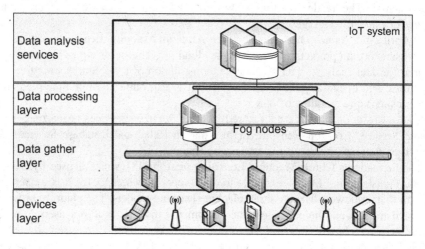

Fig. 2. IoT system with fog nodes

Such architectures are popular due to the absence of the drawbacks described earlier. However neither existing cloud analytical services nor systems that perform scalable data analysis can be used for such systems. The suggested approach and it is implementation on the actor model allow to solve this problem.

3   The Essence of the Approach

3.1   Presentation of Data Mining Algorithm as a Set of Functional Blocks

According to [12, 13], a data mining algorithm can be written as a sequence of functional blocks (based on the principle of functional programming). A data mining algorithm can be presented as a sequence of function calls:

$$\text{dma} = f_n(d, f_{n-1}(d, \ldots .f_i(d, \ldots .f_1(d, m) \ldots) \ldots)), \tag{1}$$

where f_i: is a function that analyses the input data set d of type D and changes the mining model m of type M. This function is called functional block. It is of the type:

$$FB :: D \to M \to M, \text{ where}$$

- D: is the input data set that is analyzed by functional block,
- M: is the mining model that is built by functional block.

Note that not all of the functional blocks of the data mining algorithms need to use the data:

$$f_i^c(d, m) = f_i^c(nil, m)$$

Such blocks are called calculation functional blocks. Accordingly the blocks, which use the data:

$$f_b^f(d, m) \neq f_b^f(nil, m)$$

are called processing functional block. Thus if the algorithm is represented as a set of functional blocks:

$$A = \{f_1, f_2, \ldots, f_i, \ldots, f_n\},$$

it is possible to divide the set into two subsets depending on the functional block's type:

$$A = A_c \cup A_f = \{f_1^c, f_2^c, \ldots, f_i^c, \ldots, f_v^c\} \cup \{f_1^f, f_2^f, \ldots, f_b^f, \ldots, f_w^f\}.$$

A data mining algorithm is also a functional block since according to (1) it can be presented as a composition of functional blocks:

$$dma = f_n \circ f_{n-1} \circ \ldots \circ f_i \circ \ldots \circ f_1.$$

The different flowchart structures (decisions, loops and other) can also be presented by functional blocks [14]. For example, we rewrite the 1R [15] algorithm as set of functional blocks:

- the conditional function which checks weather the current attribute is a target attribute. If so, it calls a composition of two functional blocks:
  - *addingOneRule:* adding a new rule to the mining model,
  - *incrementOneRule:* incrementing the count of vectors validated for this rule
    isCurrAttrTarget = if cf (d, m) then addingOneRule°incrementOneRule, where
    - *cf* – function to calculate the conditional expression,
- loop - function, which calls the functional block *isCurrAttrTarget* for all the attributes of the current vector,
  attrsCycle = loop'(d, f_{initA} (d, m), cf_A, f_{preA}, isCurrAttrTarget), where

- f_{initA}: the function initializes an attribute counter with the first index of the list of attributes
- cf_A: the conditional function checks whether all the attributes have been processed
- f_{prevA}: the preprocessing function changes an attribute counter assigning the index of the next vector
- the vectors cycle function calls the functional block *attrsCycle* for all vectors, *vectorsCycle* = *loop'*(*d*, f_{initW} (*d*, *m*), cf_W, f_{preW}, *attrsCycle*), *where*
  - cf_W: conditional function, that checks whether all the vectors have been processed
  - f_{initW}: the function, which initializes a vector counter with the first index of the list of vectors
  - f_{prevW}: preprocessing function, that changes the vector counter by assigning the index of the next vector
- loop - function, which for all values of the target attribute calls a functional block:
  - *selectBetterScoreRule:* selection of rule with minimal error for the current value of the target attribute
    targetsValuesCycle = *loop'*(*d*, f_{initT}(*d*, *m*), cf_T, f_{preT}, *selectBetterScoreRule*), where
  - cf_T: conditional function, that checks whether all the values of the target attribute (classes) have been processed
  - f_{initT}: the function, initializes a class counter with the first index of the list of classes
  - f_{prevT}: preprocessing function, that changes the class counter by assigning the index of the next class
- cycle function that calls functional block *targetsValuesCycle* for all rules:
  rulesCycle = *loop'* (*d*, fb_{initR} (*d*, *m*), cf_R, fb_{preR}, *targetsValuesCycle*), *where*
  - cf_R: conditional function, that checks whether all the rules have been processed
  - f_{initR}: the function initializes a rules counter with the first index of the list of rules
  - f_{prevR}: preprocessing function, that changes the rules counter by assigning the index of the next rule

So, the 1R algorithm can present as composition of two the functional blocks:

$$1R = rulesCycle \circ vectorsCycle \tag{2}$$

The functional block *rulesCycle* does not require the presence of the dataset and is a calculation functional block. The *vectorsCycle* block processes data and is a processing functional block.

3.2 Conversion of a Data Mining Algorithm into Parallel Form

According to the Church-Rosser theorem [12] the reduction (execution) of functional expressions (algorithm) can be done concurrently. The expression (1) has to be transformed into a representation, from which the functional blocks will be invoked as arguments. For this purpose a function *parallel* which takes care of data-parallelization in the algorithms has been added [16].

Using the *parallel* function, different parallelized forms of one data mining algorithm can be created. For example, the 1R algorithm can be converted into the following parallel forms:

- with parallel processing of the data sets by the vectors:

$$\text{vectorsCycleParall} = \text{parallel}(d, m, \text{vectorsCycle})$$
$$\text{1RVectorsCycleParallel} = \text{rulesCycle} \circ \text{vectorsCycleParall}. \tag{3}$$

- with parallel processing of data sets by the attributes:

$$\text{attrCycleParall} = \text{parallel}(d, m, \text{attrsCycle})$$
$$vectorsCycle = loop'(d, fb_{initW}(d, m), fb_{initW}, fb_{preW}, \text{attrCycleParall}) \tag{4}$$
$$\text{1RVectorsCycleParallel} = \text{rulesCycle} \circ vectorsCycle.$$

3.3 Mapping a Data Mining Algorithm on a IoT System with Fog Nodes

The IoT system can be represented as a union of two sets of nodes:

$$S = C \cup F = \left\{ n_0^c, n_1^c, \ldots, n_p^c, \ldots, n_u^c \right\} \cup \left\{ n_0^f, n_1^f, \ldots, n_q^f, \ldots n_z^f \right\}, \text{ where}$$

- n_p^c - computing node of a system that does not store data and is used to perform the analysis services (located at the data analysis services level on the Fig. 2),
- n_q^f - a node of a system that stores data and is used for preprocessing (fog nodes on the Fig. 2).

To execute analysis algorithms in such systems, the actor model [17] has been proposed [18]. The execution environment based on the actor model can be represented as a set of actors:

$$E = \left\{ r, a_0, a_1, a_2, \ldots, a_j, \ldots, a_g \right\}, \text{ where}$$

- r – the router, which distributes messages among actors,
- a_0 – the actor, which carries out the main algorithm sequence,
- a_1–a_g – the actors, that carry out the parallel function of the algorithm.

Actors can execute functional blocks and therefore run a distributed execution of the data mining algorithm [18]. The described approach was implemented as the data mining library DXelopes [19]. The library has adapters for the integration in the actors environments [18].

The actors environment was used to create the prototype of the distributed IoT system with fog nodes. Mapping actors to the nodes of the system divides the set of actors into two subsets: computing and processing:

$$E \rightarrow S = E \rightarrow (C \cup F) = (E_c \rightarrow C) \cup (E_f \rightarrow F) =$$
$$\{ \left(a_j^c, n_p^c \right) \mid a_j^c \in E, n_p^c \in C \} \cup \{ \left(a_r^f, n_q^f \right) \mid a_r^f \in F, n_q^f \in F \}$$

The types of the functional blocks in the Data Mining algorithm can be reconsidered when mapping them to actors. The blocks that interact with data should be located at the processing actors and the blocks that do not interact with the data should be located at the computing actors:

$$A \rightarrow E \rightarrow S = A \rightarrow E \rightarrow (C \cup F) = (A_c \rightarrow E_c \rightarrow C) \cup (A_f \rightarrow E_f \rightarrow F) =$$
$$\{ \left(f_i^c, a_j^c, n_p^c \right) \mid f_i^c \in A, a_j^c \in E, n_p^c \in C \} \cup \{ \left(f_b^f, a_r^f, n_q^f \right) \mid f_b^f \in A, a_r^f \in F, n_q^f \in F \} \quad (5)$$

Information on the fog nodes can be distributed horizontally or vertically. If the distribution is horizontal the data, that is recorded by the sensors at each node has the same metadata but is related to different objects. For example, there can be sensors for pressure, temperature, humidity etc. that would measure the parameters of similar objects but, for example, be located in different regions.

If the distribution is vertical, the data recorded at each node is related to one or several parameters. Thus the data, which is stored at each node has different meta data but is usually related to a single object. In this case, the synchronization can be achieved through comparison of timestamps.

The suggested approach allows to easily transform sequential algorithms into parallel processing on attributes or data vectors for both cases. The functional blocks of the algorithm that processes data can be moved to the fog nodes that store information according to (5).

4 Experiments

Experiments for centralized and distributed IoT systems have been carried out. The Apache Spark MLlib was used for the centralized approach. It has been deployed on high-performance servers supporting hardware virtualization and providing the possibility to perform cloud computing. The following objects of the computing cluster infrastructure were used for the experiments:

- two servers with following characteristics:
  - CPU - IntelXeon 2.9 GHz (2 CPU on 6 kernels, performance of calculations in 2 streams on a kernel, only 24 streams on the server), RAM - 128 GB,
  - CPU - Power7 3.3 GHz (2 CPU on 4 kernels, performance of calculations in 4 streams on a kernel, only 32 streams on the server), RAM - 128 GB,
- two StorageSystemStorwizev700 with 13.6 TiB.

For the distributed approach, actors with the functional blocks of the algorithm 1R were distributed between the nodes of the system. As an example, a specific parallel form of the algorithm 1R was used in each of the data distribution types (Fig. 3). Algorithm R1 was parallelized into vectors (3) for a system with a horizontal distribution (Fig. 3a) and into attributes (4) for the system with a vertical distribution (Fig. 3b).

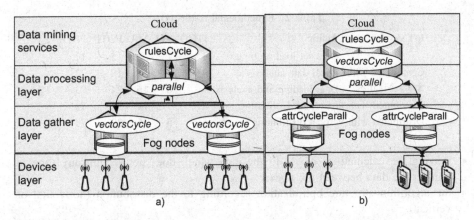

Fig. 3. Using the actors for execution of 1R algorithm in IoT system with fog nodes (a) horizontal data distribution, (b) vertical data distribution

The calculation functional blocks of the algorithm were run on the same cluster as Spark ML. Fog nodes were created as virtual machines on a high performance server with the following characteristics:

- Huawei FusionServer RH2288 V3 Rack Server with 2 Intel® Xeon® E5-2600 v3/v4 processors, the volume of random access memory - 128 GB.

We performed experiments for systems on 2, 4 and 8 fog nodes. Each of these nodes stored equal parts of the data set. The data sets from Azure ML were used for the experiments. The parameters of the data sets are presented in Table 1.

Table 1. Experimental datasets

Input data set	Number of rows	Number of attributes	Size of data (Kb)
Iris Two Class Data (ITCD)	100	4	2
Telescope data (TD)	19 020	10	1 499
Breast Cancer Info (BCI)	102 294	5	4 832
Movie Ratings (MR)	227 472	4	6 055
Flight on-time performance (Raw) (FOTP)	504 397	5	39 555
Flight Delays Data (FDD)	2 719 418	5	136 380

The experimental results are provided in Table 2. Data loading time and analysis time were measured separately for the centralized systems. Therefore the total analysis time in such systems is the sum of loading- and analysis time. The results of the experiments show that the total analysis time in such systems is higher than in distributed systems. The reasons for this are:

Table 2. Experimental results (s)

IoT system	Action	ITCD	TD	MR	FOTP	FDD
-	Data set loading time	1	1	2	5	11
Centralized with Spark MLib	Local data analysis	4	7	14	19	72
	Data loading and analysis	5	8	16	24	83
Distributed with horizontal distributing		4	7	14	20	74
Distributed with vertical distributing		5	5	15	22	77

- moving the calculations closer to the data which does not require any time to transfer the data between the nodes,
- the algorithm structure optimization according to the distribution (horizontal or vertical).

Additionally the network traffic between the end devices and the cloud has been decreased.

5 Conclusion

IoT can have a centralized or distributed architecture. Most of the existing data mining solutions for IoT can work only within a centralized architecture. However, distributed architecture (Fog computing) became more popular since it allows to decrease network traffic in a network with a large number of endpoint devices.

The paper describes an approach to building analytical services in IoT systems with distributed architecture that uses distributed data analysis which is based on an actor model. The decomposition of the algorithm into functional blocks and their mapping to actors allows to distribute the calculations between the nodes of an IoT system and observe the following advantages:

- moving computing blocks that process the data to the nodes that store information thus increasing data processing speed and decreasing network traffic,
- optimizing the structure of the algorithm depending on the data distribution (horizontal or vertical) and thus increasing data processing speed (parallelization).

The performed experiments demonstrated the efficiency of the suggested approach. The execution time for an algorithm, distributed between the nodes of an IoT system considering the data storage places and a distribution type, is less than for the IoT systems with a centralized architecture (the ones that use cloud analytical services and the ones based on scaled data analysis platforms use). In future we plan to propose automated methods for estimation and distribution the functional blocks and the actors in IoT systems with distributed architecture.

Acknowledgments. The work has been performed at the Saint Petersburg Electrotechnical University "LETI" within the scope of the contract Board of Education of Russia and science of the Russian Federation under the contract № 02.G25.31.0058 from 12.02.2013. The paper has been prepared within the scope of the state project "Organization of scientific research" of the main part of the state plan of the Board of Education of Russia, the project part of the state plan of the Board of Education of Russia (task 2.136.2014/K) as well as supported by grant of RFBR (projects 16-07-00625).

References

1. Evans, D.: The internet of things. how the next evolution of the internet is changing everything. Cisco White Paper, Cisco Systems (2011)
2. Gartner Says the Internet of Things Installed Base Will Grow to 26 Billion Units By 2020. Gartner (2014)
3. Tsai, C.-W., Lai, C.-F., Vasilakos, A.V.: Future internet of things: open issues and challenges. Wireless Netw. **20**(8), 2201–2217 (2014)
4. Chen, F., Deng, P., Wan, J., Zhang, D., Vasilakos, A.V., Rong, X.: Data mining for the internet of things: literature review and challenges. Int. J. Distrib. Sens. Netw. **2015**, 14 (2015). Article ID 431047. http://dx.doi.org/10.1155/2015/431047
5. Gubbi, J., Buyya, R., Marusic, S., Palaniswami, M.: Internet of things (IoT): a vision, architectural elements, and future directions. Future Gener. Comput. Syst. **29**, 1645–1660 (2013)
6. Gronlund, C.J.: Introduction to machine learning on Microsoft Azure, 18 April 2016. https://azure.microsoft.com/en-gb/documentation/articles/machine-learning-what-is-machine-learning/
7. Barr, J.: Amazon Machine Learning – Make Data-Driven Decisions at Scale. Amazon Machine Learning, 18 April 2016. https://aws.amazon.com/ru/blogs/aws/amazon-machine-learning-make-data-driven-decisions-at-scale/
8. Google Cloud Machine Learning at Scale, 18 April 2016. https://cloud.google.com/products/machine-learning/
9. Meng, X., Bradley, J., Yavuz, B., Sparks, E., Venkataraman, S., Liu, D., Freeman, J., Tsai, D., Amde, M., Owen, S., Xin, D.: Mllib: Machine learning in apache spark (2015). arXiv preprint arXiv:1505.06807
10. Ingersoll, G.: Introducing apache mahout. Scalable, commercialfriendly machine learning for building intelligent applications. IBM (2009)
11. Bonomi, F., Milito, R., Zhu, J., Addepalli, S.: Fog computing and its role in the internet of things. In: Processing of MCC, 17 August 2012, Helsinki, Finland, pp. 13–16 (2012)
12. Church, A., Barkley Rosser, J.: Some properties of conversion. Trans. AMS **39**, 472–482 (1936)
13. Kholod, I., Petukhov, I.: Creation of data mining algorithms as functional expression for parallel and distributed execution. In: Malyshkin, V. (ed.) PaCT 2015. LNCS, vol. 9251, pp. 62–67. Springer, Heidelberg (2015)
14. Kholod, I., Kupriyanov, M., Shorov, A.: Decomposition of data mining algorithms into unified functional blocks. Math. Probl. Eng. **2016**, 11 (2016). Article ID 8197349
15. Holte, R.C.: Very simple classification rules perform well on most commonly used datasets. Mach. Learn. **11**, 63–90 (1993)
16. Kholod, I., Kuprianov, I., Petukhov, A.: Parallel and distributed data mining in cloud. In: Perner, P. (ed.) ICDM 2016. LNCS, vol. 9728, pp. 349–362. Springer, Heidelberg (2016). doi:10.1007/978-3-319-41561-1
17. Hewitt, C., Bishop, P., Steiger, R.: A universal modular actor formalism for artificial intelligence. In: IJCAI, pp. 235–245 (1973)

18. Kholod, I., Petuhov, I., Kapustin, N.: Creation of data mining cloud service on the actor model. In: Balandin, S., Andreev, S., Koucheryavy, Y. (eds.) NEW2AN/ruSMART 2015. LNCS, vol. 9247, pp. 585–598. Springer, Heidelberg (2015)
19. Kholod, I.: Framework for multi threads execution of data mining algorithms. In: Proceeding of 2015 IEEE North West Russia Section Young Researchers in Electrical and Electronic Engineering Conference, (2015 ElConRusW), pp. 74–80. IEEE Xplore (2015)

ruSMART: Smart Services Serving Telecommunication Networks

Neural Network System for Monitoring State of a Optical Telecommunication System

Anton Saveliev[✉], Sergey Saitov, Irina Vatamaniuk, Oleg Basov, and Nikolay Shilov

SPIIRAS, 39, 14th Line, St. Petersburg 199178, Russia
saveliev@iias.spb.su

Abstract. The paper presents a methodology for the synthesis of systems for monitoring the state of a fiber-optical linear path of the optical transport networks, based on the information and measuring control system that implements neural network recognition algorithms with synthesis by dominance. The proposed information-measuring system processes levels of the average intensity of the optical signal received at various carriers in a certain retrospective over a defined period of time. It is seen that the reliability of control increases with the growth of size of the watch window.

Keywords: Optical Transport Networks · Main optical path · Neurocomputers · Neural network · Information and Measuring Control System

1 Introduction

Achievements of neuromathematics and realization of a new generation of optical neurocomputers are actively stimulating new directions for application of neural network technologies [1]. So, considering unique properties and possibilities of systems for optical parallel information processing, it can be noted that relevance of researches in science areas, aimed at improving monitoring tools for optical telecommunication systems (OTS), significantly increases.

In the last decades, the optical transport networks (OTN), serving an average of 90 % of volume of long-distance and international traffic, have become a basis of national OTS. It is possible to mark out the following features of the current state of the art in a subject domain [1]:

- productivity of the optical fiber baseband transmission paths (OFBTP) with wavelength-division multiplexing (WDM) has increased many times and reached tens of terabits per second;
- transmission distance of nonregenerative signaling exceeded 1000 km;
- synchronous transmission technologies are succeeded by the whole group of asynchronous technologies: from the known operator options *Ethernet* to perspective *OTN (Optical Transport Network)* and *GMPLS* [2]. It has become possible to transmit signals of various formats in spectral channels of one OFBTP with WDM (OFBTP have become heterogeneous).

© Springer International Publishing AG 2016
O. Galinina et al. (Eds.): NEW2AN/ruSMART 2016, LNCS 9870, pp. 39–49, 2016.
DOI: 10.1007/978-3-319-46301-8_4

These factors significantly complicate monitoring process of the baseband transmission path (BTP) state of the real time. Standard control facilities of OFBTP of von Neumann type [3] have no sufficient productivity for processing the incoming massifs of optical signal intensity measurements. Meanwhile there are results of researches that show a possibility for application of neural network approaches to formation of a system for monitoring the state of heterogeneous OFBTP with WDM [4, 5].

2 The Essence of the Approach

Let heterogeneous OFBTP with WDM be an object of control (Fig. 1), in which there are R carriers with wavelengths $\lambda_1, \ldots, \lambda_R$. Transferring optical module (TOM) switches the sources of optical radiation (SOR) and the optical multiplexer (OM), from the output of which a group optical signal is entered into the optical fiber (OF) with power $P_{inp\Sigma}$. The optical fiber amplifier (OFA) not only increases signal power, transforming pump current I_{Ampm} into amplification with a coefficient G_{Ampm}, but also introduces noise (n). From the optical fiber output, the optical power of a signal $P_{sphoto\Sigma}$ and noise $P_{nphoto\Sigma}$ arrives at the fiber optical receiving module (FORM), where after frequency selection in the optical demultiplexer (OD), signals at the corresponding wavelengths arrive at individual photodetectors (PD). In decision devices (DD), photocurrent force during a digit-time slot is estimated, and the conclusion is drawn about which symbol was accepted "0" or "1".

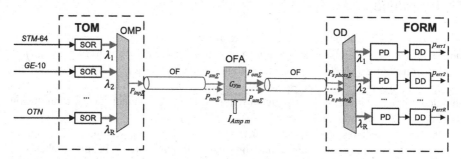

Fig. 1. The scheme of a heterogeneous OFBTP with WDM serving traffic of Synchronous Digital Hierarchy (*STM-64*), *G-Ethernet* and *OTN*

For realization of continuous control of the state of OFBTP, in subject domain it is often proposed to organize a measurement channel (Fig. 2) via which the part of group signal energy will be fed through a coupler to the Information and Measuring Control System (IMCS).

Research has shown that it is expedient to use a specialized optical neural network as a compute kernel of IMCS [3].

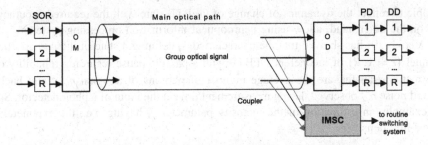

Fig. 2. The organization of a measurement channel in OFBTP

3 Estimability Analysis of Time History of the State of Optical Components and OFBTP

Let τ_{MP} be a maintenance period (MP) of OFBTP during which all meaning characteristics of all components of the main optical path (MOP) can be measured, and τ_{split} be some time period, during which there is a transition of a certain spectral channel (SC) from operating state Ω_1 to a state of parametric failure Ω_2. Then it is possible to allocate the following two groups of factors of operating conditions of MOP [3, 6].

I. Irreversible factors, for which $\tau_{split} > \tau_{MP}$ is true, connected with phenomena of wear of the OFBTP elements.

II. Factors (reversible and irreversible), causing changes of intensity parameters of an optical signal at the input of a photodetector with $\tau_{split} < \tau_{MP}$.

Processes of *the first group* are the cause of insignificant attenuation increase during a time period τ_{MP}, brought by the spectral channel elements and smooth increase of their background noise (bn) levels. The existing techniques [6] allow one to successfully predict optical components failure because of their ageing on the future time interval τ_{MP}, which gives an opportunity to replace or restore them in due time.

The second group is characterized by the following factors: fluctuations of temperature; bends of the optical fiber (reversible and irreversible), which resulted from errors of the service personnel or attempts of unauthorized access to optical fiber; thermofluctuation and corrosion growth of microcracks; radiation thickening of optical fiber, caused by influence of factors of the nuclear weapon or strokes of atmospheric electricity. These events lead to change of attenuation of MOP and the corresponding change of intensity parameters values of an optical signal, simultaneous and identical to all spectral channels, at the input of a photodetector [3].

The processes of the second group also include: redistribution of power between spectral channels, caused by nonlinear effects of optical fiber [7]; trend of the gain coefficient of OFA [8]. These situations are followed by change of transfer factor of the spectral channel, simultaneous but different as to amplitude. The period of the observed fluctuations of signal strength significantly exceeds duration of a single impulse τ_{imp} of an optical signal, i.e. $\tau_{split} \gg \tau_{imp}$. It is obvious that alterations of a state of OFBTP will generally be determined by the factors of *the second group*. Therefore, it is necessary to choose such intensity parameters of an optical signal, which, on the one hand, are capable to reflect the dynamics of change of OFBTP state with the required accuracy, and on the other hand, allow using neurooptical information processing systems.

A study [9] has showed that the characteristics, defining a state of the r-th spectral channel $(r = \overline{1, R})$ of incoherent OFBTP with a passive pause regarding reliability of information transfer, are the average number of photons of a signal n_{avgr} and background noise n_{nr} observed during measurement time at the input of a photodetector. So, for example, an estimation of the intensity parameter γ, having, i.e., the exponential probability density

$$f(\gamma) = \frac{1}{2\sigma_\gamma^2} \exp\left(-\frac{\gamma}{2\sigma_\gamma^2}\right),$$ (1)

where σ_γ is the distribution parameter, a value characterizing aprioristic intensity uncertainty, will be given by

$$\hat{\gamma} = \left(\frac{1}{n_{sig} + 1/2\sigma_\gamma^2}\right) \cdot \sum_{k=1}^{K} n_k - \frac{n_{bn}}{n_{sig}},$$ (2)

where $n_{sig} = J_{sigR} \cdot \tau_{obs}$ is a number of photons of a signal with uptake intensity J_{sigR} observed in a time period τ_{obs}, and $n_{bn} = J_{bnR} \cdot \tau_{obs}$ is a number of photons of background noise, respectively; $k = 2, \ldots, K$.

Such an assessment of the $\hat{\gamma}$ parameter γ has an average value:

$$M[\hat{\gamma}] = \left\{ M\left[\frac{\sum_{k=1}^{K} n_k}{n_{sig} + 1/2\sigma_\gamma^2}\right] - \frac{n_{bn}}{n_c} \right\} = \frac{M[\gamma] \cdot n_{sig} + n_{bn}}{n_{sig} + 1/2\sigma_\gamma^2} - \frac{n_{bn}}{n_{sig}},$$ (3)

and dispersion

$$D[\hat{\gamma}] = \frac{M[\gamma] \cdot n_{sig} + n_{bn}}{(n_{sig} + 1/2\sigma_\gamma^2)^2} = \frac{2\sigma_\gamma^2 n_{sig} + n_{bn}}{(n_{sig} + 1/2\sigma_\gamma^2)^2}.$$ (4)

As a time of supervision τ_{obs} grows, it appears that $D[\hat{\gamma}] \approx 2\sigma_\gamma^2/n_{sig}$, i.e., errors of intensity parameter measurement are generally caused by quantum noise of a signal, and dispersion of such an estimate is directly proportional to the value of aprioristic intensity uncertainty, and with growth of n_{sig} tends to the Cramér-Rao bound. Therefore, the accuracy of estimation of these parameters will be due to a number of received photons $n_{sigr}(t)$ during bit τ_{imp} and observation time τ_{obs}.

Numerical and full-scale experiments showed that for the existing OFBTP, a measurement of sample average number of photons with the given accuracy and reliable estimation $\psi = 0.99$ will require observation of several thousand impulses, and observation time τ_{obs} will be units of milliseconds.

4   Formalization of States of OFBTP and Preparation of Data Array for Processing in Neural Network Information and Measuring System

Further, it is proposed to identify the following sets of states of OFBTP that reflect the suitability of their R spectral channels [3].

Fault-free (FF) set, when all spectral channels provide transfer of a flow with the required speed B_r^{RQRD} and necessary reliability $p_{err\,r}$ of transmission:

$$Q_{FF}^{LP} = \{R, B_r^{RQRD} | p_{err\,r} \leq p_{err\,r}^{perm}\};\tag{5}$$

where $p_{err\,r}^{perm}$ is a permissible value of a bit mistake in the channel.

Operational (Op) (intermediate), when most spectral channels R_{FF} are fault-free, and in other R_{Op} spectral channels there can be a certain (limited) decrease in reliability of transfer (not worse than $p_{err\,r}^{al}$) regarding the maximum transmission speed:

$$Q_{Op}^{LP} = \{\exists r \in R_{Op}, B_r^{RQRD} | p_{err\,r}^{al} \geq p_{err\,r} > p_{err\,r}^{perm}\},\tag{6}$$

where $R_{Op} \cup R_{FF} = R$, $\forall r \in R_{FF}(B_r^{RQRD} | p_{err\,r} \leq p_{err\,r}^{perm})$; $R_{FF} \geq R_{thld}$, $R_{FF} + R_{Op} \geq R_{crit}$.

Alarmed (Al) (inoperable), when a number of fault-free spectral channels appears to be less than a threshold value R_{thld}, or the sum $R_{FF} + R_{Op}$ of fault-free and operational spectral channels becomes less than a value R_{crit}:

$$Q_{Al}^{LP} = \{\exists r \in R_{Op}, B_r^{RQRD} | R_{FF} < R_{thld} \text{ or } R_{FF} + R_{Op} < R_{crit}\}.\tag{7}$$

Requirements for values B_r^{RQRD}, $p_{err\,r}^{perm}$, $p_{err\,r}^{al}$, R_{FF}, R_{Op}, R_{thld}, R_{crit} are considered to be set by a metasystem.

For unambiguous reference of a state of OFBTP to the sets Q_{FF}^{LP}, Q_{Op}^{LP} or Q_{Al}^{LP}, during functioning of spectral channels it is necessary to have some combinations of impulses with the determined characteristics in a group signal at all bearing wavelengths [10]. Such impulses combinations (IC), having length N_{IC} of symbols can periodically be entered into TOM by methods of temporary division of channels (TDM) and branch off in IMCS (Fig. 2) together with an information optical signal.

In the developed system, by processing N_{IC} of symbols at each of the carriers a sample average value of a number of photons per bit μ_{rk} in the *r-th* spectral channel in the *k-th* moment of time will be received. It follows that IMCS for a state of OFBTP has to process the values of average intensity of an optical signal received at various carriers in a certain retrospective in some period (Fig. 3).

Suppose that in the presence of several classes of states of spectral channels of OFBTP $Y = (\Omega_z)$, $w = \overline{1, W}$ K measurements are carried out on N_{IC} impulses, and estimates $\{\mu_{rk}, r = \overline{1, R}; k = \overline{1, K}\}$ are obtained for each spectral channel. It is required to determine an estimate $y_{rK+1} = f(\mu_{rK+1})$, $y_r = 1, \ldots, W$ of processes of parameters alteration $\{\mu_{rk}\}$ according to the criterion $c_r(y_r, \dot{y}_r)$, set by a penalty function:

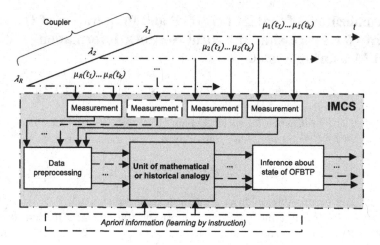

Fig. 3. A generalized functional structure of IMCS

$$c_r(y_r, \vartheta) = 1 - \delta(y_r; \dot{y}_r), \dot{y}_r = 1, \ldots, W, \tag{8}$$

where δ is the Kronecker symbol; $\dot{y}_r$ is a true value of a sought estimate.

Each image y_{rk} ($k = 1, \ldots, K, K + 1$) can be put in correspondence with its distribution $\varpi(\mu_r | \dot{y}_r)$ in space of the observed R-dimensional random process $(\mu_{r1}, \ldots, \mu_{rK+1})$. Knowing some a priori distribution $p(\dot{y}_r)$, it is possible to solve the problem by the usual Bayesian methods. However, during analysis of MOP the situation is complicated by the fact that the variables μ_{rk} are not independent, and the distributions $\varpi(\mu_r | 1), \ldots \varpi(\mu_r | W))$ and $p(\dot{y}_r)$ are a priori unknown. Their a priori knowledge can be replaced by the process of learning by instruction, i.e. the process of supervisory communication of additional information to IMCS. For this purpose, during measurements and tests of MOP a set of training examples for each IC is formed.

Everything mentioned above can be ensured with the help of means of neurooptic that allow for realization of parallelized analysis of parameters of the whole MOP on the time interval $\tau_P = [t_1, t_k]$, i.e., when an extrapolation argument is presented by a volume optical image – a parameter matrix of the form

$$X'_{\text{FF}} = \begin{pmatrix} \mu_{11} & \mu_{12} & \cdots & \mu_{1k} \\ \mu_{21} & \mu_{22} & \cdots & \mu_{2k} \\ \cdots & \cdots & \cdots & \cdots \\ \mu_{R1} & \mu_{R2} & \cdots & \mu_{Rk} \end{pmatrix}, \tag{9}$$

where μ_{rk} is the mean number of photons in the r-th spectral channel by N_{IC} in the k-th moment of time. It is necessary to create a type of a matrix-prediction according to the given argument by methods of mathematical and/or historical analogy

$$X''_{\text{Pred}} = \begin{pmatrix} \mu_{1(k+1)} & \mu_{1(k+2)} & \cdots & \mu_{1q} \\ \mu_{2(k+1)} & \mu_{2(k+2)} & \cdots & \mu_{2q} \\ \cdots & \cdots & \cdots & \cdots \\ \mu_{R(k+1)} & \mu_{R(k+2)} & \cdots & \mu_{Rq} \end{pmatrix}. \tag{10}$$

Analysis of this matrix allows us to make a conclusion about the projected state of OFBTP for taking control actions.

5 The Technique for Synthesis of Neural Network Information-Measuring Control System for OFBTP State

Let the vector X_e in the selected feature space be represented by the magnitudes and directions of changes of the matrix elements X'_p (9). A vector of the governing parameters X_{gove} of MOP combines characteristics of SOR and OFA in the e-th situation, $e = \overline{1, E}$, known to the control system by values of the corresponding pumping currents of SOR and OFA: $I_{Se}(t_k)$ and $I_{\text{Ampe}}(t_k)$ respectively.

Thus a case of the functioning of MOP can be written in the form

$$C_e = (X_e(t_k); Y_e(t_k + \tau_{\text{Cs}})) = (X_{\text{Ope}}(t_k), I_{\text{FFe}}(t_k), I_{\text{Ampe}}(t_k); Y_e(t_k + \tau_{\text{Cs}})),$$

where $X_e = (X_{\text{Pe}}(t_k), I_{Se}(t_k), I_{\text{Ampe}}(t_k))$ is the vector of causes (Cs) of the e-th situation; $Y_e(t_k + \tau_{\text{Cs}})$ is the vector taken as a consequence of the e-th situation; τ_{Cs} is a time interval between the cause and the consequence.

Obtained in the course of traffic control of MOP, a training set cannot contain the full volume of information characterizing all the possible consequences $Y(t_k + \tau_{\text{Cs}})$ of all the possible set of causes $X(t_k)$. So extrapolating functional, modeling on its basis an operator F of cause-effect relation $Y(t_k + \tau_{\text{Cs}}) = F(X(t_k))$, will not be predetermined, i.e., $Y(t_k + \tau_{\text{Cs}}) = F_M^{(1)}(X(t_k))$, where $F_M^{(1)}$ is a model of operator F of the first approximation.

Using the assumption that all tests are conducted at one point in time t_0, from the study of governing parameters in the time space we can pass to their analysis in the feature space. For this purpose, the test report can be written in the form:

$$R_{test} = \{S_{ij}\}, \tag{11}$$

where each situation C_{ij} corresponds to the expression

$$C_{ij} = (X_{\text{test } ij}; Y_{ij}) = (X_{\text{Opij}}, I_{\text{FFi}}, I_{\text{Ampi}}; Y_{ij}), \tag{12}$$

where Y_{ij} is the reaction of the state of the spectral channel of a regeneration section in the i-th situation to the j-th effect of an algorithm for MOP testing; $X_{\text{test } ij} = (X_{\text{Opij}}, I_{\text{FFi}}, I_{\text{Ampi}})$ is a set of parameters describing a situation in which this reaction to the j-th impact is carried out; $j = 1, \ldots, G$. Therefore, in interests of control system training, the information array can be obtained, where each i-th case of exploitation contains G_i observation situations by which the functionality $F_M^{(2)}$ of the operator $Y(t_i + \tau_{Cs}) = F_M^{(2)}(X'(t_k))$ of the second approximation is synthesized.

Then the task of developing the control system is reduced to a task of synthesis of a set of neural network algorithms for formation F_M by optimizing the neural network parameters allowing one to draw an inference

$$Y_M : Y = F(X) \rightarrow Y_M = F_M(X), \tag{13}$$

$$|Y - Y_M| \le \varepsilon^*, \tag{14}$$

where ε^* is the specified value, reflecting the requirements to the adequacy of a generated model. In vector view, this is consistent with the form

$$Y_{Mi} = F_M(X_{Opi}; I_{FFi}, I_{Ampi}). \tag{15}$$

The training set for formation of model structure must be presented by a set of cases of the form

$$R_a = \{C_a\} = \{(X_{Opa}, I_{FFa}, I_{Ampa}; Y_a)\}, \tag{16}$$

where $(X_{Opa}, I_{FFa}, I_{Ampa}) = X_a$ is the cause of the consequence Y_a.

The process of obtaining F_M with the known neural network structure G by available experimental data means the adjustment of transmission coefficients of interneuronal communication with the aim of minimizing the model error. However, direct measurement of this error in practice is not achievable, so we use the estimate

$$\xi_1 = \sum_{X \in X_a} |F_M(X) - Y_a|, \tag{17}$$

called a learning error, where the summation over X is carried out by the final set of parameters X_a, called a learning set, for which Y_a are known.

The unknown error ξ_2 made by the model F_M on data not previously used in training is called an error of model integration. Since the true value of the integration error ξ_2 is inaccessible, in practice its estimation is used, which is obtained from the analysis of a part of the examples X_b for which system responses Y_b are known, but were not used in the training $X_b \in C_b = \{X_b, Y_b\}$, $X_b \cap X_a = \varnothing$.

$$\xi_2 = \sum_{X \in X_b} |F_M(X) - Y_b|. \tag{18}$$

A sample $C_b = \{X_b, Y_b\}$ is hereinafter referred to as a test sample (verification).

The actual *method of the synthesis of control system* for MOP state is considered as a sequence of the following stages.

1. The synthesis of an artificial neural network that reproduces the logic of MOP functioning, i.e., the structure definition of G_l and parameters $\Phi = \{\phi_{lh}\}$ of a neural network, modeling an operator

$$F : Y_a = F(X_a), F \rightarrow F_M. \tag{19}$$

2. The choice of a learning algorithm for neural networks allowing realization of information-measuring elements transformation of a set $\{X_a\}$ to the elements of a set $\{Y_a\}$ in accordance with the criterion of suitability:

$$\Delta = \left| Y_a - F_M(X_{Pai}, X_{Sai}, X_{Ampai}) \right| \leq \Delta^*, \tag{20}$$

where Δ^* is a permissible value of error Δ of representation of a precedent C_a, i.e., recovery of the consequence Y_{Ma} of the cause X_a, where $Y_{Ma} = F_M(X_a)$.

3. Based on processing of vectors $\{X_b\}$, the search for such a set of vectors $\{Y_{Mb}\}$, which would reflect the predicted states of $\{Y_b\}$ spectral channels. The reliability criterion of prediction is the expression

$$D_{Pred} = \left| Y_b - F_M(X_{Pbi}, X_{Sbi}, X_{Ampbi}) \right| \geq D_{Pred}^*. \tag{21}$$

A neural network, synthesized in such a way after training, will ensure with the required accuracy displaying of the observed vector X_c in the inference about the state of MOP Y_{Mc}. If in the report on the current observations R_c there is a vector X_c similar to the well-known one of the neural network, $X_c = X_a = (X_{OPai}, X_{FFai}, X_{Ampai})$, then the functions of IMCS are limited to associative search (by historical analogy) of a given precedent and the restoration of its consequence as a sought-for prediction, i.e., $Y_c = Y_{Ma}$.

If X_c does not coincide with any of the causes $X_a \in R_a$, then 'intelligent' algorithms are implemented for estimation of Y_c and X_c by the report R_a. This is what distinguishes the proposed approach to recognition using a neural network from any other method of pattern recognition consisting in assigning X_c to the closest X_a based on a proximity measure (e.g., in the Hamming space).

We studied dependence of indices of control reliability on the retrospection depth for $\tau_{obs} = 1$ s and $\tau_1 = 10$ s. Based on the available statistical data, when increasing the number of "sensors" in the input layer of the IMCS and each time conducting the synthesis of the forecast model to achieve $\xi_2 = 10$ %, the expected values of D_{Pred} were obtained and summarized in Table 1.

Table 1. Dependence of the reliability of control on the size of an observation window

τ_r, sec	$MO(D_{Pred})$	$minD_{Pred}$	$maxD_{Pred}$	$\sigma(D_{Pred})$
8	47.99	42.86	54.31	3.07
10	54.90	49.61	60.27	3.56
12	62.43	56.23	63.22	1.34
14	60.73	59.52	61.67	0.77
16	65.12	63.03	65.90	0.84
18	65.88	63.88	66.04	1.01
20	65.56	63.36	66.59	0.83

It is seen that the reliability of control increases with the growth of size of the watch window. Therefore, to ensure the adequacy of IMCS functioning under these conditions, the period of retrospection $\tau_r = [t_1, t_k]$ of the monitoring data should be increased to 12–15 s.

During the research, we have developed the algorithm [3] to identify the state of the MOP on the long-living intervals with the *synthesis by dominance*. We have also received a patent [11] for invention of neurooptical controller that implements the above-mentioned approach on the components of integrated and nonlinear optics.

6   Conclusion

The presented approach to creation of the control system for a state of OFBTP based on application of a control combination entered into a group optical signal, as well as the information and measuring system realizing neural network algorithms of recognition with synthesis by dominance, allows one to monitor a state of OFBTP in real time. It can be implemented not only in high-speed OFBTP of OTN, but also in OTS used at organization of multipoint videoconferencing. At present, based on computer simulation, the following is being carried out: (1) testing of neural networks of different structures to ensure maximum reliability of OFBTP state identification and minimize training time; (2) search for ways to optimize the characteristics of the control system such as the depth of retrospection (τ_r), frequency rate of current measurements (τ_m) and a range (τ_l) of lead, which depend on the characteristics (a number of spectral channels and speed of information transmission in them) of the specific OFBTP. The developed approach for optical path monitoring is oriented to implementation in smart environments and cyberphysical systems to support safe and trust connection between distributed embedded modules, robots, cloud services, user devices and users [12–21].

Acknowledgment. This work is partially supported by the Russian Foundation for Basic Research (grants № 15-07-06774, № 16-08-00696, 16-29-04101) and the Council for Grants of the President of Russia (Projects No. MK-7925.2016.9).

References

1. Hudgings, J., Nee, J.: WDM all-optical networks. In: EE228 Project Report – Oslo, 29 p. (1996)
2. Rosen, E., Viswanathan, A., Callon, R.: Multiprotocol label switching Architecture. In: RFC 3031, 144 p., January 2001
3. Saitov, I.A., Muzalevskii, D.Y.: Continuous monitoring of a fiber-optical baseband transmission path based on intellectual optical-signal processing facilities. Telecommun. Radio Eng. **70**(16), 1501–1508 (2011). (In Russ.)
4. Khomonenko, A.D., Yakovlev, E.L.: Neural network approximation of characteristics of multi-channel non-Markovian queuing systems. SPIIRAS Proc. **41**(4), 81–93 (2015)
5. Nesteruk, P.G., Kotenko, I.V., Shorov, A.V.: Analysis of bio-inspired approaches for protection of computer systems and networks. SPIIRAS Proc. **18**(3), 19–73 (2011)

6. Toge, K., Ito, F.: Recent research and development of optical fiber monitoring in communication systems. Photonic Sens. **3**(4), 304–313 (2013)
7. Saitov, I.A., Myasin, N.I.: A model of a fiber-optical baseband transmission path with wavelength-division multiplexing and fiber-optical amplifiers. Telecommun. Radio Eng. **70** (19), 1729–1738 (2011). (In Russ.)
8. Delavaux, J.-M.P., Nagel, J.A.: Multi-stage erbium-doped fiber amplifier design. Lightwave Technol. **13**(5), 703–720 (1995)
9. Robert, M., Gagliardi, Sh.: Optical communication: transl. from Eng. / under the editorship of A.G. Sheremetyev, – M. Svyaz', 424 p. (1978). (In Russ.)
10. Maamoun, K., Mouftah, H.: Survivability Issues in Optical and Optical Wireless Access Networks: Monitoring Trail Deployment for Fault Localization in All-Optical Networks and Radio-Over-Fiber Passive Optical Networks. LAP Lambert Academic Publishing, Saarbrücken (2012). 172 p.
11. Saitov, I.A., Myasin, N.I., Muzalevskii, D.Y.: Device for continuous monitoring operating capacity of fibre-optic linear channel. Patent of RF for an invention № 2400015 from 20.09.2010. Application № 2009102711/28(003451) from 27.01.2009
12. Yusupov, R.M., Ronzhin, A.L.: From smart devices to smart space. Herald Russ, Acad. Sci. **80**(1), 45–51 (2010). MAIK Nauka
13. Budkov, V., Prischepa, M., Ronzhin, A.: Dialog model development of a mobile information and reference robot. Pattern Recogn. Image Anal. **21**(3), 458–461 (2011). Pleiades Publishing
14. Saveliev, A., Basov, O., Ronzhin, A., Ronzhin, A.: Algorithms for low bit-rate coding with adaptation to statistical characteristics of speech signal. In: Ronzhin, A., Potapova, R., Fakotakis, N. (eds.) SPECOM 2015. LNCS, vol. 9319, pp. 65–72. Springer, Heidelberg (2015)
15. Karpov, A.A., Ronzhin, A.L.: Information enquiry Kiosk with multimodal user interface. Pattern Recogn. Image Anal. **19**(3), 546–558 (2009). Moscow: MAIK Nauka/Interperiodica
16. Basov, O., Ronzhin, A., Budkov, V., Saitov, I.: Method of defining multimodal information falsity for smart telecommunication systems. In: Balandin, S., Andreev, S., Koucheryavy, Y. (eds.) NEW2AN/ruSMART 2015. LNCS, vol. 9247, pp. 163–173. Springer, Heidelberg (2015)
17. Ronzhin, A.L., Budkov, V.Y.: Multimodal interaction with intelligent meeting room facilities from inside and outside. In: Balandin, S., Moltchanov, D., Koucheryavy, Y. (eds.) ruSMART 2009. LNCS, vol. 5764, pp. 77–88. Springer, Heidelberg (2009)
18. Karpov, A., Ronzhin, A., Kipyatkova, I.: An assistive bi-modal user interface integrating multi-channel speech recognition and computer vision. In: Jacko, J.A. (ed.) Human-Computer Interaction, Part II, HCII 2011. LNCS, vol. 6762, pp. 454–463. Springer, Heidelberg (2011)
19. Ronzhin, A.L., Budkov, V.Y., Karpov, A.A.: Multichannel system of audio-visual support of remote mobile participant at e-meeting. In: Balandin, S., Dunaytsev, R., Koucheryavy, Y. (eds.) ruSMART 2010. LNCS, vol. 6294, pp. 62–71. Springer, Heidelberg (2010)
20. Budkov, V.Y., Ronzhin, A.L., Glazkov, S.V., Ronzhin, A.: Event-driven content management system for smart meeting room. In: Balandin, S., Koucheryavy, Y., Hu, H. (eds.) NEW2AN 2011 and ruSMART 2011. LNCS, vol. 6869, pp. 550–560. Springer, Heidelberg (2011)
21. Ronzhin, A., Prischepa, M., Budkov, V.: Development of means for support of comfortable conditions for human-robot interaction in domestic environments. In: Botía, J.A., et al. (eds.) Workshop Proceedings of the 8th International Conference on Intelligent Environments, pp. 221–230. IOS Press (2012)

Optimization Algorithm for an Information Graph for an Amount of Communications

Yulia Shichkina$^{(\boxtimes)}$, Mikhail Kupriyanov, and Mohammed Al-Mardi

Department of Computer Science and Engineering,
Saint Petersburg Electrotechnical University "LETI", St. Petersburg, Russia
{strange.y, almardi-md}@mail.ru,
mikhail.kupriyanov@gmail.com

Abstract. In connection with the annual increase in the volume of processed data and raising the importance of computer modeling of real objects and processes, requirement to improve the technology of parallel algorithms is increasing. Successful implementation of parallel algorithms on supercomputers depends on several parameters, one of which is the amount of inter-processor data transfers. Starting at a particular number of processors, computational speedup falls due to increased volume of data transmission. For some algorithms this dependence is a linear decreasing function. Imbalance of volume of calculations and complexity of data transmission operations increases with the rising of the number of processors. In this article we present the results of investigations of dependence of the density and algorithm execution time on the amount of interprocessor transfers. Also, we present a method of reducing interprocessor communications through more efficient distribution of operations of the algorithm by processes. This method does not account for the execution time of the operations themselves, but it is a foundation for more improved methods of multiparametric optimization of parallel algorithms.

Keywords: Algorithm · Parallel execution · Sequence list · Execution time · Operation · Process · Processor · Information dependence · Equivalent conversions · Information graph

1 Introduction

The use of distributed memory processors has become of current concern due to the wide spreading of high-performance cluster computing systems. The computing in a distributed memory is different from the computing in a shared memory, because in a distributed memory a message passing interface is used. Distributed memory systems are more architecturally complicated devices than shared memory systems.

For such systems it is necessary to have knowledge of a parallel computer general architecture and the essential for the programming topology of interprocessor communications before the creation of a parallel program. This is because of the absence of an automatic parallelization that affords to turn any sequential program into a parallel one and maintains its high performance [26]. The structure of an algorithm of the current task has to be connected explicitly with structure of a computing system and the communication among many parallel, independent processes has to be valid [1].

© Springer International Publishing AG 2016
O. Galinina et al. (Eds.): NEW2AN/ruSMART 2016, LNCS 9870, pp. 50–62, 2016.
DOI: 10.1007/978-3-319-46301-8_5

In recent times a lot of research were devoted to parallel computing. As a whole, these research may be conventionally divide into the range of main categories: research of parallel algorithms, their structure and quality [4, 6, 10]; the development of the general theory of parallel programming [3, 5, 16]; handling of applied partial problems [27]; research devoted to the development of parallel algorithms for restrictive class tasks of some area, with parallel algorithms of computing mathematics examples [11]; research those include formal models letting to describe the functioning of sequential processes executed simultaneously [2, 9, 20]; resolving problems of sequencing [15, 18, 23] etc.

As is known, the computing speed may be increased in two ways. The first way is to choose the high-speed modification of computer architecture. But this way is of the little scope because of its physical features. The second way is program. The program builder using this way has to choose the architecture model that allows the parallel algorithm realization and to settle the important issue of the creation of a parallel program.

Nowadays parallel programs builders divides into two clauses: those who thinks that a parallel program has to be built from the ground up without using of sequential analogues, and those who depends upon accumulated for decades bundle of sequential programs.

Both approaches have their advantages and disadvantages. But both approaches are united in one issue: the necessity of the analysis of the algorithm structure for the effective use of computing resources and the looking for opportunities for the speed-up of computing processes. This analysis may be conducted whether previously in the case of the creation of a parallel algorithm on the base of sequential one, or in an inter-mediate way for the obtaining of the information on the success of parallel execution, or finally for the comparison of algorithms and their implementation against each other.

In the past few decades the modeling of sequential and parallel algorithms are the object of intense interest.

As a whole, worked out nowadays methods of the building of parallel algorithms on multiprocessing systems [7, 8, 12, 14, 17, 24] do not let to built rather effective and high-performance programs, because their feature is the adaptation for concrete tasks with a concrete architecture of a computing system.

Among methods those let to get the proper idea of possible parallel branches of algorithms the following may be mentioned: the method of the search of mutually independent activities [13], the method of the definition of early and old terms of the execution of operations of an algorithm [13], methods of timetabling based on a movement list [22].

The last method is the least laborious. In according with this method, the existing algorithm has to be subdivided into operation, the information graph has to be built [25], features (height and width of algorithm (the time and the quantity of processors, used in calculations) have to be defined. As every method, the last one [22] has its disadvantages (the necessity to subdivide algorithm into separate operations and to build the graph of information dependences between operations) and its advantages: the researcher can observe the amount of locales those are necessary for the paralleling, the effectiveness of the use of locales and ultimately the possibility of the parallel real-ization of the suspected method, used in the shape of a concrete algorithm.

One of quality parameters of a parallel program is the loading density of compute nodes. Time delays at data channeling from one processor to another one leads to summarily length processors downtimes and increasing of the whole algorithm execution.

In the present article the method of the development of the effective algorithm of used processors quantity, algorithm execution time and the scope of interprocessor transitions is devised. This method can be used not only for sequential algorithms for the obtaining of their parallel analogous, but for parallel algorithms for the improvement of their quality.

2 Problem Description

Any sequential or parallel algorithm is a complicated multicoupling system with a set of parameters having an impact on the operating quality of this system. Multiparameter optimization of the performance of the algorithm is a rather complicated task, but the task that may be completed gradually.

In this article results of the first stage of the completion of the task of the obtaining of the timetable of the execution of the algorithm of an indicated information graph are presented. The algorithm has to be optimal for the interprocessor transfer size at following restrictions of optimized algorithm and computing system:

- The quantity of processors (bases) of the computing system is unrestrained;
- Input data of each operation are equal;
- All operations have the same time of execution conventionally equal to 1 item;
- Time of the data transfer between any two processors is constant and conventionally equal to 1 item.

Obviously that there are no algorithms and computing systems of such characteristics in practice, but this model of parallel algorithm is a start model for the obtaining of the method of parallel algorithms execution time optimization with account of the metadata package of the algorithm itself and the computing system.

Consider the following example. Suppose this information graph of the algorithm is defined (Fig. 1).

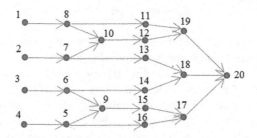

Fig. 1. Information graph of the algorithm

After the distribution of graphs points over tiers with the application of the method of the optimization of the graph for the width on the base of the connectivity matrix or adjacency lists [22] we shall following according to tiers initial groups of points of the graph (Fig. 2):

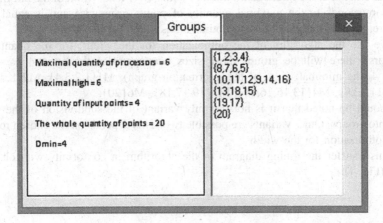

Fig. 2. Initial timetable of the algorithm

If we build the timetable of the algorithm by obtained groups on the computing system with account of interprocessor data transfer, it will be of the following shape (Fig. 3):

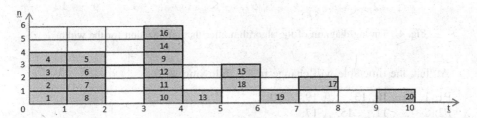

Fig. 3. Timing diagram of the algorithm with account of the duration of the interprocessor data transfer

P1: 1, 8, _, 10,13, _, 19, _, _,720;
P2: 2, 7, _, 11, _, 18, _, 17;
P3: 3, 6, _, 12, _, 15;
P4: 4, 5,_, 9;
P5: _, _, _, 14;
P6: _, _, _, 16;

where Pi is the number of the processor, i = 1,…,6; the symbol of the underlining '_' is the down time of the processor (the "bubble") waiting for obtaining of new input data from other operations.

In accordance with this timetable, the whole duration of the activity of the algorithm t = 10, the quantity of processors n = 6, total amount of down time p = 16.

For the definition of the possibility to fit the compute density of processors, we shall define hypothetically minimal width of the informational graph: Dmin = 4.

In our case, after the initial distribution of points over tiers, width of the information graph corresponds to the maximal quantity of points groupwise and is equal to 6. Therefore, there is the possibility to optimize this graph for the width.

If we use the algorithm of the optimization for the width (for the quantity of processors), there will be groups those sizes correspond to the optimal parameter Dmin = 4 (the minimal width of the information graph): M1{1,2,3,4}, M2{8,7,6,5}, M3{10,11,12,9}, M4{13,15,16,14}, M5{19,17,18}, M6{20}.

It should be noted that it is not the only variant of the subdivision of the set of points into groups. Other variants are possible two. This depends on the chosen method of the optimization for the width.

Let us transfer the timing diagram of the algorithm in conformity with obtained groups (Fig. 4):

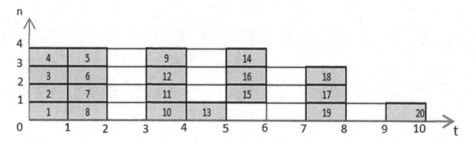

Fig. 4. Timing diagram of the algorithm after the optimization for the width

At that, the timetable will change in the following way:

P1: 1, 8, _, 10, 13, _, _, 19, _, 20;
P2: 2, 7, _, 11, _, 15, _, 17;
P3: 3, 6, _, 12, _, 16, _, 18;
P4: 4, 5, _, 9, _, 14.

According to this timetable, the whole duration of the activity of the algorithm t = 10, the quantity of processors n = 4, the whole down time p = 12.

The obtained timetable is better than initial one (ref. Fig. 3), because it lets to use computing system of less quantity of processors for the implementation of the algorithm while keeping the whole duration of the implementation of the algorithm and decreasing the downtime of processors.

3   Method of the Algorithm Optimization for the Amount of Communications

It is clear from provided diagrams that the time sent for the data transfer increases the duration of processors activity and the whole duration of the activity of the algorithm.

Provided examples are consistent with the known fact that the function of the computational speedup of the algorithm on the system of n computing devices F(n) has the following normal probability plot (Fig. 5):

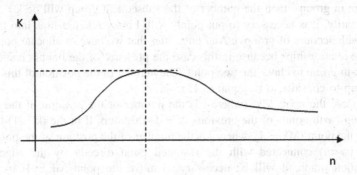

Fig. 5. Graph of the dependency of computation speedup on the quantity of processors, where K is the speedup, n is processors

The computation speedup begins from some n and droops due to the increase of the amount of data transfer. For some algorithms this dependency is the linear decreasing function; for example, the parallel version of the bubblesort acts slower than the initial sequential method because the amount of data transferred between processors is rather large and is comparable with the quantity of executable computing operations (and this unbalance of the amount of computing and complexity of data transfer operations increases with the growth of the quantity of processors) (Fig. 6) [13].

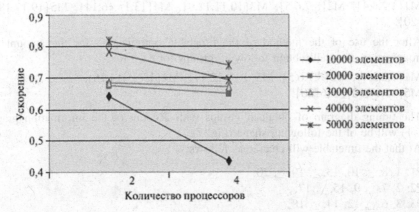

Fig. 6. Graph of the dependency of computing speedup on the quantity of processors for the bubble sort algorithm

Therefore, the next step towards the obtaining of the optimal to the execution time algorithm is the decrease of the amount of data transfers between processors.

The offered by authors method of information graph optimization for the amount of communications consists of following:

1. To place at the center the group of points corresponded to tiers, obtained in arbitrary way. It will be better if these groups are obtained in a formalized way, for example, by the method of the optimization of the information graph for the width by means of a matrix or an adjacency list.
2. The process of the replacement of points begins with the last group. Assume that there are m groups, then the number of the subsequent group will be k = m.

   Primarily, it is necessary to put points with binary relationships into the time-table (with account of groups). And only after that we have to allocate points with multiple relationships because in this case the presence of the bubble is inevitable.
3. In the k-th group to chose the first point. The number of the position of this point in the group to consider to be equal to 1: i = 1.
4. To compare the point Mki (where i is the number of the position of the point in the group) with points of the previous (k − 1)-th group. If in the (k − 1)-th group there is the point (Mk − 1j, where j is the number of the position of the point in the group, j >= i) connected with the assigned point directly by the edge in the information graph, it will be necessary to move the point Mk − 1j to the i-th position in its group. If in the (k − 1)-th group there is no point, connected with the point Mki then ref. step 6.
5. If k > 2 then k = k − 1 and ref. step 4.
6. If in the m-th group there are points those have not been analyzed, then k = m, i = i + 1 and ref. step 4.
7. If in the last group all points have been analyzed and m > 2, then m = m − 1 and ref. step 6.
8. If m = 1 then the method is ended.

Example: for the information graph of the Fig. 1 let us base on groups obtained after the optimization of the graph for the width:

M1{1,2,3,4}, M2{8,7,6,5}, M3{10,11,12,9}, M4{13,15,16,14}, M5{19,17,18}, M6{20}.

After the use of the method of the timetable optimization for the amount of communication, we shall obtain following groups for each tier:

M1{1,2,3,4}, M2{8,7,6,5}, M3{10,9,12,11}, M4{13,15,14,16}, M5{19,17,18}, M6{20}.

The timing diagram of obtained groups with account of the information graph (Fig. 1) will be of the following shape (Fig. 7):

At that the timetable will change as follows:

P1: 1, 8, _, 10, 13, _, 19, _, 20;
P2: 2, 7, _, 9, 15, _, 17;
P3: 3, 6, _, 12, 14, _, 18;
P4: 4, 5, _, 11, 16.

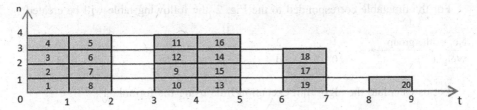

Fig. 7. Timing diagram of the algorithm optimized to the amount of communications

If we recalculate the amount of communications, we shall recognize that the lost of the time of the data transfer between processors has grown twice less and become equal to 8 items instead of initial 16 items.

The total execution time of the algorithm has grown less from 10 to 9.

4 Estimate of the Execution Time of the Algorithm with Account of the Degree of the Continuity of Information Graph Points

Any modification of the algorithm takes a time that may be spent for the solution to other, more important problems. That's why, before the algorithm optimization it would be desirable to know what shall a researcher obtain as a result of the optimization and will the new algorithm be better than previous one.

While researching communication dependencies and their influence on the algorithm whole execution time, we managed to get the estimation of the minimal algorithm whole execution time that should be calculated from the following formula:

$$t_{min} = \sum_{k=1}^{m} M + 2 \sum_{k=1}^{m} M_c. \tag{1}$$

where t_{min} is the minimal algorithm whole execution time that may be achieved by the optimizing of the timetable for the amount of communication between processors, $k = 1,...,m$, m is the quantity of groups of the graph, M is the quantity of groups containing points with input data binary relationships only, M_c is the quantity of groups, where even one point has more than one edge in it (multiple relationships).

Example: for the approximation of calculation, let us assign the weight to each point, i.e. the figure 0 for a point with only one edge in it and the Fig. 1 for a point with more edges in it. The following table will be created:

No. of the point	1	2	3	4	5	6	7	8	9	10
Weight	0	0	0	0	0	0	0	0	1	1
No. of the point	11	12	13	14	15	16	17	18	19	20
Weight	0	0	0	0	0	0	1	1	1	1

Let us assign the weight to each group: the figure 0 for a group with points of the weight 0 i.e. with only one input point in it and the Fig. 1 otherwise:

For the timetable corresponded to the Fig. 2, the following table will be created:

No. of the group	1	2	3	4	5	6
Weight	0	0	1	1	1	1

Let us calculate the algorithm execution time using the formula 1:

$$t_{min} = \sum_{k=1}^{m} M + 2 \sum_{k=1}^{m} M_{c}. = 2 + 2 \cdot 4 = 10 \qquad (2)$$

Therefore, we can make assertions about the algorithm execution time on the base of data from the information graph only, without the construction of a diagram.

For the timetable corresponded to the diagram of the Fig. 7 the time of the execution may be calculated using the following formula:

$$t_{min} = \sum_{k=1}^{m} M + 2 \sum_{k=1}^{m} M_{c} = 3 + 2 \cdot 3 = 9 \qquad (3)$$

Using the formula 1, theoretically possible algorithm execution time may be calculated. There are 6 points of the weight equal to 1. Therefore, in an ideal timetable they may be subdivided into two groups of weights equal to 1. Therefore, the rest 4 groups are of the weight equal to 0. If we substitute values M = 4 and Mc = 2 to the formulae 1, we shall discover that theoretically the algorithm execution time may be reduced to 8 items:

$$t_{min} = \sum_{k=1}^{m} M + 2 \sum_{k=1}^{m} M_{c} = 4 + 2 \cdot 2 = 8 \qquad (4)$$

Unfortunately, information dependencies between points do not let to obtain ideal algorithm execution time always.

5   Upsizing of Operations

Upsizing of operations before the optimization for the reducing of the quantity of operations of the weight equal to 0 seems to be the logical decision. This lets to escape an accidental break of a linear chain of operators between different processors and appearance of new bubbles.

Let us take as a basis the information graph of the Fig. 1 and upsize its operations by the joining of directly interrelated points of the weight of 1. As a result, we shall obtain the following collection of points:

$$1' = [4 + 5 + 16]$$
$$2' = [3 + 6 + 14]$$
$$3' = [2 + 7 + 12]$$
$$4' = [1 + 8 + 11]$$
$$5' = [10 + 13]$$

$$6' = [9 + 15]$$
$$7' = [18]$$
$$8' = [19]$$
$$9' = [17]$$
$$10' = [20]$$

where: the number of the point i' is the new number and the collection of points in square brackets is previous points those were joined into one new point.

As a result, the information graph of the Fig. 8 shall assume the following shape:

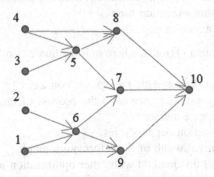

Fig. 8. Information graph of the algorithm with upsized points

If we build groups of point for this graph, we shall obtain following results: the quantity of necessary processors is 4, the quantity of groups is 4 too:

$$M1\{1, 2, 3, 4\}, \; M2\{6, 7, 5\}, \; M3\{9, 8\}, \; M4\{10\}.$$

The timing diagram will correspond to these groups (Fig. 9):

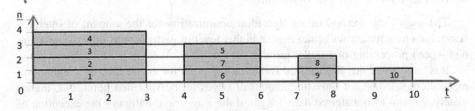

Fig. 9. Timing diagram of the algorithm after the upsize of operations.

At that the timetable shall change as follows

P1: 1, _, 6, _, 9, _, 10;
P2: 2, _, 7, _, 8;
P3: 3, _, 5;
P4: 4.

In conformity with this timetable, the whole algorithm execution time t = 10, the quantity of processors n = 4, total down time amount p = 6.

In this case, the optimization for the width and communications does not lead to the reduction of the quantity of processors and algorithm execution time but allows to decrease slightly the amount of bubbles.

6   Conclusion

The method of information graph optimization for the amount of communications allows to reduce the amount of communications between processors and, therefore, to reduce the whole algorithm execution time.

Advantages of this method are:

- Low timing labor content O (md^2), where m is the quantity of groups, d is the width of the graph.
- The possibility to work not with the sparse connectivity matrix, but with the adjacency list. This possibility speeds up the process of the timetable calculation process and economizes a memory.
- Preservation of information dependencies.
- Preservation of the initial width of the information graph.
- Possibility to combine this method with other optimization methods.

It would be effective to use the method of an algorithm optimization for the amount of data transfer between processors in accordance with the following methodology:

1. To subdivide the algorithm into operations.
2. To build an information graph of the algorithm.
3. To build the parallel form of the information graph and the timing diagram of the algorithm.
4. To realize the optimization for the width (the quantity of processors).
5. To realize the optimization for interprocessor communications.
6. To realize the upsizing of operations.

The use of the method of an algorithm optimization for the amount of interprocessor data transfer allows achieving the higher level of performance, effectiveness and high-speed processing of parallel programs.

It is to be noted that this method is the starting point for the creation of a higher-end method considering not only the amount of edges in the specified point, but, immediately, amount of transferred data, length of the way, and duration of the execution of each operation.

Acknowledgements. The paper was prepared in SPbETU and is supported by the Contract № 02.G25.31.0149 dated 01.12.2015 (Board of Education of Russia).

References

1. Abramov, O.V., Katueva, Y.: The technology of parallel computing for the analysis and optimization. In: Management Problems, № 4 (2003)
2. Ahmad, I.: A parallel algorithm for optimal task assignment in distributed systems. In: Ahmad, I., Kafil, M.: Proceedings of the 1997 Advances in Parallel and Distributed Computing Conference, p. 284 (1997)
3. Aho, A., Hopcroft, J., Ullman, J.: The Design and Analysis of Computer Algorithms. Addison-Wesley, Reading (1974)
4. Akl, S.: The Design and Analysis of Parallel Algorithms. Prentice-Hall, Englewood Cliffs (1989)
5. Andrews, G.R.: Concurrent Programming: Principles and Practice. Benjamin/Cummings, Menlo Park (1991)
6. Arvindam, S., Kumar, V., Rao, V.: Floorplan optimization on multiprocessors. In: Proceedings of 1989 International Conference on Computer Design, pp. 109–113. IEEE Computer Society (1989)
7. Boyer, L.L., Pawley, G.S.: Molecular dynamics of clusters of particles interacting with pairwise forces using a massively parallel computer. J. Comput. Phys. **78**, 405–409 (1988)
8. Brunet, J.P., Edelman, A., Mesirov, J.P.: Hypercube algorithms for direct N-body solver for different granularities. SIAM J. Sci. Stat. Comput. **14**, 1143–1149 (1993)
9. Hu, C.: MPIPP: an automatic profileguided parallel process placement toolset for SMP clusters and multiclusters. In: Proceedings of the 20th Annual International Conference on Super-Computing, pp. 353–360 (2006)
10. Coddington, P.: Random number generators for parallel computers. In: NHSE Review, № 2 (1996)
11. Drake, D.E., Hougardy, S.: A linear-time approximation algorithm for weighted matchings in graphs. ACM Trans. Algorithms **1**, 107–122 (2005)
12. Frank, J., Vuik, C., Vermolen, F.J.: Parallel deflated Krylov methods for incompressible flow. In: Parallel Computational Fluid Dynamics: Practice and Theory, pp. 381–388. Elsevier Publishing, Amsterdam (2002)
13. Gergel, V.P.: Lectures of Parallel Programming. In: Gergel, V.P., Fursov, V.A.: Proceedings of Benefit. Samara State Aerospace University Publishing House, p. 163c (2009)
14. Hu, Y.F., Emerson, D.R., Blake, R.J.: The communication performance of the cray T3D and its effect on iterative solvers. Parallel Comput. **22**, 22–32 (1993)
15. Jackson, D.B., Snell, Q.O., Clement, M.J.: Core algorithms of the Maui scheduler. In: Feitelson, D.G., Rudolph, L. (eds.) JSSPP 2001. LNCS, vol. 2221, pp. 87–102. Springer, Heidelberg (2001)
16. Jordan, H.F., Alaghband, F.: Fundamentals of Parallel Processing. Pearson Education, Inc., Upper Saddle River (2003)
17. Matrone, A., Bucchigniani, E., Stella, F.: Parallel polynomial preconditioners for the analysis of chaotic flows in Rayleigh-Benard convection. In: Keyes, D., et al. (eds.) Parallel Computational Fluid Dynamics: Practice and Theory, pp. 139–145. Elsevier Publishing, Amsterdam (2000)
18. Lifka, D.A.: The ANL/IBM SP scheduling system. In: Feitelson, D.G., Rudolph, L. (eds.) IPPS 1995. LNCS, vol. 949, pp. 295–303. Springer, Heidelberg (1995)
19. Plasmeijer, M.J., Plasmeijer, R., Eekelen, M.C.: Functional Programming and Parallel Graph Rewriting. Addison Wesley Publishing Company, Reading (1993). 592 p.
20. Rauber, N., Runger, G.: Parallel Programming: For Multicore and Cluster Systems. Springer, Heidelberg (2010)

21. Roosta, S.H.: Parallel Processing and Parallel Algorithms: Theory and Computation. Springer, New York (2000)
22. Shichkina, Y.: Reducing the height of the information graph of the parallel algorithm. Scientific and technical statements STU. Inform. Telecommun. Manag. **3**(80), S. 148–152 (2009)
23. Thain, D., Tannenbaum, T., Livny, M.: Grid Computing: Making The Global Infrastructure a Reality. Wiley, New York (2003). 1060 p.
24. Van Der Vorst, H.A.: Parallel linear systems solvers: sparse iterative methods. In: Wesseling, P. (ed.) High Performance Computing in Fluid Dynamics, pp. 173–200. Kluwer, Dordrecht (1996)
25. Voevodin, V.V., Voevodin, V.l.V.: Parallel Computing. BHV-Petersburg, St. Petersburg (2002)
26. Walshaw, C., Cross, M., Everett, M.G.: Mesh partitioning and load balancing for distributed memory parallel systems. In: Topping, B.H.V. (ed.) Advances in Computational Mechanics for Parallel & Distributed Processing, pp. 97–104. Saxe-Coburg Publications, Edinburgh (1997)
27. Wilkinson, B., Allen, M.: Parallel Programming Techniques and Applications Using Networked Workstations and Parallel Computers. Pearson Education, Upper Saddle River (2005). p. 468

Application of Fuzzy Sections for Constructing Dynamic Routing in the Network DTN

Yulia Shichkina[1(✉)], Mikhail Kupriyanov[1], Anastasia Plotnikova[1], and Yaroslav Domaratsky[2]

[1] Department of Computer Science and Engineering,
Saint Petersburg Electrotechnical University "LETI", St. Petersburg, Russia
strange.y@mail.ru, mikhail.kupriyanov@gmail.com,
nastya19922008@gmail.com
[2] Technology Office Motorola Solutions, Inc., Schaumburg, USA
yaroslav.domaratsky@motorolasolutions.com

Abstract. This article provides a methodology for construction of data transfer paths through DTN dynamic network, implemented with the devices mounted on moving objects and connected via WI-FI, Bluetooth and LTE D2D. The methodology covers all the five stages of the Knowledge Discovery in Databases technology. The stage of application of Data Mining tools was studied in more detail. It is based on the application of fuzzy logic instrument to select subset that meet the network parameters from a set of moving objects. Further application to the subset of related objects of Yen's algorithm of search for optimal paths on a weighted graph with weights of the grade of the selected subset ownership of the object allows to build the most credible data transfer path and several alternative paths, ranked by descending of data delivery probability.

Keywords: Fuzzy sets · Database query · Delay tolerance network · Route · Graph algorithm

1 Introduction

Recently, the active integration of computer networks into many areas of human activity has led to the fact that the information channels throughput, functioning and reliability are some of the main problems in the field.

The problem of quality and reliability of information transfer is particularly pressing in new generation networks which are time delays-tolerant. In practice, to refer to these networks which are tolerant to the time delays, the English abbreviation DTN (Delay Tolerance Network) is used.

It should be noted that the scope of the delay tolerance networks application is very diverse. They are used in the following spheres: at work of Ministry of Internal Affairs in violation of the communications infrastructure integrity; in remote areas with the lack of infrastructure; on airplanes; in mines; as well as in the new rapidly developing field of Internet of Things.

Internet of Things is a globally connected system of devices, objects and subjects, based on IP technology. The Internet of Things term was introduced by Kevin Ashton

© Springer International Publishing AG 2016
O. Galinina et al. (Eds.): NEW2AN/ruSMART 2016, LNCS 9870, pp. 63–75, 2016.
DOI: 10.1007/978-3-319-46301-8_6

in 2009. Internet of Things implies the formation of an environment where all objects of the world, from the transport aircrafts to pens, have access to the Internet.

With the concept of Internet of Things new methods of communication organization of D2D (device to device) devices and the mesh are intimately connected, with the help of which a large number of "smart" devices not only interact with the user, but with each other as well. Often these devices are connected to the server appliances, but can also work independently. Earlier, analysts spoke of synchronization of applications between different devices. It is expected that the level of interaction of various devices with each other will grow. In this regard, there is a number of urgent tasks from the development of models of compatibility of different systems of Internet of Things (protocols, interfaces, algorithms) to the development of new electronic components. These problems include the task of constructing specialized dynamic networks that can store data on the nodes and transmit them on the route created from the available for communication nodes.

This article proposes an approach to the calculation of the data transmission path in a dynamic network DTN on the ground of the database mining of DTN nodes movement history and the current network status. Optimal construction of the dynamic DST network will allow in the field of Internet of Things, not only with lower latency on the network to transmit information between devices, but also to improve the use of cloud computing techniques to process information on the devices themselves or the nearest to them compute nodes.

2 Construction Routing in DTN

DTN networks, originally designed for deep space communications organization, are increasingly being used in conventional telecommunication computer systems. Organization of DTN network is significantly different from the usual organization of data transmission networks. A distinctive feature is data delivering regardless of the current state of communication channels. For the "classic" data transfer protocols in the event of non-delivery of data in the "current" moment, the data are deleted. DTN is an approach to construction of network architecture, which was developed for the solution of TCP/IP protocol problems in networks with large messaging delay period. This protocol is based on the paradigm of "store data and pass them on" [4]. The feature of this approach is the delivery of data regardless of the current state of the communication channel since the data is stored, unlike with other protocols, and transmitted as soon as it becomes possible. DTN protocol uses special messages that contain the information necessary for routing as well as data for transfer.

Existing DTN-network data routing protocols can be divided into three main categories [5]: one copy (or direct transfer), multiple copies (or broadcasting) and hybrid (limited broadcasting). Protocols of the first type transfer only one copy of the message through selected route to the destination. Broadcasting protocols transfer multiple copies of the message to sensor node within the network expecting that at least one copy will reach the destination. Broadcasting protocols, such as protocols based on theory of epidemics, or even pandemic, [6], can improve the guarantee for delivery. But, obviously, at network connectivity losses and/or random connection breaks the demands for buffer storage capacity grow quickly at the increase of the network size.

Selection of direct transfer or broadcasting conditioned by the following reasons:

- Broadcasting protocols can be operated with the presence of minimal information about the parameters and status of the network, while at the use of the direct transfer protocols to evaluate the best route there is the need to have complete information about the network.
- Broadcasting protocols of reusable copies result in redundancy, but the cost of repeated transfer due to the packet loss can be sufficiently high.

Network protocols can be passively or actively adaptive, depending on the way the route is estimated. In actively adaptive network protocols all the routes are predicted and evaluated before they will be in demand.

Passively adaptive network protocols evaluate routes only in the case of the need. If in the sensor nodes the information on the network connectivity is updated regularly, the best routes in the actively adaptive protocols are evaluated fairly simply. But these network protocols require more network resources to evaluate and update the routing tables, especially when the network topology changes frequently.

On the other hand, in the passively adaptive protocols, routing tables are smaller, so their evaluation and update are easier. However, due to the need for route evaluation an extra delay before message transfer is introduced.

Hybrid network protocols combine both approaches.

In this article, we consider the case of constructing the DTN not with static objects, which can sometimes, for various reasons, leave the network, but with dynamic objects that are constantly on the move, interact with each other and participate in the process of data transfer.

The example of this can be the device mesh, built on the devices installed in the ambulances, police cars, public transport, mobile phones of supermarket customers or tourists in national parks, historic centers, museums, etc.

Dynamic routing organization technology to build DTN together with technology of global geolocation GLONASSS/GPS can solve most of the problems of the traditional management and communication systems. Furthermore, this technology can significantly improve the technical characteristics of the system by means of data preprocessing directly on the receiving and transmitting devices without sending the large amounts of information to data centers.

But, in the light of all the above, it is necessary not only to accurately evaluate technical characteristics of the specialized DTN networks, but also to organize correct data transferring.

Further it will be shown how to use fuzzy slices from the database to extracted the information in order to build the routing rules for the DTN networks with it.

3   The Methods Used in Data Processing

To implement the construction of dynamic routing technology for DTN it is proposed to use the apparatus of the two theories: Data Mining and KDD – Knowledge Discovery in Databases.

Knowledge Discovery in Databases (KDD) methodology arose in 1989. With the help of this method not specific algorithms or mathematical apparatus are described, but the sequence of actions to be performed for the detection of useful knowledge. This method does not depend on the subject area. KDD technology includes the stages of selection of informative attributes, models construction, data pre-processing, cleaning, post-processing and interpretation of the results (Fig. 1).

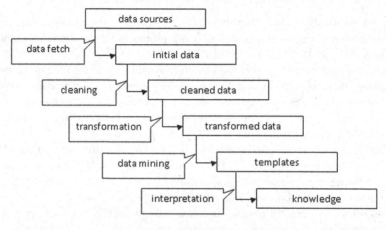

Fig. 1. Stages KDD

Considering KDD technology, it should be noted that the first stage consists of a data selection process. Here, the first step of the analysis is to obtain the original selection. At this stage the data often must be not only collected, but also to consolidated and transformed. With regard to the task of dynamic routing construction for the DTN, for example, the database on the movement of objects had to be transformed from the text format csv to the MongoDB format. Often at this stage other mechanisms are used: queries, data filtering or sampling, and as the source the specialized data storage is used, that consolidates all the information necessary for the analysis.

At the second stage of data cleaning is performed. Real data require processing to produce more useful knowledge. The need for pre-processing the data in the course of analysis arises without regard for technology and algorithms used. Data cleaning tasks include: fill-in-the-blanks, duplicates elimination etc. With regard to the problem of dynamic routing constructing for DTN at this stage the selection of relevant properties of moving objects was made.

On the third stage data transformation is considered. This step is necessary for those methods, at the use of which the raw data should be presented in some specific form. The fact is that the various analysis algorithms require specially pre-processed data. In solving the problem of dynamic routing constructing for DTN at this stage the individual attribute values conversion into new units was performed [7].

The next, the fourth stage, is Data Mining. On this stage the process of detection of previously unknown, non-trivial, practically useful knowledge, available for interpretations, is carried out in the raw data.

On the fifth, the last stage, the interpretation is carried out. In the case when the extracted dependencies and patterns are opaque to the user, there should exist the post-processing techniques that would allow to get them into the interpreted form. In the case of dynamic routing construction for DTN the mechanism of construction by the formal rules of the real route data transfer used in the DTN protocols is applied on this stage.

The main parts of this process are Data Mining techniques that allow to detect patterns and knowledge. The Data Mining term was introduced by Gregory Piatetski-Shapiro in 1989. The knowledge gained in the course of Data Mining techniques application should describe the new relations between the properties, to predict the values of some attributes on the basis of others. The basis of the Data Mining techniques are all sorts of methods of classification, modeling and prediction, based on the use of decision trees, artificial neural networks, genetic algorithms, evolutionary programming, associative memory, fuzzy logic.

With regard to the problem of dynamic routing constructing for DTN, from the Data Mining methods the decision trees and fuzzy logic are the most appropriate. In the case of decision trees construction by the values of the selection one solution is derived - plan for data transfer by the network. The methods of the decision-making theory are very laborious and belong to the class of NP-problems. The advantages of fuzzy logic for the task of constructing DTN is obvious, it is the possibility to obtain several routes, ranked by the range of parameters, rarer routes reconstruction in changing situation, less labor requiring. As the result of the fuzzy logic application in the task of the dynamic routing construction for DTN associated directed graph is constructed, the further application to which the Yen's algorithm, for example, results in a set of requested routes.

4   Construction of Fuzzy Sections in Databases

Most of the data processed in modern information systems, are of clear, numeric nature. However, database queries, which a human tries to formulate, often contain omission and uncertainties.

Fuzzy slices - is a good example of enrichment of one technology (database) with another one (fuzzy logic). Fuzzy slices are understood as filters by measurements, which involve fuzzy values, such as "all objects moving to the north of the city." In this example, the concept of "the north of the city" is not clear, and if to take into account that the objects can move to the north of the city not in a straight line, the fuzziness appears in the definition of the movement [1].

The mathematical theory of fuzzy sets and fuzzy logic are generalizations of the classical theory of sets and formal logic. These concepts were first introduced by the American scientist Lotfi Zadeh in 1965. The main reason for the emergence of the new theory was the presence of fuzzy and approximate arguments in describing processes, systems, objects by a human.

In relational databases this role is performed by fuzzy queries (flexible queries).

A characteristic of a fuzzy set is a Membership Function. Let's denote with $\mu(x)$ the degree of membership of x to the fuzzy set, which is a generalization of the concept of the characteristic function of a crisp set. Then the fuzzy set C is the set of ordered pairs of the form C = {μ(x)/x}, μ(x) here can possess any value within the interval [0, 1], x ∈ X. The value of μ(x) = 0 means the absence of membership at the set, 1 - full membership [2].

For the variables related to the continuous data class mean it is more convenient to denote the membership function with the analytic formula and represent graphically for illustrative purposes. There are over a dozen of typical curve shapes for membership functions setting.

There are over a dozen of typical curve shapes for membership functions setting. The simplest examples of representation of fuzzy sets are piecewise linear functions: V-type and trigonal. They are defined by the following formulas:

Triangular:

$$\begin{cases} \frac{x-a}{b-a}, a \leq x \leq b, \\ \frac{c-x}{c-b}, b \leq x \leq c, \\ 0. \end{cases}$$

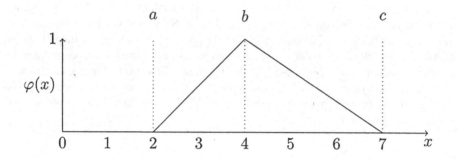

Trapezoidal:

$$\begin{cases} \frac{x-a}{b-a}, a \leq x \leq b, \\ 1, b \leq x \leq c, \\ \frac{d-x}{d-c}, c \leq x \leq d, \\ 0. \end{cases}$$

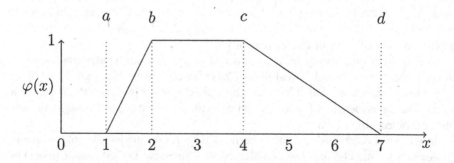

An example of more complex membership function is the following function:

$$\begin{cases} x^2, 0 \le x \le a, \\ 1, a \le x \le b, \\ \frac{c-x}{c-b}, b \le x \le c, \\ 0. \end{cases}$$

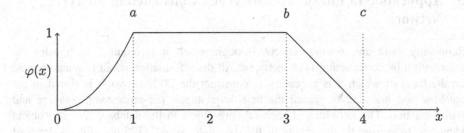

Fuzzy sets are helping to work with the fuzzy concepts, i.e. concepts that do not have an exact value. They change their values depending on the task. For example, the "situated nearby" parameter, in the framework of the problem of the availability of Wi-Fi spot, the operating range of which is 100 m, will be in the range from 0 to 20 m, and in the framework of the problem of finding the neighboring town, the parameter will be in the range of 10 to 80 km and so on.

Fuzzy sets can have different degree of fuzziness. The set, which membership function grows slowly is more exact than the set, the Membership Function of which grows more rapidly. Fuzzy degrees are important in the application of fuzzy sets theory. The degree is an indicator of the quality of different algorithms in decision-making, information search models [2]. It is possible to evaluate the degree of fuzzy sets with the help of:

- normalized entropy,
- matrix approach,
- axiomatic approach.
- for fuzzy sets the following logical operations are defined:
- intersection of two fuzzy sets A ∩ B (fuzzy "AND"):

$$\varphi(x) = \min(\varphi_A(x), \varphi_B(x))$$

- consolidation of two fuzzy sets A ∪ B (fuzzy "OR"):

$$\varphi(x) = max(\varphi_A(x), \varphi_B(x))$$

- negation of fuzzy set ¬A [3]:

$$\varphi(x) = 1 - \varphi_A(x)$$

where $\varphi_A(x), \varphi_B(x)$ are the membership functions of A and B sets respectively.

Fuzzy search in databases brings the most benefit when it is required not only extract information, operating with fuzzy concepts, but rank it somehow by descending (ascending) degree of relevance of the request [5]. It allows to answer the following questions: which data transfer path should be considered the main and which - the reserve; what information should be sent first, etc.

5 Application of Fuzzy Sections When Constructing the DTN Network

Supposing there are moving objects, through which it is required to transfer the information by constructing DTN network. All the information about moving objects, on the basis of which it is necessary to construct the DTN network, is stored in the database and has already passed the first three stages: pre-processing, cleaning and transformation. The attributes of each moving object in the database are: id - object identifier, time from the beginning of the day, date, speed, GPS coordinates, level of confidence, speed [6]. For the construction of the rules of data transfer through DTN network it is necessary to make a few slices of the database of: time, direction and distance:

1. "Distance" Slice

Let us assume that x for "distance" membership function lies in the range from 0 to 500 (the radius of equipment operation). This slice can be represented as a simple piecewise linear membership function (Fig. 2).

$$\begin{cases} 1, a \leq x \leq b, \\ \frac{c-x}{c-b}, b \leq x \leq c, \\ 0, c \leq x. \end{cases}$$

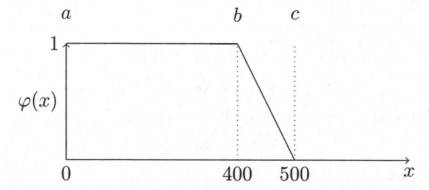

Fig. 2. Schedule the membership function "distance"

2. "Time" Slice

For the membership function of "time", let us assume that x should lie in the range from 0 to 15 (the delay time of the moving object). This slice can also be represented as a piecewise linear membership function (Fig. 3).

$$\begin{cases} 1, a \leq x \leq b, \\ \frac{c-x}{c-b}, b \leq x \leq c, \\ 0, c \leq x. \end{cases}$$

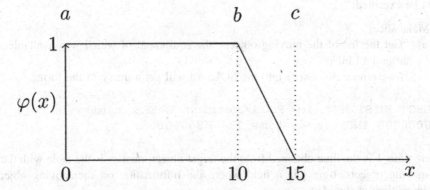

Fig. 3. Schedule the membership function "time"

3. "Direction" Slice

Assume that x of "direction" membership function, and x lies in the range from 0 to 3 (the degrees of deviation from the exact direction). Then the "direction" slice can also be represented in the form of membership functions (Fig. 4).

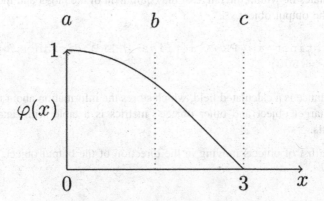

Fig. 4. Schedule the membership function "direction"

$$\begin{cases} \frac{x^2}{8}+1, a \leq x \leq b, \\ \frac{c-x}{c-b}, b \leq x \leq c, \\ 0, c \leq x. \end{cases}$$

Once membership functions are identified, it is possible to create a database query and evaluate the obtained entries. To do this, it is necessary to specify the degree of membership of 0.8. This number will mean that it is required to select all records which degree of membership is greater than 0.8.

Example. Assume that it is necessary to transfer the data from the moving object with the identifier id = 1 (input object) to the object (possibly stationary) with id = 4 (output object) at 14:00. For the construction of transfer rules the following algorithm must be executed:

1. Make slices:
   (a) "Get the list of the moving objects, the equipment of which was available at around 14:00".

   If we render this query into the SQL, we will get a query of the form:

```
SELECT DISTINCT id FROM metrics WHERE time>= time_1 -
600000000 AND time<= time_1 + 600000000;
```

where time_1 is the time obtained from the target object, metrics is the table with data on moving objects, time = is a field where the information on the moving object residence time is stored.

With this query the identifiers of moving objects that between 13–50 and 14–10 were transferring the data to the base about their location will be obtained.

   (b) "Get the list of moving objects that are located closely enough to the target object to receive the data from it."

   To do this it is required to calculate the distance between the target object and other objects. If the distance is less than 500 m, the point is located within the zone of equipment operation. After filtration of the obtained data only those entries will remain, which coordinates lie within the range of the equipment of the target and moving in the direction of the output object.

```
SELECT distinct id FROM metrics WHERE distance >= 0 AND
distance <= 500;
```

where the distance is a calculated field, which stores the information about the distance between the target object and other objects; metrics is a table with data about the moving objects.

   (c) "Get a list of objects moving in the direction of the output object."

To do this using the formula of finding the cosine of the angle between the vectors it is required to determine the angle between the target object and output object:

$$cos\ \alpha = \frac{ab}{|a||b|}$$

```
SELECT distinct id FROM metrics WHERE cosine >= 0 AND co-
sine<= 0,98;
```

where cosine is the calculated field, which stores the information about the cosine of the angle between the vectors, metrics is the table with data about the moving objects.

If the angle is less than 3 degrees, the moving object is subject to filtering.

2. After finding all three slices it is required to calculate the degree of membership of filtered entries using membership formulas and logical rules for fuzzy slices and finally select only those moving objects, the degree of membership of which is greater than 0.8.

This will provide the list of all moving objects, which were at the time of data transfer (±10 min) in the specified object operating range and were moving in the direction of the output object.

The results of the first execution of the algorithm can be schematically represented as a circle with the radius of 500 m, and certain set of points, which time and direction fulfill the specified conditions (Fig. 5).

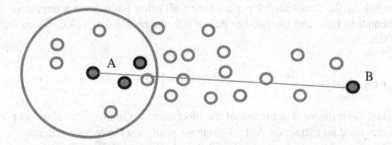

Fig. 5. The results of the first iteration of the algorithm

The filled point in the center is the output object. Two filled points closer to the circle are the moving objects with a degree of membership greater than 0.8.

Repeating this algorithm for the two found moving objects, adopting that each of them by turns is the input object, the whole set of circles will be constructed (Fig. 6).

The circles containing the output object will be the latter. The algorithm is finite, as from a certain iteration, the distance between the input and output objects will be reducing along with the number of moving objects with the degree of membership greater than 0.8.

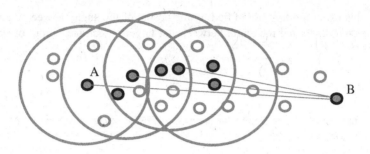

Fig. 6. Several iterations of the algorithm

As a result of the entire algorithm based on fuzzy slices, a set of associated points corresponding to the moving objects will be obtained. This set of points is substantially a connected oriented graph with one input and one output point.

At the last stage of KDD, the «interpretation», when constructing dynamic routing for DTN, to the resulting graph Yen's algorithm can be applied, which allows to select the number of shortest paths by k-e graph. The degrees of membership will be the graph scales. All paths k's will be ranked from the most to the least optimal. The number of paths k is defined by a network administrator. The same degree of membership will serve as the optimality criterion. The first data transfer route in the list of found paths will have the highest degree of membership, and thus the chance the data will be delivered on time and with no loss of communication.

For each path found it is possible to calculate the total degree of membership, such as the arithmetic middling or the minimal value of the degrees of membership of all the arcs included in the path. Actually the arc with minimal degree to membership is the weakest link in the data transfer path. Even if all other paths have a degree of membership equal to zero, and one arc has that of 0.8, the probability value of data delivery will be 0.8.

6 Conclusion

The studies have shown that the use of the mechanism of fuzzy slices allow to perform a multi-criteria data extraction from a database with fuzzy selection criteria.

Application of graph theory or decision making algorithms to the obtained set of points allows to build one or more routes ranked by descending of probability of data delivery from one object to another for construction of dynamic DTN network.

This article contains the example of relational database queries. But modern databases based on NoSQL or NoSQL technologies also support complex queries and the presented technique for routes construction for DTN dynamic network in these databases works with the same success. The only limitation in this case is that it should be considered that such databases do not have the relations and the data have to be stored in one collection. In general, the choice of database type depends on the task and sometimes one document-oriented database or relational database is not enough.

Acknowledgements. The paper was prepared in SPbETU and is supported by the Contract № 02.G25.31.0149 dated 01.12.2015 (Board of Education of Russia).

References

1. Zadeh, L.A.: Fuzzy logic, neural networks, and soft computing. Commun. ACM **37**(3), 77–84 (1994)
2. Zadeh, L.A.: Fuzzy sets. Inf. Control **8**, 338–353 (1965)
3. Sivanandam, S.N., Sumathi, S., Deepa, S.N.: Introduction to Fuzzy Logic Using MATLAB. Springer, Heidelberg (2007). 441 p.
4. Wood, L., Holliday, P., Floreani, D., Eddy, W.M: Sharing the dream. In: Workshop on the Emergence of Delay/Disruption-Tolerant Networks (E-DTN), Part of the International Conference on Ultra Modern Telecommunication (ICUMT), St. Petersburg, Russia, 14 October, 2009, pp. 1–2 (2009)
5. Tanenbaum, A.S.: Computer Networks, 5th edn. Prentice Hall, Upper Saddle River (2011). Tanenbaum, A.S., Wetherall, D.J., Cloth, 960 p.
6. Mundur, P.: Delay tolerant network routing: beyond epidemic routing. In: Mundur, P., Seligman, M. (eds.) Proceedings of Third International Symposium on Wireless Pervasive Computing, ISWPC 2008, 7–9 May 2008, pp. 550–553 (2008)
7. Batyrshin, I., Kacprzyk, J., Sheremetov, L., Zadeh, L.A. (eds.): Perception-Based Data Mining and Decision Making in Economics and Finance. Studies in Computational Intelligence, vol. 36, pp. 85–118. Springer, Heidelberg (2007)

Strategic Analysis in Telecommunication Project Management System

Ekaterina Abushova, Ekaterina Burova, and Svetlana Suloeva[✉]

Peter the Great Saint-Petersburg Polytechnic University, Saint-Petersburg, Russia
{abushova_ee, burova_ev, emm}@spbstu.ru

Abstract. A lot of issues are tackled in the project management system in the field of telecommunications. The most important one for achieving companies' strategic goals is a process of selecting an optimal project portfolio. Strategic analysis of the external and internal business environment the company operates in has an important impact when selecting such a portfolio. Since the business environment in telecommunications is developing dynamically and the competition is fierce, assessment indicators, analysis methods and tools must be carefully selected so that the costs of analysis and further application of a mechanism for selecting an optimal project portfolio would not exceed the benefits of its use.

Keywords: Strategic analysis · Portfolio analysis · PEST- analysis · SNW-analysis · Matrix set

1 Introduction

Selection of preferable projects and formation of an optimal project portfolio are some of the most important decisions to be taken by enterprises specializing in telecommunications [1]. Viability of enterprises, their future potential and life span depend on how correctly the projects will be selected. Selection of the projects and formation of the portfolio should be consistent with the strategic development plan of the enterprise. Since such a plan is strategic and drawn up for more than one year, the indicators, methods and tools, which are suggested for use in the optimal portfolio selection mechanism and are reviewed in this paper, must be strategic too [2].

2 Main Content

The mechanism of selecting an optimal project portfolio includes several stages: data acquisition and strategic analysis of the information about the business environment of the enterprise; selection of projects which can be included in the market portfolio; formation of an optimal project portfolio given the existing limitations provided the company achieves the strategic goals it sets.

Let us see into this mechanism in more detail and pay special attention to analysis of business environment. The mechanism of selecting an optimal project portfolio is presented in Fig. 1.

O. Galinina et al. (Eds.): NEW2AN/ruSMART 2016, LNCS 9870, pp. 76–84, 2016.
DOI: 10.1007/978-3-319-46301-8_7

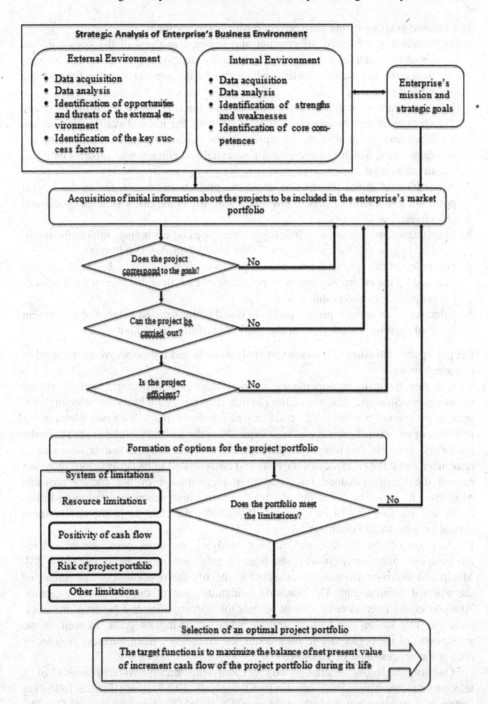

Fig. 1. Mechanism of forming an optimal project portfolio

1. Strategic analysis of the enterprise's business environment:
   (a) Acquisition of initial information and strategic analysis of the obtained data about the external environment in order to identify opportunities and threats, define the key success factors, which should be considered when forming the project portfolio.
   (b) Acquisition of initial information about the internal environment the company operates in and its analysis aimed at defining the internal competitive advantages and potential of the enterprise in competitive struggle and revealing the drawbacks, which can decrease the potential and efficiency of projects included in the portfolio.
2. Acquisition of initial information about the projects to be included in the market portfolio, evaluation of their consistency with the enterprise's strategic goals and objectives.
3. Identification of economic efficiency of the projects, excluding admittedly inefficient or impossible for carrying out in the existing market conditions.
4. Formation of the options for the project portfolio, their evaluation by the limitations imposed by the enterprise: resource, positivity of cash flow, balance of life cycles of the projects, risk acceptability, etc.
5. Selection of an optimal project portfolio based on maximized balance of net present value of increment cash flow of the market portfolio throughout its life.

Let pay special attention to the strategic analysis tools and indicators which are used in this mechanism.

The first stage of implementation of the mechanism is strategic analysis of the environment. Strategic analysis of the environment ensures a basis for selecting projects in the market portfolio. The efficiency of decisions that will be taken later about selection of an optimal project portfolio depends on the accuracy of the analysis. So the indicators, methods and tools must be selected very carefully. However, one should remember about the efficiency of analysis and costs related to its performance must not exceed the results obtained later due to implementation of the chosen projects. Moreover, it should be noted that the telecommunication sector is developing turbulently and analysis should be recurrent. If possible, the changes in the environment should be monitored continuously.

Analysis of the environment implies studying its three components: macro-environment, micro-environment and internal environment of the organization [3]. Macro and micro-environment is analyzed to identify the opportunities and threats of the external environment. The outcome is identification of the key success factors. Analysis of the internal environment reveals the opportunities and potential the company can rely on in competitive struggle when achieving its goals, as well as the weaknesses of the organization. As a result, the company's main business abilities or core competences must be identified.

One of the tools recommended for analyzing the micro-environment of a telecommunication enterprise can be PEST analysis. PEST stands for the following factors of the sector: political (P), economic (E), social (S) and technological (T). The following specification can be given to the components for the telecommunication sector:

1. Political/legal factors

In Russia the telecommunication sector is regulated by the Ministry of Telecom and Mass Communications of the Russian Federation. It performs the functions related to development of state policy and legislation in the field of information technology, telecommunications, mail communications, mass communication, mass media communication, including electronic means (among other things the Internet, systems of television (including digital) and radio broadcasting and new technologies in these fields), print, publishing and printing activities, personal data processing. The state regulates and approves the tariffs for long-distance and local phone calls. Since the tariffs are not high, communications are available practically to everyone. However, the other side of the medal is that due to this fact profitability of the national telecommunication companies is at a relatively low level.

2. Social factors

Today's life cannot be conceived without telecommunications which embrace practically all spheres of people's life. Their significance is constantly growing. Fashion for telecommunications affects the general culture of the population.

3. Economic factors

The telecommunication sector is in high demand in the global market and marked with absolutely fierce competition. At the same time, the price of telecommunication services is low, which is why virtually anyone can use them.

4. Technological factors

Today in the telecommunication factor the fastest network channels are fiber channels. Their speed is quite sufficient for transferring voice and text messages. However, this type of communications is very expensive and a lot of money has to be invested in maintaining operability of the equipment and its updating.

At the same time wireless communication systems are actively developing. Numerous satellites on the orbit ensure communications between any locations on the planet.

The factors we suggest for evaluating the macro-environment in terms of PEST analysis of a telecommunication company by relevant aspects are given in the Table 1:

To assess competitive attraction of the projects, which can be seen as strategic economic activity zones, we suggest using not only individual tools, models and methods, but them as a combination too. Thus, we recommend a complex tool for analyzing attractiveness of projects, which we call the "matrix set". It is rather difficult to evaluate the attractiveness of projects (strategic zones of business, SBU) and further formation of the optimal project portfolio and all activities of the enterprise depend a lot on this decision. That is why strategic analysis is to provide clear, objective and up-to-date information which allows evaluating the attractiveness of the project and correcting it in case of need in the future. A scheme for the use of the "matrix set" is represented in Fig. 2 [4, 5].

Designations in Fig. 2: RMS – relative market share; PLC – product life cycle; SBU – strategic business unit; US – unit size; KPA – competitive price advantage;

Table 1. Recommended factors of PEST analysis

Aspect	Factors
Political factors	Legislation, regulating the telecommunication sector
	Presence of state telecommunication companies in the sector
	Tax policy
	Trends for regulating or deregulating of the telecommunication sector
	State policy in the field of investments
Economic factors	Economic situation in the country
	Currency exchange rate dynamics
	Life cycle stage of the telecommunication sector (growth rates, profitability level trade-wise)
	Level of people's income
	Inflation and interest rates
Social factors	Population growth rates
	Lifestyle and consumer habits
	Requirement for the quality of products made by telecommunication companies and the level of service
	Gender and age of major consumers
	Absolute number of users for this type of services
Technological factors	Innovations and technological advancement of the telecommunication sector
	Legislation in the field of techniques in the sector
	Expenses on research and development in telecommunication companies
	Degree of use, introduction and transfer of technologies in the sector
	Accessibility of the latest technologies for the companies

KQA – competitive quality advantage; MSreal - real share of the market; MSequit – equitable share of the market; K – specific weight of SBU in the enterprise's total volume of sales.

It is preferable to use SNW analysis for strategic analysis of the internal business environment of telecommunication enterprises. SNW is a common abbreviation, which stands for strength, neutral and weakness.

SNW analysis of the internal business environment of an enterprise is quite an efficient way to estimate the organization's competitiveness whereby the best option is to choose the average market condition for a certain situation as neutral. SNW analysis of an enterprise studies the following aspects of the internal environment:

- Main business strategy of the organization.
- Competitiveness of the service in the relevant market.
- Availability of finances.
- Efficiency of the trademark, innovations and employees' performance.
- Marketing and production level.

We suggest examining the factors presented in the Table 2 for telecommunication companies:

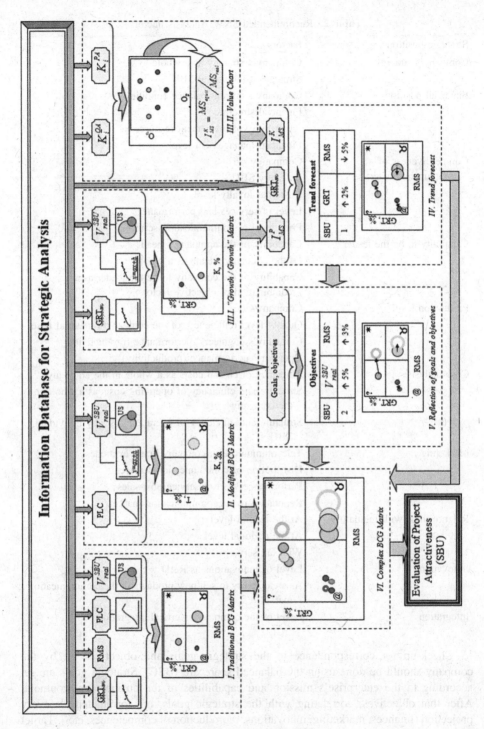

Fig. 2. Scheme for using the matrix set

Table 2. Recommended SNW Analysis Factors

Strategic position	Factors
Company's strategy	Company's strategy as a whole
	Strategy of individual SBU
Financial position	Company's current balance condition
	Financial structure
	Availability of investment resources
	Financial management level
Competitiveness of communication services	Companywide
	By every SBU
	Customer loyalty level
	Level of services and performance
	Price level of communication services
Capability to be the leader	Capability of the company's head
	Of the company's team as a whole
	Capability of the company to be the leader as a combination of objective factors
Production level	Companywide
	Quality level (efficiency) of the company's material base
	Company's engineers' performance (productivity)
	Operators' performance (productivity)
Cost analysis	Efficiency of capital costs as a whole in the company
	Structure and efficiency of operating costs as a whole in the company
	Structure and efficiency of operating costs by individual SBU
Marketing	Telecommunication company brand awareness
	Distribution of communication services
	Performance level of service post-sales
	Reputation in telecommunication market
Relationships with regulatory bodies	At the federal level
	At the regional level
	With tax services
Innovations	Level of innovations as R&D
	As a capability to sell new products in the communication market
Integration	Level of the company's vertical integration

Check-up for correspondence to the strategic goals and objectives set by the company should be done using the balanced scorecard (BSC). Strategic goals are set according to the enterprise's mission and capabilities of the business environment. After that objectives, correlating with the strategic goals, are formulated by every projection (finances, marketing, innovations, reproduction of competences, etc.). Target programs are developed to achieve every objective, which include definite measurable factors for performance evaluation: results, time schedules, resources. It is these very factors which must be used to check if the projects meet the strategic goals and objectives of the enterprise.

3   Conclusion

The technique for selecting an optimal market portfolio is the most important part of the project management process. In case the managers can skillfully manage the process of project implementation, but the projects themselves have been chosen wrongly, most probably, the enterprise is not going to have a positive result.

The paper proposes a mechanism for selecting an optimal project portfolio, whose specific feature is accommodating the projects to be selected with the strategic goals and objectives set by the enterprise and corresponding to the constraint system with maximized NPV balance of increment cash flow of the market portfolio throughout its life cycle.

The most responsible stage in this mechanism is strategic analysis of the enterprise's business environment and evaluation of the competitive position for the projects, so the main emphasis in this paper is put on recommendation of tools and factors of strategic analysis. There is no universal system of analysis tools. In each sector and at every enterprise there are factors which have an impact on project selection, so the proposed set of analysis tools and factors considers the specifics of the telecommunication sector and can be recommended, in particular, for enterprises of this sector.

The matrix set for evaluating attractiveness of projects suggested in this paper allows carrying out complex analysis of projects, given both the trends in the telecommunication market and competitive position (current and required one) of individual SBU of the company in the market. Moreover, this set makes it possible as early as at the stage of analysis to clearly understand how balanced the project portfolio under consideration is by life cycle stages.

The strategic analysis methods and factors, which are reviewed and suggested for use in the project management system, both existing and enhanced or developed by the authors meets the needs of today's business conditions of enterprises, are aimed at solving specific project management tasks in the telecommunication sector, ensure that an enterprise can adapt to the changing conditions of the external and internal environment.

References

1. Glukhov, V.V., Balashova, E.S.: Economics and Management in Info-Communications: Textbook. Piter, Saint Petersburg (2012)
2. Suloeva, S.B.: Strategic controlling in enterprise: theory, methodology, tools. Dissertation of Dr. of Sciences. SPb (2005). (rus)
3. Strategic Management Accounting at an Industrial Enterprise. Concept, Methodology and Tools. Lap Lambert Academic Publishing, 264 p. (2014)

4. Abushova, E.E., Suloeva, S.B.: Methods and models of modern strategic analysis. St. Petersb. State Polytech. Univ. J., Economics Series. The RF Ministry of Education and Science, St. Petersburg, no. 1(187), pp. 165–176 (2014). (Economic and Mathematical Methods and Models). ISSN 1994-2354, ISSN 2304-9774, References: p. 176 (10 items)
5. Fleisher, K., Bensoussan, B.: Strategic and Competitive Analysis. Methods and Techniques for Analyzing Business Competition. BINOM Knowledge Laboratory, Moscow (2005)

ruSMART: Role of Context for Smart Services

RCOS: Real Time Context Sharing Across a Fleet of Smart Mobile Devices

Julien Dhallenne[1(✉)], Prem Prakash Jayaraman[2], and Arkady Zaslavsky[3,4]

[1] Lappeenranta University of Technology, Lappeenranta, Finland
`julien.d@iki.fi`
[2] Swinburne university of technology, Melbourne, Australia
`pjayaraman@swin.edu.au`
[3] Commonwealth Scientific and Industrial Research Organisation, Data61, Clayton, VIC 3168, Australia
`arkady.zaslavsky@csiro.au`
[4] Saint Petersburg National Research University of ITMO, 49 Kronverksky Pr., St. Petersburg 197101, Russia
`arkady.zaslavsky@acm.org`

Abstract. Sharing context is a key challenge and will be a requirement of future IoT systems and services. To this end, in this paper, we propose, develop, implement and validate a Real Time Context Sharing (RCOS) system. RCOS takes advantage of the widely used publish/subscribe paradigm embedding context-awareness. We also propose a new context-aware subscription language enabling publishers to express data with sufficient contextual information and subscribers to subscribe to data by matching publisher context to subscribers contextual preferences. Finally, as a proof of concept, we extend the Apache ActiveMQ Artemis software and create a client prototype. We evaluate our proof of concept for larger scale deployment.

Keywords: Context aware publish/subscribe · Context sharing · Semantic web

1 Introduction

Today, as awareness rises on the value of data, users and machines tend to share newly discovered contexts more frequently through digital means. The publish/subscribe paradigm is widely used to exchange messages between parties according to their interest. However, most of the systems follow a topic-based model or a content-based model. This may not be sufficient in the future as the Internet evolves and the Internet of Things paradigm takes a more and more important place in our daily life. In this paper, the term context sharing is used to describe the ability of entities to share their context based on the publish/subscribe paradigm. The entities in this case could be smart mobile devices used by customers and producers, who could assume the role of publisher/subscriber. In this paper, real time stands for as quick as possible.

© Springer International Publishing AG 2016
O. Galinina et al. (Eds.): NEW2AN/ruSMART 2016, LNCS 9870, pp. 87–100, 2016.
DOI: 10.1007/978-3-319-46301-8_8

It is worth mentioning that International Erasmus Mundus Masters Program Pervasive Computing & Communications for Sustainable Development (EMM PERCCOM) [1] enabled the research reported in this paper.

Consider the following scenario example, namely "Digital fruit market". Petri is selling Pink Lady apples, for delivery or pick-up, in the Lappeenranta area. The price for a kilogram is 3.50 euros. In this given situation, the product "Pink lady apple" has the following context attributes, namely sourness, delivery location, and cost that need to be matched to an interested customer. We want the producer to be able to subscribe to a particular set of interests which will be matched in a near real-time manner when publishers publish their interests (i.e. clients and direct consumers), but it can also be the other way around.

Tero is holding a restaurant in the Lappeenranta area and he/she is looking for sour apples in this area. In this situation, the consumer Tero has a set of preferences defined by the context attributes which are the taste of the apple and the pick-up location. If this set of preferences matches with an existing product, we have a semantic match. This is the case, since Petri is producing Pink Ladies in the Lappeenranta area.

In a typical publish/subscribe system, the producer has to specify, in a single string, the product attributes he/she wishes to be part of the matching process. Moreover, it is not possible to express relationships between the entities being matched, and only a limited matching based on logical operators and string comparisons can be made. With the proposed context-based system it is possible to do semantic matching. This semantic matching can also be, for instance, a location within a certain radius of another location. In a content-based system, this could only be handled by a client tool and not by the broker of the publish/subscribe system since it cannot process contextual data. In the case of a topic based system, we would be required to have an apple category with sub-categories such as "sour apples" or "apples in Lappeenranta". In this situation, cross-matching would not be possible and the operation would imply a high computational resource need.

This paper makes the following contributions:

(a) We have conducted an extensive literature survey to identify the current state-of-the art, gaps in publish/subscribe based systems and the corresponding context-aware capable subscription languages.

(b) We have proposed and developed RCOS, a real time context sharing system based on semantic web principles. The contribution also includes a history based approach and a mobile application enabling smart mobile devices to share context seamlessly. The history is a graph, expanding as ontologies are removed from the main graph, when publications are canceled. It features contextual attributes, values and dates.

(c) We have deployed a proof of concept implementation and evaluation, to study the performance of the system and validate its efficacy.

2   Literature Review and Related Work

2.1   Publish/Subscribe Systems

The publish/subscribe paradigm is constituted of publishers publishing information, known as events, and subscribers sending subscriptions that represent their interests. The distribution of events to the corresponding subscribers is handled by a broker. Subscribers are notified of published events matching their interests. The broker can be on a server or distributed. Eugster et al. [2] differentiate the following types of subscription models.

Topic-Based Model: This model is based on topics. Notifications are transmitted to matching topics which subscribers have subscribed to. They are consequently forwarded all the messages transmitted to these particular topics. The advantage of this model is the low processing time required to answer a request. However, this approach being non-hierarchical, it is impossible for a subscriber to subscribe to a subset of events in a given topic. Eugster et al. [2] describe it as a flat approach. This issue is addressed by some implementations, such as the one from Oki et al. [3]. This model also lacks involving the content of a subscription in the process.

Content-Based Model: Rosenblum and Wolf [4] has introduced a subscription scheme based on the content of events. Through a set of operators and a specific subscription language, subscribers can also specify conditions over the content of the notification they wish to receive. We also consider XML and JSON based models to be a subset of the content-based approach in the sense that the brokers will interpret the content without relying on a defined knowledge base.

Context-Based Model: Some works also refer to this model as concept based [5]. Concept based addressing was introduced by Buchmann and Moody [6]. Tarkoma et al. [5] mention that "concept-based addressing allows to describe event schema at a higher level of abstraction by using ontologies that provide a knowledge base for an unambiguous interpretation of the event structure by using metadata and mapping functions". In our system, we propose a context-based model using ontologies as a knowledge base. These ontologies are defined in the JSON-LD format [7].

2.2   Context Awareness in Existing Publish/Subscribe Systems

Works on implementing the context awareness paradigm in publish/subscribe brokers already exists in the research community. Loke et al. [8] introduce a context-based addressing effort for Elvin. This work contributes to allowing to distribute messages to users in a chosen context according to ontologies interpretation. For this, they have also created a context-aware capable language [9]. Elvin was also included in the ECORA framework from Padovitz et al. [10], which provides a hybrid architecture for context-oriented pervasive computing.

However, Elvin as well as its open source implementation suffer from a lack of popularity nowadays, and are missing maintenance as well as cross-language support, since they handle the messaging process using its own standard that is not widely implemented.

Other recent studies such as the one realized by Tarkoma et al. [11] propose very efficient and flexible solutions. However, the concept of ontology is not fully considered and implemented. It is also worth noting the work and vision brought by Cugola et al. [12], which brings a distributed protocol to publish/subscribe systems according to their location to allow a more efficient distributed broker system. Our vision, however, places the location as a specific ontology. While their work is focused on the physical distribution of messages, our contribution is meant to bring a more generic way to embed context-awareness into publish/subscribe paradigm.

In [13], Zahariadis et al. introduce a novel context-aware publish/subscribe system. This system is developed within the effort of a single digital market in the European Union. Their context-based broker, however, does not include a subscription language, nor does it support preliminary defined ontologies as knowledge base with history.

2.3   Subscription Languages in the Existing Systems

Subscription languages used in publish/subscribe systems are more or less descriptive. The more operations are defined in a language, the higher is the complexity and processing time. Campailla et al. define three types of subscription query languages. (a) The Simple Subscription Language (SiSL) is used where all messages are total and of known format. If an attribute is not defined in the query, any value queried for it would return true; (b) the Strict Subscription Language (StSL) is an extension of SiSL where all attributes that occur in the query must be defined; and (c) in the Default Subscription Language (DeSL), all attributes are initialized to a default value, which are then updated by the message. This allows to test if the attributes are defined by a message.

To the best of our knowledge, Elvin [9] is the only publish/subscribe system in the research community with a subscription language defined so that it could be extended to incorporate a general context awareness capability considering ontologies due to its <action> <proposition> tuple integration. According to Campailla et al. [14], Elvin's and our approach are Simple Subscription Language (SiSL) since non-defined attributes results in accepting any value for them. This allows us to keep the query representation as small as possible.

The way we distinguish our approach from the one presented in Elvin is that we have a defined subscription language that does not contain operators. In our subscription language, the comparisons are done according to defined semantics from the schema.org effort [15]. JSON-LD definition allows us to directly embed in, through defined attributes, logic operators interpreted due to their involvement as attributes in a given ontology. For example, an ontology containing the properties *maxPrice* and *minPrice*, as defined in schema.org, involves that the broker compares the property price for a similar ontology, so that it defines a result for *minPrice* > *price* > *maxPrice*. Our approach also allows defining nested relations between entities. In this manner, we can express how an ontology relates to another ontology and easily integrate with current semantic web services. In the future, this could also allow publish/subscribe systems to be able to automatically match content discovered across the semantic web.

2.4   Comparison of Implemented Publish/Subscribe Systems

The information bus [3] is one of the early works on publish/subscribe systems. It uses Remote Method Invocation in order to publish and subscribe information. SCRIBE [16] and CORBA [17] are two publish/subscribe research following the information bus. SCRIBE brings reliable and scalable alternative to IP multicasting through the publish/subscribe paradigm on application level, balancing the load between nodes and being focused on a peer-to-peer configuration, while CORBA is designed to facilitate the communication of systems deployed on different platforms. The CORBA standard is implemented in C++ and Java and has standard mappings in Ada, C, C++, C++11, COBOL, Java, Lisp, PL/I, Object Pascal, Python, Ruby and Smalltalk. To the best of our knowledge, none of these systems has been designed considering smart mobile devices integration. In terms of early content-based systems, none provide a smart mobile device consideration. Rebeca [18] can, however, communicate through RMI, SNMP and HTTP, which brings a better interoperability. Elvin and Eugster et al. [19] have a specific way of subscribing and publishing, which limits their interoperability. ECA [6] transmits XML over SOAP, this brings a better interoperability with external systems since it follows the SOAP protocol. SOAP is, however, an information rich protocol, and in terms of performance, it is less oriented towards smart mobile devices than REST for example. As for the commercial publish/subscribe systems, all of them can be integrated into mobile technology, Apache ActiveMQ projects being the ones that support the biggest number of protocols. ActiveMQ Artemis distinguishes itself from ActiveMQ by supporting a more native approach of REST. FIWARE Orion brings novelty in the commercial publish/subscribe systems by being context-based. It still, however, lacks a subscription language and handles queries via REST GET methods on specific URLs.

3   RCOS Real Time Context Sharing

We created RCOS, the architecture of this system is represented in Fig. 1. RCOS can be included in existing publish/subscribe systems in order to enable context-awareness and history consideration into them. We use ontologies to semantically represent our context information. This approach allows us to provide a system that can be easily integrated with current and future semantic web applications. As represented in Fig. 1, it is composed of three modules which are the Queue management module, the Context & History Aware Broker module and

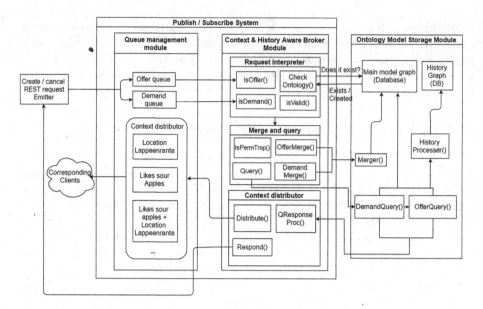

Fig. 1. RCOS overall architecture

Ontology Model Storage module. Each of these modules interact programmatically together. The Ontology Model Storage module is external to the Publish/Subscribe System in the sense that it does not interact directly with the publishing and subscribing processes, but is invoked by them for ontology modeling and storage. The following paragraphs give more details on the internals of each module.

3.1 Queue Management Module

In RCOS, an entity publishes or subscribes to information about another entity. An entity emitting an offer offers information about another entity; this can be considered as a publication. An entity emitting a demand requests information about another entity; this can be considered as a subscription.

The Queue management module handles two queues to which the offers and demands are posted (publications and subscriptions) via a REST interface. These REST requests can be of two types. Create requests imply that *the Queue management module* will request for an addition to the ontology base during the merging operation, while a cancel requests implies that *the Queue management module* will request for a deletion to the ontology base during the merging operation. The subscription language we define in RCOS brings together two ontologies on the base of information offer/demand relationship. In the Fig. 2 below, we represented our scenario where Tero wants to find sour apples in the Lappeenranta area.

```
{
    "@context":
    ["http://schema.org/",
    {
        "lfd": "http://example-localfood.org/"
    }],
    "@type": "Person",
    "@id": "http://example.org/profile/Tero5872",
    "seeks": {
        "@type": "Demand",
        "itemOffered":{
            "@type": "lfd:Apple",
            "description": "",
            "name": "",
            "lfd:species": "",
            "lfd:taste": "sour",
            "lfd:origin": "",
_ (some lines omitted)
                                "@type": "GeoCoordinates",
                                "latitude": "",
                                "longitude": ""
                        },
                        "geoRadius": ""
_ (some lines omitted)
```

Fig. 2. Subscription query for Digital fruit market

In the subscription presented in Fig. 2, all attributes of the ontologies that are to be returned from the knowledge base by the *Ontology Model Storage module*, relayed by the *Context & History Aware Broker module*, are specified. The wildcard "*" indicates to RCOS that the history for a certain attribute is requested. Such a history will be extracted from the history graph of previously canceled offers by the Ontology Model Storage. This extracted information is from the same context as the returned results for the demand itself.

3.2   Context & History Aware Broker Module

The Context & History Aware Broker module consists of three sub-modules. It serves requests from *the Queue management module*. In the "Digital fruit market" scenario, it will handle the context attributes consideration. We designed the sub-modules as follows.

Request Interpreter Sub-module: This sub-module first checks that the ontology format is valid through the isValid() function. If it is not, the request is rejected. The ontology is then handled as an offer or a demand. If it is an offer, it then queries the main model graph to know if the ontology exists through ChckOntology(), and if it does not, it is created. An offer or a demand object is then passed on to the Merge and Query sub-module.

Merge and Query Sub-module: The merge and query sub-module behaves according to the object type it obtains from the request interpreter sub-module. In the case of an offer or a demand which is permanent and marked as such by the IsPermTmp() function, it will be merged in the ontology model graph by *the Ontology Model Storage module* described in the next paragraph. A demand or an offer cancellation will then be handled by *the Query module*. This query function also relies on *the Ontology Model Storage module.*

Context Distributor Sub-module: This sub-module handles *the Ontology Model Storage module*'s response through QResponseProc(). It will also handle the context to distribute to the context distributor in the *Queue management module* via Distribute(), and if the response was for demand request, it will return currently existing results to the sender via Respond(). Otherwise, a simple acknowledgment is delivered to the sender.

3.3   Ontology Model Storage Module

The Ontology Model Storage module handles the ontologies knowledge base and history graph, which are loaded in the RAM memory, and the database to store them. In the "Digital fruit market" scenario, it will hold the information about the nearby available apple produces. It serves *the Context & History Aware Broker module* for verifying ontologies, merging them to the knowledge base graph and handling queries to issue responses.

When a merge request is issued from *the Context & History Aware Broker module*, the Merger() function will merge the ontology into the knowledge base graph. For queries, it will accordingly serve it as a demand by matching the corresponding sub-graph and returning it, or as a canceled offer by deleting the corresponding ontology from the graph, and calling the HistoryProcessor() function with this given ontology. If a demand query requests for the history of an attribute, it will also be handled by HistoryProcessor() function, which will take care of querying the history graph for the requested ontology's attributes present in the history graph linked to a date. For instance, for the history of price of a given product, each price returned from the history graph is linked to a date on which the offer started, this date being an attribute of the offer.

Schema.org's standard ontology vocabulary [15] allows modeling a product and its attributes. A product can also include nested ontologies containing their own attributes. This is the case in RCOS; an apple ontology has an offer ontology that is associated with it. The offer ontology includes aggregated and individual offers information. An individual offer includes information such as the seller, the minimum ordering price and the eligible area for the offer. The schema.org's [15] and other similar efforts are mainly joint commercial effort. The apple ontology does not yet exist as such to the extent of our knowledge. In this ontology, which we modeled, we include new attributes such as vitamins, label, species and origin of the apple.

4   Proof of Concept - Prototype Implementation

Our proof-of-concept implementation "Digital fruit market" is divided into two parts, and it is constituted of a prototype of RCOS, and a prototype client for smart mobile device communicating with RCOS. We focused our effort for this proof of concept on the possibility to emit a demand to RCOS, interpreting it and receiving the answer which we format visually in the smart mobile device client application.

4.1   The RCOS Client

We developed the RCOS client using the Ionic 2 framework, which relies on Angular technology. Our client allows us to emit demands (subscriptions) and display the ontologies resulting from the sub-graph transmitted by the RCOS server as a response. Figure 3 shows the mobile application client that we developed.

RCOS is constituted of views, controllers and providers. The *Data provider* is the core component of our application. It handles the REST requests to RCOS server and provides an SQLITE local storage that is used as local cache by the application. The *Products view and controller* are responsible for displaying the apples resulting from our subscription and stored in the local storage. The *Producers view and controller* have this responsibility for the producers and handle a map. Finally, the *Subscription view and controller* are meant to allow subscription. Since this is a proof of concept, it only handles subscriptions for sour apples in the Lappeenranta area and relies on initiated REST requests. This can be extended in a more generic and automated manner.

(a) Products view

(b) Producers view

Fig. 3. RCOS proof of concept's client

4.2 The RCOS Server

The RCOS server relies on using Apache ActiveMQ Artemis 1.3 as an external tool. However, the component we add can directly be embedded into the broker of a publish/subscribe system. Currently, our prototype implementation handles demands on an ontology knowledge base. This ontology knowledge base is stored in a JSON-LD file and is loaded in memory as a graph by RCOS. In order to create ontologies and place them in a graph for our knowledge base, we use the software tool Apache Jena 3.0.1. We have chosen this software due to its high interoperability and compatibility with the JSON-LD format. The main graph is loaded from a JSON-LD file and the extracting of the sub-graph used to answer a "demand" request is processed through the Jena SPARQL [20] interpreter. The query we use in our prototype, which includes nested elements, allows us to obtain the full ontologies that present a sour taste, and whose seller's locality is Lappeenranta. It could be made generic independently to the subscription's attributes formatting by analyzing it and identifying parent nodes. However, we limited our current proof-of-concept to a static query for simplicity reasons. We evaluate our prototype performances in the following section for larger scale deployment.

5 Evaluation of RCOS

5.1 Evaluation Setup

In this section, we evaluate our prototype in terms of performance variation for a larger scale deployment. We want to know which factors influence the processing time the most when querying our knowledge base graph. In order to do that, we continuously inject ontologies into our knowledge base graph and do measurements while querying it. We have chosen to evaluate this part of our system, because it is the one that will be the most affected as the knowledge base grows in a large deployment case. Current publish/subscribe systems are already able to handle a significant amount of requests efficiently, so it is worth measuring the knowledge base graph querying. This allows us to understand the behavior of the processing time according to the evolution of the number of ontologies in it.

5.2 Evaluation of the Matching Process

In Fig. 4, we are interested in the performance of the matching process's behavior. The matching process is handled via SPARQL [20] CONSTRUCT operation. This operation allows us to extract a sub-graph from an existing graph, according to restrictions we define in the query. We start with ten ontologies, two of which match lfd:taste = "sour" and eight lfd:taste = "sweet". The blue curve corresponds to the case where the SPARQL CONSTRUCT will create a sub-graph out of n-8 matching ontologies, n being the number of ontologies in the queried knowledge base graph. The green curve corresponds to the case where SPARQL

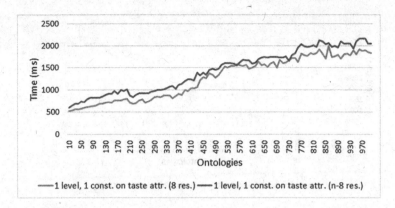

Fig. 4. Processing time and number of results against number of ontologies (Color figure online)

CONSTRUCT will create a sub-graph out of 8 matching ontologies for any size of the queried knowledge base graph.

From these results, we consider the CONSTRUCT operation processing time to be negligible against the time it takes for the SPARQL processor to go through the knowledge base graph, in order to find the right ontologies matching the query. Due to this fact, and in order to obtain clear data on the impact of the number of ontologies in the knowledge base graph, the following tests (when they have a constraint) are made with a static n-8 ontologies result matching to the query.

We are firstly interested in knowing the actual influence of introducing a single constraint into the SPARQL query that is represented in Fig. 5. In this first case, we do not involve any nested elements (ontologies included into ontologies) in our knowledge base graph. Only the first level attributes of our ontologies are queried.

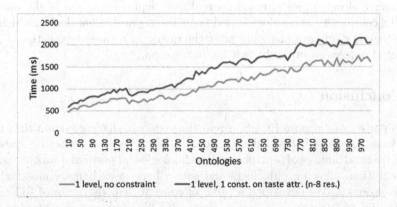

Fig. 5. Processing time for non-constrained/constrained taste attribute against number of ontologies (Color figure online)

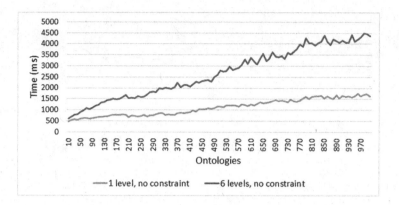

Fig. 6. Processing time for one and six levels ontologies against number of ontologies

The blue line represents the case where we do not introduce a constraint in the query (n results), and the orange line the case where we limit the results to sour apples (n-8 results). As we can observe when introducing a constraint, the processing time of the query can increase by 20 % and we can also observe that this trend is slightly increasing as the number of ontologies increases. This first test involves a single-level type of ontologies. Such ontologies are constituted of attributes, and cannot include any other ontology.

In Fig. 6, we observe the behavior of the processing time as we introduce new ontologies including six levels of complexity. They involve complex computing operations for the SPARQL engine, resulting in a bigger processing task. The processing time of an ontology with six levels can be three times the one of an ontology which has a single-level.

From these measurements, we can conclude that the processing time variation follows a linear trend against the number of ontologies in our knowledge base graph. However, it grows significantly more as we introduce complex ontologies. Ontologies complexity is the most affecting factor on the processing time as the number of ontologies grows. In a real case deployment, one might want to limit the complexity of ontologies, and favor two separated ontologies, which are linked together programmatically after the query, in a case of expecting a high number of ontologies in the knowledge base graph.

6   Conclusion

In this paper, we propose RCOS, a real time context sharing system that aims to address the gap of knowledge on context-aware publish/subscribe systems. RCOS takes advantage of semantic web technologies, in particular an ontological representation, allowing publishers and subscribers to exchange context about various products. We developed a proof of concept demonstration of RCOS for the "Digital fruit market" scenario. Using RCOS, producers can provide contextual annotations to produces. This allows subscribers to subscribe and query for relevant produces based on location-based and personalization context. We also proposed a novel history feature for attributes, which enables to enrich the contextual information, over the time, around a given produce. Finally, experimental evaluations validate the efficacy of the system and identify which criteria has the biggest impact on the performances.

Acknowledgments. Authors acknowledge the support from EMM PERCCOM, IoT EPI bIoTope Project, which is co-funded by the European Commission under H2020-ICT-2015 program, Grant Agreement 688203. The research has also been carried out with the financial support from the Ministry of Education and Science of the Russian Federation under grant agreement RFMEFI58716X0031. Julien Dhallenne would also like to thank Prof. Ahmed Seffah, Susanna Koponen, Prof. Jari Porras and Prof. Éric Rondeau for their support.

References

1. Klimova, A., Rondeau, E., Andersson, K., Porras, J., Rybin, A., Zaslavsky, A.: An international Master's program in green ICT as a contribution to sustainable development. J. Clean. Prod. **135**, 223–239 (2016)
2. Eugster, P.T., Felber, P.A., Guerraoui, R., Kermarrec, A.M.: The many faces of publish/subscribe. ACM Comput. Surv. (CSUR) **35**(2), 114–131 (2003)
3. Oki, B., Pfluegl, M., Siegel, A., Skeen, D.: The Information Bus: an architecture for extensible distributed systems. ACM SIGOPS Oper. Syst. Rev. **27**(5), 58–68 (1994)
4. Rosenblum, D.S., Wolf, A.L.: A design framework for internet-scale event observation and notification. ACM SIGSOFT Softw. Eng. Notes **22**(6), 344–360 (1997)
5. Baldoni, R., Querzoni, L., Tarkoma, S., Virgillito, A.: Distributed event routing in publish/subscribe communication systems. In: Garbinato, B., Miranda, H., Rodrigues, L. (eds.) Middleware for Network Eccentric and Mobile Applications, pp. 219–244. Springer, Heidelberg (2009)
6. Buchmann, A.P., Moody, K.: An active functionality service for open distributed heterogeneous environments. Shaker (2002)
7. Sporny, M., Longley, D., Kellogg, G., Lanthaler, M., Lindstrm, N.: JSON-LD 1.0, W3C Recommendation (2014)
8. Loke, S.W., Padovitz, A., Zaslavsky, A.: Context-based addressing: the concept and an implementation for large-scale mobile agent systems using publish-subscribe event notification. In: Stefani, J.-B., Demeure, I., Zhang, J. (eds.) DAIS 2003. LNCS, vol. 2893, pp. 274–284. Springer, Heidelberg (2003)
9. Loke, S.W., Zaslavsky, A.: Communicative acts of Elvin-enhanced mobile agents. In: IEEE/WIC International Conference on Intelligent Agent Technology, pp. 446–449. IEEE (2003)
10. Padovitz, A., Loke, S.W., Zaslavsky, A.: The ECORA framework: a hybrid architecture for context-oriented pervasive computing. Pervasive Mob. Comput. **4**(2), 182–215 (2008)
11. Tarkoma, S., Lindholm, T., Kangasharju, J.: Collection and object synchronization based on context information. In: Magedanz, T., Karmouch, A., Pierre, S., Venieris, I.S. (eds.) MATA 2005. LNCS, vol. 3744, pp. 240–251. Springer, Heidelberg (2005)

12. Cugola, G., Margara, A., Migliavacca, M.: Context-aware publish-subscribe: model, implementation, and evaluation. In: IEEE Symposium on Computers and Communications, pp. 875–881. IEEE (2009)
13. Zahariadis, T., Papadakis, A., Alvarez, F., Gonzalez, J., Lopez, F., Facca, F., Al-Hazmi, Y.: FIWARE lab: managing resources and services in a cloud federation supporting future internet applications. In: IEEE/ACM 7th International Conference on Utility and Cloud Computing (UCC), pp. 792–799. IEEE (2014)
14. Campailla, A., Chaki, S., Clarke, E., Jha, S., Veith, H.: Efficient filtering in publish-subscribe systems using binary decision diagrams. In: Proceedings of the 23rd International Conference on Software Engineering, pp. 443–452. IEEE Computer Society (2001)
15. Barker, P., Campbell, L.M.: What is schema.org? LRMI, vol. 21 (2014)
16. Castro, M., Druschel, P., Kermarrec, A.M., Rowstron, A.I.: SCRIBE: a large-scale and decentralized application-level multicast infrastructure. IEEE J. Sel. Areas Commun. **20**(8), 1489–1499 (2002)
17. Object Management Group: The Common Object Request Broker (CORBA): Architecture and Specification. Object Management Group (1995)
18. Parzyjegla, H., Graff, D., Schröter, A., Richling, J., Mühl, G.: Design and implementation of the Rebeca publish/subscribe middleware. In: Petrov, I., Guerrero, P., Sachs, K. (eds.) Buchmann Festschrift. LNCS, vol. 6462, pp. 124–140. Springer, Heidelberg (2010)
19. Eugster, P.T., Guerraoui, R., Damm, C.H.: On objects and events. ACM SIGPLAN Not. **36**(11), 254–269 (2001). ACM
20. Prud' Hommeaux, E., Seaborne, A.: SPARQL query language for RDF. W3C Recommendation 15 (2008)

Reasoning over Knowledge-Based Generation of Situations in Context Spaces to Reduce Food Waste

Niklas Kolbe[1,2], Arkady Zaslavsky[1,3(✉)], Sylvain Kubler[2], and Jérémy Robert[2]

[1] Commonwealth Scientific and Industrial Research Organisation,
Data61, Clayton, VIC 3168, Australia
{niklas.kolbe,arkady.zaslavsky}@csiro.au
[2] Interdisciplinary Center for Security, Reliability and Trust,
University of Luxembourg, 4 rue Alphonse Weicker, 2721 Luxembourg, Luxembourg
{niklas.kolbe,sylvain.kubler,jeremy.robert}@uni.lu
[3] Saint Petersburg National Research University of ITMO,
49 Kronverksky Pr., St. Petersburg 197101, Russia
arkady.zaslavsky@acm.org

Abstract. Situation awareness is a key feature of pervasive computing and requires external knowledge to interpret data. Ontology-based reasoning approaches allow for the reuse of predefined knowledge, but do not provide the best reasoning capabilities. To overcome this problem, a hybrid model for situation awareness is developed and presented in this paper, which integrates the Situation Theory Ontology into Context Space Theory for inference. Furthermore, in an effort to rely as much as possible on open IoT messaging standards, a domain-independent framework using the O-MI/O-DF standards for sensor data acquisition is developed. This framework is applied to a smart neighborhood use case to reduce food waste at the consumption stage.

Keywords: Situation awareness · Context awareness · Pervasive computing · Ontologies · Internet of Things · Context Space Theory

1 Introduction

The Food and Agriculture Organization of the United Nations (FAO) estimates that up to 50 % of produced food is wasted all over the world, which has a non-negligible impact on the society, environment and economy (e.g., starvation, carbon emission and economic cost, *etc.*) [8,10]. The information and system intelligence and analytics capabilities enabled by pervasive environments and the so-called Internet of Things (IoT), could potentially help drive innovative sustainable development and business models. The IoT offers provisions for real-time analysis on any operation or process as a game-changer when it came to creating environmental benefits. To take full advantage of the IoT, it is nonetheless crucial not to focus only on sensor data, but also on the "context" in which

© Springer International Publishing AG 2016
O. Galinina et al. (Eds.): NEW2AN/ruSMART 2016, LNCS 9870, pp. 101–114, 2016.
DOI: 10.1007/978-3-319-46301-8_9

this data was generated, monitored, and so forth. Context is any information that characterizes the environment of an entity (a person, group of person, a place or a Thing) relevant to the interaction of the application and end-users [1]. Context-awareness means understanding the whole environment and current situation of the entity. Context can be processed to developed more advanced services such as "situation-awareness", which can be seen as a course of events that evolves to more sophisticated relations between entities (or even situations).

Situations are defined as external semantic interpretation of sensor data on a higher level of abstraction than activities or context [22]. Thus, situation awareness strongly depends on expert knowledge to interpret sensed data. A situation aware approach requires modeling, reasoning, and sensor data acquisition, while considering several functional requirements for each step. Defining expert knowledge, adopting reasoning engines and integrating sensor data are extensive and error-prone tasks, complicating the development of situation aware applications. To address this problem and easily capture all domain- and application-specific dependencies, this paper investigates a general ontology-based framework for situation awareness based on standardized technologies.

The paper is structured as follows: Sects. 2 and 3 present the background and related work in the area. Section 4 proposes the core ontology and the framework architecture for general situation awareness. Section 5 presents a use case of a smart neighborhood to reduce food waste and an evaluation of the proposed framework; the conclusion follows. It is worth mentioning that the International Erasmus Mundus Masters Program "Pervasive Computing & Communications for Sustainable Development" (EMM PERCCOM) [11] enabled the research reported in this paper.

2   Background

Several theories (e.g., Context Space and Situation Theory, Semantic Sensor Network) and technological building blocks (e.g., O-MI/O-DF standards) have been considered to design the proposed framework for general situation awareness. This section therefore provides the necessary background regarding each of these theories and technologies.

First, the use of ontological approach allows situation modeling with rich semantics that can be understood and shared among humans and machines. In the life sciences community, Ontology Web Language (OWL) is extensively used and has become a de facto standard for ontology development [15]. OWL provides a vocabulary for the Resource Description Format (RDF) by extending the RDF Schema vocabulary. Given this, our framework is designed based on the OWL standard.

Another key theory considered in our framework is the Situation Theory Ontology (STO) [12], which was developed based on situation semantics referred to as Situation Theory [7]. This theory will be applied to model situations, where facts of situations are formulated as "infons", and "situations" are defined by specifying which infons they support (see Eq. 1). An infon is a relation of

n objects, whereas objects can be individuals, attributes or situations. The polarity $(0/1)$ specifies whether this relation is true or false.

$$S \models \ll relation, a_1, ..., a_n, 0/1 \gg \tag{1}$$

The Semantic Sensor Network (SSN) ontology [6] is a standard ontology to represent knowledge about a sensor network (e.g., sensing devices, measured properties and deployed platforms...), without initially taking into consideration actuators. The Semantic Actuator Network (SAN) ontology [17] was further developed as a counterpart to SSN. Both ontologies will be considered in our framework to specify the system setup.

The Context Space Theory (CST) [16] was developed – *based on a spatial representation of context* – to provide a general context model with a rich theoretical foundation. The context space is defined through context attributes and situations are modeled as subspaces. By combining specification- and learning-based techniques, and by supporting algebraic operations, CST allows for general reasoning about situations. CST-based reasoning is implemented in ECSTRA [4] with a flexible architecture for situation aware systems, which will be applied for the implementation of this study. The knowledge defined in STO will be used to generate the situation spaces in the context space.

Finally, to increase interoperability of the framework in a range of IoT settings, recent IoT messaging standards published by The Open Group, namely the O-MI (Open-Messaging Interface) and O-DF (Open-Data Format) standards, are used to enable peer-to-peer data exchange between different systems and devices [20]. O-MI messages can be exchanged on top of well-known protocols like HTTP, SOAP or SMTP, while O-DF [19] is a generic content description model for Things in the IoT, which can be extended with more specific vocabularies (e.g., using domain-specific ontology vocabularies) [9]. The knowledge defined in SSN and SAN will be used to generate context collectors based on O-MI/O-DF.

3   Related Work

Besides STO, other upper ontologies for situation awareness were developed by the research community. In the Core SAW Ontology [14], situations are represented as a set of entities with attributes, goals and foremost relations. It furthermore integrates observed sensed data in the ontology. The Situational Context Ontology [2] starts from a context perspective and adds a situational structure around it, while offering provisions for modeling imprecise sensor data (using fuzzy logic). The Situation Ontology developed in [21] is based on a context and situation layer, and allows the definition of atomic and composite situations based on context values.

Several hybrid approaches, combining ontologies with other reasoning techniques to achieve situation awareness have been proposed. In [5] the feasibility of integrating ontological knowledge into CST has been shown, based on both a context ontology and rule-based situation definitions. Situation spaces are generated by processing the rules and querying the ontology with SPARQL. The Wavellite framework [18] was proposed to achieve situation awareness in environmental monitoring. It uses upper ontologies, including STO, as a knowledge base and combines it with rules and neural networks for inference. However, the reasoning engines are application-specific. A number of approaches add rule-based reasoning around an ontology, such as BeAware! [3], which proposes a general reasoning technique by extending the Core SAW Ontologies and including relation types.

The aforementioned approaches have not been previously used in a food waste reduction or management process. We could nonetheless point out a few community-based social networks that apply pervasive computing to address this challenge, such as EUPHORIA (standing for Efficient food Use and food waste Prevention in Households through Increased Awareness) [13], which is a project that allows users to log and track their everyday food related behavior and redirect these, through social influence, towards more sustainable food related practices. Nonetheless, the project has focused on social behavior around food consumption (necessitating manual inputs via a mobile application), and has not proposed any IoT-based services to automate the discovery of food in the neighborhood that is e.g. close to its expiry date, and propose to end-users appropriate recipes. The paper investigates and develops a framework that fulfills such IoT-based services.

4   Framework Design

This section presents the core ontology for CST, generation of situation spaces and the framework architecture. In this respect, Sects. 4.1 and 4.2 respectively detail the CST Ontology, and how situations are generated based on this ontology. Section 4.3 provides an "at a glance" overview of the overall proposed framework.

4.1   CST Ontology

Situation spaces in CST are defined through a set of acceptable regions for all context attributes. Each context attribute is assigned with a relevance weight $w_i \in [0, 1]$. Furthermore acceptable regions are assigned with a contribution function η_i^S, which assigns a contribution $\in [0, 1]$ to each value within the acceptable region. The overall confidence if a situation is occurring is calculated based on the relevance and the contribution for a context state x, as shown in Eq. 2.

$$\mu_s = \sum_{i=1}^{n} w_i * \eta_i^S (x_i) \tag{2}$$

For final inference, the confidence value is compared to a threshold ε_i, as formulated in Eq. 3.

$$\gamma = (\mu_s \geq \varepsilon_i) \tag{3}$$

The specifications are the requirements for situation modeling. Thus, concepts of STO and SSN/SAN were mapped and extended with CST specific information. The core of the CST ontology is shown in Fig. 1. STO and SSN/SAN were mapped through two key connections: (i) `sto:Attribute` is defined as subclass of `ssn:Property`, and (ii) `sto:Individual` is defined as subclass of `ssn:FeatureOfInterest`. Through these definitions, sensors are observing attributes of individuals in STO situation definitions. Context attributes in CST correspond to both `sto:Attributes` and `ssn:Properties`. An acceptable region can be defined as a `sto:Value` of an attribute, whereas the value is specified as an interval. The ontology was extended to capture further CST-related concepts, which includes the `csto:ConfidenceThreshold` for a situation, `csto:Relevance` for an infon[1] and `csto:Contribution` for values of acceptable regions.

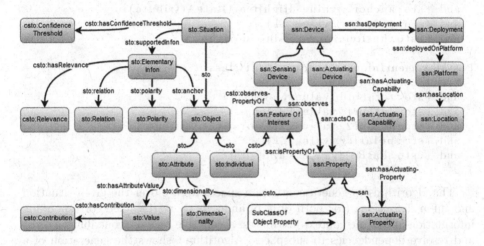

Fig. 1. Core ontology for Context Space Theory

4.2   Situation Generation

Situation generation from the core ontology can be declined into three categories, namely based on (i) situation objects, (ii) situation types, and (iii) situation objects generated by type definitions. Individual situations in STO are modeled with concrete instances for situations, infons and other objects. The objects involved in a situation may be application dependent, which makes it unable to reuse the situation specification. Instead, situation types can be defined based on OWL class axioms. Listing 1.1 presents an example for a situation type definition to infer if a person is running. The situation supports two infons, which are based on the movement speed and the heart rate of a person. The complete definition

---

[1] The relevance belongs to context attributes, but an attribute can have a different weight for different facts about situations.

(not shown in the listing for simplicity) further includes the specification of the individuals, attributes, acceptable regions and contribution.

Listing 1.1. Example for Situation Type Definition in the Ontology

```
PersonRunning  owl:equivalentClass  (
    sto:Situation
    and (sto:supportedInfon value HighHeartRateInfon)
    and (sto:supportedInfon value FastMovementInfon)
    and (sto:relevantIndividual value Person)
    and (csto:hasConfidenceThreshold value 0.8)
)
HighHeartRateInfon  owl:equivalentClass  (
    sto:ElementaryInfon
    and (sto:relation value Heartrate)
    and (sto:anchor1 value Person)
    and (sto:anchor2 value HighHeartRateAttribute)
    and (sto:polarity value _1)
    and (csto:hasRelevance value 0.6)
)
FastMovementInfon  owl:equivalentClass  (
    sto:ElementaryInfon      ,
    and (sto:relation value Movement)
    and (sto:anchor1 value Person)
    and (sto:anchor2 value FastMovementAttribute)
    and (sto:polarity value _1)
    and (csto:hasRelevance value 0.4)
)
```

The algorithms to generate situation spaces iterate over the given situation and infon definitions (for both objects and types), retrieve the corresponding information about the context attributes, acceptable regions, contribution, *etc.*, and resolve dependencies to subspaces. Algorithm 1 shows the generation of a situation space for one situation definition. Each generated situation space is then added with its confidence threshold to the context space.

Situation types ease the modeling process because situation definitions do not depend on application specific objects. In CST, objects can share the same situation space, while each object maintains a different state in the context space. The origin of the state may come from different sensors for different objects. This is captured and maintained through the integration of the SSN ontology. If it is not desired to resolve these dependencies via the ontology, separate situation spaces can be generated for each relevant individual involved in the situation (case (iii)).

4.3   Framework Architecture

Overall, the architecture needs to integrate the following major building blocks for a situation aware system:

– Knowledge base (CST ontology)

Algorithm 1. Generation of Situation Space

1: **function** GENERATESITUATIONSPACE(situation)
2:    $situationSpace \leftarrow new\ SituationSpace(situation.name);$
3:    **for all** $situation.getInfons()$ **do**
4:        **for all** $infons.getAnchors()$ **do**
5:            **if** $anchor.type() == attribute$ **then**
6:                $axis \leftarrow newAxis(attribute.name)$
7:                **for all** $attribute.getAcceptableRegions()$ **do**
8:                    **if** $infon.polarity() == 1$ **then**
9:                        $axis \leftarrow addRegion(value, contribution)$
10:                   **else**
11:                       $axis \leftarrow addAsymmetricRegion(value, contribution)$
12:               $situationSpace \leftarrow addAxis(axis, infon.getRelevance())$
13:           **else if** $anchor.type() == situation$ **then**
14:               $SubSpace \leftarrow$ GENERATESITUATIONSPACE($anchor$)
15:               $situationSpace \leftarrow addAxisSubSpace(SubSpace)$
16:           **else if** $anchor.type() == individual$ **then**
17:   ▷ Not considered in CST Situation Spaces
18:    **return** $situationSpace$

- Ontology management (OWL API, SPARQL-DL API, Pellet, Protégé)
- CST-based reasoning (ECSTRA)
- Sensor data acquisition (IoT Data Server for O-MI agents)
- Client application

Figure 2 illustrates the complete architecture designed in our study to integrate those building blocks. The knowledge base consists of CSTO-based application ontologies. Multiple ontologies with different situation specifications and the application setup can be provided for the system (ontology editors like Protégé can be used in this respect). The ontology management component is responsible for the programmatic access and manipulation of the knowledge base. Since the algorithm needs to access the TBox axioms (terminology) of the ontology, an OWL-centric approach is preferred over a RDF-centric approach. Tools used in our framework include OWL API, SPARQL-DL API and the Pellet reasoner. The ECSTRA implementation is used for CST-based reasoning. Finally, the O-MI/O-DF standards are integrated, meaning that instead of subscribing to a central publish/subscribe engine, the context collectors subscribe directly to one or more O-MI nodes and receive the notifications in an O-DF payload format. The central manager (*cf.* Fig. 2) forms an interface to integrate and coordinate all these components, while providing a facade to client applications to initialize the system and send enhanced reasoning requests. At a more concrete level, the tasks of the manager are:

1. Loading and merging given ontologies.
2. Initializing the ontology reasoner.
3. Generating situation spaces based on object and type definitions.

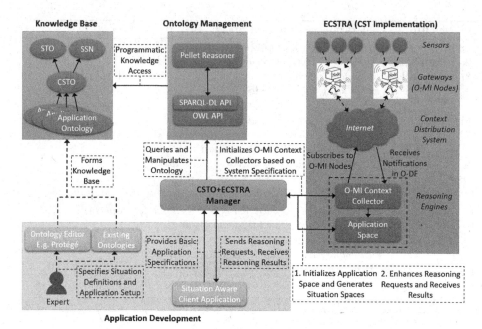

Fig. 2. Framework architecture for situation awareness

4. Generating O-MI context collectors based on the given specifications.
5. Initialize the application space.
6. Resolving dependencies to individuals and sensors from reasoning requests.
7. Distributing reasoning results.

From an operational perspective, the overall architecture has been implemented as a JVM-based library, which can be deployed in an agent-based architecture. Context collectors can potentially be added manually to access other types of information sources.

5   Use Case: Reducing Food Waste

This section describes a proof-of-concept and an evaluation of the proposed framework. Firstly, the situation awareness framework is applied to a use case to reduce food waste in a smart neighborhood in Sect. 5.1. Subsequently a discussion of the features of the framework and a performance evaluation follows in Sect. 5.2.

5.1   Use Case Scenario and Implementation

The overall scenario is depicted in Fig. 3, which considers a connected neighborhood and exploits situation awareness to give best recommendations about the consumption of food items (e.g. relevant recipes, incentives for food sharing...).

The application is developed as a JVM-based web application, where recipes are requested from an open REST API[2].

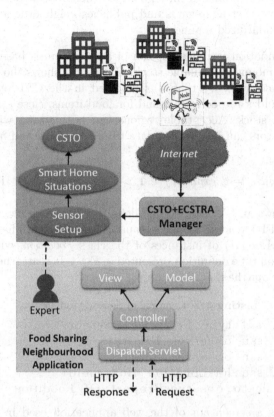

Fig. 3. Use case architecture

It should be noted that our implementation is based on the following assumptions:

- The implementation is based on a simulated smart neighborhood, composed of three households.
- Each household generates (simulated) sensor data values. To sense information about food items, it is assumed that each item is labeled with an RFID tag and smart fridges are equipped with RFID readers, to read these tags when items are placed inside the fridge.
- Information stored in the RFID tags includes the available amount of items and related expiration date.
- Sensor data providing information about when and how to access food items is simulated. For example, this input can be simulated based on human being's

---

[2] Yummly Recipe API: https://developer.yummly.com/.

activity in the household or on the availability of smart access devices (e.g., smart locks like slock.it[3]).

- All sensor values are published through an IoT/neighborhood avatar (an O-MI node in our case) that aggregates and publishes neighborhood-related information in a standardized manner.

The recommendation for consumption of a food item is based on the shelf life and relative amount of available stock. Equation 4 shows the situation type definition in situation theory that was modeled in the CST ontology, where parameters $\dot{f}$, $\dot{e}$ and $\dot{s}$ respectively stand for food items, close expiration dates and relative high stock. Acceptable regions can be modeled with fuzzy sets, e.g. context collectors can fuse sensed data to *low*, *medium* and *high* amount of available stock.

$$\left[\dot{S}_R | \dot{S}_R \models \ll expires, \dot{f}, \dot{e}, 1 \gg \land \ll stock, \dot{f}, \dot{s}, 1 \gg \right] \tag{4}$$

Listing 1.2 shows an OWL individual definition of the ontology for a sensor. It specifies an RFID reader, which is capable to observe different attributes (*Expiration* and *Amount*) of instances of the class *Fooditem*, which in turn is part of the situation type definition presented in Eq. 4. It is attached to a specific household via the ssn:hasLocation object property.

Listing 1.2. Example of Sensor Modeling

```
RFIDSensor001  rdf:type  FridgeRFIDSensor
RFIDSensor001  ssn:observes  Expiration
RFIDSensor001  ssn:observes  Amount
RFIDSensor001  ssn:hasLocation  Household1
RFIDSensor001  csto:observesPropertyOf  Fooditem
```

Figure 4 shows a screenshot of the web application used in our neighborhood waste management system, along with the recommendation outputs. The system identifies the items recommended for usage and displays the available amount, location and current accessibility. Furthermore it shows recipes that can be cooked with the available food items. A sustainability index is calculated for each recipe, based on the consideration of ingredients that are recommended for usage and the amount that will be prevented from being wasted. The index takes into account the environmental impact of the commodity group (carbon footprint, blue water footprint, economic cost [8]) of each ingredient.

5.2   Evaluation and Ontology Performance

The framework presented in this paper is based on a rich foundation for both situation modeling and situation reasoning, whose key functionalities are:

---

[3] Smart Locks slock.it: https://slock.it.

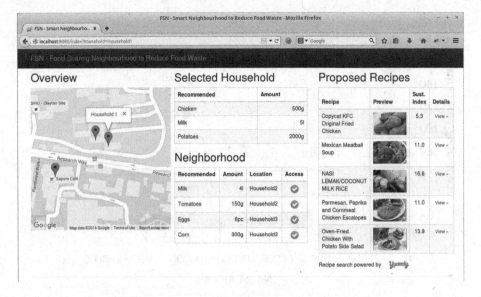

Fig. 4. Food sharing neighborhood application

- **Situation modeling.** Space and time aspects, situation types, roles of objects, relations.
- **Knowledge.** Integrating knowledge about situations and systems, allowing reuse and sharing with semantic web technologies.
- **Reasoning.** General applicability, uncertainty and temporal aspects, can be extended with prediction and proactive adaption. Enhanced through consideration of involved individuals via STO.
- **Application development.** Automated integration of sensor data acquisition, less complex for deployment.

After the automated initialization of the system, situation reasoning can be performed directly with ECSTRA or with enhanced requests involving access to the ontology. The proposed architecture does not the capabilities of the existing ECSTRA implementation. To perform (optional) enhanced reasoning requests, access to the ontology during run-time is required. Figure 5 shows the added computation time to resolve dependencies to individuals, attributes, situations and sensors for reasoning requests via the ontology.

The ontology was populated with test data. The largest data set consisted of 7500 situation definitions with attached infons, attributes etc. which corresponded to 137400 axioms in the ontology. The figure shows the average computing time with standard derivation for 1000 test runs per data set.

The test indicates a complexity of $O(n)$. However, ontology reasoners demand high memory. With further increased testing data the available heap space (6 GB) was not sufficient. This might be an issue for very large-scale systems. In this case run-time reasoning requests should be sent directly to individual specific generated situation spaces.

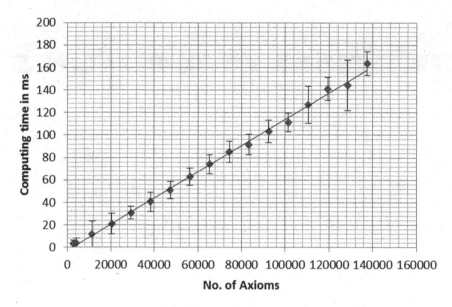

Fig. 5. Performance results for ontology access during reasoning requests

6   Conclusion and Future Work

The selected approach was based on an intensive study of situation aware approaches. As a result of this discussion, the combination of Situation Theory and Context Space Theory was motivated by automatic generation of Situation Spaces in CST with knowledge specified in an ontology. By identifying requirements for a holistic framework, STO, SSN and SAN were combined and extended to serve as a core ontology for a CST-based system. Algorithms to extract the knowledge from the ontology and initialize the application space were proposed.

Further discussion led to a design of an overall framework based on the proposed core ontology and ECSTRA for CST-based reasoning. In order to meet the requirements for platform independent sensor data acquisition, the IoT standards O-MI/O-DF were integrated into the system. The contribution of this work is a Java library which was designed to allow an efficient use of these components to develop situation aware applications.

As a proof-of-concept the framework was applied to a use case, which validated the feasibility of the approach. By showcasing a system to reduce food waste at consumption stage the use case demonstrated the enabling effects of situation aware systems regarding the contribution to sustainability.

Further work identified includes the consideration of a dynamic environment (joining and leaving objects, discovery of new sensor sources), validation of situation occurrences based on OWL axioms specified in the ontology, generating situation spaces with incomplete knowledge and adding actuation to the situation aware framework.

Acknowledgments. Authors acknowledge support from EMM PERCCOM, IoT EPI bIoTope Project, which is co-funded by the European Commission under H2020-ICT-2015 program, Grant Agreement 688203, as well as financial support from Ministry of Science & Education of Russian Federation, Grant RFMEFI58716X0031.

References

1. Abowd, G.D., Dey, A.K.: Towards a better understanding of context and context-awareness. In: Gellersen, H.-W. (ed.) HUC 1999. LNCS, vol. 1707, pp. 304–307. Springer, Heidelberg (1999)
2. Anagnostopoulos, C.B., Ntarladimas, Y., Hadjiefthymiades, S.: Situational computing: an innovative architecture with imprecise reasoning. J. Syst. Softw. **80**(12), 1993–2014 (2007)
3. Baumgartner, N., Gottesheim, W., Mitsch, S., Retschitzegger, W., Schwinger, W.: BeAware! - situation awareness, the ontology-driven way. Data Knowl. Eng. **69**(11), 1181–1193 (2010)
4. Boytsov, A., Zaslavsky, A.: ECSTRA – distributed context reasoning framework for pervasive computing systems. In: Balandin, S., Koucheryavy, Y., Hu, H. (eds.) NEW2AN 2011 and ruSMART 2011. LNCS, vol. 6869, pp. 1–13. Springer, Heidelberg (2011)
5. Boytsov, A., Zaslavsky, A., Eryilmaz, E., Albayrak, S.: Situation awareness meets ontologies: a context spaces case study. In: Christiansen, H., Stojanovic, I., Papadopoulos, G.A. (eds.) CONTEXT 2015. LNCS (LNAI), vol. 9405, pp. 3–17. Springer, Heidelberg (2015)
6. Compton, M., Barnaghi, P., Bermudez, L., GarcA-Castro, R., Corcho, O., Cox, S., Graybeal, J., Hauswirth, M., Henson, C., Herzog, A.: The SSN ontology of the W3C semantic sensor network incubator group. Web Semant. Sci. Serv. Agents World Wide Web **17**, 25–32 (2012)
7. Devlin, K.: Situation theory and situation semantics. Handb. Hist. Log. **7**, 601–664 (2006)
8. FAO: Food wastage footprint: impacts on natural resources. Technical report, FAO (2013). http://www.fao.org/docrep/018/i3347e/i3347e.pdf
9. Framling, K., Kubler, S., Buda, A.: Universal messaging standards for the IoT from a lifecycle management perspective. IEEE Internet Things J. **1**(4), 319–327 (2014)
10. Gustavsson, J., Cederberg, C., Sonesson, U., Otterdijk, R.V., Meybeck, A.: Global Food Losses and Food Waste. Food and Agriculture Organization of the United Nations, Rome (2011)
11. Klimova, A., Rondeau, E., Andersson, K., Porras, J., Rybin, A., Zaslavsky, A.: An international Master's program in green ICT as a contribution to sustainable development. J. Clean. Prod. **135**, 223–239 (2016)
12. Kokar, M.M., Matheus, C.J., Baclawski, K.: Ontology-based situation awareness. Inf. Fusion **10**(1), 83–98 (2009)
13. Lim, V., Yalva, F., Funk, M., Hu, J., Rauterberg, M.: Can we reduce waste and waist together through EUPHORIA? In: 2014 IEEE International Conference on Pervasive Computing and Communications Workshops (PERCOM Workshops), pp. 382–387. IEEE (2014)

14. Matheus, C.J., Kokar, M.M., Baclawski, K., Letkowski, J.A., Call, C., Hinman, M.L., Salerno, J.J., Boulware, D.M.: SAWA: an assistant for higher-level fusion and situation awareness. In: Defense and Security, pp. 75–85. International Society for Optics and Photonics (2005)

15. McGuinness, D.L., Harmelen, F.V.: OWL web ontology language overview. W3C Recommendation 10(10), 2004 (2004)

16. Padovitz, A., Loke, S.W., Zaslavsky, A.: Towards a theory of context spaces. In: Proceedings of the Second IEEE Annual Conference on Pervasive Computing and Communications Workshops, 2004, pp. 38–42. IEEE (2004)

17. Seydoux, N., Alaya, M.B., Drira, K., Hernandez, N., Monteil, T.: San (semantic actuator network). https://www.irit.fr/recherches/MELODI/ontologies/SAN.html

18. Stocker, M., Ronkko, M., Kolehmainen, M.: Situational knowledge representation for traffic observed by a pavement vibration sensor network. IEEE Trans. Intell. Transp. Syst. 15(4), 1441–1450 (2014)

19. The Open Group: Open data format standard (O-DF) (Open Group Standard) (2014). https://www2.opengroup.org/ogsys/catalog/C14A

20. The Open Group: Open messaging interface technical standard (O-MI) (Open Group Standard) (2014). https://www2.opengroup.org/ogsys/catalog/C14B

21. Yau, S.S., Liu, J.: Hierarchical situation modeling and reasoning for pervasive computing. In: The Fourth IEEE Workshop on Software Technologies for Future Embedded and Ubiquitous Systems, 2006 and the 2006 Second International Workshop on Collaborative Computing, Integration, and Assurance, SEUS 2006/WC-CIA 2006, p. 6. IEEE (2006)

22. Ye, J., Dobson, S., McKeever, S.: Situation identification techniques in pervasive computing: a review. Pervasive Mob. Comput. 8(1), 36–66 (2012)

Storing and Indexing IoT Context for Smart City Applications

Alexey Medvedev[1(✉)], Arkady Zaslavsky[2,3],
Maria Indrawan-Santiago[1], Pari Delir Haghighi[1],
and Alireza Hassani[1]

[1] Faculty of Information Technology, Monash University, Melbourne, Australia
{alexey.medvedev,maria.indrawan,pari.delir.haghighi,
ali.hassani}@monash.edu
[2] CSIRO, Data61, Melbourne, Australia
arkady.zaslavsky@csiro.au
[3] ITMO University, St. Petersburg, Russia
arkady.zaslavsky@acm.org

Abstract. IoT system interoperability, data fusion, data discovery and access control for providing Context-as-a-Service as well as tools for building context-aware smart city applications are all significant research challenges for IoT-enabled smart cities. These middleware platforms have to cope with potentially big data generated from millions of devices in large cities. The amount of context, metadata, annotations in IoT ecosystems equals and may even exceed the amount of raw data. This paper discusses the challenges of context storage, retrieval and indexing for smart city applications. We analyse, compare and categorise existing approaches, tools and technologies relevant to the identified challenges. The paper proposes a conceptual architecture of a hybrid context storage and indexing mechanism that enables and supports the Context Spaces theory based representation of context for large-scale smart city applications. We illustrate the proposed approach using solid waste management system with adaptive on-demand garbage collection from IoT-enabled garbage bins.

Keywords: Smart city · Context · Storage · Internet of Things (IoT) · Waste management

1 Introduction

Research and development projects in Smart Cities are actively pursued by the ICT research community. The number of systems and applications in such areas as Intelligent Transportation Systems (ITS), smart buildings and homes, city security and smart metering continues to grow by virtue of significant achievements in proliferation of various Internet enabled devices, sensor manufacturing and wireless communication services [1].

While some parts of the Smart city infrastructure are already implemented, the problem of providing relevant and reliable data to applications has not yet been solved. We refer to relevant metadata and annotations that describe the raw data as context. Applications that consume context are referred as context-aware applications. We

© Springer International Publishing AG 2016
O. Galinina et al. (Eds.): NEW2AN/ruSMART 2016, LNCS 9870, pp. 115–128, 2016.
DOI: 10.1007/978-3-319-46301-8_10

define both these terms in more detail in the following section. Existing context-aware systems are incompatible in the ways they represent and store context. To develop smarter applications we need to provide methods for acquiring context from context producers without being dependent on their particular properties or domains [2].

Data decentralisation is one of the main features of the Internet that adds complexity to the context incompatibility problem. Smart city applications need to integrate data from different sources like information systems, users' histories and sensory data. While being used, these systems and applications generate vast amounts of data, which is transferred, processed and stored by service providers and users' endpoints. This Big Data is constantly changing and the volume of data streams is growing, raising the questions of scalability, performance, interoperability, data format conversion, data discovery and security.

Mostly, modern Smart City systems work in an isolated fashion, having low level of interoperability that limits their functionality [2]. This leads to the concept of deploying Smart city scale middleware systems that will serve a number of agents (city authorities, enterprises, users, sensors, actuators) and provide them with transparent access to all needed context by communicating with similar middleware systems of other companies or organisations.

The idea of an IoT platform, a middleware system that enables various systems to interchange data for mutual benefits, is widely discussed in academic and research community [3–5] and a number of problems/questions are raised. One of these emerging questions is finding ways for structuring, modelling, storing, indexing and retrieving context from large datasets that are generated by different kinds of virtual and physical sensors and used for analytical, predictive and other purposes.

The process of storing and retrieving large datasets is not new and is well studied in both relational and NoSQL paradigms. As we move from internal systems owned by one organisation to a system of systems (SoS) [3], we see that each system, even in one domain of knowledge, uses different structures for representing information. Integrating exterior services into large systems usually lead to the development of drivers, protocols, parsers, ETL procedures and tests. Another problem is that the information is often needed in a higher level contextual form, rather than in a raw form. For example, a smart home system needs to be notified that the user left work and is driving home, so it is time to switch on the air-conditioning system. Notification with prediction of users' arrival time is high-level context in this case. In contrast, if the smart home system would receive GPS data directly from the user's smartphone instead of high-level context, the system will have to incorporate the whole stack of sophisticated data acquisition and reasoning algorithms. Such an approach would significantly increase the complexity and cost of the smart home system. At the same time, raw context can be used for a number of other applications like building a dynamic map of city traffic, timetable management etc. The solution is providing Context-as-a-Service [4, 6] by doing reasoning in a cloud system and delivering the context in interoperable form and at the level of the abstraction that is needed by different applications.

Development of novel efficient scalable methods for reasoning, aggregating, representing, storing and retrieving context on middleware side can bring us closer to seamless system interoperability, enabling the possibility for faster creation of more context-aware Smart City applications.

This paper addressed the challenges of context storage, retrieval and indexing for smart city applications. We analyse, compare and categorise existing approaches, tools and technologies relevant to the identified challenges. The paper proposes a conceptual architecture of a hybrid context storage and indexing mechanism that enables and supports the Context Spaces theory (CST) based representation of context for large-scale Smart City applications. The proposed approach is illustrated using a solid waste management system scenario with adaptive garbage collection from IoT enabled garbage bins.

The paper is organised in the following way: in Sect. 2, we provide information about related work in the area of context-awareness and its connection with storage systems for IoT middleware. In Sect. 3, we discuss requirements to storage systems that will be used on the persistence level of Smart City scale IoT platform. In Sect. 4, we describe existing software solutions that are suitable for large-scale context storage and introduce the Context Spaces theory as an efficient instrument for dealing with situation awareness. In Sect. 5, we describe how the proposed storage can be used in a Smart City usage scenario. Section 6 concludes the paper and defines directions for future work.

2   Related Work

The term "context" is well studied in literature and has a number of definitions. In this paper we will use the definition provided by Dey in [7]: "Context is any information that can be used to characterise the situation of an entity. An entity is a person, place, or object that is considered relevant to the interaction between a user and an application, including the user and applications themselves." Dey also provides a definition for context-aware computing: "A system is context-aware if it uses context to provide relevant information and/or services to the user, where relevancy depends on the user's task".

Several context representation standards (e.g. ContextML [8] and SensorML [9]) have been proposed. Antunes el al. in [2] suggest that none of these standards are widely used and the lack of compatibility between context-aware platforms forces to deal with various context representation forms.

Bazire et al. stated that "it was not possible to develop in isolation a model of context because context, knowledge and reasoning are strongly intertwined" [10]. Brezillon also suggested that "the notion of context can take on different meanings, depending on, well. . . . the context" [11]. Dourish stated that there is no need in deciding what is context and what is not in general. He defines contextuality as "a relational property that holds between objects or activities" [12].

While the concept of context is still debated, from the middleware perspective context is any data and metadata that can be queried for making decisions about an entity's situation. Unstructured information, like images, video streams or files, sound or natural language text are not considered context, as any direct precise queries on such data are impossible. Video streams, images and sound can be processed by recognition software for situation-awareness. The outputs of such processing, if they are presented in any structured form, can become part of contextual information.

There are several dimensions for characterizing context. First of all, context can be classified by as sensed (e.g. current GPS position), static (e.g. map of the location), derived (e.g. address of presence) and profiled (slowly changing) [13]. All these types can be used as context by different applications. Secondly, context can be current (e.g. GPS coordinates), historical (set of points representing users track), aggregated and compressed (most common tracks of user represented by critical points only) etc. Historical and aggregated context plays significant role for any machine-learning algorithms, which are used for making reasoning and predictions. Thirdly, context can be used by different types of applications ranging from one person's needs in managing any smart space to city authorities needs for making tactical or even strategic decisions about infrastructural management. All of these significant differences indicate that trying to model or represent all the possible variants of context using one approach can be a serious challenge.

Wagner et al. [4] analyse requirements for the Context-as-a-Service middleware platform. The defined requirements related to storage part of the platform are (i) possibility to exchange context information that is heterogeneously and (ii) consumption of resources used by context services should be minimised. Hong and Landay [14] advocate advantages of an infrastructure approach to context aware computing which include (i) system interoperability, (ii) loose coupling and independence of systems and (iii) simpler mobile devices with less power consumption. They also declare five challenges for context-aware infrastructure which are (i) simple but expressive data formats for context data representation, (ii) building discovery services, (iii) finding balance between smart infrastructure and smart devices, dividing their responsibilities, (iv) defining scopes for dealing with security and privacy of data and (v) building scalable infrastructures for dealing with large number of sensors and devices.

There has been a number of academy and industry projects in the field of IoT [15], but providing scalable hybrid storage solution for middleware support of context-aware applications, to the best of our knowledge, is not well introduced yet. A survey of context modelling techniques is presented in [16]. The authors mention all the popular methods for modelling context but do not concentrate on storage or representation approaches for large-scale environments. Another important aspect for most storage solutions is building indexes. Approaches for Big data indexing are analysed in [17].

A novel approach for building semantically rich interoperable services is JavaScript Object Notation for Linked Data (JSON-LD) [18, 19]. While being a valid JSON, that allows processing it with all existing techniques, it adds possibilities for building RESTful services with Semantic web data integration. This approach promises to seriously increase loose coupling of systems and self-descriptiveness of context. JSON was traditionally used for serialising data and exchanging messages in JavaScript applications. The design of JSON-LD targets easy integration of existing deployed JSON-based systems with Linked Data approach and enables development of interoperable Web services as well as using document-oriented datastores for storing Linked Data [20].

One of the main concept of JSON-LD is also called "context". Here this term means something different and enables linking of JSON documents with RDF model that represents an ontology. JSON-LD context can be described together with the main document, but it can also be contained in another document to which the main document is linking. The developers of JSON-LD were trying to preserve the concepts that are known to developers and the decision was made to use entity-centric approach rather than triple-centric approach [21].

Some projects based on JSON-LD are already implemented. For example, in [22] Szekely et al. describe the development of a system for preventing human trafficking. The project involved building a knowledge graph based on indexing JSON-LD documents by ElasticSearch instead of using triple store and SPARQL. The collection of JSON-LD documents was generated by text mining techniques and represented in denormalized way.

3   Context Storage Requirements

In this section, we identify and discuss requirements of a context storage middleware. These requirements can have sufficient differences with context representation that is optimal in ubiquitous/mobile computing scenarios because of the significant difference in computing power, value and veracity of data streams and durability expectations.

Endpoint applications need to acquire context from various sources. The only way for these application to get context about the outside world is to communicate with some middleware, as communication with enormous numbers of sensors is not feasible due to many restrictions, such as network bandwidth, energy efficiency, access control and complexity of task. The middleware receives requests for context from clients and tries to fulfil these requests. For this, the middleware platform should either store all the information inside or query some other systems for retrieving the needed data. The first approach is disc space consuming, but it can improve performance. For example, modern search engines use indexed information for providing search results. The second approach is time-consuming, as querying other systems and especially mobile sensors and devices can be a time-consuming process due to networks delays and slow response time or inaccessibility of mobile data sources. The IoT middleware may combine both approaches that will result in a better balance of disc space consumption, performance and data relevance.

We have identified the following requirements of a context storage middleware:

Disk Based – although in-memory systems are getting more attention nowadays, the amount and variety of data make processing not possible without keeping data persistently on disk.

Scalability – it is hard to predict the amount of stored information, but in case of a Smart City it would not be possible to provide the storage service by one server node. This means that proposed solution must be horizontally scalable.

High Availability – the storage should not have a single point of failure (SPoF).

Structural Freedom – storage must be able to store structured data without applying restrictions on its structure.

Interconnected Entities – in some cases storage must facilitate the means for storing highly interconnected data (e.g. relations of people, organisations, transport, infrastructure etc.) and effectively running queries over such data.

Veracity – different sources can supply information that can be conflicting or uncertain and there should be a way to store all variants of incoming data with annotations about the identity and trust level of the originator and rank of the suggestion. Context of the querying side must be treated respectively during responding to the query.

Large Amounts of Sensory Data – sensors and other Internet enabled devices generate large number of time series events of similar but not the same structure.

Ontology Support – a number of research projects [15] model data using ontological principles as it is a good way for modelling the domain interconnections and facilitating reasoning over data. However, this approach does not seem to be suitable for storing large amounts of raw data and low-level context.

Fast Information Retrieval and Rich Indexing Capabilities – performance is the key requirement for context delivery in smart cities applications. This highlights the need for efficient indexing of stored context.

Fast Writes – streams of sensor readings must be written on disk without long queues and expensive rebuilding of indexes.

Geospatial Data – many of Smart City applications are highly dependable of geospatial context, so the middleware storage must be able to provide effective indexing possibilities for this type of context.

Various Approaches to CAP Theorem. Traditionally, one of the main principles of database management systems is ACID – Atomicity, Consistency, Isolation and Durability. According to the CAP theorem, we cannot have consistency, availability and partitioning tolerance in one system at the same time. As mentioned, the context coming from different sources can already be uncertain and conflicting. That means the middleware solution in some cases can afford lack of transactional support and consistency in favour of high availability and partitioning, as the requirements for horizontal scalability and availability have higher priority. At the same time some parts of middleware system can have strong requirements for consistency and these requirements must be satisfied. After analysing the requirements for the middleware storage system it becomes clear that fulfilling all the requirements with one existing solution is not feasible. The variety of data processing approaches leads us to the idea of hybrid storage architecture.

One of the recent trends in software development is polyglot persistence [23, 24]. It means systems no longer try to accomplish all tasks using one data storage, but rather use different technologies to store data where each technology provides certain capabilities. In [23] a use case of PolygotHIS, a health information system using three different databases is presented. The system uses relational database (PostgreSQL) for storing structured transactional data, document-oriented datastore (MongoDB) for storing schemaless documents and graph datastore (Neo4J) for storing data containing relationships. PolygotHIS implements various software agents to achieve interoperation between involved data stores. In [24] polyglot persistence approach is used to Enhance Performance of the Energy Data Management System (EDMS). EDMS uses MySQL, MongoDB and OpenTSDB [25] that runs over HBase.

In the next section we briefly describe the most popular approaches for context modelling and representation with respect to storage considerations. We also mention the most popular open-source software projects in each area.

4 Existing Context Representation and Storage Technologies

4.1 Modelling Techniques

Key-Value is a popular NoSQL modelling and storage technique that represents any information with a key association and retrieves it by the given key effortlessly and quickly. The key-value modelling is the fastest, easiest and noticeably scalable way of retrieving information from storage. However, standards, schema, verification and relations between entities are not offered. The most important point from our per-spective is the absence of means for searching inside values, making it possible to request data only by key.

Document-Oriented or Mark-up scheme tagged encoding is another NoSQL tech-nique for representing context. It is still very flexible and scalable, but allows to organise data in structures, usually using JSON as serialization format. Documents are organised in collections and the most important – there are ways to organise different types of indexes over collections, making fast queries possible. Data denormalization is a strong and at the same time weak point of this approach. It is fast to retrieve and write, but the data can easily become inconsistent. Furthermore, document-oriented approaches consume more disk space in comparison with the relational approaches due to applying data denormalization as a main data modelling technique. Organizing relations between documents is possible, but document storage engines usually do not support joins, as it assumes that this work should be done by higher-level software components. The most widely used document-oriented stores are MongoDB [26] and CouchDB. JSON-LD fits naturally with MongoDB document model. Another example of document-oriented datastore is ElasticSearch [27]. It is a multi-tenant search engine based on Lucene. The difference between ElasticSearch and other document-oriented datastores is its ability to automatically create mappings and index documents of structure that was not defined in advance. ElasticSearch indexes data using inverted lists and wide-columns based on the type of incoming data. ElasticSearch uses a specialised JSON-based query language called Query DSL. Another distinguishing feature of Query DSL is the presence of scoring function that enables data search based on unprecise queries with computing the relevance of returned data.

Relational Database is another way of context storage. Relational database manage-ment systems (RDBMS) technology is one of the most well established technologies and have been used as a main approach for data management for more than 40 years. Allowing an excellent level of stability, functional richness, knowledge base and other benefits, relational model has a serious disadvantage for modelling context – it is the rigid schema that makes it hard to store any information that is not structured in the way that is defined by relational schema. Another problem is the expensiveness of joins between tables. Most well-known open-source relational databases are PostgreSQL and MySQL.

Ontology Based Modelling is a way of organizing context into ontologies using semantic technologies like RDF or OWL. A large number of development tools, reasoners, standards and storage engines [28] are available. Ontologies give capabilities for defining entities and expressing relations between them. However, when dealing with Big Data, retrieval of context can be resource consuming and issues with scalability may arise. Besides, ontologies are not recommended for representing streams of sensor data. Examples of RDF storage engines are Jena2, Sesame, AllegroGraph, Virtuoso, etc. [28] Most popular serialization formats are Turtle, N-Triples, N-Quads, N3, RDF/XML and JSON-LD.

Graph-Based Modelling is a natural way for representing entities and interconnections between them. They are ideal for representing unstructured information and information that has ambiguity. Graphs are typeless and, schemaless, and there are no constraints on relations. This structure is ideal for representing social networks and is recommended for read-mostly requirements. Graph databases have a lot in common with RDF storages but use different languages for querying data. Some graph databases can be used as RDF storages with special plugins applied. According to [29] the popularity of graph databases has increased by 500 % within 2014–2015 years period. Most popular graph Databases are Neo4J, Titan and OrientDB.

Object Based Modelling. Numerous projects focus on context-awareness common object-oriented programming languages technique of modelling context as objects [30, 31]. These projects deliver huge theoretic base and numerous advanced features for context processing without focusing on the persistence problem that makes them hard to use in a large-scale environment. Though numerous attempts were taken to develop object storage, the industry standard is still mapping objects to a relational database schema. This is usually done manually or with a special object/relational framework facilitating automatic process of mapping entities and hiding the persistence level under ORM abstractions [32]. The main problem of this approach is called object-relational impedance mismatch [33], which represents a set of difficulties while transferring data from object model with polymorphism, inheritance and encapsulation to the denormalized table-based database approach.

Our research of context representation approaches is summarised by providing quantitate analysis in Table 1. We use the following designations: Disk based (D); Relations (R); Veracity (C); Geospatial data indexing (GSI); Storage of Sensory Data

Table 1. Summary of context representation approaches and their intersections with Smart City platform storage requirements

	D	R	V	GSI	SD	SL	HS	FW
Relational	+	+	−	+	+/−	−	−	+/−
Ontology	+	+	+	−	−	+	−	−
Key-value	+	−	+	−	+/−	++	++	++
Document	+	+/−	+	+	++	++	++	+
Wide-column	+	−	+	+	+	+	++	+
Graph	+	++	+	+	−	++	+/−	−
Object	−	+	−	−	−	++	−	+

(SD); Schemaless/Structural data freedom (SL); Horizontal Scalability (HS); Fast Writes (FW); Strong/native support (++); Supported (+); Limited support (+/−); Not supported (−).

According to the analysis of context representation and storage techniques we identify the document-oriented approach as the most suitable for our purposes.

4.2 Context Spaces Theory

One of the above mentioned theoretical foundations for context representation and reasoning about situation awareness is the CST [34]. It uses geometric metaphors for representing context attributes and building multidimensional spaces. Special context situations algebra is used for situation detection and prediction.

The visualization of a situation subspace and context-situation pyramid [35] in CST is presented in Fig. 1. CST proposes steps to a generic framework for context-aware applications and provides a model and concepts for context description and operations over context. This theory is implemented in two frameworks ECORA [31] and ECSTRA [30] and has been extended in Fuzzy Situation Inference (FSI) [27] for situation modelling and reasoning under uncertainty and other advanced reasoning capabilities. These frameworks use the aforementioned object-based modelling approach and do not focus on issues such as scalability or persistence. Developing methods for mapping context spaces theory approaches to scalable and efficient hybrid storage can help to implement these methods in large-scale Smart City middleware usage scenarios.

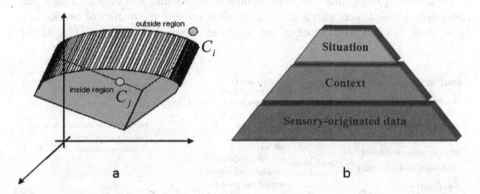

Fig. 1. Visualization of situation subspace in context spaces theory (a) and context-situation pyramid (b) [35]

5 Smart City Use Case

5.1 Waste Collection Scenario

One of the possible use cases of IoT platforms is the management of solid waste collection in Smart Cities [36, 37]. The high level view of the whole data exchange

Fig. 2. The big picture of data exchange between stakeholders in solid waste management domain [36]

process and its stakeholders is shown in Fig. 2. The IoT platform is a meeting point for interests of all the stakeholders including citizens, municipalities, truck owning companies, recycling factories, city administration and others. Smart city invests in installing wireless level sensors on waste bins, enabling provisioning of context information to truck owning companies that are responsible for collecting solid waste.

By consuming the above-mentioned context information, decision support systems of truck owning companies can build optimised routes that will help to reduce fuel consumption and drivers efforts at the same time keeping the quality of service at a high- level. General architecture of the proposed approach is shown in Fig. 3. The

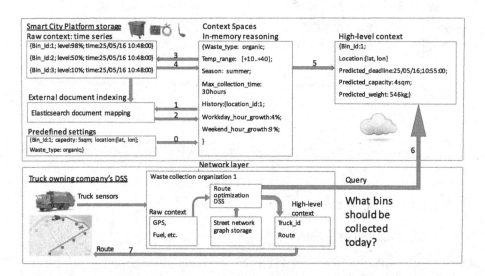

Fig. 3. Smart City platform storage and requests from context consumers

storage part of the IoT platform consists of raw context and high level context parts. Sensors generate raw context. The parameters are sensor id, time, location, level, air pollution etc. Using raw context for building routes is possible, but there are a number of problems that need to be solved. The hardest problem is reasoning over all available sensory information on the fly. The only information that the truck owning company's system needs for route optimisation is the list of bins that need to be collected in the nearest time. After making reasoning on raw context we receive only the ID of bin, waste capacity and the deadline for collecting the waste from the current bin. We propose to store this high level context in the second layer of the storage. This high level context gives the route planner precise information and is easy to process. The estimated deadline can be reasoned using historical information from the raw context (how fast the bin is being filled depending on part of the day, day of week etc.) and special reasoning rules that specify behaviour for different types of waste (organic, plastic etc.), rules for different temperatures, seasons and locations. These attributes are closely related to context attributes of the CST making the formal situation prediction possible. One of the responsibilities of the CTS reasoners is to check the correctness of situation recognitions or predictions and to refresh the high-level context if needed. This is done by background scanning of indexed raw context.

We propose a multi-level context storage architecture with background near real time index scanning for mapping CST notions on persistence layer for solving scalability and performance issues in Smart City-scale deployments on commodity hardware.

5.2   Choice of Technologies

According to our findings and [38], document-oriented storage is the most suitable technology for storing raw context as the "nature" of raw context is based on XML or JSON formats. Moreover, JSON documents are the most straightforward way for serializing in-memory objects, which makes document-oriented approach the most suitable for facilitating scalable datastore for CST reasoning algorithms.

Although wide-column storage engines like Apache Cassandra or Apache HBase perform better on some benchmarks [39], we have chosen MongoDB [26] as a primary raw context storage because of its horizontal scalability and native compatibility with JSON-LD. MongoDB offers a wide range of indexing capabilities, but issues with indexing nested objects can significantly spoil the performance [40]. As mentioned before, there is no information at the system design time as to what document structures exactly will be stored causing a lack of indexes on unexpected fields. To encompass this problem, we propose to use an external search engine based in full text search technology. We chose Elasticsearch [27] for this purpose. Although some projects propose using Elasticsearch as a main datastore [41], we prefer to have a more durable solution for storing master data. Usage of external indexing engine allows to reduce performance decrease of main storage that could be caused by rebuilding indexes while doing frequent inserts of sensory data.

6 Conclusion and Future Work

In this paper we briefly reviewed the definition of context from the IoT middleware viewpoint. We have found out that basic qualities of context are its searchability, presence of structure but absence of fixed schema, and semantic interoperability for system integration. We discussed the main approaches for context representation with respect to existing storage technologies and have found that in general context is closer to document-oriented representation with addition of semantic mark-up for storage purposes.

We propose a novel context storage approach based on polyglot persistence methods with addition of external indexing engine. This architecture is based on the presented analysis of context-modelling approaches and existing software solutions. Our solution adopts JSON-LD for semantically rich context modelling. In addition, we implement a multi-level context storage approach that is powered by utilizing CST algorithms. The proposed middleware storage solution opens perspectives for using the CST in the Smart City-scale environment for alleviating system interoperability and higher situational awareness in applications.

CST offers a number of techniques for context reasoning. In the future, we plan to perform large-scale tests of different storage solutions to discover the most optimal way for storing context. Another target is combining our proposed storage solution with efficient access mechanisms. Having a proper approach for context provisioning and acquisition is crucial for IoT applications. To ease the process of application development for IoT, it is vital to have a proper language to query context. Our research group is currently developing Context Definition and Query Language (CDQL), a convenient and flexible query language to express context information requirements without considering details of the underlying structure. An important feature of the query language is to make it possible to query entities in IoT environment based on their situation in a fully dynamic manner where users can define the situations and context entities as a part of the query.

Acknowledgements. Part of this work has been carried out in the scope of the project bIoTope which is co-funded by the European Commission under Horizon-2020 program, contract number H2020-ICT-2015/688203 – bIoTope. The research has been carried out with the financial support of the Ministry of Education and Science of the Russian Federation under grant agreement RFMEFI58716X0031.

References

1. Petrolo, R., Loscrí, V., Mitton, N.: Towards a smart city based on cloud of things. In: Proceedings of the 2014 ACM International Workshop on Wireless and Mobile Technologies for Smart Cities - WiMobCity 2014, pp. 61–66. ACM Press, New York (2014)
2. Antunes, M., Gomes, D., Aguiar, R.: Semantic-based publish/subscribe for M2M. In: Proceedings - 2014 International Conference on Cyber-Enabled Distributed Computing Knowledge Discovery CyberC 2014, pp. 256–263 (2014)

3. Maia, P., Cavalcante, E., Gomes, P., Batista, T., Delicato, F.C., Pires, P.F.: On the development of systems-of-systems based on the Internet of Things. In: Proceedings of the 2014 European Conference on Software Architecture Workshops - ECSAW 2014, pp. 1–8. ACM Press, New York (2007)
4. Wagner, M., Reichle, R., Geihs, K.: Context as a service - requirements, design and middleware support. In: 2011 IEEE International Conference on Pervasive Computing and Communications Workshops (PERCOM Workshops), pp. 220–225. IEEE (2011)
5. Balandina, E., Balandin, S., Koucheryavy, Y., Balandin, S., Mouromtsev, D.: Innovative e-tourism services on top of Geo2Tag LBS platform. In: 2015 11th International Conference on Signal-Image Technology and Internet-Based Systems (SITIS), pp. 752–759. IEEE (2015)
6. Moore, P., Xhafa, F., Barolli, L.: Context-as-a-Service: a service model for cloud-based systems. In: 2014 Eighth International Conference on Complex, Intelligent and Software Intensive Systems, pp. 379–385. IEEE (2014)
7. Dey, A.K.: Understanding and using context. Pers. Ubiquit. Comput. **5**, 4–7 (2001)
8. Knappmeyer, M., Kiani, S.L., Frà, C., Moltchanov, B., Baker, N.: ContextML: a light-weight context representation and context management schema. In: ISWPC 2010 - IEEE 5th International Symposium Wireless Pervasive Computing 2010, pp. 367–372 (2010)
9. Sensor Model Language (SensorML). http://www.opengeospatial.org/standards/sensorml
10. Bazire, M., Brézillon, P.: Understanding context before using it. In: Dey, A.K., Kokinov, B., Leake, D.B., Turner, R. (eds.) CONTEXT 2005. LNCS (LNAI), vol. 3554, pp. 29–40. Springer, Heidelberg (2005)
11. Brézillon, P., Gonzalez, A.J.: Context in computing. In: Igarss 2014, pp. 1–571 (2014)
12. Dourish, P.: What we talk about when we talk about context. Pers. Ubiquit. Comput. **8**, 19–30 (2004)
13. Henricksen, K., Indulska, J.: A software engineering framework for context-aware pervasive computing. In: Proceedings of the Second IEEE Annual Conference on Pervasive Computing and Communications, 2004, pp. 77–86. IEEE (2004)
14. Hong, J., Landay, J.: An infrastructure approach to context-aware computing. Hum.-Comput. Interact. **16**, 287–303 (2001)
15. Perera, C., Zaslavsky, A., Christen, P., Georgakopoulos, D.: Context aware computing for the internet of things: a survey. IEEE Commun. Surv. Tutor. **16**, 414–454 (2014)
16. Strang, T., Linnhoff-Popien, C.: A Context Modeling Survey
17. Adamu, F.B., Habbal, A., Hassan, S., Cottrell, R.L., White, B., Abdullahi, I.: A Survey On Big Data Indexing Strategies. In: 4th International Conference on Internet Applications, Protocol and Services (NETAPPS2015). Cyberjaya, Malaysia (2015)
18. JSON for Linking Data. http://json-ld.org/
19. Lanthaler, M., Gütl, C.: On using JSON-LD to create evolvable RESTful services. In: Proceedings of the Third International Workshop on RESTful Design - WS-REST 2012. p. 25. ACM Press, New York (2012)
20. JSON-LD 1.0, A JSON-based Serialization for Linked Data. https://www.w3.org/TR/json-ld/
21. Lanthaler, M.: Creating 3rd generation web APIs with hydra. In: Proceedings of the 22nd International World Wide Web Conference (WWW 2013), pp. 35–37. ACM Press, Rio de Janeiro (2013)
22. Szekely, P., et al.: Building and using a knowledge graph to combat human trafficking. In: Arenas, M., et al. (eds.) ISWC 2015. LNCS, vol. 9367, pp. 205–221. Springer, Heidelberg (2015). doi:10.1007/978-3-319-25010-6_12

23. Kaur, K., Rani, R.: A smart polyglot solution for big data in healthcare. IT Prof. **17**, 48–55 (2015)
24. Prasad, S., Avinash, S.B.: Application of polyglot persistence to enhance performance of the energy data management systems. In: 2014 International Conference on Advances in Electronics Computers and Communications, pp. 1–6. IEEE (2014)
25. OpenTSDB. http://opentsdb.net/index.html
26. MongoDB. https://www.mongodb.com/
27. Elasticsearch. https://www.elastic.co/
28. Faye, D.C., Cure, O., Blin, G.: A survey of RDF storage approaches (2016)
29. Andlinger, P.: Graph DBMS increased their popularity by 500 % within the last 2 years. http://db-engines.com/en/blog_post/43
30. Boytsov, A., Zaslavsky, A.: ECSTRA – distributed context reasoning framework for pervasive computing systems. In: Balandin, S., Koucheryavy, Y., Hu, H. (eds.) NEW2AN 2011 and ruSMART 2011. LNCS, vol. 6869, pp. 1–13. Springer, Heidelberg (2011)
31. Padovitz, A., Loke, S.W., Zaslavsky, A.: The ECORA framework: a hybrid architecture for context-oriented pervasive computing. Pervasive Mob. Comput. **4**, 182–215 (2008)
32. Hybernate. http://hibernate.org/
33. Ireland, C., Bowers, D., Newton, M., Waugh, K.: A classification of object-relational impedance mismatch. In: Proceeding - 2009 1st International Conference on Advance Databases, Knowledge, Data Applications DBKDA 2009, pp. 36–43 (2009)
34. Tuesday, O., Prime, B., Tony, M.: 1 Towards a theory of context, pp. 1–27 (2010)
35. Padovitz, A., Zaslavsky, A., Loke, S.W.: A unifying model for representing and reasoning about context under uncertainty
36. Medvedev, A., Fedchenkov, P., Zaslavsky, A., Anagnostopoulos, T., Khoruzhnikov, S.: Waste management as an IoT-enabled service in smart cities. In: Balandin, S., Andreev, S., Koucheryavy, Y. (eds.) NEW2AN/ruSMART 2015. LNCS, vol. 9247, pp. 104–115. Springer, Heidelberg (2015)
37. Anagnostopoulos, T.: Robust Waste Collection Exploiting Cost Efficiency of IoT Potentiality in Smart Cities, pp. 7–9 (2015)
38. Santos, N., Pereira, O.M., Gomes, D.: Context storage using NoSQL. In: Conferência sobre Redes Computadores (2011)
39. Abramova, V., Bernardino, J.: NoSQL databases. In: Proceedings of the International C* Conference on Computer Science and Software Engineering - C3S2E 2013, pp. 14–22. ACM Press, New York (2013)
40. MongoDB vs. Elasticsearch: The Quest of the Holy Performances. http://blog.quarkslab.com/mongodb-vs-elasticsearch-the-quest-of-the-holy-performances.html
41. Elasticsearch as a Time Series Data Store. https://www.elastic.co/blog/elasticsearch-as-a-time-series-data-store

ruSMART: Smart Services in Automotive Industry

"Connected Car"-Based Customised On-Demand Tours: The Concept and Underlying Technologies

Alexander Smirnov[1,2], Nikolay Shilov[1,2(✉)], and Oleg Gusikhin[3]

[1] SPIIRAS, St. Petersburg, Russia
{smir,nick}@iias.spb.su
[2] ITMO University, St. Petersburg, Russia
[3] Ford Motor Company, Dearborn, MI, USA
ogusikhi@ford.com

Abstract. Wide spreading of such concepts as ubiquitous computing, connectivity, cyberphysical systems, Internet of Things opens various possibilities both in increasing the human productivity in various tasks and developing new business models that allow companies to transform from product suppliers to service providers or even to virtual companies acting as brokers. The paper proposes a concept of customised on-demand tours by cars or minivans as one of the phenomena of the above transformation. It also describes the technological basis underlying the concept and explains the proposed scenario via an illustrative case study.

Keywords: Connected car · On-demand tour · Service · Technological basis

1 Introduction

Such concepts as ubiquitous computing, connectivity, cyberphysical systems, Internet of things have been deeply penetrating into our lives. This trend doesn't only help to increase the productivity in various tasks but also opens a whole new world of business models allowing companies to transform from product suppliers to service providers or even to virtual companies acting as brokers.

For example, Rolls-Royce instead of selling aircraft engines now charges companies for hours that engines run and takes care of servicing the engines [1]. Another famous example is Uber, that does not only provides taxi services, but it does this without actually owning cars and acts just as a connecting link between the taxi drivers and passengers. Timely changed business model can provide for a significant competitive advantage (e.g., the current capitalisation of Uber is about $68 billion, which is $20 billion higher than that of GM [2]).

As a result, all significant players of the global markets are searching for various ways to extend and update their businesses in order to keep pace with the changing markets (e.g., Toyota is investing into Uber and promoting its cars for Uber drivers [2]).

O. Galinina et al. (Eds.): NEW2AN/ruSMART 2016, LNCS 9870, pp. 131–140, 2016.
DOI: 10.1007/978-3-319-46301-8_11

In this paper we propose a concept and a technological basis for customised on-demand tours by cars or minivans. The concept integrates the ideas of e-tourism and on-demand taxi ride.

The reminder of the paper is structured as follows. Section 2 describes the current trends, achievements and challenges in the e-tourism area. Section 3 introduces the connected car phenomenon and some related innovative mobility models. The concept of the customized on-demand tours is proposed in Sect. 4. It is followed by the technological basis. The main results are summarised in the conclusion.

2   e-Tourism

The ubiquitous world in which we live is characterized by a high mobility of individuals, most of them wearing devices capable of geo-localization (smartphones or GPS-equipped cars) [3]. Mobiquitous environment is a next generation of ubiquitous environment, which supports adaptation to mobility of people and applications, and changes in devices state. In other words, mobiquitous environment has a mobile and ubiquitous nature. This is also a step towards the "infomobility" infrastructure, i.e. towards operation and service provision schemes whereby the use and distribution of dynamic and selected multi-modal information to the users, both pre-trip and, more importantly, on-trip, play a fundamental role in attaining higher traffic and transport efficiency as well as higher quality levels in travel experience by the users [4]. It is a new way of service organization appeared together with the development of personal mobile and wearable devices capable to present user multimodal information at any time. Infomobility plays an important role in the development of efficient transportation systems, as well as in the improvement of the user support quality. In accordance with the forecast of [5], the market of such technologies as mobile Internet, automation of knowledge work, and Internet of Things by 2025 can increase 20 trillion USA dollars.

Development of tourist services and apps (mobile device applications) has got popularity recently [e.g., 6]. "In a field trial in Görlitz (Germany), 421 tourists explored the city with one of two different mobile information systems, a proactive recommender of personalized tours and a pull service presenting context-based information on demand. A third group of tourists was tracked by GPS receivers during their exploration of the destination relying on traditional means of information. Results point out that both mobile applications gained a high level of acceptance by providing an experience very similar to a traditional guided tour. Compared to the group tracked by GPS loggers, tourists using a mobile information system discovered four times more sights and stayed at them twice as long" [7].

"The findings of the evaluation carried out have demonstrated that the widget-based solution is better than the notification-based solution. Despite the fact that both options are considered good solutions to achieve proactivity, the second one is considered by the users more annoying. <...> We can state that the "time pressure" factor is a good indicator to know when a proactive recommendation is reasonable or not, because in these situations users give less feedback" [8].

Analysis of existing at the moment apps [9, 10] in the market shows that there is a trend towards providing proactive tourist support based on his/her location, preferences, and current situation in the area (weather, traffic jams, and etc.) [11]. Development of such systems is still an actual task that attracts researchers from all over the world [e.g., 12–14]. Such systems are aimed to solve the following tasks:

- generate recommended attractions and their visiting schedule based on the tourist and region contexts and attraction estimations of other tourists; tourist context characterizes the situation of the tourist, it includes his/her location, co-travelers, and preferences; region context characterizes the current situation of tourist location area, it includes his/her location, co-travelers, and preferences; region context includes such information as weather, traffic jams, closed attraction, etc.
- collect information about attractions from different sources and recommend the tourist the best for him/her attraction images and descriptions;
- propose different transportation means for reaching the attraction;
- update the attraction visiting schedule based on the development of the current situation.

3   Connected Cars and Innovative Mobility Models

"Connected car" or "connected vehicle" is a relatively new term originating from the Internet-of-Things vision standing for the vehicle's connectivity with the around on a real time basis for providing the safety and expedience to the driver [15].

Car manufacturers are continuously developing the in-vehicle electronic systems that have made a significant step forward recently. Such systems have transformed from simple audio players to complex solutions (referred to as "infotainment systems") that enable communication with smartphones, sharing information from different vehicle sensors, information delivery through in-vehicle screen or stereo system (e.g., Ford SYNC[1], GM OnStar MyLink[TM2], Chrysler UConnect[®3], Honda HomeLink[4], Kia UVO[5], Hyundai Blue Link[6], MINI Connected[7], Totyota Entune[8], BMW ConnectedDrive[9], Apple CarPlay[10], Google's Auto Link[11], etc.). A detailed review can be found in [16].

---

[1] http://www.ford.com/technology/sync/.

[2] https://www.onstar.com.

[3] http://www.chryslergroupllc.com/innovation/pages/uconnect.aspx.

[4] http://www.homelink.com/.

[5] https://www.myuvo.com/.

[6] https://www.hyundaiusa.com/technology/bluelink/.

[7] http://www.mini.com/connectivity/.

[8] http://www.toyota.com/entune/.

[9] http://www.bmw.com/com/en/insights/technology/connecteddrive/2013/-index.html.

[10] https://www.apple.com/ios/carplay/.

[11] http://www.motorauthority.com/news/1092768_googles-auto-link-in-car-system-to-rival-apple-carplay.

Such systems do not only address the driver's experience, but also are aimed at entertaining the passengers of the car, e.g. engaging the video subsystems for rear seat passengers, multi-zone climate control, etc.

Integration of different mobile apps with in-vehicle system is a promising task related to the "connected car" concept. Integration of on-board infotainment systems with various cloud services can help in creating various intelligent decision support systems capable of providing a richer driving experience and seamless integration of information from various sources. Recent advances in car on-board infotainment systems make it possible to organize the mentioned above infomobile support for the driver and passengers.

Besides the mentioned above innovative taxi services (such as Uber, Gett, Yandex Taxi, Lyft, etc.) there could be mentioned several more interesting innovative mobility models. There is a number of initiatives related to the car sharing concept (BMW's DriveNow, Zipcar, Anytime, enjoy, etc.). The key idea of such services is that the user has an app in his/her mobile phone that is used to locate the car, reserve and enter it (no need for keys). So, everything is done in an automated way "on the go".

Another direction of the developing mobility services is parking support. The app would connect to a cloud service with information on parking spot availability, reserve a spot and update the route destination in the cars' navigation system to lead the driver to the reserved parking spot (BMW's ParkNow, Ford's Parking Spotter[12], some research efforts [17]).

One more model to be mentioned is Uber Tour taking advantage of the good knowledge of cities by Uber drivers. One can order one of the predefined tours with a certain duration, and the driver will pick the traveler up and taking him/her through a number of points of interests.

In this paper we have tried to achieve a synergy between the connected car and e-tourism ideas.

4 Customized On-Demand Tours: The Concept

Imagine the following scenario. You are about to leave a foreign city but your flight is in the evening and the hotel check out is at 11 am. Wouldn't it be nice to have a tour for few hours? There are a number of existing tourist support systems and services and some more intelligent ones are being developed [18] that can advice some tourist routes and attractions based on your preferences. You even can have a mobile app that analyses your schedule and suggests the tour automatically.

You can modify the suggested tour and reserve a car or (in case of a bigger travelling company) a comfortable minivan with the driver is reserved for you. The driver will pick you up at the predefined time and will follow the route loaded into the car's navigation system automatically (Fig. 1).

---

[12] https://media.ford.com/content/fordmedia/fna/us/en/news/2015/01/06/mobility-experiment-parking-spotter-atlanta.html.

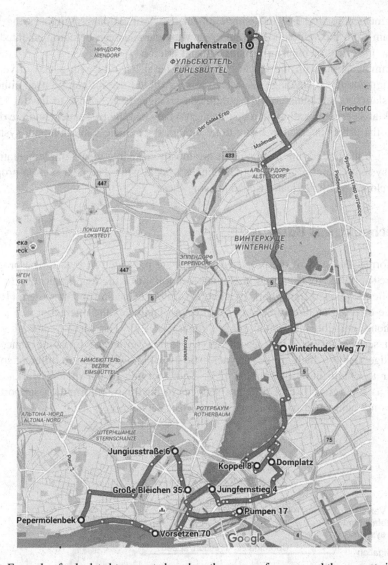

Fig. 1. Example of calculated tour route based on the user preferences and the current situation

Another important aspect of the concept is the usage of the passengers' personal electronic devices (such as tablets) during the tours. These devices can be used for the following purposes:

- Tour guidance. The passengers can read, view and listen about the attractions they are passing by.
- Vehicle systems control. Though for safety reasons it is not allowed to switch or on off various systems and subsystems of the connected car, it is still possible to adjust some settings such as climate control, switch on/off interior lights, open/close windows, etc.

- Augmented reality. For the entertainment and tour guidance purposes the video stream from the vehicle's front and rear view cameras can be complemented with some historical or other informative visual artefacts.
- Communication with the driver. In case of having a ride with a taxi driver in a foreign country a language issues can arise. A cloud service can be used either for translation or for transferring pre-defined messages (e.g., a wish to have a 15 min walk at a certain location) can be transferred between the passengers and the driver, with the driver interaction being done though the vehicle's infotainment system.

Though the underlying technologies are mostly available today, implementation of such a system creates a number of challenges to deal with. The next section proposes the developed technological basis addressing these.

5 Technological Basis

The information flows of the proposed concept are presented in Fig. 2. The proposed ideas are supported by advanced intelligent technologies with their application to Web. The earlier developed framework of a context-driven decision support system [19] has a service-oriented architecture. Such architecture facilitates the interactions of service components and the integration of new ones [20–22]. The services are integrated through service fusion. The idea of service fusion originates from the concept of knowledge fusion, which implies a synergistic use of knowledge from different sources in order to obtain new information [23]. Thus, service fusion in this work can be defined as synergistic use of different services to have new information support possibilities not achievable via usage of the services separately.

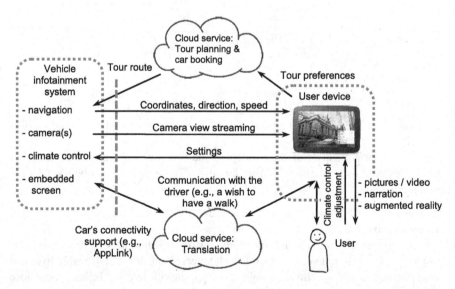

Fig. 2. Information flows

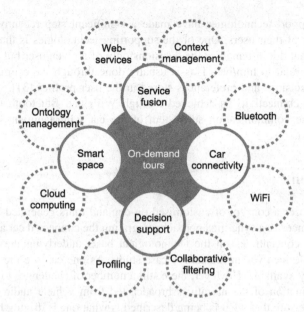

Fig. 3. Technological basis

The proposed technological basis is presented in Fig. 3. The solid circles denote higher level technologies, while dashed circles denote lower-level technologies, supporting the former.

The synergic integration of different services (or service fusion [24]) is the basis for intelligent usage of information from various sources. Context-based service fusion can provide a new, previously unavailable level of personalised on-board information support via finding compromise decisions taking into account possibilities of various Web-services and the current situation.

Due to the mobile device restrictions (limited computational capacities and power consumption), it is not reasonable to perform a complex computations in a mobile device. In this case, an infrastructure is needed that allows different devices to interact with each other for delegation of computations to a cloud during solving their tasks.

The smart spaces technology [25–29] aims at the seamless integration of different devices by developing ubiquitous computing environments, where different services can share information with each other, perform computations, and interact with each other for joint task solving.

The open source Smart-M3 platform [30] is one of the platforms that can be used for implementation of the smart space. The platform aims at providing a Semantic Web information sharing between software entities and devices. Usage of this platform makes it possible to significantly simplify further development of the system, include new information sources and services, and to make the system highly scalable due to the usage of the common protocols and semantics. The semantics is supported by the common ontology. The key idea of this platform is that the formed smart space is device-, domain-, and vendor-independent. Smart-M3 assumes that devices and software entities can publish their embedded information for other devices and software entities through simple, shared information brokers.

Decision support technologies have made a significant step recently to the customized support of their users. One of the supporting technologies is that of collaborative filtering that doesn't analyze the decisions only of a given user but also of other users that are similar to him/her. This is usually done through the estimation of previously made decisions and preferences stored in the user profiles [31].

The vehicle connectivity is achieved through WiFi and Bluetooth technologies, supported by the communication subsystem of the car's infotainment system (e.g., Ford's AppLink).

6 Conclusion

The paper proposes a concept of customised on-demand tours generated based on the personal preferences as a synergic approach integrating the connected car and e-tourism technologies. It concentrates on the technological basis underlying the concept and explains to whole idea via an illustrative case study. As one can see the technologies above are mostly available. However, there still a number of challenges to address. For example, coordination of the narratives broadcasted from vehicle audio system with navigation (to be sure that what is being described is what one is driving by); dynamic tour adjustment based on the traffic/weather changes; on-the-go tour update by the passenger; multimodal tours assuming not only taxi-like rides but also including some public traffic routes or bicycling; and many others.

The goal of the future research is to implement a prototype modelling the major functions of the proposed concept and to develop underlying models and methods aimed to solve the above challenges.

Acknowledgements. The presented results are part of the research carried out within the project funded by grants #15-07-08092, 15-07-08391, 16-07-00375, 16-07-00375 of the Russian Foundation for Basic Research. The work has been partially financially supported by the Government of Russian Federation, Grant 074-U01.

References

1. Bryson, J.R., Daniels, P.W. (eds.): Handbook of Service Business: Management, Marketing, Innovation and Internationalisation. Edward Elgar Publishing, Cheltenham (2015)
2. The Wall Street Journal (2016). http://www.wsj.com/
3. Artigues, C., Deswarte, Y., Guiochet, J., et al.: AMORES: an architecture for mobiquitous resilient systems. In: ARMOR 2012 Proceedings of the 1st European Workshop on AppRoaches to MObiquiTous Resilience, pp. 7:1–7:6. ACM (2012)
4. Ambrosino, G., Nelson, J.D., Bastogi, B., Viti, A., Romazzotti, D., Ercoli, E.: The role and perspectives of the large-scale Flexible Transport Agency in the management of public transport in urban areas. In: Ambrosino, G., Boero, M., Nelson, J.D., Romanazzo, M. (eds.) Infomobility Systems and Sustainable Transport Services, ENEA 2010, pp. 156–165 (2010)
5. Manyika, J., Chui, M., Bughin, J., Dobbs, R., Bisson, P., Marrs, A.: Disruptive Technologies: Advances that will Transform Life, Business, and the Global Economy: Executive Summary. McKinsey Global Institute, San Francisco (2013). 22 p. http://www.mckinsey.com/

6. Balandina, E., Balandin, S., Koucheryavy, Y., Mouromtsev, D.: Innovative e-tourism services on top of Geo2Tag LBS platform. In: The 11th International Conference on Signal Image Technology and Internet Systems (SITIS 2015), Bangkok, Thailand, pp. 752–759 (2015)
7. Modsching, M., Kramer, R., ten Hagen, K., Gretzel, U.: Effectiveness of mobile recommender systems for tourist destinations: a user evaluation. In: Yorke-Smith, N. (ed.) Interaction Challenges for Intelligent Assistants: Papers from the AAAI Spring Symposium, Technical report SS-07-04, pp. 88–89 (2007)
8. Bader, R., Siegmund, O., Woerndl, W.: A study on user acceptance of proactive in-vehicle recommender systems. In: 3rd International Conference on Automotive User Interfaces and Interactive Vehicular Applications (AutomotiveUI 2011), pp. 47–54 (2011)
9. Price, E.: Travel apps that can replace your tour guide (2015). http://www.cntraveler.com/stories/2015-02-24/travel-apps-that-can-replace-your-tour-guide
10. Cowen, B.: A personal tour guide – almost everywhere – for $9.99 or less! (2015). http://www.johnnyjet.com/2015/01/a-personal-tour-guide-almost-everywhere-for-9-99-or-less/
11. Smirnov, A., Kashevnik, A., Ponomarev, A., Shilov, N., Teslya, N.: Proactive recommendation system for m-tourism application. In: Johansson, B., Andersson, B., Holmberg, N. (eds.) BIR 2014. LNBIP, vol. 194, pp. 113–127. Springer, Heidelberg (2014)
12. Gerhardt, T.: 3 Ways Multi-Modal Travel is Tricky for App Developers (2015). http://mobilitylab.org/2015/03/10/3-ways-multi-modal-travel-is-tricky-for-app-developers/
13. Staab, S., Werthner, H., Ricci, F., Zipf, A., Gretzel, U., Fesenmaier, D.R., et al.: Intelligent systems for tourism. IEEE Intell. Syst. 17(6), 53–64 (2002)
14. Hasuike, T., Katagiri, H., Tsubaki, H., Tsuda, H.: Interactive approaches for sightseeing route planning under uncertain traffic and ambiguous tourist's satisfaction. In: Eto, H. (ed.) New Business Opportunities in the Growing e-Tourism Industry, pp. 75–96. Business Science Reference, Hershey (2015)
15. Shim, H.B.: The technology of connected car. J. Korea Inst. Inf. Commun. Eng. 20(3), 590–598 (2016)
16. Smirnov, A., Kashevnik, A., Shilov, N., Ponomarev, A.: Smart space-based in-vehicle application for e-tourism: technological framework and implementation for Ford SYNC. In: Balandin, S., Andreev, S., Koucheryavy, Y. (eds.) NEW2AN/ruSMART 2014. LNCS, vol. 8638, pp. 52–61. Springer, Heidelberg (2014)
17. Smirnov, A., Shilov, N., Gusikhin, O.: Socio-cyberphysical system for parking support. In: Proceedings of the 5th International Conference on Information and Electronics Engineering (ICIEE 2015), pp. 145–151 (2015)
18. Smirnov, A., Kashevnik, A., Ponomarev, A., Teslya, N., Shchekotov, M., Balandin, S.I.: Smart space-based tourist recommendation system: application for mobile devices. In: Balandin, S., Andreev, S., Koucheryavy, Y. (eds.) NEW2AN/ruSMART 2014. LNCS, vol. 8638, pp. 40–51. Springer, Heidelberg (2014)
19. Smirnov, A., Kashevnik, A., Shilov, N., Gusikhin, O.: Context-driven on-board information support: smart space-based architecture. In: The 6th International Congress on Ultra Modern Telecommunications and Control Systems (ICUMT 2014), pp. 195–200. IEEE (2014)
20. Web Services Architecture, W3C Working Group Note, Web (2014). http://www.w3.org/TR/ws-arch/
21. Alonso, G., Casati, F., Kuno, H.A., Machiraju, V.: Web Services. Concepts, Architectures and Applications. Springer, Heidelberg (2004)
22. Papazoglou, M.P., van den Heuvel, W.-J.: Service oriented architectures: approaches, technologies and research issues. VLDB J. 16(3), 389–415 (2007)
23. Smirnov, A., Levashova, T., Shilov, N.: Patterns for context-based knowledge fusion in decision support. Inf. Fusion 21, 114–129 (2015)

24. Smirnov, A., Shilov, N., Makklya, A., Gusikhin, O.: Context-based service fusion for personalized on-board information support. In: Fischer-Wolfarth, J., Meyer, G. (eds.) Advanced Microsystems for Automotive Applications 2014. LNMOB, vol. 1, pp. 111–120. Springer, Heidelberg (2014)

25. Cook, D.J., Das, S.K.: How smart are our environments? An updated look at the state of the art. Pervasive Mob. Comput. 3(2), 53–73 (2007)

26. Balandin, S., Waris, H.: Key properties in the development of smart spaces. In: Stephanidis, C. (ed.) UAHCI 2009, Part II. LNCS, vol. 5615, pp. 3–12. Springer, Heidelberg (2009)

27. Kiljander, J., Ylisaukko-oja, A., Takalo-Mattila, J., Eteläperä, M., Soininen, J.-P.: Enabling semantic technology empowered smart spaces. Comput. Netw. Commun. 2012, 1–14 (2012)

28. Ovaska, E., Cinotti, T.S., Toninelli, A.: The design principles and practices of interoperable smart spaces. In: Advanced Design Approaches to Emerging Software Systems: Principles, Methodologies, and Tools, pp. 18–47. IGI Global (2012)

29. Korzun, D.G., Balandin, S.I., Gurtov, A.V.: Deployment of smart spaces in Internet of Things: overview of the design challenges. In: Balandin, S., Andreev, S., Koucheryavy, Y. (eds.) NEW2AN 2013 and ruSMART 2013. LNCS, vol. 8121, pp. 48–59. Springer, Heidelberg (2013)

30. Honkola, J., Laine, H., Brown, R., Tyrkko, O.: Smart-M3 information sharing platform. In: Proceedings of ISCC 2010, pp. 1041–1046. IEEE Computer Society (2010)

31. Smirnov, A., Shilov, N.: Ontology matching in collaborative recommendation system for PLM. Int. J. Prod. Lifecycle Manag. 6(4), 322–338 (2013)

Smart Driving: Influence of Context and Behavioral Data on Driving Style

Mikhail Sysoev[✉], Andrej Kos, and Matevž Pogačnik

Laboratory for Telecommunications, Faculty of Electrical Engineering,
University of Ljubljana, Tržaška 25, 1000 Ljubljana, Slovenia
mikhail.sysoev@ltfe.org

Abstract. In this article, we present an approach to determine stress level in a non-invasive way using a smartphone as the only and sufficient source of data. We also present the idea of how to partly transfer such approach to the determination of the driving style, as aggressive driving is one of the causes of car accidents. For determination of the driving style a variety of methods are used including the preparation movements before maneuvers, identification of steering wheel angle, accelerator and brake pedal pressures, glance locations, facial expressions, speed, medical examinations before driving as well as filling out of the questionnaires after the journey. In our paper we present a methodology for estimation of potentially unsafe driving (in the meaning of more intensive acceleration and braking compared to average driving) and discuss how to estimate such unsafe driving before it actually takes place. We present sensors and data which can be used for these purposes. Such data include heart rate variability from chest belt sensor, behavioral and contextual data from smartphone, STAI short questionnaire to assess personal anxiety and anxiety as a state at certain moment, and initial interaction with car during opening and closing of the car doors. To determine intensive acceleration and braking we analyzed GPS data like speed, acceleration and also data from accelerometer inside the car to avoid interference in GPS-signal. Actually, our long term goals are to provide feedback about potentially unsafe driving in advance and thus strengthening driver's attention on the driving process before the start.

Keywords: Sensors · Driving style · Context · GPS · Stress

1 Introduction

Stress is undoubtedly an important factor in our lives. Problems related to mental health are gradually moving to priority positions in the structure of public health of today's world. Stress is one of the main reasons for this, as it causes activation of the sympathetic division of the autonomic nervous and the hypothalamo–hypophyseal portal systems. The result of this reaction is secretion of hormones, including cortisol which is responsible for neurotoxic damages, emotional and vegetative reactions, and ultimately, behavioral and mental disorders and somatic diseases.

Stress can be both a brief reaction to some events, and it can also have a prolonged/chronic effect on the body. The most common stress recognition methods are connected with determination of physical and physiological responses of the body to stress.

© Springer International Publishing AG 2016
O. Galinina et al. (Eds.): NEW2AN/ruSMART 2016, LNCS 9870, pp. 141–151, 2016.
DOI: 10.1007/978-3-319-46301-8_12

According to researchers [1], the following body features can be used for these purposes: heart rate (HR), heart rate variability (HRV), electrodermal activity (EDA), electrocardiogram (ECG), electromyogram (EMG), skin temperature (ST), pupil dilation (PD), blood volume pulse (BVP), respiration, voice features, facial expression, eye gaze, and blink rates. Despite the progress and the potential of physiological and physical stress recognition, these methods also have its disadvantages: In order to gather data on body (physiological data) parameters, it is necessary to wear various wearable sensors (chest belts, wrist bands, head wearing devices, skin patches, and others).

Rapid growth of smartphone use, development of its technical capabilities, and an increasing number of sensors built into them allow us to analyze various stress situations using the data received from smartphones as the only and sufficient source (behavioral pattern and contextual data). A smartphone is a device that we always carry with us. Smartphone data analysis method for stress recognition can replace the physical and physiological data analysis method and consequently reach maximal noninvasiveness and unobtrusiveness in the cases where it is necessary [1]. Stress can also influence our driving style in bad way.

Why do we want to relate our stress detection approach to the driving style? The current status of world road safety remains an ongoing concern. The Global Status Report on Road Safety 2013 [2], the UN World Health Organization (WHO) declares that:

- the total number of road traffic deaths remains unacceptably high at 1.24 million per year;
- 20 and 50 million drivers and passengers injured in the accidents;
- $518 billion globally can be estimated losses from the accidents.

The main causes of accidents are, in addition to drunk driving, running of red lights, speeding, reckless driving and weather conditions, also aggressive driving styles, like intensive acceleration and braking. One approach to reducing the number of accidents is the identification of potentially unsafe driving before it actually takes place. Determination of reasons for changes in driving style (e.g. from calm to aggressive) typically requires collection of driver and context related data before the journey, such as driver's activity, walking style, driver's behavior and contextual data, physiological features. And also collecting data during driving: HRV, light, which is related to weather conditions, sound level, phone calls etc.

For determination of the driving style a variety of methods are used including the preparation movement before maneuvers, identification of steering wheel angle [3], accelerator and brake pedal pressures, glance locations, facial expressions, speed [4–6], medical examinations before driving (especially for public transport drivers) as well as filling out of the questionnaires after the journey. Additional information can be obtained by measuring changes in the drivers' behavior such as deviation in the manner of opening and closing car door (different intensity of opening and closing the car door) [7, 8] and collecting context data before the journey to know what has happened.

Information about driving style can be used in car safety systems, corporate systems of the driving safety assessment, parental control systems for young drivers, driver engagement and coaching, eco-driving system, because aggressive driving also leads to higher fuel consumption and emissions of the entire cycle [9].

The remainder of the paper presents a methodology of stress determination using only smartphone and partially, based on this approach, methodology of driving style estimation, using contextual data. It includes sensors review, methodology, experiment conditions, data review, discussion and conclusions.

2 The Proposed Approach for Stress Determination

Next we describe an approach for perceived stress recognition using data collected from a smartphone. Aggressive driving, if it is not a feature/habit of person, could be caused by stressful situations/conditions, which could be determined using above mentioned data (we assume that aggressive driving can be caused by stress or some situation), for this reason we provide this example. A smartphone Nexus 5 was applied (behavioral and contextual data collection, provision of current stress level self-assessments) for the purpose of developing a solution for stress determination in an indirect way (noninvasive). No additional wearable sensors were applied.

The information collected includes audio, gyroscope and accelerometer features, light condition, screen mode (on/off), current stress level self-assessment (by providing 7-scale self-report of current stress level every hour) and the current activity type.

Three stress analysis models have been built: two with the consideration of current activities of a participant and one without those. Classification of low- and high-stress conditions, which was executed for a separate model for a certain kind of activity only, enabled us to achieve approximately 4 % higher accuracy than under the conditions when those activities were neglected [1]. Results of applying different analysis algorithms with model considering the activity type are presented in Table 1. Also, an

Table 1. Classification results using various models and algorithms.

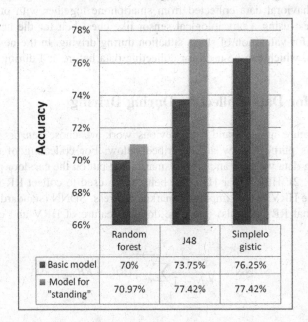

	Random forest	J48	Simplelo gistic
■ Basic model	70%	73.75%	76.25%
■ Model for "standing"	70.97%	77.42%	77.42%

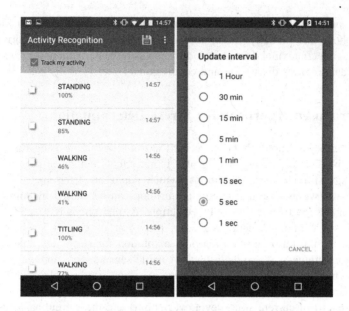

Fig. 1. Android application for activity recognition.

Android application was developed as a means for the current activity-type identification based on Google activity recognition. The application interface is presented in Fig. 1.

Although the results are preliminary they show, with conjunction with others researches in this field, that data from smartphone can be applied for stress determination. Because of this result, we are going to partly transfer such approach to the driving activity. To provide methodology for estimation of driving style using contextual and behavioral data collected from smartphone together with other kinds of data. We suggest using a physiological sensor like chest belt for the heart rate variability (HRV), for validation of stress situation during driving. In the next section we discuss sensors, which can be used for collecting data before and during driving.

3   Sensors for Data Collection During Driving

Based on literature reviews and our previous work we chose sensors which seem suitable for this purpose, they are described below. For collecting of the car door opening/closing data we used an android smartphone put on the car door pocket inside the holder (Fig. 2). HRV Polar H7 chest belt can be used to collect RR-intervals and then to calculate HRV - as an important marker of stress. SDNN (standard deviation of normal to normal RR intervals) is a time domain feature of HRV and can be easily calculated:

$$SDNN = \sqrt{\frac{1}{N-1}\sum_{n-2}^{N}\left(RR_j - RR_{avgj}\right)^2} \tag{1}$$

Fig. 2. Measuring of car door acceleration

It is established by physicians that low HRV leads to increase feelings of fatigue and is associated with stress [10]. For GPS data we used two sensors; one is a built-in smartphone sensor (smartphone was placed on the dashboard) and second one is a U-blox GSP receiver + external antenna. It is important to note that most smartphones provide only 1 Hz sampling of GPS data, because of hardware limits, while the U-blox receiver is supposed to provide 18 Hz sampling. Actually, during real-life test U-blox provided only 10 Hz. In practice it is advisable to check all sensors in real-life scenarios and not rely only on documentation, because in different conditions sensors can provide different resolution.

In our previous work [1] we developed an Android application based on framework Funf [11] to collect contextual data during working days. Then we extended our app, called Sensoric, which is now able to collect:

- contextual data like activity level, calendar entries, screen on/off, audio features/ ambient noise, external light in lux, gyroscope and accelerometer data, call and SMSlog;
- questionnaires, the Spielberger State-Trait Anxiety Inventory (STAI) before driving (Fig. 3). Test users are able to note if there was anything unusual during the journey and leave a comment, assess traffic conditions and assess how aggressive was their driving in their opinion;
- recognition of current activity type based on Google activity recognition.

Also we collected accelerometer data during driving with 40 Hz resolution to compare it with GPS data. It is depicted in Fig. 4.

It should be noted that if there are no places on the way with poor GPS signal, the acceleration calculated from GPS with 1 Hz sampling and acceleration provided by

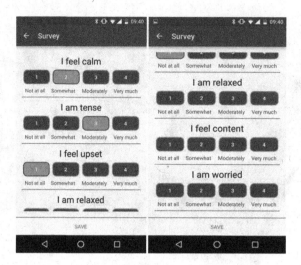

Fig. 3. 6-items STAI questionnaire.

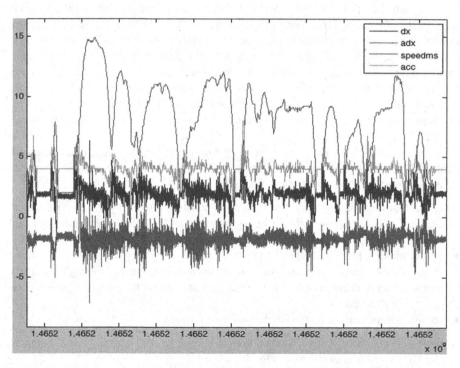

Fig. 4. Correlation between acceleration calculated from GPS NMEA data and from smartphone accelerometer data (speed, acceleration, value of X and Y axes).

accelerometer inside the smartphone with 40 Hz sampling, are very similar, especially if the data was smoothed before, for example by moving average. Both these data channels can be used to correct each other, especially when GPS receiver loses the signal.

4   Methodology and Experiment Scenarios

In this chapter we explain the experiment scenarios, which we used for test data collection among 6 test participants. The application Sensoric starts collecting contextual data and activity type data approximately 1 h before going from work, while it is also possible to switch it on manually, if necessary. The participant puts on a heart rate chest belt and starts recording HRV data 10–20 min before driving to get adapted to the chest belt. Just before driving (leaving work), the participant takes the first survey (the Spielberger State-Trait Anxiety Inventory), while the system switches on the accelerometer on the car door through an SMS. In the car we use a smartphone on the dashboard (LG Nexus 5, Nexus 6, Sony Xperia, Nexus 4 were tested) to collect GPS NMEA and Sensoric data. After driving, the participant takes the second survey and also assesses traffic and his aggressive driving, which leads to automatic switch off of the accelerometer on the car door by sending another SMS and stopping collection of Sensoric data.

Because of different contextual data influences on the driving style it is important to set or at least take into account the context of the experiment. These are the participant's driving experience, sex, age group, weather, traffic, route, number of passengers, time of the day and others. The best way, if it is possible, is to set up all conditions of experiment and use them during all journeys (for example, run the experiment only in sunny weather, on the same route and passengers amount etc.).

4.1   Opening/Closing of the Car Door

The door of the car is the first "system", which the driver interacts with. In our opinion, this is an interesting opportunity to collect contextual data in an unobtrusive manner through opening and closing of the car door.

The update rate of the accelerometer on car door is 40 Hz. It is not enough for building a pattern of opening and closing a car door, but it is enough for getting certain features from this data. These features are:

- Time between opening and closing car door;
- Time between opening car door and starting the engine;
- Intensity of car door movement (how fast one opens/closes the car door).

In Fig. 5 a raw data set of the car door opening and closing is depicted. And these features could be added to the analysis model as contextual data.

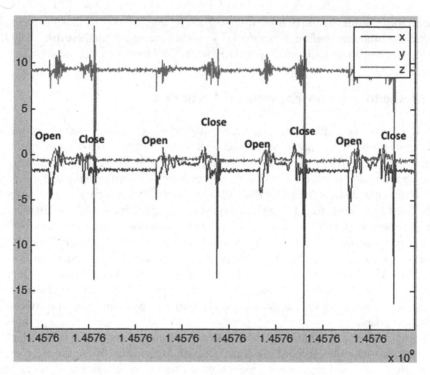

Fig. 5. Car door opening and closing data, 10 Hz. (X, Y, Z axis).

4.2 STAI Questionnaire - Self-assessment of Current Anxiety

The Spielberger State-Trait Anxiety Inventory (STAI) [12] is a reliable and sensitive method for assessment of anxiety level as anxiety in a given moment (reactive anxiety as a state) and personal anxiety (as a stable characteristic of the person or a trait). Full STAI questionnaire includes 20 questions. In case of time-limited studies one can use the short version of STAI questionnaire (6 questions instead of 20) and it still provides sufficient results with correlation coefficient 0.9 (as compared to the full form) [13].

Description of short STAI questionnaire: For each statement in the questionnaire the participant chooses the appropriate value to indicate how he/she feels right at that moment. There are no right or wrong answers, just what seems to describe his/her present feelings best. One shouldn't spend too much time on any one statement - Fig. 3.

Description of STAI form Y-2 (20 questions should be filled out only once) - read each statement and then select the appropriate value of the statement to indicate how one generally feels. There are no right or wrong answers, which describe your present feelings best. One should not spend too much time on any one statement, but give the answer which seems to describe how one generally feels [12].

5   Identification of an Aggressive Driving Style

There are several definitions of aggressive driving style but all of these are described in terms of speed, acceleration, braking, lane changes. GPS data or accelerometer data inside the car easily provide data like speed, acceleration and braking manner, so we are going to define aggressive driving in this way. In research [14] the acceleration values above 2 m/s2 were set as critical values considering the safety aspect. In another study [15] authors claim that the critical value for acceleration is 1.25 m/s2. Other approaches determine the critical value of acceleration based on current speed. Researchers in [16] derived acceleration and deceleration from the speed model and defined critical values for 20 km/h as 2.16 m/s2 and for up to 80 km/h as 1.27 m/s2. Road profile and drivers experience have also influenced the determination of critical values. So, in this case it could be possible to define aggressive driving style as compared to average driving.

In Figs. 6 and 7 a sample of calm driving and a sample of aggressive driving are presented (speed, acceleration from GPS and X, Y axis values from accelerometer). In Fig. 6 one can see higher acceleration and braking peaks (Y axe) and higher peaks during turning (X axe). The last feature is an interesting parameter, as the largest Slovenian insurance company Triglav uses similar features in its application "Drive" [17] to assess safety driving and to provide special discounts for good drivers. Also Russians insurance companies, like AlfaStrakhovanie Group, use similar approach to promote new insurance products based on safe driving.

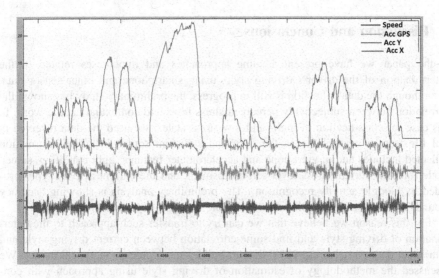

Fig. 6. More calm driving. (Speed, acceleration, value of X, Y axis).

There is another option for assessment of the driving style - using a subjective self-report after driving about how the driver him/herself was satisfied with his/her

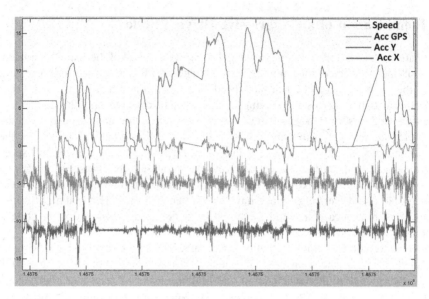

Fig. 7. More aggressive driving. (Speed, acceleration, value of X, Y axis).

driving and give basis for comparison of the current manner of acceleration and braking (how they are intensively) to driver's average manner. Drivers can fill out short STAI questionnaire after the journey, in order to catch the changes in his/her behavior.

6 Discussion and Conclusions

In the paper we have presented some approaches and hypotheses related to the determination of the driver's driving style, using smartphone and other sensor data. Even though the data collection is still in progress, the preliminary study has shown the correlation between subjectively perceived stress level and contextual data at work. In this case for classification of high- and low-stress states, we used the data received in real life from a smartphone as the only and sufficient source [8]. The information collected includes audio, gyroscope and accelerometer features, light condition, screen mode (on/off), current stress level self-assessment, and the current activity type (provided by Google activity recognition). The preliminary analysis is showing accuracy around 77 % in binary classification using the decision tree algorithms.

For this reason we believe that we can try to transfer such approach to the determination of driving style and find some correlation between current driving style and contextual data and also use HRV data to build a more precise analytical model. We discussed the methodology of estimation of driving style using approach with contextual, driving and physiological data and suggest appropriate sensors for it.

Acknowledgments. The work was supported by the Ministry of Education, Science and Sport of Slovenia, and the Slovenian Research Agency.

References

1. Sysoev, M., Kos, A., Pogačnik, M.: Noninvasive stress recognition considering the current activity. Pers. Ubiquit. Comput. **19**(7), 1045–1052 (2015)
2. Global status report on road safety (2013). http://www.who.int/violence_injury_prevention/road_safety_status/2013/en/
3. Pentland, A., Liu, A.: Modeling and prediction of human behavior. Neural Comput. **11**(1), 229–242 (1999)
4. Hong, J.-H., Margines, B., Dey, A.K.: A smartphone-based sensing platform to model aggressive driving behaviors. In: Proceedings of the SIGCHI Conference on Human Factors in Computing Systems. ACM (2014)
5. Oguchi, K., Weir, D.: System for predicting driver behavior. U.S. Patent Application 11/968,864
6. Liu, L., et al.: Toward detection of unsafe driving with wearables. In: Proceedings of the 2015 Workshop on Wearable Systems and Applications. ACM (2015)
7. Kavthekar, N., Badadhe, A.: Numerical analysis of door closing velocity for a passenger car (2015)
8. Wang, L., et al.: Silhouette analysis-based gait recognition for human identification. IEEE Trans. Pattern Anal. Mach. Intell. **25**(12), 1505–1518 (2003)
9. Pelkmans, L., et al.: Influence of vehicle test cycle characteristics on fuel consumption and emissions of city buses. No. 2001-01-2002. SAE Technical paper (2001)
10. QMedical. http://www.qmedical.com/dt_hrv_stress_relationship.htm
11. FUNF, Open Sensing Framework. http://funf.org
12. Spielberger, C.D., et al.: Manual for the state-trait anxiety inventory (1983)
13. Marteau, T.M., Bekker, H.: The development of a six-item short-form of the state scale of the Spielberger State-Trait Anxiety Inventory (STAI). Brit. J. Clin. Psychol. **31**(3), 301–306 (1992)
14. Son, Y.-T., Kim, C.-K.: Research for the method of design consistency evaluation using individual driving behavior. J. Korean Soc. Civ. Eng. **28**(6D), 767–774 (2008)
15. Fitzpatrick, K., et al.: Speed prediction for two-lane rural highways. No. FHWA-RD-99-171 (2000)
16. Han, I.-H., Yang, G.-S.: Recognition of dangerous driving using automobile black boxes. J. Korean Soc. Transp. **25**(5), 149–160 (2007)
17. Triglav. https://www.triglav.si/mobilno/mobilna-aplikacija-drajv

NEW2AN: Cooperative Communications

A Source Prioritizing Scheme for Relay Cooperative Networking

Ioannis Giannoulakis[1(✉)], Emmanouil Kafetzakis[2], and Anastasios Kourtis[1]

[1] National Centre for Scientific Research "Demokritos",
Patr. Gregoriou & Neapoleos, 15310 Agia Paraskevi, Greece
{giannoul,kourtis}@iit.demokritos.gr
[2] ORION Innovations, Ameinokleous 43, 11744 Athens, Greece
mkafetz@orioninnovations.gr

Abstract. This paper elaborates on the rising important paradigm of cooperative communications and suggests a novel algorithm that enhances performance at low complexity. Our motivation emerges from a channel constrained context, like e.g., cognitive networks, where communication channel distribution, as well as relay assignment, play a pivotal role for the total network throughput and the overall performance. The proposed scheduling scheme attempts to maximize the overall throughput by maximizing the number of transmission sources. Moreover, simulation results for the impact of the number of relays and the available communication channels are provided. The results show that the use of a suitable number of relays produces a severe improvement in the network's overall throughput. On the other hand, applying more communication channels tends to remove these benefits.

Keywords: Device-to-Device · Cooperative commounications · Relays

1 Introduction

Wireless connectivity becomes increasingly powerful in order to meet the demands of ubiquitous mobile services. Portable devices enriched with processing and storage capabilities are called to serve resource demanding applications in the wireless domain, introducing significant changes to mobile networking. As a consequence, next generation communications should be able to deliver crucial improvements in many key aspects of wireless communications, including performance, coverage, and reliability.

The term Device-to-Device (D2D) commonly refers to technologies that empower wireless devices with direct communication and data exchange capabilities, without the need of fixed infrastructure. However, the latter may still be responsible for centralized management and allocation of resources, radio link control and other important tasks, as in the case considered here. With the rise of context-aware software and the advance of location-based applications, cooperative and D2D communications play an increasingly important role due to

© Springer International Publishing AG 2016
O. Galinina et al. (Eds.): NEW2AN/ruSMART 2016, LNCS 9870, pp. 155–165, 2016.
DOI: 10.1007/978-3-319-46301-8_13

the ability to use relays in order to optimize the received quality of experience. With all the above in mind, D2D communication capabilities have the potential to introduce added value for existing wireless networks.

This work studies a relay cooperative communication mechanism which attempts to keep at high level the total throughput of the network. More precisely, the followed approach handles the potential participation of relay nodes in the transmission procedure according to the Decode and Forward (DF) mode. The adopted approach targets at environments with volatile wireless channel conditions. Under this point of view, our study manages to capture a straightforward representation, that provides stability and overall throughput optimization.

The work presented here has been inspired by [8], where a similar model was presented. The proposed strategy of [8] attempts to maximize the minimum throughput of the source and relay nodes collectively. Differently from that, the scheduler proposed in this paper targets to environments where apart from maximal throughput, fairness among all sources is also of equal importance. This may occur due to e.g., large differences in average Signal to Noise Ratio (SNR) and throughput in some of the source-destination pairs, handling of applications featuring low latency constraints, potentially increased interference at the relays, and so on. Hence, our scheduler adopts a selection framework that facilitates the overall network throughput and at the same time does not promote always the transmission of sources with high signal quality. Moreover, when relays are present, in order to increase the overall throughput, a scheme for allocating them to low signal quality source-destination pairs is proposed.

Cooperative communications have been introduced to enhance cellular networks with relaying. Following this strategy, some works (see, e.g., [2,4,9]) proposed the application of D2D communications in order to exploit more efficiently the actual cellular infrastructure. Furthermore, several other use cases based on D2D notion have come into play, handling a variety of issues like e.g., peer-to-peer communications, multicasting, machine-to-machine communications, and so on (see, e.g. [3,6]). A good survey can be found in [1]. In the field of cooperative networking some important studies have also emerged, targeting to relaying and traffic offloading. For example, mobile nodes are allowed to communicate directly with each other, and at the same time, to operate within the conventional cellular infrastructure, contributing important enhancements to the cellular network's performance and to users' satisfaction level [2].

A common categorization of cooperative relay systems is between pure relay systems and hybrid relay systems, i.e., cases where relay nodes play only the role of supporting the communication between the source and the destination and when nodes perform as sources and as relays simultaneously [7]. While in the relay operation, data can be retransmitted through an intermediate relay or a mobile station that has been dedicated to act as relay. Infrastructure based relays have been analyzed in [12] and this is also the case adopted in our framework. Besides, relevant techniques to increase capacity have appeared in [10,11].

Simulation results have been included in the paper in order to verify and evaluate the proposed technique. The conducted tests indicate that our method outperforms direct transmission schemes. The intuition that attaching relays to low quality communication links enhances the overall network's throughput is verified.

The main contributions of this paper can be outlined in the following aspects. To begin with, we consider the case where reallocation of communication channels to both source and destination nodes takes place at the beginning of each timeslot. Therefore, this model suits well to highly volatile wireless communication conditions, where scheduling decisions need to be repeated on a timeslot basis. Secondly, a new scheduler which attempts to maximize the overall network throughput and at the same time the number of transmitting nodes, is proposed. Finally, by conducting simulation experiments, we are in position to provide useful results with respect to the enhancement produced by relays and with regard to the number of channels used in the network. Our results disprove the intuitive reasoning that relays always strengthen network performance. In addition, it is shown that the number of channels plays critical role in the overall throughput.

This paper is structured as follows. Section 2 introduces our model, the basic assumptions and gives insight to common scheduling procedures within the context of cooperative communications. Section 3 presents the proposed scheme in the scheduling domain and provides some further intuition on the expected performance outcomes that rise from its adoption. In Sect. 4, the simulations results that support the proposed technique are provided. Section 5 concludes the paper.

2   System Model

In this study, we consider a wireless network with N source-destination communication links (s_i, d_i), $s_i \in S = \{s_1, \ldots, s_N\}$ and $d_i \in D = \{d_1, \ldots, d_N\}$. Each D2D link can be uniquely identified by the pair (s_i, d_i), $i = 1, \ldots, N$ and let $\mathcal{N}$ be the set containing all such links. Source and destination nodes are assumed to participate in D2D communication in half-duplex mode. Information originates at the source node and needs to be transmitted at the destination for each link. Also, without loss of generality, it will be implied that source nodes have an infinite queue of data to transmit and always compete for network resources.

In this model it is assumed that there exist a finite number k, $k \in \mathbb{N}$ of separate communication channels (depending on the specific wireless technology applied) and they will be denoted by $c_i \in C$, where $C = \{c_1, \ldots, c_k\}$. Time is separated into short slots. At the beginning of every timeslot, each source and destination node s_i, d_i, $i \in \mathcal{N}$ is associated with one or more communication channels c_i. This scenario is widely applied in cases like e.g., cognitive networks, where the available channels can change dynamically. In our case, in order to add a source-destination pair in the scheduling process, both the source and the destination of each pair should be assigned at least one common communication channel. The channels' distribution is assumed identical for all nodes and the channels are allocated according to a p-persistent approach. For an example we refer to Fig. 1.

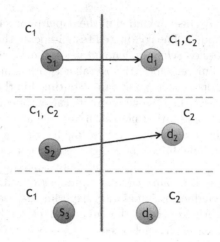

Fig. 1. Communication links with limited channel availability

Usually, under the DF scheme which is also assumed here, a relay node can be used to forward the received signal towards the destination on the same channel with the source. In this study we leverage on the prospect of deploying relays that can be used to enhance performance by decoding and retransmitting the received data at the same channel. The relays are assumed capable to communicate in all communication channels $c_i \in C$, and a scheduling rule that allows to forward the received signal conformally to the transmissions of the rest source nodes applies. This topic will be discussed in further detail in the next section.

As a common practice, cooperative communications with the use of relays follow a reception and forward policy at the same channel. Leveraging on the possibility of exploiting channel diversity, one may improve significantly the system's performance by relaxing this practice. For example, we refer to Fig. 2. Therein, both sources can transmit only in c_1, but the relays may use both c_1 and c_2. According to [8], performance can be improved by following a more flexible scheme of channel allocation, see, e.g., Fig. 3. The model presented here enhances this practice by offering an additional scheme for efficient node and relay scheduling.

With respect to the achievable transmission rate, when only a direct transmissions exist, we follow the well established relation from Shannon's law, i.e.,

$$\mathcal{Q}(p,q) = W \log_2(1 + \mathrm{SNR}_{pq}).$$

In the above equation, W denotes the available channel bandwidth and SNR_{pq} the SNR from source p to destination q.

When relays exist, a well accepted model for DF mode (see, e.g., [5]) is adopted. According to this, under DF mode, the transmission rate can be expressed as

$$\mathcal{Q}(s,r,d) = \frac{W}{2} \min\{\log_2(1 + \mathrm{SNR}_{sr}),$$

$$\log_2(1 + \mathrm{SNR}_{sd} + \mathrm{SNR}_{rd})\},$$

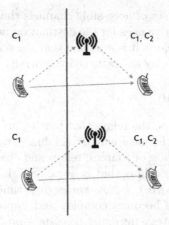

Fig. 2. Cooperative links where channel diversity can be applied

	t_1	t_2	t_3	t_4	t_5	t_6	
c_1	s_1	r_1	s_2	r_2	s_1	r_1	...
c_2	-	-	-	-	-	-	...

(a) Traditional channel allocation

c_1	s_1	s_2	s_1	s_2	s_1	s_2	...
c_2	-	r_1	r_2	r_1	r_2	r_1	...

(b)   Refined channel allocation

Fig. 3. Modes of channel allocation

where r indicates the relay node. Furthermore,

$$\text{SNR}_{pq} = \frac{P_p x_{pq}^{-\delta}}{\sigma^2}$$

applies, where P_p is the transmission power of source p, x_{pq} is the Euclidean distance between the source p and the destination q, and δ stands for the path loss exponent. Finally, σ^2 provides the variance of the background noise at destination node q.

With respect to the network resource allocation, it should be mentioned that we assume that a central entity exists and is responsible for control and management of the cooperative transmission pairs. When a communication pair joins or leaves the network, the set of accessible channels that is attainable from the source (and also the respective ones for the destination node), are communicated to the control entity. Therefore, it is assumed that the scheduling decisions along with the channel and the relay allocations are centrally concluded.

3   Proposed Scheme

The source-destination pairs, the relays that participate in the network and the available communication channels can be combined to obtain the scheduling decision, i.e., the combination of sources, relays and channels that will apply in the following timeslot. By recalling that in our model the availability of channels in each cooperative link (s_i, d_i), $i \in \mathcal{N}$ varies over time, the problem of channel selection for each node becomes complex and important as well. Actually, the channel allocation strategy normally consists a problem featuring NP-hard complexity. A usual choice is to attempt to maximize some of the key performance metrics, as for example, throughput, latency, and so on, throughout all source-destination links.

At first, we examine the direct transmission case, i.e., without the participation of relay nodes. The key concept of our technique resides in scheduling the active nodes, so as to maximize the number of transmitting nodes scheduled in each timeslot. This way, our method differs from [8] in that scheduling does not aim to maximize the minimum throughput, but it attempts to keep the overall network throughput as high as possible.

In order to illustrate better our idea, an example is presented in Fig. 4. The upper part of it refers to a simplistic setting without installation of relays in the network. Three cooperative links are active and their respective allocated channels in each timeslot have been included inside the parenthesis. According to the proposed direct transmission scheduling, in timeslot 1 the source nodes s_1 and s_2 will compete for c_1 and the ambiguity is resolved by a random selection. At the same time, c_3 is assigned to s_3. However, the merits of the proposed scheduling schema become more evident in timeslot 2. For that case, although all three pairs are eligible for transmission in c_2, s_2 is preferred since this allocation allows both s_1 and s_3 to transmit. The lower part of Fig. 4 presents the channel allocation strategy which is proposed.

As a second step, the inclusion of a number of relays at arbitrary locations within the network's area is considered. In case that some devices can also act as relays, the model presented so far can be easily modified to include that case. Differently from this situation, we consider here the context that a fixed number of relays have been deployed at constant (but random) points in the network and that they are able to receive and to transmit on any available communication channel, $c_i \in C$.

Relays here act according to the DF mode. In this case, each timeslot is further split into two equal frames. In the first frame only the source transmits, while during the second frame, the source and the relay retransmit the data of the first frame to the destination. Although approximately half of the normally data transmission time is involved in this mode, the gains due to the higher SNR ratios can prove much worthier.

t_1	t_2	t_3	
$s_1(c_1,c_3)$ $d_1(c_1,c_2)$	$s_1(c_1,c_2)$ $d_1(c_1,c_2)$	$s_1(c_1,c_2,c_3)$ $d_1(c_1,c_2)$	...
$s_2(c_1)$ $d_2(c_1,c_3)$	$s_2(c_2)$ $d_2(c_2)$	$s_2(c_1,c_3)$ $d_2(c_3)$	
$s_3(c_2,c_3)$ $d_3(c_3)$	$s_3(c_2,c_3)$ $d_3(c_2,c_3)$	$s_3(c_1,c_2)$ $d_3(c_1,c_2)$	

(a)   Communication channel distribution on a time slot basis

	t_1	t_2	t_3	
c_1	S_1	S_1	S_3	...
c_2	-	S_2	S_1	
c_3	S_3	S_3	S_2	

(b)   Scheduling decisions according to the proposed algorithm

Fig. 4. Channel allocation and scheduling decisions

Since relays usually consist a way for improving the quality of communication links, here they are used to improve the links featuring low SNR. Therefore, if there exists a number m of relays, the m links with the lowest SNR are selected to feature the cooperative scheme respectively. Moreover, although the initial distribution of relays in the area is random, the assignment of relays to links is not random. On the contrary, for each link, the relay which is nearest to the middle between the source and the destination is assigned, beginning form the link featuring the lowest SNR.

4   Simulation Results

In this section we present the simulation results that are used to strengthen the arguments presented already. Our purpose is to describe the network topology and configuration details and then to present our simulation findings.

In the first place, we consider a square area with side 750 m. Inside that area, $n = 20$ source-destination pairs, along with $r = 5$ relays, have been placed according to a uniform distribution. The power of transmitting nodes is equal to 500 mW and it is assumed that the range of transmission is large enough to cover the whole network area. Also, the value of the background noise is $\sigma^2 = 10^{-10}$ W and the path loss exponent is $\delta = 4$. The total available bandwidth is $W = 22$ MHz and this is equally divided to $k = 10$ communication channels. A channel assignment takes place at the beginning of each time slot and a time interval of 100 timeslots is examined. To keep visualization complexity low, a similar context with $n = 10$ source - destination pairs has been presented in Fig. 5.

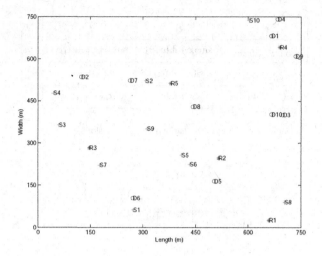

Fig. 5. Sample topology of the cooperative network

We address a setup with varying number of relays r, $1 \leq r \leq 10$, in order to assess the impact of the number of relays to overall network throughput. In a similar fashion to the assumptions of Sects. 2 and 3, the relays may work on any communication channel (one in every timeslot) and they are assigned to the pairs with minimum SNR at a descending order, i.e., the pair with the lowest throughput is assigned the relay that can improve it most, and so on. Figure 6 shows the average network throughput for the above configurations. As a remark, we observe that in the context of the throughput maximizing scheduler presented in Sect. 3, indeed the use of relays greatly enhances overall performance. In addition, as reflected by Fig. 6, even a small number of relays, i.e., $r = 2$ or $r = 3$, can produce important benefits.

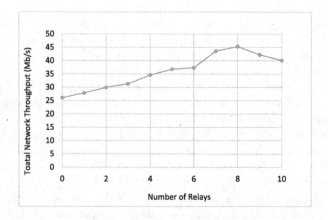

Fig. 6. Total network throughput versus number of relays

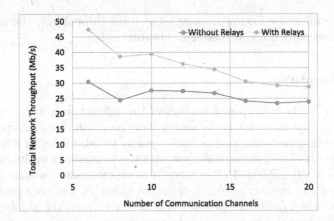

Fig. 7. Total network throughput versus number of communication channels

As a second step, we keep the configuration with $r = 5$ relays introduced earlier, but we change the number of available communication channels in the given bandwidth. With this strategy, we attempt to provide greater insight in the way the structural networking attributes affect the average throughput. To that end, we refer to Fig. 7. As it may be seen, when no relays are present, it is more convenient to keep the number of channels at an adequately high level, i.e., $k = 6$. Nevertheless, here the presence of relays poses several implications. Indeed the upper curve of Fig. 7 reveals that increasing the number of communication channels removes to some extent the benefits introduced by the relays.

As a final experiment, we provide here the case where the number of cooperative pairs i is variable, i.e., $2 \leq i \leq 20$. The number of relays remains $r = 5$ and the available communication channels are $k = 10$. For that configuration, the resulted network throughput is shown in Fig. 8. As we observe, although

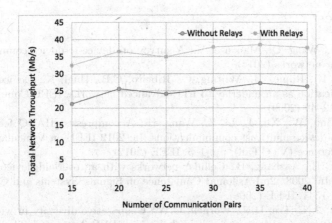

Fig. 8. Total network throughput versus number of communication pairs

the existence of relays always improves the total throughput, the number of communication pairs does not play an important role.

5   Conclusions

In this paper we have proposed an overall throughput maximization scheme, applicable to cooperative networks that feature communication channel allocation constraints. Simulation results have been included, taking as reference the proposed schema. The outcomes suggest that the throughput optimizing scheduler can be severely enhanced when relays are present in the network. Moreover, the impact of the number of communication channels has been investigated, showing that when relays are present, it is better to keep the number of communication channels limited.

As a future work, the proposed methodology can be elaborated to include a limited range of transmission and reception for all participating nodes. Effectively, this would give rise to the possibility to reuse some of the already allocated communication channels inside the network's area without causing severe interference. Another potential direction can be the generalization of the model to include flow-level performance metrics, relaxing the assumption that all sources have an infinite queue of data to transmit. In that case, in each timeslot only the sources that have data to transmit would participate in the channel allocation. Finally, one can also consider the case where the relays are not installed in the network, but on the contrary, an arbitrary node plays that role. The framework presented here can be readily adapted to handle such cases, by allocating also to relays a subset of the communication channels at the beginning of each timeslot.

Acknowledgement. This research has received funding from the European Union's H2020 Research and Innovation Action under Grant Agreement No. 671596 (SESAME project).

References

1. Asadi, A., Wang, Q., Mancuso, V.: A survey on device-to-device communication in cellular networks (2014)
2. Doppler, K., Rinne, M., Wijting, C., Ribeiro, C.B., Hugl, K.: Device-to-device communication as an underlay to LTE-advanced networks. IEEE Commun. Mag. **47**(12), 42–49 (2009)
3. Du, J., Zhu, W., Xu, J., Li, Z., Wang, H.: A compressed HARQ feedback for device-to-device multicast communications. In: 2012 IEEE on Vehicular Technology Conference (VTC Fall), pp. 1–5. IEEE (2012)
4. Kaufman, B., Aazhang, B.: Cellular networks with an overlaid device to device network. In: 2008 42nd Asilomar Conference on Signals, Systems and Computers, pp. 1537–1541. IEEE (2008)
5. Laneman, J.N., Tse, D.N., Wornell, G.W.: Cooperative diversity in wireless networks: efficient protocols and outage behavior. IEEE Trans. Inf. Theor. **50**(12), 3062–3080 (2004)

6. Lei, L., Zhong, Z., Lin, C., Shen, X.: Operator controlled device-to-device communications in LTE-advanced networks. IEEE Wirel. Commun. **19**(3), 96 (2012)
7. Li, G.Y., Xu, Z., Xiong, C., Yang, C., Zhang, S., Chen, Y., Xu, S.: Energy-efficient wireless communications: tutorial, survey, and open issues. IEEE Wirel. Commun. **18**(6), 28–35 (2011)
8. Li, P., Guo, S., Leung, V.C.: Improving throughput by fine-grained channel allocation in cooperative wireless networks. In: 2012 IEEE on Global Communications Conference (GLOBECOM), pp. 5740–5744. IEEE (2012)
9. Peng, T., Lu, Q., Wang, H., Xu, S., Wang, W.: Interference avoidance mechanisms in the hybrid cellular and device-to-device systems. In: 2009 IEEE 20th International Symposium on Personal, Indoor and Mobile Radio Communications, pp. 617–621. IEEE (2009)
10. Vanganuru, K., Puzio, M., Sternberg, G., Fan, Y., Kaur, S.: Downlink system capacity of a cellular network with cooperative mobile relay. In: 25th Wireless World Research Forum (2010)
11. Xiao, L., Fuja, T.E., Costello, D.J.: Mobile relaying: coverage extension and throughput enhancement. IEEE Trans. Commun. **58**(9), 2709–2717 (2010)
12. Yang, Y., Hu, H., Xu, J., Mao, G.: Relay technologies for WiMAX and LTE-advanced mobile systems. IEEE Commun. Mag. **47**(10), 100–105 (2009)

QoS Aware Admission and Power Control for Two-Tier Cognitive Femtocell Networks

Jerzy Martyna$^{(\boxtimes)}$

Institute of Computer Science, Faculty of Mathematics and Computer Science,
Jagiellonian University, ul. Prof. S. Łojasiewicza 6, 30-348 Cracow, Poland
martyna@ii.uj.edu.pl

Abstract. In two-tier femtocell networks, the femtocell users (FUs) can
be admitted to the femto base stations (FBSs) provided that the interference to the primary users (PUs) is no higher than the defined thresholds.
Moreover, several FBSs may require different quality of service (QoS)
with different payments. This paper proposes a method to obtain from
maximally achievable revenue from FBSs to the required QoS. It also
proposes an efficient QoS aware admission algorithm which is compared
with other schemes, such as minimum signal-to-interference-plus-noise
ratio (SINR) removal algorithm (MSRA) and random SU removal algorithm (RSRA). Presented QoS aware admission algorithm can achieve
a much higher revenue with required QoS. In addition, the proposed
algorithm is evaluated by the simulation experiments.

Keywords: Cognitive femtocell networks · QoS aware admission ·
Power control

1 Introduction

Cognitive radio (CR) networks are seen as a key solution to meet the FCC (e.g.
Federal Communications Commission) policy and to build the future generation of wireless networks [1,16]. A cognitive radio must periodically perform
spectrum sensing and operate at any unused frequency in the licensed and unlicensed band, regardless of whether the frequency is devoted to licensed services
or not. However, with the spectrum usage being both space and time dependent,
there is a great amount of "white space" (unused bands) available sparsely that
can potentially be used for both licensed (primary users, PUs) and unlicensed
(secondary users, SUs) users.

Two-tier femto networks, consisting of macrocell overlaid with femtocells in
co-channel deployment, have increased attention in recent years because of providing in building for indoor users femtocell base stations (femto-BSs). These
femto-BSs are complemented by the poor signal from the macrocell BS (macro-
BS). It is obvious that the femtocells also enable to concurrent transmissions
that can be accommodated in the network. It improves spatial reuse, but makes
interference as a challenging issue. These problems have been studied in several papers. Among others, Giivenc in the paper [8] considered the impact of

© Springer International Publishing AG 2016
O. Galinina et al. (Eds.): NEW2AN/ruSMART 2016, LNCS 9870, pp. 166–178, 2016.
DOI: 10.1007/978-3-319-46301-8_14

spreading on the capacity neighbouring femtocells. The feasibility of coexistence of femto-macrocells in the same frequency bands has been studied in [3]. The uplink capacity and interference avoidance for two-tier femtocell networks has been investigated by Chandrasekhar *et al.* [4] in which an exact outage probability at a macrocell and tight lower bounds on the femtocell outage probability have been derived. By employing stochastic geometry model, bounds on the distribution of aggregated interference from two-tier spatial point processes have been successfully analyzed by Huang *et al.* [10] and Law [12]. With these models the maximum femto-BS density and thus maximum overall capacity satisfying a per-tier outage constraint have been formulated. The methods to interference avoidance in femtocell networks, based on the OFDMA method, were presented by Lopez-Perez *et al.* [14]. The decentralized strategies for interference management in femto-cell networks are presented in the paper Bharucha *et al.* [2].

The integration of cognitive radio and femtocells provided a new quality of these technologies. The sensed information provided by the CR devices allows changing the communication parameters of the secondary users. On the other hand, femtocells give the limit of interference, thereby, increasing network capacity, in a small area. In the literature, there have been a many works addressing different aspects of two-tier cognitive femtocell networks. Among others, cognitive interference management in heterogeneous two-layers femtocell networks has been studied by Kaimaletu *et al.* [11]. Authors have been presented the spectrum allocation scheme to improve cognitive interference for these networks. The downlink capacity of two-tier cognitive OFDMA-based femto networks was studied by Cheng *et al.* [5]. A cost-effective scheme to manage the downlink interference from user-deployed femto-cells to macrocell user was proposed by Li and Sousa *et al.* [13]. An integrated architecture and a multiobjective optimization problem for the joint power control, base station assignment and channel assignment scheme was formulated by Torregoza *et al.* [19]. The same issues, the resource allocation problem as a joint relay, subcarrier and power allocation problem with the objective of maximizing the sum of the weighted rates of the femtocell system subject to protecting the macrocell network's communication, have been analyzed by Gamage *et al.* [7]. Recently, a pricing power control with statistical delay QoS provisioning in uplink of two-tier OFDMA femtocell network have been studied by He *et al.* [9]. However, none of these studies have attemted to analyze the QoS requirements with simultaneous optimization in the two-tier cognitive femtocell networks.

This paper provides a formulated optimization problem of the maximization of the secondary revenue in terms of required QoS parameters and permissible interference. The efficient solution of this problem has been obtained using the presented admission control algorithm. The effective power control algorithm with demanded QoS parameters and minimizing the interference was achieved as a solution. Moreover, the presented algorithm has properties comparable to well-known admission control algorithms, such as minimal SINR removal algorithm (MSRA) and random SU removal algorithm (RSRA).

The paper is organized as follows. In Sect. 2, the system model is presented. In Sect. 3, the optimization problem of secondary revenue in terms of QoS requirements is formulated. Section 4 provides an admission control algorithm as an efficient heuristic solution of optimization problem. In Sect. 5, some simulation results are presented. Section 6 concludes the paper.

2   System Model

In this section, the system model being applied is presented.

As shown in Fig. 1, the downlink of an orthogonal frequency division multiple access (OFDMA) based system is composed of two-tier macro- and femto-cells. Primary user (PU) (or licensed user) has a license to operate in a certain spectrum band. This access can only be controlled by the Primary Base Station (Primary-BS) (or licensed base station). In principle, the Primary-BS is a fixed infrastructure network component. Denote $K =| \mathcal{K} |$ as the number of macro users (MUs), $F =| \mathcal{F} |$ as femto-BS and $M =| \mathcal{M} |$ as femtocell users (FUs), which are randomly located inside the coverage area of the macrocell. The femto-BS, located at the center of each femtocell, provides services for a set of femtocell users. A set of macro users is serviced by both the cognitive base station and also by femto-BS. We assume that the macro user MU can only be served by cognitive base station even it locates within the femtocell coverage. All femto-BS, macro users, femtocell users are operate as secondary users (SUs). Additionally, we consider a set of PUs which are staying in receiving mode. Thus, all SUs are

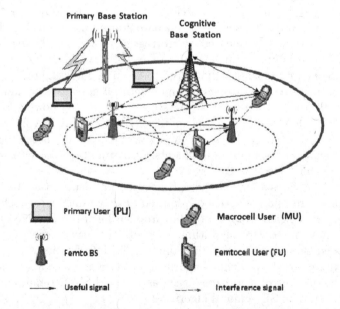

Fig. 1. An example of interference model in the cognitive femtocell network.

trying to transmit their data in the uplink to their femto- or macro-BSs. It is associated with the interferences which are received by PUs.

In the system model, the total bandwidth is divided into N subcarriers with two of them being grouped into one subchannel. Let all femtocells and macrocells operate in the same frequency band and have the same number of subcarriers. We assume that the transmission in two-tier cognitive femtocell network may occur as long as the aggregated interference incurred by the femto-BS is below some acceptable constraint. Thus, the total transmit power of all femtocell users in each channel is no more than the interference power threshold P_{int}^{max}.

Let P_{max} be the maximum transmit power of PU. The maximum transmit power of all M femtocell user FU over the subchannel n is given by

$$\sum_{m=1}^{M} P_m^n G_{m,k}^n \leq P_{int}^{max}, 1 \leq k \leq K \tag{1}$$

where $G_{m,k}^n$ is the channel gain between m and all k macrouser MU.

For each transmit power P_m^n must be satisfied the following condition

$$P_{min} \leq P_m^n \leq P_{max}, 1 \leq m \leq M \tag{2}$$

where p_{min} is the minimum of transmit power of femtocell.

While femto SU share the spectrum with PUs, femtocell users will cause interference to the PUs. Let T_j^I be the interference power received by the j-th PU, namely

$$T_j^I = \sum_{i=1}^{n_s} h_{ij}^{sp} P_i^s x_i, 1 \leq j \leq N_p \tag{3}$$

where n_s is the number of all femtocell users, N_p is the number of PUs, h_{ij}^{sp} is the power attenuation from the femtocell user i to j-th PU, x_i indicates whether i-th FU is admitted or not. We assume that $x_i = 1$ shows that i-th femtocell user is admitted, zero otherwise. The transmit power of i-th femtocell user is equal to P_i^s and the transmit power over all n channels is given by $P_i = \sum_{i=1}^{n} P_i^n$, h_{ij}^{sp} denotes the power attenuation from i-th femtocell user to j-th PU and is given by

$$h_{ij}^{sp} = \frac{G_i^s G_j^p}{(d_{ij}^{sp})^\eta}, 1 \leq i \leq n_s, 1 \leq j \leq N_p \tag{4}$$

where d_{ij}^{sp} denotes the distance from the femtocell user i to j-th PU. The exponent η is the path fading factor. G_i^s and G_j^p denote the antenna gains of i-th femtocell user and j-th PU, respectively. The interference power caused by i-th FU is defined by

$$\tau_{ij} = h_{ij}^{sp} P_i^s = \frac{G_i^s G_j^p P_i^s}{(d_{ij}^{sp})^\eta}, 1 \leq i \leq n_s, 1 \leq j \leq N_p \tag{5}$$

The interference power caused by i-th FU and K macro cell users, which are located outside the coverage area of the macrocells is given by

$$\tau_{mj} = h_{mj}^{sp} \cdot P_m^s = \frac{G_m^s \cdot G_j^p \cdot P_m^s}{(d_{mj}^{sp})^\eta}, 1 \le m \le M \tag{6}$$

where G_m^s and G_j^p denote the antenna gains of m-th macro cell user and j-th PU, respectively; P_m^s is the transmit power of m-th macro cell user, d_{mj}^{sp} denotes the distance from m-th macro user to j-th PU.

3 Optimization Problem of Secondary Revenue

This section analyzes the impact of received interference in two-tier cognitive femto network on the revenue of secondary users.

The uplink maximum data transmission rate λ_l from the l-th femto-BS, macro users and femtocell users to BS is given by [15,18]

$$\lambda_l = B \cdot \log_2(1 + \xi_l^\eta), 1 \le l \le F + K + M \tag{7}$$

where B is the uplink bandwidth, ξ_l^η is the uplink SINR of the l-th femtocell user or femto-BS or macro user measured by BS. We assume that the minimal value the required SINR for given SU is given by ξ_l^{min}. Thus, the demanded SINR for the l-th femtocell user or femto-BS or macro user is expressed as

$$\xi_l^{min} = 2^{\frac{\lambda_l^{min}}{B}} - 1 \tag{8}$$

where λ_l^{min} is the minimum uplink data transmission rate for the l-th SU.

The downlink SINR of an active femto-BS or macro user or femtocell user l is defined as

$$
\begin{aligned}
\xi_l^q &= \frac{H_l^{sq} P_l^q}{N_0 + \sum_{j=1, j \neq l}^{n_s} H_j^{sq} P_l^q \cdot x_j} \\
&= \frac{H_l^{sq} P_l^q}{N_0 + I_q - H_{ij}^{sq} \cdot P_i^s - H_{mj}^{sq} \cdot P_m^s}
\end{aligned}
\tag{9}
$$

where H_l^{sq} denotes the power attenuation from l-th SU, I_q denotes the accumulated interference of the BS caused by all active femto- and macro-BS, i.e. $I_q = \sum_{l=1}^{n_s} H_l^{sq} P_l^q x_l$.

We are interested in finding the optimal admitted femtocell user and macro user and femtocell-BS subject to constraints such that the secondary revenue is maximized. Specifically, we will solve the following optimization problem

$$\arg_{P_l^s, x_l} \max \sum_{l=1}^{n_s} \left(\sum_{f=1}^{F} \tau_{fj} + \sum_{i=1}^{K} \tau_{ij} + \sum_{m=1}^{M} \tau_{mj} \right) x_l \tag{10}$$

subject to

$$\sum_{l=1}^{n_s} \left(\sum_{f=1}^{F} \tau_{fj} + \sum_{i=1}^{K} \tau_{ij} + \sum_{m=1}^{M} \tau_{mj} \right) x_l \leq \Gamma_j,$$
$$j \in N_p, x_l \in \{0,1\}, l \in N_s \tag{11}$$

$$\xi_l^s \geq \xi_l^{sq}, \quad \text{if } x_l = 1, \quad l \in \{\mathcal{F} \cup \mathcal{K} \cup \mathcal{M}\} \tag{12}$$

$$P_l^s \in [0, P_{max}^s], \quad l \in \{\mathcal{F} \cup \mathcal{K} \cup \mathcal{M}\} \tag{13}$$

where N_s is the number of all SUs, N_p is the number of all PUs. The first constraint, see Eq. (11), shows that the interference to PUs can not exceed the interference threshold given by Γ_j. The constraint given by Eq. (12) represents that the SINR requirement of active SUs should be satisfied. The third constraint, given by Eq. (13) indicates that the power limitation of SUs and P_{max}^s denotes the maximum transmission power of SU.

4   An Admission and Power Control Algorithm for Two-Tier Cognitive Femtocell Networks

In this section, an admission and power control algorithm is presented, which can be used in the two-tier cognitive femtocell network. It can be shown, in the previous section, that the optimization problem of admission and power control in two-tier femtocell cognitive network is nonconvex due to the mixed integer programming and high order objectives. These problems are known as NP-hard problems. However, for a fixed number of variables, it is possible to use the heuristic method. Using the proposed method, the below presented algorithm of admission and power control in two-tier cognitive femtocell networks is formulated. The presented method is show by Algorithm 1 (Fig. 2).

In the proposed method applied to solving our problem for given power of admitted PUs, it is firstly assumed that an initial set of possible solutions for the uplink SINR of an active set of SUs is given and called *core_set*. Initially, this set is without the constraints presented in the conditions (11), (12) and (13). Thus, the integer variables are released and considered to be continuous variables. The lower bound for the initial set *core_set* can be found by solving the iteration. The first step adds to *core_set* a new SU with minimal SINR. If there are all SUs and the QoS parameters are satisfied, then the is obtained the feasible solution. It does not, it randomly removes SU in each iteration. In consequence, the set of SUs will be updated by randomly removing in each iteration. As solution, the proposed method finds the feasible solution from the set of SUs by given power of PUs. By iterative calculation of given procedure for admitted PUs with various values of power control, we can obtain suitable value of power of each PU.

```
procedure QoS_aware_admission_power_control;
begin
  k := 0;  l := 1;
  let {S_i}, i = 1, …, N_p be a power for admitted PUs;
  set L_0 - lower bound of interference for admitted PUs;
  set U_0 - upper bound of interference for admitted PUs;
  set core_set - set of all SUs;
  while (U_k − L_k ≤ Γ_{N_p}) do
    begin
      add a new SU with minimal SINR to core_set;
      if l = N_s and Eqs. (12 - 13) and QoS parameters are satisfied
      then  solution := x;
        goto label;
      else
        randomly remove the SU from core_set;
      end if;
      k := k + 1;
    end while;
    label: feasible_solution := solution;
end;
```

Fig. 2. A procedure for QoS aware admission and power control in two-tier cognitive femtocell network.

5   Simulation Results

In order to evaluate the performance of the proposed algorithm, we developed a femtocell OFDM simulator. The core of this tool is based on a system radio network simulator called the Rudimentory Network Simulator (RUNE) [20]. This simulator has been extended to include numerous features of a two-tier cognitive femtocell network, including channel models, the SIR and the outage probability, QoS admission control, femto-BS location algorithm, etc. The simulator generates a Rayleigh fading map by filtering white noise through a Bessel function. It allows to calculate the SINR estimation of the downlink channels. In the simulation a scenario was used with one Primary-BS located at the center of cell with radius equal to 400 m. The number of PUs was varied from 1 to 10. The distance between PUs and Primary-BS was settled as 200 m. Additionally, was assumed that the bandwidth is set as 10 MHz. Ihe uplink data transmission was chosen from the 32–128 kbps.

The main simulation parameters are summarized in Table 1.

In Fig. 3, we plot the obtained outage probability versus SIR and compare it with the Monte Carlo simulation. It can be seen that at the low SIR values and the high SIR values, the obtained outage probability and the simulated are similar to each other.

Table 1. Summary of main simulation parameters

Parameters	Values
Maximum Primary BS	1 W
Maximum Femto-BS	0.125 W
Femto SINR threshold	3.2 dB
Femtocell coverage radius	15 m
Indoor pathloss exponent	2
Pathloss exponent α_1	3
Pathloss exponent α_2	5
Interference zone for Femto-BS	50 m
FU shadowing standard deviation	3 dB
MU shadowing standard deviation	6 dB

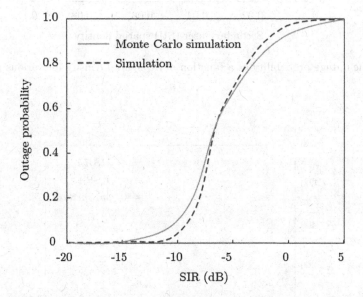

Fig. 3. The outage probability as a function of SIR.

Figure 4 illustrates the outage probability as a function of SUs (FUs and MUs taken together) spatial density for different number of PUs. The obtained curve shows the number of active SUs that could be accepted to meet a desired value of the outage probability. This curve is necessary for investigation of admission control mechanism. For instance, in order to maintain the outage probability less than 0.02, the spatial density of active SUs showed not exceed 0.04 for 4 PUs.

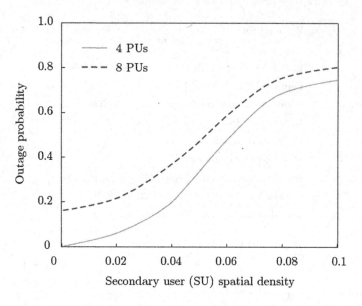

Fig. 4. The outage probability as a function of SU spatial density for various number of PUs.

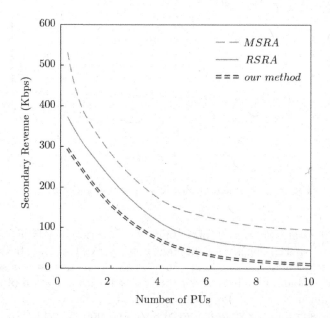

Fig. 5. The revenue of secondary users in dependence of the number of PUs.

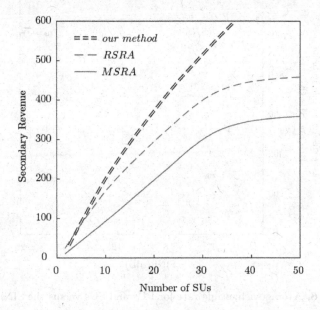

Fig. 6. The revenue of secondary users in dependence of the number of SUs.

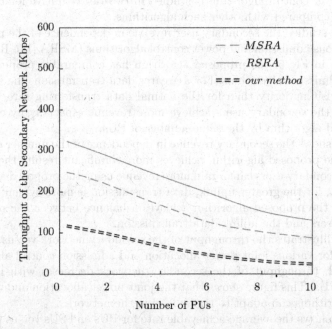

Fig. 7. Throughput of the secondary network versus the number of PUs.

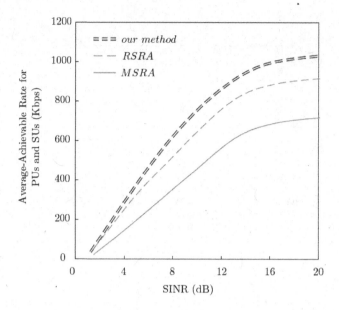

Fig. 8. Average-achievable rate for PUs and SUs versus the SINR.

In the simulation, the performance of the above presented algorithm of admission and power control algorithm is studied in two-tier cognitive femtocell network and is compared with other such algorithms.

First, we studied the secondary user revenue in dependence of the number of PUs for various admission and power control algorithms (MSRA [17], RSRA [6]). As is shown in Fig. 5, the proposed algorithm has comparable properties with the other admission algorithms. For a greater data transmission, the secondary revenue is visit majority than for the normal data transferring rate. It is also visible that the secondary users achieve more revenue especially by employing the proposed algorithm by the same number of PUs.

Figure 6 shows the secondary revenue in dependence of the number of SUs. In this case, the proposed algorithm achieves more significant results than others. Thus, the secondary users can obtain more revenue using the proposed algorithm. Additionally, for the greater uplink, data transmission is more evident there. In summation, the proposed algorithm achieves a balance between the revenue of secondary users and the uplink data transmission.

Figure 7 illustrates the throughput of the secondary network versus the number of PUs for various joint power allocation and admission control algorithms. As expected, throughput of the secondary network decreases with increasing number of PUs. This figure shows that the joint power allocation and admission control algorithms can adapt to traffic load in the network.

Figure 8 shows the average-achievable rate for PUs and SUs versus the signal interference noise ratio (SINR) for different joint power allocation and admission control algorithms. It is to be noted that the proposed algorithm acts like the

MSRA [17] and the RSRA [6] algorithms. As it was expected, the increase of the SINR induces the growth of the average-achievable rate for PUs and SUs.

6   Conclusion

In this paper, the problem of maximizing the secondary revenue in terms of QoS requirements in the two-tier cognitive femtocell networks has been investigated. To solve this optimization problem, a methodology to search the subset of SUs with defined constraints has been proposed. Using the presented admission control algorithm, the effective power control with required QoS and minimized interference has been obtained. Finally, the simulation results showed that the introduced admission control algorithm, in comparison to minimal SINR removal algorithm (MSRA) and random SU removal algorithm (RSRA), possesses good properties. Among others, the revenue of all secondary users is higher than in the above-mentioned admission control algorithm.

Further work includes investigation with the increase in any of the following cases: the number of PUs increases, the number of femto and macro users increases, and the value of interference increases.

References

1. Akyildiz, I.F., Lee, W.-Y., Vuran, M.C., Mohantry, S.: Next generation/dynamic spectrum access/cognitive radio wireless networks: a survey. Comput. Netw. J. **50**(13), 2127–2159 (2006)
2. Bharucha, Z., Haas, H., Auer, G., Cosovic, I.: Femto-cell resource partitioning. In: Proceedings of IEEE Globecom 2009, pp. 1–6 (2009)
3. Chandrasekhar, V., Kountouris, M., Andrews, J.G.: Coverage in multi-antenna two-tier networks. IEEE Trans. Wirel. Commun. **8**(10), 5314–5327 (2009)
4. Chandrasekhar, V., Kountouris, M., Andrews, J.G.: Uplink capacity and interference avoidance for two-tier femtocell networks. IEEE Trans. Wirel. Commun. **8**(7), 3498–3509 (2009)
5. Cheng, S.-M., Ao, W., Chen, K.-C.: Downlink capacity of two-tier cognitive femto networks. In: IEEE 21st International Symposium on Personal, Indoor and Mobile Radio Communications (PIMRC), 26–30 September, pp. 1303–1308 (2010)
6. Kim, D.I., et al.: Joint rate and power allocation for cognitive radios in dynamic spectrum access environment. IEEE Trans. Wirel. Commun. **7**(12), 5517–5528 (2008)
7. Gamage, A.T., Alam, M.S., Shen, X.S., Mark, J.W.: Joint relay, subcarrier and power allocation for OFDMA-based femtocell networks. In: IEEE Wireless Communications and Networking Conference (WCNC), pp. 679–684 (2013)
8. Giivenc, I.: Impact of spreading on the capacity of neighbouring femtocells. In: IEEE 20th International Symposium on Personal, Indoor and Mobile Radio Communications (PIMRC), 26–30 September, pp. 1814–1818 (2009)
9. He, S., Lu, Z., Wen, X., Zhang, Z., Zhao, J., Jing, W.: A pricing power control scheme with statistical delay QoS provisioning in uplink of two-tier OFDMA femtocell networks. Mob. Netw. Appl. **20**(4), 413–423 (2015)

10. Huang, K., Lau, V., Krishnamurthy, S.V., Faloutsos, M.: Downlink ad hoc networks: transmission-capacity trade-off. IEEE J. Sel. Areas Commun. **27**(7), 1256–1267 (2009)
11. Kaimaletu, S., Krishnan, R., Kalyani, S., Akhtar, N., Ramamurthi, B.: Cognitive interference management in heterogeneous femto-macro cell networks. In: 2011 IEEE International Conference on Communications (ICC), pp. 1–6 (2011)
12. Law, L.K., Pelechrinis, K., Krishnamurthy, S.V., Paloutsos, M.: Downlink capacity of hybrid cellular ad hoc networks. IEEE/ACM Trans. Netw. **18**(1), 243–256 (2010)
13. Li, Y.-Y., Sousa, E.S.: Cognitive femtocell: a cost-effective approach towards 4G autonomous infrastructure networks. Wirel. Pers. Commun. **64**(1), 65–78 (2012)
14. Lopez-Perez, D., Valcarce, A., de la Roche, G., Zhang, J.: OFDMA femtocells: a roadmap on interference avoidance. IEEE Commun. Mag. **47**(9), 41–48 (2009)
15. MacKay, D.J.C.: Information Theory, Interferences, and Learning Algorithms. Cambridge University Press, Cambridge (2003)
16. Mitola, J., Maguire Jr., G.Q.: Cognitive radio: making software radios more personal. IEEE Pers. Commun. **6**(4), 13–18 (1999)
17. Ngo, D.T., et al.: Distributed interference management in femtocell networks. In: IEEE Vehicular Technology Conference, pp. 1–5 (2011)
18. Proakis, J.G.: Digital Communications, 4th edn. McGraw-Hill, New York (2000)
19. Torregoza, J.P.M., Enkhbat, R., Hwang, W.-J.: Joint power control, base station assignment, and channel assignment in cognitive femtocell networks. EURASIP J. Wirel. Commun. Netw. **2010**. Article ID 285714
20. Zander, J., Kim, S.-L.: Radio Resource Management for Wireless Networks. Artech House, Boston (2001)

NEW2AN: Wireless Networks

Improving Efficiency of Heterogeneous Wi-Fi Networks with Energy-Limited Devices

Dmitry Bankov, Evgeny Khorov$^{(\boxtimes)}$, Aleksey Kureev, and Andrey Lyakhov

Institute for Information Transmission Problems,
Russian Academy of Sciences, Moscow, Russia
{bankov,khorov,kureev,lyakhov}@iitp.ru

Abstract. A heterogeneous wireless network consists of various devices that generate different types of traffic with heterogeneous requirements for bandwidth, maximal delay and energy consumption. An example of such networks is a Wi-Fi HaLow network that serves a big number of Machine Type Communication battery-powered devices and several offloading client stations. The first type of devices requires an energy-efficient data transmission protocol, while the second one demands high throughput. In this paper, we consider a mechanism that allocates a special time interval (Protected Interval) inside of which only battery powered-powered devices can transmit. We show that appropriate selection of the Protected Interval duration allows battery-powered devices to consume almost the minimal possible amount of energy on the one hand, and to provide almost the maximal throughput for offloading stations on the other hand. To find such duration, we develop a mathematical model of data transmission in a heterogeneous Wi-Fi network.

Keywords: Internet of Things · Machine Type Communications · Restricted Access Window · Traffic Indication Map · Power save

1 Introduction

The number of devices connected to the Internet grows every day. According to Cisco prognosis [1], the number of connected devices, both Human and Machine Type Communications, will exceed 50 billion by 2020, most of them being wireless. Along with the number of devices, grows the variety of device and traffic types, increasing the heterogeneity. Wi-Fi networks are perfectly fit for such a challenge, which can be illustrated by the new amendment of the Wi-Fi standard, IEEE 802.11ah. On one hand, it is designed for the Internet of Things (IoT) scenarios with a swarm of energy-limited sensors that rarely transmit data. On the other hand, one of its use cases is cellular data offloading, which includes user devices that transmit saturated data flows.

The reported study was partially supported by RFBR, research project No. 15-07-09350 a.

O. Galinina et al. (Eds.): NEW2AN/ruSMART 2016, LNCS 9870, pp. 181–192, 2016.
DOI: 10.1007/978-3-319-46301-8_15

Let us consider data transmission in an infrastructure heterogeneous Wi-Fi network with two types of devices. The first type, active stations (STAs), transmit or receive heavy data streams and are not sensitive to energy consumption. The second type, power-saving (PS) STAs, rarely transmit or receive single data frames from the Access Point (AP). To minimize energy consumption, the Wi-Fi standard has a palette of methods which are based on the following idea.

In the PS mode, a STA switches between the awake and the doze states. In the awake state, the STA can receive and transmit frames, while in the doze state the STA consumes minimal energy and neither receives, nor transmits any information. A PS STA is mainly in the doze state and awakes only to transmit a message or to receive one from the AP. The latter works as follows. The AP buffers data targeted for the PS STAs. To inform the STAs about buffered data, the AP periodically includes a Traffic Indication Map (TIM) in broadcast beacon frames. From time to time, the PS STA awakes to check the TIM in the next beacon for the pending data. When a STA receives a beacon with a TIM indicating that it has data to receive, the STA sends a PS-poll frame (after contending for the channel), and the AP sends buffered data as a response to the successfully received PS-poll frame.

Since STAs can sleep during some beacons, i.e., the period between two consequent awakenings can be greater than the beacon period, the set of STAs that are awake after a beacon is unknown to the AP. So the AP cannot schedule transmissions for the PS STAs and they use random channel access to transmit PS-polls. However the performance of random channel access degrades with the increase of the number of contending STAs, and it is an important issue for dense IoT networks. To limit the contention and to prioritize the PS STAs, it is reasonable to protect their transmissions from the active STAs, dividing each Beacon Interval (BI) into two parts: the Protected Interval (PI) and the Shared Interval (SI). During the PI, only PS STAs are allowed to retrieve their frames, while in the SI all STAs can transmit their data. The Wi-Fi standard specifies several ways how to organize the PI, one of which is the new Restricted Access Window mechanism introduced in 802.11ah [2,3]. However, at this point a problem arises: *how to choose a proper duration of the PI?* On one hand, it shall be long enough to let the PS STAs receive the buffered data. On the other hand, the longer is the PI, the less time the active STAs have to transmit their data, therefore their throughput degrades. In addition, the PS STAs can finish their transmissions during the SI, but the contention with active STAs increases energy consumption.

In this paper, we study this problem and propose a solution based on a mathematical model of such heterogeneous data transmission. Our model takes into account the fact that during the PI, devices operate in non-saturated mode, i.e., they simultaneously start contending for the channel to transmit a frame to the AP, while in SI, we combine the transmission of non-saturated and saturated flows.

The rest of the paper is organized as follows. Section 3 reviews related papers. Section 2 contains the formal problem statement. The developed analytical model is described in Sect. 4. Section 5 shows the numerical results. The conclusion is given in Sect. 6.

2   Problem Statement

Consider a network with an AP, PS STAs and $M - 1$ active STAs (PS STAs and active STAs are different devices) being in the transmission range of each other. Active STAs work in saturated mode, i.e., they always have packets for transmission. The AP queue of packets destined to active STAs is saturated too. For shortness, considering saturated data transmission, we do not distinguish the AP and $M - 1$ active STAs, assuming that we have M active STAs.

As for the PS STAs, they rarely receive single packets from the AP. Specifically, the AP periodically broadcasts beacons containing TIM information element, which indicates the STAs for which the AP has buffered data. We consider a situation, when the AP has data for several, say, K, PS STAs. As PS STAs do not need to awake before every beacon, not all K PS STAs receive the TIM. For clarification, we introduce probability p that a PS STA awakes to receive a particular beacon and receives the TIM.

When a PS STA receives a TIM which indicates that no data is buffered for the STA, it switches to the doze state. Otherwise, the PS STA transmits PS-poll frame to retrieve the buffered packet from the AP. To protect PS-poll frames from collisions with active STAs' packets, the AP establishes the PI of duration τ_{PI} which should, obviously, depend on K and p. If some PS STAs do not successfully retrieve the data from the AP during τ_{PI}, they can try again in the SI, contending for the channel with M active STAs. After a successful data reception from the AP, each PS STA turns to the doze state.

To contend for the channel, both PS STAs and active STAs use the default Wi-Fi channel access — called the Distributed Coordination Function (DCF) — which works in the following way. When a STA has a frame for transmission, it waits random backoff time. Specifically, it initializes backoff counter with a random integer value uniformly drawn from interval $[0, CW - 1]$, where CW is the contention window. It equals $CW_{min} = 16$ for the first packet transmission attempt and doubles after every unsuccessful transmission, until it reaches $CW_{max} = 1024$ (we use the default CW_{min} and CW_{max} values for 802.11ah STAs, provided in [2]). Being in the awake state, STAs continuously sense the channel. While the channel is idle, STAs decrement the backoff counter every T_e seconds. A STA suspends its backoff counters when the channel is busy, and resumes when the channel becomes idle and DCF InterFrame Space (DIFS) passes. Finally, when the backoff counter reaches zero, the STA transmits a packet.

If a transmission is successful, a Short InterFrame Space (SIFS) after that the receiver sends an acknowledgement (ACK). If the STA does not receive the ACK frame within T_{ACK}, it retries, unless the packet retry counter reaches retry limit RL. In this case, the packet is dropped.

Depending on the channel state — idle, STA's transmission or reception — the STA consumes power N_{TX}, N_{RX} and N_{IDLE}, respectively.

To achieve the best performance, the AP has to estimate the number of active STAs each BI and to re-select the duration τ_{PI}^* of the PI in such manner that (i) energy consumption by PS STAs is minimal while (ii) the active STAs' throughput is maximal. Note that generally speaking, such a value may not exist. However in this paper, we show that there is a range of values which provide suboptimal results for both performance indices.

To find such values, we need to develop a mathematical model of the described transmission process. Specifically, given τ_{PI}, the model shall allow obtaining

- the probability P that a chosen awaken PS STA retrieves its buffered data from the AP till the next beacon (i.e. within BI);
- the average energy $\overline{E}$ that this PS STA consumes when retrieving its packet from the AP;
- aggregated throughput S of M active STAs.

3   Related Work

A process of energy efficient transmission of single packets by a large number of stations has been studied in many papers.

Specifically, in [4], the authors consider a system, where multiple RFID tags transmit data to a single reader. The reader periodically allocates to the tags a frame that consists of several slots. The tags use slotted Aloha to choose the slots when they transmit data. The authors solve the problem of choosing such a frame duration, that the ratio of the expected number of successful slots duration to the total frame duration is maximal. This is done by estimating the number of tags, initially unknown to the reader, using the Maximum Likelihood approach. Unfortunately, the results of this paper cannot be applied to select the PI duration in our case, for two reasons. First, the medium access used in Wi-Fi is significantly different from the one used for RFID. Second, the number of contending PS STAs may significantly vary from a BI to a BI, as well as inside a BI.

In [5], similarly to our paper, the authors consider a 802.11ah network consisting of an AP and several associated STAs. The AP periodically allocates a Restricted Access Window to the STAs to protect their transmission, thus implementing the PI. The authors devise a way for the AP to estimate the number of STAs and describe a model of data transmission during the PI that can be used to find the probability of the successful transmission. However, the authors consider a case when the STAs transmit in saturated mode, i.e., in their case the number of contending STAs does not decrease during the PI, therefore their model is inapplicable in our case.

Non-saturated transmission is considered in [6], the authors of which model a network of PS STAs connected to an AP. The STAs awake every BI to receive their data from the AP. The authors develop a model that can be used to find the average packet delay and the average energy consumption per packet. However, this model is inapplicable to solve our problem, because it does not consider the existence of active STAs that transmit their data along with the PS STAs.

To address the issues of modeling the transmission of non-saturated data flows along with background saturated traffic, in our paper we develop a new mathematical model, which is described in Sect. 4.

4   Analytical Model

4.1   General Description

To estimate throughput of active STAs, we assume that the probability that a PS STA succeeds to retrieve its data from the AP during the PI is close to 1. With this assumption we, firstly, can consider each BI separately, i.e., there are no PS STAs that had started to retrieve their data in the previous BI and did not succeed until the considered BI. Secondly, the number of PS STAs that retrieve their data during the SI is low and these PS STAs do not affect the probability that an active STA transmits in a slot during the SI. Note that this assumption likely holds in the range of suboptimal τ_{PI} values, as we show in Sect. 5. We also assume that the duration of SI is much longer than the duration of a packet. Under such assumptions the active STAs throughput in the SI can be estimated with well-known Bianchi's model [7]. Since active STAs cannot transmit outside the SI, the average throughput can be found as follows:

$$S = S_{Bianchi}(1 - \frac{\tau_{PI}}{\tau_{BI}}),\tag{1}$$

where $S_{Bianchi}$ is the throughput found with Bianchi's model.

Let us find P and $\overline{E}$. For that, we consider a BI. As a PS STA wakes up with probability p, the total number k of awaken PS STAs among the K STAs for which the AP has buffered data in the beginning of a BI is a binomially distributed random number. In Sects. 4.2 and 4.3, we obtain probability P' that a chosen PS STA succeeds to retrieve the data till the end of the BI and the average energy $\overline{E'}$ consumed by the PS STA to retrieve a data packet provided that the number k of awaken PS STAs is given. Obviously, sought-for values of P and $\overline{E}$ can be expressed as follows

$$P = \sum_{k=0}^{K-1} \binom{K-1}{k} p^k (1-p)^{K-k-1} P'(k, M),\tag{2}$$

$$\overline{E} = \sum_{k=1}^{K} \binom{K}{k} p^k (1-p)^{K-k} \overline{E'}(k, M).\tag{3}$$

4.2   Process of Retrieving Packets

Since all STAs and the AP are located within transmission range of each other, they count their backoffs synchronously. Similar to [7], we denote a time interval between two consequent backoff countdowns as a *slot*. We distinguish the following types of slots.

- In an *empty* (e) slot, no STAs transmit.
- In a *successful* (s) slot, only one PS STA transmits its PS-poll and the AP replies with a data frame. Then the PS STA acknowledges reception of the frame.

- In a *collision* (c) slot, more than one PS STAs transmit a PS-poll, while active STAs do not transmit.
- In an *active* (a) slot, at least one active STA transmits. The transmission may be successful or not, however we do not distinguish these cases. Note that active slots can be present only in SI.

The durations of empty, successful, collision and active slots are T_e, T_s, T_c and T_a, respectively.

Let t be a slot number starting from the beginning of the BI. We choose an arbitrary PS STA and describe the evolution of the network from the beginning of the BI with a Markov process I_t with state $I = (c, s, a, r)$, where c, s and a are the numbers of collision, successful and active slots, respectively, and r is the number of retries for the chosen PS STA. Note that unlike Bianchi's model, we consider not a steady-state distribution of the process I_t, but its evolution during the BI.

In each state we can easily find the number of empty slots as $t - c - s - a$. So the real time T can be obtained from the model time t and the process state parameters as

$$T(t, I = (s, c, a, r)) = T_e(t - c - s - a) + T_s s + T_c c + T_a a. \tag{4}$$

For the defined process we introduce successful A^S and unsuccessful A^U absorbing states. A transition to A^S occurs when the chosen STA successfully transmits its data. The process transits to A^U when the chosen STA reaches its retry limit RL or the end of the BI is reached.

Let Q_a be the probability that at least one active STA transmits in a given slot. Obviously, during the PI, i.e. when $T < \tau_{PI}$, $Q_a(T) = 0$. In the SI $Q_a > 0$ and under aforementioned assumption can be estimated with the Bianchi's model [7]:

$$Q_a(T) = \begin{cases} 0, & T < \tau_{PI}, \\ 1 - (1 - \pi)^M, & otherwise, \end{cases} \tag{5}$$

where π is the probability that a given active STA transmits in a slot, obtained using Bianchi's model.

Figure 1 shows possible transitions from state (c, s, r, a). We denote a transition probability as P_X^Y, where X represents the type of slot t, i.e. X is e, s, c or a, and Y is either $+$ or $-$ depending on whether the chosen PS STA transmits or not. Since by definition, in an empty slot no STAs transmit, we simply write P_e, omitting the upper index.

To find these probabilities, we introduce $P_{TX|t,r}$ which is the conditional probability that the chosen PS STA transmits in slot t, provided that by that time it has made r unsuccessful transmission attempts. We also introduce $P_{e|-}, P_{s|-}, P_{c|-}$ which are the conditional probabilities that the slot is empty, successful or collision, respectively, provided that the chosen STA does not transmit, and express transition probabilities as

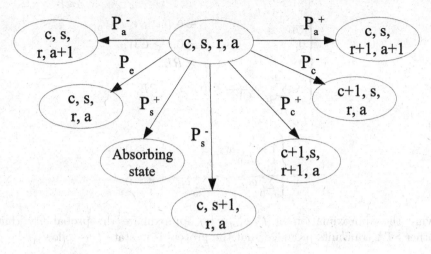

Fig. 1. Transitions for process I_t.

$$P_e = (1 - P_{TX|t,r})\, P_{e|-},$$
$$P_s^+ = P_{TX|t,r} P_{e|-},$$
$$P_s^- = (1 - P_{TX|t,r}) P_{s|-},$$
$$P_c^+ = P_{TX|t,r}(1 - P_{e|-} - Q_a(T)),$$
$$P_c^- = (1 - P_{TX|t,r}) P_{c|-},$$
$$P_a^+ = P_{TX|t,r} Q_a(T),$$
$$P_a^- = (1 - P_{TX|t,r}) Q_a(T).$$

Given $P_{TX|t,I}$, the probability of a PS STA to transmit if the process is in state I at time instant t, these probabilities can be found as follows:

$$P_{e|-} = (1 - P_{TX|t,I})^{k-s-1}(1 - Q_a(T)),$$
$$P_{s|-} = (k - s - 1)P_{TX|t,I}(1 - P_{TX|t,I})^{k-s-2}(1 - Q_a(T)),$$
$$P_{c|-} = 1 - P_{e|-} - P_{s|-} - Q_a(T).$$

Using the approach from [8], we estimate $P_{TX|t,r}$ as follows:

$$P_{TX|t,r} = \frac{a_{t,r}}{b_{t,r}},$$

where $a_{t,r}$ and $b_{t,r}$ approximate the unconditional probability of the chosen STA transmitting in slot t with retry counter r, and the probability of the STA having retry counter r in time slot t, respectively:

$$
a_{t,r} = \begin{cases}
\frac{1}{CW_0}, & r = 0, 0 \le t < CW_0, \\
0, & r = 0, t \ge CW_0, \\
0, & r \ge RL, \\
\sum\limits_{i=t-CW_r}^{t-1} \frac{a_{i,r-1}}{CW_r}, & 0 < r < RL,
\end{cases}
$$

and

$$
b_{t,r} = \begin{cases}
1 - \sum\limits_{i=0}^{t-1} b_{i,r}, & r = 0, \\
\sum\limits_{i=0}^{t-1} b_{i,r-1} - \sum\limits_{t=0}^{t-1} b_{i,r}, & r > 0.
\end{cases}
$$

Having the approximation of $P_{TX|t,r}$, we approximate the probability that another STA transmits provided that the process is in state I as follows:

$$
P_{TX|t,I=(s,c,a,r)} = \frac{\sum\limits_{\hat{r}=0}^{\min(c+a,RL-1)} P_{TX|t,\hat{r}} \Pr\left(t, I = (s,c,a,\hat{r})\right)}{\sum\limits_{\hat{r}=0}^{\min(c+a,RL-1)} \Pr\left(t, I = (s,c,a,\hat{r})\right)}
$$

Now we consider the evolution of the process. The process starts in time 0 in state $(0,0,0,0)$. With the described transitions of the process, we can iteratively find its state probability distribution $\Pr(t, I)$ at each time slot t. Moreover using (4), we can find the probability that the chosen STA successfully retrieves its data during the BI with given k:

$$
P'(\tau_{BI}) = \sum_{t,I:T(t,I)<\tau_{BI}} \Pr(t,I)P_s^+. \tag{6}
$$

4.3   Calculating Average Energy $\overline{E}'$

Let $E(I,t)$ be the average energy that the chosen STA has consumed by time instant t when its state is I. It equals 0 for $t = 0$, otherwise it is calculated according to the following equation:

$$
E(I,t) = \sum_{I' \overset{X,Y}{\to} I} \Pr(I'|t-1)P_X^Y(E(I',t-1) + E_X^Y), \tag{7}
$$

where we sum over all possible states I' at time $t - 1$ and corresponding transitions to state I at time t. Similar to transition probabilities, we denote the energy consumed by such transitions as E_X^Y. This energy depends on N_{tx}, N_{rx} and N_{idle}, which are the power consumed in transmission, reception or idle state, as follows:

$$E_e = N_{idle}\sigma,$$
$$E_s^- = N_{rx}(T_{PS} + T_D + T_{ACK}) + N_{idle}(2SIFS + DIFS),$$
$$E_s^+ = N_{tx}(T_{PS} + T_{ACK}) + N_{rx}T_D + N_{idle}(2SIFS + DIFS),$$
$$E_c^- = N_{rx}T_{PS} + N_{idle}(SIFS + T_{ACK} + DIFS),$$
$$E_c^+ = N_{tx}T_{PS} + N_{idle}(SIFS + T_{ACK} + DIFS),$$
$$E_a^- = N_{rx}(T_{AD} + T_{ACK}) + N_{idle}(SIFS + DIFS),$$
$$E_a^+ = N_{tx}T_{PS} + N_{rx}(T_{AD} - T_{PS}) + N_{idle}(SIFS + T_{ACK} + DIFS).$$

where $T_{PS}, T_{ACK}, T_D, T_{AD}$ are the durations of PS-poll, ACK, data frames transmitted for PS STAs and data frames transmitted by active STAs, respectively.

The average energy that the chosen STA consumes to successfully retrieve a packet equals

$$\overline{E'} = \frac{\sum_t E(A^S, t) + E(A^U, t)}{P(\tau_{PI})}. \tag{8}$$

5   Numerical Results

To validate our model, we consider an 802.11ah network with K PS STAs for which the AP has buffered data at every BI and M active STAs, all STAs being located 10 meters from the AP. Such a short distance guarantees that transmission errors are caused only by collisions. Active STAs transmit and receive data frames 1500 bytes long and PS STAs retrieve data frames 100 bytes long. A PS STA awakes with probability $p = 0.5$. We assume that active STAs and PS STAs operate in a 2 MHz channel. Active STAs use MCS5 while PS STAs use the most reliable modulation coding scheme (MCS0)[2]. Table 1 shows the scenario parameters. Transmit, receive and idle power are derived from voltage and current values given in the IEEE simulation scenario recommendations [9].

Table 1. Scenario parameters

Parameter	Value	Parameter	Value	Parameter	Value
T_e	52 µs	CW_{min}	16	T_{PS}	320 µs
T_s	1560 µs	RL	7	$SIFS$	160 µs
T_c	988 µs	Transmit power N_{TX}	308 mW	$DIFS$	264 µs
T_a	3276 µs	Receive power N_{RX}	11 mW	T_{ACK}	240 µs
τ_{BI}	100 ms	Idle power N_{IDLE}	5.5 mW	$ACKTimeout$	400 µs

Although an 802.11ah AP can support up to 8191 connected STAs, we consider only small numbers of simultaneously operating STAs, assuming that the AP uses some standard mechanisms to decrease the contention between the

STAs. One of such mechanisms — TIM segmentation — is described in 802.11ah amendment [2]. It allows the AP to divide the TIM into several parts and broadcast only a single part of the TIM in a beacon. The AP thus divides the STAs into groups, e.g. with a clusterization method from [10], and services each group in a separate BI. Even if the total number of STAs is big, only a reasonable number of them will retrieve their data during the BI, while the rest of the STAs will wait for a beacon containing their part of the TIM.

At first, we find the average energy $\overline{E}$ that the chosen PS STA consumes to retrieve a packet from the AP. Figure 2 shows that the results obtained with the developed analytical model almost coincide with those obtained with simulation, except for short PIs. If the PI is short, the PS STAs mostly finish their

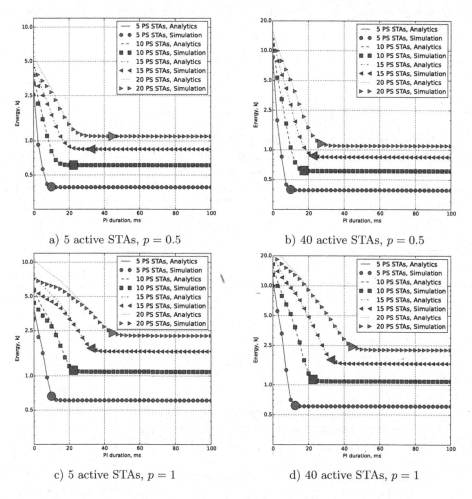

a) 5 active STAs, $p = 0.5$ b) 40 active STAs, $p = 0.5$

c) 5 active STAs, $p = 1$ d) 40 active STAs, $p = 1$

Fig. 2. Dependency of the chosen STA average energy consumption on the PI duration, big markers indicate the τ^* value.

transmissions during the SI, contending for the channel with active STAs, so only in this area our assumption that PS STAs do not affect the performance of active STAs does not hold. Small PI duration yields high energy consumption, and by increasing the PI duration we can manifold reduce it down to some value at τ^* (indicated with big markers), where it reaches a plateau that corresponds to the case when the PI is long enough for PS STAs to receive their data during it. More accurately, we can state that τ^* is the minimal τ_{PI} for which power consumption differs from the optimal one not more than by 10 %.

Increasing the PI duration beyond τ^* does not significantly affect power consumption, but reduces the time available for active STAs, decreasing their throughput. In more detail, such an effect is shown in Fig. 3. An important result is that initially, i.e. when PI is small, the throughput of active STAs grows with the PI. It means that in addition to PS STAs, active STAs benefit from the introduction of the PI, too, since they do not suffer from contention for the channel with a high number of PS STAs. However, at some point the throughput reaches the maximal value, after which it goes down, since the amount of the time available for active STAs decreases.

An important fact is that although choosing $\tau_{PI} = \tau^*$ does not maximize throughput, it provides suboptimal performance. Indeed, in the worst considered case, with 40 active STAs and 20 PS STAs, the throughput at τ^* differs from the maximal one by less than 30 %, see Fig. 3, but for smaller number of STAs the difference decreases. Note that the maximal throughput can only be achieved with manifold increase of energy consumption.

Another important fact is that the throughput obtained analytically at this point is close to the value obtained with simulation.

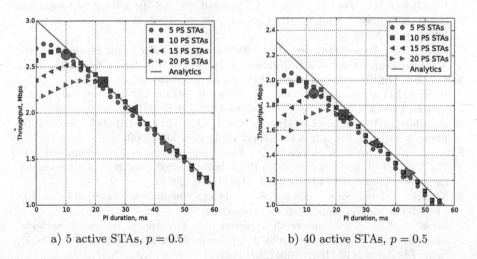

a) 5 active STAs, $p = 0.5$ b) 40 active STAs, $p = 0.5$

Fig. 3. Dependency of the active STA throughput on the PI duration, big markers indicate the τ^* value.

6 Conclusion

In this paper, we have studied efficiency of the Wi-Fi power saving mechanism in an heterogeneous Wi-Fi network with energy-limited devices. To improve performance of the network, the AP protects transmission of PS-polls from collisions with transmission of active STAs. To model operation of the network, we have developed a mathematical model, which provides us such performance indices as throughput and power consumption. With this model, we show that the duration of the protection interval can be chosen in such a way that provides suboptimal results for both the throughput and power consumption, which is important for implementation.

The future research is connected with considering scenarios with hidden STAs and TIM segmentation, a novel mechanism introduced in IEEE 802.11ah.

References

1. Evans, D.: The internet of things: how the next evolution of the internet is changing everything. CISCO white paper, San Jose, vol. 1, pp. 1–11 (2011)
2. IEEE P802.11ahTM, D5.0 Draft Standard for Information technology – Telecommunications, information exchange between systems Local, metropolitan area networks Specific requirements - Part 11: Wireless LAN Medium Access Control (MAC) and Physical Layer (PHY) Specifications - Amendment 6: Sub 1 GHz License Exempt Operation, March 2015
3. Khorov, E., Lyakhov, A., Krotov, A., Guschin, A.: A survey on IEEE 802.11 ah: an enabling networking technology for smart cities. Comput. Commun. **58**, 53–69 (2015)
4. Khandelwal, G., Lee, K., Yener, A., Serbetli, S.: ASAP: a MAC protocol for dense and time-constrained RFID systems. EURASIP J. Wirel. Commun. Netw. **2007**(2), 3 (2007)
5. Park, C.W., Hwang, D., Lee, T.-J.: Enhancement of IEEE 802.11 ah MAC for M2M communications. IEEE Commun. Lett. **18**(7), 1151–1154 (2014)
6. Liu, R.P., Sutton, G.J., Collings, I.B.: Power save with offset listen interval for IEEE 802.11 ah smart grid communications. In: 2013 IEEE International Conference on Communications (ICC), pp. 4488-4492. IEEE (2013)
7. Bianchi, G.: Performance analysis of the IEEE 802.11 distributed coordination function. IEEE J. Sel. Areas Commun. **18**(3), 535–547 (2000)
8. Khorov, E., Krotov, A., Lyakhov, A.: Modelling machine type communication in IEEE 802.11 ah networks. In: 2015 IEEE International Conference on Communication Workshop (ICCW), pp. 1149–1154. IEEE (2015)
9. TGax Simulation Scenarios. https://mentor.ieee.org/802.11/dcn/14/11-14-0980-10-00ax-simulation-scenarios.docx
10. Levin, M.S.: Combinatorial clustering: literature review, methods, examples. J. Commun. Technol. Electron. **60**(12), 1403–1428 (2015). http://dx.doi.org/10.1134/S1064226915120177

Mathematical Model of QoS-Aware Streaming with Heterogeneous Channel Access in Wi-Fi Networks

Alexander Ivanov, Evgeny Khorov[✉], Andrey Lyakhov, and Ilya Solomatin

Institute for Information Transmission Problems, Moscow, Russia
{a.ivanov,khorov,lyakhov,solomatin}@iitp.ru

Abstract. In Wi-Fi networks, preliminary channel reservation protects transmissions in reserved time intervals from collisions with neighboring stations. However, making changes in established reservations takes long time spent on negotiating changes with neighboring stations and dissemination of information about these changes. This complicates serving of Variable Bit Rate (VBR) flows which intensity varies with time, what leaves no choice but to reserve some additional time for handling data bursts and packet retransmissions (caused by random noise and interference from remote stations). In the paper, we consider a more flexible approach when bursts and retransmissions are handled by some random access method while a constant part of an input flow is served in preliminarily reserved intervals. We build a mathematical model of a VBR flow transmission process with this heterogeneous access method and use the model to find transmission parameters which guarantee that Quality of Service requirements of the flow are satisfied at the minimal amount of used channel time.

Keywords: Wi-Fi · VBR · QoS · Heterogeneous method · Mathematical model

1 Introduction

The users' desire for better Quality of Service (QoS) drives development of new QoS-aware protocols. In wireless technologies, special attention is paid to channel access mechanisms which to a considerable degree influence the ability to provide QoS. A very good example is the Wi-Fi technology where the IEEE 802.11e amendment, which was the first one to introduce QoS support into Wi-Fi, defined two QoS-aware channel access mechanisms: Enhanced Distributed Channel Access (EDCA) and Hybrid coordination function Controlled Channel Access (HCCA). The former is a *random* (contention-based) channel

The research was done in IITP RAS and supported by the Russian Science Foundation (agreement No 16-19-10687).

access mechanism providing prioritized QoS. The latter is a *deterministic* channel access mechanism providing parameterized QoS by means of contention-free polling.

Both EDCA and HCCA were suitable for QoS provisioning at the moment of their development. But constantly increasing density of Wi-Fi networks has made these mechanisms barely applicable to QoS provisioning in modern scenarios, where multiple Wi-Fi networks usually coexist in one area. Because of a low number of available frequency channels, access points (APs) have to choose channels which are already occupied. It makes EDCA suffer from frequent collisions and severe interference. Using HCCA, an AP can slightly relieve the situation by protecting data transmissions from interference with its own stations (STAs), but not from STAs of associated with other APs. Thus, both EDCA and HCCA cannot be used to provide QoS in dense Wi-Fi networks.

Reliable data transmission in dense Wi-Fi networks requires some sort of coordination between STAs in order to reduce interference. In recent Wi-Fi amendments, such coordination has been provided by means of new deterministic channel access mechanisms based on *preliminarily channel time reservation*. For example, the IEEE 802.11aa amendment developed for robust audio and video streaming has appended HCCA with the HCCA Negotiation mechanism, which allows an AP to reserve time intervals during which this AP can serve its STAs while the neighboring APs and their STAs do not transmit. This is achieved through information dissemination about the reserved time intervals. To reduce the overhead caused by such dissemination, time intervals are reserved not individually but in sequences: an AP reserves *a sequence of periodic time intervals of equal duration*. Next, we refer to such a sequence simply as a *(periodic) reservation*. Thanks to the periodicity, a reservation can be described only by three parameters: the *period* of the reserved time intervals, their *duration* and the *beginning of the first interval* (see Fig. 1). The same approach is used in the IEEE 802.11s amendment (Wi-Fi Mesh technology) where deterministic Mesh coordination function Controlled Channel Access (MCCA) is defined. Using MCCA, a STA can set up a periodic reservation in order to protect its data transmissions from interference with neighboring STAs.

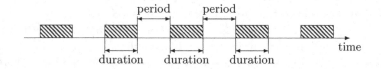

Fig. 1. Periodic reservation

Generally speaking, deterministic channel access mechanisms provide higher reliability as compared with random access ones and, thus, seem to be more desirable when serving data with strict QoS requirements. However, preliminarily channel reservation requires some negotiation between neighboring STAs and dissemination of information about established reservations. On the one hand, this ensures that reservations do not overlap, thus, increasing reliability. On the other hand, the negotiation require both additional signaling and time to be performed. As for the dissemination, it takes even longer time since it is performed via special management frames — beacons — which are transmitted quite rarely. Thus, quick changes in already established reservations are not possible. While it is not a problem when serving Constant Bit Rate (CBR) flows [2–5], it comes to the fore in case of Variable Bit Rate (VBR) flows, which intensity vary with time. Along with packets retransmissions, caused by inevitable transmission failures, it turns the choice of appropriate reservation parameters into a non-trivial problem, especially when a served VBR flow imposes strict QoS requirements. One approach here consists in reserving a redundant amount of channel time, accounting for the worst possible situation. In this case, the flow can be transmitted with required QoS though at cost of overall network degradation since too much time is reserved in anticipation of the flow bursts.

To improve the situation, a random access method (like EDCA) can be used jointly with channel reservations. The idea is to handle flow bursts and retransmissions by means of the random access method while serving the constant component of the flow intensity inside periodic intervals. Indeed, random access mechanisms do not need to wait for the nearest reserved interval to transmit data and can start to contend for the channel after it remains idle for a specified duration. Thus, though experiencing backoff-induced delays and frequent collisions, such mechanisms are able to handle variations of a VBR flow intensity.

Standardization activity of Wi-Fi community has resulted in emergence of *a heterogeneous channel access method* in Wi-Fi. This method based on EDCA and HCCA is not described very well and only few studies [8–11] consider its usage for QoS provisioning. In this paper, we propose how to use the heterogeneous access method to transmit a VBR flow with QoS requirements and build a mathematical model of this transmission process. The model can be used to find such parameters of the method which guarantee QoS requirements satisfaction at the minimal amount of used channel time. Moreover, we demonstrate gains which the heterogeneous access method achieves over the deterministic one.

The rest of the paper is organized as follows. In Sect. 2, we briefly review the existing studies on heterogeneous access methods. We formulate the problem of the paper in Sect. 3. In Sect. 4, we develop the mathematical model of the considered transmission process. In Sect. 5, we use the model to select appropriate transmission parameters and demonstrate gains of the heterogeneous channel access. Finally, Sect. 6 concludes the paper.

2   Related Papers

A number of papers study how to share resources between random and deterministic access methods in Wi-Fi networks. For example, [12] investigates how to share time between EDCA and HCCA to achieve the maximum overall throughput. [7] considers coexistence of EDCA and MCCA in Wi-Fi Mesh networks and shows how EDCA throughput depends on the percentage of time reserved by MCCA. As for the heterogeneous access, a few existing studies like [8] and [10] mainly consider a heterogeneous access method defined in the Wi-Fi standard. In [8], the authors consider streaming of saturated data and show that the more data is transmitted with HCCA, the higher is ratio of successful transmissions. The authors of [10] propose to use a Markov decision process to coordinate simultaneous usage of HCCA and EDCA to achieve maximum channel utilization. However, none of the mentioned papers consider streaming of data with QoS requirements other than throughput. The usage of heterogeneous access for transmitting data with QoS requirements is considered in [11] where a good description/analysis of EDCA, HCCA as well as the heterogeneous access method is presented. [11] proposes an enhanced admission control algorithm which can be used to improve performance of almost any HCCA scheduler as shown by simulation. However, so far no papers have studied how to transmit unsaturated data with QoS requirements by means of a heterogeneous access method, i.e., how to choose its transmission parameters to satisfy QoS. In this paper, we fill this gap by developing a mathematical model of a VBR flow transmission in the presence of noise with such an access method. We exploit the mathematical approach from [6], which considers transmission of a VBR flow with QoS requirements (delivery delay and packet loss ratio) inside periodic time intervals.

3   Problem Statement

We consider transmission of a VBR flow between two stations. Packets of the flow arrive strictly periodically (with period T_{in}) in batches of random size (the batch size takes value j with probability p_j^{in}, $j \in \{1, \ldots, \mathcal{M}\}$). Such structure of an input flow corresponds to transmission of a video flow with RTP [1]. The QoS requirements are represented by a) the bound on packet delivery time D_{QoS} and b) the bound on packet loss ratio PLR_{QoS}, so that the sender drops any flow packet standing in the queue longer than D_{QoS}.

To transmit the flow, the sender uses a heterogeneous access method. For that, the sender and receiver set up a periodic reservation with period T_{res} and duration D_{det} of the reserved intervals which is enough only for a single packet transmission attempt. The sender transmits packets in the reserved time intervals as well as outside them. The sender always transmits the oldest packet in the queue. This packet is transmitted until it is successfully delivered or its lifetime exceeds D_{QoS}. In the latter case, the packet is discarded and the sender starts serving the next packet in the queue. Additionally, to control the number of transmission attempts of a packet, the sender maintains a retry counter: each packet can be transmitted no more than $\mathcal{R}$ times.

The probability of unsuccessful packet transmission equals q_{det} in reserved time intervals and q_{ran} outside them. Since generally $q_{ran} > q_{det}$, the sender uses the random access only for packets which will become outdated before the next reserved time interval and, thus, will be dropped. We assume that the time

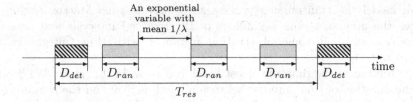

Fig. 2. Time parameters

needed to get an access to the channel between two consecutive reserved time intervals, i.e. in a *contention-based interval*, is an exponential random variable with mean $1/\lambda$. The duration of the contention-based interval equals $T_{res} - D_{det}$, while the time needed to transmit a packet with the random access equals D_{ran} (see Fig. 2).

In the described transmission process, we need to choose such T_{res} and $\mathcal{R}$ that guarantee that the QoS requirements are satisfied while keeping the amount of the used channel time as low as possible. To solve this problem, we develop a mathematical model of the considered process which can be used to find PLR as a function of T_{res} and $\mathcal{R}$.

4 Mathematical Model

4.1 Markov Chain

Further we assume that $T_{res} \leq T_{in} \leq D_{QoS}$. First, we split the time into slots of duration $\tau = \gcd(T_{res}, T_{in})$, so that the beginnings of the reserved intervals coincide with the beginnings of some slots. We express all time values in slots (Fig. 3):

$$t_{res} = \frac{T_{res}}{\tau}, t_{in} = \frac{T_{in}}{\tau}, \ t_{res}, t_{in} \in \mathbb{N}.$$

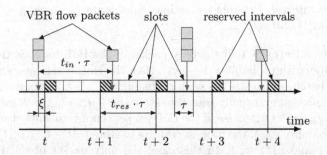

Fig. 3. Packet arrivals and periodic reservation

We model the transmission process as a discrete time Markov chain: we observe the process at the beginnings of the reserved intervals and describe the process state at some observation instant t with three integer values $(h(t), m(t), r(t))$.

$h(t)$ represents the time the oldest batch has spent in the queue. We denote ξ as the duration of time interval between a batch arrival and the beginning of the next slot (ξ is the same for all batches). Since nothing happens with batches during time ξ after their arrivals, we virtually shift all batch arrivals up to the beginnings of the next slots. To keep the process behavior unchanged, we decrease the bound D_{QoS} by ξ. We define the age of a batch as the difference between the current time and the batch arrival time. This definition is valid even for a batch which has not arrived yet, though in this case its age is negative. Next, we define the *Head of Line (HoL) batch* as the batch with the highest age among all batches which are currently in the queue and batches which have not arrived yet. We refer to the first packet of the HoL batch as the *HoL packet*. Finally, $h(t)$ is the age of the HoL packet (batch) expressed in slots. Thanks to the time shift we made above, the age of the HoL packet at any observation instant equals an integer number of slots ($h(t) \in \mathbb{Z}$). $m(t)$ is the remaining number of packets in the HoL batch and $r(t)$ is the remaining number of transmission attempts of the HoL packet.

Due to the definition of $h(t)$, it is always higher than $-t_{in}$ since the next batch arrives not later than $(t_{in} - 1)$ slots. Since the age of the HoL batch cannot exceed D_{QoS}, $h \leq d = \lfloor \frac{D_{QoS}}{\tau} \rfloor$.

Let the system be in state $(h(t), m(t), r(t))$ at instance t. Further we describe all possible transitions from this state.

$h(t) < 0$. In this case, the queue is empty and the next batch of size $m(t)$ arrives only $|h(t)|$ slots later. By the next reserved interval the age of this batch simply increases by t_{res} slots. Thus, the process transits to state $(h(t)+t_{res}, m(t), r(t) = \mathcal{R})$ with probability 1.

$0 \leq h \leq d-t_{res}$. In this case, the queue is not empty and the sender transmits the HoL packet in the current reserved interval. If not transmitted successfully, the HoL packet remains in the queue since it does not become outdated by the next reserved interval. Possible transitions from state $(h(t), m(t), r(t))$ depend on values of $m(t)$ and $r(t)$:

- $m(t) = 1$ and $r(t) > 1$. If the only packet of the HoL batch is transmitted successfully (with probability $1 - q_{det}$), then the HoL batch leaves the queue. The next batch, which size is j with probability p_j^{in}, is t_{in} slots younger, what makes the process eventually transit to state $(h(t)-t_{in}+t_{res}, j, \mathcal{R})$ with probability $(1-q_{det})p_j^{in}$. Otherwise, if the HoL packet is not transmitted successfully (with probability q_{det}), the process transits to state $(h(t)+t_{res}, m(t), r(t)-1)$.
- $m(t) = 1$ and $r(t) = 1$. In this case, the only packet of the HoL batch anyway leaves the queue because of the retry limit. So, the process transits to state $(h(t) - t_{in} + t_{res}, j, \mathcal{R})$ with probability p_j^{in}.

- $m(t) > 1$ and $r(t) > 1$. The process transits to state $(h(t), m(t)-1, \mathcal{R})$ with probability $1 - q_{det}$ and to state $(h(t), m(t), r(t) - 1)$ with probability q_{det}.
- $m(t) > 1$ and $r(t) = 1$. Since the HoL packet anyway leaves the queue because of the retry limit, the process transits to state $(h(t)+t_{res}, m(t)-1, \mathcal{R})$ with probability 1.

$d - t_{res} < h \le d$. In this case, the HoL batch becomes outdated by the next reserved time interval and leaves the queue by the beginning of the next interval. Thus, the process transits to state $(h(t) - t_{in} + t_{res}, j, \mathcal{R})$ with probability p_j^{in}.

Having described all possible transitions, we are able to build the transition matrix and find the steady-state distribution $\boldsymbol{\pi}(h, m, r)$ of the process states.

4.2 PLR Calculation

Given the average number of packets I_{in} arriving into the queue during one transition and the average number of packets I_{dis} discarded during one transition, we can find PLR as follows:

$$PLR = \frac{I_{dis}}{I_{in}}. \tag{1}$$

I_{in} can be easily found as

$$I_{in} = \frac{T_{res} \sum j p_j^{in}}{T_{in}}. \tag{2}$$

The calculation of I_{dis} is more complicated and presented below.

Packets can be discarded only during transitions from the following two types of states: (a) states with $r = 1$ and $0 \le h \le d - t_{res}$ and (b) states with $h > d - t_{res}$. Let $N_{dis}(h, m, r)$ be the average number of packets discarded during transition from state (h, m, r). If the state does not belong to the mentioned types, then $N_{dis}(h, m, r) = 0$. If the state belongs to type (a), then only the HoL packet can be discarded because of the retry limit if not transmitted successfully, thus, $N_{dis}(h, m, r) = q_{det}$. In case (b), the entire HoL batch becomes outdated before the next reserved interval. It occurs $D_{QoS} - h \cdot \tau$ s after the beginning of the current one. Outdating packets are transmitted with the random access method with all transmissions performed within $D_{QoS} - h \cdot \tau - D_{det} + D_{ran}$ s following the end of the current interval (see Fig. 4). This time, which we denote as $T(h)$, cannot be higher than $T_{res} - D_{det}$, thus, $T(h)$ can be accurately calculated as follows:

$$T(h) = \min\{D_{QoS} - h \cdot \tau - D_{det} + D_{ran}, T_{res} - D_{det}\}. \tag{3}$$

To find $N_{dis}(h, m, r)$, we need to find the probability distribution of the number of transmission attempts, which can be performed in an interval of duration $T(h)$. Let $\Lambda(w; T)$ be the probability of exactly w transmission attempts being possible inside a contention-based interval of duration T. The number of transmission in an interval of duration T cannot exceed $W(T) = \lfloor T/D_{ran} \rfloor + 1$. Thus, if $w > W(T)$, then $\Lambda(w, T) = 0$. Otherwise, let t_i^{wait} be the duration of a time interval between the end of transmission $i - 1$ and the beginning of

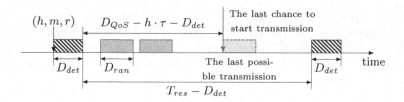

Fig. 4. Usage of random access

transmission i, $1 \leq i \leq \mathcal{W}(T)$. t_i^{wait} is an exponential random variable with mean $1/\lambda$. Thus, $\Lambda(w;T)$ can be calculated as follows:

$$\Lambda(w;T) = \mathbb{P}(\sum_{i=1}^{w} t_i^{wait} \leq T - wD_{ran}, \sum_{i=1}^{w+1} t_i^{wait} > T - (w+1)D_{ran}) = \quad (4)$$

$$= \mathbb{P}(\sum_{i=1}^{w} t_i^{wait} \leq T - wD_{ran}) - \mathbb{P}(\sum_{i=1}^{w+1} t_i^{wait} \leq T - (w+1)D_{ran}),$$

where

$$\mathbb{P}(\sum_{i=1}^{w} t_i^{wait} \leq X) = e^{-\lambda X} \sum_{i=w}^{+\infty} \frac{\lambda^i X^i}{i!} = 1 - e^{-\lambda X} \sum_{i=0}^{w-1} \frac{\lambda^i X^i}{i!}.$$

Now, we are able to calculate $N_{dis}(h,m,r)$ for transition from state with $h > d - t_{res}$. With probability $1 - q_{det}$ the HoL packet is successfully transmitted and $m - 1$ remaining packets of the HoL batch are transmitted during the contention interval. Since none of these packets has been transmitted before even once, all of them have $\mathcal{R}$ transmission attempts. Otherwise, with probability q_{det} all the m packets are transmitted with the random access where the remaining number of transmission attempts of the first packet equals $r - 1 \geq 0$, while other packets have $\mathcal{R}$ transmission attempts. Let the maximum number of transmission attempts in the contention interval be equal to w (with probability $\Lambda(w;T(h))$). We consider transmission of $n \geq 0$ packets in a *transmission window* of size $0 \leq w \leq \mathcal{W}(T(h))$. Let $U(r,n,w)$ be the average number of packets not transmitted successfully in this window, where $r \geq 0$ is the remaining number of transmission attempts of the first packet. If $n = 0$, then $U(r,n,w) = 0$. For $n \geq 1$, the values of $U(r,n,w)$ can be found recurrently:

$$U(r,n,w) = \sum_{t=1}^{\min\{w,r\}} q_{ran}^{t-1}(1 - q_{ran})U(\mathcal{R}, n-1, w-t) +$$

$$q_{ran}^{\min\{w,r\}}(1 + U(\mathcal{R}, n-1, w - \min\{w,r\})). \quad (5)$$

Combining (3), (4) and (5) we can calculate $N_{dis}(h,m,r)$ for $h > d - t_{res}$:

$$N_{dis}(h,m,r) = \sum_{w=0}^{\mathcal{W}(T(h))} \Lambda(w;T(h)) \left(q_{det}U(r-1,m,w) + (1 - q_{det})U(\mathcal{R}, m-1, w) \right).$$

Finally, I_{dis} can be calculated as follows:

$$I_{dis} = \sum_{h,m,r} \pi(h,m,r)N_{dis}(h,m,r). \tag{6}$$

4.3 Channel Time Consumption

We define channel consumption C as the percentage of the channel time occupied by a flow transmission with the heterogeneous channel access method. C is composed of two components: $C = C_{det} + C_{ran}$. The former is the percentage of the reserved channel time, and the latter is the percentage of the channel time occupied by transmissions in contention-based intervals. Evidently,

$$C_{det}(T_{res}) = \frac{D_{det}}{T_{res}}. \tag{7}$$

To find C_{ran}, we denote $N_{tx}(h,m,r)$ as the average number of transmission attempts performed with the random access during transition from state (h,m,r). If $h \le d - t_{res}$, then $N_{tx}(h,m,r) = 0$. For $h > d - t_{res}$, values of $N_{tx}(h,m,r)$ can be calculated in a similar way as values of $N_{dis}(h,m,r)$ are calculated in Sect. 4.2. Let $V(r,n,w)$ be the average number of transmission attempts performed in a contention-based interval to transmit $n \ge 0$ packets. Here $r \ge 0$ is the remaining number of transmission attempts of the first packet, while other packets have $\mathcal{R}$ transmission attempts, and $w \ge 0$ is the maximum possible number of transmission attempts in a contention-based interval. If $n = 0$, then $N_{tx}(r,n,w) = 0$. For $n \ge 1$, the following recurrent formula can be used:

$$V(r,n,w) = \sum_{t=1}^{\min\{r,w\}} q_{ran}^{t-1}(1 - q_{ran})(t + V(\mathcal{R}, n-1, w-t)) +$$
$$q_{ran}^{\min\{r,w\}}(\min\{r,w\} + V(\mathcal{R}, n-1, w - \min\{r,w\})).$$

Similar to (4.2), for $N_{tx}(h,m,r)$ we obtain

$$N_{tx}(h,m,r) = \sum_{w=0}^{\mathcal{W}(T(h))} \Lambda(w; T(h))(q_{det}V(r-1,m,w) + (1 - q_{det})V(\mathcal{R}, m-1, w)).$$

Finally, C_{ran} can be calculated as follows:

$$C_{ran}(T_{res}, \mathcal{R}) = \frac{D_{ran}}{T_{res}} \sum_{h,m,r} \pi(h,m,r)N_{tx}(h,m,r).$$

5 Numerical Results

Let us show how to use the model to find appropriate parameters in case of a video flow transmission over a Wi-Fi network with IEEE 802.11a PHY. We

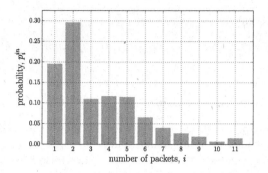

Fig. 5. The batch size distribution of the VBR flow

consider a VBR flow which batches arrive each $T_{in} = 40$ ms and have the size distribution shown in Fig. 5. The delay bound D_{QoS} equals 150 ms. The sender transmits data packets of size 1500 bytes at the rate of 54 Mbps and ACKs are transmitted at the rate of 6 Mbps, so that $D_{det} = 312$ μs and $D_{ran} = 346$ μs. The probabilities of unsuccessful transmissions are the following: $q_{det} = 0.05$ and $q_{ran} = 0.2$. The mean access time $1/\lambda$ equals 171 μs. This value is obtained experimentally in a saturated Wi-Fi network with 10 stations all of which use the EDCA random access mechanism and transmit best effort data at the rate of 54 Mbps. Given all these parameters, we use the model to find PLR as a function of $\mathcal{R}$ and T_{res}, which is shown in Fig. 6. The same function but obtained by simulation negligibly differs from the analytical one, that is why it is not explicitly shown in Fig. 6.

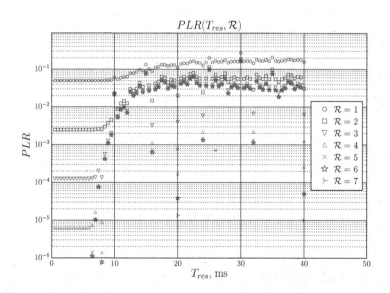

Fig. 6. PLR as a function of T_{res} and $\mathcal{R}$

Function $PLR(T_{res}, \mathcal{R})$ significantly drops at one points while raising at others. To explain this, let us consider a reservation with period $T_{res} = 20$ ms. In this case, packets can become outdated only in the middle between two consecutive reserved intervals, thus, having approximately $D_{det} \bmod T_{res} = 10$ ms to be transmitted with the random access method. Since 10 ms $\gg 1/\lambda$, multiple transmissions in a contention-based interval are possible. That is why, the higher is value of $\mathcal{R}$, the lower is the value of PLR. Another situation can be observed if $T_{res} = 30$ ms. In this case, outdating packets are never transmitted with the random access since $D_{QoS} \bmod T_{res} = 0$ ms, what increases PLR values in comparison with neighboring values of T_{res}.

It is worth to mention, that the developed model can be used to analyze transmission with the deterministic access method: PLR can be found as for the heterogeneous method simply by putting $q_{ran} = 1$, while channel consumption is given by (7). We use this ability of the model to compare the heterogeneous and deterministic access methods in terms of the minimal amount of the channel time needed to satisfy the QoS requirements. Given a value of PLR_{QoS}, we find such parameters $\tilde{T}_{res}$ and $\tilde{\mathcal{R}}$, which guarantee that the QoS requirements are satisfied at the minimal value of channel consumption C. We denote this value as $\tilde{C}(PLR_{QoS})$:

$$\tilde{C}(PLR_{QoS}) = \min_{\substack{T_{res}, \mathcal{R}: \\ PLR(T_{res}, \mathcal{R}) \leq PLR_{QoS}}} C(T_{res}, \mathcal{R}) = C(\tilde{T}_{res}, \tilde{\mathcal{R}})$$

Functions $\tilde{C}(PLR_{QoS})$ for the both considered methods are demonstrated in Fig. 7. It shows that the heterogeneous access method outperforms the deterministic one for all relevant values of PLR_{QoS} ($10^{-5} \ldots 10^{-2}$) by reducing channel consumption by 25–35 %. Though achieving this gains requires adjustment of the method parameters, this task can be successfully solved by means of the developed mathematical model, what finally turns the proposed heterogeneous method intro a more preferable choice for serving VBR flows.

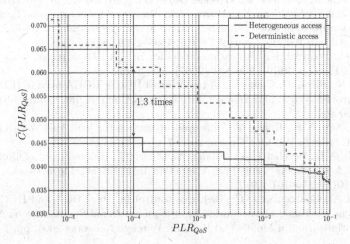

Fig. 7. $\tilde{C}(PLR_{QoS})$ for heterogeneous and deterministic access methods

6   Conclusion

In the paper, we have considered transmission of a VBR flow with a hetero-geneous channel access method being a combination of the random and deter-ministic. With this method, packets of the flow are transmitted in preliminarily reserved periodic time intervals as well as outside them, so that retransmissions and data bursts can be handled by means of the random access, while a constant component of the flow can be served in reserved interval. We have developed a mathematical model of this transmission process which can be used to find such parameters of the heterogeneous method, which guarantee that QoS require-ments of the flow are satisfied at the minimal amount of the used channel time. It is shown that the heterogeneous method generally occupies less channel time to satisfy QoS requirements then the deterministic one. The future work includes further improvement of the model to make it account for possibility of erroneous estimation of the transmission failure probabilities.

References

1. RFC 3550. http://www.ietf.org/rfc/rfc3550.txt
2. Ansel, P., Ni, Q., Turletti, T.: FHCF: a fair scheduling scheme for 802.11e WLAN (2003)
3. Cecchetti, G., Ruscelli, A.L.: Real-time support for HCCA function in IEEE 802.11e networks: a performance evaluation. Secur. Commun. Netw. **4**(3), 299–315 (2011)
4. Cowling, J., Selvakennedy, S.: A detailed investigation of the IEEE 802.11e HCF reference scheduler for VBR traffic. In: 13th International Conference on Computer Communications and Networks (ICCCN 2004). Citeseer (2004)
5. Grilo, A., Nunes, M.: Performance evaluation of IEEE 802.11e. In: The 13th IEEE International Symposium on Personal, Indoor and Mobile Radio Communications 2002, vol. 1, pp. 511–517. IEEE (2002)
6. Ivanov, A., Khorov, E., Lyakhov, A.: QoS support for bursty traffic in noisy channel via periodic reservations. In: Wireless Days (WD), 2014 IFIP, pp. 1–6. IEEE (2014)
7. Krasilov, A.N., Lyakhov, A.I., Moroz, Y.I.: Analytical model of interaction between contention-based and deterministic channel access mechanisms in Wi-Fi Mesh net-works. Autom. Remote Control **74**(10), 1696–1709 (2013)
8. Kuan, C., Dimyati, K.: Utilization model for HCCA EDCA mixed mode in IEEE 802.11e. ETRI J. **29**(6), 829–831 (2007)
9. Lai, W.K., Shien, C., Jiang, C.: Adaptation of HCCA/EDCA ratio in IEEE 802.11 for improved system performance. Int. J. Innov. Comput. Inf. Contr. **5**(11), 4177–4188 (2009)
10. Ng, B., Tan, Y.F., Roger, Y.: Improved utilization for joint HCCA-EDCA access in IEEE 802.11e WLANs. Optim. Lett. **7**(8), 1711–1724 (2012)
11. Ruscelli, A.L., Cecchetti, G., Alifano, A., Lipari, G.: Enhancement of QoS support of HCCA schedulers using EDCA function in IEEE 802.11e networks. Ad Hoc Netw. **10**(2), 147–161 (2012)
12. Siris, V.A., Courcoubetis, C.: Resource control for the EDCA and HCCA mech-anisms in IEEE 802.11e networks. In: 4th International Symposium on Modeling and Optimization in Mobile, Ad Hoc and Wireless Networks 2006, pp. 1–6. IEEE (2006)

Mobility Load Balancing Enhancement for Self-Organizing Network over LTE System

Sangchul Oh[1]([⊠]), Hongsoog Kim[1], Jeehyeon Na[1], Yeongjin Kim[1],
and Sungoh Kwon[2]

[1] 5G Giga Communication Research Laboratory,
Electronics and Telecommunications Research Institute (ETRI),
Daejeon 305-700, Korea
{scoh,kimkk,jhna,yjkim}@etri.re.kr
[2] School of Electrical Engineering, University of Ulsan, Ulsan 680-749, Korea
sungoh@ulsan.ac.kr

Abstract. The objective of mobility load balancing (MLB) is to intelligently spread user traffic load out on a network in order to avoid degradation of end-user experience and performance due to overloaded or congested cells. The load standard deviation (LSD) as a new key performance indicator (KPI) for MLB performance evaluation has been proposed in the paper. The paper aims to minimize the LSD over network level for equally spreading cell load out on a network with low radio resource control (RRC) signaling load. To support the MLB function enhancement for self-organizing network (SON), the novel MLB algorithm has been proposed in this paper. The performance of the algorithm has been analyzed and compared through computer simulations as well.

According to the results, we found that the proposed MLB algorithm can reduce the LSD from 7.48 % to 60.74 %. On the other hand, we observed that the proposed MLB algorithm required 10.79 % more handovers than the non-MLB operation.

Moreover, the overall number of RLF was produced as many as 140 from the proposed MLB algorithm operation. This information indicates that MLB and mobility robustness optimization (MRO) coordination is needed for reducing the number of RLF. Furthermore, looking at the impacts of RLF, we can conclude that MRO should have higher priority than MLB.

Keywords: LTE · Self-organizing networks · Mobility load balancing

1 Introduction

A small cell access point (AP) is a nomadic or mobile access point with small size supporting cheap wired and wireless convergence services by connecting a mobile phone with the Internet in indoor environments such as home and office. Although it is similar to a Wi-Fi access point in a functional aspect point of view, it is different that a primary role of a small cell access point is to relay a mobile phone call unlike a Wi-Fi AP.

© Springer International Publishing AG 2016
O. Galinina et al. (Eds.): NEW2AN/ruSMART 2016, LNCS 9870, pp. 205–216, 2016.
DOI: 10.1007/978-3-319-46301-8_17

Small cell deployments are expected to occur in different scenarios, such as venues, enterprise, dense urban residential and business areas. These scenarios include both uncoordinated deployments with customer and coordinated deployments with the operator, or small cell vendors.

The self-organizing network (SON) function includes both network self-configuration and self-optimization functions. In a SON architecture aspect, SON functionalities are divided into two types: a distributed SON (dSON) and a centralized SON (cSON).

The dSON algorithm is executed locally at the individual small cell to get a fast interaction with the direct neighborhood. UE specific information and rich input data sets are easier implemented locally by dSON. The dSON entities autonomously adjust local parameter settings within each Home eNB (HeNB), or interacting with neighbor HeNBs over X2 interface. On the other hand, the cSON algorithm requiring wide area visibility and parameter settings, without the need for fast response times is performed at a central server (a.k.a. SON server).

Therefore, a hybrid architecture for SON combines with interacting cSON and dSON components. A cSON entity collecting information like performance counters retrieved from the small cell could be seen as a part of the network manager (NM) and/or of the element manager (EM) and provides guidelines and parameter ranges to dSON functions.

The mobility load balancing (MLB) by cell reselection in radio resource control (RRC) idle mode has been studied in [1], and the MLB for RRC connected mode has been proposed in [2–4]. Although a few problems have been addressed in these papers, the MLB problems when both serving cell and neighbor cell were in overload status have not been addressed, yet. Moreover, the additional RRC signaling load regarding early HO should be considered in MLB algorithm in order to apply for a live LTE network. The objective of our MLB approach is to intelligently spread a load of cells out on a network with low RRC signaling load nevertheless in this severe overload status. The load standard deviation (LSD) as a new key performance indicator (KPI) for MLB performance evaluation has been proposed as well.

The rest of this paper is organized as follows. Section 2 describes the system model and assumptions considered in this paper. Details of the proposed MLB algorithm are presented in Sect. 3. We then provide the simulation environments and results in Sect. 4, and offer some concluding remarks in Sect. 5.

2   System Model

There are two load balancing (LB) strategies to perform load balancing by SON function over LTE system level. One is to modify downlink (DL) power such as reference signal power or antenna tilt. However, degradation of coverage in reduced power cells can be archived by this strategy, and over-provisioning of power amplifiers in increased power cells can be required. This can be one of the reasons that the cost of small cell AP is increased. The other is to modify handover (HO) parameter such as cell individual offset (CIO). The CIO is a target parameter for mobility load balancing (MLB). Although this strategy can overcome the cons of DL power modification, free resource of neighbor cell is needed to archive the pros of load balancing.

With regard to MLB, early HO is needed in case of overloaded cell, and delayed HO is needed in case of normal loaded cell for MLB optimization.

The handover procedure to provide mobility management is one of the main features in the 3rd generation partnership project long term evolution (3GPP LTE). The HO procedure in 3GPP LTE is started with measurement report (MR) message transmission from a UE to its serving cell in a first stage. The UE monitors the reference signal received power (RSRP) of reference symbols or reference signal received quality (RSRQ) on downlink channel based on cell-specific reference signals (C-RS) periodically. If the conditions for MR transmission is satisfied with the signal strength, the UE sends the corresponding MR indicating the triggered event. In this paper, the event A3 and MR messages in [6] are used for our MLB optimization. Equation (1) shows the entering condition of event A3 in case that a neighbouring cell becomes offset better than PCell.

$$M_n + O_{fn} + O_{cn} - Hys > M_p + O_{fp} + O_{cp} + Off, \tag{1}$$

where M_n is the RSRP(or RSRQ) of the neighbouring cell. O_{fn} is the frequency specific offset of the frequency of the neighbour cell. O_{cn} is the cell specific offset of the neighbour cell. M_p is the RSRP(or RSRQ) of the primary cell (PCell). O_{fp} is the frequency specific offset of the primary frequency. O_{cp} is the cell specific offset of the PCell. Hys is the hysteresis parameter for this event. Off is the offset parameter for this event (i.e. a3-Offset).

In this paper O_{fn} and O_{fp} are ignored, since inter-frequency scenario is excluded. Off is also removed, since target parameters for our MLB algorithm are O_{cn} and O_{cp} only. If there are multiple neighbors along with a UE, a UE has several M_ns, O_{fn}s, and O_{cn}s. They are considered as candidate target cells in this paper.

Equations (2) and (3) show the resource block utilization ratio (RBUR) over a HeNB level and a network level respectively.

$$\rho_{h_k}(t) = \frac{\sum_{j=1}^{m} N_{UE_j}(t)}{C \times B}, \tag{2}$$

where $\rho_{h_k}(t)$ indicates the RBUR on tth time slot over kth HeNB. $N_{UE_j}(t)$ indicates the number of the physical resource block allocated to jth UE connected to kth HeNB on tth time slot. m is the number of UE connected to kth HeNB. C is the total number of carrier per a HeNB. B is the total number of physical resource block (PRB) per carrier.

$$\rho(t) = \sum_{k=1}^{K} \rho_{h_k}(t) = \frac{\sum_{i=1}^{M} N_{UE_i}(t)}{C \times B \times K}, \tag{3}$$

where $\rho(t)$ indicates the RBUR on tth time slot over overall network level. $N_{UE_i}(t)$ indicates the number of the physical resource block allocated to ith UE on tth time slot over overall network level. M is the total number of UE over overall network level. K is the total number of HeNB over overall network level.

The LSD is defined as

$$\sigma = \sqrt{\frac{\sum_{t=1}^{n}(\rho(t) - \bar{\rho}(t))^2}{n}}, \tag{4}$$

where $\bar{\rho}(t) = \frac{\sum_{t=1}^{n}\rho(t)}{n}$ and n is the total number of time slot t.

Our proposed MLB algorithm aims to minimize σ over network level in order to spread their load equally out with low RRC signaling load.

3   Proposed MLB Algorithm

As stated before, the LSD in Eq. (4) is defined newly as a key performance indicator (KPI) in terms of MLB algorithm performance evaluation. An appropriate UE for early HO should be selected to satisfying this goal in the first stage, and then CIO values of each UE should be adjusted within safty range to avoid a radio link failure (RLF) for early HO.

Table 1 shows Q-OffsetRange (a.k.a. cell individual offset CIO), defined in [6]. The values are in dB. Value dB-24 corresponds to -24 dB, dB-22 corresponds to -22 dB and so on. The Q-OffsetRange was used to indicate a cell specific offset such as O_{cn} or O_{cp} to be applied when evaluating triggering conditions for event A3 measurement reporting.

Table 1. Q-OffsetRange

Q-OffsetRange ::=	ENUMERATED {dB-24, dB-22, dB-20, dB-18, dB-16, dB-14, dB-12, dB-10, dB-8, dB-6, dB-5, dB-4, dB-3, dB-2, dB-1, dB0, dB1, dB2, dB3, dB4, dB5, dB6, dB8, dB10, dB12, dB14, dB16, dB18, dB20, dB22, dB24}

Figure 1 explains the basic concept and procedure of the MLB algorithm. Given that five UEs are connected to the HeNB1, three UEs are located at cell edge area and two UEs are located at the center of cell area. When the RBUR ρ_{HeNB1} of HeNB1 is more than overload threshold $TH_{overload}$, the proposed MLB Algorithm 1 is performed over the HeNB1. If $RRCConnectionReconfiguration$ in order to adjust O_{cn} or O_{cp} of event A3 should be sent to all UEs, heavy RRC signaling load is generated. To avoid this drawback, the appropriate user equipment (UE) selection procedure located in cell edge by periodic MR was added into our MLB algorithm. The UEs at cell edge are with RSRP values less

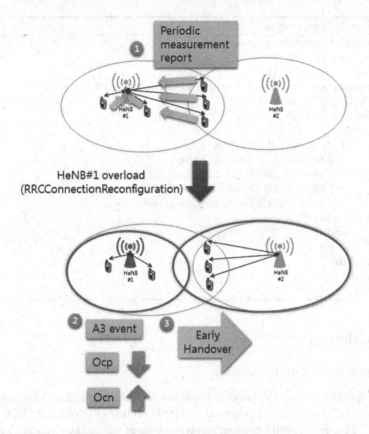

Fig. 1. The proposed MLB algorithm concept

than $TH_{serving}$ and more than TH_{target}. Moreover, if both the serving cell and the neighboring cell are in a similar overloaded status, the conventional MLB operation make unnecessary HO or ping-pong HO as well as signaling overload. Therefore, we took $TH_{loadgap}$ into account to resolve this defect. O_{cn} of event A3 over the target cell is increased and O_{cp} of event A3 over the target cell is decreased. Consequently early HO has been made forcibly.

Where $\mathbb{H}$ indicates the set of HeNBs. $\mathbb{H} = \{h_1, h_2, ..., h_K\}$, where K is the number of the HeNBs. $\mathbb{U}$ indicates the set of UEs located at cell edge with RSRP values less than $TH_{serving}$ and more than TH_{target}. $\mathbb{U} = \{u_1, u_2, ..., u_M\}$, where M is the number of the UEs. $\mathbb{CT}$ indicates the set of candidate target cells belong to $\mathbb{U}$. $\mathbb{CT} = \{ct_1, ct_2, ..., ct_L\}$, where L is the number of the candidate target cells belong to $\mathbb{U}$. ρ_s is the load of serving cell. ρ_{ct_l} is the load of lth candidate target cell. $TH_{overload}$ is the threshold value with respect to overload. $TH_{loadgap}$ is the threshold value with respect to load differentiation between source cell and target cell.

Algorithm 1. The proposed MLB algorithm

1: Collecting the resource usage information of $\mathbb{H}$;
2: **for** h_1 to h_K **do**
3:    **if** $\rho_{h_k} > TH_{overload}$ **then**
4:       Finding current active $\mathbb{U}$ located at cell edge and belong to h_k;
5:       Finding $\mathbb{CT}$ belong to $\mathbb{U}$;
6:       **for** u_1 to u_M **do**
7:          **for** ct_1 to ct_L **do**
8:             **if** $ct_l = $ max. RSRP **then**
9:                **if** $(\rho_s - \rho_{ct_l}) > TH_{loadgap}$ **then**
10:                   O_{cn} += ΔQ-$OffsetRange$ of A3 event over ct_l;
11:                   O_{cp} −= ΔQ-$OffsetRange$ of A3 event over ct_l;
12:                   send $RRCConnectionReconfiguration$ to u_m
13:                **end if**
14:             **end if**
15:          **end for**
16:       **end for**
17:    **end if**
18: **end for**

---

4   Simulation

4.1   Simulation Environments

The simulations assume the small cell parameter values modified from scenario #3 (dense) in [7] to consider practical smartphone test environment (LTE FDD Band 3). This paper simulates an urban environment where the users are walking at 1 m/sec with circular footprint. Table 2 describes the simulation environments using the detailed parameters and assumptions in this paper. In this work, we implemented the MLB algorithm with an LTE model of the NS-3 simulation environment in [8]. Moreover, the RLF in [5] was implemented, since there was no RLF implementation in current version of NS-3 simulation environment.

Figure 2 shows the cell layout topology for the simulation. The load status of HeNBs is expressed by green (normal status) and red (overload status) colors. The overload threshold was 0.7. This means that HeNB goes into the overload status if it is spending PRB over 70 % comparing to total own PRB. The evolved packet core (EPC) includes core network interfaces, protocols and entities such as the mobility management entity (MME), the serving gateway (S-GW), and the packet data network gateway (PDN-GW). the application server provide UEs with LTE application service. Video service regarding QCI 2 (GBR video call) is implemented in this simulation.

4.2   Simulation Results

Figure 3 presents RBUR $\rho(t)$ per HeNB without an MLB algorithm. As we can see in the figure, the fluctuation of RBUR per HeNB is kept on very low status,

Table 2. Simulation Parameters and Assumptions

Parameters	Settings
System bandwidth per carrier	20MHz, Number of PRB (Physical Resource Block) = 100

Carrier frequency	1.8 GHz,
	    [Band 3 frequency (fc)]
	    DL = 1842.5 MHz
	    UL = 1747.5 MHz
	    [Band 3 EARFCN]
	    DL = 1575
	    UL = 19575

Carrier number	1
Total BS TX power	24 dBm
Number of cells	10
Inter-site distance (ISD)	30 m
Antenna pattern	Omni-directional
Number of UEs	80

Path loss (PL)
    (non-line-of-sight (NLOS))
	$147.4 + 43.3 \log_{10}(R)$,
	    distance R in km

Shadowing	standard deviation $(\sigma) = 4$ dB
UE mobility model	circular way (radius = 10m, speed = 1 m/sec)
deployment scenario	urban
MAC scheduler	channel and qos aware (CQA)

UE dropping	* 50 % mobile UE: Randomly and uniformly
	    distributed over whole area.
	* 50 % fixed UE: Randomly and uniformly
	    distributed over limited handover available area
	    (cell edge) only.

$TH_{overload}$	0.7
$TH_{loadgap}$	0.1
Hysteresis	2dB
time-to-trigger (TTT)	256 msec
Simulation time	605 sec

since there is no MLB operation as well as identical UE circular mobility pattern when time goes by.

However, we can see the big fluctuation of each RBUR in Fig. 4 due to the effect of MLB operation. The overload compensation at network scope is performed by the forced early HO depending on MLB algorithm.

Figure 5 describes average RBURs $\rho(t)$ per HeNB with MLB and without MLB. The overall average values of RBUR with MLB operation and without

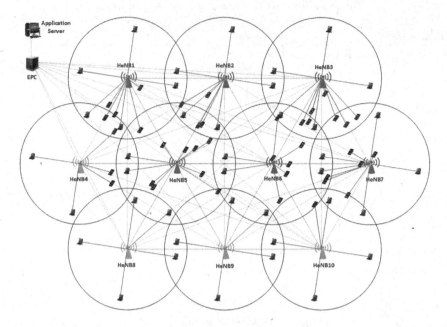

Fig. 2. Simulation cell layout topology

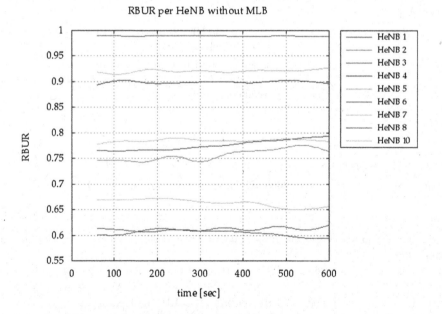

Fig. 3. RBUR $\rho(t)$ per HeNB without MLB

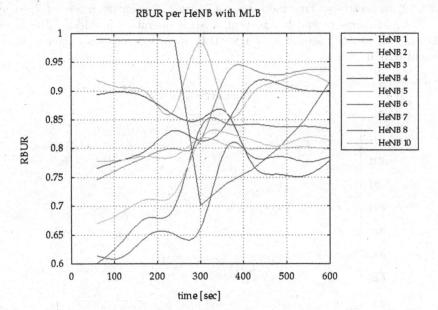

Fig. 4. RBUR $\rho(t)$ per HeNB with MLB

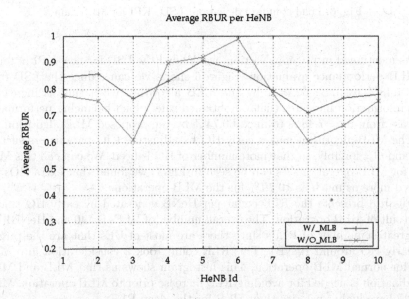

Fig. 5. Average RBUR $\rho(\bar{t})$ per HeNB

MLB operation have 8.1 and 7.7. While, their standard deviation values are 0.05 and 0.12, respectively. This indicates we can get smaller deviation of average RBUR by MLB operation. It means that MLB operation provides safety load balancing on network scope, although MLB operation requires bigger average value of RBUR.

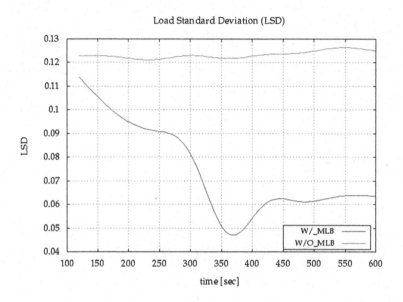

Fig. 6. Load standard deviation (LSD) KPI for MLB gain

As mentioned in previous section, we defined the LSD for new KPI in terms of MLB performance evaluation. Figure 6 shows we can reduce the LSD from min. 7.48 % to max. 60.74 % by the MLB algorithm when time is increased. This information indicates that we obtained safety load balancing performance increase from min. 7.48 % to max. 60.74 % by the proposed MLB algorithm.

The MLB algorithm makes early HO by adjusting CIO parameters such as O_{cn} and O_{cp} forcibly, so that more number of HO is need. We can call this MLB cost for MLB operation. As can be seen in Fig. 7, we found that extra HOs are needed more as much as 10.79 % for the MLB operation.

Figure 8 presents the RLF count per HeNB generated by early HO that is the result of MLB operation. The overall number of RLFs is 140 and HeNB1 has the greatest number of RLFs since there are most of UEs that are the targets for early HO around HeNB1. This RLF count took a consideration into MLB cost for normal MLB operation. This histogram shows us that MLB and MRO coordination is needed for avoiding RLF increase prior to MLB operation. MRO should have higher priority than MLB for the clean RLF.

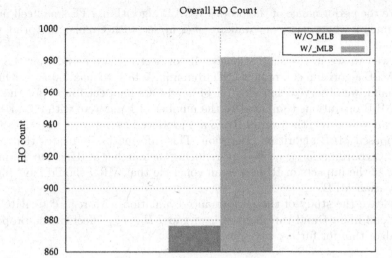

Fig. 7. Overall handover count for MLB cost

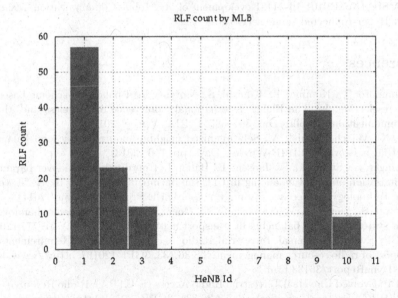

Fig. 8. RLF count for MLB cost

5   Conclusions

In this paper, we discussed important properties such as MLB gains and costs that take place in the proposed MLB algorithm operation over LTE system. The objective of this paper was to minimize LSD over network level in order to spread their load equally out with low RRC signaling load. To analyze and evaluate the performance of the proposed MLB algorithm, LTE small cell urban environments where users are walking at 1 m/sec with circular footprint were simulated.

As we can see in the simulation results of Sect. 4.2, we found that the proposed MLB algorithm can reduce LSD from min. 7.48 % to max. 60.74 %. On the other hand, we observed that 10.79 % of handovers happened more for the proposed MLB operations compared to the number of handovers without an MLB algorithm. Moreover, the overall RLFs are produced as many as 140 times from the proposed MLB algorithm operation. This information indicates that MLB and MRO coordination is needed for reducing the number of RLF. Furthermore, looking at the impacts of RLF, we can conclude that MRO should have higher priority than MLB.

We leave the study of the performance evaluation with regard to RRC signaling load impact and threshold of load gap $TH_{loadgap}$ effect in the proposed MLB algorithm for further research.

Acknowledgments. This work was supported by Institute for Information & communications Technology Promotion(IITP) grant funded by the Korea government(MSIP)(No. R0101-16-244, Development of 5G Mobile Communication Technologies for Hyper-connected smart services.

References

1. Yamamoto, T., Komine, T., Konishi, S.: Mobility load balancing scheme based on cell reselection. In: The Eighth International Conference on Wireless and Mobile Communications Mobility, pp. 381–387. IARIA, Venice (2012)
2. Zia, N., Mitschele-Thiel, A.: Self-organized neighborhood mobility load balancing for LTE networks. In: IFIP Wireless Days, pp. 1–6 (2013)
3. Lobinger, A., Stefanski, S., Jansen, T., Balan, I.: Coordinating handover parameter optimization and load balancing in LTE self-optimizing networks. In: IEEE Vehicular Technology Conference Spring, pp. 1–5. IEEE Press, Yokohama (2011)
4. Marwat, S., Meyer, S., Weerawardane, T., Goerg, C.: Congestion-aware handover in LTE systems for load balancing in transport network. ETRI J. **36**, 761–771 (2014)
5. 3GPP: Evolved Universal Terrestrial Radio Access (E-UTRA); Requirements for support of radio resource management. TS 36.133, 3GPP (2010). http://www.3gpp. org/DynaReport/36133.htm
6. 3GPP: Evolved Universal Terrestrial Radio Access (E-UTRA); Radio Resource Control (RRC); Protocol specification. TS 36.331, 3GPP (2010). http://www.3gpp.org/ DynaReport/36331.htm
7. 3GPP: Small cell enhancements for E-UTRA and E-UTRAN - Physical layer aspects. TS 36.872, 3GPP. (2013). http://www.3gpp.org/DynaReport/36872.htm
8. NS3: discrete-event network simulator 3. version 3.23, NS3 Consortium (2015). https://www.nsnam.org/

Optimizing Network-Assisted WLAN Systems with Aggressive Channel Utilization

Aleksandr Ometov[1(✉)], Sergey Andreev[1], Alla Levina[2], and Sergey Bezzateev[3]

[1] Tampere University of Technology, Korkeakoulunkatu 1, 33720 Tampere, Finland
{aleksandr.ometov,sergey.andreev}@tut.fi
[2] Saint Petersburg National Research University of Information Technologies, Mechanics and Optics (ITMO University),
Lomonosova St., 9, 191002 St. Petersburg, Russia
levina@cit.ifmo.ru
[3] Saint Petersburg University of Aerospace Instrumentation,
Bolshaya Morskaya St., 67, 190000 St. Petersburg, Russia
bsv@aanet.ru

Abstract. Cellular network assistance over unlicensed spectrum technologies is a promising approach to improve the average system throughput and achieve better trade-off between latency and energy-efficiency in Wireless Local Area Networks (WLANs). However, the extent of ultimate user gains under network-assisted WLAN operation has not been explored sufficiently. In this paper, an analytical model for user-centric performance evaluation in such a system is presented. The model captures the throughput, energy efficiency, and access delay assuming aggressive WLAN channel utilization. In the second part of the paper, our formulations are validated with system-level simulations. Finally, the cases of possible unfair spectrum use are also discussed.

1 Introduction and Motivation

Today, modern Wireless Local Area Networks (WLANs) are widely utilized all over the world. This is due to their low deployment and service costs, relatively simple channel access protocols, and ubiquitous availability of radio interfaces on most of the contemporary user devices. Being a major technology trend, IEEE 802.11 (WiFi) took its niche as one of the most popular wireless communication solutions [1]. It utilizes unlicensed bands (2.4, 5 GHz), thus allowing for unrestricted high-speed connectivity between users and network infrastructure. Based on its previous editions, the current protocol version [2] introduces advanced Medium Access Control (MAC) mechanisms built over a wide range of Physical (PHY)-layer features that altogether support the steadily growing numbers of networked devices[1].

---

[1] See Ericsson mobility report: On the pulse of the Networked Society, http://www.ericsson.com/mobility-report.

© Springer International Publishing AG 2016
O. Galinina et al. (Eds.): NEW2AN/ruSMART 2016, LNCS 9870, pp. 217–229, 2016.
DOI: 10.1007/978-3-319-46301-8_18

On the other hand, cellular network operators are increasingly willing to offload their excess traffic demand onto unlicensed bands[2] by e.g., leveraging direct connectivity between user devices in license-exempt spectrum. With such cellular network assistance, there exists an opportunity to manage and improve performance of the conventional WLAN deployments [3,4]. Similar approaches have been investigated from various perspectives and shown to offer throughput and energy efficiency benefits [5]. However, we believe that the ultimate capabilities of network-assisted WLAN operation have not been studied conclusively. For instance, the potential of novel MAC algorithms and "adaptive scheduling" [6] remains largely unexplored in this context.

Fig. 1. Topology of the considered scenario

More specifically, the tentative performance gains with network-assisted traffic offloading have been considered in the past by relying on WiFi-Direct connectivity [7], as well as employing anchor access points as part of the cellular infrastructure [8]. We expect that the practical advantages of integrating the WLAN connectivity with system-wide cellular network management would grow further over the following years [9,10]. This should result in generally improved levels of performance that current IEEE 802.11 system deployments would gain with added cellular network assistance.

The above considerations call for revisiting the existing WiFi-specific performance evaluation models to determine the optimal network-assisted operation of real-life WLANs. In this work, a model for aggressive channel utilization based on regenerative analysis is discussed. It allows to quantify system performance with high scalability by operating with only a small number of parameters and thus may be preferred over traditional Markov chain based approaches [11,12].

---

[2] See Cisco Visual Networking Index: Global Mobile Data Traffic Forecast, http:// www.cisco.com/c/en/us/solutions/collateral/service-provider/visual-networking-index-vni/mobile-white-paper-c11-520862.html.

To illustrate its capabilities, we model the network operation for static and dynamic user arrivals. Finally, we overview our developed technology prototype that continues the authors' previous work in [13]. The described demonstrator features a conventional Linux laptop with a custom-compiled kernel WiFi module and a traffic generation software in order to mimic the case of aggressive channel utilization, as it is demonstrated in Fig. 1.

The rest of this text is organized as following. In the following section, the WiFi-specific Binary Exponential Backoff protocol operation and the corresponding analytical model are briefly reviewed. The numerical results and the model validation are elaborated in Sect. 3. The final section summarizes the work-in-progress and concludes this paper.

2   Case Description and Analysis

In this work, we consider the conventional IEEE 802.11n MAC operation, which utilizes the so-called Distributed Coordination Function (DCF) based on Binary Exponential Backoff (BEB) protocol for collision resolution and Carrier Sense Multiple Access with Collision Avoidance (CSMA/CA) mechanism [14] operating in the Request-To-Send/Clear-To-Send (RTS/CTS) mode.

For the sake of simplicity, we consider a single Access Point (AP) in our network together with M stationary devices connected to it. We also assume that all of the WiFi devices are operating in the "aggressive" saturation mode, that is, their traffic queues are always full whenever they initiate a transmission. Another important assumption is that the co-located cellular network always has an up-to-date knowledge on the number of devices connected to the AP or residing in its range, since we only consider multi-radio devices with two radio interfaces, WLAN and cellular.

The RTS/CTS based channel reservation mechanism – employed whenever a particular device seizes a channel for transmission – operates as a four-way handshake. The system in question is primarily controlled by three parameters: the initial backoff window W_0, the backoff stage m, and the retransmission counter K. If the channel is not reserved during the Arbitration Inter-Frame Spacing ($AIFS$) interval, the tagged device is applying a backoff procedure before commencing its data transmission attempt.

Hence, the Backoff Counter (BC) value is selected uniformly from the interval 0 to $W_0 - 1$, and the current contention window value is denoted as W_i. After each idle slot, the BC value is decremented. Whenever it reaches zero, a transmission attempt is initiated. In case there are two or more simultaneously transmitting users, a *collision* is detected at the AP side. As long as the packet is not discarded (K still allows to retransmit), the CW is doubled ($W_i = 2W_i - 1$) to reduce the collision probability and the BC is generated again.

It is important to note that equipment vendors set the maximum limit for the CW as $CW_{max} = W_0^m$. However, if there have been more than K unsuccessful transmissions of one packet, it is discarded and the corresponding data is lost. A diversity of WiFi chipsets available on the market calls for harmonization of the utilized BEB parameters to achieve efficient and fair medium access. To this end, our considered analytical model may be applied to select the corresponding set of MAC parameters in a close-to-optimal manner.

Over the recent years, multiple research papers use Markov chain based techniques to analyze the BEB scheme [15,16]. However, it may be difficult to scale such models to study all the system parameters of interest due to the fast state space growth. Fortunately, there are alternative analytical approaches to evaluate the performance indicators in case of saturation, which are based on regenerative analysis. We believe the latter may be scaled better and capture the system behavior more efficiently [17,18].

In what follows, we discuss an analytical model to optimize the BEB parameters for network-assisted WLAN operation with aggressive channel utilization. First, we characterize the number of successful transmissions and the corresponding transmission attempts. Accordingly, the well-known collision probability expression can be written as follows

$$p_c = 1 - (1 - p_t)^{M-1}, \tag{1}$$

where $M-1$ is the number of contending devices that may collide with the tagged user during a time slot and p_t is the transmission probability for a device, i.e., the probability for a user to start its transmission in a randomly chosen time slot. This value may be obtained as

$$p_t = \lim_{n \to \infty} \frac{\sum_{i=1}^{n} B^{(i)}}{\sum_{i=1}^{n} D^{(i)}} = \frac{E[B]}{E[D]}, \tag{2}$$

where $B^{(i)}$ is the number of packet transmission attempts for a given regeneration cycle i^{th} related to the duration of the ith cycle in slots $D^{(i)}$.

Therefore, the probability of a successful transmission can be derived as

$$p_s = \frac{M p_t (1 - p_t)^{M-1}}{1 - (1 - p_t)^M}. \tag{3}$$

Further, we may develop the discussed analytical approach for a case when some of the packets may be discarded based on the number of retransmission attempts K and the backoff stage value m. In order to produce the respective transmission probabilities, we need to evaluate the average number of transmission attempts $E[B]$ during the ith cycle as

$$E[B] = \sum_{i=1}^{K+1} i \Pr\{B = i\} = (1 - p_c) \sum_{i=1}^{K+1} i p_c^{i-1} + (K + 1) p_c^{K+1} = \frac{1 - p_c^{K+1}}{1 - p_c}. \tag{4}$$

Additionally, we have to take into account the number of transmission attempts $E[D^{1,2}]$ as

$$
\begin{cases}
E[D^1] = & (1-p_c)\left[\sum_{i=1}^{K+1}\left(2^{i-1}W_0 - \frac{W_0-i}{2}\right)p_c^{i-1}\right] + \\
& +p_c^{K+1}\left(2^K W_0 - \frac{W_0-(K+1)}{2}\right), \\
E[D^2] = & (1-p_c)\left[\sum_{i=1}^{m+1}\left(2^{i-1}W_0 - \frac{W_0-i}{2}\right)p_c^{i-1} + \right. \\
& \left. + \sum_{i=m+2}^{K+1}\left(2^{m-1}W_0(i-m+1) - \frac{W_0-i}{2}\right)p_c^{i-1}\right] + \\
& +p_c^{K+1}\left(2^{m-1}W_0(K-m+2) - \frac{W_0-(K+1)}{2}\right).
\end{cases}
\tag{5}
$$

Finally, we determine the sought transmission probabilities by calculating (4) for both (5) and (2). After straightforward technical transformations, we arrive at

$$
\begin{cases}
p_t^1 = \frac{2(1-2p_c)(1-p_c^{K+1})}{W_0(1-p_c)(1-(2p_c)^{K+1})+(1-2p_c)(1-p_c^{K+1})}, \\
p_t^2 = \frac{2(1-2p_c)(1-p_c^{K+1})}{(1-2p_c)(W_0(1-2^m p_c^{K+1})+(1-p_c^{K+1}))+p_c W_0(1-(2p_c)^m)},
\end{cases}
\tag{6}
$$

where p_t^1 is for the case when $K \leq m$ and p_t^2 stands for $K > m$. As the last step, by taking into account the results from (6), we produce the probability of a successful transmission with (3).

3   Numerical Results

In this section, we evaluate the considered analytical approach by utilizing a simple MAC-layer WiFi simulator. Correspondingly, the main operating parameters are summarized in Table 1. Our subsequent results are divided into two groups: the overview of possible BEB optimization opportunities for network-assisted WLAN deployments; and the case when the number of saturated multi-radio devices varies uniformly.

Table 1. Core system parameters

Parameter	Value
Packet size	1500 bytes
PHY data rate	65.0 Mbps
Number of users	5 to 100
Initial backoff window W_0	2 to 1024
Backoff stage R	2 to 14
Short retry limit K	7, ∞
Maximum simulation duration	30 min or 10^6 slots

Based on a live trial reported in [19] and executed in Mountain View, California, the average number of devices connected to one AP is fluctuating below 5 during the day. In this work, as the worst case, we assume 5 times more devices that have an active saturated connection to a single AP. Along these lines, we study a set of cases in order to obtain insights into the best initial Contention Window and Retransmission Counter values based on the successful transmission probability. Accordingly, we estimate the probability that a packet has been successfully delivered if its transmission is attempted by the tagged user.

The results for the actual average number of devices is shown in Fig. 2(a). Here, the horizontal line represents a suboptimal algorithm based on the approach from a well-known work in [20]. In this case, the system has knowledge of the total number of users (via the cellular network assistance) and each of the user devices is only allowed to utilize the corresponding channel access probability (which can be easily recalculated into the initial CW and RC values).

Further, our simulation tool was calibrated with the discussed analytical framework and the example for $W_0 = 16$, $R = 6$, and $K = 7$ is shown in Table 2. However, Jain's Fairness Index ($J = (\sum_{i=1}^{n} x_i)^2 / n \sum_{i=1}^{n} x_j^2$) [21] demonstrates that the numbers of successful transmissions across the users are not equal. This crucial BEB operation feature is related to the Channel Capture Effect issues [22].

Table 2. Calibration between simulation and analysis

Number of users	p_s, analysis	p_s, simulation	Jain's index
10	0.32792	0.32563	0.90665
20	0.35368	0.35822	0.94976
30	0.36366	0.36833	0.96009
50	0.37025	0.37706	0.97315

As shown in Fig. 2(a), there is only one optimal point (see the peak in the plot) with the backoff parameters of $W_0 = 5$ and $R = 2$. In order to optimize the system operation, the values need to be updated accordingly for each of the devices through the AP and in coordination with the cellular network assistance function. The step-wise behavior in the right side of the plot is due to a decreased saturation in the channel, i.e., the initial contention builds up as the channel access time increases.

The corresponding results for 10 users are illustrated in Fig. 2(b). Clearly, there is more than one peak, that is, the best successful transmission probability may be achieved in a number of ways by selecting the alternative pairs of BEB parameters. Here, to consider only one of those, the middle part of this cluster of points may become an adequate option ($W_0 = 4, R = 4$). This is due to a lower influence of the capture effect, and the initial CW value does not affect the operation in terms of the initial transmission delay either.

Finally, we increase the number of contending devices to a larger value of 100 in Fig. 2(c). We see that the number of peaks also increases and make a similar decision as in the case for 10 users to choose the suboptimal operating point. Therefore, for 100 devices, the BEB parameters may be chosen as $W_0 = 4$ and $R = 4$.

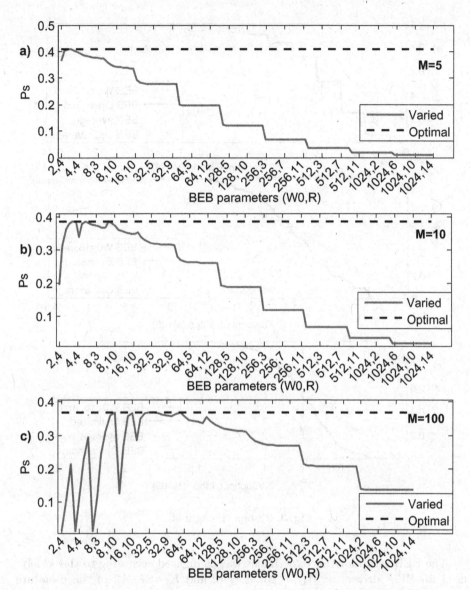

Fig. 2. Successful transmission probability for different numbers of users

We continue by modeling dynamic user arrivals and thus utilize the following setup. The devices attempt to access the channel over an interval of time equal to 5,000 slots. The randomly chosen 10 % of the maximum number of users are inactive, while others keep transmitting or activate. If a user was applying the backoff procedure in the previous operating interval, the BEB parameters remain the same in the current one. In case a device has just activated, the BEB parameters are set to the initial values.

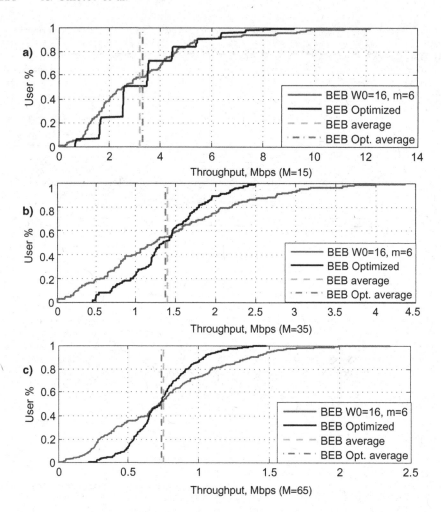

Fig. 3. System throughput

The BEB setup for this experiment was configured according to the widely-used *MadWifi* driver[3] as $W_0 = 16$, $R = 6$, and $K = 7$. All of the users are operating in the "lossy" mode [23], that is, packet drops are possible according to the Short Retry Limit value. The results for the system throughput (Fig. 3) fully support our previous discussion by indicating the same average throughput value for all the algorithms, while fairness may be optimized for higher numbers of devices.

The system operation from the delay perspective is illustrated in Fig. 4. The optimized solution shows better results even for the small number of devices (Fig. 4(a)). This is generally due to shorter channel access times. The standard

---

[3] See ath9k, https://wireless.kernel.org/en/users/Drivers/ath9k/.

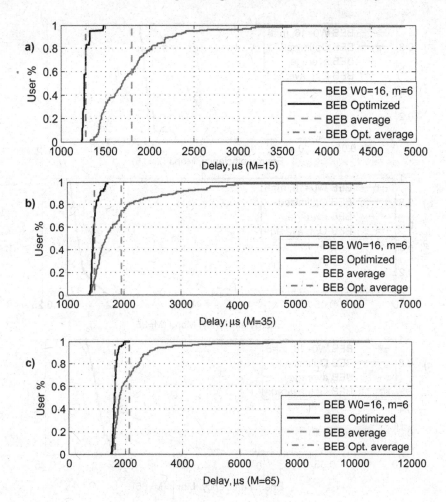

Fig. 4. Packet transmission delay

backoff procedure with the RTS/CTS mechanism requires each device to wait for at least the first *CW* interval that may be significantly longer than the one with the optimized BEB parameters. Additionally, the default parameters have more impact on collision resolution time in case of higher numbers of users (Fig. 4(b) and (c)).

The delay performance is directly connected to that of energy efficiency [24], which is shown in Fig. 5. We quantify the energy efficiency based on 100 mW idling power, 200 mW RX, and constant 100 mW + the transmit power for TX [7]. The optimized solution enjoys faster collision resolution and thus the users are attempting to transmit more frequently when their number is low (Fig. 5(a)). On the contrary, as the number of devices grows, the optimized parameters make users backoff for longer time intervals (Fig. 5(b) and (c)).

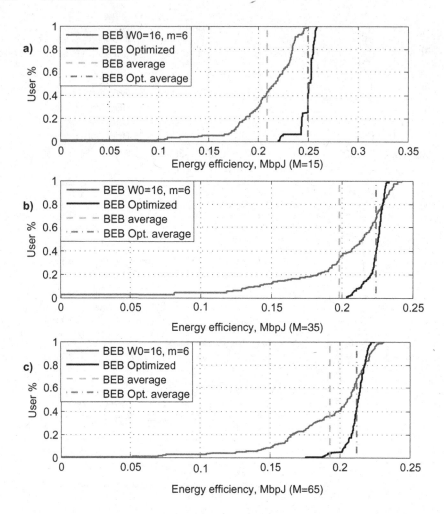

Fig. 5. Energy efficiency

Due to a lower impact of idling and RX power, the reduced collision rate leads to smaller energy consumption.

Summarizing the above, we verified our custom-made modeling tool with the developed analytical model in the saturation regime. Further, we studied the dynamics of user arrivals in the network-assisted WLAN system and can conclude that cellular assistance may bring significant benefits for all the considered performance indicators.

4   Current Work and Conclusions

In this section, we briefly discuss the system prototype under development and the available options for software-based traffic load generators. In a nutshell, our prototype operates by utilizing a conventional Linux laptop equipped with an open-source WiFi driver. It is running an *iperf server*[4] in order to mimic the intended aggressive channel utilization. The main feature of this testbed is in its ability to generate *artificial load*, i.e., it is possible to emulate the needed number of users by only modifying the BEB operation of one user. We note that more expensive APs on the market can also update the MAC parameters on the devices dynamically.

Importantly, as Linux systems allow their users to recompile the core modules, a malicious person may modify the BEB parameters in an offensive way, e.g., by setting CW, BC, and K to near-zero values. Should such a person be located in a public hot-spot (see Fig. 1), the attacker's device would transmit almost immediately without the initial channel sensing and/or waiting intervals used by others. Said attacker would then achieve a better channel utilization, while the remaining "fair" users would continue resolving their increased collisions as it was shown in the previous section. Preventive measures may thus be necessary to detect and protect from this and similar types of attacks.

Summarizing, in this work we studied the relations between the number of devices, the backoff parameters, the access delay, and the energy efficiency for cellular-assisted IEEE 802.11-based networks. Our results were obtained for both static and dynamic user arrival models. The considered optimization procedure was shown to provide improved system-wide fairness as well as generally better performance. Finally, we presented a capable testbed that may be employed to study more subtle real-world effects of network-assisted WLAN operation.

References

1. Doppler, K., Rinne, M., Wijting, C., Ribeiro, C., Hugl, K.: Device-to-device communication as an underlay to LTE-advanced networks. IEEE Commun. Mag. **47**(12), 42–49 (2009)
2. IEEE: Wireless LAN medium access control (MAC) and physical layer (PHY) specifications, IEEE standard 802.11 (2014)
3. Andreev, S., Moltchanov, D., Galinina, O., Pyattaev, A., Ometov, A., Koucheryavy, Y.: Network-assisted device-to-device connectivity: contemporary vision and open challenges. In: Proceedings of 21th European Wireless Conference, VDE, pp. 1–8 (2015)
4. Andreev, S., Gonchukov, P., Himayat, N., Koucheryavy, Y., Turlikov, A.: Energy efficient communications for future broadband cellular networks. Comput. Commun. **35**(14), 1662–1671 (2012)
5. Masek, P., Zeman, K., Hosek, J., Tinka, Z., Makhlouf, N., Muthanna, A., Herencsar, N., Novotny, V.: User performance gains by data offloading of LTE mobile traffic onto unlicensed IEEE 802.11 links. In: Proceedings of 38th International Conference on Telecommunications and Signal Processing (TSP), pp. 117–121. IEEE (2015)

---

[4] See iperf3: A TCP, UDP, and SCTP network bandwidth measurement tool, https:// github.com/esnet/iperf/.

6. Bilgir Yetim, O., Martonosi, M.: Adaptive usage of cellular and WiFi bandwidth: an optimal scheduling formulation. In: Proceedins of the seventh ACM international workshop on Challenged networks, pp. 69–72. ACM (2012)

7. Pyattaev, A., Johnsson, K., Andreev, S., Koucheryavy, Y.: 3GPP. LTE traffic offloading onto WiFi direct. In: Proceedings of Wireless Communications and Networking Conference Workshops (WCNCW), pp. 135–140. IEEE, April 2013

8. Fodor, G., Dahlman, E., Mildh, G., Parkvall, S., Reider, N., Miklós, G., Turányi, Z.: Design aspects of network assisted device-to-device communications. IEEE Commun. Mag. 50(3), 170–177 (2012)

9. Araniti, G., Calabro, F., Iera, A., Molinaro, A., Pulitano, S.: Differentiated services QoS issues in next generation radio access network: a new management policy for expedited forwarding per-hop behaviour. In: Proceedings of 60th Vehicular Technology Conference (VTC2004-Fall), vol. 4, pp. 2693–2697. IEEE (2004)

10. Petrov, V., Moltchanov, D., Koucheryavy, Y.: Applicability assessment of terahertz information showers for next-generation wireless networks. In: IEEE International Conference on Communications (ICC), pp. 1–7, May 2016

11. Kwak, B.J., Song, N., Miller, L.E.: Performance analysis of exponential backoff. IEEE Trans. Netw. 13, 343–355 (2005)

12. Tinnirello, I., Bianchi, G., Xiao, Y.: Refinements on IEEE 802.11 distributed coordination function modeling approaches. IEEE Trans. Veh. Technol. 59, 1055–1067 (2010)

13. Ometov, A., Andreev, S., Turlikov, A., Koucheryavy, Y.: Characterizing the effect of packet losses in current WLAN deployments. In: Proceedings of 13th International Conference on ITS Telecommunications (ITST), pp. 331–336. IEEE (2013)

14. Malone, D., Dangerfield, I., Leith, D.: Verification of common 802.11 MAC model assumptions. In: Uhlig, S., Papagiannaki, K., Bonaventure, O. (eds.) PAM 2007. LNCS, vol. 4427, pp. 63–72. Springer, Heidelberg (2007)

15. Bianchi, G.: Performance analysis of the IEEE 802.11 distributed coordination function. IEEE J. Sel. Areas Commun. 18(3), 535–547 (2000)

16. Bianchi, G., Tinnirello, I.: Remarks on IEEE 802.11 DCF performance analysis. IEEE Commun. Lett. 9(8), 765–767 (2005)

17. Andreev, S., Koucheryavy, Y., de Sousa, L.F.D.: Calculation of transmission probability in heterogeneous ad hoc networks. In: Proceedings of Internet Communications (BCFIC Riga), 2011 Baltic Congress on Future, pp. 75–82. IEEE (2011)

18. Wu, H., Peng, Y., Long, K., Cheng, S., Ma, J.: Performance of reliable transport protocol over IEEE 802.11 wireless LAN: analysis and enhancement. In: Proceedings of INFOCOM 2002 Twenty-First Annual Joint Conference of the IEEE Computer and Communications Societies, vol. 2, pp. 599–607. IEEE (2002)

19. Afanasyev, M., Chen, T., Voelker, G.M., Snoeren, A.C.: Usage patterns in an urban WiFi network. IEEE/ACM Trans. Netw. 18(5), 1359–1372 (2010)

20. Abramson, N.: The ALOHA system - another alternative for computer communications. In: Proceedings of AFIPS, Fall Joint Computer Conference, vol. 37, pp. 281–285 (1970)

21. Jain, R., Chiu, D., Hawe, W.: A quantitative measure of fairness and discrimination for resource allocation in shared computer systems. DEC Research Report TR-301 (1985)

22. Ometov, A.: Fairness characterization in contemporary IEEE 802.11 deployments with saturated traffic load. In: Proceedings of The 15th Conference of FRUCT assosiation (2014)

23. Moltchanov, D., Koucheryavy, Y., Harju, J.: Loss performance model for wireless channels with autocorrelated arrivals and losses. Comput. Commun. **29**(13), 2646–2660 (2006)
24. Andreev, S., Koucheryavy, Y., Himayat, N., Gonchukov, P., Turlikov, A.: Active-mode power optimization in OFDMA-based wireless networks. In: Proceedings of IEEE Globecom Workshops, pp. 799–803. IEEE (2010)

Randomized Priorities in Queuing System with Randomized Push-Out Mechanism

Alexander Ilyashenko[(✉)], Oleg Zayats, and Vladimir Muliukha

Russian State Scientific Center for Robotics and Technical Cybernetics,
Peter the Great St. Petersburg Polytechnic University, St. Petersburg, Russia
ilyashenko.alex@gmail.com, zay.oleg@gmail.com,
vladimir@mail.neva.ru

Abstract. For queuing systems with two incoming flows and single channel with randomized priority and push-out mechanism were obtained analytical expressions of generating function and loss probabilities. As general parameters for model control were used push-out probability $0 \leq \alpha \leq 1$ and probability of selecting service packets from first flow rather than second $0 \leq \beta \leq 1$. Theoretically and experimentally found that in certain combinations of incoming flows load factors loss probabilities dependence may be linear. This behavior is called "law of linear losses". Areas where this effect is shown are called "areas of linear behavior". In article shown such areas for considered randomized priorities system.

Keywords: Priority queuing · Randomized push-out mechanism · Randomized priority

1 Introduction

Simplest queuing systems considered enough in details. These models are getting complicated by increasing the amount of incoming flows or service channels and limiting system buffer size. However, there are several ways more to change behavior of queuing models. One of them is introduction of push-out mechanism. In classical literature, considered cases of his absence and deterministic pushing out. In [1–3] considered a generalization of these two cases as a randomized push-out mechanism. This method of model characteristics control was shown to be effective and allowed to change the loss probabilities in the tens of percent.

It is also possible to add a priority on service in the system, i.e. clearly indicate which flow prevails over others. Such cases are considered for the preemptive and relative priorities in [1–3]. In works related to priority queuing models also considered two more kinds of priorities: alternating and randomized [4]. Alternating priority was studied by authors of this article in [6]. Just switching priority between two type of incoming packets had very strong influence on model behavior.

For classification of priority queuing models usually used Kendall's notation [5] with extension introduced by Basharin [4]. For a description of the priority system has been used a symbol f_i^j. Originally, i takes the values: $i = 0$ (no priority), $i = 1$ o (non-preemptive priority), $i = 2$ (preemptive priority). The superscript j chosen out of

© Springer International Publishing AG 2016
O. Galinina et al. (Eds.): NEW2AN/ruSMART 2016, LNCS 9870, pp. 230–237, 2016.
DOI: 10.1007/978-3-319-46301-8_19

the two values: $j = 0$ (no push-out), $j = 2$ (deterministic push-out). Following Vilchevsky recommendations, lets use $j = 1$ for randomized push-out. Value $i = 3$ was used for alternating priorities [6]. Furthermore, in this paper we propose to reserve $i = 4$ for randomized priority. Model which will be studied in this article can be classified using described notation as where $i = 4$.

In this paper, proposed to investigate the combination of randomized priority and randomized push-out mechanism when before every stage of requirements selection out of storage takes place a random selection of requirements with specified probability β. And proposed model can be described using extended notation as $\bar{M}_2/M/1/k/f_4^1$.

2   Building Kolmogorov's System of Linear Equations

Studying queuing systems starts with consideration of phase space. For model with two incoming flows randomized push-out mechanism and randomized priority, where before packet from each flow is taken for service with specified probability phase space can be specified as

$$\Omega = \{O\} \cup \{(n_1, n_2) : n_1 = \overline{0, k - 1}, n_2 = \overline{0, k - i - n_1}\}. \tag{1}$$

State $\{O\}$ means totally empty system, when buffer and service channel are empty. Other states (n_1, n_2) mean that model has busy service channel and n_1 packets from first flow in buffer and n_2 from second. Sum of them can't be greater than buffer size $(k - 1)$.

According to phase space, probabilities of different model states expressed as

$$P_O(t), \ P_{n_1 n_2}(t) = P\{N_1(t) = n, N_2(t) = n_2\}.$$

All these probabilities satisfy normalization condition

$$P_O(t) + \sum_{n_1=0}^{k-1} \sum_{n_2=0}^{k-1-n_1} P_{n_1, n_2}(t) = 1.$$

According to Markov's theorem [5] process in this model is ergodic. So can be introduced final probabilities for model states

$$P_O = \lim_{t \to \infty} P_O(t), P_{n_1, n_2} = \lim_{t \to \infty} P_{n_1, n_2}(t), (n_1 = \overline{0, k - 1}, n_2 = \overline{0, k - 1 - n_1}.)$$

In this model is used randomized priority. As a main parameter used $\beta = \mathfrak{x}_1$ as probability of choosing packet from first flow on service rather than from second. $\mathfrak{x}_2 = 1 - \beta$ is probability of packet from second flow being chosen.

State graph of model with phase space (1) presented on Fig. 1. Using this graph was built system of balance equations

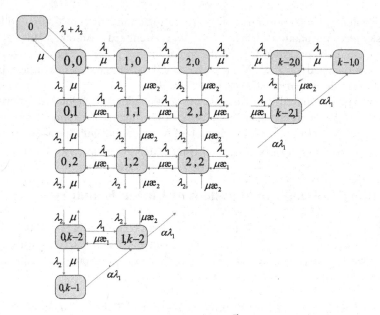

Fig. 1. State graph for system $\vec{M}_2/M/1/k/f_4^1$

Balance equation for state $\{O\}$ will have only two unknown probabilities

$$-(\lambda_1 + \lambda_2)P_O + \mu P_{0,0} = 0.$$

Also probability P_O will be in equation for state $(0,0)$

$$-(\lambda_1 + \lambda_2 + \mu)P_{0,0} + (\lambda_1 + \lambda_2)P_O + \mu(P_{0,1} + P_{1,0}) = 0.$$

By substituting P_O obtain following balance equation

$$-(\lambda_1 + \lambda_2)P_{0,0} + \mu(P_{0,1} + P_{1,0}) = 0.$$

Then we can modify state graph (Fig. 1) and remove state $\{O\}$ and use following equation for getting P_o.

$$P_O = \frac{1}{\rho}P_{0,0}$$

Balance equations will not change for both state graphs and looks like

$$\begin{cases}
-(\lambda_1 + \lambda_2)P_{0,0} + \mu(P_{1,0} + P_{0,1}) = 0, (i = 0, j = 0), \\
-\mu P_{k-1,0} + \lambda_1 P_{k-2,1} + \alpha\lambda_1 P_{k-2,0} = 0, (i = k-1, j = 0), \\
-(\alpha\lambda_1 + \mu)P_{0,k-1} + \lambda_2 P_{0,k-2} = 0, (i = 0, j = k-1), \\
-(\lambda_1 + \lambda_2 + \mu)P_{0,j} + \text{æ}_1\mu P_{1,j} + \mu P_{0,j+1} + \lambda_2 P_{0,j-1} = 0, (i = 0, 1 \leq j \leq k-2), \\
-(\lambda_1 + \lambda_2 + \mu)P_{i,0} + \mu P_{i+1,0} + \text{æ}_2\mu P_{i,1} + \lambda_1 P_{i-1,0} = 0, (1 \leq i \leq k-2, j = 0), \\
-(\alpha\lambda_1 + \mu)P_{i,k-1-i} + \lambda_2 P_{i,k-i} + \lambda_1 P_{i-1,k-1-i} + \alpha\lambda_1 P_{i-1,k-i} = 0, (1 \leq i \leq k-2, i+j = k-1) \\
-(\lambda_1 + \lambda_2 + \mu)P_{i,j} + \text{æ}_1\mu P_{i+1,j} + \text{æ}_2\mu P_{i,j+1} + \lambda_1 P_{i-1,j} + \lambda_2 P_{i,j-1} = 0, (1 \leq i \leq k-2, 1 \leq i+j \leq k-2),
\end{cases} \quad (2)$$

3 Generating Function Method Application

Application of method from [2, 3, 6] to any model starts from generalization of (2). It can be generalized using Kronecker's delta-symbol in following way:

$$
\begin{aligned}
&((\lambda_1 + \lambda_2)(1 - \delta_{i+j,k-1}) + \alpha\lambda_1\delta_{i+j,k-1}(1 - \delta_{i,k-1}) + \mu)P_{ij} = \lambda\delta_{i,0}\delta_{j,0}P_\emptyset + \\
&+ \lambda_1(1 - \delta_{i,0})P_{i-1,j} + \lambda_2(1 - \delta_{j,0})P_{i,j-1} + \ae_1\mu(1 - \delta_{j,0})(1 - \delta_{i+j,k-1})P_{i+1,j} + \\
&+ \ae_2\mu(1 - \delta_{i,0})(1 - \delta_{i+j,k-1})P_{i,j+1} + \mu\delta_{j,0}(1 - \delta_{i,k})P_{i+1,0} + \\
&\mu\delta_{i,0}(1 - \delta_{j,k-1})P_{0,j+1} + \alpha\lambda_1\delta_{i+j,k-1}(1 - \delta_{i,0})P_{i-1,j+1}
\end{aligned}
\tag{3}
$$

For getting a solution for this system of equations will be used generating functions method. Generating function of phase state (1) is:

$$
G(u,v) = \sum_{i=0}^{k}\sum_{j=0}^{k-i} P_{ij}u^i v^j
\tag{4}
$$

Multiplying left and right part of (3) on $u^i v^j$ and summarizing by all integer values of (i,j), satisfying $i+j \le k-1$, obtain equation for generating function:

$$
\begin{aligned}
&'((\lambda_1 + \lambda_2 + \mu) - (\lambda_1 u + \lambda_2 v) - (\frac{\ae_1\mu}{u} + \frac{\ae_2\mu}{v}))G(u,v) = \\
&= \lambda P_\emptyset + ((\lambda_1 + \lambda_2) - \alpha\lambda_1 - (\lambda_1 u + \lambda_2 v) + \alpha\lambda_1\frac{u}{v})\Sigma_1 + \alpha\lambda_1 P_{k-1,0}u^{k-1}(1 - \frac{u}{v}) + \\
&+ G(u,0)(\frac{\mu}{u} - \frac{\ae_1\mu}{u} - \frac{\ae_1\mu}{v}) + G(0,v)(\frac{\mu}{v} - \frac{\ae_1\mu}{u} - \frac{\ae_2\mu}{v}) + P_{0,0}(\frac{\ae_1\mu}{u} + \frac{\ae_2\mu}{u} - \frac{\mu}{u} - \frac{\mu}{v})
\end{aligned}
\tag{5}
$$

We can check correctness of (5) by comparison with equation for model $\vec{M}_2/M/1/k/f_1^1$ from [1, 2], with $\ae_1 = 1, \ae_2 = 0$. Also we can check it by getting distribution for total amount of all types packets in system buffer (like in model $M/M/1/k$). This model has one Poisson incoming flow with intensity $\lambda = \lambda_1 + \lambda_2$ and service rate μ, like in considered model $\overrightarrow{M_2}/M/1/k/f_4^1$.

Well known fact that N_Σ in $M/M/1/k$ model is distributed as truncated geometric law [3]:

$$
P\{N_\Sigma = n\} = \frac{1-\rho}{1-\rho^{k+1}}\rho^n, (n = \overline{0, k-1})
\tag{6}
$$

So, according to (6) probabilities that model has n packets in buffer are equal to

$$
\begin{aligned}
r_n &= P\{N = n\} = \frac{1-\rho}{1-\rho^{k+1}} \cdot \rho^{n+1}, (n = \overline{0, k-1}), P_O = P\{N_\Sigma = 0\} = \frac{r_O}{\rho} \\
&= \frac{1-\rho}{1-\rho^{k+1}}.
\end{aligned}
$$

Generating function for total amount of packets in model is

$$G_\Sigma(u) = G(u, u), G_\Sigma(u) = \sum_{n=0}^{k-1} r_n u^n, G_\Sigma(u) = \frac{\rho(1-\rho)}{1-\rho^{k+1}} \cdot \frac{1-(\rho u)^k}{1-(\rho u)}.$$

Then using geometric progression could be proven that this generating function describes truncated geometric law distribution.

Applying technique from [2, 3, 6] to generating function Eq. (5) can be obtained series expansion by degrees of u and v, which allows to get expression for probabilities of all model states through two sets of probabilities: "diagonal" $P_{k-1-i,i}, (i = \overline{0, k-1})$ and "boundary" $P_{k-1-i,0}, (i = \overline{0, k-1})$:

$$P_{k-1-j,l} = \alpha P_{k-1,0}(\theta_{j+1,l+1}^{(+)} - \theta_{j,l}^{(+)}) + \frac{æ_2}{\rho_1} \sum_{i=0}^{j-2-l} P_{k-1-i,0}(\theta_{j-i,l+1}^{(+)} - \theta_{j-i-1,l}^{(+)}) +$$

$$+ \sum_{i=0}^{l} P_{k-1-i,i} \theta_{j-i+1,l-i}^{(+)} + \sum_{i=l+1}^{\frac{l+l}{2}} P_{k-1-i,i} \theta_{j-i+1,-l+i}^{(-)} - \alpha \sum_{i=0}^{l+1} P_{k-1-i,i} \theta_{j-i+1,l+1-i}^{(+)} -$$

$$- \alpha \sum_{i=l+2}^{\frac{l+l+1}{2}} \theta_{j-i+1,-l-1+i}^{(-)} P_{k-1-i,i} + ((\alpha-1) - \varepsilon) \sum_{i=0}^{l} P_{k-1-i,i} \theta_{j-l,-l+i}^{(-)} + \varepsilon \sum_{i=0}^{l-1} P_{k-1-i,i} \theta_{j-l,l-i-1}^{(+)} +$$

$$+ \varepsilon \sum_{i=l}^{\frac{l+l-2}{2}} \theta_{j-i,-l+i+1}^{(-)} P_{k-1-i,i}$$

where

$$\theta_{j,l}^{(+)} = (\frac{\rho_1}{æ_1})^{1-j}/2 \sum_{m=l}^{\frac{j-1+l}{2}} \frac{\rho_2^m æ_2^{m-l}(-1)^l}{2^{2m-l}(\rho_1 æ_1)^{\frac{2m-l}{2}} m!(m-l)!} P_{j-1}^{(2m-l)}(c)$$

$$\theta_{j,l}^{(-)} = (\frac{\rho_1}{æ_1})^{1-j}/2 \sum_{m=0}^{j-1-l} \frac{\rho_2^m æ_2^{m+l}(-1)^l}{2^{2m+l}(\rho_1 æ_1)^{\frac{2m+l}{2}} m!(m+l)!} P_{j-1}^{(2m+l)}(c)$$

So, all probabilities for model $\overrightarrow{M_2}/M/1/k/f_4^1$ can be calculated using only $2k - 1$ unknown probability.

To complete studying considered model we have to build smaller system of linear equation only for «diagonal» and «boundary» probabilities. First set of equations can be obtained by summation all the diagonals of the graph, in the same manner as it was done in [2]. The second set of equations is obtained using analyticity conditions for generating function at the origin in the same manner as was done for a system with alternating priority [6]. To do this, go to the limit $v \to 0$ in the expression for the generating function (5) and using L'Hopital's rule and equating to zero coefficients of series expansion obtain k more equations:

$$P_{k-1,0} - \rho_1 P_{k-2,0} - \alpha\rho_1 P_{k-2,1} = 0, (j = k-1)$$
$$(\rho+1)P_{i,0} - \rho_1 P_{i-1,0} - P_{i+1,0} - \text{æ}_2 P_{i,1} = 0, (j = \overline{1, k-2})$$
$$(\rho+1)P_{0,0} - P_{1,0} - P_{0,0} - P_{0,1} = 0, (j = 0)$$

4 Computational Results

In [3] were defined areas of linear behavior and closing. Using results from previous section these areas could be found for model with randomized priority and presented on Figs. 2, 3, 4, 5, 6.

One of the special cases in randomized priority model is system with non-preemptive priority when $\text{æ}_1 = 1$ and $\text{æ}_2 = 0$, considered in [1, 2]. Obtained areas of linear behavior and closing for this case coincided with the results of [2] and weren't included in this article. Figure 2 shows a case of equiprobable packets selection from

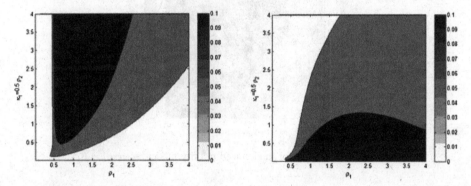

Fig. 2. Areas of linear behavior for randomized priority model with $\text{æ}_1 = \text{æ}_2 = 0.5$: (a) first flow; (b) second flow packets.

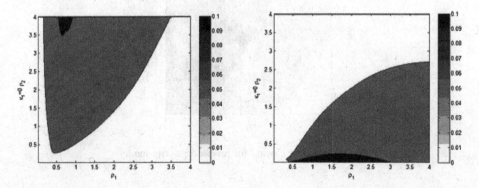

Fig. 3. Areas of linear behavior for reversed priority model where $\text{æ}_1 = 0$ and $\text{æ}_2 = 1$: (a) first flow; (b) second flow packets.

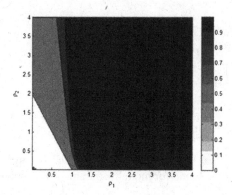

Fig. 4. Areas of closing for preemptive priority model

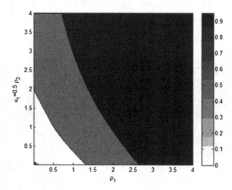

Fig. 5. Areas of closing for randomized priority model $\ae_1 = \ae_2 = 0.5$.

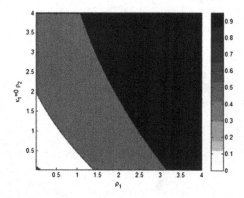

Fig. 6. Areas of closing for reversed priority model

model buffer for service with $æ_1 = æ_2 = 0.5$. Figure 3 shows areas for case of opposing priorities when flow with priority for the pushing out of the buffer has lower priority for queuing than another flow (the case of $æ_1 = 0$ and $æ_2 = 1$).

Figures 4, 5, 6 show closing areas (when changing probability of pushing out makes possible to achieve the loss probability of non-priority requirements close to 1) for the same cases, as discussed above.

5 Conclusion

In this article was considered model with two Poisson incoming flows, one service channel, randomized priorities and randomized push-out mechanism. Was shown application of generating functions method to obtain simplifies system of balance linear equations and shown method of getting loss probabilities for such models. Also were build areas of linear behavior and closing for three cases: non-preemptive, equiprobable and reversed priorities. In all considered cases, it is clear that a change of parameters $æ_1$ and $æ_2$ significantly change areas of linear behavior and closing in wide range. Using these parameters in combination with changing probability of pushing out allows at first make raw model configuration, and then more accurate with the push of probability.

Acknowledgements. The reported study was partially supported by RFBR, research project No. 15-29-07131 ofi_m.

References

1. Ilyashenko, A., Zayats, O., Muliukha, V., Laboshin, L.: Further investigations of the priority queuing system with preemptive priority and randomized push-out mechanism. In: Balandin, S., Andreev, S., Koucheryavy, Y. (eds.) NEW2AN/ruSMART 2014. LNCS, vol. 8638, pp. 433–443. Springer, Heidelberg (2014)
2. Avrachenkov, K.E., Shevlyakov, G.L., Vilchevsky, N.O.: Randomized push-out disciplines in priority queueing. J. Math. Sci. **122**(4), 3336–3342 (2004)
3. Muliukha, V., Ilyashenko, A., Zayats, O., Zaborovsky, V.: Preemptive queuing system with randomized push-out mechanism. Commun. Nonlinear Sci. Numer. Simul. **21**(1–3), 147–158 (2015)
4. Gnedenko, B.V., Danielyan, E.A., Dimitrov, B.N., Klimov, G.P., Matveev, V.F.: Priority Queueing Systems. Moscow State University, Moscow (1973). (in Russian)
5. Kleinrock, L.: Queueing Systems. Wiley, New York (1975)
6. Ilyashenko, A., Zayats, O., Muliukha, V., Lukashin, A.: Alternating priorities queueing system with randomized push-out mechanism. In: Balandin, S., Andreev, S., Koucheryavy, Y. (eds.) NEW2AN/ruSMART 2015. LNCS, vol. 9247, pp. 436–445. Springer, Heidelberg (2015)

NEW2AN: Wireless Sensor Networks

Improving Energy-Awareness in Selective Reprogramming of WSNs

Hadeel Abdah[1], Emanuel Lima[2], and Paulo Carvalho[1(✉)]

[1] Centro Algoritmi, Departamento de Informática, Universidade do Minho,
Braga, Portugal
pg28633@alunos.uminho.pt, pmc@di.uminho.pt
[2] Instituto de Telecomunicações and Department of Electrical and Computer
Engineering, University of Porto, Porto, Portugal
emanuellima@fe.up.pt

Abstract. Saving energy is considered one of the main challenges in wireless sensor networks (WSNs), being radio activities such as message transmission/reception and idle listening the main factors of energy consumption in the nodes. These activities increase with the increase of reliability level required, which is usually achieved through flooding strategies. Procedures such as remote WSNs reprogramming require high-level of reliability leading to an increase in radio activity and, consequently, waste of energy. This energy waste is magnified when dealing with selective reprogramming where only few nodes need to receive the code updates. The main focus of this paper is on improving energy efficiency during selective reprogramming of WSNs, taking advantage of wise routing, decreasing the nodes' idle listening periods and using multiple cooperative senders instead of a single one. The proposed strategies are a contribution toward deploying energy-aware selective reprogramming in WSNs.

Keywords: WSNs · Selective reprogramming · Energy-aware strategies

1 Introduction

Wireless sensors networks consist of large numbers of small, resource-constrained, self-organizing, low cost, computing motes. Nowadays, these networks are considered ideal candidates for a wide range of applications such as monitoring environmental issues, military operations and other application fields where it is hard to maintain a continuous presence of human beings [1].

After deploying these networks, it might be necessary to update the code running on nodes due to factors such as changes in the environment, software updates or changes in the application goals. However, this type of networks is often deployed in environments with harsh conditions where reprogramming the nodes manually may be a cumbersome or even an impossible task. Therefore, remote reprogramming is suggested as the best solution to achieve such modifications on the nodes [1]. Remote reprogramming can be applied to the whole network or just to some specific nodes (selective reprogramming), either way, it is crucial to provide reliability for such procedure. Unfortunately, most of the approaches oriented to remote WSNs reprogramming resort to network flooding,

© Springer International Publishing AG 2016
O. Galinina et al. (Eds.): NEW2AN/ruSMART 2016, LNCS 9870, pp. 241–253, 2016.
DOI: 10.1007/978-3-319-46301-8_20

leading to a major waste of energy in network nodes. When dealing with selective reprogramming, the waste of energy increases steeply specially when just a small number of nodes need to get update messages, as these messages may still be received and retransmitted from all network nodes.

This work is focused on considering multiple scenarios of selective reprogramming and proposing an effective energy-aware approach for each one. The proposed approaches aim at reducing energy consumption in the network by taking advantage of multiple and complementary solutions such as wise routing, clustering and the ability to manage nodes sleeping time instead of using typical flooding approaches. In this paper, we also propose enhancements to the Deluge [2] extension proposed in [3] in order to reduce the overhead of selective reprogramming in WSNs, making this process more energy efficient.

This paper is structured as follows: related work on WSNs remote reprogramming is discussed in Sect. 2; different scenarios for selective reprogramming and the proposals for improving it are debated in Sect. 3; the obtained results are included as proof-of-concept in Sect. 4; and the final conclusions are presented in Sect. 5.

2   Related Work

Designing and implementing a protocol for remote reprogramming of WSNs faces many challenges due to tight constraints in network nodes regarding processing and communication activities, directly impacting energy consumption. The main concern in this paper is related to energy consumption in remote reprogramming, especially in selective reprogramming. This section presents an overview of the most popular remote reprogramming protocols and the mechanisms proposed in the literature to reduce energy consumption during the reprogramming process.

2.1   Remote Reprogramming in WSNs

Many remote reprogramming approaches were proposed having reliability of code dissemination as main concern, leading to solutions that still evince limitations regarding energy efficiency. Deluge [2] uses a three-stage handshaking protocol consisting of advertisement, request and data, where updated nodes advertise their code version and outdated nodes request these nodes for the new code version. Deluge provides reliability, robustness, and support for multi-hop network reprogramming while being simple to implement. However, it requires the nodes to be always in idle listening mode during the reprogramming process, increasing considerably the amount of energy waste, since idle listening is one of the major sources of energy consumption in WSNs [4]. MDeluge [5] tries to reduce energy consumption by disseminating the code image to a designated subnet of the WSN using a distribution tree, which is formed when the sensor nodes send code request messages. A micro server keeps the code and sends it based on the requests received. In [5], simulations results show that MDeluge performs better than Deluge when disseminating the new code to designated sensor nodes. However, MDeluge is not suitable for many real network scenarios, as the authors establish excessive operational assumptions.

Other reprogramming protocols, such as Stream [6], try to conserve energy by reducing the size of transmitted code. This is accomplished by installing the reprogramming protocol on each node beforehand through the segmentation of the flash memory. Therefore, unlike Deluge where both the code image and the reprogramming protocol are disseminated, Stream only sends the code image and the information about the reprogramming protocol to be used. This reduces the amount of data sent during the reprogramming process. However, the dissemination process is the same as in Deluge; thus, suffering from the same problems of energy waste due to long periods of idle listening. Other protocols also try to reduce the amount of data transmitted by sending the differences between code versions (delta). The receiving nodes use this delta to rebuild the new code image before reprogramming. Zephyr [7] is presented as an incremental reprogramming protocol where a delta is sent to each node whenever an update is required. Hermes [8], an improvement of Zephyr, reduces the delta using techniques to mitigate the effect of changes in functions and global variables caused by differences in the code. However, these two protocols, being based on Stream regarding the dissemination process, have the same problems concerning long idle listening periods.

Some remote reprogramming protocols try to reduce the idle listening periods to preserve energy, such as MNP [9]. This protocol attempts to guarantee that in a neighborhood there is at most one source transmitting the new code at a time and tries to select the sender that is expected to have the most impact. MNP reduces the problems of collision, hidden nodes and long periods of idle listening by putting the node into a "sleep" state whenever its neighbors are transmitting an uninteresting segment. Although MNP saves energy, the dissemination process takes long [10]. Freshet [11] also conserves energy by putting nodes to sleep. It operates in three phases for each new code image: blitzkrieg, distribution, and quiescent. Nodes are put to sleep between the blitzkrieg and the distribution phases as well as in the quiescent phase. Infuse [12] disseminates data in an energy-efficient. Since Infuse uses a TDMA-based MAC protocol, sensors only need to listen to the radio in the slots assigned to their neighbors. In the remaining slots, sensors can turn their radio off.

2.2   Selective Remote Reprogramming in WSNs

Despite the undeniable relevance of performing selective reprogramming in WSNs, only few protocols were proposed covering this facility.

The first protocol is the Socially-aware Dissemination of Code Updates [13], which was applied in a real-world scenario with animals and humans taking advantage of their social behavior. Code updates are relayed opportunistically from one animal to another upon contact. Unlike existing approaches that propagate updates to the entire network, the authors limit dissemination as much as possible to the target nodes, taking advantage of a characteristic common to many mobile WSNs scenarios, namely, the fact that the monitored individuals exhibit social behavior. The implicit structure of social interactions, once elicited, provides an effective tool for steering efficient routing decisions. This approach is able to reduce the network overhead but it is only applied in limited scenarios.

The second protocol is a Deluge extension for selective reprogramming of WSNs [3], which allows reprogramming the network in two modes: (i) reprogramming the entire network (as in Deluge); and (ii) reprogramming nodes selectively according to the type of platform in use or to a list of node identifiers. The protocol extension is fully compatible with the default Deluge operations [2], has minimal impact on network traffic and offers a high-degree of reliability. Despite the new features the protocol extension brings in, the dissemination process is still based on Deluge, therefore, reducing overhead and optimizing energy are still open design issues. As an example, the packets used in reprogramming (DE and DF packets) introduce an additional overhead, which leads to additional power consumption.

3 Strategies for Enhancing Selective Reprogramming

When performing selective reprogramming in WSNs, two scenarios can be considered depending on the location in the network of the new code image to be disseminated. The first scenario is when the user wants to reprogram a subset of nodes for the first time, so it is necessary to send the new code image from the base station (BS) to the selected nodes. In this scenario, flooding the whole network with code messages should be avoided. Instead, the code image should be routed from the BS to the selected nodes through the path with the highest energy levels. During this process, other nodes not involved in routing should turn their radios off to save their energy.

The second scenario is when some nodes in the network already have the new code image. In this case, the code image needs to be routed from these nodes to the selected nodes for reprogramming. In this scenario, we discuss two approaches: (i) when clustering is used in the wireless sensor network; and (ii) when the network is flat.

3.1   Revisiting Selective Reprogramming Within Deluge

For the strategies discussed in this section, we will take advantage of Deluge extension for selective reprogramming [3]. In this protocol, the authors propose two types of packets (DE, DF) in order to combine the normal operation of Deluge [2] with selective reprogramming. A DE packet corresponds to the packet originally sent in reprogramming, including minor changes to specify the type of reprogramming, the cardinality of the set of nodes to be reprogrammed using that packet, and the corresponding Node IDs. A DF packet, oriented to selective reprogramming, is designed to carry Nodes IDs in excess, i.e., Node IDs that cannot be transported in a DE packet.

As shown in Fig. 1, the field Reprogram Type in DE packets determines whether the selective reprogramming is carried out through Node ID or platform type. Although these features increase the flexibility of selective reprogramming, the format of DE and DF packets do not optimize the way node IDs are handled. In fact, the protocol reserves two bytes to identify each node ID, which leads to an increase in packet overhead and, consequently, to an increase in the number of DE and DF packets required for selective reprogramming. This increase is more significant when the list of node IDs to reprogram is long, urging for a more efficient handling of node IDs.

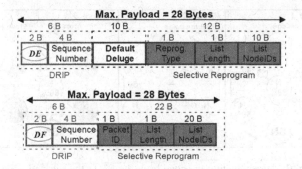

Fig. 1. Structure of DE and DF packets [3].

In order to make this procedure more energy-efficient, we suggest making the number of bits reserved to represent a node ID adjustable. This is accomplished by defining a 4-bit long packet field called BPN (bits per node), allowing a maximum length of 16 bits to represent node IDs. This allows reducing DE and DF packets overhead in scenarios where only a small number of nodes with limited range of IDs is present. In practice, the number of DE and DF packets exchanged in the network will be reduced. The structure of the modified version of a DE packet is illustrated in Fig. 2.

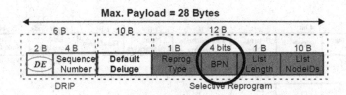

Fig. 2. DE packet modified structure.

In WSNs several metrics are commonly matter of concern, namely, the energy consumption in the nodes, the network lifetime, and the number of messages received and transmitted per node. The following subsections will detail the two reprogramming scenarios mentioned above, assuming the use of Deluge extension for Selective Reprogramming [3] due to the flexibility it brings to Deluge.

3.2 Selective Reprogramming Using the BS

In this scenario, we assume that the selected node (SN) might be several hops away from the base station (BS) and the network is flat (no hierarchy). During the process of forwarding the code image from the BS to the SN, we propose an energy-aware algorithm for performing selective reprogramming divided into three phases: (i) Discovering possible paths; (ii) Choosing the best path; and (iii) Disseminating code, as shown in Fig. 3.

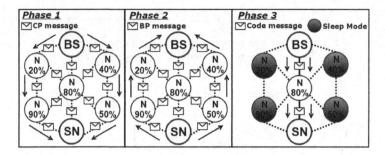

Fig. 3. Disseminating the code from the BS to the selected node.

Discovering Possible Paths (Phase 1) - In the first phase, the BS will sense the network to discover possible delivery paths up to the selected node. For this, each node in a path shall report its current energy to assist the routing decision. Thus, initially, the BS sends a specific message type named CP (Check Path), with the structure illustrated in Fig. 4. CP messages, apart from allowing sensor nodes to perform energy and route pinning, allow the BS to inform the network on the number of packets required to send the new code image (code packets). This information is relevant to evaluate the sleeping time that some nodes may undergo for saving energy.

	2 B	4 bits	1 B	1 B	1 B	1 B	1 B per Node	1 B per Node
CP msg	Message ID	NOH	Reprog. Type	SelectedNode/ Platform	Num. CodePackets	TimeStamp	List NodeIDs	List NodesEnergy
							Path Hops	Path Energy

	2 B	1 B	1 B per Node
BP msg	Message ID	TTS	List NodeIDs
			Path Hops

Fig. 4. CP and BP message structure.

As shown in Fig. 4, the first two bytes in a CP message are used to define the message identifier. The NOH field is used to specify the maximum allowed number of hops that a path may have. This allows controlling the dissemination scope, i.e., any path exceeding NOH will be ignored. As mentioned before, the RepType field identifies the type of reprogramming, namely, if reprogramming will be based on the node identifier or on the platform type. When the BS sends a CP message, both the list of hops in the path and corresponding list of energy levels are empty. One byte is reserved for each element in these lists, allowing a total of ten NodeIDs as hops in the path. The number of CodePackets identifies the number of code packets that the BS should send to reprogram the node, while the TimeStamp field indicates when the CP message was originated. These fields are used in the second phase of the algorithm.

The first phase involves checking all possible paths in the network; thus, CP messages are flooded. When a node receives a CP message, it checks if it matches either the SelectedNode field containing the NodeID of the node to be reprogrammed or the platform type. If the node is not the selected one, then it checks the list of NodeIDs to verify if its NodeID is already in the list of hops in the path, before inserting it. The node also inserts its residual energy in the list of NodesEnergy field (Path energy) to allow for subsequent path decision. After that, the node will broadcast the modified CP message to its neighbors, which will handle the message in the same way. This procedure will be repeated in every hop so that nodes in the path to the selected node indicate their NodeIDs and residual energy. A node will drop a CP message if its own NodeID is already in the path to the selected node.

Choosing the Best Path (Phase 2) - The selected node will receive multiple CP messages reporting different paths and corresponding energy status. Several decision-making rules can be defined for helping the selected node to choose the best path. A straightforward approach is to establish a minimum energy threshold, below which a path is not eligible. For instance, the selected node may eliminate paths containing at least a hop with residual energy below 20 %, and then choose the path with the maximum percentage of energy, on average. This avoids the selection of a path including energy-constrained nodes. After choosing the best path, the selected node is expected to send a BP message (Best Path message) reporting that choice, see Fig. 4. A BP message is also designed to convey relevant timing information so that nodes outside the best path enter into sleeping mode for a correct amount of time.

In more detail, as shown in Fig. 4, the first two bytes of a BP message are used to identify the message type. The selected node can then set the sleeping time (TTS field) for the nodes outside the best path. This value is evaluated based on the number of code packets that the BS is expected to send, previously announced in the CP message. Thus, the nodes are able to turn their radio off for a period of time that corresponds to the number of code packets sent by the BS times the time a message takes to reach the selected node. This amount of time is calculated using TimeStamp value in the CP message, and corresponds to the time elapsed from sending a CP message and receiving it in the selected node.

When a node receives a BP message it checks if its NodeID is in the list of NodeIDs of the message. If this is the case, the node knows it is in the best path, therefore, it will forward the BP message to its neighbors and will keep its radio on to forward the code packets to the SN. Nodes that do not belong to the chosen path will forward the message and immediately turn their radio off according to TTS.

Disseminating Code (Phase 3) - This phase starts as soon as the BS receives the BP message. Although the code dissemination can be performed using any dissemination protocol such as Deluge, the energy costs will be significantly reduced as the nodes that do not belong to the best path do not receive or transmit any messages.

3.3 Selective Reprogramming Using Updated Nodes

Selective Reprogramming with Clustering. In large scale WSNs, some level of hierarchy is expected to be present, being data aggregation the most common mechanism used for this purpose. In data aggregation, the network is divided into groups or clusters, and instead of making each node forward its data to the BS, data is sent to a group leader node, usually named the cluster head (CH). The selection of a CH is dependent on the clustering protocol that is being used. Clustering protocols such as HEED [14] and Dynamic Multi Level Hierarchical Clustering [15] choose a CH depending on its residual energy and number of neighbors. In our approach, these protocols are preferable due to their efficiency, although, choosing any other clustering protocol is also acceptable. Our objective is to take advantage of clustering to enhance selective reprogramming. In the clustering scenario, we assume that the code to be disseminated already resides in other nodes within the same cluster as the SN. This assumption reflects a scenario which may well occur in practice, therefore, selective reprogramming should take advantage of existing up-to-date code versions distributed in the WSN. We advocate the use of multiple senders instead of only one to perform code dissemination so that each sender can contribute by sending a number of code messages according to its residual energy. The proposed algorithm is divided into three phases: (i) Selective Reprogramming Setup; (ii) Selection of Senders; and (iii) Code Dissemination, as illustrated in Fig. 5.

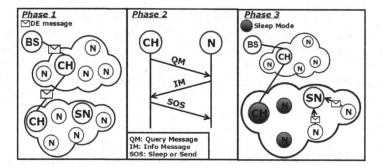

Fig. 5. Code dissemination in a WSN with clustering.

Selective Reprogramming Setup (Phase 1) - In this phase, the BS sends packets to the CH in order to identify in which clusters the selected nodes for reprogramming are located. This can be accomplished resorting to DE and DF packets [3].

Selection of Senders (Phase 2) - When a CH receives a DE or DF packet it will send a Query message to the nodes in its cluster asking for their identifiers, residual energy and the version of code image they are running. This control message has a minimal structure, as only a message type, and packet identifier are required. In response to Query messages, each node will send an Info message to the CH. Based on the received messages; the CH determines which nodes in the cluster require reprogramming and which nodes can be selected as senders. If reprogramming is needed inside the cluster, the CH will divide the cost of updating the selected node by multiple senders. For this process, we propose a mechanism where the CH starts to eliminate potential senders with energy less than a specific threshold if better candidate senders are in place. This extends the lifetime of nodes that have a small amount of energy, as they will not be involved in the updating process. Then, the CH calculates the number of packets containing the new image that each sender is expected to send during the updating process. For this calculation the CH takes in consideration the energy of each sender and the number of packets that each sender should sent, i.e.:

$$\frac{\textbf{Sender residual energy} \times \textbf{Number of code packets needed for reprogramming}}{\textbf{Summation of the potential senders energy}}$$

The total number of code packets that need to be sent for updating a selected node can be taken from the DE packet, where the field "DefaultDeluge" (see Fig. 1) contains information concerning the structure of the code image. Using the proposed mechanism, the number of code packets that each sender has to send is proportional to its remaining energy. Figure 6 depicts the structure of Info and Sleep or Send (SOS) messages that are used by the CH to control the updating process of the selected node.

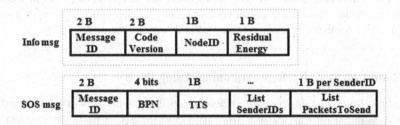

Fig. 6. Info and SOS message structure.

The TTS (Time To Sleep) field in SOS messages is used to inform passive nodes, i.e., nodes neither being selected as senders nor undergoing update, about the amount of time they should turn their radio off. The CH can determine this time by multiplying the number of code packets needed to update the node by their delivery time to the selected node (an approximation of the time between sending the Query message and receiving the last Info message). Each sender knows the number and order of packets to be sent depending of its NodeID position in the List of SenderIDs of the SOS message. For instance, if the first NodeID in the List of SenderIDs is 23 and the number of packets in the first position of the List of PacketsToSend is 10, then node 23 must send the first 10 packets of the code image, i.e. both lists are co-indexed.

Code Dissemination (Phase 3) - In the third and last phase, the designated senders will start using default Deluge [2] to disseminate the code image while the nodes that are not participating in the updating process will be in sleeping mode.

In presence of WSN clustering, the strategy proposed above is expected to improve the energy efficiency during selective reprogramming. The proposed solution takes advantage of: (i) using multiple collaborative senders, allowing the nodes to share the energy cost of updating a selected node; and (ii) putting the nodes that are not involved in reprogramming in sleeping mode during selective reprogramming.

Selective Reprogramming in Flat Networks. In the previous scenario, we handled selective reprogramming in presence of clustering, assuming that the update code version resides in other nodes of the same cluster as of the SN. A similar approach can be used when the SNs in need of reprogramming are in a flat network, i.e., without clustering. In this case, the SN can play the same role as the CH in the previous scenario. When the SN node receives the modified version of a DE or DF packet, it will send a Query message to its neighbors, and waits for Info responses. The SN can then determine the best senders and the number of packets each one is expected to send, and will send SOS messages to its neighbors asking them for either turning their radio off or playing the role of senders.

4 Proof-of-Concept: Simulation Results

To evaluate the performance of the proposed strategies, they were implemented and compared with both flooding and Deluge using OMNeT++ [16]. In order to facilitate the comparison among these different approaches, the sensor nodes used in the simulations are static. The metric considered is related to the radio activity in a node, expressing the number of messages received and sent in that node.

4.1 Results on Selective Reprogramming Using the BS

For the first strategy - *selective reprogramming using BS* - the network is flat, and the code is transmitted for the first time from the BS to the SN. A multihop scenario was considered including fifteen static sensor nodes with different percentages of residual energy. The nodes were programmed to forward only three CP messages with a maximum number of hops (NOH) set to six hops. Note that these values vary extensively according to the type of application where the WSN is used.

The radio activity in nodes was analyzed for: the selected node (SN); a neighboring node that is outside the estimated best path (NPN); and finally, a node that belongs to the best path (PN). We studied the radio activity in these nodes while varying the number of code packets needed to be transferred to the selected node. The mean values resulting from ten simulation runs were considered in order to obtain accurate results. The number of messages sent and received in the SN, NPN and PN are illustrated respectively in Figs. 7a, b, c. The simulation results show a significant improvement when applying our strategy over both flooding and Deluge in terms of reducing radio activity in all three nodes, thus, reducing power consumption in them. This improvement is mostly evident in the node that is not included in the best path (NPN) node. As illustrated in Fig. 7b, the radio activity in NPN node increases rapidly with the increasing number of code packets in both flooding and Deluge, whereas it is put to sleep in our approach, so its energy is saved.

As expected, for the SN and PN nodes, the activity increases with the increase of code packets, however, the increase is still smaller in comparison to flooding and Deluge.

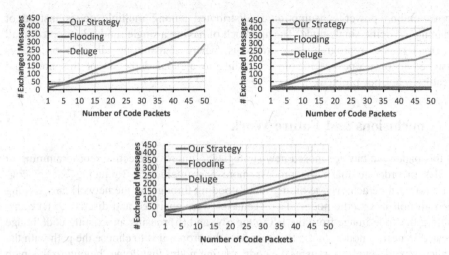

Fig. 7. (a) Messages sent and received at SN. (b) Messages sent and received at NPN. (c) Messages sent and received at PN

4.2   Results on Selective Reprogramming Using Updated Nodes

For the second strategy - *selective reprogramming using update nodes* - the network is clustered and some neighboring nodes already have the code to be transferred to the SN. The initial simulation scenario includes one network cluster, comprising five cluster members and one CH. These nodes are in bidirectional communication with each other. The results below are for two cluster nodes, the selected node SN and a cluster member node (CN) with residual energy of 25 %, making of it a potential sender. Figure 8 illustrates the number of messages sent and received at both SN and CN while varying the number of code packets.

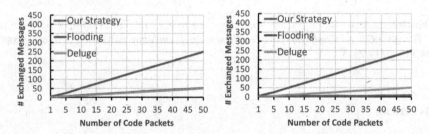

Fig. 8. Messages sent and received at SN (left) and CN (right)

The results show that when applying the strategy based on cooperative senders, the radio activity of SN, in terms of the number of messages sent and received, is similar to the radio activity of this node in Deluge. However, the radio activity in other cluster nodes is reduced sizably when this strategy is implemented. It must be also noted that using our approach, where the code packets are sent via multiple senders, the corresponding power consumption is distributed among multiple nodes instead of draining a specific single node. The approach of having a single sender is adopted in all previous remote reprogramming protocols, where transmitting all code packets is assigned to only one specific node, leading to a rapid depletion of this node and creating a hotspot problem.

5 Conclusions and Future Work

In this paper, we have proposed new strategies to improve selective reprogramming of WSNs in order to make the process more energy-efficient. Reducing energy consumption can be achieved avoiding blind flooding throughout the network and turning the radio off in specific nodes. Two scenarios were analyzed and discussed: (i) transmitting the code image from the BS to the SN; and (ii) reusing an existing code image located in nearby nodes. In the first scenario, the proposal is to choose the path with the highest energy levels to transfer the code, putting nodes that do not belong to this path into sleep for the whole reprogramming period. The second scenario is analyzed both for cluster-based and flat WSNs. We proposed the use of multiple senders to transmit the code to the SN eliminating single sender exhaustion, while forcing other nodes to sleep during the reprogramming process, thus, avoiding unnecessary reception of code messages in these nodes. We tested these two approaches and compared them with typical flooding and Deluge solutions. The results show a significant reduction in the number of messages received and sent in the nodes, leading to reduction of the power consumption and making selective reprogramming more energy-efficient.

Although Deluge [2] has become a standard reprogramming protocol for WSNs, the presented proposals can be understood as contributions toward an energy-aware deployment of Deluge or Deluge extension for selective reprogramming [3].

As future work, we plan to extend the study to networks with larger number of nodes, and to analyze the impact of applying these strategies on the time needed to accomplish the reprogramming process.

Acknowledgements. This work has been supported by COMPETE: POCI-01-0145-FEDER-007043 and FCT – *Fundação para a Ciência e Tecnologia* within the Project Scope: UID/CEC/00319/2013.

References

1. Brown, S., Sreenan, C.J.: Software update recovery for wireless sensor networks. In: Komninos, N. (ed.) SENSAPPEAL 2009. LNICST, vol. 29, pp. 107–125. Springer, Heidelberg (2010)
2. Chlipala, A., Hui, J., Tolle, G.: Deluge: data dissemination for network reprogramming at scale. Technical report, University of California, Berkeley (2004)
3. Lima, E., Carvalho, P., Gama, O.: A protocol extension for selective reprogramming of WSNs. In: 2015 23rd International Conference on Software, Telecommunications and Computer Networks (SoftCOM), pp. 280–284. IEEE (2015)

4. Zheng, X.-L., Wan, M.: A survey on data dissemination in wireless sensor networks. J. Comput. Sci. Technol. **29**, 470–486 (2014)
5. Zheng, X., Sarikaya, B.: Code dissemination in sensor networks with MDeluge. In: 2006 3rd Annual IEEE Communications Society on Sensor and Ad Hoc Communications and Networks, SECON 2006, pp. 661–666. IEEE (2006)
6. Panta, R.K., Khalil, I., Bagchi, S.: Stream: low overhead wireless reprogramming for sensor networks. In: IEEE INFOCOM 2007, 26th IEEE International Conference on Computer Communications, pp. 928–936. IEEE (2007)
7. Panta, R.K., Bagchi, S., Midkiff, S.P.: Efficient incremental code update for sensor networks. ACM Trans. Sens. Netw. **7**, 1–32 (2011)
8. Panta, R.K., Bagchi, S.: Hermes: fast and energy efficient incremental code updates for wireless sensor networks. In: IEEE INFOCOM 2009, pp. 639–647. IEEE (2009)
9. Kulkarni, S.S., Wang, L.: MNP: multihop network reprogramming service for sensor networks. In: Proceedings of 25th IEEE International Conference on Distributed Computing Systems, ICDCS 2005, pp. 7–16. IEEE (2005)
10. De, P., Liu, Y., Das, S.K.: Energy-efficient reprogramming of a swarm of mobile sensors. IEEE Trans. Mob. Comput. **9**, 703–718 (2010)
11. Krasniewski, M.D., Panta, R.K., Bagchi, S., Yang, C.-L., Chappell, W.J.: Energy-efficient on-demand reprogramming of large-scale sensor networks. ACM Trans. Sens. Netw. **4**, 2 (2008)
12. Kulkarni, S.S., Arumugam, M.: Infuse: a TDMA based data dissemination protocol for sensor networks. Int. J. Distrib. Sens. Netw. **2**, 55–78 (2006)
13. Pásztor, B., Mottola, L., Mascolo, C., Picco, G.P., Ellwood, S., Macdonald, D.: Selective reprogramming of mobile sensor networks through social community detection. In: Silva, J.S., Krishnamachari, B., Boavida, F. (eds.) EWSN 2010. LNCS, vol. 5970, pp. 178–193. Springer, Heidelberg (2010)
14. Younis, O., Fahmy, S.: HEED: a hybrid, energy-efficient, distributed clustering approach for ad hoc sensor networks. IEEE Trans. Mob. Comput. **3**, 366–379 (2004)
15. Tomar, G.S., Verma, S.: Dynamic multi-level hierarchal clustering approach for wireless sensor networks (2009)
16. OMNeT++. http://www.omnetpp.org. Accessed May 2016

Modified Elastic Routing to Support Sink Mobility Characteristics in Wireless Sensor Networks

Imane Benkhelifa[1,2]([⊠]), Nassim Belmouloud[2], Yasmine Tabia[2],
and Samira Moussaoui[2]

[1] CERIST Research Center, Algiers, Algeria
i.benkhelifa@cerist.dz
[2] Department of Computer Science, USTHB University, Algiers, Algeria
{nbelmouloud,ytabia,smoussaoui}@usthb.dz

Abstract. This paper presents improvements for the geographic routing protocol Elastic so to support the different sink mobility characteristics. We have proposed a strategy to support multiple mobile sinks; tested Elastic under high speeds of the mobile sink; proposed two strategies in case of the sink temporary absence and finally proposed to predict the sink location by the source node and then by all the nodes. Simulation results show that our propositions improve much the delivery ratio and reduce the delivery delay.

Keywords: Elastic protocol · Geographic routing · Sink mobility · Mobility management · Wireless Sensor Networks

1 Introduction

Geographic protocols are currently being thoroughly studied due to their application potential in networks. They are very efficient in wireless networks for several reasons. First, nodes need to know only the location information of their direct neighbors in order to forward packets and hence the stored state is minimal. Therefore they can achieve high scalability with reasonable memory requirements [1, 2]. Second, such protocols conserve energy and bandwidth since discovery floods and state propagation are not required beyond a single hop. Third, in mobile networks with frequent topology changes, geographic routing has fast response and can find new routes quickly by using only local topology information [3, 4]. Therefore, geographic routing is generally considered as an attractive routing method for both mobile wireless ad-hoc and sensor networks [5].

It is well known that in the case of a static sink the energy consumption of individual nodes varies strongly across the WSN, since the nodes close to the sink are much more heavily burdened than those farther away from the sink due to relay operations [6]. Also, many applications such as target tracking, emergency response and smart cities need the design of a routing protocol that considers mobile elements as part of the design. Some geographic routing protocols with mobile sinks [2, 7, 19] have been proposed in order to respond to some specific applications as well as for improving the network performance.

O. Galinina et al. (Eds.): NEW2AN/ruSMART 2016, LNCS 9870, pp. 254–268, 2016.
DOI: 10.1007/978-3-319-46301-8_21

In this paper, we study the sink mobility characteristics and propose improvements for one the geographic routing protocols namely Elastic protocol. We call the new protocol M-Elastic for Modified-Elastic. The paper is organized as follows: Sect. 2 presents some related work. Section 3 highlights the main characteristics of sink mobility. A brief presentation of Elastic routing is presented in Sect. 4. Our improvement propositions are detailed in Sect. 5. Simulation assumptions and results are discussed in Sect. 6. Finally we conclude with a conclusion and future work in Sect. 7.

2 Related Work

In [8], authors evaluated the ability of data transmission and reception of WSN to a mobile sink on the basis of its speed; they conclude that the maximum data delivery depends upon this parameter. However, authors didn't mention which routing protocol was used; hence, the results cannot give a clear conclusion.

Stojmenovic et al. in [9] discussed the sink mobility in WSNs focusing on delay-tolerant networks and real-time networks. They investigated the theoretical aspects of the uneven energy depletion phenomenon around static sinks and addressed the problem of energy-efficient data gathering by mobile sinks.

Authors in [10] highlighted the importance of multiple mobile sinks in WSNs and proposed a new geo-casting protocol that allows the dissemination to multiple mobile sinks.

In [11], authors observed though simulation the impact of a single mobile sink in WSN. They employed a mobile sink to a multi-hop routing platform namely the connected K-neighbors (CKN) sleep algorithm. The first scenario considers a mobile sink that moves randomly within a rectangular area and then another within a restricted circular area and finally, an event-driven sink. These scenarios were compared with the results obtained when considering a stationary sink. Authors concluded that mobility maximizes the network lifetime especially in the case of event-driven where the sink moves towards the source to get the packet.

Authors of [12] identified and highlighted the interactions between the controlled mobility and the layers of the control stack in self-organizing wireless networks and came up with a case study in which they show how controlled mobility can be exploited practically. Their advantages and limitations were also presented.

In the following section, we will discuss the sink mobility characteristics and their impact on the network.

3 Mobile Sinks

The sink mobility assumption can be imposed by the application nature. For example, in security constrained scenario, if a static sink is located, it can be easily compromised and damaged by malicious users, causing disconnection between sensors and the end-user [13]. Hence, the use of a mobile sink makes harder the damage of such component.

It is commonly agreed that a sink node is a powerful device with unconstrained supply energy and computing capacity. In addition, the following characteristics of the sink can critically influence the communication operations in sensor networks.

The Number. Even though the typical number of sinks is "one", in most practical applications such as Emergency Response, the increase in the number of sinks provides more robust data collection and helps to increase the network lifetime and reduces the communication overhead within the network. In addition, it alleviates the uneven energy depletion problem of a single-sink deployment and can bring more uniform energy dissipation, therefore, the possibility of energy hole will be reduced and network coverage will be improved [14].

The Mobility. During the lifetime of the network, the sink can be stationary or mobile. In some cases, the mobility is derived by the application. For example, sinks are integrated in mobile devices such as mobile phones carried by mobile users or attached to animals or vehicles equipped with radio devices. The mobile sink could provide the ability to closely monitor the objects that we want to guard in the WSN and to look at the events as smaller granularity than static sinks [15]. For delay tolerant applications, single mobile sink in fact equals virtually multiple static sinks at different positions [14]. To support the mobility of sink, it is essential to manage the relationship between the moving speed of the sink and the tolerable delay associated with data transmission.

A mobile sink can either move at fixed or variable speed and can be considered as slow (up to 1 m/s), moderate (1 to 20 m/s), or fast (greater than 20 m/s) [21].

As frequent updates of the position of the mobile sink can generate excessive energy consumption of sensors, routing strategies manipulating mobile sinks should provide effective means for monitoring sinks to keep all (or some) sensor nodes updated for further data reports.

The Presence. The sink can be continuously or partially present during the lifetime of the network. In the latter case, the routing protocol must support the temporary absence of a sink. Instead of dropping messages during the absence of the sink, messages can be buffered in source nodes or other predefined locations (i.e. a set of sensors near the sink) to send them to the sink when it is available again.

The Trajectory. When considering a mobile sink, the sink trajectory can be arbitrary or predefined by the network administrator in order to cover the whole network. For example, the sink can be mounted on a helicopter or on a fire truck monitoring a disaster area. In the case where the sink moves arbitrary, some sensor nodes may never be visited or the sink may go always far from the events, hence, not all events will be reported. In addition, if the sensors can predict the mobile sink's movement, the energy consumption would be greatly reduced and data packets handoff would be smoother [16].

4   Geographic Routing – Case Elastic

The geographic routing assumes that the sensor node has information about its location in the network. The packets contain the locations of the source node and the destination node. With the use of intermediate nodes, routing decisions are taken. In general, the node holding a packet chooses the closest node to the destination as the next hop. This forwarding process is called Greedy Forwarding [17] and is repeated until the packet reaches its destination. Some geographic protocols consider other network metrics, such as the available energy, the level of congestion, load balancing or time constraints.

If the sink is mobile, it is constantly in motion; its location information must be regularly updated to the source node [18]. Further, since a rate of communication overhead is created during the update location information of the Sink node, the power consumption increases. Therefore, it is necessary to find an efficient method to update the location information of the mobile sink. Among the protocols that consider the mobility of sink and the efficiency of location updates, we are interested in Elastic Protocol.

In Elastic [19], a source node uses the Greedy forwarding before transmitting data to a mobile sink, and the location information update of the mobile sink is transmitted to the source node in the opposite direction along the same path used for data transmission. Data are transmitted to the new location of the mobile sink when its location information is found in the way of the data transmission, i.e. one the nodes in the route knows the current location of the sink.

The location service is executed in the order A–B–C, as shown in Fig. 1 (a). However, when the Sink node approaches to node B, as shown in Fig. 1(b), the service is performed in the order Sink-B–C since B is a neighbor of the sink and can directly know the sink position. When the sink node escapes the transmission limit of A, as indicated in Fig. 1(c), the sink node transmits the location information to A (its last hop forwarder) via unicast. Then, A backups information for the new location of Sink and expects the new package and changes the destination of the packet with the new position and sends the next packet to the sink via Greedy Forwarding. In Fig. 1(d–f), while node A is sending this packet to the sink, node B can overhear this transmission and derives the new position of the sink and finds another route via Greedy forwarding to send the next packet. This process is repeated for each packet until the source knows the new position of the sink.

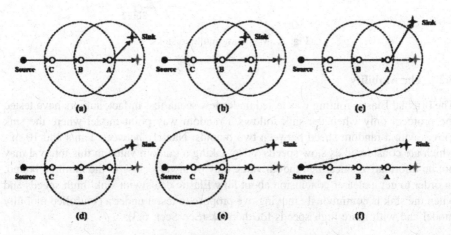

Fig. 1. Location process in Elastic (Yu et al. [19]).

5 Modified Elastic to Support Sink Mobility Characteristics

To support applications such as emergency response or smart cities where there is a great need to deploy multiple sinks, each one has its own mobility model with different speeds and trajectories, we propose in this section, improvements and modifications to Elastic to support such sink mobility characteristics.

5.1 The Number

We suppose that multiple sinks are deployed. At the beginning the source node determines to whom the message is destined. This information is included in the packet with an additional list of the other sinks in the case where the first sink is not available. This list is considered as a reserve list. The order of this list is determined by the source node. If the last forwarder notices the absence of the main sink or fails to transmit the message to this sink, it checks the sinks list and changes the packet destination with the first sink of the list. This strategy allows avoiding discarding messages in the case of the sink's failure and even avoiding using face routing [17]. Indeed, if the last forwarder is faced to the local minima problem, and instead of using the planarization, it simply chooses another sink from the reserve list and then sends the message to this new destination (sink 2 in Fig. 2).

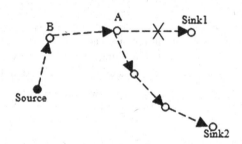

Fig. 2. Managing multiple sinks.

5.2 The Mobility

The original Elastic routing was tested under few scenarios. In fact, authors have tested the protocol only when the sink follows a random way point model where the sink moves with a random speed between two bounds. Namely between 1 m/s and 10 m/s which are considered as slow speeds. Also, taking a random value in this interval may not be informative since the random value may be always close to the minimal bound. In order to get a clearer conclusion about how Elastic deals with sink' high speeds and when the sink is continuously moving, we propose to test it under a controlled mobility model and with more high speeds for the sink (See Sect. 6.3).

5.3 The Presence

During the lifetime of the network, a sink can be temporary absent. For example, a sink mounted on a helicopter. When the helicopter goes far away from the sensor area, nodes cannot transmit their messages. An efficient routing protocol has to support such kind of situation. However, the original Elastic protocol does not treat this problem at all. For that, we propose to improve Elastic routing so to support the temporary absence of the sink.

The deployment of multiple sinks can be of a huge benefit in the case of the absence of one sink (as proposed in Sect. 5.1), because there is a high probability to find another sink to receive messages, even with longer routes.

The real problem occurs when there is only one sink deployed in the sensor area. In the following, we explain our proposition to improve Elastic so to support the sink absence.

Note that the last hop forwarder is the first who notices the sink absence.

- When the sink is absent, the last forwarder (node A in Fig. 3) buffers the received messages. If its buffer is full, it asks its neighbors (neighbors of A) to buffer the next received messages.

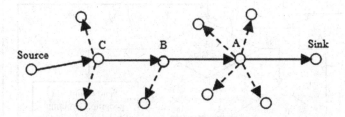

Fig. 3. Buffering messages during the sink's absence.

- If all the buffers of all the neighbors (neighbors of A) become full, the last forwarder delegates its last forwarder (node B in Fig. 3) to receive and buffer new messages. Remark: If the buffer of an intermediate node is empty, this means that the node still believes that the sink is available. At the contrary, if the node buffer is not empty, the node concludes that the sink is not available and it's its task to ensure the buffering process.
- Then, in turn, the second last forwarder (node B in Fig. 3) executes the same process with its neighbors following the reverse path of the geographic routing during the data delivery, until arriving at the source node. From there, the source node understands that the sink is not available and all the buffers of the route nodes are full and can no longer transmit messages. We propose two solutions to face this problem at the level of the source node:
  - Not sending messages anymore until the sink becomes available again. But, what if there are more important messages than previous ones to be sent? The source node has to find another solution. We propose to prioritize messages. When an intermediate node receives a message with higher priority than the

priority of the message it holds, it discards the old message and buffers the new one. Note that prioritized messages will replace old messages beginning from the last hop forwarder following the reverse path of greedy forwarding. That is A–B–C in Fig. 3.

- Finding another route. However, if the absence period is very long, there is a high risk that the entire network faces the congestion problem because whenever a route is full, the source looks for another until exhausting all its neighbors. Since managing congestion is out of our scope, we omit this solution.

Note that nodes buffer packets for a known period of time (buffering-time). If the buffering-time is finished and yet the sink is absent, nodes are then obliged to discard the packet to save their memory and energy. Diagram in Fig. 4 explains the algorithm managing the temporary absence of the sink by buffering messages in intermediate nodes.

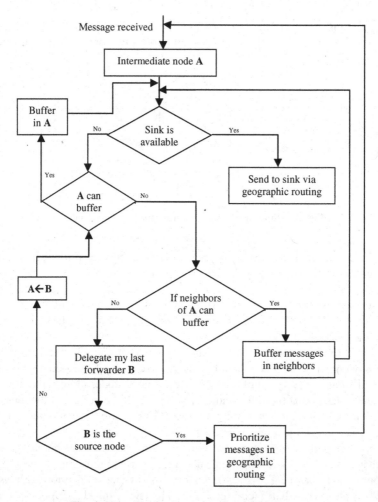

Fig. 4. Diagram for managing the sink's absence.

When the sink comes back, we use the same strategy as the sink location service, which is the overhearing concept. The sink informs by unicast its last hop forwarder (node A in Fig. 3) about its availability and its new location if it has changed. Node A begins to send its buffered messages to the sink. Neighbors of A overhear these transmissions (including node B) and notice the availability of the sink and its location. In turn, they transmit their buffered messages via greedy forwarding. Neighbors of B do the same thing and so forth until transmitting all the buffered messages and reaching the source node. From now onwards, the source is aware about the availability of the sink and transmit messages normally.

5.4 The Trajectory

In the original Elastic, before obtaining the new location of a mobile sink, the source still encapsulates the known original location of the sink in each data packet, and this invalid location information may lead to unsuccessful data delivery [19]. If the source can predict the new location of the sink, the delay of finding a shorter route can be highly reduced. However, authors didn't mention whether the sink path is arbitrary or predefined, but in their tests, they supposed that the sink follows the random way point model, which leads us to deduce that the sink trajectory is arbitrary. This makes it hard to predict.

First Improvement. We suppose that the sink trajectory is predefined and the source knows the sink's trajectory and speed. This allows the source to predict easily the sink location at any time. Before sending a packet, the source encapsulates its belief about the sink location (the predicted location) on the data packet. This allows finding a shorter route to the sink's new location instead of sending the packet through the old path until arriving at a node that knows the new location of the sink. The predicted location by the source may not be the exact location of the sink but at least the packet will be sent in the right direction towards the sink and will meet surly one of the sink' neighbors since anyway the sink communicates its current location to its nearby sensors periodically. Despite this, we keep the overhearing principle so that the source node can update the sink real location since the location got by overhearing is more credible that the predicted one.

Second Improvement. In the original Elastic, with a higher moving speed, the sink may move out of the radio range of the last hop forwarding node with a higher probability. In this case, the sink has to inform its location to the last hop forwarding node by unicasting, and the data packets forwarded by the last hop forwarding node during this period are all dropped [19]. If the last hop forwarder can predict the new location of the sink, the number of dropped messages can be reduced. However, the last forwarder is not always the same, so we propose that all nodes know the sink's trajectory and speed. This allows nodes to predict easily the sink location at any time. While waiting the unicat message from the sink, the last forwarder encapsulates their belief about the sink's new location. Now, the sink location extracted from overhearing messages is no longer the only information about the new location of the sink but it is considered as a trusted update helping as a basic reference for nodes to predict the sink location after certain time.

6 Performance Evaluation

In this paper, our goal is to observe the impact of sink characteristics on Elastic geographic routing protocol and to improve it so to support these mobility character-istics. Simulating our improvement proposals using NS2 simulator, we analyze the performance results by means of calculating the delivery ratio of the transmitted packets, as well as the mean delay of transmission and the number of hops necessary to transmit successfully a packet from the source to the sink.

We generate a sensor network with 100 nodes distributed uniformly in an area of 200×200 m^2 with a transmission range of 30 m. This gives a density of 7.07 fol-lowing the formula (1):

$$D = \frac{\pi N R^2}{A} \tag{1}$$

Where N is the total number of nodes, A is the sensor area and R is the transmission range.

The sink (s) is (are) continuously moving without pause time, In Elastic routing, the sink sends location announcement messages once it moves 1 m away from its previous location. Each second, the source node (which is determined beforehand) sends 10 packets to the sinks (at the position that it holds about the sinks). Note that the simulation results are averaged over 10 runs. Table 1 summarizes the default simula-tion parameters. We suppose that all the nodes except sinks have fixed positions during all the simulation time and have the same energy of transmission and reception. Note that the rate packet loss is not due to collision but to the different scenarios of sink mobility.

Table 1. Simulation parameters.

Parameter	Default value
Number of nodes	100
Area size	200×200 m^2
Mac layer	800.11
Communication range	30 m
Antenna	Omni Antenna
Propagation	TwoRayGround
Packet size	512 bytes
Sink speed	20 m/s
Sink trajectory	arbitrary
Simulation time	200 s

We compare M-Elastic with Elastic [19] and GPSR [17] as it is considered a reference for geographic routing in Ad Hoc networks and sensor networks.

6.1 M-Elastic Considering Multiple Mobile Sinks

To support multiple sinks, we have proposed that the source node determines to whom the packet is destined with a reserve list. In this test, we vary the number of mobile sinks and we study its effect on the delivery ratio and the average delay of transmission. After 50 % of simulation time, we provoke intentionally the failure of the main sink and after 70 % of simulation time; we provoke the failure of the first sink in the reserve list.

As shown in Fig. 5(a), with the increase of the number of the mobile sinks, the success delivery ratio increases too. Indeed, the more the number of sinks increases, the more a packet has the chance to be received by one of the sinks of the reserve list in case of delivery failure to some sinks.

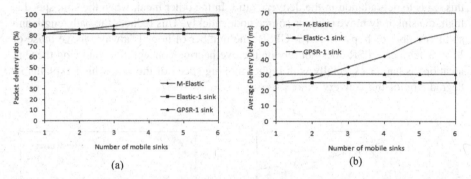

Fig. 5. (a): Delivery ratio with multiple mobile sinks; (b): delivery delay with multiple mobile sinks.

Now, we measure the average delivery delay. We notice according to Fig. 5(b) that the delay increases when the number of sinks increases. This can be explained by two factors: first, the deployment strategy where in order to have a good coverage; sinks are deployed initially in relatively distant areas (otherwise, no significant advantage can be derived) while the source remains static. Then, they move arbitrary and may become more and more distant from the source. The number of hops increases and so the delay. The second factor is the failure of some sinks. This obliged the last hop forwarder to choose a sink from the reserve list and begin another route discovery towards this new sink which needs additional time. Note that the transmission delay is proportional to the number of hops. Indeed, whenever, the path length increases, the delivery delay increases too.

6.2 M-Elastic Considering Sink High Speeds

In the original paper of Elastic [19], authors tested their protocol under low speeds (maximum speed is from 1 m/s to 10 m/s) and the sink moves with a random speed between 1 m/s and 10 m/s. However, authors did not mention if there is a pause time or not. In addition, the real speed of the sink is not known. The scope of this paper is to test Elastic under more different sink characteristics. To do so, we consider, in this test, one single mobile sink and we vary its speed between 5 m/s and 35 m/s. the sink's speed is known and constant during a test. The choice of this interval is motivated by the different applications that can benefit from mobile sinks. For example, in the case of an emergency situation, the sink can be mounted on a fire truck or even a civil helicopter (The speed can be higher than 20 m/s). In smart cities, the sink can be attached to persons or vehicles (the speed can vary between 5 m/s and 20 m/s). Figure 6(a) shows the success delivery ratio and Fig. 6(b) shows the mean number of hops. For M-Elastic, with a higher moving speed, the sink may move quickly out of the radio range of the last hop forwarding node with a higher probability. In this case, the sink has to inform its location to the last hop forwarding node by unicast, and the data packets forwarded by the last hop forwarding node during this period are all dropped, this leads to degradation in the delivery ratio. In the other hand, when the sink speed is high, the sink may move towards source node quickly. This reduces the path length and so the number of hops while for GPSR, the number of hops is always above those of M-Elastic since GPSR does not rely on the overhearing concept, thus knowing the sink locations is harder especially with higher moving speed of the sink which explains its degradation of the delivery ratio.

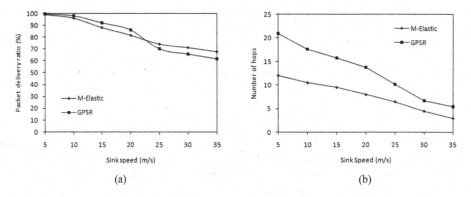

Fig. 6. (a) Delivery ratio vs sink speed; (b) number of hops vs sink speed.

6.3   M-Elastic Considering Sink Absence

In this test, we vary the sink absence duration between 10 % and 50 % of the simulation time. We assume that each node can buffer up to 2 packets. The sink speed was set to 20 m/s and the buffering-time was set to 5 s. The packet generation rate was set to 10 packets/s. We suppose that during the absence of the sink, generated packets have more and more priority.

Figure 7(a) shows the percentage of the packets delivered successfully with respect to packets transmitted during the absence of the sink. For M-Elastic, the results reveal that the delivery ratio remains constant until 20 %. Indeed, during the absence of the sink, intermediate nodes buffer the packets, each one up to 2 packets in our case. If their buffers become full, packets will be buffered in their neighbors. When the sink is

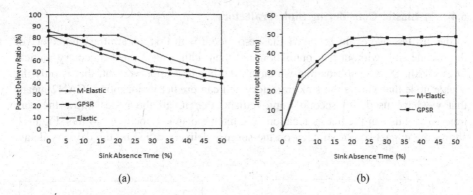

Fig. 7. (a) Impact of the sink absence time on delivery ratio; (b) delivery delay vs sink absence time.

available again, nodes holding packets send them to the sink via geographic routing along the initial path. All packets reach the destination; packets were just blocked for moments. In our case, the density is 7, meaning that each intermediate node has around 7 neighbors. After 20 % of absence, the route path becomes full (all intermediate nodes and their neighbors are buffering packets) because the average path length is 7–8 hops as deduced from Fig. 6(b), source node begins to prioritize packets and old packets will be dropped. In addition, after 25 % absence, nodes begin to discard packets to save their energy and memory. All these factors explain the degradation of the delivery ratio but still it is a gain instead of dropping them all with the original Elastic and GPSR. Figure 7(b) shows the average interrupt latency of a packet. That is, the transmission time from the comeback of the sink until arriving at destination. Obviously the end-to-end delay is calculated following this formula:

$$\text{End-to-End Delay} = \text{greedy forwarding Time} + \text{sink absence Time} + \text{Interrupt Latency} \quad (2)$$

Note that we applied the same buffering strategy to GPSR to compare it with M-Elastic.

From Fig. 7(b), we notice that the interrupt latency increases gradually until 20 % of absence. This is explained by the fact that when the sink is back, buffered packets will be released in their order of arrival to be sent to the sink. Thus, newest packets will wait the transmission of oldest ones. After 25 % of sink absence, the route becomes full (intermediate nodes and their neighbors are all buffering packets) and prioritized packets will replace oldest ones. When the sink is back again, packets will be sent in the same order as before prioritization. Note that after prioritization, the number of buffered packets is the same since the route is full. What happened is only replacement of packets, this explained the almost stability in interrupt latency. We notice from Fig. 7 that the degradations follow Exponential Distribution for both M-Elastic and GPSR.

6.4   M-Elastic Considering Sink Trajectory

In the last test, we evaluate the M-Elastic protocol with first, an arbitrary trajectory of the mobile sink without any prediction and then with a predefined trajectory namely SCAN [20], SCAN1 means that we have used our first improvement, that is only the source node that knows the sink trajectory and can predict its location. SCAN2 means that we have used our second improvement, that it, all the nodes know the sink trajectory, thus can predict its location. We fixed the sink velocity to 20 m/s. Figure 8 shows the impact of our strategy on the success delivery ratio and the delivery delay.

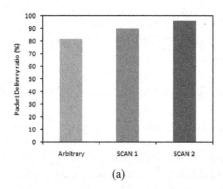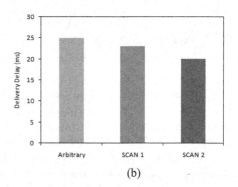

(a)   (b)

Fig. 8. (a) Impact of the sink trajectory with location prediction on delivery ratio; (b) on delivery delay.

Clearly, Fig. 8 shows that a mobile sink following a predefined trajectory gives better results in terms of packet delivery and average delay compared to an arbitrary one. In fact, with a predefined trajectory, source node can predict the new location of the sink and encapsulates this information on the packet. This allows finding quickly a shorter route to the sink's new location. With the second improvement, all nodes can predict the sink location, and while waiting the sink unicast, packets will not be dropped but sent to the predicted location. This will reduce the number of dropped packets as well as reducing finding another route, thus reducing the transmission delay.

7   Conclusion

In this paper, we have proposed improvements for the geographic routing protocol Elastic to support different sink mobility characteristics namely the multiplicity of mobile sinks, the temporary absence of the sink and higher speeds of the mobile sink. The results reveal that with multiple mobile sinks and thanks to our strategy of reserve list, M-Elastic gives better results in terms of success delivery ratio. However, whenever the sink speed is high, the delivery ratio becomes lower but with the advantage of having less number of hops, that means less delivery delay. Thus in real-time applications we suggest to deploy a mobile sink with high speed. The major improvement we have done is modifying Elastic so to support the absence of the sink. Instead of dropping packets during the absence time and thanks to our strategy of buffering and prioritizing packets, M-Elastic saves up to 25 % of packets sent and important packets will be sent even after the fullness of the route. Finally, predicting the sink location improves the delivery ratio and reduces the delivery delay.

Still remain some challenges to complete this work. For example, what is the adequate percentage of mobile sinks that should be deployed? Even if we believe it is a NP-hard problem. What is the best trajectory that should be taken by the sink? What it the best trade-off between the buffer size, the buffering-time and the packet generation rate? All these questions and others are under consideration in our ongoing work.

Acknowledgment. This work is part of the National Research Project "The Contribution of Vehicular and Sensor Networks in Risk Management" funded from the Algerian National Direction of Scientific Research.

References

1. Liu, Y., Shi, S., Zhang, X.: Balance-aware energy-efficient geographic routing for wireless sensor networks. In: Proceedings of the 8th International Conference on Wireless Communications, Networking and Mobile Computing (WiCom), pp. 1–4 (2012)
2. Li, X., Yang, J., Nayak, A., Stojmenovic, I.: Localized geographic routing to a mobile sink with guaranteed delivery in sensor networks. IEEE J. Sel. Areas Commun. **30**(9), 1719–1729 (2012)
3. Seada, K., Helmy, A.: Geographic routing in sensor networks. In: the Encyclopedia of Sensors, vol. 4, pp. 193–204. American Scientific Publishers (2006)
4. Al-Karaki, J.N., Kamal, A.: Routing techniques in wireless sensor networks: a survey. IEEE Commun. Mag. **11**(6), 6–28 (2006)
5. Popescu, A.M., Tudorache, I.G., Peng, B., Kemp, A.H.: Surveying position based routing protocols for wireless sensor and ad-hoc networks. Int. J. Commun. Netw. Inf. Sec. (IJCNIS) **4**(1), 41–67 (2012)
6. Khan, M., Gansterer, W.N., Haring, G.: Static vs. mobile sink: the influence of basic parameters on energy efficiency in wireless sensor networks. Comput. Commun. **36**, 965–978 (2013)
7. Ma, C., Wang, L., Xu, J., Qin, Z., Zhu, M., Shu, L.: A geographic routing algorithm in duty-cycled sensor networks with mobile sinks. In: Proceedings of the 7th International Conference on Mobile Ad-hoc and Sensor Networks (MSN), pp. 343–344 (2011)
8. Khalid, Z., Ahmed, G., Khan, N.M.: Impact of mobile sink speed on the performance of wireless sensor networks. J. Inf. Commun. Technol. **1**(2), 49–55 (2007)
9. Li, X., Nayak, A., Stojmenovic, I.: Sink Mobility in Wireless Sensor Networks, in Wireless Sensor and Actuator Networks: Algorithms and Protocols for Scalable Coordination and Data Communication. Wiley Inc., Hoboken (2010)
10. Erman, A.T., Dilo, A., van Hoesel, L., Havinga, P.: On mobility management in multi-sink sensor networks for geocasting of queries. Sensors. **11**, 11415–11446 (2011)
11. Jordan, E., Baek, J., Kanampiu, W.: Impact of mobile sink for wireless sensor network. In: Proceedings of the 49th Annual ACM Southeast Regional Conference, pp. 338–339 (2011)
12. Natalizio, E., Loscrì, V.: Controlled mobility in mobile sensor networks: advantages issues and challenges. Telecommun. Syst. **52**(4), 2411–2418 (2013)

13. Hamida, E.B., Chelius, G.: Strategies for data dissemination to mobile sinks in wireless sensor networks. IEEE Wirel. Commun. **15**(6), 31–37 (2008)
14. Munir, S.A., Dongliang, X., Canfeng, C., Ma, J. (eds.): Mobile Wireless Sensor Networks: Architects for Pervasive Computing. InTech, Rijeka (2011). ISBN 978-953-307-325-5
15. Erman, A.T., van Hoesel, L., Wu, J., Havinga, P.: Enabling mobility in heterogeneous wireless sensor networks cooperating with UAVs for mission-critical management. Technical report TR-CTIT-08-14, University of Twente, The Netherlands (2008)
16. Liu, X., Zhao, H., Yang, X., Li, X., Wang, N.: Trailing mobile sinks: a proactive data reporting protocol for wireless sensor networks. In: Proceedings of the IEEE 7th International Conference on Mobile Adhoc and Sensor Systems (MASS), pp. 214–223 (2010)
17. Karp, B., Kung, H.T.: GPSR: greedy perimeter stateless routing for wireless networks. In: Proceedings of the Sixth Annual International Conference on Mobile Computing and Networking, pp. 243–254 (2000)
18. Yu, J., Jeong, E., Jeon, G., Seo, D.-Y., Park, K.: A dynamic multiagent-based local update strategy for mobile sinks in wireless sensor networks. In: Murgante, B., Gervasi, O., Iglesias, A., Taniar, D., Apduhan, B.O. (eds.) ICCSA 2011, Part IV. LNCS, vol. 6785, pp. 185–196. Springer, Heidelberg (2011)
19. Yu, F., Park, S., Lee, E., Kim, S.-H.: Elastic routing: a novel geographic routing for mobile sinks in wireless sensor networks. IET Commun. **4**(6), 716–727 (2010)
20. Koutsonikolas, D., Das, S.M., Hu, Y.C.: Path planning of mobile landmarks for localization in wireless sensor networks. In: Proceedings of IEEE Distributed Computing Systems Workshops, p. 86 (2006)
21. Khan, A.W., Abdullah, A.H., Anisi, M.H., Bangash, J.I.: A comprehensive study of data collection schemes using mobile sinks in wireless sensor networks. Sensors **14**, 2510–2548 (2014)

Connectivity Estimation in Wireless Sensor Networks

Ilhom Nurilloev, Alexander Paramonov$^{(\boxtimes)}$, and Andrey Koucheryavy

Saint-Petersburg State University of Telecommunications,
Saint Petersburg, Russia
{ilhom_nurulloev,akouch}@mail.ru,
alex-in-spb@yandex.ru

Abstract. The connectivity of a wireless sensor network is one of the most important indicators of the network capabilities. This article describes the characteristics of connectivity and proposed a method of its estimation for wireless sensor networks. We use the Erdos-Renyi's model for random graphs as the connectivity model. The applicability of this model to connectivity estimation of the network with different number of nodes was investigated. A possibility of using UAVs for improving the connectivity in wireless sensor network was also considered.

Keywords: WSN · Connectivity · Random graphs · Erdos-Renyi model · Flying sensor networks · Energy efficiency · UAV

1 Introduction

Number of the applications for the wireless sensor network (WSN) is continuously expanding [1–3]. Currently, these technologies have been used in household, transport, logistics, housing and utilities, security, military affairs, medicine and many other areas. With their help, and their interaction with other networks, the monitoring and control tasks are solved. Using wireless technology and self-organization functions in many cases they allow obtaining quick and efficient solutions of information delivering tasks, which their decision using traditional techniques is impossible, ineffective or requires a significant investment of time [4–6]. As a continuation of the evolution of wireless sensor networks the development of the concepts of ubiquitous sensor networks and the Internet of Things (IoT) can be considered. The variety of applications of these technologies requires the determination of the main indicators of their performance, and evaluation methods. The various of target destination of wireless sensor networks have different requirements for certain qualities, for which methods and models is necessary that establish their connection with the technical parameters of the network and its elements [7–9]. In this paper we study the connectivity of network, as one of the main indicators of performance. The choice of this indicator for the study is due to the fact that namely connectivity is characterized the possibility of the service delivery of information, i.e. the availability of service.

O. Galinina et al. (Eds.): NEW2AN/ruSMART 2016, LNCS 9870, pp. 269–277, 2016.
DOI: 10.1007/978-3-319-46301-8_22

2   Formulation of the Problem

The main function, which performs the WSN, is a data delivery. Depending on the final service, received by the user, it can be telemetry data or transfer the data of streaming services. Telemetry services cover a wide range of applications such as the control of environmental parameters, processes, geographical coordinates, meter of testimony electricity consumption or heat, state of the human or animal body and many others. By streaming data transmission services include sound and video, which can also be realized on the basis of wireless sensor networks, at the choice of appropriate technologies of organization of radio channels, providing the required bandwidth. Depending on the purpose and services, requirements to the network can vary significantly. In terms of probability-time performance is accepted to allocate network tolerant to delays and a network tolerant to losses. Tolerance to the size of a network indicator says that it is not of paramount importance for a particular purpose. Basic quality indicators of the WSN, as well as other networks indicators can be divided into three main groups: the availability indicators, reliability and time indices. By the time indices should include data delivery time, variation time and delivery (jitter), bandwidth. For reliability parameters - probability of data loss (loss coefficient), the probability of errors in the data delivered. For availability indices should include the probability of availability of data delivery services, which in WSN is largely determined by the relative position of nodes and is characterized by connectivity (the probability of connectivity) [10]. Also, to the last group of indicators should include the time of life, which in many cases is limited due to the use of non-renewable energy sources. Depending on the purpose, some of the above indicators of network quality becomes dominant. For example, to monitoring of the fire safety a message delivery time is critical, and for the transmission data of electricity consumption (hot and cold water), the delivery time is not critical. If the network is used to monitor a certain area, than covering should be a paramount component.

For almost all the applications, connectivity plays a significant role. Connectivity is the property of network which does mean to be able to establish a connection between network elements such as gateway and any of the network nodes. In sensor networks is often meets heterogeneous connectivity, i.e., sensor node may be have multiple independent pathways connecting it to the gateway. Network connectivity is an important parameter which largely determines the vitality of the network. When the network connectivity is lost, the nodes cease to perform its functions. Calculate the connectivity parameter of a real network with a large number of nodes is difficult. For estimation of connectivity certain models can be used.

Since variants of the network design may be different, then it is advisable to characterize the connectivity by probability of connectivity. The probability of connectivity is equal to one, when any node on the network can be connected with any other nodes on the network.

3   Selecting the Model of Connectivity

To represent the network structure model graph is frequently used, in which the nodes are represented by vertices and connection lines (channels) by edges. In the wireless network model, existence an edge between vertices determined by the location of nodes and characteristics of their radio zones. In the simulation of WSN, we use a circle of radius R centered at the node location as a radio zone model. Because of the random location of network nodes, the link between pairs of vertices is also random and can be described by a probability of falling of vertex to the circle of radius R. The random character of existence of the link between the nodes of WSN allows assuming a choice a random graph as its model.

There are various models to describe of random graphs [11]. In graph theory, the Erdos–Renyi model is one of main models for generating random graphs. They are named after Paul Erdos and Alfred Renyi, who first introduced one of the models in 1959. This model is described as follows. Given a set $Vn = \{1,\ldots,n\}$, whose elements are called vertices and on this set constructed random graph with random edges. Potential edges of the graph no more than C_n^2 pieces. Any two vertices i and j are connected by an edge with some probability $p \in [0, 1]$ which does not depend from other C_n^2 pairs of vertices. In other words, the edges appear in accordance with a standard Bernoulli scheme in which C_n^2 trials and "the probability of success" p. Denote by E a random set of edges that occurs as a result of the implementation of such scheme. The random graph we denote by $G = (Vn, E)$. This is a random graph in Erdos - Renyi model [12, 16].

There are several theorems for this model. One of them exactly describes a method which estimates a connectivity of the graph.

Theorem: Consider the model $G(n, p)$. Let the $p = c\frac{\ln(n)}{n}$. If $c > 1$, it is almost always a random graph is connected. If $c < 1$, it is almost always a random graph is not connected. The main purpose of this theorem in our case is that when $c = 1$, the probability of connectivity Pc of the graph is equal to a threshold value. At $c < 1$, the probability of connectivity of graph less than this value, and for $c > 1$, the probability of connectivity over it. There are many publications dedicated to using of random graph model for wireless networks [17–24]. From the point of view of the tasks of constructing WSN, by largest of c one can judge on what extent the problem of connectivity is solved. In this model, there are no restrictions related to the length of the edges, while in the WSN maximum possible length of the edges is limited by the R. For example, for the most common building technology of sensor networks, such as ZigBee (IEEE 802.15.4) and Bluetooth (IEEE 802.15.1) radius of the nodes is from 10 to 100 m. And for the technology Wi-Fi (IEEE 802.11b) it is from 20 to 300 m. Along with the restriction to length of edges in WSN model there is a limitation to field of nodes location.

4   The Simulation Model and Modeling Results

To study the connectivity of the considered network and check of this theorem applicability we create the simulation model. It generates the given number of nodes with random coordinates in the restricted 3D area, and then searches the shortest paths between all pairs of nodes using Floyd algorithm and then estimates the part of founded paths from all possible paths. This estimation represents the connectivity probability. The considered area of nodes placement is cubic ($V = 250 \times 250 \times 250$ m^3), and the

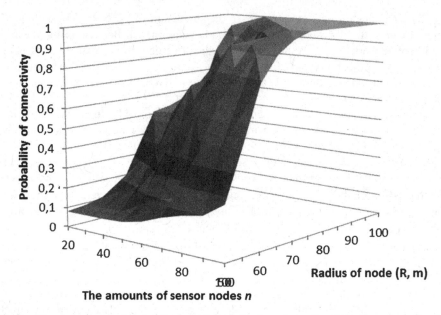

Fig. 1. The dependence of probability of connectivity from the radius (R) and number of nodes (n) in the cube

nodes are distributed randomly, i.e., form the Poisson field. We changed the amount of nodes in the cube from 20 to 100. Taking into account the characteristics of ZigBee and Bluetooth we selected the communication range from 50 to 100 m. The results of simulation are shown in Fig. 1.

Since the network nodes form a Poisson field, then probability of existence of communication (the existence of an edge) is described by the probability of falling a random point (node) in the area limited by a ball of radius R (Fig. 2).

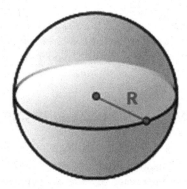

Fig. 2. Communication zone of node in the three-dimensional space

$$p = \frac{V_{ball}\,\rho}{n} \tag{1}$$

Where p the probability of falling a nodes in the node action radius,
ρ – the density of the network nodes (nodes/m^3);
n - the total number of nodes in the network;

$$V_{ball} = \frac{4\pi R^3}{3}\,\mathrm{m}^3,$$

R - radius of the node connection.

Comparing the p value with threshold probability that from theorem of Erdos-Renyi can be defined as

$$p_0 = \ln n / n \tag{2}$$

It is possible to estimate the probability of network connectivity under given parameters R and n.

Approximate value of connectivity probability may be obtained by [16]

$$p_C = e^{-e^{-c}} \tag{3}$$

Were c – is the constant value from the equation $p_0 = \frac{\ln n + c_0}{n}$. For $c_0 = 0$ $p_C = e^{-1} \approx 0.37$.

We estimate dependence of the network connectivity from radius of the nodes for a fixed number of nodes. Figure 3 shows the dependence of probability of network

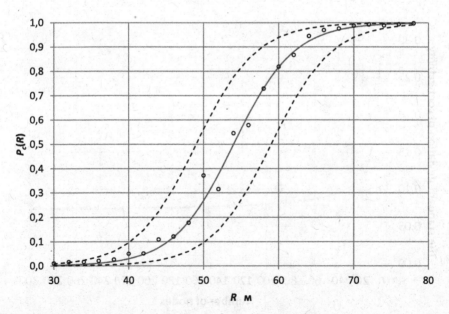

Fig. 3. The dependence of the probability of connectivity from radius of nodes (R) in a cube at $V = 250^3$ m^3, $n = 100$ m.

connectivity, which is calculated by simulation, from radius of the network nodes for $n = 100$. According to the results of simulation the probability of connectivity is equal 0.34 at $R = 52$ and 0.55 at $R = 54$. At a linear extrapolation for probability of connectivity 0.37 $R = 52.4$. According to (2) the threshold probability $p_0 = 0.046$ and the probability of falling to ball for this radius by the formula (1) is equal to 0.039. Difference between these values is 15 %. Simulation results was approximated by S-curve

$$\tilde{p}_C(R) = \frac{1}{1 + e^{-\frac{R - r_0}{b}}} \tag{4}$$

where R and r_0 are parameters, obtained by numerical feting of the curve to the simulation data. In this picture the upper and lower bounds of the confidence interval also given for significance level of 5 %. The confidence interval covers theoretical point p_0 given by theorem and approximation (3).

Thus, these results show that the random graph model and Erdos-Renyi theorem may be used to estimate parameters of the network for given requirements of connectivity or for connectivity estimation for given network parameters.

It is obvious that the use of a random graph model for WSN is expedient in case the number of nodes is large enough. When a small number of nodes is considered, in most cases the network structure can be described geometrically and for it is not required use of probabilistic methods. For the study of the applicability of random graph model to describe the connectivity of network simulation was conducted in which standard deviation of connectivity was estimated. Figure 4 shows the standard deviation of the probability of network connectivity from amounts of nodes in the network.

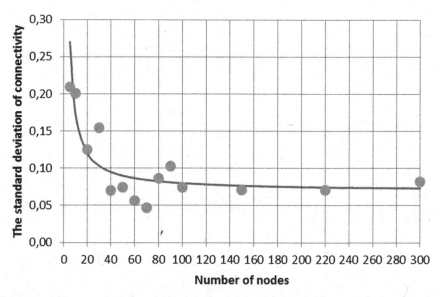

Fig. 4. The dependence of the standard deviation in the evaluation of the probability of network connectivity on the number of network nodes

As can be seen from the Fig. 4 the standard deviation, and hence the error of connectivity estimation decreases as $1/n$. It is the maximum for a small n, and stabilized at values $n > 50$ at 20 % of the estimated quantity. Thus, at a relatively small number of network nodes a connectivity estimation error using the model of a random graph may be too great. If the number of nodes is 50 or more, error is small enough (less than 20 %) for practical calculations.

When the network connectivity probability p is less than the threshold probability p_0, network is divided into clusters. To ensure the connectivity of clusters, we need to increase the communication range R of the nodes; this sometimes cannot be done. In this case, it is advisable to use unmanned aerial vehicles (UAVs), for which R is more because of the fact that it is located high above the ground (greater area of the line of sight, better antenna and more energy) [13–15]. Thus it is possible to provide network connectivity.

5 Conclusion

1. When organizing wireless sensor network on a set of randomly positioned nodes, it is advisable to apply the model of a random graph.
2. When simulating WSN in a random graph, the probability of the existence of edges is defined as the probability of falling a node in a communication range and depends on characteristic of nodes placement in a service zone.
3. Results of the WSN simulation with random placement of nodes (Poisson field) showed that the probability of network connectivity depends on the number of nodes and communication range and can be described by theorem of Erdos-Renyi.
4. Study of a dependence of connectivity probability of the WSN in three-dimensional space from a communication range and number of nodes showed statistical equality to the results obtained from the Erdos-Renyi model.
5. A study of the applicability of a random graph model from the number of nodes in the network showed that the error of connectivity estimation decreases with increasing n by law that is closed to $1/n$. At $n > 50$ error of connectivity estimation is less than 20 %, which is acceptable in most practical cases.
6. To ensure network connectivity when the probability of connectivity is less than p0 it is advisable to use the UAV.

Acknowledgment. The reported study was supported by RFBR, research project No. 15 07-09431a "Development of the principles of construction and methods of self-organization for Flying Ubiquitous Sensor Networks".

References

1. Andreev, S., Gerasimenko, M., Galinina, O., Koucheryavy, Y., Himayat, N., Yeh, S.-P., Talwar, S.: Intelligent access network selection in converged multi-radio heterogeneous networks. IEEE Wirel. Commun. **21**(6), 86–96 (2014). Art. no. A18
2. Galinina, O., Andreev, S., Gerasimenko, M., Koucheryavy, Y., Himayat, N., Yeh, S.-P., Talwar, S.: Capturing spatial randomness of heterogeneous cellular/WLAN deployments with dynamic traffic. IEEE J. Sel. Areas Commun. **32**(6), 1083–1099 (2014). Art. no. 6824742
3. Andreev, S., Larmo, A., Gerasimenko, M., Petrov, V., Galinina, O., Tirronen, T., Torsner, J., Koucheryavy, Y.: Efficient small data access for machine-type communications in LTE. In: IEEE International Conference on Communications, pp. 3569–3574 (2013). Art. no. 6655105
4. Akyildiz, I.F., Vuran, M.C., Akan, O.B., Su, W.: Wireless sensor networks: a survey revisited. Comput. Netw. J. (2005)
5. Koucheryavy, A., Al-Naggar, Y.: The QoS estimation for physiological monitoring service in the M2M network. In: Goldstein, B., Koucheryavy, A. (eds.) Proceeding of the Internet of Things and Its Enablers (INTHITEN), Saint-Petersburg State University of Telecommunications, SUT, St. Petersburg, pp. 133–139 (2013)
6. Abakumov, P., Koucheryavy, A.: Clustering algorithm for 3D wireless mobile sensor network. In: Balandin, S., Andreev, S., Koucheryavy, Y. (eds.) NEW2AN/ruSMART 2015. LNCS, vol. 9247, pp. 343–351. Springer, Heidelberg (2015)
7. Mašek, P., Zeman, K., Hošek, J., Tinka, Z., Makhlouf, N., Muthanna, A., Herencsár, N., Novotný, V.: User performance gains by data offloading of LTE mobile traffic onto unlicensed IEEE 802.11 links. In: Proceedings of the 38th International Conference on Telecommunication and Signal Processing, TSP 2015, pp. 1–5. AsszisztenciaSzervezoKft, Prague (2015). ISBN: 978-1-4799-8497- 8
8. Vybornova, A., Koucheryavy, A.: Traffic analysis in target tracking ubiquitous sensor networks. In: Balandin, S., Andreev, S., Koucheryavy, Y. (eds.) NEW2AN/ruSMART 2014. LNCS, vol. 8638, pp. 389–398. Springer, Heidelberg (2014)
9. Futahi, A., Paramonov, A., Koucheryavy, A.: Wireless sensor networks with temporary cluster head nodes. In: 18th International Conference on Advanced Communication Technology (ICACT), Phoenix Park, Korea, pp. 283–288. IEEE (2016)
10. Al-Qadami, N., Koucheryavy, A.: Coverage and connectivity and density criteria in 2D and 3D Wireless sensor networks. In: Balandin, S., Andreev, S., Koucheryavy, Y. (eds.) NEW2AN/ruSMART 2015. LNCS, vol. 9247, pp. 319–328. Springer, Heidelberg (2015)
11. Bollobas, B.: Random graphs. In: Bollobas, B., Fulton, W., Katok, A., Kirwan, F., Sarnak, P. (eds.) Cambridge University Press (2001)
12. Erdos, P., Renyi, A.: On random graphs I. Publ. Math. Debrecen. **6**, 290–297 (1959)
13. Kirichek, R., Paramonov, A., Koucheryavy, A.: Swarm of public unmanned aerial vehicles as a queuing network. In: Vishnevsky, V., et al. (eds.) DCCN 2015. CCIS, vol. 601, pp. 111–120. Springer, Heidelberg (2016). doi:10.1007/978-3-319-30843-2_12
14. Dao, N., Koucheryavy, A., Paramonov, A.: Analysis of routes in the network based on a swarm of UAVs. In: Kim, J.K., Joukov, N. (eds.) Information Science and Applications (ICISA) 2016. Lecture Notes in Electrical Engineering, vol. 376, pp. 1261–1271. Springer, Heidelberg (2016)
15. Kirichek, R., Paramonov, A., Koucheryavy, A.: Flying ubiquitous sensor networks as a quening system. In: Proceedings of the International Conference on Advanced Communication Technology, ICACT 2015, 01–03 July 2015, Phoenix Park, Korea (2015)

16. Erdos, P., Renyi, A.: On the evolution of random graphs. In: Publication of the Mathematical Institute of the Hungarian Academy of Sciences (1960)
17. Newman, M.E.J.: Random graphs as models of networks. Cornell University Library. http://arxiv.org/abs/cond-mat/0202208. Accessed 12 Feb 2002
18. Kawahigashi, H., Terashima, Y., Miyauchi, N., Nakakawaji, T.: Modeling ad hoc sensor networks using random graph theory. In: Second IEEE Consumer Communications and Networking Conference, CCNC 2005 (2005)
19. Dong, J., Chen, Q., Niu, Z.: Random graph theory based connectivity analysis in wireless sensor networks with Rayleigh fading channels. In: 2007 Asia-Pacific Conference on Communications (2007)
20. Ding, L., Guan, Z.H.: Modeling wireless sensor networks using random graph theory. Physica A **387**, 3008–3016 (2008). www.elsevier.com/locate/physa, 2007
21. Ambekar, C., Agrawal, A., Bhanushali, R., Deshpande, A.: Energy efficient modeling of wireless sensor networks using random graph theory. In: International Conference on Issues and Challenges in Intelligent Computing Techniques (ICICT) (2014)
22. Vu, T.M., Safavi-Naini, R., Williamson, C.: On applicability of random graphs for modeling random key predistribution for wireless sensor networks. In: Dolev, S., Cobb, J., Fischer, M., Yung, M. (eds.) SSS 2010. LNCS, vol. 6366, pp. 159–175. Springer, Heidelberg (2010)
23. Haenggi, M., Andrews, J.G., Baccelli, F., Dousse, O.: Stochastic geometry and random graphs for the analysis and design of wireless networks. IEEE J. Sel. Areas Commun. **27**(7), 1029–1046 (2009)
24. Ling, Q., Tian, Z.: Minimum node degree and k-connectivity of a wireless multihop network in bounded area. In: IEEE Global Telecommunications Conference, IEEE GLOBECOM 2007 (2007)

NEW2AN: Security Issues

Survey: Intrusion Detection Systems in Encrypted Traffic

Tiina Kovanen[(✉)], Gil David, and Timo Hämäläinen

University of Jyväskylä, Jyväskylä, Finland
{tiina.r.j.kovanen, gil.david, timo.t.hamalainen}@jyu.fi

Abstract. Intrusion detection system, IDS, traditionally inspects the payload information of packets. This approach is not valid in encrypted traffic as the payload information is not available. There are two approaches, with different detection capabilities, to overcome the challenges of encryption: traffic decryption or traffic analysis. This paper presents a comprehensive survey of the research related to the IDSs in encrypted traffic. The focus is on traffic analysis, which does not need traffic decryption. One of the major limitations of the surveyed researches is that most of them are concentrating in detecting the same limited type of attacks, such as brute force or scanning attacks. Both the security enhancements to be derived from using the IDS and the security challenges introduced by the encrypted traffic are discussed. By categorizing the existing work, a set of conclusions and proposals for future research directions are presented.

Keywords: Intrusion detection system · Encrypted traffic · Traffic analysis

1 Basics of Intrusion Detection Systems

Intrusion detection system, IDS, is used to examine network traffic and to detect malicious activities. One classification basis of IDSs is the location of the sensor. It can be local on the protected machine or reside at some point in the network protecting a larger group of machines. The local IDS is known as Host-based Intrusion Detection System, HIDS, and the one residing in the network is Network Intrusion Detection System, NIDS. As for encrypted traffic IDS, it is easier for a HIDS to have the encryption keys and thus be able to analyze decrypted data. On the other hand, HIDS is unable to detect attacks that spread in the network as the HIDS is limited to the local view. Another security tool to be noted is application firewall which has similar detection capabilities as an IDS. Encryption is a challenge that is often dealt with by sharing decryption keys.

Another way to classify IDS is based on the attack detection methodology. Usually IDSs are separated into two different categories, signature based or anomaly detection based (see e.g. [1]). Signature based detection is done by inspecting the payload and comparing the findings to known attacks. This method is generic since the same signatures should apply to any known network. Signature based method is accurate for known attacks, but is unable to handle new threats. Anomaly detection based method compares the traffic to known normal traffic patterns and if big enough deviation is

© Springer International Publishing AG 2016
O. Galinina et al. (Eds.): NEW2AN/ruSMART 2016, LNCS 9870, pp. 281–293, 2016.
DOI: 10.1007/978-3-319-46301-8_23

detected, the traffic is classified as malicious. This method is less generic since it adapts the normal traffic model to the learnt network, thus providing an adaptive model that captures the actual behavior. Anomaly detection based methods are less accurate than signature based methods but they are able to detect new types of attacks also called as zero-day attacks.

Koch [1] stated that the rise of encrypted traffic is one of the challenges that future IDS have to be able to cope with. As encrypted traffic comes more and more common, the payload inspection becomes less valid approach. Nonetheless malicious activities are still in the networks and need to be detected. There are three different approaches in handling encrypted traffic inspection and all of them have their unique disadvantages [1, 2]. The first one is protocol based detection, but it only detects misuse of encryption protocol itself. Attacks done using an encrypted channel remain unnoticed. The second option modifies network infrastructure and needs decryption. Last option is based on traffic analysis, which extracts information from the traffic flow. In this survey the protocol based detection is included in traffic analysis category if it does not require decryption. This leaves us two categories that are separated by the need for decryption.

As most IDS detection methods rely on inspecting the payload of the messages, traffic decryption becomes a natural solution for encrypted data. This reduces the problems of encrypted traffic IDS back to the traditional challenges of any IDS. Accessing decrypted data can be done by several different methods varying from shared encryption keys to reverse engineering applications. However, requiring decryption has several drawbacks. The first one is obvious, decryption is not possible nor allowed in all network environments. The second one is ethical and has to do with user privacy. The third one is the concern for security. As end-to-end encryption is broken, this gives malicious adversaries another attack point. The third concern is addressed in some publications but none of them can render the first concern obsolete. Thus we have to take a look at what traffic analysis methods can provide.

Traffic analysis methods have the advantage that they do not require decryption. This allows their implementation in all the same places as traditional IDSs. Traffic analysis methods aim to define characteristics of normal network traffic and often focus on the network flow. Different aspects are taken into consideration such as timing and size of the packets. After the features are extracted different anomaly detection and machine learning methods are applied. The major drawback of this approach is a lower detection accuracy than a traditional IDS. This is due to the fact that payload data is encrypted, hence less information is available for the analysis.

Dyer et al. [3] report that encryption protocols such as TLS, SSH and IPsec were all susceptible to traffic analysis attacks. This means that information can be extracted from different kinds of encrypted traffic.

2   Attacks

Encryption of traffic limits heavily the availability of different features in data. Therefore, it is tempting to focus on attacks that presumably are visible through increased traffic. These attacks include scanning, brute force and dictionary attacks and DoS/DDoS. It is shown in many articles that such attacks are visible from network flows [4]. However, detection of other types of attacks in encrypted traffic using methods that do not require decryption is much less studied. To be visible without decryption, the attack needs to fulfill few requirements. Firstly, the detection of attacks is only possible if the traffic goes through the IDS. Some attacks can be locally executed or use alternative route to target. In these cases, IDS cannot detect the direct attack but might detect side effects of the attack. One example of this type of an attack is virus on a USB drive, where the virus activates locally on the machine. On the other hand, some more advanced IDSs are able to detect further stages of the attack such as contacting C&C server and possible exfiltration of files. Secondly, the attack has to change some of the network traffic features. Detection of zero-days attacks and targeted attacks is possible by focusing on the network traffic behavior by modelling various detectable attack features, without relying on known attack signatures. We present a set of detectable attack features in Table 1.

Table 1. Detectable attack features.

Detectable attack features
1. Frequency of sent packets
2. Frequency of received packets
3. Ratios between sent / received packets
4. Ratios between sent / received packet sizes
5. Time between sent / received packets
6. Number of packets
7. Size of packets
8. Session duration
9. Changed request – reply sequences
10. Endpoint identity
11. Connection hour and day
12. Response time

IDS has to identify malicious activities that have varying features depending on the phase and type of the attack. According to Engen [5] the phases of the attack can be divided into four sections: surveillance, exploitation, mark and masquerade. The first phase includes scanning and probing activities to gather information of the target system. This phase includes possible password cracking by brute force or dictionary attacks. The second phase includes using suitable exploit to the system to gain access with administrator privileges. This phase may contain Denial of Service (DoS) attacks, which usually try to flood the system with requests so that it is unable to respond. The third phase includes the malicious activity of the attacker in the system. This phase may consist of stealing or destroying data, planting malicious software on the system or using it for other attacks. The last phase involves hiding evidence of successful intrusion. This may be done by deleting activity log entries and removing executed malware files.

Table 2. Attack phases, corresponding attack types and examples.

Phase	Attack type	Examples
Surveillance	Scanning (Targeted)	Targeted against a specific computer or vulnerability
	Scanning (Mass)	Large scale scanning to find any target
	Password cracking	Dictionary or brute force attack
Exploitation	DoS / DDoS flooding type	Server resources are depleted by numerous requests
	DoS / DDoS vulnerability type	Server resources are depleted by targeting known vulnerability
	Remote-to-Local	SQL injection
	User-to-Root	Buffer overflow
	Zero-day attacks	Previously unknown attacks
Mark	Data exfiltration / deletion / alteration	Insider action or after a breach
	Spyware	Keylogger
	Other malware execution	Spying or causing harm
Masquerade	Log entry, malware trace etc. deletion	Possible backdoor left open

Table 2 presents the four phases of an attack according to Engen [5], with corresponding types of attacks and examples. By using a taxonomy for attacks it is easier to cover more different types of attacks. All the attacks that have detectable features might be detectable with traffic analysis. Many of the attack types can belong to several phases depending on the motivation of the attacker. For example, DoS can be seen as an exploitation phase action, where it enables other attack vectors to succeed. It can also be seen as the actual attack belonging to the mark phase. In this case the motivation is just to shut down the target system with no further intentions. Also most of the Mark phase attacks are exploits too but are here seen as the end goal of an attack and thus categorized to Mark phase.

Another aspect is the location of the IDS. This affects, for example, the endpoint identity feature's usability. If the protected machine is a web server, all the connections are accepted unless black listing is used. This means that every endpoint is regarded benign unless an attack is discovered by other means. On the other hand, if the protected machine is in limited network, the approach adds a valuable feature in the case when not all of the machines are allowed to contact each other. This also requires identification of endpoints. If NAT connections are used, simple identification based on IP addresses fails. Also spoofing IP addresses and port information is possible.

3 Literature

During recent years, the percentage of encrypted traffic is constantly increasing, and the number of articles discussing the encrypted traffic IDS domain is increasing respec-tively. In this survey, we selected research papers that discuss the detection of attacks from encrypted traffic. Snowball search is technique which searches more publications from the citations of referred publications [6]. This was used to retrieve the central citations of the found publications. These citing articles were reviewed by title. The number of found articles was limited but more insight was found from articles dis-cussing traffic analysis attacks and encrypted traffic classification. Also most of the encrypted traffic IDS papers refer to these near field topics as the basis of their own research. The approaches in traffic analysis attacks and encrypted traffic classification research are slightly different than in encrypted traffic IDS but the methods can be useful in all of these research activities.

Traffic analysis attack studies discuss different methods of extracting information from encrypted traffic based on information available without decryption [3], [7–9]. The aim is not to the detect intruders but rather to point out that encryption has limitation in hiding sensitive information. For example, revealing the identity of visited web pages in encrypted web traffic is possible without decryption [9]. Encryption protocols use packet padding to obfuscate packet size information and thus attempt to prevent traffic analysis attacks. It is shown that padding encrypted traffic is not enough to prevent traffic analysis attacks [3, 10, 11]. In encrypted traffic IDS the ideas of traffic analysis attacks are used to reveal malicious activities rather than gaining sensitive information from encrypted traffic.

Encrypted traffic classification aims to identify applications sending the encrypted traffic. The research question is not focused in finding malicious activities but aims to produce viable information for quality of service decisions. For example, VoIP calls or SSH connections are identifiable without decryption [12–21]. Deep packet inspection and port numbers are not needed to identify various application and protocols. This enables the creation of normal traffic pattern from which the deviations can be detected. However, the results are often too coarse to meet the expectations of traffic managers [19]. The publication on encrypted traffic classification enforce the view that encryption does not hide the typical features of an application. The more information it is possible to gather from encrypted traffic without decryption, the more possible it becomes to detect attacks from the features they have.

3.1 Encrypted Traffic IDS

Some of the articles discussing encrypted traffic IDS use an approach that requires decrypting the traffic before IDS analysis is done. Decrypted traffic can be obtained from target's protocol stack [22] or by reverse engineering applications [23]. Cen-tral IDS, CIDS, approach was presented by Goh et al. [24–26]. Their solution mirrored the traffic to a CIDS, which was able to decrypt the traffic and perform deep packet inspection to the decrypted traffic. They used Shamir's secret sharing scheme and VPN. The compromised host problem is addressed. However, this solution required decryption and therefore is only suitable to limited range of network configurations. The encrypted traffic IDS solutions, which require decrypting can use the same detection approaches as traditional IDS. Therefore, the accuracy of their solutions is not the main interest here as the accuracy of the IDS is not related to encryption.

From the literature discussed so far it is shown that different types of actions and patterns can be identified from encrypted traffic without decryption. The solutions that do not require decryption are more interesting as they could use the methods used in traffic analysis attacks and encrypted traffic classification. The suitability of traffic analysis methods for detection operations in high speed networks have been addressed by Hellemons et al. [27] and Amoli et al. [28, 29].

One of the earliest publications on encrypted traffic IDS was made by Joglekar and Tate [30]. Their solution ProtoMon was based on detection of protocol misuse. Even though this approach is limited to detection of protocol violations only, it formed a basis for many of the other studies discussed in this survey.

Yamada et al. [31] proposed an approach, which only uses data size and timing information without decrypting the traffic. By comparing client's access frequency to the characteristic of normal accesses it was determined whether the access was malicious or not. They tested the method by using an actual dataset gathered at a network gateway and DARPA dataset. For DARPA dataset they added random padding for each data size to simulate the encryption. They tested three different attack classes: Scanning attacks, scripting vulnerabilities and buffer overflows. The results for the actual dataset were good with low false alarm and low false negative rates. Different types of attacks were distinguished from normal accesses. However, they were not able to detect all attacks from the DARPA dataset mainly because some of the attacks did not include a scanning phase before the intrusion. They state that this situation differs from an actual network attack scenario and therefore future work should focus on different datasets.

Foroushani, Adibnia, and Hojati [32] also used traffic analysis methods to detect intrusions without decrypting the traffic. They focused on detecting the attacks from accesses with SSH2 protocol to network public servers. The method was implemented on Snort IDS and evaluated using DARPA dataset. In scanning attacks, the requests are similar to normal requests but responses are smaller than normal. Script language attacks are similar to normal HTTP traffic when the attack is successful (small request, large response). However, when the attacker is looking for vulnerable applications, the attack evokes small responses with error messages. This pattern can be used to detect abnormal activities. Buffer overflow attacks send large requests in order to overflow the vulnerable buffer. They show that their method is able to detect intrusions with false alarm rate of about 15 % and scanning, script and buffer overflow attacks are detected with high accuracy. Numerical results for accuracy are not given. The reasoning for detecting different types of attacks is clearly stated but the results for different types of attacks are not separated in the analysis. Therefore, it is impossible to tell if the presented approach is working as intended.

Koch and Rodosek [33, 34] explored a security system for encrypted environments. Their solution was based on multiple analysis blocks, namely: Command evaluation, Strategy analysis, User identification and Policy conformity. They were able to identify limited range of commands such as 'ls–l' and login sequences from SSH-traffic. The analysis was based on the sizes of input packet series, size of answer packet series, divergences in packet sizes, server delays arising from system access and split answer packet series. The analysis is based on both the sender packets and the server answers.

Augustin and Balaz [10] proposed IDS architecture that combined encrypted application recognition to the anomaly detection based pattern identification in SSL traffic. This dual approach gives more accurate information for threat classification. One of the most cited result was that encryption does not hide size information completely. No numerical test results were given.

In 2012, Hellemons et al. [27] published a flow-based SSH intrusion detection system SSHCure. It was based on a three phase state machine, which monitored packets-per-flow and minimum number of flow records. The three phases were scanning, brute force and die-off phase. Every attack had to include either scanning or brute force phase. Each phase had different threshold values for monitored features. The method was evaluated with two datasets recorded at University of Twente's campus in 2008 and 2012. They manually inspected the dataset and found 29 (in year 2008) and 101 (in year 2012) incident in the scanning phase. Their method found correctly 28 and 100 incidents respectively and had 1 false positive in both sets. No false positives were recorded. They used to their algorithm to see how many of these attacks progressed to further phases. From the 2008 dataset, 17 attacks reached brute-force phase and 16 reached die-off phase. From the 2012 dataset, 58 attacks reached brute-force phase and 25 reached die-off phase. Correctness of these later classifications was not presented.

In 2013, Barati et al. [2] proposed a data mining solution for the encrypted traffic IDS problem and used flow-based features instead of packet features. They presented a hybrid model of Genetic Algorithm and Bayesian Network classifier for finding the best subset of features. The model was tested by trying to detect brute force attacks from SSH traffic. Their model extracted 12 most efficient features from the original 42. In classification phase, with the selected features, they received average ROC area value of 0.983 and false positive rate of 0.015. The results were promising but they state that different types of attacks need to be tested with larger dataset. In 2014, Barati, Abdullah, Udzir, Behzadi, Mahmod and Mustapha [35] published an article on SSH IDS in cloud environment. The method extracted most representative features and classified them by using the Multi-Layer Perceptron model of Artificial Neural Network. It was evaluated against brute force attacks. Their method was able to classify correctly 94 % of the instances. The ROC area value was 0.978 and False Positive Rate was 1.6 %.

Amoli and Hämäläinen published in 2013 [28] an article on detecting zero-days attacks and encrypted network attacks in high speed networks. High speed network requires too much resources that a deep packet inspection based IDS would be feasible. Amoli's and Hämäläinen's work is suitable for both normal and encrypted networks as it only uses network flows for analysis. The real-time detection model they suggest is based on two engines. The first engine is aimed to find attacks that increase network traffic (e.g. DoS and scanning). The second engine is designed to find out botnet's master in DDoS attack. Implementation and testing the model was presented later [29]. For evaluating the first engine they extracted fast network intrusions in DoS, probes and DDoS from DARPA dataset. They received 100 % Recall, 98.39 % Accuracy and False Positive Rate of 3.61 %.

In 2014, Koch et al. [36] wrote a more comprehensive article on their ideas. Their solution is based on multiple modules. They aim to detect attacks from network traffic and to identify insider threat and extrusion activities. In the case of network attacks the detection is done based on a similarity measure. This method assumes that there are more normal events than malicious. Therefore, a normal connection has high correlation to the majority of connections. Malicious connections are rare and have lower correlation to the other connection. The more similar the event is to the majority of events, the more likely it is normal. If attacker tries to influence the detection system by flooding malicious traffic, the correlations of normal connections drop. However, this behavior can be detected as well. In the case of small amounts of connections, this similarity measure is not applicable. Their proposed method uses intra-session correlation, which uses segments of one connection for correlations. Extrusion- and insider detection uses command identification and attack trees to identify possible attacks. Sequence evaluator is used to analyze if the used sequence of commands is a part of known attacks. Personal typing characteristics are used to confirm the identity of the user. Then the authorization verification module verifies if the action is allowed for this user. Once all the evaluation results are accomplished the Action Selection can form automatic firewall rules when needed. The evaluation was done on a HTTPS webshop. Users and normal traffic were simulated using Tsung benchmarking tool. Brute force login attempts and SQL injections were added to the traffic. Up to 63 parallel user connections were simulated and malicious traffic varied between 1 % and 2.7 % of the connections. Network attacks were identified with accuracy of 72.05–74.39 % with false alarm rates of 27.80–25.92 %. The article combines the detection results of both SQL injections and brute force attempts, therefore it is not possible to evaluate the detection performance for each attack by itself.

In 2015, Zolotukhin et al. [37] presented data mining based solution to detect DoS attacks. They implemented DBSCAN algorithm and compared it to other well-known algorithms: K-means, K-Nearest Neighbors, Support Vector Data Description (SVDD) and Self-Organizing Map. All but SVDD performed with accuracy over 99.99 %. SVDD achieved an accuracy of 99.94 %. The false alarm rate for the DBSCAN approach was lowest being 0.0697 %. This confirms that detection of DoS attack is relatively accurate even in encrypted traffic.

Most of the encrypted traffic IDS papers using the traffic analysis approach focus on relatively small amount of different attacks. The most frequent ones are different DoS attack scenarios and scanning attacks. Traffic analysis fares well against this type of threats but only few articles state the performance against more difficult types of attacks such as SQL injection. Another common feature is that only successful detections are reported. Articles do not evaluate detections methods across various scenarios but focus on few relatively easy use cases. In Table 3 are presented the articles discussing possible solutions for encrypted traffic IDS. The solutions are based on either protocol analysis or traffic analysis and do not require decryption.

The majority of articles presented in Table 3 do only limited testing on mass attack types such as scanning, brute force and DoS. Few list other types of attacks in their test sets but the results have limitations. First limitation is that attacks have to have some sort of mass attack component. Either scanning or brute force phase is required before subtler attacks can be detected. Another type of limitation is in presenting the results.

Table 3. Intrusion detection solutions that do not require decryption.

Article	Method	Attack types	Cons
[30]	Protocol misuse detection	Protocol misuse	Only protocol misuse detection
[31]	Data size, timing. Datasets: Darpa and live recording	Scanning, scripting language vulnerabilities, buffer overflow	Attacks must contain scanning phase
[32]	Data size, time interval. Dataset: Darpa	Scanning, script, buffer overflow	Attacks are not separated in analysis
[33]	Statistical command evaluation	Identifies commands	
[34]	Command evaluation, Strategy analysis, User identification and policy conformity. No testing.		
[36]	Similarity measurements	Brute force, SQL injection	Attacks are not separated in analysis
[10]	Application identification. No testing.	DoS	
[27]	3 state machine, Packets-per-flow and minimum number of flow records	SSH: Scanning, brute-force, exploit	Attacks must contain scanning or brute force phase
[28]	DBSCAN No testing.	DoS, DDoS, scanning, zero days, botmaster	
[2]	Choosing most effective features with Genetic Algorithm	Brute force	
[35]	Artificial Neural Network (Multi-Layer Perceptron)	Brute force	
[37]	DBSCAN	DDoS	
[29]	DBSCAN	DoS, DDoS, scanning, zero days, botmaster	

The attack traffic contains subtler attacks but results are only reported for the whole attack data. This leaves the possibility that the detection rate is based on the noisier attacks alone.

4 Discussion and Conclusions

Although several papers discussing encrypted traffic IDSs have been published, only few of them challenge the current detection boundaries. Even negative results would increase the valid information available. Currently it seems that attacks that distinctly change the normal traffic pattern, can be distinguished with relatively high accuracy. This is enough to detect various DoS/DDoS, brute force and scanning attacks that are based on the amount of messages.

Future research should be made systematically. Detection features and attacks in Tables 1 and 2 give a basis for creating test sets. Testing should be conducted with large dataset consisting of multiple types of attacks. Thorough consideration should be used while choosing the dataset. For example, the DARPA dataset might give too promising results if both the training and testing are done only using it because the DARPA dataset has documented disadvantages (see e.g. [38, 39]). To some extent, using the same dataset for both training and testing can cause problems, such as overfitting, in all datasets.

The reporting of results should also be made systematically and the results should include negative findings when the approach is unable to detect certain types of attacks. Using clear numerical results instead of descriptions makes comparison possible. Recommended values include at least accuracy and false alarm rate. For more detailed analysis, true positive, false positive, true negative and false negative values, Receiver Operating Characteristic curve (ROC) and area under curve (AUC) should be presented. By testing attack types that at first seem hard to detect and reporting even the negative results, it is possible to realistically evaluate the limits of detection.

We acknowledge wholeheartedly that this research is far from trivial. The environment is complex and changing from scenario to other. Still we see that the detection of subtler attacks than scanning, brute force and DoS is possible. This is based on detection results on near field research on traffic analysis attacks and encrypted traffic classification. In this article, we have analyzed encrypted traffic security challenges and presented a comprehensive review of the research work on encrypted traffic IDSs. Our analysis identifies that regardless of the way encrypted traffic is analyzed, there is yet more to be done; more untapped potential and more unresolved challenges.

References

1. Koch, R.: Towards next-generation intrusion detection. In: 2011 3rd International Conference on Cyber Conflict (ICCC), pp. 1–18 (2011)
2. Barati, M., Abdullah, A., Mahmod, R., Mustapha, N., Udzir, N.I.: Feature selection for IDS in encrypted traffic using genetic algorithm. In: Proceedings of the 4th International Conference on Computing and Informatics, (ICCI 2013), pp. 279–285 (2013)

3. Dyer, K.P., Coull, S.E., Ristenpart, T., Shrimpton, T.: Peek-a-Boo, i still see you: why efficient traffic analysis countermeasures fail. In: 2012 IEEE Symposium on Security and Privacy (SP), pp. 332–346 (2012)
4. Sperotto, A., Schaffrath, G., Sadre, R., Morariu, C., Pras, A., Stiller, B.: An overview of IP flow-based intrusion detection. IEEE Commun. Surv. Tutor. **12**(3), 343–356 (2010)
5. Engen, V.: Machine learning for network based intrusion detection. Bournemouth University (2010)
6. Paradis, J.G., Zimmerman, M.L.: The MIT Guide to Science and Engineering Communication. MIT Press, Cambridge (2002)
7. Liberatore, M., Levine, B.N.: Inferring the source of encrypted HTTP connections. In: Proceedings of the 13th ACM Conference on Computer and Communications Security, pp. 255–263 (2006)
8. Hintz, A.: Fingerprinting websites using traffic analysis. In: Dingledine, R., Syverson, P.F. (eds.) PET 2002. LNCS, vol. 2482, pp. 171–178. Springer, Heidelberg (2003)
9. Bissias, G.D., Liberatore, M., Jensen, D., Levine, B.N.: Privacy vulnerabilities in encrypted HTTP streams. In: Danezis, G., Martin, D. (eds.) PET 2005. LNCS, vol. 3856, pp. 1–11. Springer, Heidelberg (2006)
10. Augustin, M., Balaz, A.: Intrusion detection with early recognition of encrypted application. In: 2011 15th IEEE International Conference on Intelligent Engineering Systems (INES), pp. 245–247 (2011)
11. Raymond, J.-F.: Traffic analysis: protocols, attacks, design issues, and open problems. In: Federrath, H. (ed.) Designing Privacy Enhancing Technologies. LNCS, vol. 2009, pp. 10–29. Springer, Heidelberg (2001)
12. Alshammari, R., Lichodzijewski, P.I., Heywood, M., Zincir-Heywood, A.N.: Classifying SSH encrypted traffic with minimum packet header features using genetic programming. In: Proceedings of the 11th Annual Conference Companion on Genetic and Evolutionary Computation Conference: Late Breaking Papers, New York, NY, USA, pp. 2539–2546 (2009)
13. Alshammari, R., Zincir-Heywood, A.N.: A flow based approach for SSH traffic detection. In: IEEE International Conference on Systems, Man and Cybernetics, ISIC, pp. 296–301 (2007)
14. Alshammari, R., Zincir-Heywood, A.N.: Investigating two different approaches for encrypted traffic classification. In: Sixth Annual Conference on Privacy, Security and Trust, PST 2008, pp. 156–166 (2008)
15. Alshammari, R., Zincir-Heywood, A.N.: Machine learning based encrypted traffic classification: identifying SSH and skype. In: IEEE Symposium on Computational Intelligence for Security and Defense Applications, CISDA 2009, pp. 1–8 (2009)
16. Alshammari, R., Zincir-Heywood, A.N.: Can encrypted traffic be identified without port numbers, IP addresses and payload inspection? Comput. Netw. **55**(6), 1326–1350 (2011)
17. Arndt, D.J., Zincir-Heywood, A.N.: A comparison of three machine learning techniques for encrypted network traffic analysis. In: 2011 IEEE Symposium on Computational Intelligence for Security and Defense Applications (CISDA), pp. 107–114 (2011)
18. Bacquet, C., Gumus, K., Tizer, D., Zincir-Heywood, A.N., Heywood, M.I.: A comparison of unsupervised learning techniques for encrypted traffic identification. J. Inf. Assur. Secur. **5**, 464–472 (2010)
19. Cao, Z., Cao, S., Xiong, G., Guo, L.: Progress in study of encrypted traffic classification. In: Yuan, Y., Wu, X., Lu, Y. (eds.) Trustworthy Computing and Services, pp. 78–86. Springer, Berlin Heidelberg (2012)
20. Erman, J., Mahanti, A., Arlitt, M., Cohen, I., Williamson, C.: Offline/realtime traffic classification using semi-supervised learning. Perform. Eval. **64**(9–12), 1194–1213 (2007)

21. Maiolini, G., Baiocchi, A., Rizzi, A., Di Iollo, C.: Statistical classification of services tunneled into SSH connections by a K-means based learning algorithm. In: Proceedings of the 6th International Wireless Communications and Mobile Computing Conference, New York, NY, USA, pp. 742–746 (2010)

22. Abimbola, A.A., Munoz, J.M., Buchanan, W.J.: NetHost-Sensor: investigating the capture of end-to-end encrypted intrusive data. Comput. Secur. **25**(6), 445–451 (2006)

23. Kilic, F., et al.: iDeFEND: intrusion detection framework for encrypted network data. In: Reiter, M. (ed.) CANS 2015. LNCS, vol. 9476, pp. 111–118. Springer, Heidelberg (2015). doi:10.1007/978-3-319-26823-1_8

24. Goh, V.T., Zimmermann, J., Looi, M.: Towards intrusion detection for encrypted networks. In: International Conference on Availability, Reliability and Security, ARES 2009, pp. 540–545 (2009)

25. Goh, V.T., Zimmermann, J., Looi, M.: Experimenting with an intrusion detection system for encrypted networks. Int. J. Bus. Intell. Data Min. **5**(2), 172–191 (2010)

26. Goh, V.T., Zimmermann, J., Looi, M.: Intrusion detection system for encrypted networks using secret-sharing schemes. In: International Journal of Cryptology Research, Hotel Equatorial, Melaka, Malaysia (2010)

27. Hellemons, L., Hendriks, L., Hofstede, R., Sperotto, A., Sadre, R., Pras, A.: SSHCure: a flow-based SSH intrusion detection system. In: Sadre, R., Novotný, J., Čeleda, P., Waldburger, M., Stiller, B. (eds.) AIMS 2012. LNCS, vol. 7279, pp. 86–97. Springer, Heidelberg (2012)

28. Amoli, P.V., Hämäläinen, T.: A real time unsupervised NIDS for detecting unknown and encrypted network attacks in high speed network. In: 2013 IEEE International Workshop on Measurements and Networking Proceedings (M N), pp. 149–154 (2013)

29. Amoli, P.V., Hämäläinen, T., David, G., Zolotukhin, M., Mirzamohammad, M.: Unsupervised network intrusion detection systems for zero-day fast-spreading attacks and botnets. Int. J. Digit. Content Technol. Its Appl. **10**(2), 1–13 (2016)

30. Joglekar, S.P., Tate, S.R.: ProtoMon: embedded monitors for cryptographic protocol intrusion detection and prevention. In: Proceedings of the International Conference on Information Technology: Coding and Computing, ITCC 2004, vol. 1, pp. 81–88 (2004)

31. Yamada, A., Miyake, Y., Takemori, K., Studer, A., Perrig, A.: Intrusion detection for encrypted web accesses. In: 21st International Conference on Advanced Information Networking and Applications Workshops, AINAW 2007, vol. 1, pp. 569–576 (2007)

32. Foroushani, V.A., Adibnia, F., Hojati, E.: Intrusion detection in encrypted accesses with SSH protocol to network public servers. In: International Conference on Computer and Communication Engineering, ICCCE 2008, pp. 314–318 (2008)

33. Koch, R., Rodosek, G.D.: Command evaluation in encrypted remote sessions. In: 2010 4th International Conference on Network and System Security (NSS), pp. 299–305 (2010)

34. Koch, R., Rodosek, G.D.: Security system for encrypted environments (S2E2). In: Jha, S., Sommer, R., Kreibich, C. (eds.) RAID 2010. LNCS, vol. 6307, pp. 505–507. Springer, Heidelberg (2010)

35. Barati, M., Abdullah, A., Udzir, N., Behzadi, M., Mahmod, R., Mustapha, N.: Intrusion detection system in secure shell traffic in cloud environment. J. Comput. Sci. **10**(10), 2029 (2014)

36. Koch, R., Golling, M., Rodosek, G.D.: Behavior-based intrusion detection in encrypted environments. Commun. Mag. IEEE **52**(7), 124–131 (2014)

37. Zolotukhin, M., Hämäläinen, T., Kokkonen, T., Niemelä, A., Siltanen, J.: Data mining approach for detection of DDoS attacks utilizing SSL/TLS protocol. In: Balandin, S., Andreev, S., Koucheryavy, Y. (eds.) NEW2AN/ruSMART 2015. LNCS, vol. 9247, pp. 274–285. Springer, Heidelberg (2015)

38. McHugh, J.: Testing intrusion detection systems: a critique of the 1998 and 1999 DARPA intrusion detection system evaluations as performed by lincoln laboratory. ACM Trans. Inf. Syst. Secur. 3(4), 262–294 (2000)
39. Mahoney, M.V., Chan, P.K.: An analysis of the 1999 DARPA/Lincoln laboratory evaluation data for network anomaly detection. In: Vigna, G., Kruegel, C., Jonsson, E. (eds.) RAID 2003. LNCS, vol. 2820, pp. 220–237. Springer, Heidelberg (2003)

Architecture for the Cyber Security Situational Awareness System

Tero Kokkonen[(✉)]

Institute of Information Technology, JAMK University of Applied Sciences,
Jyväskylä, Finland
tero.kokkonen@jamk.fi

Abstract. Networked software systems have a remarkable and critical role in
the modern society. There are critical software systems in every business area.
At the same time, the amount of cyber-attacks against those critical networked
software systems has increased in large measures. Because of that, the cyber
security situational awareness of the own assets plays an important role in the
business continuity. It should be known what is the current status of the cyber
security infrastructure and own assets and what it will be in the near future. For
achieving such cyber security situational awareness there is need for the Cyber
Security Situational Awareness System. This study presents the novel architecture of the Cyber Security Situational Awareness System. The study also
presents the use case of threat mitigation process for such Cyber Security Situational Awareness System.

Keywords: Cyber security · Situational awareness · Multi sensor data fusion ·
Situational awareness information sharing · Early warning

1 Introduction

Situational awareness and early warning capability is extremely important for command and control of the own assets or making decisions related to the mission or
business. Military aviation has a long history of using command and control systems
with situational awareness generated by multi sensor information that could also be
shared from the systems of other organisations. There are similar requirements for
situational awareness in the cyber domain. Sensor feed from multiple different sensors
should be fused automatically and visualised for the decision maker. Additionally, the
information of known cyber threats should be shared with other organisations.

The terms situational awareness and situation awareness are mixed in the literature
and used for describing the same phenomenon. In this paper the term situational
awareness is used because situational awareness is considered to describe the phenomenon more accurately.

As stated in [1] real time cyber security situational awareness and data exchange are
required in several strategic guidelines of different countries, for example in Finland's
Cyber Security Strategy [1, 2]. A systematic literature review [1] indicates that there are
several studies related to situational awareness in cyber domain; however, it is still
stated in [3] that there is no solution for Cyber Common Operating Picture (CCOP).

© Springer International Publishing AG 2016
O. Galinina et al. (Eds.): NEW2AN/ruSMART 2016, LNCS 9870, pp. 294–302, 2016.
DOI: 10.1007/978-3-319-46301-8_24

This paper proposes state of the art architecture for the Cyber Security Situational Awareness System including a multi sensor data fusion component and data exchange with trusted partner organisations. The paper also presents the use case process for the Cyber Security Situational Awareness System and threat mitigation. The paper consists of a comprehensive set of reference literature and research papers as the background of the study. First, the Cyber Security Situational Awareness is discussed and the Data Fusion process is described. Also, the interfaces are presented, and the requirements for Human Machine Interface and data visualisation are analysed, followed by the description of the proposed architecture and finally, the conclusion with proposed items for further work is presented.

2 Cyber Security Situational Awareness

Endsley specifies one of the most used definitions of situational awareness (or as stated in the original reference situation awareness) as in the volume of time and space gathering information and elaborating understanding of what is happening and prediction of what will happen in the near future [1, 4]. From the point of view of Cyber Security Situational Awareness System, it means that there is multi sensor information available indicating what is happening, there is the capability for analysing such information, and there is also capability for making predictions what will happen in the near future.

As stated in [5] there are three types of information needed for situational aware- ness in cyber security: information of computing and network components (own assets), threat information, and information of mission dependencies. According to [6] there are four components of situational awareness: Identity (organisation's goals, structure, decisions making processes and capabilities), Inventory (hardware and software components), Activity (past and present activity of own cyber assets), and Sharing (both inbound and outbound). Paper [7] proposes a framework that consists of real-time monitoring, anomaly detection, impact analysis, and mitigation strategies (RAIM). The U. S. Army Innovation Challenge for Cyber Situational Awareness covers analytics, data storage, and visualisation of networks, assets, open-source information, user activity, and threats [8].

It is important to notice that there is a large and increasing number of systems, devices and cyber security applications or sensors in the organisation network pro- viding data to be analysed. Analysing that increasing amount of information requires high computational power [9]. Data fusion is a recognised technique in surveillance and the security systems used for merging the scattered surveillance and status infor- mation as integrated totality. For example, paper [10] introduces data fusion for intrusion detection information.

3 Multi Sensor Data Fusion

The data fusion is defined as *"the process of combining data to refine state estimates and predictions"* [11]. The dominant data fusion model is JDL model by the US Joint Directors of Laboratories Data Fusion Sub-Group. In the JDL model the fusion process is divided into different levels. Originally, there were levels 0–4. Nowadays, there are levels 0–6 which can be described for the cyber domain as follows [11–16]:

- Level 0 (Data Assessment). Cyber security sensor feed to the system.
- Level 1 (Object Assessment). Identification of cyber entities for example services, devices, physical network connections or information flows and the properties of those entities.
- Level 2 (Situation Assessment). State of the systems in cyber domain. Combining, for example information of software versions, vulnerabilities or patches installed.
- Level 3 (Impact Assessment). Information related to an ongoing attack or threat, indicating the damage and mitigation actions or incident response required or already done.
- Level 4 (Process Refinement/Resource Management). Management of cyber sensors. Selection of used sensors, configuration of sensor settings and definition of the reliability score of each sensor.
- Level 5 (User Refinement/Knowledge Management). Human Machine Interface (HMI) providing access to control each layer of fusion. An important part of that level is effective visualisation of information to the user.
- Level 6 (Mission Management). Determination of mission objectives and policy for supporting decision making.

Giacobe presents an application of the JDL data fusion process model for cyber security utilising JDL levels 0–5 [14], and paper [15] introduces adapted national level JDL data fusion model for levels 0–5.

Paper [16] divides multi sensor data fusion algorithms under four main categories: Fusion of imperfect data, Fusion of correlated data, Fusion of inconsistent data, and Fusion of disparate data. There are several mathematical algorithms under those four categories. For example, [17] utilises Support Vector Machines (SVMs) as the fusion algorithm for network security situational awareness, and paper [18] proposes a Hierarchical Network Security Situation Assessment Model (HNSSAM) with DS data fusion for cyber security. Spatiotemporal event correlation is used for anomaly detection and for network forensics in study [19].

4 Interfaces

The proposed architecture includes several types of input information for data fusion supporting all the levels of JDL Data Fusion process. Because of that, the data fusion engine should implement several different data fusion algorithms chosen to support data fusion of such data. Following interfaces are proposed for the architecture.

4.1 Sensor Information

Input interfaces for the information from the cyber security sensor feeds such as information from anomaly based or signature based Intrusion Detection Systems (IDS), Intrusion Prevention Systems (IPS), firewalls, antivirus systems, log file analyser, authentication alarms etc.

4.2   Own Assets Status Information

Input interfaces for the information of the systems in the cyber domain. All the entities and their properties should be identified as well as their status and configuration information. Includes also the information of the sensors with their status and configuration information. Some of the systems are able to automatically inform their status and configuration information. Otherwise, the user will update the status information using HMI. If the service is under attack, the impact assessment status information is most likely input to the system by user. Additionally, the spare parts of the physical devices should be input to the system.

4.3   Analysis Information

The analysed impact assessment information about an ongoing attack or threat; caused damage, information of attacker, what are the used attack methods, what are the countermeasures, present and past mitigation activities or incident response activities, and the result of those activities. The analysis information also consists of Indicators Of Compromise (IOC) information and open source intelligence information originated, for example from social media, news or CERT-bulletins concerning systems in the use or the business area represented. Such open source intelligence information might offer early warning information about incoming threats or information needed for incident response. Paper [20] states that pure technical data is just a part of bigger situational awareness fused with intelligence information.

Certain policies or objectives that should be noticed as part of the Situational Awareness and decision-making information are input as part of the analysis information. The analysis information is input to the system both automatically and using HMI.

4.4   Sharing the Information

Information sharing is one of the most critical elements in cyber security. If there is a trusted network of other organisations and there is the capability to share information with those organisations, there is much more information available for the data fusion. With shared information there are requirements for filtering the information before sharing it according to the company policy. All the information cannot be shared because of the confidentiality of the security information. Inbound data should also be analysed and the reliability score assigned.

In the case of simultaneously ongoing data fusion and data sharing processed the origin of the information should be indicated because of the data-loops. If the information is shared (outbound) to any organisation of the information sharing community and after while the same information is shared back (inbound) from any organisation of the information sharing community, there is a data-loop. Data-loops produce problems with the data fusion algorithms. If the origin of the information is indicated and data fusion algorithm notices that inbound information originates from itself, such information should be perceived in the fusion process.

There are standards called Structured Threat Information eXpression (STIX™) [21] and Trusted Automated eXchange of Indicator Information (TAXII™) [22] for exchanging cyber threat information. The information sharing community using such standards could be formed as described in paper [23].

5 HMI and Visualisation Layer

HMI should propose access to modify and add information for all the layers of fused data as described earlier in 3 and 4. It should also visualise the information efficiently for the user to obtain the scattered information more understandable format.

The visualisation part of the Cyber Security Situational Awareness System offers the Cyber Common Operating Picture to the user. The required data is in the system, the question is how to visualise that data to the user, especially for the decision maker who might not have deep technical background and knowhow. Many tools in cyber security are only for special purpose and certain data, not for integration of several types of data and without interoperability with other tools [24]. The authors of paper [25] used attack graphs for visualisation and ArcSight was used for visualisation in [26]. The paper [27] focuses on visualisation of threat and impact assessment.

The cyber domain is complex and there is plenty of different information available. The main conclusion for the visualisation problem is that there should be different visualisation tools and techniques for different purposes and for different user roles. Visualisation tools for high level decision makers are totally different to the tools for the analyst. Using case studies, the authors of paper [28] emphasise the potential for several different visualisation tools.

A solution for visualisation problem would be the usage of common symbols. Paper [29] suggests usage of military symbols, for example defined in standard MIL-STD-2525 [30]. Such standards should be extended for cyber domain, for example a military symbol for pending identity could mean a new incident in cyber domain. The common symbols should be defined and adopted as global standard for cyber security.

6 Proposed Architecture

The proposed novel architecture for the Cyber Security Situational Awareness System includes data fusion engine according to 3, interfaces described in 4, as well as HMI and Visualisation layer described in 5. Because there is plenty of different information from different sources the information needs to be normalised. The blog diagram of the proposed architecture is presented in Fig. 1.

The ultimate goal for such systems is that described functionalities are as automatic as possible; however, there is analyst operator required for controlling the data fusion, controlling the sensors, and adding analysis information to the system. For example, cyber security sensors might produce false alarms and the data fusion might help with the false alarms by fusing the information from multiple sources; however, the analyst operator is required to analyse the sensor feed and maybe configure the sensors or indicating to the system that false alarms are occurring. Also, if there is a real incident ongoing, the analyst operator is capable of inputting the case related additional information to the system. The possible process for situational awareness and threat mitigation using the proposed architecture is presented in Fig. 2.

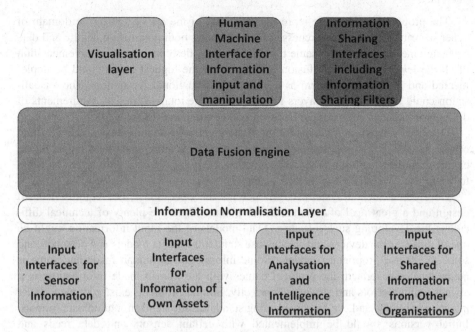

Fig. 1. Blog diagram of proposed architecture

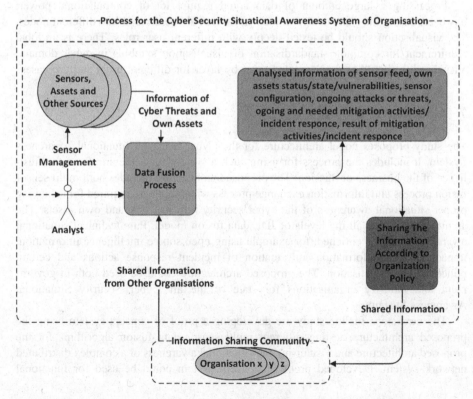

Fig. 2. Use case process for the cyber security situational awareness system

The proposed architecture represents the state of the art system in the domain of cyber security situational awareness systems utilising both data fusion engine and data exchange mechanisms at the same time. It also provides capability for implementation of all the levels of JDL data fusion process. Even the highest levels could be implemented and input to the system as a part of the Situational Awareness. The Visualisation could be deployed in layers for supporting the totally different requirements of different user roles, for example decision maker compared to analyst.

Detailed system requirements for the Cyber Security Situational Awareness System can be derived using proposed architecture. There are requirements for the data fusion engine according to 3, interfaces according to 4, HMI and Visualisation layer according to 5 and use case process presented in Fig. 2.

Developing such a system as product for the operational use requires detailed design and a great deal of software development. There are plenty of technical difficulties for developing such a system. Data models of the input information might be one of those. Some devices and sensors use standardised data models and protocols and some might use proprietary models. Some information is human made and some is automatically generated, the problem comes with the human made information, is it always without errors and structured correctly. Similar problems exist with many other integrated systems and can be solved using standardisation and structured data formats. Initial versions should be implemented with certain sensors and data feeds and extended gradually.

Processing a large amount of data could require lot of computational power; however, during the exact design of the system it could be divided into different nodes. The visualisation should be tested deeply with different user roles. There is a global requirement for common standardisation of visualisation symbols in cyber domain. Visualisation should be implemented layer by layer for different users and use cases.

7 Conclusion

The study proposes novel architecture for the Cyber Security Situational Awareness System. It includes the process for using such a system for achieving the cyber resilience of the business or mission. The proposed architecture includes both multi sensor fusion process and information exchange process which both are required for achieving proper situational awareness of the cyber security infrastructure and own assets. The architecture utilises all the levels of JDL data fusion model. Pure technical situational awareness could be enriched, for example using open source intelligence information, impact analysis information, information of incident response actions and certain polices of the organisation. The proposed architecture could be used both in government and industry organisations for state of the art Cyber Security Situational Awareness System.

The next steps for the study are developing a proof of concept system using the proposed architecture, testing different multi sensor data fusion algorithms for the proposed architecture and visualising the situational awareness of a complex distributed network system. Developed proof of concept system could be used for functional evaluation of the theoretical architecture proposed in this study. Additionally, automatic threat mitigation based on situational awareness would be an interesting domain of research and development.

Acknowledgment. This work was funded by the Regional Council of Central Finland/Council of Tampere Region and European Regional Development Fund/Leverage from the EU 2014–2020 as part of the JYVSECTEC Center project of JAMK University of Applied Sciences Institute of Information Technology.

References

1. Franke, U., Brynielsson, J.: Cyber situational awareness - a systematic review of the literature. Comput. Secur. **46**, 18–31 (2014)
2. Secretariat of the Security Committee: Finland's Cyber Security Strategy. Government Resolution, 24 January 2013
3. Conti, G., Nelson, J., Raymond, D.: Towards a cyber common operating picture. In: Proceedings of the 5th International Conference on Cyber Conflict (CyCon). NATO CCDCOE Publications, Tallinn (2013)
4. Endsley, M.: Toward a theory of situation awareness in dynamic systems. Hum. Factors: J. Hum. Factors Ergon. Soc. **37**(1), 32–64 (1995)
5. The MITRE Corporation: Cybersecurity, Situation Awareness. https://www.mitre.org/capabilities/cybersecurity/situation-awareness/. Accessed 23 May 2016
6. The Industrial Control System Information Sharing and Analysis Center (ICS-ISAC): Situational Awareness Reference Architecture (SARA). http://ics-isac.org/blog/sara/. Accessed 23 May 2016
7. Ten, C.W., Manimaran, G., Liu, C.C.: Cybersecurity for critical infrastructures: attack and defense modeling. IEEE Trans. Syst. Man Cybern. - Part A: Syst. Hum. **40**(4), 853–865 (2010)
8. Keller, J.: Army cyber situational awareness innovation challenge focuses on cyber threats at brigade level. In: Military & Aerospace Electronics, 18 November 2015. http://www.militaryaerospace.com/articles/2015/11/army-cyber-threats.html. Accessed 23 May 2016
9. Yu, W., Xu, G., Chen, Z., Moulema, P.: A cloud computing based architecture for cyber security situation awareness. In: Proceedings of the IEEE Conference on Communications and Network Security (CNS), National Harbor, MD, pp. 488–492 (2013)
10. Bass, T.: Intrusion detection systems and multisensor data fusion. Commun. ACM Mag. **43**(4), 99–105 (2000)
11. Steinberg, A., Bowman, C., White, F.: Revisions to the JDL data fusion model. In: SPIE Proceedings, Sensor Fusion: Architectures, Algorithms, and Applications III, vol. 3719, pp. 430–441 (1999)
12. Azimirad, E., Haddadnia, J.: The comprehensive review on JDL model in data fusion networks: techniques and methods. Int. J. Comput. Sci. Inf. Secur. (IJCSIS) **13**(1) (2015)
13. Blasch, E., Steinberg, A., Das, S., Llinas, J., Chong, C., Kessler, O., Waltz, E., White, F.: Revisiting the JDL model for information exploitation. In: Proceedings of the 16th International Conference on Information Fusion (FUSION), Istanbul, pp. 129–136 (2013)
14. Giacobe, N.: Application of the JDL data fusion process model for cyber security. In: SPIE Proceedings, Multisensor, Multisource Information Fusion: Architectures, Algorithms, and Applications, vol. 7710, p. 77100R, 28 April 2010

15. Swart, I., Irwin, B., Grobler, M.: MultiSensor national cyber security data fusion. In: Proceedings of the 10th International Conference on Cyber Warfare and Security (ICCWS), pp. 320–328 (2015)
16. Khaleghi, B., Khamis, A., Karray, F.O., Razavi, S.N.: Multisensor data fusion: a review of the state-of-the-art. Inf. Fusion **14**(1), 28–44 (2013)
17. Liu, X., Wang, H., Liang, Y., Lai, J.: Heterogeneous multi-sensor data fusion with multi-class support vector machines: creating network security situation awareness. In: Proceedings of the Sixth International Conference on Machine Learning and Cybernetics, Hong Kong, pp. 2689–2694 (2007)
18. Zhanga, Y., Huanga, S., Guob, S., Zhu, J.: Multi-sensor data fusion for cyber security situation awareness. In: Proceedings of the 3rd International Conference on Environmental Science and Information Application Technology (ESIAT 2011). Procedia Environ. Sci. **10**, 1029–1034 (2011)
19. Xie, Y.: A spatiotemporal event correlation approach to computer security. Doctoral Dissertation, Carnegie Mellon University, School of Computer Science, Pittsburgh, PA, USA (2005)
20. Kornmaier, A., Jaouën, F.: Beyond technical data - a more comprehensive situational awareness fed by available Intelligence Information. In: Proceedings of the 6th International Conference on Cyber Conflict (CyCon). NATO CCDCOE Publications, Tallinn (2014)
21. Barnum, S.: Structured Threat Information eXpression (STIX™). Version 1.1, Revision 1, 20 February 2014. http://stixproject.github.io/getting-started/whitepaper/. Accessed 24 May 2016
22. Connolly, J., Davidson, M., Schmidt, C.: Trusted Automated eXchange of Indicator Information (TAXII™), 2 May 2014. http://taxiiproject.github.io/getting-started/whitepaper/. Accessed 24 May 2016
23. Kokkonen, T., Hautamäki, J., Siltanen, J., Hämäläinen, T.: Model for sharing the information of cyber security situation awareness between organizations. In: Proceedings of the 23rd International Conference on Telecommunications (ICT), Thessaloniki, Greece (2016)
24. Fink, G., North, C., Endert, A., Rose, S.: Visualizing cyber security: usable workspaces. In: Proceedings of the 6th International Workshop on Visualization for Cyber Security (VizSec), Atlantic City, NJ, pp. 45–56 (2009)
25. Jajodia, S., Noel, S., Kalapa, P., Albanese, M., Williams, J.: Cauldron mission-centric cyber situational awareness with defense in depth. In: Proceedings of the Military Communications Conference (MILCOM), Baltimore, MD, pp. 1339–1344 (2011)
26. Briesemeister, L., Cheung, S., Lindqvist U., Valdes, A.: Detection, correlation, and visualization of attacks against critical infrastructure systems. In: Proceedings of the 8th Annual Conference on Privacy, Security and Trust, Ottawa, Canada (2010)
27. Nusinov, M.: Visualizing threat and impact assessment to improve situation awareness. Thesis, Rochester Institute of Technology (2009)
28. Hall, P., Heath, C., Coles-Kemp, L.: Critical visualization: a case for rethinking how we visualize risk and security. J. Cybersecur. **1**(1), 93–108 (2015)
29. Grégoire, M., Beaudoin, L.: Visualisation for network situational awareness in computer network defence. In: Visualisation and the Common Operational Picture. RTO-MP-IST-043 (2005)
30. U.S Department of Defence Interface Standard, Joint Military Symbology: MIL-STD-2525D, 10 June 2014

Detecting the Origin of DDoS Attacks in OpenStack Cloud Platform Using Data Mining Techniques

Konstantin Borisenko[1], Andrey Rukavitsyn[1], Andrei Gurtov[2,3],
and Andrey Shorov[1(✉)]

[1] Department of Computer Science and Engineering, Saint-Petersburg
Electrotechnical University "LETI", Professora Popova str. 5,
Saint-Petersburg, Russia
{borisenkoforleti,ashxz}@mail.ru, rkvtsn@gmail.com
[2] Department of Computer and Information Science, Linköping University,
581 83 Linköping, Sweden
gurtov@acm.org
[3] SCA Research Lab, ITMO University, 49 Kronverkskiy pr.,
Saint-Petersburg, Russia

Abstract. The paper presents the results of the design and implementation of detection system against DDoS attacks for OpenStack cloud computing platform. Proposed system uses data mining techniques to detect malicious traffic. Formal models of detecting components are described. To train data mining models real legitimate traffic was combined with modelled malicious one. Paper presents results of detecting the origin of DDoS attacks on cloud instances.

Keywords: Cloud security · DDoS attacks · Cloud security components · Data mining

1 Introduction

Cloud computing systems are rapidly developing. 30 years ago there was nothing comparable to a cloud. However, according to [1], 47 billion dollars have been globally spent on clouds only in 2013. And the sum is expected to be doubled by 2017, as companies invest in cloud services to create new competitive services.

The international scientific group in the cloud computing security field published 2013 a "threats report" [2]. Accordingly, cloud infrastructure attacks are placed 5th in the list of threats to clouds. Infrastructure attacks, such as "distributed denial of service" (DDoS attacks) represent a huge threat to any representative of the cloud computing service standard model (Infrastructure as a Service - IaaS, Platform as a Service - PaaS, Software as a Service - SaaS). Interestingly, in the 2010 s report [3] these attacks were not mentioned in the list of notable threats to clouds.

DDoS attacks are especially harmful to the companies, which provide cloud services for customers. It is essential for service providers to protect themselves against DDoS attacks, since successful attacks can cause to an essential economical damage [4].

© Springer International Publishing AG 2016
O. Galinina et al. (Eds.): NEW2AN/ruSMART 2016, LNCS 9870, pp. 303–315, 2016.
DOI: 10.1007/978-3-319-46301-8_25

In this paper, the authors present a novel approach to protect cloud computing systems against DDoS attacks. We distinguish external and internal DDoS attacks, depending on the location of the attacking source relative to the cloud infrastructure. Such an attack classification allows us to select correct counter measures.

For example, rigorously blocking all network traffic from an internal virtual node can affect business services. In case, the detection module detects that the attack originates from within the cloud, the countermeasure module will try to block network traffic coming from the specific ports of the suspicious node only. We also propose a security module architecture that is able to detect both kinds of attacks. The architecture does not require the installation of any sensors on the client-side and thus all processes in the cloud are kept confidential. At the same time it analyzes not only incoming external traffic, like it is done in some commercial tools [5], but also internal traffic.

We also created detection techniques that are based on data mining and machine learning methods, including self-learning models. We use supervised learning models in order to classify network traffic. Experiments showed, that the developed algorithms are fast and malicious traffic is noticed within five seconds after the attack starts. We collected and analyzed traffic to train the data mining models using the Netflow [6] protocol. In addition, the usage of self-learning algorithms makes it easier to maintain cloud security, because models are learning to adapt to new types and scenarios of DDoS attacks.

All modules of the security module are flexible and can be deployed on those nodes, where the cloud platform is installed. The module prototype was implemented in the OpenStack cloud computing platform.

2   Related Work

Nowadays, researchers are developing and implementing different defensive techniques to detect malicious traffic and protect cloud computing platforms against DDoS attacks.

Security Solutions for Clouds. Elastic Cloud Security System (ECS^2) [7] provides complex security methods against malicious traffic. ECS^2 has antibot IP reputation tables and antivirus engines. Firewalls are rule-based and working in real time. The disadvantage of using this approach is the required time to update the filter tables and antivirus signatures. Delayed table updating can cause the acceptance of malicious traffic from new, unlabeled IP addresses.

In [8], the authors propose a defense method that places detection processes in virtual machines (VMs). The authors tested their security methods on 108 services, which were launched on different VMs. Those methods are based on data mining techniques and the launched applications were analyzed on VMs. The drawback is the placement of the detection system. Some customers do not want to have background security processes inside their VMs due to specific security regulations.

A confidence-based filtering method is presented in [9]. This method monitors the transport and network layers and creates correlation characteristics of attributes in the

IP- and TCP header. Fewer attack types were tested. The authors conclude that the model has a high efficiency and low storage requirements, when operating within high-workload networks.

In [10] the authors present a neural network model for malicious traffic detection in cloud networks. Their model searches anomalies in traffic flows and creates alerts for administrators to prevent damage. The authors note, that increasing the sample period improves the results. Therefore, higher accuracy forces an increase of the reaction time and therefore minors the defenses effectiveness.

Paper [4] presents a new, flow-based anomaly detection scheme using the K-mean clustering algorithm. Training data, containing unlabeled flow records, are separated into clusters of normal and anomalous traffic. The corresponding cluster centroids are used as patterns to efficiently detect anomalies in real-time data. The authors state that applying the clustering algorithm separately for different services improves the detection quality.

The authors of [11] propose an algorithm to detect Denial of Service attacks that utilizes the SSL/TLS protocol. The algorithm is based on filtering noise data. Clustering detects malicious traffic. They trained models on the data obtained from realistic environments. The authors conclude, that the proposed model allows detecting all intrusive flows with a very small number of alarms.

In this paper, a novel approach is described. To understand the differences between existing defense solutions Table 1 is presented.

Table 1. Solutions comparison

	Hardware solutions	Redirecting traffic	Host-based solutions	Authors approach
Automatic defense	+	+	−	+
Traffic redirection	−	+	−	−
Capability to detect new attack scenarios	−	±	±	+
Keeping data in cloud confidential	+	?	?	+
Possibility to detect internal malicious traffic	−	−	+	+
Possibility to organize defense for concrete virtual machines in cloud computing platform	−	−	+	+

Table 1 presents three different most often used defense approaches. Among hardware solutions are Arbor Pravail, NSFOCUS, DefensePro. QRator and Kaspersky DDoS Protection are using redirecting traffic approach to filter malicious traffic. Snort, Suricata are host-based solutions. All of them except host-based defense solution provides automatic defense against DDoS attacks. Sometimes traffic redirection does not meet policy requirements of companies, using the cloud. That happens, because redirecting traffic can be analyzed, so data and process in the cloud can be compromised. The same issue has host-based solution, when defense process is situated inside virtual machines.

Ability to automatically reflect new attack scenarios and sources is a significant advantage of defense systems. Hardware solutions are not able to defend against new attacks. Host-based and redirecting traffic approaches are not able to reflect zero-day attacks, but companies, providing this defense, try to update system to be able to reflect new malicious patterns as fast as possible. Sometimes it can take a couple days to update defense system.

Cloud computing technology provides the ability to perform attacks inside the cloud. Hardware and redirecting traffic approaches are not able to deal with these attacks. In addition, they cannot provide defense for concrete virtual machine in cloud.

To summarize, presented in this paper approach has significant advantages comparing to other solutions. Organizing automatic defense of internal and external DDoS attacks will make it easier to cloud administrators to keep cloud resources online without compromising data and processes in cloud computing platforms.

3   Cloud Security Component Architecture

The analysis of the common security problems in cloud computing infrastructures showed that a security system should meet the following requirements: (1) be capable to serve high-workload networks; (2) mitigate attacks with high accuracy as soon as possible, preferable in real-time; (3) meet the requirements of the customer's security policies; (4) should not consume notable resources of the cloud platforms performance.

According to these requirements, the following security architecture is proposed. The key parts of the security module are the gate sensor and the security controller that consists of a collector, and analyzer and a countermeasure module. They are shown on the Fig. 1. All external and internal traffic passes through the gate. This means that every instance communicates with every other inside the cloud network through the gate.

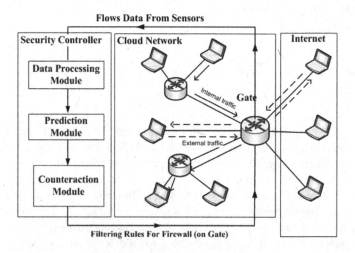

Fig. 1. Defense system architecture

The proposed architecture consists of the following modules: sensor, collector, analyzer, countermeasure, and firewall.

The sensor collects data about the traffic flow values and sends it to the collector. It stores information in the database and sends data to the data-processing module, that prepares vectors for the analyzer. This module consists of two layers. The first layer searches for malicious traffic inside the cloud network.

The second layer uses five steps to find victims as well as malicious source and their ports. The sources can be detected as in- or external. As a result the module generates data and sends it to the countermeasure module to choose the optimal way to block malicious activities.

We now consider the main models and algorithms of the proposed defense system.

The sensor probe works based on the Netflow protocol. In the OpenStack environment, the Open Virtual Switch uses a Netflow probe v5. We configured it to send data to the collector.

The collector receives raw data and stores it in the database. The collector's model will be represented as follows:

$NetflowCollector = \langle NetflowPort, MessageHandler, DataBase \rangle$, where:

$NetflowPort$ — listening port for sensor's data; $MessageHandler$ — packet handler to store them in database; $DataBase$ — connection settings to database.

The MessageHandler's algorithm:

```
While TRUE:
    Data = Receive(Netflow_packet);
    Prepair(Data, Time);
    SendToDatabase(Data);
    Each 5 seconds do:
    SendAllDataToAnalyzator();
```

The collector can be implemented in any cloud- or external node. The Netflow sensor sends data about the traffic every five seconds. The collector receives this data and sends it to the database, adding a timestamp to the received packet. This enables deep time analysis. In addition, every five seconds, the collector prepares a data vector and sends it to the first layer of the analyzer module.

$Analyzer = \langle CollectorPort, FirstLayer, SecondLayer, Database, CAPort \rangle$, where:

$CollectorPort$ — listening port for collector's data; $FirstLayer$ — first layer component; $SecondLayer$ — second layer component; $Database$ — connection settings to database; $CAPort$ — settings to countermeasure module connection.

The analyzer is divided into two layers to reduce resource consumption. In 48 h, the first layer processes about 31500 vectors. During the same time, the second layer processes over 600 million vectors. The prediction models are based on a supervised learning classification. Data vectors are composed of the following attributes: the amount of packets, the amount of bytes and the amount of unique pairs of IP addresses and ports. If the first layer predicts malicious traffic it generates an alert for the second layer. When the second layer receives the alert, it requests data from the database and groups data using the following filters: destination IP addresses, source and destination IP addresses, source IP address and the port with the destination IP address, and both IP addresses and ports. Such a composition allows the correct prediction of the traffic type. Moreover, it allows the defense systems to defend specific cloud instances instead of blocking all instances at once.

The second layer writes an information table about the ports of the malicious sources and their IP addresses, which are sorted regarding to the victims' addresses. The table is structured as follows:

$$Answer = Victim_{1-n} : \{Ip_{1-n} : \{ScType, VctAddr_{1-m}, MalSrcPort_{0-k}, SrcDstPort_{0..l}\}\}$$

The table stores information about every attacked instance *Victim*. Every *Victim* has a table with lists of IP addresses, which are suspected to generate malicious data *Ip*.

That list contains information about the predicted attack scenario type *ScType*, the victims addresses list *VctAddr*, its malicious ports list *MalSrcPort* and the source and destination ports list *SrcDstPort*. The last one is needed to measure the attack power using vital necessity ports that are required to be active according to the clients' requirements. In case of an internal attack, the countermeasure module will decide if it is able to close those ports or not.

Figure 2 shows the second layer of the algorithm. Every time when the second layer module receives an alert from the first layer, it executes all five steps. Malicious sources and victims are detected using Data Mining classification models.

The countermeasure component receives the table from the analyzer component and chooses the optimal reaction to block malicious traffic.

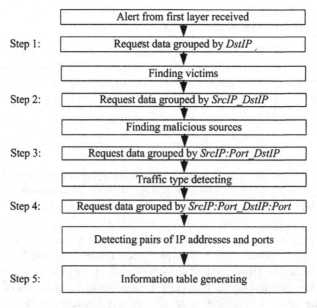

Fig. 2. Second layer working algorithm.

4   Data Mining Process and Experiment Results

Dataset Collection. To test the first and second layer we used the previously developed frameworks RSVNET [12] and IDACF [13]. They were used to model DDoS attacks on cloud instances.

Flow data was collected using the collector and afterwards used to train the data mining models. It has proven difficult to simulate legitimate traffic, since a vast amount of uncertainties takes place here. A frequently used approach to generate traffic is to use low power settings of attack scenarios. Such an approach can lead to a loss of accuracy due to disparities between legitimate traffic characteristics and scenarios. Therefore, we created an approach of combining data of real legitimate traffic with generic, malicious traffic using our frameworks.

To create legitimate datasets, the proposed modules were installed on the gateway of the Second Saint Petersburg Gymnasium. Its corporate network consists of 200 computers. A sensor was installed to mirror traffic from the WAN port to the collector component. The connection scheme is shown in Fig. 3:

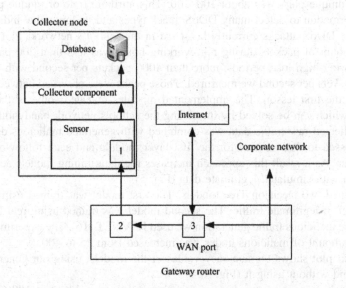

Fig. 3. Connection scheme to collect real legitimate data

The gateway router has a WAN port (3), which connects all computers to the Internet. The gateway has also several ports connected to local area routers, spread over the school. Our collector node is connected to the gateway port using port (2). All traffic goes through the WAN port and is mirrored to port (2). The sensor is set up to sniff data coming through port (2). The collector module stores the raw Netflow packets in the database. The results are presented in Fig. 4.

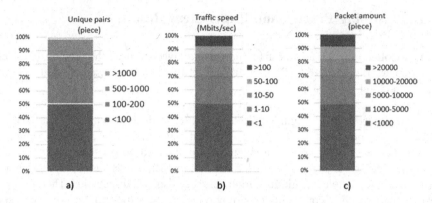

Fig. 4. Real legitimate traffic analysis

Figure 4b shows power dependencies in relation to the traffic in the WAN port. Half of the time the traffic was below 1 Mbit/s. This is due to the schools timetables peculiarities. Figure 4a shows the amount of unique pairs and ports. About half of the time, the amount of such pairs fell below 100. During a working day the average amount of unique pairs was about 100–200. This attribute (rate of unique pairs) is a crucial information to detect many DDoS attack types and its low value indicates that there are no DDoS attacks currently launched in the school's network. Plot c shows, that the amount of packets during non-working hours was less than 200 packets per second. During high load periods, more than 4000 packets per second with a peak of 100 000 packets per second we measured. Those peaks were observed between lessons and before the first lesson. The implemented modules revealed some of the schools problems, which can be solved by increasing the schools network bandwidth.

The collected raw traffic data was combined with generated malicious data. They were processed to data vectors for the first layer module and each following step of second layer. As a result this approach increases the data mining models accuracy in comparison with simulated legitimate data (Fig. 5).

We trained two Decision Tree models. The first model was trained using a simulated dataset of legitimate traffic. The second model was trained using real legitimate data. For the malicious traffic generation we used RSVNET [16]. The experiment lasted 10 h. The amount of malicious nodes was increased from 25 to 400.

The next plot shows the accuracy values of the models, using our generic-traffic-approach and without using it (Fig. 5).

For a high amount of attacking nodes, the accuracy is almost at 100 %. With a smaller amount of attackers, the accuracy of the simulated approach declines. That is due to the legitimate modeling scenario and becomes similar to the attacking scenarios. However, using our approach, the model trained on real legitimate traffic showed very high accuracy results. Which legitimates its use in the subsequent experiments.

Experiment Results. The experiments were conducted using a framework for modeling internal attacks on cloud instances (IDACF). Training and testing datasets were produced using 65 different scenarios of malicious client's behavior. Five victims were

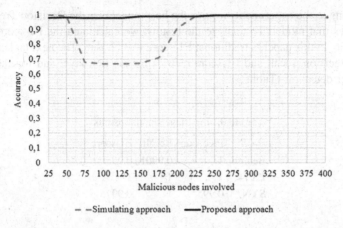

Fig. 5. Accuracy value comparison between the proposed approach and simulated legitimate traffic

attacked by 104 instances. The experiment lasted 48 h in total: 20 h were used to collect the training dataset, 28 h to collect the testing dataset. We used the restriction, that certain attack types can only be executed one at a time: SYN Flooding or HTTP Flooding. The dataset, that we collected from Second Saint Petersburg School was added to the generated datasets. All in all 6.5 GB Netflow data were collected. The dataset was divided into training and testing subsets: 46 % for training and 54 % were used for testing.

After finishing the collection phase, the data processing module used the data to create data vectors for every data-mining model of the first and the second layer. The Number of vectors for each model is shown in Table 2.

Table 2. Number of vectors for every data mining model (training dataset)

	1st layer	2nd layer			
		1 step	2 step	3 step	4 step
Number of vectors	31500	1.75 million	5.67 million	283.82 million	454.23 million

Each step in the second layer of the data processing model generates a lot more data than the previous step. Lots of resources are needed to process the amount of data. That's why, the second layer will only start to work, if the first layer sends an alert. In online testing for each 5 s the input dataset will contain around 25000 vectors. This is an average value estimated during legitimate traffic monitoring.

Previous articles were devoted to the comparison of different data mining models for first layer malicious traffic detection. We measured Naive Bayes, SVM, kNN and Decision Tree models.

According to the experiments results, the best model was the Decision Tree. Further research uses that model. To analyze the results, we calculated the F1-score for each type of traffic. In this paper, a traditional F1-score was chosen. The next tables show the results of precision, recall and F1-score metrics for the first layer and for each step of second layer detection (Table 3).

Table 3. First layer model results

	Precision	Recall	F1-score
Benign	1	0.9901	0.9950
HTTP	1	1	1
SYN	0.995	1	0.9975

The false positive rate equals 0 %. The false negative rate is 0.03 %. The use of real legitimate traffic data significantly increases the effectiveness of the predicting model. The precision of detecting benign traffic is 100 %. Therefore, all legitimate traffic was predict correctly. Little amount of SYN Flooding traffic was predicted as benign. That happened only for traffic at the beginning of the attack. The attacks impact is not enough to cause damage at this time.

In case that malicious traffic is detected by the first layer, the first step in the second layer searches for victims. The results are shown in Table 4.

In the 1 step the amount of false positive errors slightly increases. This happens due to peculiarities of the captured legitimate traffic. During breaks between lessons the network load that can be interpreted similar to DDoS attacks. All metrics have high

Table 4. Second layer model results

	Precision	Recall	F1-score
1 step results			
Benign	0.9791	0.9936	0.9863
HTTP	0.9946	0.9905	0.9926
SYN	0.9932	0.9896	0.9914
2 step results			
Benign	0.9632	0.9646	0.9639
HTTP	0.9832	0.9797	0.9814
SYN	0.9967	0.9984	0.9975
3 step results			
Benign	0.9757	0.9493	0.9623
HTTP	0.9986	0.9990	0.9988
SYN	0.9999	1	0.9999
4 step results			
Benign	0.9796	0.9528	0.966
HTTP	0.9992	0.9991	0.9991
SYN	0.9998	1	0.9999

values. After identifying the victim nodes, the second step finds the malicious sources. Vectors are grouped by source and destination IP addresses. In the 2 step the results of this step are similar to the first step. After finding malicious sources, the 2 step finds malicious ports, which generate attacks. In the 3 step the amount of false positives errors slightly increased. However, the amount of false negative errors significantly reduced. Also the percentage of correctly predicted HTTP and SYN Flooding reaches almost 100 %. So, most of malicious ports were predicted correctly. The step detects if malicious traffic impacts vital necessary victim ports. In the 4 step the amount of false positive errors slightly reduced. The accuracy of the predicted attack scenarios is almost 100 %.

In the experiments the effectiveness of five data mining models was measured. The false positive rate is less than 6 % for all of them. The smaller this value, the less amount of traffic will be interpreted as malicious and therefore blocked. To reduce the value of its metric we made a hypothesis: not considering the type of attack reduces the amount of false positives errors. The experiments were repeated with the possibility to execute both attack types at one time. The attack traffic was marked as "malicious". The same models were trained on new data. The results are shown in Table 5.

Table 5. Comparison between results of models with separate attacks types and with attacks marked malicious.

		1st layer	2nd layer			
			1 step	2 step	3 step	4 step
SYN, HTTP, Benign	FPR, %	*0*	3.78	*3.77*	5.45	4.95
	FNR, %	0.03	0.87	0.86	0.01	0.007
MALICIOUS, Benign	FPR, %	2.04	*1.11*	3.99	*4.49*	*3.98*
	FNR, %	0	*0.54*	0.802	*0.01*	*0.007*

We chose the metric to be the main measure for the effectiveness of models to be the false positive rate (FPR). The smaller it is, the better the model works. For the first layer model, the hypothesis could not be proved. For the second layers steps 1, 2 and 3, the results were better using the hypothesis. The error-rate was reduced by 25 %. The hypothesis improved the outcome in 3 out of 5 times.

5   Conclusion

In this paper, we presented an approach to detect internal and external DDoS attacks in cloud computing services using data mining techniques. We also proposed an architecture of a cloud security system to detect DDoS attacks. The architecture consists of a sensor, a controller storing information about traffic flow, an analyzer processing data for the data mining model and a countermeasure module sending commands to the firewall to hamstring attacks. The proposed architecture does not affect the customer's data and therefore is not in conflict with the requirements of the company's security policies and can be used to protect services in the cloud computing platforms from

DDoS attacks. This paper presents formal models of a defense system. The authors implemented the prototype within an OpenStack cloud computing system and carried out a set of experiments towards DDoS attacks.

In addition, the authors presented an approach to combine real legitimate and generated malicious traffic. This approach was used to increase the effectiveness of the data mining models. To collect real legitimate traffic we implemented the sensor and collector modules in the corporate network of the Second Saint Petersburg School. We merged legitimate and generated traffic into one dataset to train and test our data mining models.

The first layer's prediction component predicts legitimate traffic with 100 % accuracy. Previous researches outlined models with a significantly smaller effectiveness. The second layer's detecting component is divided into five steps. Each of the first four steps uses different data mining models. However, the false positive rate raised to 6 %. To improve the precision, all attack types were named as "malicious". In result, the new models showed a reduction of the false positive rate by 25 %.

Future works will be devoted to test the detection system online in order to measure the response time of the models and the whole system. Also we are planning to improve the second layer data mining models.

Acknowledgements. The authors want to gratefully thank the director of Second Saint Petersburg Highschool Marder Ludmila Maratovna and system administrator Shilnikov Denis Evgenievich for providing us the possibility to implement our components in schools corporate network and monitoring the traffic flows values.

The paper has been prepared within the scope of the state project "Organization of scientific research" of the main part of the state plan of the Board of Education of Russia, the project part of the state plan of the Board of Education of Russia (task 2.136.2014/K), supported by grant of RFBR #16-07-00625, supported by Russian President's fellowship, as well as with the financial support of the Foundation for Assistance to Small Innovative Enterprises in the scientific and technical spheres #10134gu2015.

References

1. Salesforce.com: What is Cloud Computing? - Salesforce UK. http://www.salesforce.com/uk/cloudcomputing/#where
2. Secucloud web-site: Secucloud. https://secucloud.com/en/company/about-us
3. Weins, K.: RightScale State of the Cloud 2013: A New Industry Survey. http://www.rightscale.com/blog/cloud-industry-insights/rightscale-state-cloud-2013-new-industry-survey
4. Munz, G., Li, S., Carle, G.: Traffic anomaly detection using k-means clustering. In: GI/ITG Workshop MMBnet (2007)
5. Docs.openstack.org: OpenStack Docs: Scenario: Legacy with Open vSwitch. http://docs.openstack.org/networking-guide/scenario_legacy_ovs.html
6. Michael Scheck. Netflow For Incident Detection/Cisco CSIRT. https://www.first.org/global/practices/Netflow.pdf
7. Oracle web-site: Oracle Exalogic Elastic Cloud: System Overview. http://www.oracle.com/us/products/middleware/exalogic/exalogic-system-overview-1724075.pdf

8. Delimitrou, C., Kozyrakis, C.: Security Implications of Data Mining in Cloud Scheduling. IEEE Comput. Arch. Lett. 1–1 (2015)
9. Dou, W., Chen, Q., Chen, J.: A confidence-based filtering method for DDoS attack defense in cloud environment. Future Gen. Comput. Syst. **29**, 1838–1850 (2013)
10. Vieira, K., Schulter, A., Westphall, C., Westphall, C.: Intrusion detection for grid and cloud computing. IT Prof. **12**, 38–43 (2010)
11. Zolotukhin, M., Hamalainen, T., Kokkonen, T., et al.: Data mining approach for detection of DDoS attacks utilizing SSL/TLS protocol. In: 15th International Conference, NEW2AN 2015, St. Petersburg, Russia, pp. 274–285 (2015)
12. Bekeneva, Y., Borisenko, K., Shorov, A., Kotenko, I.: Investigation of DDoS attacks by hybrid simulation. In: Khalil, I., et al. (eds.) ICT-EurAsia 2015 and CONFENIS 2015. LNCS, vol. 9357, pp. 179–189. Springer, Heidelberg (2015). doi:10.1007/978-3-319-24315-3_18
13. Borisenko, K., Smirnov, A., Novikova, E., Shorov, A.: DDoS attacks detection in cloud computing using data mining techniques. In: Perner, P. (ed.) ICDM 2016. LNCS (LNAI), vol. 9728, pp. 197–211. Springer, Heidelberg (2016). doi:10.1007/978-3-319-41561-1_15

Investigation of Protection Mechanisms Against DRDoS Attacks Using a Simulation Approach

Yana Bekeneva, Nikolay Shipilov, and Andrey Shorov$^{(\boxtimes)}$

Department of Computer Science and Engineering,
Saint Petersburg Electrotechnical University "LETI",
Professora Popova Str. 5, Saint Petersburg, Russia
{yana.barc,ashxz}@mail.ru, nikshipilov@gmail.com

Abstract. Nowadays, permanent availability is crucial for a growing number of computer services. An increasing quantity and power of DoS attacks frequently disrupts online network communication. Therefore it is important to create new effective defense methods for networks. In this paper we outline a programming library for the simulation of distributed reflected denial of service attacks and security mechanisms against them. Using this framework, a protection mechanism to detect and mitigate DRDoS attacks based on DNS and NTP protocols is developed. To evaluate the effectiveness of the proposed protection mechanism, a series of experiments was conducted. Also a comparison between the proposed protection mechanism and the protection mechanisms proposed by other researchers was carried out.

Keywords: DDoS reflection · DDoS amplification · DRDoS detection · DRDoS protection

1 Introduction

Lately, computer networks have been exposed to powerful DDoS attacks based on network traffic reflection and amplification – so called Distributed reflection DoS (DRDoS attacks) [1].

Those kind of attacks are executed as follows. The attacking nodes generate some requests where the source IP address is replaced by the IP address of the attacked host. These requests are sent to servers or other devices that may be used to reflect network traffic. The replies to these requests are sent to the target node. The mechanism of traffic reflection increases the complexity to identify the real source of the attack.

Therefore, common protection mechanisms against traditional DDoS attacks are unable to detect or even to identify the source of this attack. Usually traffic reflection does not provide the high power of attacks, but some vulnerabilities of protocols allow to amplify the reflected traffic, thereby significantly increasing the attack power [2].

The size of some types of replies is several times larger than the requests [3]. Thus, attackers do not need to send a large amount of traffic to successfully execute the attack. That is why, the strongest observed attacks were DRDoS attacks [1].

© Springer International Publishing AG 2016
O. Galinina et al. (Eds.): NEW2AN/ruSMART 2016, LNCS 9870, pp. 316–325, 2016.
DOI: 10.1007/978-3-319-46301-8_26

2 Related Work

Nowadays, many groups are doing research related to the investigation of DRDoS attacks and protection mechanisms against them.

After reviewing a number of publications devoted to DRDoS attacks, we concluded that today there is no clear consence how DRDoS attacks are defined. Therefore, in this article we distinguish reflected attack, amplification attack and its combination that we call DRDoS attack.

As a reflected attack, we define the use of legitimate nodes to execute DoS or DDoS attacks by sending them a large number of requests, specifying the IP address of the victim, which will function as the sender of these requests. As a result, when a node receives the requests it will send all replies to the IP address of the victim. Amplification attacks aim to overload the victim node. The attack is based on sending large numbers of special requests to obtain a very large response. That leads to the overload of the output of the DNS server and eventually to the denial of service.

Next, we consider the scientific works that are devoted to the protection against DRDoS like attacks. In [4] the authors describe existing methods of protection against DRDoS attacks and identify their strengths and weaknesses. The survey shows that most of the protection mechanisms against DRDoS attacks have been developed for the DNS protocol.

In [5] the authors propose a protection method, that is based on the fact that the victim receives more replies than requests were sent. Their protection mechanism is called DNS Amplification Attacks Detector (DAAD). The authors propose to monitor to which DNS servers the requests were sent from each node and store the information in a database. All replies from a DNS server are checked and if the incoming packet is really a reply to the request it will be accepted. If the node did not send the DNS request, the response is rejected.

The authors of [6] propose to install a preliminary DNS resolver and create a tunnel using IPSec or the SSL protocol between the preliminary resolver and the DNS resolver on the client side. All external DNS requests arrive at the preliminary DNS resolver and cannot directly enter the client's DNS server from external sources.

In [7] a mechanism called Response Rate Limiting (RRL) is outlined. It is designed to limit the number of unique responses from the DNS server. This protection mechanism is used on the DNS server side and analyzes outgoing traffic only. It completely ignores the incoming traffic. The method bases on the fact that the addresses to which the replies were sent are recorded. The number of replies from the server to each address is limited. If this limit is exceeded, no answers will be sent.

In [8] a method based on the detection of traffic deviation with respect to a template is proposed. The authors analyzed the number of incoming packets of the DNS protocol as well as their size. If the number and size of the packets exceed a predetermined value, the situation is recognized as an attack.

In this paper, we compare the performance of some of the protection mechanisms against DRDoS attacks using the developed simulation library. We also offer a protection mechanism against DRDoS attacks which has been designed using our simulation library. Experiments have shown that the developed protection mechanism is quite versatile and an effective method to protect against DRDoS like attacks.

3   Simulation Environment

To execute the experiments we used a simulation system, developed by us and described in [9]. This simulation system is based on the discrete-event simulation platform OMNeT++ as well as the INET library. We designed the DRDoS library for different DRDoS attack simulations and to test corresponding protection mechanisms. The DRDoS library contains models for the attacking nodes, models of legitimate nodes that are used as traffic reflectors and models of routers that allow to install different protection mechanisms on them. Models of legitimate traffic were based on monitoring our university's network. Our library allows to configure and simulate different DRDoS attack scenarios.

Modeling the attack as well as the reflected and amplified traffic is more compli-cated than modeling direct DDoS attacks [2]. Models of malicious traffic include models of malicious requests as well as reflected traffic models. Depending on the vulnerability, the reflected traffic can be similar to malicious requests or reinforced by varying degrees. Amplification can be determined in two ways: the ratio of response and the request size in bytes or the number of packets. These parameters are taken into account when creating models.

The developed system allows to connect real nodes to the simulated network. We have created a model of the router, which has specific parameters: the number of external connections and configuration of network interfaces. Each external connection is represented as a sub-module of the router and has a set of parameters characterizing this connection. Within the simulated system, the external connection is represented by a network router interface. Thus the interface can be connected to any simulated node. In this simulation, the system converts the transmitted packets of the models into real packages, and also performs reverse transformations.

The software tcpdump is used to debug experiments for monitoring network activity in real time. To record all network packets to the real server and real nodes during the experiments, the software tshark is used.

4   Proposed Protection Mechanism and Its Realization

In this section, a new mechanism for DRDoS attack detection is described. It is based on the fact that the protocols and their vulnerabilities, which are used for reflection are well known. Usually a packet's characteristics of legitimate and malicious traffic differ from each other. If an attacker uses a vulnerability, he uses some specific mode or commands with specific code. Usually such fields in the packet header are different for legitimate and malicious traffic.

It is important to know that malicious, reflected traffic always is a response to a request. The attack prevention mechanism is based on traffic symmetry.

Our approach is based on traffic symmetry and some differences in the packet's headers.

Since legitimate requests may have some deviations from the pattern, they can be recognized as an attack and blocked, which is not desirable. Therefore, legitimate requests coming from the same local network, must be analyzed and stored until a response is received in order to avoid blocking legitimate answers. The proposed method analyzes outgoing requests and incoming responses. For each vulnerable protocol a count is created (Count).

The algorithm of outgoing traffic analysis is shown in Fig. 1.

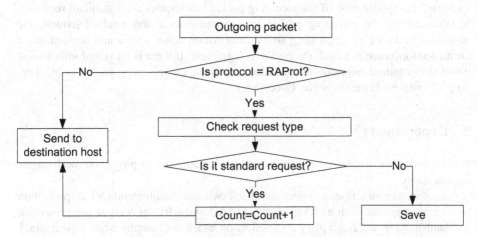

Fig. 1. Algorithm of outgoing traffic analysis for proposed mechanism

If any outgoing packet is sent by a vulnerable protocol, the mechanism checks if it is a standard request or not. If it is a non-standard request, the information about the source and destination is stored. Any outgoing packet increases the Count.

For incoming packets the analysis algorithm is shown in Fig. 2.

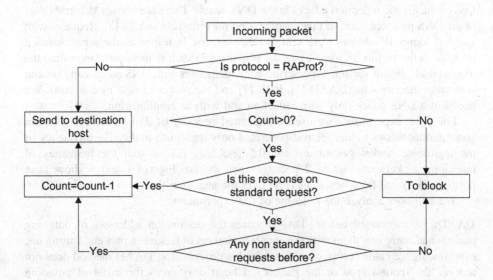

Fig. 2. Algorithm of incoming traffic analysis for proposed mechanism

If an incoming packet is sent by a vulnerable protocol, our algorithm checks if Count > 0. If Count = 0, there were no requests sent by this protocol from this local network and this packet should be blocked. This is the first step and it helps to detect the attacks in this network.

If Count > 0, there were some requests based on this protocol. The next step is to analyze the response type. If the incoming packet is a response to a standard request it is legitimate. If the incoming packet is a response to a non-standard request, the algorithm looks for corresponding stored information about source and destination. If such an information is found, the packet is legitimate. If there is no stored information about non-standard requests from this local network the incoming packet is blocked. Any allowed packet reduces the Count by 1.

5   Experiments

In this section we discuss experiments on reflection attacks and protection mechanisms against them.

At first, we describe a reflection attack without amplification. The protection mechanisms against such attacks, namely DAAD and RRL, as well as our protection mechanism were run. Exemplary for a reflection attack with amplification a NTP attack has been chosen. The option to apply DAAD, RRL and our protection mechanism against NTP attacks were evaluated. Since we used real servers for our experiments, the attack was comparably small.

5.1   Experiments with DNS Attacks and Protection Mechanisms Against Them

One of the classic reflection attacks is the DNS attack. There are known vulnerabilities of the DNS protocol, each of them can be used for reflection attacks [2]. Requests with spoofed source IP-addresses are sent to DNS servers. In response, the server sends a package to the victim. In some cases the responses to such requests are larger than the request itself. In this section we describe our experiments with DNS attacks and several protection mechanisms (DAAD [5], RRL [7] and the proposed new mechanism). We executed a DNS attack only with reflection and without amplification.

For these experiments we used a simulated network of 250 hosts, 150 of them generated malicious traffic, the rest generated only legitimate traffic. The frequency of the legitimate packet generation was 0.5 packages per second; the frequency of malicious packets was set to 2 packages per second. Figures 3 and 4 show false positives (FP) and false negatives (FN) over the attacking time.

We will now analyze the outcome of the experiments.

DAAD: As mentioned before, DAAD stores the destination addresses of outgoing packets and compares them with the source addresses of incoming packets. During our experiments we came across some flaws in this process. The DAAD method does not analyze the request type or the packet's data, it only notes the event of outgoing

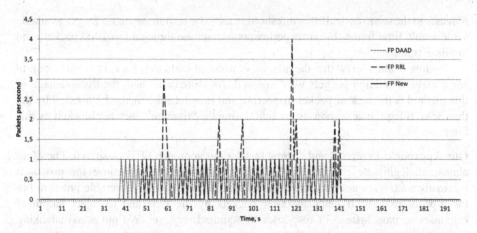

Fig. 3. Amount of FP for DAAD, RRL and proposed approach against DNS based reflected attack

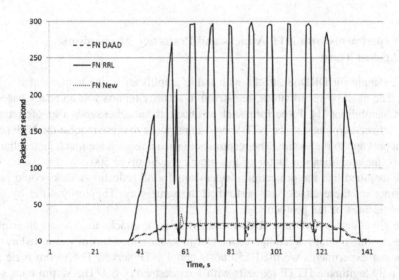

Fig. 4. Amount of FN for DAAD, RRL and proposed approach against DNS based reflected attack

requests. It also does not analyze incoming packets regarding to their the response type and allows any packet from source IP-addresses to be stored in the database.

This method leads to false positive and false negative errors and the error rate increases proportionally with the traffic intensity. If the victim node sends some requests to a server and the attackers use the same server to reflect malicious responses, those could pass through the filter and some legitimate responses could be blocked.

RRL: This method is used on the server side and analyzes only outgoing traffic. The destination addresses are stored and a threshold prevents uncontrolled outgoing requests to the same node. If this threshold is met, the responses are no longer available for a certain time frame. In our experiments, we set the threshold to 1000 packets and the time frame to 5 s.

Figures 3 and 4 show that the amount of allowed malicious packets is rather big; in some cases all malign packets were allowed. As aforementioned, the disadvantage of this method is the lack of packet inspection and therefore the method is limited by the threshold. If legitimate responses are sent during the "timeout", they are blocked by this filter.

Our Approach: Figures 3 and 4 show that for our approach FP is close to 0. Therefore almost all legitimate packets reached their destination node because the proposed mechanism analyses not only the event of sending packets by vulnerable protocol but the type of request. It helps to differ packets by the type and to reject amplified responses to pass instead of responses to standard requests. We put some attacking nodes in the same network as the victim server, the spoofed requests of such nodes were rated as legitimate and the responses to them were rated legitimate as well. Therefore it caused the FN error.

5.2   Experiments with NTP Attacks and Protection Mechanisms Against Them

As an example for DRDoS attacks with traffic amplification, we implemented a NTP attack. The most powerful attack, registered in security reports was an attack based on NTP vulnerabilities [1]. To perform such an attack, the attacker sends a spoofed request with a defined command to the NTP server. In response to these request the NTP server sends a package to the victim. The responses to such a request are much larger than the requests, the amplification factor of this attack can be up to 500.

We assumed that the protection mechanisms for all reflection attacks could be the same since all these attacks are performed the same way. They only differ by some specific features of the protocols.

We simulated DAAD and RRL methods for NTP attacks to evaluate their effectiveness towards other reflection attacks. Our protection mechanism was tested as well.

For our experiments we used 250 hosts and 1 NTP server. The victim node generated only legitimate HTTP requests with 1 request every 6 s. The victim node starts to send packets at random points in time between 0 to 170 s. Each other host sends a legitimate NTP request once at a random point of time between 0 to 170 s. 125 hosts are generating legitimate HTTP requests starting between 0 and 5 s and finishing between 185 and 190 s. The interval between the requests is randomly distributed within 10 to 20 s.

We now discuss the results of our experiments with NTP attacks.

Figures 5 and 6 show FP and FN for each protection mechanism.

DAAD: This method has proven to be also suitable to deflect NTP based reflection attacks. Since legitimate NTP traffic usually comes in a rather small amount of packets, the number of allowed incoming packets is also supposed to be small. However, there

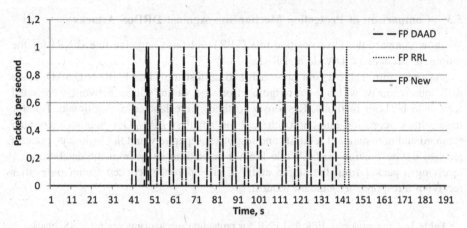

Fig. 5. Amount of FP for DAAD, RRL and proposed approach against NTP reflection attack

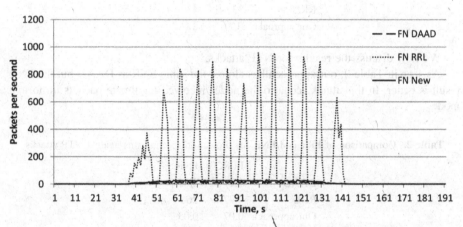

Fig. 6. Amount of FN for DAAD, RRL and proposed approach against NTP reflection attack

were some false positives and false negative errors. The reason is that the method does not pay attention to the type of request.

RRL: This protection mechanism has been implemented on the router on the NTP server side. As for DNS attacks, this method does not analyze the packets type nor content. All malicious packets were rated "legitimate" and passed the filter until the threshold was met. During the timeout also legitimate traffic has been blocked.

Our Approach: The false positive rate was also marginal. Almost all packets reached the destination node. The method counted standard requests and did not allow any malicious packets to pass through.

5.3   Comparison of Protection Mechanisms Against DRDoS Attacks

We now compare the false positive rate (FPR) and false negative rate (FNR) for the proposed mechanism, DAAD and RRL.

As seen from Table 1, our approach had almost the same FNR as DAAD, since malicious requests, which were outgoing from the victim's local network were rated legitimate by both methods. Responses to them were labeled as legitimate. But for these attack scenarios our approach had the smallest FPR. Our approach counted standard and non-standard requests and responses separately and the legitimate packets reached the destination hosts. While DAAD deleted an entry from the database after receiving a packet from the stored address, our approach reduced Count only after receiving a response to standard request.

Table 1. Comparison of FPR and FNR for protection mechanisms against DNS attacks

	FPR, %	FNR, %
DAAD	40,3	7,5
RRL	33	43
Our approach	0,7	8

We now discuss the results for NTP attacks.

As seen in Table 2, our approach has almost the same FNR as DAAD but the FNR result is better. In this attack scenario the nodes received legitimate packets in normal mode.

Table 2. Comparison of FPR and FNR for protection mechanisms against NTP attacks

	FPR, %	FNR, %
DAAD	1,1	0,02
RRL	0,77	26
Our approach	0,07	0,03

6   Conclusion

We introduced experiments devoted to the simulation of DDoS attacks based on traffic reflection and amplification as well as protection methods against them. Experiments devoted to the simulation of methods, designed for DNS attacks were executed as well as evaluation of their effectiveness towards different types of attacks (NTP). We designed a new attack detection mechanism for attacks based on traffic reflection and amplification. We assume it is possible to apply this detection mechanism to any type of attacks based on traffic reflection and amplification. Experiments devoted to evaluation of our method's effectiveness were executed, errors of all protection mechanisms, which were discussed in this paper, were compared.

In future we plan to test these protection mechanisms in other attack scenarios to evaluate the effectiveness of the proposed method. We plan as well to improve our detection algorithm to reduce the false negative rate.

Acknowledgments. The paper has been prepared within the scope of the state project "Organization of scientific research" of the main part of the state plan of the Board of Education of Russia, the project part of the state plan of the Board of Education of Russia (task 2.136.2014/K), supported by grant of RFBR # 16-07-00625, supported by Russian President's fellowship, as well as with the financial support of the Foundation for Assistance to Small Innovative Enterprises in the scientific and technical spheres.

References

1. ARBOR Networks Security Reports. http://www.arbornetworks.com/resources
2. Rossow, C.: Amplification hell: revisiting network protocols for DDoS abuse. In: Symposium on Network and Distributed System Security (NDSS) (2014). http://www.internetsociety.org/sites/default/files/01_5.pdf
3. Shankesi, R., AlTurki, M., Sasse, R., Gunter, C.A., Meseguer, J.: Model-checking DoS amplification for VoIP session Initiation. In: Backes, M., Ning, P. (eds.) ESORICS 2009. LNCS, vol. 5789, pp. 390–405. Springer, Heidelberg (2009)
4. Bekeneva, Ya., Shipilov N., Borisenko K., Shorov A.: Simulation of DDoS-attacks and protection mechanisms against them. In: Proceedings of the 2015 IEEE North West Russia Section Young Researches in Electrical and Electronic Engineering Conference, pp. 50–56 (2015)
5. Kambourakis, G., Moschos, T., Geneiatakis, D., Gritzalis, S.: Detecting DNS amplification attacks. In: Lopez, J., Hämmerli, B.M. (eds.) CRITIS 2007. LNCS, vol. 5141, pp. 185–196. Springer, Heidelberg (2008)
6. MacFarland, D.C., Shue, C.A., Kalafut, A.J.: Characterizing optimal DNS amplification attacks and effective mitigation. In: Mirkovic, J., Liu, Y. (eds.) PAM 2015. LNCS, vol. 8995, pp. 15–27. Springer, Heidelberg (2015)
7. Rozekrans, T., Mekking, M., de Koning, J.: Defending against DNS reflection amplification attacks. Technical report, University of Amsterdam (2013). https://nlnetlabs.nl/downloads/publications/report-rrl-dekoning-rozekrans.pdf
8. Huistra, D.: Detecting reflection attacks in DNS flows. In: 19th Twente Student Conference on IT (2013). http://referaat.cs.utwente.nl/conference/19/paper/7409/detecting-reflection-attacks-in-dns-flows.pdf
9. Bekeneva, Y., Borisenko, K., Shorov, A., Kotenko, I.: Investigation of DDoS attacks by hybrid simulation. In: Khalil, I., et al. (eds.) ICT-EurAsia 2015 and CONFENIS 2015. LNCS, vol. 9357, pp. 179–189. Springer, Heidelberg (2015). doi:10.1007/978-3-319-24315-3_18

Weighted Fuzzy Clustering for Online Detection of Application DDoS Attacks in Encrypted Network Traffic

Mikhail Zolotukhin[1(✉)], Tero Kokkonen[2], Timo Hämäläinen[1],
and Jarmo Siltanen[2]

[1] Department of Mathematical Information Technology, University of Jyväskylä,
Jyväskylä, Finland
{mikhail.m.zolotukhin,timo.t.hamalainen}@jyu.fi
[2] Institute of Information Technology, JAMK University of Applied Sciences,
Jyväskylä, Finland
{tero.kokkonen,jarmo.siltanen}@jamk.fi

Abstract. Distributed denial-of-service (DDoS) attacks are one of the
most serious threats to today's high-speed networks. These attacks
can quickly incapacitate a targeted business, costing victims millions
of dollars in lost revenue and productivity. In this paper, we present a
novel method which allows us to timely detect application-layer DDoS
attacks that utilize encrypted protocols by applying an anomaly-based
approach to statistics extracted from network packets. The method
involves construction of a model of normal user behavior with the help
of weighted fuzzy clustering. The construction algorithm is self-adaptive
and allows one to update the model every time when a new portion of
network traffic data is available for the analysis. The proposed technique
is tested with realistic end user network traffic generated in the RGCE
Cyber Range.

Keywords: Network security · Intrusion detection · Denial-of-service ·
Anomaly detection · Fuzzy clustering

1 Introduction

Distributed denial-of-service (DDoS) attacks have become frighteningly common
in modern high-speed networks. These attacks can force the victim to significantly downgrade its service performance or even stop delivering any service
by using lots of messages which need responses consuming the bandwidth or
other resources of the system [1]. Since it is difficult for an attacker to overload
the target's resources from a single computer, DDoS attacks are launched via a
large number of attacking hosts distributed in the Internet. Although traditional

© Springer International Publishing AG 2016
O. Galinina et al. (Eds.): NEW2AN/ruSMART 2016, LNCS 9870, pp. 326–338, 2016.
DOI: 10.1007/978-3-319-46301-8_27

network-based DDoS attacks have been well studied recently [2–4], there is an emerging issue of detection of DDoS attacks that are carried out on the application layer [5–9]. Unlike network-layer DDoS attacks, application-layer attacks can be performed by using legitimate requests from legitimately connected network machines which makes these attacks undetectable for signature-based intrusion detection systems (IDSs). Moreover, DDoS attacks may utilize protocols that encrypt the data of network connections in the application layer. In this case, it is hard to detect attacker's activity without decrypting web users network traffic and, therefore, violating their privacy.

Anomaly-based approach is probably the most promising solution for detecting and preventing application-layer DDoS attacks because this approach is based on discovering normal user behavioral patterns and allows one to detect even zero-day attacks. Attacks that involve the use of HTTP protocol is the most prevalent application-layer DDoS attack type nowadays [7] due to the protocol popularity and high number of vulnerabilities in this protocol. For these reasons, many recent studies have been devoted to the problem of anomaly-based detection of application-layer DDoS attacks that utilize HTTP protocol. For example, paper [5] analyzes application-layer DDoS attacks against a HTTP server with the help of hierarchical clustering of user sessions. Study [6] shows a novel detection technique against HTTP-GET attacks, based on Bayes factor analysis and using entropy-minimization for clustering. In [7], authors propose the next-generation application-layer DDoS defense system based on modeling network traffic to dynamic web-domains as a data stream with concept drift. In [8], authors detect application-layer DDoS attacks by constructing a random walk graph based on sequences of web pages requested by each user. In [13], we propose an algorithm for detection of trivial and intermediate application-layer DoS attacks in encrypted network traffic based on clustering conversations between a web server and its clients and analysis of how conversations initiated by one client during some short time interval are distributed among these clusters.

In this study, we improve our technique for anomaly-based detection of application-layer DDoS attacks that utilize encrypted protocols proposed in [13]. The technique relies on the extraction of normal behavioral patterns and detection of samples that significantly deviate from these patterns. For this purpose, we analyze the traffic captured in the network under inspection. It is assumed that the most part of the traffic captured during the system training is legitimate. This can be achieved by filtering the traffic with the help of a signature-based intrusion detection system. Unfortunately, the method presented in [13] requires the entire training dataset to be stored in memory, and, therefore, it cannot be applied when the amount of network traffic is really huge. The improvement proposed in this study allows one to update the normal user behavior model every time when a new portion of network traffic is available for the analysis. As a result, it requires significantly less amounts of computing and memory resources and can be successfully employed for prevention of DDoS attacks in high-speed networks.

The rest of the paper is organized as follows. Extraction of the most relevant feature vectors from network traffic is considered in Sect. 2. Fuzzy clustering algorithms are presented in Sect. 3. Section 4 describes the DDoS attack detection technique based on weighted fuzzy clustering. In Sect. 5, we evaluate the performance of the technique proposed. Finally, Sect. 6 draws the conclusions and outlines future work.

2   Feature Extraction

We concentrate on the analysis of network traffic flows in SSl/TLS traffic transferred over TCP protocol as the most popular reliable stream transport protocol. A flow is a group of IP packets with some common properties passing a monitoring point in a specified time interval. In this study, we assume that these common properties include IP address and port of the source and IP address and port of the destination. The length of each such time interval should be picked in such a way that allows one to detect attacks timely. When analyzing a traffic flow extracted in some time interval, we also take into account all packets of this flow transfered during previous time intervals. Resulting flow measurements provide us an aggregated view of traffic information and drastically reduce the amount of data to be analyzed. After that, two flows such as the source socket of one of these flows is equal to the destination socket of another flow and vice versa are found and combined together. This combination is considered as one conversation between a client and a server.

A conversation can be characterized by following four parameters: source IP address, source port, destination IP address and destination port. For each such conversation at each time interval, we extract the following information:

1. duration of the conversation
2. number of packets sent in 1 second
3. number of bytes sent in 1 second
4. maximal, minimal and average packet size
5. maximal, minimal and average size of TCP window
6. maximal, minimal and average time to live (TTL)
7. percentage of packets with different TCP flags: FIN, SYN, RST, etc.

Features of types 2–7 are extracted separately for packets sent from the client to the server and from the server to the client. It is worth to mention that here we do not take into account time intervals between subsequent packets of the same flow. Despite the fact, that increasing of these time intervals is a good sign of a DDoS attack, taking them into consideration leads to the significant increasing of the number of false alarms. It is caused by the fact, that when the server is under attack it cannot reply to legitimate clients timely as well, and, therefore, legitimate clients look like attackers from this point of view.

Values of the extracted feature vectors can have different scales. In order to standardize the feature vectors, max-min normalization is used. Extracted feature vectors are recalculated as follows:

$$x_{ij} = \frac{x_{ij} - \min_{1 \leq i \leq n} x_{ij}}{\max_{1 \leq i \leq n} x_{ij} - \min_{1 \leq i \leq n} x_{ij}}. \tag{1}$$

As a result, new value of feature x_{ij} is in range $[0, 1]$. We denote vectors that contain minimum and maximum feature values as x_{min} and x_{max} respectively.

3   Theoretical Background

In this section, we present theoretical background for the DDoS attack detection algorithm presented in the next section.

3.1   Fuzzy C-Means

Let us consider data set $X = \{x_1, x_2, \ldots, x_n\}$ where x_j is a feature vector of length d. We are aiming to divide these vectors into c clusters. Fuzzy c-means is a method of clustering which allows one vector x_j to belong to two or more clusters [10]. It is based on minimization of the following objective function:

$$J = \sum_{i=1}^{c} \sum_{j=1}^{n} u_{ij}^m \|v_i - x_j\|^2, \tag{2}$$

where $m > 1$ is a fuzzification coefficient, u_{ij} is the degree of membership of the j-th feature vector x_j to the i-th cluster and v_i is the center of this cluster. This objective function can be minimized by iteratively calculating the cluster centers as follows:

$$v_i = \frac{\sum_{j=1}^{n} u_{ij}^m x_j}{\sum_{j=1}^{n} u_{ij}^m}, \tag{3}$$

where

$$u_{ij} = \frac{1}{\sum_{k=1}^{c} \left(\frac{\|v_i - x_j\|}{\|v_k - x_j\|}\right)^{2/(m-1)}}. \tag{4}$$

The clustering algorithm can be described as follows:

1. Initialize the membership matrix $U = \{u_{ij}\}$ in such a way that $u_{ij} \in (0,1)$ for $\forall i \in \{1, \ldots, c\}$ and $\forall j \in \{1, \ldots, n\}$ and $\sum_{i=1}^{c} u_{ij} = 1$.
2. Calculate fuzzy cluster centers v_i for $i \in \{1, \ldots, c\}$ using (3).
3. Compute the objective function with (2).
4. Compute new U using (4) and go back to step 2.

The algorithm stops if at some iteration the improvement of the objective function over previous iteration is below a certain threshold, or the maximum number of iterations is reached.

3.2   Weighted Fuzzy C-Means

One of the biggest drawbacks of the fuzzy c-means algorithm is that it requires to store all feature vectors in memory before the clustering algorithm starts. For this reason, this algorithm cannot be applied in the case when the dataset to be clustered is huge. One solution of this problem is dividing the dataset into subsets of data so that the size of each subset does not exceed the amount of memory resources of the processing system. In this case, the clustering result of the first subset can be summarized as weighted centroids. These centroids can be then used as weighted points when clustering vectors of the second and subsequent subsets. This technique is often used for clustering data streams. In this case, each subset consists of data vectors that have arrived for processing at the recent time window.

Weighted fuzzy c-means [11,12] relies on the principle described above. Let us consider data set $X = \{x_1, x_2, \ldots, x_n\}$. In addition, there is a set of weighted centroids $\{v_1, \ldots, v_c\}$ defined based on previous subsets. We are aiming to update these centroids based on vectors of subset X. The objective function minimized by the weighted fuzzy c-means can be defined as follows:

$$J = \sum_{i=1}^{c} \sum_{j=1}^{c+n} u_{ij}^m w_j \|v_i - x_j\|^2. \tag{5}$$

where w_j is the weight of the j-th point. Cluster centers can be found as

$$v_i = \frac{\sum_{j=1}^{c+n} w_j u_{ij}^m x_j}{\sum_{j=1}^{c+n} w_j u_{ij}^m}. \tag{6}$$

Thus, the weighted fuzzy c-means algorithm for clustering a dataset that consists of T subsets X^t, $t \in \{1, \ldots, T\}$ can be formulated as follows:

1. Apply fuzzy c-means to vectors of the first subset X^1 to find centroids v_i and calculate weights for the resulting centroids:

$$w_i = \sum_{j=1}^{n} u_{ij} \tag{7}$$

2. Import vectors of the next subset and calculate membership matrix $U = \{u_{ij}\}$ where $i \in \{1, \ldots, c\}$ and $j \in \{1, \ldots, c+n\}$ as follows:

$$u_{ij} = \begin{cases} 1, & \text{if } i = j, \\ 0, & \text{if } i \neq j \text{ and } j \in \{1, \ldots, c\}, \\ \dfrac{1}{\sum_{k=1}^{c} \left(\frac{\|v_i - x_j\|}{\|v_k - x_j\|}\right)^{2/(m-1)}}, & \text{if } j \in \{c+1, \ldots, c+n\}. \end{cases} \tag{8}$$

3. For each vector x_j of the new subset, assign weight equal to 1 and recalculate centroid weights as follows:

$$w_i' = \begin{cases} \sum_{j=1}^{c+n} u_{ij} w_j, & \text{if } i \in \{1, \ldots, c\}, \\ 1, & \text{if } i \in \{c+1, \ldots, c+n\}. \end{cases} \tag{9}$$

where $w_j = 1$, $\forall j \in \{c+1, \ldots, n\}$.

4. Update cluster centroids v_i, $i \in \{1, \ldots, c\}$ by substituting weight values w_i in (6) with new weights w'_i.
5. Compute the objective function by substituting new weights and centroids in (5).
6. Compute new U using (4) and go back to step 3.
7. If, at some iteration, the improvement of the objective function (5) over previous iteration is below a certain threshold, or the maximum number of iterations is reached, stop modifying centroids and move to step 2.

The algorithm stops once all subsets have been imported and analyzed.

4   DDoS Attack Detection Algorithm

In this section, we formulate our DDoS attack anomaly-based detection algorithm. The algorithm relies on constructing a normal user behavior model and detecting patterns that deviate from the expected norms. We consider two versions of obtaining the normal user behavior model: offline and online. Offline training algorithm can only be applied to a small training dataset that can be stored in memory with the resulting model that cannot be modified afterwards. Online version of the training algorithm allows one to rebuild the normal user behavior model based on feature vectors arrived in the recent time interval. As a result, it requires significantly less amounts of memory since there is no need to store all feature vectors extracted.

4.1   Offline Training Algorithm

In study [13], we presented the general description of our DDoS attack detection approach. This approach relies on the training stage that can be divided into two main parts. First, all conversations extracted from the network traffic are clustered. After that, for each particular user, distributions of conversations among the resulting clusters are analyzed. In this subsection, we formulate the training algorithm for the case of fuzzy clustering in more details.

Let us consider the set of standardized feature vectors $X = \{x_1, \ldots, x_n\}$ extracted from a training set of network traffic conversations. We apply fuzzy c-means clustering algorithm to obtain cluster centroids $V^x = \{v_1^x, \ldots, v_c^x\}$ and partition matrix $U^x = \{u_{ij}^x\}$. Clustering allows us to discover hidden patterns presented in the dataset to represent a data structure in a unsupervised way. Each cluster centroid calculated represents a specific class of traffic in the network system under inspection. For example, one such class can include conversations between a web server and clients which request some web page of this server. Since the traffic may be encrypted it is not always possible to define what web page these clients request. However, since it is assumed that traffic being clustered is mostly legitimate, we can state that each cluster centroid describes a normal user behavior pattern.

To evaluate how much a feature vector is similar to the revealed normal patterns or their combination a reconstruction criterion can be considered [14]. The reconstruction of x_j is defined as follows:

$$\bar{x}_j = \frac{\sum_{i=1}^{c} (u_{ij}^x)^m v_i^x}{\sum_{i=1}^{c} (u_{ij}^x)^m}. \tag{10}$$

We can calculate the reconstruction error E of this vector as

$$E(x_j) = ||x_j - \bar{x}_j||^2. \tag{11}$$

For the detection purposes, we store the average value μ^x of reconstruction errors calculated for feature vectors contained in the training set X as well as the standard deviation of these error values σ^x.

After that, we group all conversations which are extracted in certain time interval and have the same source IP address, destination IP address and destination port together and analyze each such group separately. Such approach is in-line with other studies devoted to the problem of application-based DDoS attacks detection [5,6,8]. Those studies analyze sequences of conversations (requests) belonging to one HTTP session. In our case, since the session ID cannot be extracted from encrypted payload, we focus on conversations initiated by one client to the destination socket during some short time interval. We can interpret a group of such conversations as a rough approximation of the user session.

Let us consider these user session approximations $S = \{s_1, \ldots, s_N\}$ found in the training set. In [13], we introduced histogram vectors each element of which was equal to the percentage of feature vectors contained in each particular conversation cluster. In this study, we extend this approach for the case of fuzzy clustering by introducing session membership matrix $A = \{a_{ij}\}$ with $i \in \{1, \ldots, N\}$ and $j \in \{1, \ldots, n\}$ such that

$$a_{ij} = \begin{cases} 1, & \text{if } x_j \in s_i, \\ 0, & \text{if } x_j \notin s_i \end{cases} \tag{12}$$

We consider new feature matrix

$$Y = A(U^x)^T. \tag{13}$$

Element y_{ij} of this matrix is directly proportional to the average probability that a conversation from the i-th "session" belongs to the j-th cluster of conversations.

Once new feature vectors have been extracted, fuzzy c-means clustering algorithm is applied to obtain cluster centroids $V^y = \{v_1^y, \ldots, v_C^y\}$ and partition matrix $U^y = \{u_{ij}^y\}$. Similarly to the clusters of conversations, each new cluster centroid represents a specific class of traffic in the network under inspection. For example, one such centroid can relate to clients that use some web service in similar manner. As previously, we consider each resulting cluster centroid as a normal user behavior pattern, because it is assumed that traffic captured during the training is mostly legitimate. We calculate the reconstruction error for each of these new feature vectors and store its average value μ^y and standard deviation σ^y.

As a result, once the training has been completed, the following model of normal user behavior is obtained:

$$M_{offline} = \{x_{min}, x_{max}, V^x, \mu^x, \sigma^x, V^y, \mu^y, \sigma^y\}. \tag{14}$$

4.2 Online Training Algorithm

The biggest drawback of the algorithm described above is that it requires to store all feature vectors X and Y in memory until the moment the clustering is completed. In this subsection, we improve our technique by proposing an online training algorithm that allows to rebuild the model of normal user behavior every time when a new portion of network traffic is available for the analysis.

Let us assume that the network traffic is captured during several short time intervals and, for each time interval, there is a set of feature vectors extracted from conversations between users. We are aiming to build the model of normal user behavior under the condition that there is only room for storing one such set of vectors in memory. For this purpose, we consider feature vectors extracted during the first time interval. The offline version of the training algorithm can be applied to these vectors to obtain x_{min}, x_{max}, V^x, μ^x, σ^x, V^y, μ^y and σ^y. In addition, we calculate weights $w^x = (w_1^x, \ldots, w_c^x)$ and $w^y = (w_1^y, \ldots, w_C^y)$ of cluster centroids V^x and V^y by using partition matrices U^x and U^y respectively as shown in (7). Moreover, we calculate the following matrix H:

$$H = V^y(e^T w^x)^{-1}, \tag{15}$$

where e is vector of length c each element of which is equal to 1. The following model of normal user behavior is obtained after the traffic captured during the first time interval is analyzed:

$$M_{online} = \{x_{min}, x_{max}, V^x, w^x, \nu^x, \mu^x, \sigma^x, H, V^y, w^y, \nu^y, \mu^y, \sigma^y\}, \tag{16}$$

where ν^x and ν^y are correspondingly the total number of conversations and the total number of user sessions analyzed.

Once a new portion of network traffic has been captured, we extract necessary features from conversations to form new set $X = \{x_1, \ldots, x_n\}$. Let us denote vectors x_{min} and x_{max} obtained during the previous training iteration as x_{min}^{old} and x_{max}^{old}. New vectors x_{min} and x_{max} can be found as follows:

$$x_{min,j} = \min\{x_{min,j}^{old}, \min_{1 \le i \le n} x_{ij}\},$$
$$x_{max,j} = \max\{x_{max,j}^{old}, \max_{1 \le i \le n} x_{ij}\}. \tag{17}$$

After that, cluster centroids V^x are recalculated:

$$v_{ij}^x = \frac{(x_{max,j}^{old} - x_{min,j}^{old})v_{ij}^x + x_{min,j}^{old} - x_{min,j}}{x_{max,j} - x_{min,j}}. \tag{18}$$

Next, we apply weighted fuzzy c-means (starting from step 2) to vectors from X taking into consideration recalculated cluster centroids V^x and their weights w^x. Once new cluster centroids and their weights have been defined, we calculate reconstruction error for each vector in X as shown in (11) and find its mean value $\bar{\mu}^x$ and standard deviation $\bar{\sigma}^x$. New values of μ^x and σ^x can be found as follows:

$$\mu^x = \frac{\nu^x \bar{\mu}^x + n\mu^{x,old}}{\nu^x + n},$$

$$\sigma^x = \sqrt{\frac{\nu^x(\bar{\sigma}^x)^2 + n(\sigma^{x,old})^2 + \nu^x(\mu^{x,old} - \mu^x)^2 + n(\mu^{x,old} - \mu^x)^2}{\nu^x + n}}. \tag{19}$$

After that, new value ν^x is recalculated as

$$\nu^x = \nu^{x,old} + n. \tag{20}$$

The second part of the model is also updated. First, new cluster centroids V^y are obtained as

$$V^y = Hw^x. \tag{21}$$

After that, feature matrix $Y = \{y_1, \ldots, y_N\}$ for user session approximations is found (13). Weighted fuzzy c-means is applied to matrix Y taking into account recalculated centroids V^y and their weights w^y. As a result, new cluster centroids with updated weights are obtained. Average value μ^y and standard deviation σ^y of reconstruction errors for vectors in Y are recalculated in the same manner as it is shown in (19). The total number of user sessions is updated as

$$\nu^y = \nu^{y,old} + N. \tag{22}$$

Finally, new matrix H is calculated as shown in (15).

As a result, the entire model of normal user behavior is updated based on new vectors from X. The training is finished, once the traffic captured during each time interval has been taken into account.

4.3   Attack Detection

To detect a trivial DoS attack we extract necessary features from a new conversation and calculate the reconstruction error e^x for the resulting feature vector x according to centroids V^x. We classify this conversation as an intrusion if

$$e^x > \mu^x + \alpha^x \sigma^x, \tag{23}$$

where α^x is the parameter that is configured during tunning the detection system.

If a client initiates several connections during the recent time interval, we calculate a partition matrix for these connections and find the corresponding vector y. After that, the reconstruction error e^y is defined based on centroids V^y. This client is classified as an attacker if

$$e^y > \mu^y + \alpha^y \sigma^y. \tag{24}$$

As previously, parameter α^y is supposed to be configured during the system validation. Although, in this case, we cannot define which connections of the client are normal and which connections have bad intent, we are able to find the attacker and what web service he attempts to attack. Moreover, this technique allows one to detect more sophisticated types of DDoS attacks.

5   Numerical Simulations

The proposed technique is tested with network traffic generated in Realistic Global Cyber Environment (RGCE) [15]. A web shop server is implemented in RGCE to serve as a target of three different DDoS attacks carried out by several attackers. Communication between the server and its clients is carried out with encrypted HTTPS protocol. An IDS prototype that relies on the proposed technique is implemented in Python. The resulting program analyzes arriving raw packets, combines them to conversations, extracts necessary features from them, adds the resulting feature vectors to the model in the training mode and classifies those vectors in the detection mode. The IDS is trained with the traffic that does not contain attacks by using offline and online training algorithms proposed. Once the training has been completed, several attacks are performed to evaluate true positive rate (TPR), false positive rate (FPR) and accuracy of the algorithms.

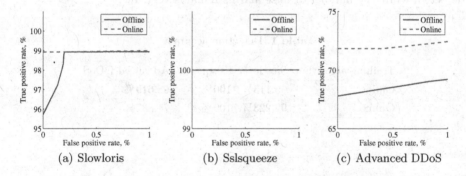

(a) Slowloris   (b) Sslsqueeze   (c) Advanced DDoS

Fig. 1. ROC curves for detection of different DDoS attacks.

First DDoS attack tested is Slowloris that is usually classified as a trivial application-layer DDoS attack. During this attack, each attacker tries to hold its connections with the server open as long as possible by periodically sending subsequent HTTP headers, adding to-but never completing-the requests. As a result, the web server keeps these connections open, filling its maximum concurrent connection pool, eventually denying additional connection attempts from clients. Figure 1(a) shows how TPR depends on FPR when detecting Slowloris for different values of parameter α^x. As one can notice, for low values of FPR both variants of the algorithm are able to detect almost all conversations related to Slowloris (TPR $\approx$ 99%).

However, if conversations related to a DDoS attack are close enough to the cluster centroids V^x of the normal behavior model, they will remain undetected. In this case, vectors of feature matrix Y should be taken into consideration. In order to test the second part of the model, Sslsqueeze attack is performed. In this case, attackers send some bogus data to the server instead of encrypted client key exchange messages. By the moment the server completes the decryption of the bogus data and understands that the data decrypted is not a valid key exchange message, the purpose of overloading the server with the cryptographically expensive decryption has been already achieved. As can be seen at Fig. 1(b) this attack is detected with 100 % accuracy.

Finally, we carry out a more advanced DDoS attack with the attackers that try to mimic the browsing behavior of a regular human user. During this attack several bots request a random sequence of web resources from the server. Since all those requests are legitimate, the corresponding conversations are classified as normal. However, the analysis of feature vectors Y calculated for user sessions allow us to detect the most part of the attacking attempts. Figure 1(c) shows how TPR depends on FPR for different values of parameter α^y when we try to detect this advanced DDoS attack. As one can see, for low values of FPR about 70 % of attacking attempts can be detected.

Table 1 shows the accuracy of detection of these three DDoS attacks for the cases when parameters α^x and α^y are selected in an optimal way, i.e. when the detection accuracy is maximal. As one can see, almost all attacks can be properly detected, while the number of false alarms remains very low.

Table 1. Detection accuracy

Training algorithm	Slowloris	Sslsqueeze	Advanced DDoS
Offline	99.113 %	100 %	68.619 %
Online	99.223 %	100 %	71.837 %

6   Conclusion

In this research, we aimed to timely detect different sorts of application-layer DDoS attacks in encrypted network traffic by applying an anomaly-detection-based approach to statistics extracted from network packets. Our method relies on the construction of the normal user behavior model by applying weighted fuzzy clustering. The proposed online training algorithm allows one to rebuild this model every time when a new portion of network traffic is available for the analysis. Moreover, it does not require a lot of computing and memory resources to be able to work even in the case of high-speed networks. We tested our technique on the data obtained with the help of RGCE Cyber Range that generated realistic traffic patterns of end users. As a result, almost all DDoS attacks were properly detected, while the number of false alarms remained very low. In the future, we are planning to improve the algorithm in terms of the detection accuracy, and test it with a bigger dataset which contains real end user traffic captured during several days. In addition, we will focus on the simulation of more advanced DDoS attacks and detection of these attacks by applying our anomaly-based approach.

Acknowledgments. This work is partially funded by the Regional Council of Central Finland/Council of Tampere Region and European Regional Development Fund/Leverage from the EU 2014–2020 as part of the JYVSECTEC Center project of JAMK University of Applied Sciences Institute of Information Technology.

References

1. Durcekova, V., Schwartz, L., Shahmehri, N.: Sophisticated denial of service attacks aimed at application layer. In: ELEKTRO, pp. 55–60 (2012)
2. Yuan, J., Mills, K.: Monitoring the macroscopic effect of DDoS flooding attacks. IEEE Trans. Dependable Secure Comput. **2**(4), 324–335 (2005)
3. Chen, R., Wei, J.-Y., Yu, H.: An improved grey self-organizing map based DOS detection. In: Proceedings of IEEE Conference on Cybernetics and Intelligent Systems, pp. 497–502 (2008)
4. Ke-xin, Y., Jian-qi, Z.: A novel DoS detection mechanism. In: Proceedings of International Conference on Mechatronic Science, Electric Engineering and Computer (MEC), pp. 296–298 (2011)
5. Ye, C., Zheng, K., She, C.: Application layer ddos detection using clustering analysis. In: Proceedings of the 2nd International Conference on Computer Science and Network Technology (ICCSNT), pp. 1038–1041 (2012)
6. Chwalinski, P., Belavkin, R.R., Cheng, X.: Detection of application layer DDoS attacks with clustering and bayes factors. In: Proceedings of IEEE International Conference on Systems, Man, and Cybernetics (SMC), pp. 156–161 (2013)
7. Stevanovic, D., Vlajic, N.: Next generation application-layer DDoS defences: applying the concepts of outlier detection in data streams with concept drift. In: Proceedings of the 13th International Conference on Machine Learning and Applications, pp. 456–462 (2014)
8. Xu, C., Zhao, G., Xie, G., Yu, S.: Detection on application layer DDoS using random walk model. In: Proceedings of IEEE International Conference on Communications (ICC), pp. 707–712 (2014)
9. Ndibwile, J., Govardhan, A., Okada, K., Kadobayashi, Y.: Web Server protection against application layer DDoS attacks using machine learning and traffic authentication. In: Proceedings of IEEE 39th Annual Computer Software and Applications Conference (COMPSAC), vol. 3, pp. 261–267 (2015)
10. Dunn, J.: A fuzzy relative of the ISODATA process, and its use in detecting compact well-separated clusters. J. Cybern. **3**(3), 32–57 (1973)
11. Hore, P., Hall, L., Goldgof, D.: A fuzzy c means variant for clustering evolving data streams. In: Proceedings of IEEE International Conference on Systems, Man and Cybernetics, pp. 360–365 (2007)
12. Wan, R., Yan, X., Su, X.: A weighted fuzzy clustering algorithm for data stream. In: Proceedings of ISECS International Colloquium on Computing, Communication, Control, and Management, vol. 1, pp. 360–364 (2008)

13. Zolotukhin, M., Hämäläinen, T., Kokkonen, T., Siltanen, J.: Increasing web service availability by detecting application-layer DDoS attacks in encrypted traffic. In: Proceedings of the 23rd International Conference on Telecommunications (ICT) (2016)
14. Izakian, H., Pedrycz, W.: Anomaly detection in time series data using a fuzzy c-means clustering. In: Proceedings of Joint IFSA World Congress and NAFIPS (IFSA/NAFIPS) Annual Meeting, pp. 1513–1518 (2013)
15. Kokkonen, T., Hämäläinen, T., Silokunnas, M., Siltanen, J., Neijonen, M.: Analysis of approaches to internet traffic generation for cyber security research and exercise. In: Proceedings of the 15th International Conference on Next Generation Wired/Wireless Networking (NEW2AN), pp. 254–267 (2015)

Dynamic Trust Management Framework
for Robotic Multi-Agent Systems

Igor Zikratov[1], Oleg Maslennikov[1], Ilya Lebedev[1], Aleksandr Ometov[2(✉)],
and Sergey Andreev[2]

[1] Saint Petersburg National Research University of Information Technologies,
Mechanics and Optics (ITMO University), St. Petersburg, Russia
[2] Tampere University of Technology, Korkeakoulunkatu 10, 33720 Tampere, Finland
`aleksandr.ometov@tut.fi`

Abstract. A lot of research attention has recently been dedicated to
multi-agent systems, such as autonomous robots that demonstrate proactive and dynamic problem-solving behavior. Over the recent decades,
there has been enormous development in various agent technologies,
which enabled efficient provisioning of useful and convenient services
across a multitude of fields. In many of these services, it is required that
information security is guaranteed reliably. Unless there are certain guarantees, such services might observe significant deployment issues. In this
paper, a novel trust management framework for multi-agent systems is
developed that focuses on access control and node reputation management. It is further analyzed by utilizing a compromised device attack,
which proves its suitability for practical utilization.

Keywords: Multi-agent systems · Information security · Access
control · Trust management

1 Introduction

Today, swarm multi-agent robotics is one of the most significant and complex
fields of research, especially in light of the fact that only under 5 % of our planet,
both land and oceanic, has been explored so far[1]. Modern robots employed for
surface research, protection, and monitoring are extremely complicated devices
equipped with a variety of sensing mechanisms [1]. Therefore, it is important to
keep them operational autonomously for as long as possible[2].

For example, wildfire fighting remains one of the most physically challenging
tasks faced by human-workers today. Autonomous machines can contribute significantly to facilitate this hard, dirty, exhausting, and dangerous job [2]. Robotic
devices can often operate faster and more efficiently while keeping people away

---

[1] See: NOAA National Ocean Service: How much of the ocean have we explored? 2014.
http://oceanservice.noaa.gov/facts/exploration.html.

[2] See: Autonomous Fire Guard (AFG) concept. 2009. http://www.yankodesign.com/
2009/08/21/firefighters-best-friend/.

© Springer International Publishing AG 2016
O. Galinina et al. (Eds.): NEW2AN/ruSMART 2016, LNCS 9870, pp. 339–348, 2016.
DOI: 10.1007/978-3-319-46301-8_28

from unsafe locations[3]. Conventionally, such devices are expected to cooperate with each other in order to reach common "targets" in remote locations [3].

Such distributed coordination has many advantages in reaching common group goals, lower operating costs, control system requirements, improve robustness, and achieve better scalability [4,5]. The related aspects have been widely recognized and well-studied over past years [6,7]. In multi-agent systems, the actual network topology connecting all the devices in operation plays a crucial role in determining consensus. Typically, the objective is to explicitly identify necessary and sufficient conditions for a particular network topology, such that a common agreement could be reached under specifically designed algorithms.

One of the most promising trends in modern robotics is the development of management tools that allow for intelligent Multi-Agent System (MAS) group control. In particular, considerable research attention has been paid to the collaborative planning frameworks, which operate in decentralized, "ad hoc" regime by forming a coalition [8]. This is to achieve better scalability, operational coverage, and network availability in cases of weak connectivity to the control unit. A significant body of research literature in this field has been dedicated to dynamic redistribution of goals between the operating nodes in case of their unpredictable breakdowns [9].

Due to the "ad hoc" nature of the considered networks, which often operate in a dynamic mesh fashion, the MAS becomes an attractive field for a wide range of attacks, such as: message capture and retransmission, violation of integrity, unauthorized data access, denial of service, etc. [10,11]. Hence, the currently utilized trust management schemes may be very limited due to discretionary distinction and mandate-based behavior [12,13]. We subdivide the possible attacks on a MAS into the following main categories [14]: (i) network-layer related attacks; (ii) attacks on identification and authentication of agents in the system [15]; and (iii) *compromised device* intrusion [16]. The main goal of this work is thus to develop a trust management framework suitable for resisting the harmful compromised device attack.

This paper is organized as follows. In Sect. 2, we review the decentralized MASs and the corresponding attacks. Further, in Sect. 3, we introduce a trust model resistant to the discussed type of attacks. Then, in Sect. 4, the compromised device intrusion attack detection in detailed. The last section describes our future work and offers some conclusions.

2   Modeling Background

In this section, we consider a MAS operating in a decentralized fashion [17,18]. We focus on a group of N robots targeting a common collaborative goal. During the initialization phase, each of the devices receives its utility function (goal) related data. The overall framework operation model is captured in Fig. 1.

---

[3] See: The National Interagency Fire Center (NIFC): Incident Management Situation Report. 2016. http://www.nifc.gov/nicc/sitreprt.pdf.

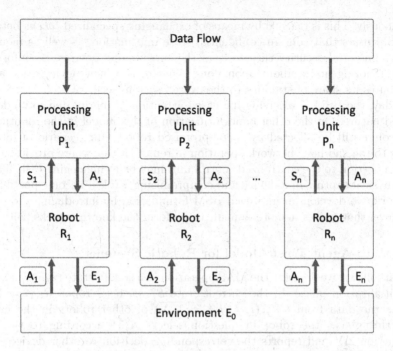

Fig. 1. Simplified decentralized MAS management framework.

2.1 Preliminary Assumptions and Definitions

Each processing unit $P_i, (i = \overline{1, N})$ of every device R_i consists of the corresponding computing unit CU_i, the data transmitting unit DT_i, the data receiving unit DR_i, the current state determination unit CS_i, and a set of sensing modules SD_i. The CU_i is communicating with other CU_j by transmitting the system state information S_i^0 and the respective operating decisions $A_i^{k+1}, (k = 0, 1, 2, \ldots, N)$. In addition, each CU_i has the knowledge of the environmental data E_i^0 and its own state S_i^0 to continuously update the utility function ΔY for any possible operating decisions in the current state. We select $max(\Delta Y)$ to be our utility function.

As an attack, we consider any malicious activity of the compromised device on the kth iteration of the system operation [19]. As a result, the following device decision A_i^{k+1} might not be selected according to the utility function. We also consider the attack with message capture and retransmission, as well as the attack on the environment estimation and the attacks targeted to affect the group decision making protocols [20].

Conventionally, a MAS utilizes the following techniques to enable secure "ad hoc" communication: a state-estimation function [21]; the lightweight cryptography solutions [22]; the time-limiting techniques [23]; and the Buddy Security Model (BSM) [24,25], among others. Interestingly, the neighboring nodes in e.g., BSM are responsible for each other's security by monitoring their environment continuously. This is reached by means of exchanging specialized *tokens* between the BSM users that contain confidential state information, as well as monitoring any potential security threats from the surrounding devices. By informing the neighboring nodes about anomalous behavior of a new device, each agent contributes its share of stability to the overall system security.

Today, the BSM is receiving increased attention primarily due to its decentralized nature. On the other hand, utilization of this model in the robotic systems could still be affected by a compromised robot. Our scenario of interest relates to the "ad hoc" network operation in remote areas, where providing reliable connection to the centralized control unit may be a challenging task. Therefore, physical capture of a device and compromising a token become possible. In this work, we develop an improved BSM formulation by introducing a device's trust level that makes it more difficult to perform the known attacks [26].

2.2   Multi-Agent Trust Model for Robotic Systems

In what follows, we describe the MAS operation in a steady-state regime i.e., past the initialization phase. In the current state S_i^0, each ith robot $R_i, (i = \overline{1, N})$ collects the data from $CU_j, (i \neq j, j = \overline{1, N})$ of other robots in the group. After this phase, the robot in question selects A_j^{k+1} according to the utility function ΔY and reports the corresponding decision to other devices with $w(A_i^{k+1})$ to $CU_j, (i \neq j, j = \overline{1, N})$. This message is based on the received information $S_1^0, S_2^0, \ldots, S_i^0 - 1, S_i^0 + 1, \ldots, S_N^0$ and the possible current decisions $A_1^{k+1}, A_2^{k+1}, \ldots, A_i^{k+1} - 1, A_i^{k+1} + 1, \ldots, A_N^{k+1}$. After receiving the message, other agents are validating the data with respect to the decision made by the ith node.

In case the above check by the jth device resulted in $\Delta Y_j, (i \neq j) \neq \Delta Y_i$, the trust level of the ith device is increased. We define the trust level as *willingness* of a particular node to report valid information to others. Alternatively, if a device has reported a faulty ΔY_i, its level of trust is decreased. As a result, we may now define a new parameter, which is a set of "steps" l required to estimate the *long-term* level of trust per device A_i^{l+1}. Therefore, when calculating ΔY_i^l at the following system iteration, the devices would rely more on nodes with higher trust levels.

In summary, the lower levels of trust would not allow the compromised robots affect the system operation in a destructive way even when sending a valid-like token. Thereby, potential malicious nodes will have to behave similar to the trusted ones for the interval of time that is necessary to build sufficient trust, which improves system robustness to the considered type of attacks.

3   Trust Model Development

In this section, we first discuss the notation related to our security mechanism formulation and then outline its implementation possibilities. Our developed trust model is illustrated in Fig. 2, where arrows represent multi-agent connections over alternative channels, such as visual, NFC, etc. Further, the wireless radio links are indicated with the dashed lines.

To ease the exposition, we introduce the notation used in this section as follows: $A = \{A_1, A_2, \ldots, A_N\}$ are the operations that may be performed by an agent; $S = \{s_1, s_2, \ldots, s_N\}$ is the set of states during the communication phase; $V = \{F, T\}$ is the set of results given by the report validation, where F corresponds to a faulty reply and T stands for a valid one; and r_l^m is the trust level determined by the mth agent for the lth device ($l \neq m$). There are different options to implement this model according to Fig. 2.

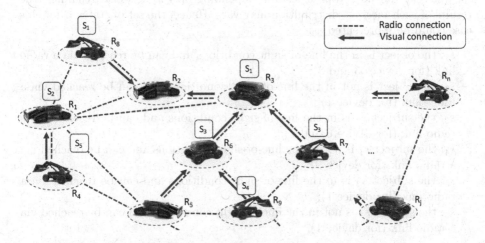

Fig. 2. Proposed trust model for interacting agents.

As an example, we assume that the system is in the state k and focus on the $2nd$ device in the network. It has reported the message $2A_i$ to the rest of the users. As shown in the figure, the $2nd$ robot has a wireless connection to the devices 1, 3, and 8, and all of these users have received the corresponding message. Further, the devices 3 and 8 have a visual proof that the $2nd$ device has indeed performed the reported action. The set of possible system states S could then be represented as follows:

- s_1: the object is in the line-of-sight conditions and can be reached via a radio link (for devices 3 and 8);
- s_2: the object is not in the line-of-sight conditions and can be reached via a radio link (for device 1).

If the $2nd$ device is acting normally, its actual operation A_2 is the same as the reported one i.e., the devices 3 and 8 should increase their levels of trust for the $2nd$ robot by $\Delta r_2^3 = \Delta r_2^8$ due to having a visual confirmation of the action. The device 1 has also received the report from the $2nd$ device, but it is only utilizing a wireless channel. In this case, the corresponding level of trust should be updated less significantly than for other nodes $\Delta r_2^1 < \Delta r_2^{3,8}$. Here, only the devices having a direct connection to the robot in question are evaluating its level of trust, but for the rest it remains unchanged.

Second, the devices having a direct connection to the robot in question may forward their knowledge to the neighbors. They would then transmit a message of type $i : A_i s_k^j v$, where i is the identifier of a reporting device, A_i is the value reported by the ith device, s_k^j is the jth device's state, and v is the validation result (either *true* or *false*). For our example, the corresponding messages should be: from the robot 1, 2 : $A_i s_2^1 T$; from the robot 3, 2 : $A_i s_1^3 T$; from the robot 8, 2 : $A_i s_1^8 T$. These messages are delivered to the agents $4, 5, 6, 7, 9$. Note that for different "ad hoc" topologies and depending on the network dynamics, the results of such message distribution may vary. Hence, the set of states S for this case could be represented as:

- s_1: the object is in the line-of-sight conditions and can be reached via a radio link (for devices 3 and 8);
- s_2: the object is not in the line-of-sight conditions and can be reached via a radio link (for device 1);
- s_3: the subject s_1 is in the line-of-sight conditions and can be reached via a radio link (for device 6 and 7);
- s_4: the subject s_1 is not in the line-of-sight conditions and can be reached via a radio link (for device 9);
- s_5: the subject s_2 is in the line-of-sight conditions and can be reached via a radio link (for device 1);
- s_6: the subject s_2 is not in the line-of-sight conditions and can be reached via a radio link (for device 1).

The subjects with different states $s_1, s_2, \ldots, s_6$ would obtain different trust levels towards the same subject based on the possibility to evaluate the device by themselves. An increment-based trust scale for the ith object may then be introduced as:

$$\Delta r_{s_1}^i > \Delta r_{s_2}^i > \cdots > \Delta r_{s_6}^i. \tag{1}$$

The main target of the indirectly connected agents is to determine their own state and select the objective level of trust based on it. Correspondingly, if the value of $v = T$, the level of trust is increased by $\Delta r_{s_i}^n$. Similarly, if $v = F$, the level of trust is decreased by the same value. Importantly, the reception of faulty data from different nodes may be caused by a variety of factors, including uncontrollable interference, severe weather conditions, and deliberate distortion of the message by one of the relaying devices. In the latter case, the collective goal may be achieved by utilizing the discussed MAS information security framework.

In summary, the second implementation option for the proposed trust model should be more efficient in practice. It allows delivering the actual trust information to a higher number of agents, but at the same time may lead to the increased amounts of signaling.

4   Detecting Attacks on a MAS

The trust level management mechanism proposed in the previous section allows controlling such threats as: capture, modification, and retransmission of messages

A_i and S_i, as well as compromising the system operation by an attacker node, which may attempt to influence the utility function ΔY. On the other hand, some MAS management protocols enable to select a *leading* device from within the group, which becomes responsible for handling system-wide decision-making functionality [27,28]. To achieve this, said device updates the utility function for the entire network.

As in any deployment with a single-node failure possibility, should an attacker seize this role, it may disrupt the operation of the entire system. Such an attack may also be executed by a set of devices within the group. In this case, the only way to detect the malicious behavior is by monitoring ΔY by all the system agents, and at each iteration of the network operation. In order to perform this monitoring, all the agents (except for the reporting one) need to recalculate the potential ΔY_{t+1} of the ith node at the lth step. In case $\Delta Y_{t+1} < \Delta Y_t$, the receiving device decreases the level of trust for the reporting robot by $\Delta r_{Y_l}^i$ and vice versa. The resulting level of trust for the ith device could be calculated as:

$$\Delta \mu^i > \alpha \Delta r_{s_i}^i + \beta \Delta r_{Y_l}^i, \tag{2}$$

where α and β are the weights corresponding to the reported agent, and the utility of its decision is selected according to the information security policy of the MAS.

5   Selected Numerical Results

In order to validate the usability of our proposed model, we conduct a set of simulation runs utilizing the V-REP robotics framework[4]. To prove the effectiveness, we compare the changes of ΔY_i^l throughout the framework operation according to Sect. 3.

The initial system setup is described as follows. Each agent has a complete knowledge of the MAS goals, the corresponding distances between the agents, and the number required to reach the goal per each target. After all the agents have exchanged the relevant data, each of them is selecting the closest target by comparing its R and the corresponding other agents' R_{min} distances to it. If $\Delta Y_i^l = A_i(min) - A_i$ is positive, the agent in question reports on its decision to proceed with the current target; otherwise, it waits.

For the sake of our experiment, we employ 50 agents uniformly distributed across the circular area with the radius of 50 m. The number of targets is three, with 5, 3, and 2 required agents, respectively. Each agent has the radio coverage of 30 m and the line-of-sight distance of 7 m. We vary the number of compromised nodes in the system to validate our framework operation.

Regardless of the system type, the utilization of the proposed trust model reduces the impact of attacks on the system efficiency (see Figs. 3 and 4). For the second option, where each agent has at least one neighboring node with the visual contact, the benefits are even more significant (Fig. 4). The developed

---

[4] See: V-REP http://www.k-team.com/mobile-robotics-products/v-rep.

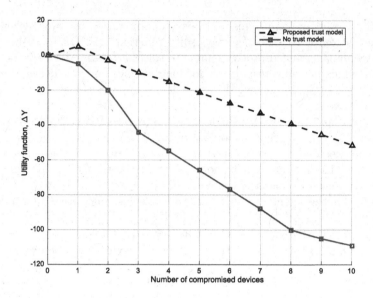

Fig. 3. Dependence of utility function on the number of compromised devices (uniform distribution of agents).

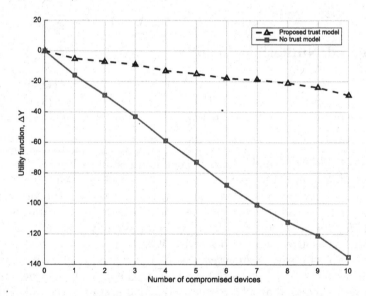

Fig. 4. Dependence of utility function on the number of compromised devices (each agent has at least one visual neighbor).

model may, however, have a negative impact on the system efficiency e.g., when the compromised node has been the best possible selection for the target.

6 Conclusions

In this work, we developed a trust model for the decentralized robotic MAS networks. Our framework provides group access based on the device-centric level of trust, which is selected and dynamically updated over time. Our formulation was successfully evaluated with simulations and may be utilized in modern MAS deployments. The main advantage of the proposed approach is in that it allows for continuous and secure communication in the robotic "ad hoc" networks that face the lack of reliable connectivity to the centralized control unit. The second benefit is in the time-driven dynamic trust level updates that determine the trust level actuality i.e., a newly-joined device would have to operate trustfully for a considerably long time in order to achieve any significant decision-making privileges.

References

1. Hernandez, L., Baladron, C., Aguiar, J.M., Carro, B., Sanchez-Esguevillas, A.J., Lloret, J., Chinarro, D., Gomez-Sanz, J.J., Cook, D.: A multi-agent system architecture for smart grid management and forecasting of energy demand in virtual power plants. IEEE Commun. Mag. **51**(1), 106–113 (2013)
2. Andreev, S., Larmo, A., Gerasimenko, M., Petrov, V., Galinina, O., Tirronen, T., Torsner, J., Koucheryavy, Y.: Efficient small data access for machine-type communications in LTE. In: Proceedings of the IEEE International Conference on Communications (ICC), pp. 3569–3574. IEEE (2013)
3. Cao, Y., Yu, W., Ren, W., Chen, G.: An overview of recent progress in the study of distributed multi-agent coordination. IEEE Trans. Ind. Inf. **9**(1), 427–438 (2013)
4. Militano, L., Fitzek, F., Iera, A., Molinaro, A.: On the beneficial effects of cooperative wireless peer-to-peer networking. In: Pupolin, S. (ed.) Wireless Communications 2007 CNIT Thyrrenian Symposium, pp. 97–109. Springer, Heidelberg (2007)
5. Petrov, V., Andreev, S., Turlikov, A., Koucheryavy, Y.: On IEEE 802.16m overload control for smart grid deployments. In: Andreev, S., Balandin, S., Koucheryavy, Y. (eds.) NEW2AN/ruSMART 2012. LNCS, vol. 7469, pp. 86–94. Springer, Heidelberg (2012)
6. Ren, W., Beard, R.W., Atkins, E.M.: A survey of consensus problems in multi-agent coordination. In: Proceedings of the 2005 American Control Conference, pp. 1859–1864. IEEE (2005)
7. Lesser, V.R.: Reflections on the nature of multi-agent coordination and its implications for an agent architecture. Auton. Agent. Multi-Agent Syst. **1**(1), 89–111 (1998)
8. Shehory, O.M., Sycara, K., Jha, S.: Multi-agent coordination through coalition formation. In: Singh, M.P., Rao, A., Wooldridge, M.J. (eds.) ATAL 1997. LNCS, vol. 1365, pp. 143–154. Springer, Heidelberg (1998)
9. Brambilla, M., Ferrante, E., Birattari, M., Dorigo, M.: Swarm robotics: a review from the swarm engineering perspective. Swarm Intell. **7**(1), 1–41 (2013)
10. Jung, Y., Kim, M., Masoumzadeh, A., Joshi, J.B.: A survey of security issue in multi-agent systems. Artif. Intell. Rev. **37**(3), 239–260 (2012)
11. Araniti, G., Calabro, F., Iera, A., Molinaro, A., Pulitano, S.: Differentiated services QoS issues in next generation radio access network: a new management policy for expedited forwarding per-hop behaviour. In: Proceedings of the IEEE 60th Vehicular Technology Conference (VTC2004-Fall), vol. 4, pp. 2693–2697. IEEE (2004)

12. Bell, D., LaPadula, L.: Secure Computer Systems: Unified Exposition and Multics Interpretation, vol. MTR-2997 R. MITRE Corp., Bedford (1976)
13. Harrison, M.A., Ruzzo, W.L., Ullman, J.D.: Protection in operating systems. Commun. ACM **19**(8), 461–471 (1976)
14. Higgins, F., Tomlinson, A., Martin, K.M.: Threats to the swarm: security considerations for swarm robotics. Int. J. Adv. Secur. **2**(2&3), 1–10 (2009)
15. Petrov, V., Edelev, S., Komar, M., Koucheryavy, Y.: Towards the era of wireless keys: how the IoT can change authentication paradigm. In: Proceedings of the IEEE World Forum on Internet of Things (WF-IoT), pp. 51–56, March 2014
16. Weis, S.A., Sarma, S.E., Rivest, R.L., Engels, D.W.: Security and privacy aspects of low-cost radio frequency identification systems. In: Hutter, D., Müller, G., Stephan, W., Ullmann, M. (eds.) Security in Pervasive Computing. LNCS, vol. 2802, pp. 201–212. Springer, Heidelberg (2004)
17. Kachirski, O., Guha, R.: Effective intrusion detection using multiple sensors in wireless ad hoc networks. In: Proceedings of the 36th Annual Hawaii International Conference on System Sciences, 8 p. IEEE (2003)
18. Mishra, A., Nadkarni, K., Patcha, A.: Intrusion detection in wireless ad hoc networks. IEEE Wirel. Commun. **11**(1), 48–60 (2004)
19. Pelechrinis, K., Iliofotou, M., Krishnamurthy, S.V.: Denial of service attacks in wireless networks: the case of jammers. IEEE Commun. Surv. Tutor. **13**(2), 245–257 (2011)
20. Basagni, S.: Distributed clustering for ad hoc networks. In: Proceedings of the Fourth International Symposium on Parallel Architectures, Algorithms, and Networks (I-SPAN 1999), pp. 310–315. IEEE (1999)
21. Karnik, N.M., Tripathi, A.R.: Security in the Ajanta mobile agent system. Softw. Pract. Exp. **31**(4), 301–329 (2001)
22. Sander, T., Tschudin, C.F.: Protecting mobile agents against malicious hosts. In: Vigna, G. (ed.) Mobile Agents and Security. LNCS, vol. 1419, pp. 44–60. Springer, Heidelberg (1998)
23. Hohl, F.: Time limited blackbox security: protecting mobile agents from malicious hosts. In: Vigna, G. (ed.) Mobile Agents and Security. LNCS, vol. 1419, pp. 92–113. Springer, Heidelberg (1998)
24. Page, J., Zaslavsky, A., Indrawan, M.: A buddy model of security for mobile agent communities operating in pervasive scenarios. In: Proceedings of the Second Workshop on Australasian Information Security, Data Mining and Web Intelligence, and Software Internationalisation, vol. 32, pp. 17–25. Australian Computer Society, Inc. (2004)
25. Page, J., Zaslavsky, A., Indrawan, M.: Countering security vulnerabilities using a shared security buddy model schema in mobile agent communities. In: Proceedings of the First International Workshop on Safety and Security in Multi-Agent Systems (SASEMAS 2004), pp. 85–101 (2004)
26. Zikratov, I.A., Lebedev, I.S., Gurtov, A.V.: Trust and reputation mechanisms for multi-agent robotic systems. In: Balandin, S., Andreev, S., Koucheryavy, Y. (eds.) NEW2AN/ruSMART 2014. LNCS, vol. 8638, pp. 106–120. Springer, Heidelberg (2014)
27. Hong, Y., Hu, J., Gao, L.: Tracking control for multi-agent consensus with an active leader and variable topology. Automatica **42**(7), 1177–1182 (2006)
28. Ni, W., Cheng, D.: Leader-following consensus of multi-agent systems under fixed and switching topologies. Syst. Control Lett. **59**(3), 209–217 (2010)

NEW2AN: IoT and Industrial IoT

Supernodes-Based Solution for Terrestrial Segment of Flying Ubiquitous Sensor Network Under Intentional Electromagnetic Interference

Trung Hoang[✉], Ruslan Kirichek, Alexander Paramonov,
and Andrey Koucheryavy

The Bonch-Bruevich Saint-Petersburg State University of Telecommunication,
Saint-Petersburg, Russia
hoangtrung.telecom@gmail.com, kirichek@sut.ru,
alex-in-spb@yandex.ru, akouch@mail.ru

Abstract. This paper is dedicated to the study of the functions of the terrestrial segment of flying ubiquitous sensor network under intentional electromagnetic interference conditions. The main feature of this influence is the violation of the integrity of the WSN and a temporary failure of nodes located in the influence zone. The aim of the study is to prove that, under IEMI conditions, there is a supernodes-based solution to ensure the functioning of the WSN. The obtained results can be used in selecting models of changing network functionality under the influence of interference so as to maintain the network stability.

Keywords: Flying ubiquitous sensor network · Terrestrial segment · Intentional electromagnetic interference · Supernode · Model reaction · Functionality · Network stability

1 Introduction

One of the most cited terms in IT publications today is the Internet of Things (IoT) [1]. Internet of Things is new technology civilization, allowing you to create real intelligent network binding billions of objects and devices and providing status information and changing commuting facilities. In the transition from the study of the characteristics on the plane to the models in three-dimensional space a new IoT- direction has been formed - Flying Ubiquitous Sensor Network (FUSN) [2]. FUSN include two network segments: terrestrial and flying. Technological base of the terrestrial segment are Wireless Sensor Networks (WSN), which belong to the class of Ubiquitous Sensor Networks (USN) [3–5]. Wireless sensor networks have a number of distinctive features compared to existing networks, the key of which are self-organization and low power consumption [6]. Great interest to the study of such networks caused first wide possibilities of their application: environment monitoring, industrial plant monitoring, transport monitoring, intrusion detection and target tracking, fire security, automobile production, medicine and others. Since the terrestrial segment is responsible for the collection, storage and transmission of data in the flying segment of the FUSN, it is

© Springer International Publishing AG 2016
O. Galinina et al. (Eds.): NEW2AN/ruSMART 2016, LNCS 9870, pp. 351–359, 2016.
DOI: 10.1007/978-3-319-46301-8_29

necessary to provide the ability to perform specified functions USN in various conditions, including the presence of external destabilizing factors, otherwise the practical application of the FUSN would be impossible.

In the field of network security, as determined in accordance with the recommendations ITU-T X.1311 "Network security structure for pervasive sensor networks" [7], there is a significant role of attacks on differences types of relationships in the USN. It is necessary to bear in mind that in the world today there is a real threat of the impact on WSN various destructive electromagnetic impacts, one of which is intentional electromagnetic interference (IEMI) [8]. There are many works devoted to this type of impact affecting on different types of networks [9–14]. Studies have been conducted to identify the sustainability and development of new standards [15], but they do not fully address the problem of protection WSN from IEMI [16, 17]. In [18] the author analyzed the influence of intentional electromagnetic effects on the functioning of the WSN. It is proved that the diagnosis of intentional electromagnetic effects on the WSN can be performed by analysis data on the network connectivity and confirmed that if there is the data on the coordinates of the nodes, the connectivity analysis allows locating the influence source.

However, the impact of IEMI in the operating channel bandwidth and out of band interference generally leads to an increase of the number of transmission errors of packets up to 100 %, i.e. inability to transfer a certain amount of nodes. At the same time the functionality of the network level can be achieved by alternative route packets through the nodes which are not subjected to (largely) destructive impact. It is obvious that, in a homogeneous network structure, when all nodes have the same functionality opportunities, the degree of influence on the network will be characterized by shares underwent the influence of nodes, and will not depend on the location exposure. In practice, WSN often has heterogeneous structure [19]. For example, if the data collection network has only one coordinator through which the data is transmitted to the user, the network routes will form a star or a tree structure, in which the amount of traffic served by different nodes is not the same. Transit nodes located in the route "closer" to the gateway will serve a greater volume of traffic than the nodes by arranging position on the periphery. Consequently, the degree of influence will depend on its localization in the service area. With the localization of impaction in the gateway area, in this example, the operation of the network is impossible. Thus, the heterogeneous network resistance to electromagnetic interference depends on the selection of the structural parameters.

2   Intentional Electromagnetic Interference

To date, almost all resources in broadband communications (Wi-Fi, Bluetooth, ZigBee, 6LoWPan, etc.) are operating in the frequency, which ranges from 300 MHz to 3 GHz.

Along with the development of wireless data transmission technology is the development of equipment, which creates electromagnetic impacts on them. Such agents in a number of cases can be used legally in order to ensure information security, as well as illegally to commit destructive actions (illegal methods of competition, cyber-terrorism and et al.). One of the common methods is the use of such effects generators of short pulses to interfere the broadband communications equipment's in the frequency range from 300 MHz to 12 GHz. Then we consider intentional electromagnetic interference as electromagnetic effect, which is accomplished by the use of portable generators moat of the electromagnetic field and leads to destruction, distortion and blocks transmitted information.

Distortion in the transmitted data structure occurs in the physical level resulting from interference generated in the transmission medium. As a result of impaction, signals with parameters comparable with the parameters of useful transmission signals leads to the violation of the normal functioning of the communication network and the distortion of the transmitted information. The main feature of IEMI is that the impulse spectrum is similar to the spectrum broadband signal. Therefore, such effects are difficult to detect with conventional methods. In addition, they can be created by using social generators from a great distance and masquerade as usual electromagnetic interference. As a result of such exposure, it violates not only the functioning of the individual sensor nodes, but also the integrity of the wireless sensor network situated in the electromagnetic impact zone.

3 Statement of the Research Problem

WSN network parameters (the number of nodes, communication range, capacity) may have different values, depending on the purpose of the network. The impact on the network parameters can also be quite varied. Therefore, the choice of model for describing the operation of the network in terms of IEMI. It presents some difficulties. If the network is composed of a small number of narrow fishing (e.g., less than a dozen) and the service area is limited to tens of meters and the exposure parameters of the same order as the transmission device node that will be exposed to nearly all the network nodes and we should expect the full network functionality loss. In the case where the number of nodes in hundreds and hundreds of service area dimensions - thousands of meters, at the same parameters impact the network can save a significant proportion of the functionality.

Considering a network consisting of three types of nodes: sensor nodes – nodes capable of performing the functions of collection, data transmission and transit; coordination – nodes - the focal point, as well as gateway network and supernodes - nodes that can perform the functions of their coordinator (gateway) if needed, we assume that for network operation we only need one coordinator (gateway). We estimate the functional relationship of network parameters by influencing it.

4 The Response Model of Wireless Sensor Networks Under Intentional Electromagnetic Interference Conditions

As noted above, there are two main scenarios for the IEMI attack on the network, which will be considered in this article (Fig. 1):

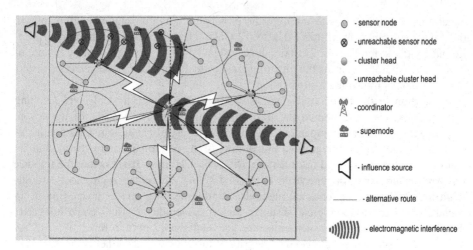

Fig. 1. The model reaction of wireless sensor networks under intentional electromagnetic interference conditions.

1. We assume, that IEMI affects only part of the sensor nodes network that performs the function of collecting and transferring data, does not affect the coordinator (gateway) of the network. At the same time, the functionality of the network much can be achieved by alternative route packets through the nodes which are not subjected to (largely) destructive impact.
2. We assume, that IEMI affects the coordinator (gateway) wherein the network functionality provided in such a way that supernode will be the coordinate (gateway), and performs its function.

Assuming that the service zone is a flat surface, and the coordinator is located in a random point, then the probability of getting the coordinates in the target area can be defined geometrically as the ratio of the area of influence zone to the area of the service zone:

$$p_G = \begin{cases} \frac{s_E}{S} \text{ at } 0 \leq s_E < S \\ 1 \text{ at } s_E \geq S \end{cases} \tag{1}$$

where s_E is the area of the influence zone;
S is the area of the service zone.
Probability functioning of coordinator can be defined as:

$$p_n = 1 - p_G \tag{2}$$

It should be noted that the expression (2) represents a resistance network in terms of the coordinator function. Assume that the network has a large number of supernodes, which are capable to do the role of coordinator, and then a complete loss of functionality would be the case when they all fall into the impact zone. For k these nodes, this probability will be determined as:

$$p_{nk} = 1 - p_G^k \tag{3}$$

Probability (3) describes the operation of at least one of the supernodes (coordinator). The functionality of the network also depends on the connectivity of the existing coordinators with other nodes in this network.

When random distribution network nodes can be conveniently described by a Poisson model of the field [20], in which the probability of getting k nodes in the region the impact of a Poisson distribution and depends on the area of impaction Se and the density of supernodes ρ_s. Then the probability that no more than $k - 1$ supernodes will be in the target area can be defined as:

$$\tilde{p}_{nk} = \sum_{i=1}^{k-1} \frac{(\rho_S S_E)^i}{i!} e^{-\rho_S S_E} \tag{4}$$

Figure 2 shows a comparison of the probabilities of dependencies (3) and (4) of the area of influence zone. The figure shows that the probability of operation at least one of k supernodes determined according to (3) decreases with increasing the area of influence zone affected from 1 to 0. When the area of influence zone equals the area of service zone, the probability of network operation is 0, which is consistent with this model. According to (4) the probability network operation is also reduced, but does not become equal to 0. This is due to the fact that the Poisson field model number of supernodes and the area of service zone is not limited to (tends to infinity).

Therefore, the smaller difference between the probability values (3) and (4) is, the bigger k is, and the ratio between the area of the influence zone and the service zone is

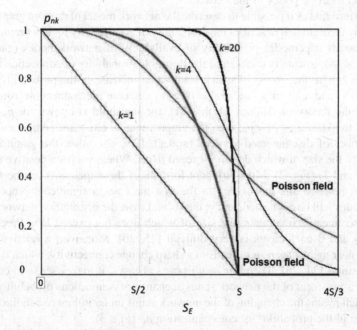

Fig. 2. The hitting probability of coordinators into the interference zone.

also smaller. From dependency analysis (Fig. 1) it can be seen that error does not exceed 10 %, with $k = 4$ or more, a square exposure region is less than 50 % of the area of service zone.

It should be noted that in the general case, the impact on the network may cause total failure due to a part of the nodes of the degradation of the quality of communication channels, maintaining the functionality of the network nodes. In this case, the model is equivalent to the loss of the components, that means decrease n. Degraded channel quality does not always result in a complete loss of their functionality [23]. Quality degradation leads to a reduction in channel resources and the appearance of changes in the settings such as the achievable data rate, error rate and communication range. If the description of the network model can be assumed, the degradation of connection channels can be the result of the impact on the network, which is equivalent to a decrease of communication range (radius connection R) of sensor nodes. Thus it can be assumed that the number of nodes in the network is large enough, and the zone area of impaction is considerably less than the area of the service zone $s_E \ll S$. In these conditions, it might be convenient to use the model of a Poisson field. It provides an opportunity to describe the probability of the existence of links between network nodes via the density of nodes or the ratio of the area of the node connection to a common area of the service zone:

$$p_r = \frac{1}{n}\rho \cdot S_C = \frac{S_C}{S} \tag{5}$$

where ρ is density of nodes;

S_C is the area of the node connection;

S is the area of the service zone;

n is a number of nodes in the network.

This in turn makes it possible to describe the network model of random graph, which is known by theorem that describes the probability of connectivity p_C in a graph [21, 22, 24], that means supernodes probability of availability of network nodes (gateways). According to (4), is usually considered the threshold probability of connection between nodes $p_0 = \ln n/n$, the excess of which leads to an increase in the probability of connectivity to 1, and a lower value, respectively to decrease the connection from 1 to 0. Also from the theory of random graphs [21], the threshold is known for $p_* = 1/n$,. According to (3), if $p_* \leq p_r < p_0$, then the graph contains one giant component size γ_n and a number of disconnected vertices (nodes). If $p_r < p_*$, then the graph contains components, the size of which does not exceed $\beta \ln n$. Where γ и β are positive numbers between 0 and 1, (Fig. 2). More stringent formula of these theorems can be found in [24–26]. In practice this means that in the first the case, a significant proportion of network nodes will be connected, and in the second case, the distribution network breaks up into disconnected fragments, the size of which does not exceed $\ln n$. Precise definition of γ and β coefficients is very difficult [25, 26]. Moreover, a relatively small change p_C near the values p_0 и p_* leads to a sharp change connectivity, which is called a phase transition [24, 26]. Therefore, with practical tasks, it makes sense to conduct a qualitative assessment of the network status at change of connections probability among nodes, which means the changing of the network status under influence conditions leads to a change in the probability of communication p_C (Fig. 3).

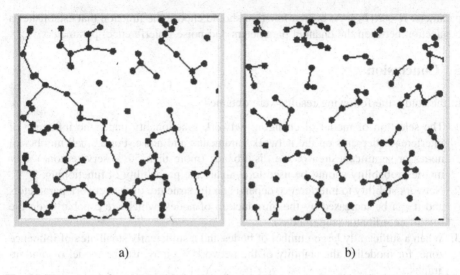

a)   b)

Fig. 3. The network fragments with the size of n (giant component) and $\ln n$ in the network with different probabilities of connection between nodes.

In other words, if $p_* \leq p_r < p_0$, then a significant part of the network nodes (giant component) is connected and the network can save a significant share functionality. The functionality of the network in this case is saved if when at least one coordinator or supernode is included in the set of connected nodes. When $p_r < p_0$ the probability of connectivity p_C is $< 0,5$. The probability of connectivity is, in fact, equal to the probability that a node comes in a variety of connected nodes. Then the probability of operation of the network can be described as the probability of maintaining connectivity with a variety of connected nodes:

$$\hat{P}nk = 1 - (P_C)^k \tag{6}$$

For sufficiently large n the connectivity probability can be estimated as it is shown in [20]

$$P_C = e^{-e^{-c}} \tag{7}$$

where C defined from $P_r = \frac{\ln n + C + 0(1)}{n}$.

If $p_C < p_*$ the network elements remain only as a connectivity insulating Rowan fragments, the number of nodes that is of the order of $\ln n$, Fig. 3b. To preserve the functionality of a particular fragment of the network, it is necessary that at least one of its nodes was supernodes (performing the function of coordinator). Then the required number of supernodes, at least, should not be less than the number of isolated fragments of the network, that means $k > n / \ln n$.

To determine the probability of connection between the nodes p_C we can use (5). To calculate the area S_C we can select corresponding model of the communication channel. In the simplest case, the model the node connection may be described as a circle with a radius R, т.е. $S_C = \pi R^2$. In describing the impaction through the channel parameters, communication range can be described as some function of the impact parameter $R = R(P_E)$. As such a function should choose the function that establishes a connection between the channel parameters and noise (interference) parameters.

5 Conclusion

In the study, the following results were obtained:

1. The selection of model of changing network functionality under the influence of interference depends on the network parameters and noise. For a small number of nodes or significant area of the disturbance (more than 50 % service zone) geometric probability should be used to evaluate the probability of functioning.
2. Network stability to interference depends on the structure of the network parameters and it can be increased by the introduction of nodes, which are capable to do the role of coordinator (supernodes).
3. When a sufficiently large number of nodes and a sufficiently small area of influence zone, for modeling the stability of the network we may use the model of random graphs.
4. The exposure model on the network can be described as the loss of the nodes, which falls within the influence zone or the decrease of the resource of connection channels.
5. The required number of supernodes can be defined from considerations of adequacy needed when changing network structure as a result of the impact.

Acknowledgment. The reported study was supported by RFBR, research project No. 15 07-09431a "Development of the principles of construction and methods of self-organization for Flying Ubiquitous Sensor Networks".

References

1. Recommendation Y.2060. Overview of Internet of Things. ITU-T, Geneva, February 2012
2. Kirichek, R., Paramonov, A., Koucheryavy, A.: Flying ubiquitous sensor networks as a queueing system. In: 2015 17th International Conference on Advanced Communication Technology (ICACT). IEEE (2015)
3. Koucheryavy, A., Salim, A.: Cluster-based perimeter-coverage technique for heterogeneous wireless sensor networks. In: 2009 International Conference on Ultra Modern Telecommunications and Workshops, ICUMT 2009. IEEE (2009)
4. Al-Qadami, N., Laila, I., Koucheryavy, A., Ahmad, A.S.: Mobility adaptive clustering algorithm for wireless sensor networks with mobile nodes. In: 2015 17th International Conference on Advanced Communication Technology (ICACT), pp. 121–126. IEEE, July 2015
5. Abakumov, P., Koucheryavy, A.: The cluster head selection algorithm in the 3D USN. In: Proceedings of International Conference on Advanced Communication Technology, ICACT 2014. Phoenix Park, Korea (2014)
6. Andreev, S., Gonchukov, P., Himayat, N., Koucheryavy, Y., Turlikov, A.: Energy efficient communications for future broadband cellular networks. Comput. Commun. **35**(14), 1662–1671 (2012)

7. Recommendation X.1311 Security Framework for Ubiquitous Sensor Networks. ITU-T, Geneva, February 2011
8. Savage, E., Radasky, W.: Overview of the threat of IEMI (intentional electromagnetic interference). In: 2012 IEEE International Symposium on Electromagnetic Compatibility (EMC), pp. 317–322. IEEE (2012)
9. Palisek, L., Suchy, L.: High power microwave effects on computer networks. In: EMC Europe 2011, York, pp. 18–21. IEEE (2011)
10. Homma, N., Hayashi, Y., Aoki, T.: Electromagnetic information leakage from cryptographic devices. In: 2013 International Symposium on Electromagnetic Compatibility (EMC EUROPE), pp. 401–404. IEEE (2013)
11. Tanuhardja, R.R., et al.: Vulnerability of terrestrial-trunked radio to intelligent intentional electromagnetic interference (2015)
12. Tientcheu, R.T., Pouhe, D.: Analysis of methods for classification of intentional electromagnetic environments. In: 2015 International Conference on Electromagnetics in Advanced Applications (ICEAA). IEEE (2015)
13. Mora, N., et al.: Experimental characterization of the response of an electrical and communication raceway to IEMI. IEEE Trans. Electromagn. Compat. **58**, 494–505 (2016)
14. Zhukovsky, M., Kirichek, R., Larionov, S., Chvanov, V.: Testing of technical security equipment for stability to intentional electromagnetic interference. In: Proceedings of EMC Europe 2011 York - 10th International Symposium on Electromagnetic Compatibility, pp. 820–823 (2011)
15. Kirichek, R.V., Chvanov, V.P.: Improvement of Russian regulatory system on protection against electromagnetic attacks. In: Proceedings of Symposium 9th International Symposium on Electromagnetic Compatibility Joint with the 20th International Wroclaw Symposium on Electromagnetic Compatibility (EMC EUROPE 2010), pp. 26–29 (2010)
16. Panasik, C.M., Siep, T.M.: Wireless network circuits, systems, and methods for frequency hopping with reduced packet interference. U.S. Patent No. 6,643,278, 4 November 2003
17. Jeong, J., Ee, C.T.: Forward error correction in sensor networks. University of California at Berkeley (2003)
18. Hoang, T., Kirichek, R., Paramonov, A., Koucheryavy, A.: Influence of intentional electromagnetic interference on the functioning of the terrestrial segment of flying ubiquitous sensor network. In: Kim, K.J., Joukov, N. (eds.) Information Science and Applications (ICISA) 2016. Lecture Notes in Electrical Engineering, vol. 376, pp. 1249–1259. (2016)
19. Galinina, O., Andreev, S., Gerasimenko, M., Koucheryavy, Y., Himayat, N., Yeh, S.-P., Talwar, S.: Capturing spatial randomness of heterogeneous cellular/WLAN deployments with dynamic traffic. IEEE J. Sel. Areas Commun. **32**(6), 1083–1099 (2014). Article No. 6824742
20. Ventcel', E.S.: Probability theory. Nauka, M. (1959). (in Russian)
21. Amit, A., Linial, N.: Random graph coverings I: general theory and graph connectivity. Combinatorica **22**(1), 1–18 (2002)
22. Janson, S., Luczak, T., Rucinski, A.: Random graphs, vol. 45. Wiley, London (2011)
23. Moltchanov, D., Koucheryavy, Y., Harju, J.: Loss performance model for wireless channels with autocorrelated arrivals and losses. Comput. Commun. **29**(13–14), 2646–2660 (2006)
24. Cannings, C., Penman, D.B.: Chap. 2. Models of random graphs and their applications. Handb. Stat. **21**, 51–91 (2003)
25. Feller, W.: An Introduction to Probability Theory and Its Applications, vol. I. Wiley, London (1968)
26. Bollobás, B., Janson, S., Riordan, O.: The phase transition in inhomogeneous random graphs. Random Struct. Algorithms **31**(1), 3–122 (2007)

A New Centralized Link Scheduling for 6TiSCH Wireless Industrial Networks

Kang-Hoon Choi and Sang-Hwa Chung[✉]

Department of Electrical, Electronics and Computer Engineering,
Pusan National University, Busan, South Korea
long567890@gmail.com, shchung@pusan.ac.kr

Abstract. Industrial wireless sensor networks require high reliability, low power usage, and timely exchange of information. To meet these requirements, the IEEE 802.15.4e time slotted channel hopping (TSCH) medium access control (MAC) protocol was developed. The recently created IETF 6TiSCH working group (WG) is implementing a protocol stack based on the Internet of Things (IoT) standard, such as the RPL routing protocol, 6LoWPAN, and CoAP, in order to enable IPv6 on IEEE 802.15.4e TSCH. This paper presents a new Centralized Link Scheduling (CLS) algorithm for operation in 6TiSCH networks. The CLS algorithm constructs efficient multi-hop schedules using a minimal number of centralized control messages, because it allocates and deallocates slots without rescheduling the entire schedule. Simulation results show that our CLS algorithm requires smaller control messages for scheduling formation than the Decentralized Traffic Aware Scheduling (DeTAS) algorithm, implemented in 6TiSCH networks, does.

Keywords: 6TiSCH · IEEE 802.15.4e · TSCH scheduling · Industrial wireless sensor network · Internet of Things

1 Introduction

The Internet of Things (IoT) is a world-wide network of interconnected, uniquely addressable objects based on standard communication protocols [1]. It is expected that the number of devices connected to the internet will reach 24 billion by 2020 [2]. The IPv4 address system is limited to uniquely assigning addresses to billions of devices, so IP version 6 (IPv6) [3] is the de-facto standard for the IoT. The IETF has defined the MAC layer of IoT protocol stacks, such as 6LoWPAN [4] for IPv6 header compression, IPv6 routing protocol for low-power and lossy networks (RPL) [5], and the constrained application protocol (CoAP) [6]. Recently, in industrial wireless sensor network devices have adopted the IoT stack to combine information technology (IT) and operational technology (OT) [7]. However, the IETF does not define the MAC layer for industrial networks. In 2012, IEEE published the IEEE 802.15.4e [8] standard, an enhanced version of IEEE 802.15.4, for industrial wireless sensor networks. TSCH mode of IEEE 802.15.4e [8] operates similarly to the Time Synchronized Mesh Protocol (TSMP) [9], which communicates according to a schedule while frequency hopping. TSCH is a MAC protocol that ensures high reliability and low power consumption to meet the needs of industrial networks. In October 2013, the IETF 6TiSCH

© Springer International Publishing AG 2016
O. Galinina et al. (Eds.): NEW2AN/ruSMART 2016, LNCS 9870, pp. 360–371, 2016.
DOI: 10.1007/978-3-319-46301-8_30

WG was created to link IEEE 802.15.4e TSCH MAC and the IETF IoT protocol stack. Since TSCH is a TDMA-based MAC protocol, it should do the scheduling. The IEEE 802.15.4e [8] standard describes how to operate a schedule in the MAC layer, but it does not specify how to build a schedule. To solve this problem, several scheduling algorithms have been proposed. The Traffic Aware Scheduling Algorithm (TASA) [10] and Decentralized Traffic Aware Scheduling (DeTAS) [11, 12] are representative examples of centralized and distributed scheduling, respectively. We propose a new Centralized Link Scheduling algorithm that minimizes the number of control packet for scheduling through allocating and deallocating slots without rescheduling the entire schedule. Section 2 briefly introduces 6TiSCH networks, and then explains not only TSCH and RPL [5] of 6TiSCH architecture [13], but also DeTAS, which is a comparison target of CLS. Section 3 describes the details of CLS. Section 4 evaluates CLS compared with DeTAS and shows that it performs better. Finally, in Sect. 5, we derive conclusions and discuss future work.

2 Related Work

2.1 6TiSCH Networks

Industrial sensor networks have long used wired networks for their high reliability, durability, and safety. However, it is very difficult to add new sensor nodes and the installation and maintenance of wired networks are expensive. To solve this problem, industrial wireless sensor network have emerged, such as WirelessHart [14], ISA100.11a [15], and IEEE 802.15.4e [8]. Unlike WirelessHart and ISA100.11a, which provide the entire communication stack, IEEE 802.15.4e [8] defined only the MAC layer. The IETF 6TiSCH WG has been working to link the IPv6-enabled [3] IoT protocol stack to the IEEE 802.15.4e TSCH MAC layer. In 6TiSCH networks, each node uses 6LowPAN header compression to transmit IPv6 packets and 6LowPAN Neighbor Discovery to search for neighbors. 6TiSCH uses the Routing Protocol for Low power and Lossy Networks (RPL) [5], so that each node adjusts its synchronization time based on the RPL root and sets a route path through the Destination Oriented Directed Acyclic Graph (DODAG).

2.2 IEEE 802.15.4e TSCH

IEEE 802.15.4e [8] is a standard that improves the MAC layer of the IEEE 802.15.4 standard to meet industrial wireless sensor network needs. The IEEE 802.15.4e [8] standard added three MAC modes, DSME (Distributed Synchronous Multi-channel Extension), LL (Low Latency), and TSCH (Time Slotted Channel Hopping), in order to guarantee low-delay, real-time transmission. IEEE 802.15.4e TSCH mode operates with high reliability and low power usage and is suitable for industrial wireless sensor networks. All nodes on the IEEE 802.15.4e TSCH network are synchronized. In TSCH, time is split into slots and a slotframe consists of a group of time slots. Each node repeats the scheduled slotframe and operates one action (transmit, receive, or sleep) in the time slot. The channel-hopping feature of TSCH reduces the problem of external interference and multi-path fading. Since TSCH is a TDMA-based MAC protocol, it should schedule when a node will communicate with another and in which time slot. The schedule is represented as a two-dimensional matrix in the form (slotOffset, channelOffset); these matrix elements are referred to as a cell. If the cell is scheduled to transmit or receive, the node communicates with a neighbor node. The IEEE802.15.4e [8] standard specifies how to execute a schedule, but has not defined the scheduling method. In this paper, we bridge this gap by proposing Centralized Link Scheduling (CLS) built to use a minimal number of control messages.

2.3 Routing Protocol for Low Power and Lossy Networks (RPL)

In 6TiSCH networks, the role of RPL is to form the network. RPL [5] is a routing protocol to find a path to the sink through a Destination Oriented Directed Acyclic Graph (DODAG), which is formed using one or more routing metrics. Each node receives a DODAG Information Object (DIO) message that is periodically broadcast from neighbors. Through DIO messages, each node selects its preferred parent. If a node receives a DIO message, it calculates its rank through an objective function (OF). Rank indicates the relative distance from the sink, so a node selects a preferred parent with the lowest rank among neighbor nodes.

2.4 Decentralized Traffic Aware Scheduling (DeTAS)

Accettura *et al.* proposed Traffic Aware Scheduling Algorithm (TASA) that is an optimal time/frequency scheduling in a centralized manner. TASA is a greedy algorithm for allocating cells, so that it does not take into account queue congestion. To solve this problem, Accettura *et al.* proposed the Decentralized Traffic Aware Scheduling (DeTAS) algorithm for 6TiSCH networks. DeTAS is able to construct optimum multi-hop schedules like TASA, in a distributed fashion. In DeTAS, to set up the schedule, each node needs to know the amount of traffic for transmission to its parent and for reception to its children. Each node sends traffic information to its parent. If the parent node receives it, it updates the traffic information, and then forwards the information to its parent to reach the sink node. The sink node running DeTAS must divide its children into even- or odd-scheduled. If a node is even-scheduled, its *transmit slot* is allocated an even slotOffset and its *receive slot* is allocated an odd slotOffset. If a node is odd-scheduled, its transmit slot is allocated an odd slotOffset and receive slot is allocated an even slotOffset. Consequently, in the same DODAG rank, even-scheduled nodes and odd-scheduled nodes are completely independent. If the sink properly divides its children into even- or odd-scheduled, DeTAS is able to perform optimal scheduling. One of the features of DeTAS is that no buffer overflow occurs because the transmit and receive slots are allocated alternately. Another feature of DeTAS is that interference does not occur. DeTAS does not allow nodes of the same DODAG rank to transmit packets in the same time slot. DeTAS calculates channel offset by $[(DODAGrank - 2) \bmod W$ $(W \geq 3)]$ in transmit slots and $[(DODAGrank - 1) \bmod W (W \geq 3)]$ in receive slots. Through this mechanism, DeTAS achieves collision-free scheduling. However, although DeTAS performs distributed scheduling, each node updates its traffic information and transmits it to the sink node for optimal scheduling, which requires significant overhead. Additionally, when a new node joins the network, the preferred parent change, or the traffic load of a node changes, DeTAS requires rescheduling for optimal performance.

3   Scheduling Description

In this section, we detail the CLS (Centralized Link Scheduling) algorithm in 6TiSCH networks based on IEEE 802.15.4e [8]. CLS is a traffic-aware algorithm that guarantees that data generated by each source node reaches the sink in a slotframe. CLS computes the scheduling in a centralized manner to avoid collision. CLS operates 6TiSCH networks organized by RPL. When a new node joins a network, CLS assigns it a proper slot. When a node's preferred parent is changed, CLS deletes the relevant slot and assigns a proper slot. CLS is the following operations: CLS Allocation Processing and CLS Deallocation Processing. CLS is designed to meet the following three goals.

(1) In order to reach the sink in the same slotframe, CLS assigns proper slots to transmit its own packets and to forward packets received from its children.
(2) CLS is built to ensure that no collision occurs in the network, resulting in high reliability and low latency.
(3) CLS allocates and deallocates slots without rescheduling the entire schedule when a new node joins the network, a preferred parent has changed, or the traffic load of a node has changed. Consequently, the number of control messages needed for scheduling is minimized.

3.1   CLS Allocation Processing

If a node is synchronized, it selects a preferred parent among the neighbors based on RPL DIO messages received from the neighbors. When a node selects a preferred parent, it sends a CLS allocation request message to its parent. A node also sends this message when its traffic load has changed. This message includes a linkQueue: the queue of links, which is consist of its address, its parent's address, its DODAG rank, and a demand slot for transmission data. When a node receives a CLS allocation request message, if it is not the sink node, it sends a linkQueue by adding a new link that consists of its address, its parent's address, its DODAG rank, and the child's demand slot. Consequently, slots are assigned to forward data generated by child nodes to the sink in a slotframe. If the sink node receives the message, Centralized Link Scheduling is initialized. Scheduling proceeds until the linkQueue is empty. First, a link is obtained from the linkQueue. If the link's $DODAGrank = 1$, the start time slotOffset is 0. If link's $DODAGrank > 1$, the start time slotOffset = (last assigned time slot of the link's parent) + 1. The channelOffset is the link's $DODAGrank - 1$, so that a node uses a channelOffset different from that of its parent, to avoid collision between node of different DODAG ranks. Second, if the cell (slotOffset, channelOffset) does not incur a collision, the schedule table is inserted. Finally, if linkQueue is empty, the slot assign message is transmitted to the requesting nodes. The algorithm for the CLS allocation request process is given in Algorithm 1.

Algorithm 1. CLS Allocation Processing

1:   **Input** : $lnkQueue$: a queue of links, which is consist of $(u, Parent(u), DODAGrank(u), Trans(u))$

     u : the local node

     $Parent(u)$: the parent node of u

     $Trans(u)$: number of slots for transmission in slotframe

     $DODAGrank(u)$: DODAGrank of u

2:   **Output** : $ScheduleTable$: 2-D matrix of (time slot, available channel)

3:   **Function** :

     $startTs (u, ScheduleTable)$: last assigned time slot of $Parent(u)$

     $Collision (ts, link, ScheduleTable)$: check collision in ts

4:   **Procedure in sink node**

5:   **while** ($lnkQueue \neq empty$)

6:     $link \leftarrow lnkQueue.pop()$

7:     **if** $link.DODAGrank > 1$ **then**

8:       $ts \leftarrow startTs(link.u, ScheduleTable) + 1$

9:     **end if**

10:    **else**

11:      $ts \leftarrow 0$

12:    $ch \leftarrow link.DODAGrank - 1$

13:    **if**   **not**

     $(ScheduleTable(ts, ch)$ **and** $Collision(ts, link, ScheduleTable))$ **then**

14:      **Insert** $ScheduleTable (ts, ch) \leftarrow link$

15:      $link.Trans \leftarrow link.Trans - 1$

16:    **end if**

17:    **else**

18:      $ts \leftarrow ts + 1$

19:    **if** $link.Trans > 0$ **then**

20:      $lnkQueue.push(link)$

21: **end while**

22: Transmit the slot assignment message to requesting nodes

23: **Procedure not in sink node** u

24: $link \leftarrow lnkQueue.pop()$

25: $link \leftarrow (u, Parent(u), DODAGrank(u), link.Trans)$

26: $lnkQueue.push(link)$

27: Transmit $lnkQueue$ to parent node

3.2 CLS Allocation Illustrative Example

Figure 1 shows a routing graph and Fig. 2 shows the schedule result of CLS Allocation processing. In Fig. 1a, when node n_1 joins the network, it sends a CLS Allocation

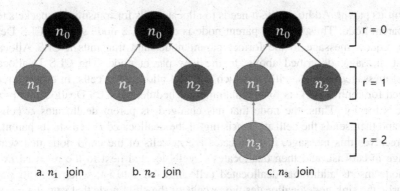

a. n_1 join   b. n_2 join   c. n_3 join

Fig. 1. CLS allocation processing example: routing graph

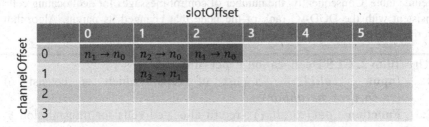

Fig. 2. CLS allocation processing example: schedule table

request message that includes the link $(n_1, n_0, DODAGrank = 1, \text{Demand Slot} = 1)$ to its parent. If the sink node n_0 receives the message, it assigns a link in the schedule table. The link's $DODAGrank = 1$; therefore, the start time slotOffset is 0 and channelOffset is also 0. The sink node n_0 assigns a link in the cell $(0, 0)$ and then sends the assigned link information to n_1. Figure 1b shows node n_2 joining the network, similar to node n_1. In Fig. 1c, when node n_3 joins the network, it sends a CLS Allocation request message including the link $(n_3, n_1, DODAGrank = 2, \text{Demand Slot} = 1)$ to its parent n_1. If node n_1 receives the message, it inserts the link $(n_1, n_0, DODAGrank = 1, \text{Demand Slot} = 1)$ into the linkQueue, and then sends it to the parent node n_0. The sink node n_0 assigns two links in schedule table, because the linkQueue's size is 2. First, take the link $(n_3, n_1, DODAGrank = 2, \text{Demand Slot} = 1)$ in the linkQueue. The link's $DODAGrank = 2$, so the start time slotOffset is the last assigned time slot $(n_1) + 1 = 1$, and the channelOffset is 1. If the cell $(1, 1)$ is available, the sink node n_0 assigns the link in the schedule table. Secondly, the Sink node takes the link $(n_1, n_0, DODAGrank = 1, \text{Demand Slot} = 1)$ in the linkQueue, and then assigns it in the same way as in Fig. 1a.

3.3   CLS Deallocation Processing

In 6TiSCH networks, the nodes update their neighbors' rank information through the receiving RPL DIO messages. If a node has a neighbor node with a smaller rank value than its preferred parent's rank, it changes preferred parent to that neighbor node. If a node changes its preferred parent, it should deallocate cells that are allocated to transmit its parent. Additionally, it needs to allocate cells for transmitting packets to the new parent node. Thus, when a parent node is changed, a node sends a CLS Deallocation request message to the former parent node, and transmits a CLS Allocation request message, described above, to the new parent node. The CLS Deallocation request message includes cellList, which is list of allocated *tx* cells. In order to reduce the need for control packets for maintaining a schedule, the CLS Deallocation process proceeds locally. Thus, the node that has changed its parent deallocates *tx* cells for itself, and then sends the cellList consisting of the deallocated *tx* cells to its parent. If a node receives this message, it deallocates the *rx* cells of the child node that sent the message in cellList, and then deallocates *tx* cells located next to the released *rx* cell. The node inserts additional deallocated cells into the cellList, sends it to its parent. Similarly, the sink node deallocates the *rx* cells of the child node that sent the message. In order to update the entire schedule, the cells of the cellList are removed from the schedule table. Consequently, the number of control messages for deallocating cells is consistent with the DODAG rank of the node that changed its parent. Algorithm 2 describes the process of CLS Deallocation.

Algorithm 2. CLS Deallocation Processing

1: **Input** : *cellList* : a list of cells, which is consist of $(ts, ch, txrx, neighbor)$
2: **Function** : $getTxCells()$: return allocated cells for transmission to parent
3: **Procedure in sink node**
4: **for** *cell* in *cellList*
5:      **Delete** *cell* **in** *ScheduleTable*
6:          **if** $cell.txrx = rx$ **and** $cell.neighbor = u$ **then**
7:              Deallocate the cell in slotframe
8:          **end if**
9: **end for**
10: **Procedure not in sink node** *u*
11: **for** *cell* in *cellList*
12:     **if** $cell.txrx = rx$ **and** $cell.neighbor = u$ **then**
13:         Deallocate the cell in slotframe
14:         **for** *txCell* **in** $getTxCells()$
15:             **if** $txCell.tx > cell.ts$ **then**
16:                 Deallocate the cell in slotframe
17:                 **Insert** *txCell* **into** *cellList*
18:                 **break**
19:             **end if**
20:         **end for**
21:     **end if**
22: **end for**
23: Transmit *cellList* to parent node

3.4 CLS Deallocation Illustrative Example

Figure 4 shows an example of CLS Deallocation processing as the routing graph changes from Fig. 3a to 3b. When node n_3 receives an RPL DIO message from the sink node n_0, it changes its preferred parent, node n_1, to the sink node n_0. Node n_3 deallocates its tx cell $(1, 1)$ and inserts the cell into cellList. Node n_3 sends a CLS Deallocation request message, including cellist, to previous parent n_1 and a CLS Allocation request message to new preferred parent n_0. If node n_1 receives the message, it deallocates the cell $(1, 1)$, which is the rx cell of node n_3, and then deallocates tx cell $(2, 0)$, which has a larger slotOffset than the released rx cell $(1, 1)$. Deallocated cells are inserted into cellList and then node n_1 sends them to the sink node n_0. The sink deallocates rx cell $(2, 0)$, deletes all cells on the cellList that are on the schedule table, and then allocates cell $(2, 0)$ through processing the CLS allocation request received from the node in the manner described in Sect. 3.1. Sink nodes should handle the allocation request after the deallocation request is processed. In this way, when a parent node is changed, nodes not related to this path maintain their schedules, so that there is no need to reschedule an entire node. Therefore, CLS can provide a flexible and efficient scheduling a change for a routing graph.

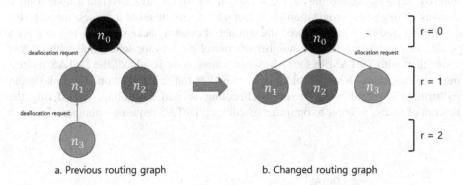

a. Previous routing graph   b. Changed routing graph

Fig. 3. CLS deallocation processing example: routing graph

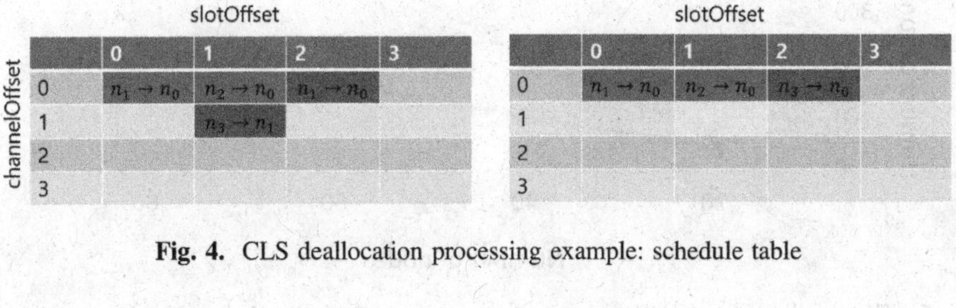

Fig. 4. CLS deallocation processing example: schedule table

4 Performance Evaluation

In this section, we evaluate the performance of CLS using a 6TiSCH simulator, which is written in the Python language and is an open-source project developed by members of the 6TiSCH WG. In the 6TiSCH simulator, the IEEE 802.15.4e TSCH MAC layer and RPL protocols have been implemented. The 6TiSCH simulator can measure the number of scheduled cells, end-to-end latency, end-to-end reliability, power consumed by adjusting various parameters such as the number of nodes, packet period, slotframe length, and deployment area. We have implemented the well-known distributed scheduling algorithm DeTAS and CLS, which we suggest in this paper. We compared the number of control packets necessary for scheduling before the routing graph stabilized to a static network. In addition, when a parent node is changed, we compared the number of control packets required to reschedule. Finally, the scheduling is completed and end-to-end latency is compared. The simulation setup is as follows: the number of nodes varies from 20 to 80, the location of the node is randomly arranged in the center of the sink within 2 km × 2 km, each node generates a packet at 1 s intervals, the available channels are 16 IEEE 802.15.4 channels. Each result was obtained through the average of 10 simulation runs in various topologies.

Figure 5 shows the number of control packets necessary for scheduling in four different networks consisting of 20, 40, 60, or 80 source nodes. When a node joins a network through synchronization from around the sink, it sends a request message for scheduling, and then we measured the number of control packets for this process. As a result, we can be sure that the number of control packets for scheduling with CLS is lower than with DeTAS. In DeTAS, when a new node is added, the DeTAS request message should arrive at the sink in order to update traffic information. The sink should redistribute its child nodes to even-scheduling or odd-scheduling, considering the amount of traffic, in order to optimize scheduling. DeTAS requires a number of control

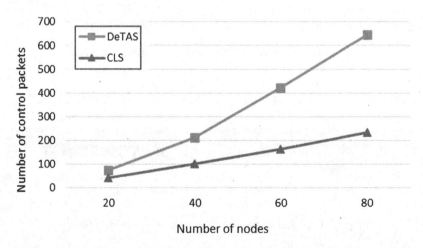

Fig. 5. The number of control packets necessary for scheduling before the routing graph stabilized with source nodes [20, 80]

packets because of the need to reschedule all nodes in the network. On the other hand, CLS requires a small number of control packets because it maintains the schedule of nodes that exist in the path from the newly joined node to the sink. In DeTAS, the more nodes there are, the more control packets are required; but in CLS, the number of control packets increases linearly because CLS does not require rescheduling. When the number of nodes is 80, CLS shows 275 % better performance than DeTAS. Also, the schedule of CLS is completed in 2.3 s, while the schedule of DeTAS is completed in 6.4 s.

Figure 6 shows the average number of control packets necessary for scheduling when a parent node is changed. Although the number of nodes increases, CLS requires a nearly constant number of control packets. Industrial wireless sensor networks are dense; most of nodes are connected in 3-hops from the sink even if the number of nodes increases. CLS deallocates cells or allocates cells without rescheduling, so that the number of control messages is measured as previous $DODAGrank+$ present $DODAGrank * 2$. On the other hand, DeTAS requires many control messages for rescheduling when a parent node is changed. When the number of nodes is 80, CLS shows 351 % better performance than DeTAS. Also, the schedule of CLS is completed in 55.2 ms, while the schedule of DeTAS is completed in 193.9 ms.

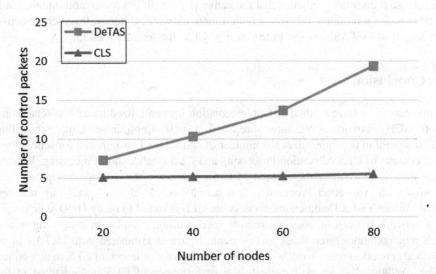

Fig. 6. The number of control packets for scheduling when a parent node is changed with the source node [20, 80]

Figure 7 shows a comparison of end-to-end latencies of CLS and DeTAS in a network with 50 source nodes. Each node begins to transmit the packet at 1 s intervals from the 10th cycle. Latency values between cycles differ due to stochastic transmission failures based on the PDR between nodes. In general, DeTAS and CLS end-to-end latencies are similar, but DeTAS seems to have a lower latency than CLS. Because DeTAS scheduling minimizes the active slot length to an optimum value, packets can

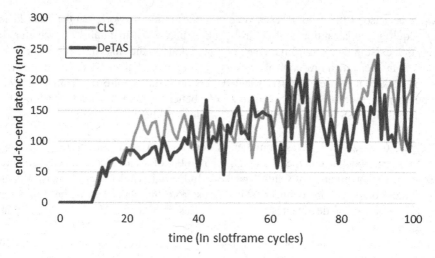

Fig. 7. End-to-end latency with 50 source nodes

reach the sink slightly faster. CLS is scheduling that minimizes the number of control packets, so it does not guarantee that the active slot length is kept to a minimum, but it is close to the minimum. When test with a network with 50 nodes, the average active slot length of DeTAS was 49; in the case of CLS, the lengths was 53.

5 Conclusion

In this paper, we have studied an implementation for centralized multi-hop scheduling in 6TiSCH networks. We have proposed a new Centralized Link Scheduling (CLS) algorithm that minimizes the number of control packets required for scheduling. CLS consists of CLS Allocation Processing and CLS Deallocation Processing. When a node joins a network or the amount of traffic increases, the sink assigns proper slot through CLS Allocation Processing. If a parent node is changed, cells are released locally through CLS Deallocation Processing. CLS is suited to many DODAG changes in a network because it does not require rescheduling. Simulation results show that CLS requires about three times as few control packets compared with DeTAS in an 80-node network. Future work will improve the active slot length of CLS to an optimal value. Additionally, we will evaluate the performance of CLS implemented in the OpenWSN [16] project, which is an open-source implementation of an IoT stack, using multiple hardware platforms.

Acknowledgments. This work was supported by the Energy Efficiency & Resources Core Technology Program of the Korea Institute of Energy Technology Evaluation and Planning (KETEP) granted financial resource from the Ministry of Trade, Industry & Energy, Republic of Korea (No. 20151110200040). This research was supported by the National Research Foundation of Korea (NRF) grant funded by the Korea government (MSIP) (NRF-2014R1A2A1A 11053047).

References

1. Atzori, L., Iera, A., Morabito, G.: The internet of things: a survey. Comput. Netw. **54**(15), 2787–2805 (2010)
2. Gubbi, J., et al.: Internet of Things (IoT): a vision, architectural elements, and future directions. Future Gener. Comput. Syst. **29**(7), 1645–1660 (2013)
3. Deering, S.E.: Internet protocol, version 6 (IPv6) specification (1998)
4. Kushalnagar, N., Montenegro, G., Schumacher, C.: IPv6 over low-power wireless personal area networks (6LoWPANs): overview, assumptions, problem statement, and goals. No. RFC 4919. RFC 4919 (Informational), Internet Engineering Task Force (2007)
5. Winter, T.: RPL: IPv6 routing protocol for low-power and lossy networks (2012)
6. Shelby, Z., Hartke, K., Bormann, C.: The constrained application protocol (CoAP) (2014)
7. Palattella, M.R., et al.: 6TiSCH wireless industrial networks: determinism meets IPV6. In: Mukhopadhyay, S.C. (ed.) Internet of Things, pp. 111–141. Springer International Publishing, Switzerland (2014)
8. IEEE Standard for Local and Metropolitan Area Networks—Part 15.4: Low-Rate Wireless Personal Area Networks (LR-WPANs) Amendment 1: MAC Sublayer, IEEE Standard 802.15.4e, 16 April 2012
9. Pister, K., Doherty, L.: TSMP: time synchronized mesh protocol. In: IASTED Distributed Sensor Networks, pp. 391–398 (2008)
10. Palattella, M.R., et al.: Traffic aware scheduling algorithm for reliable low-power multi-hop IEEE 802.15. 4e networks. In: 2012 IEEE 23rd International Symposium on Personal Indoor and Mobile Radio Communications (PIMRC). IEEE (2012)
11. Accettura, N., et al.: Decentralized traffic aware scheduling for multi-hop low power lossy networks in the internet of things. In: 2013 IEEE 14th International Symposium and Workshops on a World of Wireless, Mobile and Multimedia Networks (WoWMoM). IEEE (2013)
12. Accettura, N., et al.: Decentralized Traffic aware scheduling in 6TiSCH networks: design and experimental evaluation. IEEE Internet Things J. **2**(6), 455–470 (2015)
13. Thubert, P., Assimiti, R., Watteyne, T.: An Architecture for IPv6 over the TSCH mode of IEEE 802.15. 4. draft-ietf-6tisch-architecture-08 (work in progress) (2015)
14. Song, J., et al.: WirelessHART: applying wireless technology in real-time industrial process control. In: IEEE 2008 Real-Time and Embedded Technology and Applications Symposium, RTAS 2008. IEEE (2008)
15. Standard, I.S.A.: Wireless systems for industrial automation: process control and related applications. ISA-100.11 a-2009 (2009)
16. Watteyne, T., et al.: OpenWSN: a standards-based low-power wireless development environment. Trans. Emerg. Telecommun. Technol. **23**(5), 480–493 (2012)

Coverage and Network Requirements of a "Big Data" Flash Crowd Monitoring System Using Users' Devices

An Nguyen[1], Mikhail Komarov[1,2], and Dmitri Moltchanov[1(✉)]

[1] Department of Electronics and Communications Engineering,
Tampere University of Technology, Tampere, Finland
{truong.nguyen,mikhail.komarov}@student.tut.fi, dmitri.moltchanov@tut.fi
[2] Faculty of Business and Management, School of Business Informatics,
National Research University Higher School of Economics, Moscow, Russia

Abstract. Over the last decade aural and visual monitoring of massive people gatherings has become a critical problem of national security. Whenever possible a fixed infrastructure is used for this purpose. However, in case of spontaneous gatherings the infrastructure may not be available. In this paper, we propose the system for spontaneous "flash crowd" monitoring in areas with no fixed infrastructure. The basic concept is to engage users with their mobile devices to participate in the monitoring process. The system takes on characteristics of "big data" generators. We analyze the proposed system for coverage metrics and estimate the rate imposed on the wireless network. Our results show that given a certain level of participation the LTE network can support aural monitoring with prescribed guarantees. However, the modern LTE system cannot fully support visual monitoring as much more capacity is required. This capacity may potentially be provided by forthcoming millimeter wave and terahertz communications systems.

Keywords: Flash crowds · Monitoring · Visual and aural information

1 Introduction

The question of real-time monitoring of massive people gatherings has always been of critical nature for national security. Conventionally, aural and visual information monitoring is performed using pre-installed infrastructure, e.g., via cameras mounted on lampposts, buildings' walls, etc., connected to the Internet access points using wired or wireless technology. The advantages of this approach include feasibility of optimal coverage planning for both visual and aural information. On of the other side, it brings additional costs of infrastructure and requires apriori knowledge of areas of interest making it not suitable for monitoring of spontaneous gatherings, so-called "flash crowds".

In those areas, where no fixed infrastructure is available and/or in case of spontaneous gatherings, helicopters are conventionally used for crowd monitoring [13]. Such an approach is limited to visual information only and due to rather

© Springer International Publishing AG 2016
O. Galinina et al. (Eds.): NEW2AN/ruSMART 2016, LNCS 9870, pp. 372–382, 2016.
DOI: 10.1007/978-3-319-46301-8_31

high flying altitudes may not provide detailed information even when advanced cameras are used. To alleviate this shortcoming, recently, unmanned aerial vehicles (UAV), particularly, quadrocopters, have been proposed for flash crowds monitoring [5]. In spite of much lower cost of use compared to helicopters and potentially lower altitude allowing to achieve better resolution, such systems are still not suitable for aural information monitoring even though the generated noise is much lower than that of helicopters. The limited flying time requiring frequent and automatic recharge as well as the need for manual navigation adding to the operational costs are additional shortcomings of the system.

Neither pre-installed media capturing infrastructure nor UAV- or helicopter-based units are able to capture minor details of events at micro-scales, especially, in highly-dense environments. Furthermore, these systems are not capable of aural monitoring of the environment without the use of highly expensive directional detectors. In this paper we propose a new monitoring system for both aural and visual information for spontaneous flash-crowd environments in areas having no fixed infrastructure. Recalling that most modern handheld devices are equipped with relatively sensitive microphones and high-resolution cameras, the idea behind the proposed system is to explicitly or implicitly engage the users to participate in the monitoring process with their user devices. By downloading and installing an application a user may explicitly engage himself to the monitoring process. Assuming the uniform distribution of users over the monitored area, the coverage metrics are obtained including the cumulative distribution function (CDF) of the covered area, mean and quantiles for both aural and visual information. For visual information we explicitly take into account blocking of camera view by humans located in the area. We then translate these metrics into the rate required from the network for various audio and video codecs and compare the proposed system to that optimal infrastructure-based deployment. Our numerical results allow to make the following conclusions:

- the capacity of modern LTE system is sufficient to provide audio monitoring of the areas of interest with prescribed coverage metrics;
- novel wireless communications systems, such as those operating in millimeter wave or terahertz frequency bands, are needed for visual monitoring.

The rest of the paper is organized as follows. First, in Sect. 2, we describe the system concept, introduce the metrics of interest and review the related work. The system model and performance modeling environment is introduced next in Sect. 3. The numerical results for both coverage metrics and network requirements are provided in Sect. 4. Conclusions are drawn in the last section.

2   System Design

In this section we first present the concept of the proposed flash crowds monitoring system. We then proceed defining the metrics of interest including both coverage and network rate requirements. Finally, we formalize the problem and review the related work pertaining to the subject of interest.

2.1   The Concept

Nowadays, a high percentage of handheld mobile devices are equipped with integrated media capturing equipment including microphones and cameras. We propose to use these devices for flash crowd monitoring. Depending on implementation users can be explicitly or implicitly engaged into the monitoring process. Explicit engagement presumes an application that users download and run on their mobile devices. In implicit engagement scenario users are not notified about their involvement in the monitoring process. For obvious ethical reasons we will not consider the latter as a viable solution in this paper.

The benefits of the proposed system compared to infrastructure-, helicopter-, or UAV-based monitoring systems are that media is captured at much closer distances inside the flash crowds and that there are potentially a large number of devices providing the coverage. In addition to aural and visual information, modern smartphones equipped with numerous advanced sensing capabilities can also provide other types of information including remote sensing and telemetry. The same principle can be used for environmental monitoring applications.

There are two ways of gathering information from devices participating in environment monitoring. A smartphone-based application could itself provide the logic for information analysis gathered by the audio and video sensors. There are a number of shortcomings associated with this approach. First, the devices shall be extremely powerful as in most case the information need to be processed in real-time. The question of the use of resources not only concerns the processing power and memory but also be related to the high battery usage by applications performing real-time data processing. Although there might be additional incentives to participate in the monitoring campaign except for the "good will" of a user, the aggressive use of limited resources may prohibit the widespread use of the application. Further, there are security concerns as smartphone-cased information processing requires that the knowledge of the monitoring task to be available at user devices. Finally, local information processing may not be useful as a single node may not have enough of data to make conclusive decisions. Indeed, the strength of the proposed system is in the ability to get information from many sources located nearby. Thus, the information shall be delivered first to the certain remote server for further centralized data processing.

The devices participating in the monitoring process are expected to use the resources of cellular system uploading the data to the remote server. For specific applications such as flash crowd monitoring the density of nodes willing to simultaneously use the cellular connectivity can be extremely high and may easily overload the network not only preventing it from handling the data of security application but serving normal connections too. At the same time, in certain cases we do not need more than few nodes to monitor a certain point in space simultaneously. As a result, the external monitoring system shall be capable to turn off remote sensing capability of some nodes.

In this paper we concentrate on the flash crowd monitoring systems. In this context the problem is formalized as following: *for a random placement of users on the landscape what should be the density of nodes providing coverage for a certain type of media such that a percentage of area is covered with probability of x.* Once this question is answered we are interested in the *amount of wireless network resources needed to monitor the area of certain dimensions.*

2.2 Related Work

Based on the description of the system and metrics of interest one could notice that the problem at hand reduces to finding coverage of a space in $\Re^2$ by sets of special configuration. This problem has been extensively studied in the literature. Recent advances in this area are mostly associated with coverage of wireless sensor networks (WSN), where the set of interest is a communications range of a node having circular form. Several advanced results has been reported so far, including simple and elegant solution proposed by Lazos and Poovendran in [12], where they use the integral geometry, particularly, the notion of kinematic density, to provide simple closed-form results for k-coverage problem in WSNs under any distributions of audio sensors. Recalling the system model one may observe that this methodology is directly applicable for aural information.

Assessing performance of visual coverage in stochastic deployments is much more complicated. The reason is that individual objects (humans) block the viewing field of cameras. Figure 1 provides a simple illustration of the complex region visible from three cameras in presence of a single blocker. It also highlights the redundancy associated with the monitoring process. The problem of visibility in the random field of blockers has been addressed in the context of search in forests and, more recently, in context of extremely/tremendous high frequency electromagnetic wave propagation (EHF/THF) in crowded environments [2,8]. However, in all those studies the metric of interest was the probability that a certain point in a field of blockers is visible, not the total visible area.

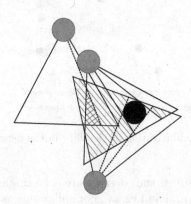

Fig. 1. An illustration of the complex visibility region in presence of a blocker.

3 Performance Modeling

In this section we first introduce the system model using visual information are the media of interest. Further, we describe the proposed simulation environment for performance analysis of the considered monitoring process.

3.1 System Model

Consider the process of visual flash crowd monitoring system. Since the height of user devices is assumed to be comparable with the height of blockers (humans) we can limit our interest to two-dimensional scenario. Fixing a certain time instant t we have a snapshot of a system as illustrated in Fig. 2. The area being covered is assumed to be 100 by 100 m. The humans are represented by circles on the landscape of diameter d. There are overall $N + M$ humans in the area comprising a crowd to be monitored. N humans are assumed to follow a conditional Mattern process with parameter d in the area [6,18]. In other words, no two users could be closer than at the distance $2d$ to each other as in practice human bodies do not overlap. M additional humans participate in the monitoring process and they also follow conditional Mattern process with parameter d. Thus, the overall number of potential blockers for viewing field of cameras is $M + N$.

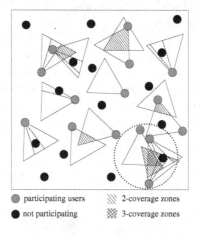

Fig. 2. The illustration of visibility in the dense crowd.

The type of a sensor affects the coverage area of a single node. We concentrate on two types of sensors, aural and visual. For audio sensors, such as microphones, the assumption of circular coverage with radius r_A around a users is taken as humans do not block acoustic waves propagation significantly [17]. For visual sensors, such as cameras capturing video or still images, the field view is by default of sectoral shape with radius r_V. For convenience we model it as an isosceles triangle with the height to the base r_V and apex angle α. To include a random orientation of cameras we assume that the bisect of the apex angle is uniformly distributed in $(0, 2\pi)$. Humans that fall into coverage field including those participating in the monitoring process block view as shown in Fig. 1.

3.2 Simulation Environment

To analyze the formalized problem we have used our own custom-build simulation environment written in C. The choice of high-performance programming language is dictated by the complexity of the coverage area estimation.

Modeling stochastic patterns of humans in the monitored area is critical for accurate performance assessment. To construct a conditional Mattern process with $N + M$ users we first checked the condition $N + M < N^\star$, where $N^\star$ is the number of humans corresponding to dense circle packing [11]. Further, for each individual human we first generated its (x, y) coordinates and checked the condition of non-overlapping. If a newly generated human overlaps with already existing ones coordinates are re-drawn and the process continues up until all the humans are generated. Once $M + N$ humans are generated, M of those are chosen as the ones participating in the monitoring process. They are further assigned audio or video sensor coverage.

Coverage analysis is the most time-consuming procedure. We use the grid method consisting in division of the area of interest into the lattice grid and checking whether nodes of a grid are covered or not [4,14,15,19]. The step of the grid is the parameter severely affecting the trade-off between accuracy of analysis and performance of the simulation framework. In our simulations it was set to 0.1 of a meter.

4 Numerical Results

In this section we first present coverage metrics including CDF, mean and quantile of the coverage process. We then demonstrate the wireless network rate requirements associated with certain coverages. The area of interest for all experiments is set to 100×100 m.

4.1 Coverage Metrics

Coverage CDFs for different number of participating users and different coverage radius of a single user are show in Fig. 3. The number of non-participating humans was kept constant and equal to 1000. Note that instead of the absolute values we plot the percentage of the covered area in OX axis. Expectedly, for the same number of participating users better coverage is provided for larger coverage radius of a single node. Furthermore, increasing the number of participating users provides better coverage. However, as one may observe, even for extremely large number of participating users (e.g., 1000 nodes) full coverage is provided with negligible probability for aural information. Thus, to reliably

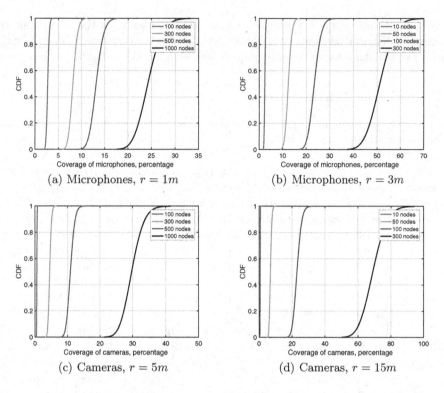

(a) Microphones, $r = 1m$ (b) Microphones, $r = 3m$

(c) Cameras, $r = 5m$ (d) Cameras, $r = 15m$

Fig. 3. Cumulative distribution functions of coverage for microphones and cameras.

cover 100×100 m area one needs to significantly more users than 1000 which might be problematic.

Cameras are characterized by significantly larger coverage radius. Thus, as one may observe already 300 participating users each with coverage radius of 15 m provide non-negligible probability of 80 % coverage of the considered area. We also highlight that the form of CDFs for both visual and aural information are highly-peaked (see, e.g., visual information for 500 nodes) meaning that they should provide rather strict guaranteed of coverage in the considered random deployment scenario. Finally, we stress that blocking of visibility field in visual information scenario does not qualitatively affect the form of CDFs compared to non-blocked aural information scenarios.

The mean values of the area coverage percentage as a function of the number of participating users and different coverage radii of a single user are shown in Fig. 4. The number of non-participating users is set to 1000. One important behavior of this metric is that it does not approach 100 % even for extremely high number of users and rather large coverage of a single users (e.g., 25 m for visual information). This behavior is attributed to completely random choice of the participating users (uniform distribution over the area). Thus, to provide the mean coverage with close to 100 % value is almost impossible for the proposed

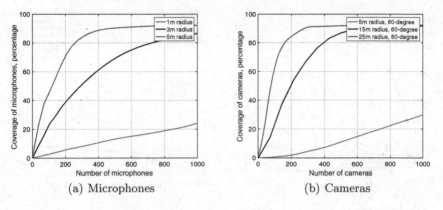

(a) Microphones   (b) Cameras

Fig. 4. Mean coverage by microphones and cameras.

system and can only be achieved using either infrastructure nodes placed in predefined places or drones/helicopters. Another option is to provide a wise choice of participating users selecting those that are located in favorable places.

4.2 Network Requirements

Let us now consider the rate requirements imposed on the wireless networks by the proposed monitoring system. Note that depending on the quality of the codec the coverage area may in generally vary for both aural and visual information. However, this effect is expected to be of minor importance and thus neglected here. Further, parameters such as resolution and compression rate may affect the performance of the automated signal processing algorithms applied to the received information and thus, smaller compression rates and higher resolution are generally preferable. Audio and video codecs we use and their parameters are listed in Table 1 [9,10,20], where MOS stands for mean opinion score.

Table 1. Parameters of audio and video codecs.

Audio codecs				Video codecs			
Type	MOS	Raw rate	LTE rate	Type	Resolution	Raw rate	LTE rate
G.732.1	3.8	5.3 kbps	6.625 kbps	H.264	LD 360p	0.7 Mbps	0.875 Mbps
G.726	3.85	32 kbps	40 kbps	H.264	SD 480p	1.2 Mbps	1.5 Mbps
G.711.1	4.1	64 kbps	80 kbps	H.264	HD 720p	2.5 Mbps	3.125 Mbps

The network requirements in terms of the bitrate needed from the network as well as 0.7 and 0.9 quantiles of the coverage process are plotted in Fig. 5 as a function of the number of participating users for different types of codecs and different coverage radius of a single node. As one may observe, the network

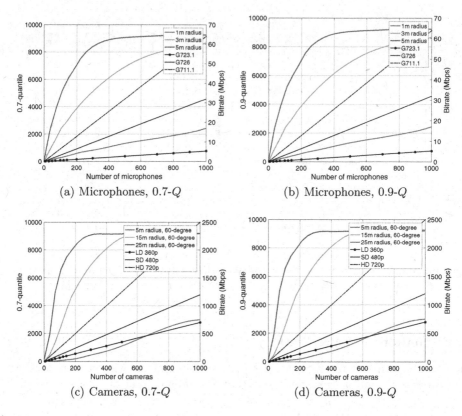

(a) Microphones, 0.7-Q (b) Microphones, 0.9-Q

(c) Cameras, 0.7-Q (d) Cameras, 0.9-Q

Fig. 5. Coverage quantiles and network requirements.

requirements for aural information approaches the value of 60 Mbps for G.711.1 codec (raw rate 64 Kbps). For G.723.1 type of a codec the aggregated rate from the nodes is just 5 Mbps which can be easily handled by the modern LTE systems. Note that even 60 Mbps can be supported by the LTE system.

Expectedly, the bitrates required by the video information are much higher. Even the lowest considered quality LD 360p requires the rate of 500 Mbps to satisfy the 0.7-quantile of the area coverage. Such rate cannot be supported by the modern mobile systems [1, 7, 16]. The millimeter wave systems operating at 28, 60 and 72 GHz and offering the effective rate of up to 7 Gbps are sufficient for visual monitoring even with HD 720p quality. Sub-terahertz systems operating even higher in the frequency band can also be considered as a way for heavy traffic traffic offloading from flash-crowds monitoring systems [3].

5   Conclusions

In this paper we proposed and analyzed the flash crowd monitoring system. The basic principle is to engage a subset of users in the crowd explicitly or implicitly to participate in the monitoring process. Assuming random positions of participating users we analyzed the system for coverage metrics of interest for both aural and visual information. We further derived wireless network requirements in terms of the bitrate for different type of codecs imposed by the proposed monitoring system.

Our numerical results indicate that the required density of the participating users needs to be exceptionally high to achieve "almost full" coverage (e.g., 0.9 quantile) for both audio and video sensors. Although the associated network requirements are exceptionally high they can be supported by the forthcoming millimeter wave or terahertz systems offering substantial rate boost at the air interface. In spite of these pessimistic conclusions we would like to note that the proposed system is the only viable option for detailed monitoring of in-crowd events. Taken altogether, we claim that the proposed system can be effectively used in conjunction with infrastructure-, helicopter- or UAV-based monitoring systems providing detailed information about the area of interest inside a flash crowd in either manual or unmanned manner.

References

1. Andreev, S., Gerasimenko, M., Galinina, O., Koucheryavy, Y., Himayat, N., Yeh, S.P., Talwar, S.: Intelligent access network selection in converged multi-radio heterogeneous networks. IEEE Wirel. Commun. **21**(6), 86–96 (2014)
2. Bai, T., Vaze, R., Heath Jr., R.W.: Analysis of blockage effects on urban cellular networks. IEEE Trans. Wirel. Commun. **13**, 5070–5083 (2014)
3. Boronin, P., Petrov, V., Moltchanov, D., Koucheryavy, Y., Jornet, J.M.: Capacity and throughput analysis of nanoscale machine communication through transparency windows in the terahertz band. Nano Commun. Netw. **5**(3), 72–82 (2014)
4. Brualdi, R., Shanny, R.: A set intersection problem. Linear Algebra Appl. **9**, 143–147 (1974)
5. Burkert, F., Fraundorfer, F.: UAV-based monitoring of pedestrian groups. Int. Arch. Photogramm. Remote Sens. Spatial Inf. Sci. **XL–1**(W2), 67–72 (2013)
6. Chiu, S., Stoyan, D., Kendall, W., Mecke, J.: Stochastic Geometry and Its Applications. Wiley, Hoboken (2013)
7. Galinina, O., Andreev, S., Gerasimenko, M., Koucheryavy, Y., Himayat, N., Yeh, S.P., Talwar, S.: Capturing spatial randomness of heterogeneous cellular/WLAN deployments with dynamic traffic. IEEE J. Sel. Areas Commun. **32**(6), 1083–1099 (2014)
8. Gapeyenko, M., Samuylov, A., Gerasimenko, M., Moltchanov, D., Singh, S., Aryafar, E., Yeh, S., Himayat, N., Andreev, S., Koucheryavy, Y.: Analysis of human body blockage in millimeter-wave wireless communications systems. In: Proceedings of the IEEE ICC, May 2016
9. Radnosrati, K., Moltchanov, D., Koucheryavy, Y.: The choice of VoIP codec for mobile devices. In: Proceedings of the ICN 2014, pp. 1–6 (2014)
10. Radnosrati, K., Moltchanov, D., Koucheryavy, Y.: Trade-offs between compression, energy and quality of video streaming applications in wireless networks. In: Proceedings of the IEEE ICC, pp. 1100–1005 (2014)
11. Kennedy, T.: Compact packings of the plane with two sizes of discs. Discrete Comput. Geom. **35**(2), 255–267 (2006)

12. Lazos, L., Poovendran, R.: Stochastic coverage in heterogeneous sensor networks. ACM Trans. Sensor Netw. **2**(3), 325–358 (2006)
13. Lee, Y., Park, Z., Bunkin, A., Nunes, R., Pershin, A., Voliak, K.: Helicopter-based lidar system for monitoring the upper ocean and terrain surface. Appl. Opt. **41**(3), 401 (2002)
14. Liu, X., Yang, B., Chen, G.: Barrier coverage in mobile camera sensor networks with grid-based deployment. J. Inf. Eng. 210–222 (2015)
15. Pullman, H.: An elementary proof of Pick's theorem. Sch. Sci. Math. **79**(1), 7–12 (1979)
16. Pyattaev, A., Johnsson, K., Andreev, S., Koucheryavy, Y.: Communication challenges in high-density deployments of wearable wireless devices. IEEE Wirel. Commun. **22**(1), 12–18 (2015)
17. Singal, S.: Radio wave propagation and acoustic sounding. Atmos. Res. **20**(2–4), 235–256 (1986)
18. Teichmann, J., Ballani, F., van de Boogaart, K.: Generalizations of matrns hardcore point processes. Spat. Stat. **3**, 33–53 (2013)
19. de Vries, P.: Area estimation with systematic dot grids. In: de Vries, P. (ed.) Sampling Theory for Forest Inventory, pp. 204–211. Springer, Heidelberg (1986)
20. Wiegand, T., Sullivan, G., Bjontegaard, G., Luthra, A.: Overview of the H.264/AVC video coding standard. IEEE Trans. Circuits Syst. Video Technol. **13**(7), 560–576 (2003)

Innovation Radar as a Tool of 5G Development Analysis

Valery Tikhvinskiy[1,2], Grigory Bochechka[1,2(✉)], Alexander Minov[1], and Andrey Gryazev[3]

[1] Icominvest, Moscow, Russian Federation
{v.tikhvinskiy, g.bochechka, a.minov}@icominvest.ru
[2] Moscow Technical University of Communications and Informatics, Moscow, Russian Federation
[3] Federal State Unitary Enterprise Central Science Research Telecommunication Institute, Moscow, Russian Federation
agryazev@zniis.ru

Abstract. 5G Technologies and network development requires all players in the mobile market to participate in the formation of long-term strategy. This includes regulatory authorities, network equipment manufacturers, manufacturers of user equipment, application developers and operators. At the current stage of development of 5G technologies and service applications, a comprehensive analysis of the innovative solutions already offered by major world vendors, research branches of telecom operators, universities and start-up companies should form the basis for long-term strategy. This paper considers an approach to the analysis of innovation by the creation of an information product - 5G Innovation Radar. The product includes the results of a global scouting of 5G innovations, selection and international expertise of the most promising solutions, ranking of 5G innovations by selected criteria of importance and the formation of individual development scenarios, taking into account the activity of each player on the national telecom market. The paper submitted also includes a comparative analysis of innovation assessment results at various stages of the development of 5G Innovation radar.

Keywords: 5G · Innovation · Radar · Mobile communications · Life cycle

1 Introduction

One major task facing the development of a strategy for 5G technologies and services is the systematization of innovative solutions already present on the world market. This will provide evidence of their priority and ranking for a national innovation strategy for 5G. It can be done by the systematization of information available on 5G innovation solutions and built by an information product able to estimate the readiness of innovations for market and the potential economic effect of implementation of such innovative solutions.

Thus, the designing of a new information product such as «5G Innovation Radar» and the solving of such tasks as those listed above can provide for the dynamic development of future 5G networks and the strengthening of competitive mobile

© Springer International Publishing AG 2016
O. Galinina et al. (Eds.): NEW2AN/ruSMART 2016, LNCS 9870, pp. 383–394, 2016.
DOI: 10.1007/978-3-319-46301-8_32

operator positions, as well as rapidly growing demand for 5G equipment and services. This is important at the present stage of national telecom market development.

"5G Innovation Radar" is intended to provide information support for a company's development strategy (vendors and operators) that form the 5G ecosystem at national level. It is also intended for information support of Telecommunication Authorities (the Regulator) that will form 5G national regulatory frameworks. Innovative superiority in 5G has been announced as one of the main objectives of the European Union [1] in implementing the Horizon 2020 program [2].

2   Methodic Approach to the Creation of 5G Innovation Radar

A methodic approach selected for the creation of 5G Innovation Radar is the approach used by one of the largest European telecommunication operators [3, 4]. This approach is called "The Technology Radar" and it includes five major stages of creating an information product and its promotion (Fig. 1):

- Searching and gathering information about the innovative solutions;
- Selection of the most important innovations;
- Ranking of innovation for the selected criteria;
- Identifying of trends and design of innovative scenarios;
- Transferring of the results (Diffusion of innovation) at the level of company management.

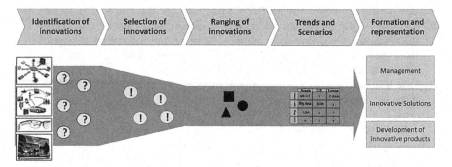

Fig. 1. Methodic approach to the creation of the information product

Stage of searching and collection of 5G innovation solutions that appeared on telecom and related markets in 2014–2016 (the study period) has included:

- The use of the international expert network providing information files with new themes and new solutions descriptions for 5G which appear in the field of view of experts;
- Formation of a complete list of innovation solutions of 5G Innovation Radar (so-called "long list"), including both technology and services for the preliminary evaluation stage for international experts.

At this stage sources of innovative solutions were not only the international network of experts, but also the information materials of numerous telecommunications 5G exhibitions, conferences, symposia, etc. [5–9].

Stage of selection of the most important innovations for 5G Innovation Radar includes:

- assessment of full list of innovations on the scale of importance by the working group of experts;
- selection of 60–70 most important 5G innovations (so-called "short list") as result of expert assessment which used for generating the demo-visual model for 5G Innovation Radar.

This stage also includes the formation of an expert group, which provides qualified expertise of the "long list" of innovation solutions generated on the first stage that is based on development strategy and the level of development of national telecom market as a whole. The international expert group involving in to 5G Innovation Radar project is composed of experts from five countries, representing Russia, Kazakhstan, France, Finland and Germany, working in the international standardization organizations, universities, mobile carrier and vendor companies and consulting companies. Stage of ranking for 5G innovations by selected criteria includes:

- Preparing a summary for each innovation included in the "short list";
- Ranking of innovations based on experts' opinion about their importance for telecom market at the national level;
- Identifying innovations on the screen of 5G Innovation radar with the color indication of innovation importance.

Stage of determination of trends and design of innovative scenarios includes:

- Preparation of the descriptions of innovation profiles;
- Identifying trends in the considered segments of the telecommunications market on the basis of the analysis of innovation;
- Clustering and determination of basic trends in the innovative development of 5G technologies and services on telecom market;
- Design of actual innovative scenarios.

The last stage of 5G Innovation Radar implementation is the most difficult, since it involves a transfer of the achieved research results on levels: Telecom Authorities, National mobile operators and vendors of 5G solutions.

5G Innovation radar methods used to solve the tasks of concerned research stages are:

- Selection and justification of clusters for the classification of several innovative solutions in: technological segment of business, stages of life cycle for innovative solutions, economic feasibility of innovative solutions implementing;
- Formation of the maintenance of inquiry for expert interviews on selected clusters of innovative solution;
- Procedure selection, and then smoothing procedure of survey results of experts;

- Analysis of the survey's results of experts by type of innovational solutions in the telecommunications market;
- Developing the criteria for classification (ranking) of innovation solutions based on technological business segment, the lifecycle of solutions, the economic feasibility of introducing of selected solutions;
- Selection justification of experts and analysis of the reliability of the classification of technological solutions in accordance with the level of experts' qualification.

3   The Choice of Clusters for Classification of 5G Innovations

Classification and further positioning of innovative solutions on a demo-visual model of 5G Innovation Radar are required for a selection and justification of appropriate clusters, which will identify the segments of the telecommunications business.

Three planes for displaying of innovations evaluation were proposed during the designing of a demo-visual model:

- Technological segments of business;
- Life cycle stages of innovation;
- economic feasibility of implementing innovative solutions.

More detailed justification of necessity of these planes is given below.

The Plane "Technological Segment of the Business". Analysis of international experience in the development of information products in the form of "Technological Radar" shows that the clustering is covered by business activities of telecommunications operators on national market. Also the objectives and tasks of the implementation of such information products [3, 4] are covered.

Therefore, clusters in the plane "technological segment of the business" cover all the innovative solutions that are relevant to the national telecom market and have an importance in a particular segment of the telecommunications business. In general cases, an innovation must meet to one of the following segments of its application on national telecom market: services, Radio network and core network, technological aspects of network entities, user equipment. Some clusters for classification of innovative solutions in the plane "technological segment of the business" can be identified by analyzing the activities of the mobile operator.

The Plane "Life Cycle Stages of Innovation". The life cycle of innovation is a certain period of time from the moment of basic research to the moment of decrease of its relevance (replacement an old innovation by new innovation), during of which the innovation has a "vital force" and generates income or other tangible benefits [3, 4]. The life cycles of innovation are different. Primarily these differences consist of the total duration of the cycle, duration of each stage of the cycle, features of the cycle development, and quantities of stages.

Analysis of typical stages of the life cycles of innovations analysis has revealed that 5G Innovation radar should reflect the following stages of the life cycle:

- **Research** - modeling, design studies (beginning of applied research);
- **Conceptual** - iron bird tests and field tests (ending of applied research);
- **Pre-Market** - creation of the prototype (ending of applied research);
- **Market Ready** - pre-series production, refinement, production testing.

The Plane "Economic Feasibility of Innovative Solutions". The economic feasibility of the introduction of innovation should take into account the following factors:

- Business - factor (market aspect);
- Technological factor (technological complexity of innovations' implementation).

Market aspect is a complex assessment which reflecting the potential market volume (revenue from the introduction of innovations), the receipt of income due to reducing the cost of services, developing potential (the ability to move to a new and perspective technologies/services).

The technological complexity of implementation of innovation is also a complex assessment reflecting the implementation complexity, the execution risk, the economic cost of implementation.

Group of international experts involved in 5G Innovation Radar design has classified all innovations in depends on the market aspect and technological implementation complexity based on three gradations: high, medium, low.

The economic feasibility of innovations was conditioned by the predominance of market aspect vs. relative implementation complexity. At the same time, the economic feasibility is ranked as "high - very high", "medium - high" and "low - medium" based on building an innovative matrix shown in Fig. 2 [3].

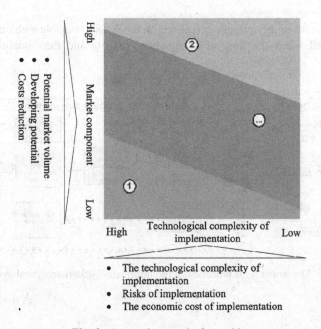

Fig. 2. Innovative matrix for ranking

When an innovation corresponds to the top area of innovation matrix, it has been classified on economic viability as a "high - very high", if an innovation corresponds average part of innovation matrix - the "medium - high" and in last case an economic viability has classified as "low - average".

4   The Structure of Processes for the Development of 5G Innovation Radar

Development of 5G Innovation radar includes a series of related processes that provide collecting of information, questioning of experts and calculation statistical metrics of the expertise, such as sample representativeness, reliability, consistency of expert opinions, assessing the quality of the selected expert.

Generalized process structure of 5G Innovation radar development is shown on Fig. 3. It includes the following stages of research and expertise:

- formation of innovative themes;
- formation graphic demonstration model;
- formation of rank estimations for the stages and business segments;
- assessment of the estimates reliability;
- assessment of the expert opinions consistency;
- formation of innovation annotations;
- formation of innovation profiles;
- economic evaluation of innovations;
- formation of innovative scenarios.

The interaction of these stages is realized on the basis of flexible web interfaces that can be adapted when the flow of information, experts and their qualifications are changing.

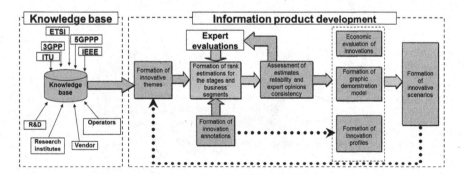

Fig. 3. The structure of processes in the information expert-analytical system

5   The Demo-Visual Model of 5G Innovation Radar

5G Innovation Radar should allow to analyze a very broad range of innovative technologies and services in the telecommunications sector that can be perspective for the National telecommunications authorities, operators and vendors at the national level. Since 5G Innovation radar can be used in different organizations and at different levels of the organization it is important to provide visualization and clarity presentation of analysis materials.

Positioning of innovations is performed on a flat demo-visual model which has a form of traditional radar screen divided into sectors. The sectors are chosen according to the segments that characterize innovative products, technologies and services in the network infrastructure and business in general, namely: user equipment, radio access network, core network, services, related technologies (Fig. 4).

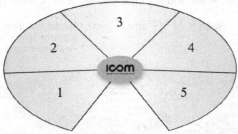

1. User equipment
2. Radio access network
3. Services
4. Core network
5. Related technologies

Fig. 4. Segmentation of innovations on business segments

The proposed classification for the "technological business segments" is sufficiently broad and covers all possible innovative technologies and services related to mobile networks.

The development of innovation is characterized by stages of its life cycle, from the moment of birth (studies in research organizations) until its implementation and use in operator network. The life cycle stages of innovations are positioned on the screen of demo-visual model of 5G Innovation Radar in the form of oval segments. They include: research & development, concept, market prototype, market presence (Fig. 5).

Thus, the demo-visual model of 5G Innovation Radar has up to 20 (5 × 4) different segments for the graphic classification of differences and features 5G innovations. Each displayed innovation can be characterized by three parameters: market significance for the national market, stage of the life cycle and business segment.

As a rule, high visual obviousness on the demo-visual model is achieved when the number of innovations equals 60–70 innovations that are the most relevant for the development of 5G in the national market.

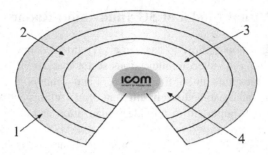

1. Research & development
2. Concept
3. Market prototype
4. Market presence

Fig. 5. Segmentation of innovations on life cycle stages

The results of the research of 5G innovations ranked by their significance degree are shown in Fig. 6.

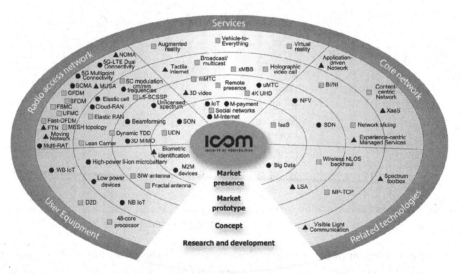

Fig. 6. Demo-visual model of 5G Innovation Radar

The following icons for different relevance of innovations are used:

● - «high to very high»,
▢ - «medium to high»,
▲ - «low to medium».

Demo-visual form of presentation of research results on positioning 5G innovations allows professionals effectively use these results as an information resource for planning company's development strategy or innovative strategy.

Results of 5G innovations analysis presented in Fig. 6 show:

- 23 innovations have a "high to very high" relevance, 32 innovations have a "medium to high" relevance, 13 innovations have a "low to medium" relevance;
- Most of innovations selected for consideration have not yet represented in the telecommunications market, 10 are in the «Market presence» stage, 15 are in the «Market prototype» stage, 18 are in «Concept» stage and 25 are in «Research & Development» stage;
- 27 innovations are in segment of Radio access network, 16 are in the segment of Services, 10 are in the segment of User equipment, 9 are in the segment of Core network, 6 are in the segment Related technologies.

Sector of Radio access network includes the largest number of innovations (40 %), underlining the importance of creating a competitive air-interface at this 5G development stage, which will be capable to implement the basic technical requirements specified by ITU and 5G PPP for ITM-2020 technology (Fig. 7) [5].

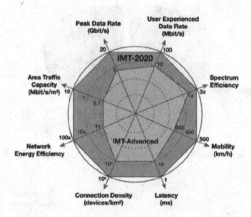

Fig. 7. Enhancement of key capabilities from IMT-Advanced to IMT-2020 (Source: ITU)

The conducted studies have shown that the number of innovations in 5G sphere, pre-selected by scouters for primary research is more than 250 titles.

Figure 8 shows the distribution of innovations in the full list of innovations before the stage of innovations ranking on segments.

Most of all 5G innovations relates to radio access network (31 %), services (21 %) and related technologies (22 %). This distribution of innovations is due to high competition of different solutions for radio access networks, due to the integration of communication networks and ICT and emergence of related technologies, as well as the desire of operators to increase ARPU through new services in existing and new market segments like M2M, IoT and Mobile Internet.

Accounting of expert opinions at the stage of ranking of innovations allows to select the most relevant innovations which leads to a change in the distribution of innovations (Figs. 9 and 10).

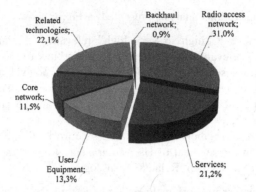

Fig. 8. An example of the distribution of pre-selected innovations

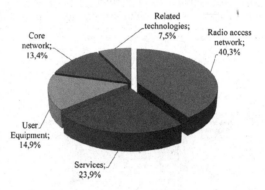

Fig. 9. An example of the distribution of innovation in clusters

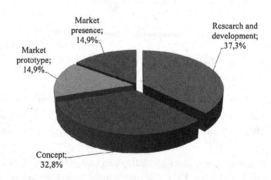

Fig. 10. An example of the distribution of innovations on the stages of the life cycle

Analysis of the distribution of innovations in Figs. 9 and 10 shows that most innovations are located on «Research & Development» and «Concept» stages, in segment "Radio access network". This is due to the current stage of 5G infrastructure development and specificity of telecommunication venders activities who receive income from the sale of such products and who are the main generators of such innovations.

Innovations that are in the stage of "Research and development", have less interest for investors. This is due to the large time interval between the step of "research and development" and "market presence" step, during which the innovation may lose its relevance given the high dynamics of development of the telecommunications market.

Most of the 5G innovations with the relevance of "high - very high" involve the utilization of cloud computing, network virtualization and realization of «All IP» concept in all network elements.

Innovations with "medium to high" relevance reflect the degree of development of advanced technological solutions are already implemented in the 3GPP technical specifications for generation 4, 5G (Release 13 and beyond). Innovations having "low to medium" relevance according to experts' opinion, have not ability to significantly increase the income of the operators, but can maintain steady growth in these revenues. These innovations have either indirectly related to the activities of operators, or are innovation, the utilization of which cannot be controlled by the telecom operator (for example, IoT/M2M, licensed spectrum sharing, augmented and virtual reality). The introduction of these innovations can stimulate the growth of traffic in mobile networks.

6   Conclusion

Information product such as Innovation radar 5G can be one of the tools of innovation management used by Communications Administration, operators and venders for 5G development at the national and corporative levels.

This product can provide the timely establishment of the regulatory framework for implementation of 5G communications, the formation of investment and technology policies of venders, development and strengthening of the competitive position of operators in order to achieve innovation excellence of the company.

Research at development of 5G Innovation Radar allowed to select the most important innovations from more than 250 innovative technologies and services on the market of 5G. These innovative solutions have been positioned at a flat demo-visual model of 5G Innovative radar which allows positioning and analyzing of the innovations based on their importance, their life cycle stage and their economic efficiency. 5G Innovation Radar is a good tool for early stage prioritization of 5G innovations to give an approximate value judgment even without detailed return on investment justifications.

International experts from 5 countries were involved in research for the creation of 5G Innovation Radar to improve its validity and reliability. These experts represent different research organizations from various segments of the telecommunications market.

Reached results reflect the period of 5G innovative activities from 2014 to 2016 and can help in decision-making at the level of the Administration of communications, national operators and 5G equipment developers.

Analysis of 5G innovations conducted for the Russian telecom market and can be applied to other national markets, with a similar level of market development.

Acknowledgements. The authors would like to express thanks to Igor Minaev and Susan Wood for their comments and corrections during the preparation of this paper.

References

1. The 5G Infrastructure Public Private Partnership: the next generation of communication networks and services, 5G PPP (2015). https://5g-ppp.eu/
2. COM/2011/0808 Horizon 2020 - The Framework Programme for Research and Innovation - Communication from the Commission. http://ec.europa.eu/
3. Golovatchev, J., Budde, O.: Technology and innovation radars. Effective instruments for the development of a sustainable innovation strategy and successful product launch. J. Innov. Technol. Manag. 7(3 & 4), 229–236 (2010)
4. Rohrberk, R., Heuer, J., Arnold, H.: The technology radar - an instrument of technology intelligence and innovation strategy. In: IEEE International Conference on Management of Innovation and Technology, pp. 978–983 (2006)
5. Rec. ITU-R M.2083-0 (09/2015). IMT Vision – Framework and overall objectives of the future development of IMT for 2020 and beyond. https://www.itu.int/
6. 3GPP RAN 5G Workshop Proceedings, Phoenix, AZ, USA, 19 September 2015. http://www.3gpp.org/
7. ETSI Workshop 5G from myth to reality, 21 April 2016. http://www.etsi.org/
8. ETSI Future Radio Technologies workshop, 27 January 2016. http://www.etsi.org/
9. Hu, F.: Opportunities in 5G Networks: A Research and Development Perspective. CRC Press, Boca Raton (2016)

Analytical Evaluation of D2D Connectivity Potential in 5G Wireless Systems

Ammar Muthanna[1(✉)], Pavel Masek[2,3], Jiri Hosek[2,3],
Radek Fujdiak[2], Oshdi Hussein[1], Alexander Paramonov[1],
and Andrey Koucheryavy[1]

[1] State University of Telecommunication,
Pr. Bolshevikov 22, St. Petersburg, Russia
ammarexpress@gmail.com
[2] Brno University of Technology,
Technicka 3082/12, 61600 Brno, Czech Republic
[3] RUDN University, 6 Miklukho-Maklaya St., Moscow 117198, Russia

Abstract. Constantly growing number of wireless devices leads to the increasing complexity of maintenance and requirements of mobile access services. Following this, the paper discusses perspectives, challenges and services of 5th generation wireless systems, as well as direct device-to-device communication technology, which can provide energy efficient, high throughput and low latency transmission services between end-users. Due to these expected benefits, the part of network traffic between mobile terminals can be transmitted directly between the terminals via established D2D connection without utilizing an infrastructure link. In order to analyse how frequently can be such direct connectivity implemented, it is important to estimate the probability of D2D communication for arbitrary pair of mobile nodes. In this paper, we present the results of the network modelling when the random graph model is used. The model was implemented as a simulation program in C# which generates random graph with a given number of vertexes and creates the minimal spanning tree (mst) by using the Prime's algorithm. All our result and practical findings are summarized at the end of this manuscript.

Keywords: 5G · Connectivity · D2D communication · Minimal spanning tree · Traffic modelling

1 Introduction

Mobile operators [1, 2] are accepting D2D (Device-to-Device) as a part of the fourth generation (4G) Long Term Evolution (LTE)-Advanced [3] standard described in 3rd Generation Partnership Project (3GPP) Release 12 [4]. As 4G wireless systems are reaching their maturity and getting to be massively deployed [10], the future fifth generation (5G) cellular networks have drawn great attention from researchers and engineers around the world. 5G networks are envisioned to attain 1.000 times higher mobile data volume per unit area, 10–100 times higher number of connecting devices and user data rate, 10 times longer battery life, and 5 times reduced latency. As a vital

© Springer International Publishing AG 2016
O. Galinina et al. (Eds.): NEW2AN/ruSMART 2016, LNCS 9870, pp. 395–403, 2016.
DOI: 10.1007/978-3-319-46301-8_33

component of future wireless networks, D2D communication refers to a radio technology that enables devices to communicate directly with each other, that is without routing the data paths through a network infrastructure. Potential application scenarios include, among others, proximity-based services where devices detect their proximity and subsequently trigger different services (such as social applications triggered by user proximity, advertisements, local exchange of information, smart communication between vehicles, etc.). Other applications include public safety support, where devices provide at least local connectivity even in case of damage to the radio infrastructure [14].

During the last several years, we have been witnessing dramatic growth of wireless network traffic and as well as a number of communicating devices. This progression is expected to continue even with faster pace in the near future. In 2020 traffic volumes are envisioned to be 1000 times higher than traffic volumes of today. As declared by many, the layout of 5G cellular system will be denser with many small cells supporting large amount of devices while the energy efficiency will be higher compared to current systems [2].

D2D communication, as one of promising answers to aforementioned significant escalation of mobile traffic, plays an increasingly important role for whole community. It facilitates the discovery of geographically close devices, and enables direct communications between these proximate nodes, which improves communication capability and reduces communication latency and power consumption. Currently, there are two main operation modes of D2D communication – with network assistance or without (e.g. when a user is out-of-coverage) [3]. It is expected that D2D will (re)use the same licensed spectrum as the cellular links.

The rest of the paper is organized as follows. Section 2 provides the description of different direct communication scenarios. Further, in Sect. 3 we discuss in detail traffic pattern expected in 5G networks showing that a part of traffic between mobile terminals (MT) can be served without network infrastructure. Created simulation model and results are provided in Sect. 4. Finally, the conclusions and future work are summarized in Sect. 5.

2 D2D Communication Scenarios

Based on the considered business models, operator can have different levels of control of D2D communication link or prefers not to have any [11, 12]. When having control, the operator can either exercise full/partial control over the resource allocation among source, destination, and relaying devices [9, 15, 16]. D2D is a communication between two or more MTs in proximity that is enabled by means of user plane (U-plane) transmission using E-UTRA technology via a path not traversing any network node, see Fig. 1. The scenarios for device discovery and direct communication within network coverage and outside are depicted in Fig. 2. There can be different situations when MT is out-of-coverage, however, we consider it when the average SINR is less than-6 dB [11].

There are multiple different D2D communication scenarios when the registered Public Land Mobile Network (PLMN), direct communication path and coverage status (in coverage or out of coverage) are taken into the account [5].

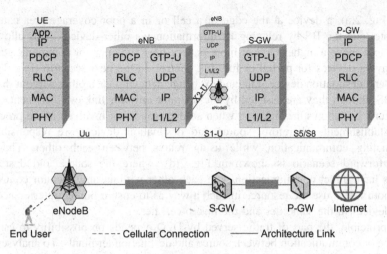

Fig. 1. User plane protocol stack between the eNodeB and UE

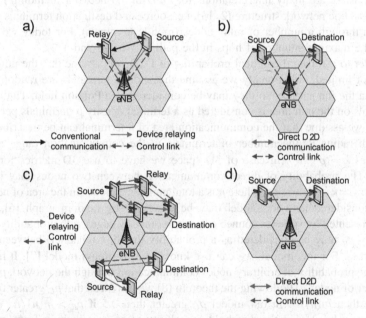

Fig. 2. Possible traffic service types in 5G network

3 Communication Service Types in 5G Network

As we mentioned before, 5G network concept includes D2D communication principles. It means that a part of traffic between mobile terminals (users) can be served without utilizing whole network infrastructure (base stations, switching centers, etc.), but only by means of mobile terminals (with or without partial participation of a network, see Fig. 2).

In Fig. 2(a), a device at the edge of a cell or in a poor coverage area can communicate with the BS by relaying its information via other devices. This allows the device to achieve a higher QoS or more battery life. The operator communicates with the relaying devices for partial or full link control. In the next scenario (Fig. 2b), the source and destination devices talk and exchange data with each other without the need for a BS, though they are assisted by the operator only for link establishment.

Figure 2(c) shows the situation when an operator is not involved in the process of link establishment. Therefore, source and destination devices are responsible for coordinating communication while using relays between each other. The most straightforward scenario is shown in Fig. 2(d), where the source and destination devices have direct communication with each other without any operator control. As such, both nodes use the resource in such a way as to ensure limited interference with other devices in the same tier and the macro-cell tier.

In principle, the part of traffic served by D2D depends on possibility to establish this type of communication between source and destination terminals. To analyse it, we have to estimate the probability of D2D communication for arbitrary pair of mobile terminals. There are many interpretations for D2D in 5G network including the possibility of ad hoc network structure [2, 13], i.e. source and destination terminals may be connected through a number of relay nodes (mobile terminals). For today we cannot propose the maximal number of hops in the path between S and T.

In order to estimate theoretical probability of D2D, we assume that the number of hops is not limited. Additionally we assume that mobile terminals are randomly distributed on the flat surface, so they may be considered as a Poisson field. The number of terminals on the unit area is considered as a terminal density ρ (terminals per square meter). If we assume that the communication area of a terminal can be described by a circle with radius R, the number of terminals in the communication range may be obtained as $k = \rho \cdot \pi R^2$ (in case of 3D space we have to use 3D metrics for ρ and volume). The probability of direct communication between two nodes may be estimated as $p_R = k/n$ $p_R = k/n$, where n is a total number of nodes in the area of network.

The considered network model may be described by random graph [6], where nodes represented by vertexes connected each to other by edges with probability p. The probability p_{ij} may be considered as a probability that link (channel) between nodes i and j exists. For this model, we can use known Erdős–Rényi model [7]. It helps to express the probability of arbitrary nodes communication through the network p_C from the number of nodes n. Following the theorem [8] we can mean that p_C greater then 0.5 if p grate then $ln(n)/n$. In our model p_C greater then 0.5 if $p_R > ln(n)/n$. We can consider $p_0 = ln(n)/n$ as the threshold probability of phase transition from not connected phase to connected phase. To illustrate this model we have done simulation of the network structure including the above described assumptions.

4   Created Model and Simulation Results

In Fig. 3, we present the results of simulation of the network structure with 300 nodes (terminals) for two communication ranges: R = 60 m (Fig. 4a) and R = 35 m (Fig. 4b). It is evident from these pictures, that in the first case the network is fully connected,

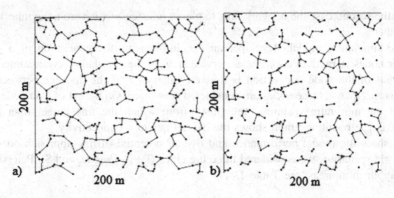

Fig. 3. Network connectivity with (a) R = 60 m, (b) R = 30 m

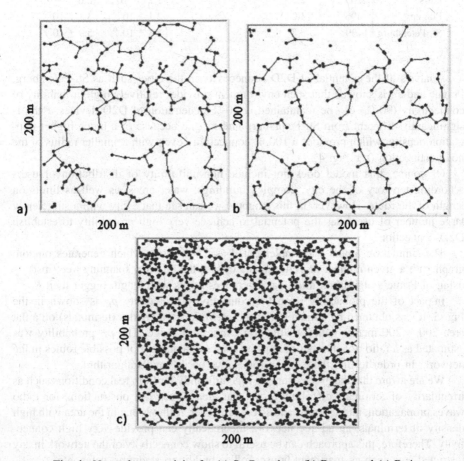

Fig. 4. Network connectivity for (a) St. Petersburg (b) Prague and (c) Paris

but in the second case the network falls to pieces (clusters) which are not connected to each other.

The simulation model, developed in C#, implements the generation of a given number nodes in random coordinates service area. In the simulation, we assumed that the radius of the node connection is a given value of r. If the distance between the nodes is less than r, these nodes are considered connected. On the set of data nodes, we applied the minimum spanning tree (mst), from which the fins longer than r are removed. The resulting graph shows the overall network connectivity.

To show the model performance and overall potential of this approach on some real-world scenarios, we considered three big cities (Paris, Prague, and St. Petersburg) as an application areas, see Table 1.

Table 1. Connectivity by D2D communication for different cities

	Square/km^2	Population [million citizens]	$p_0 = \ln(n)/n$	$p_R = \pi R^2/n$
Paris	105.4	2.2	$1.27 * 10^{-5}$	$2.68 * 10^{-5}$
Prague	496	1.2	$2.1 * 10^{-5}$	$3.1 * 10^{-5}$
St. Petersburg	1439	5.8	$3.8 * 10^{-6}$	$5.9 * 10^{-6}$

Analysis of the potential of D2D connectivity in the cities such as St. Petersburg, Prague and Paris showed that such environment provides relatively high probability of connectivity (90 %) can be maintained at 50 % penetration of D2D devices. Paris is significantly different from St. Petersburg and Prague because of a much higher population density, which provides a 100 % connectivity even with a smaller radius of the node connection (see Fig. 4).

Of course, this model does not include non-uniformity of distribution of users according territory of the city, complex buildings, water areas, as well as limits on length of the route. However, in this example we can see that a city with a sufficiently large number of users has the potential to achieve very high probability to establish D2D connection.

The simulation model was implemented as C# program which generates random graph with a given number of vertexes and creates the minimal spanning tree (mst) by using of Prime's algorithm, after that it removes edges with length bigger then R.

Impact of the probability p on the connectivity probability p_C is shown in the Fig. 5. It was obtained by simulation of the network of 100 nodes (terminals) on a flat area 200×200 meters with variable communication range. The p_C probability was estimated as a ratio of a number of existing routes to a number of possible routes in the network. In order to find the existing routes, we used Floyd's algorithm.

We are aware that the given example is significantly far from real conditions such as irregularity of terminals distribution, not flat area of service, obstructions for radio waves propagation, interferences and so on. However, it shows that in the area with high density of terminals, an ad-hoc network theoretically can provide very high connectivity. Therefore, this approach can be useful to show connectivity of the network in any restricted area such as apartment houses, office buildings, stadiums, cafes, etc.

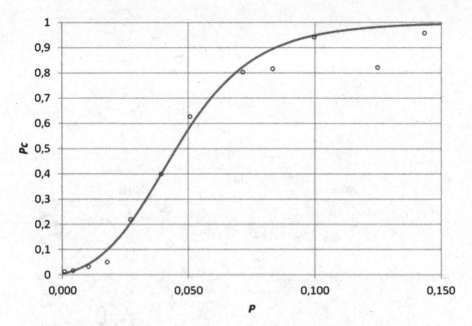

Fig. 5. Impact of the p on the connectivity probability.

We have been also analysing the network connectivity level for different densities (see the results in Fig. 6). It is clear that as the network becomes more densified, the connectivity rate grows as well. As a consequence, significant part of traffic may be served by D2D communications potentially.

In practice, to establish connection in an ad hoc network it is very important to implement self-organizing protocols. However, complex self-organizing network with a big number of terminals and not stable structure becomes difficult to control and moreover, it generates a lot of signalling which negatively impacts the channel capacity distributed between the terminals. So, in general, increasing the number of terminals leads to an increase in D2D probability, but also to decrease in overall network efficiency.

This example illustrates the network structure for different densities of communication nodes. While connectivity with high-density holds close to 100 % (Fig. 6d), for the low density, network graph is divided into a set of disconnected clusters (Fig. 6a). If the potential link between nodes is more than $1/n$, the network comprises a gigantic component with the number of nodes around n (Fig. 6b). If the probability of connection between the nodes is less than $1/n$, the network contains disconnected clusters with number of nodes around $ln(n)$ (Fig. 6a).

While analysing Fig. 6a we can propose using of any additional nodes which can increase the connectivity potential between different unconnected groups. Additional nodes may be temporary located in the area between groups by using of Unmanned Aerial Vehicles (UAVs) for example.

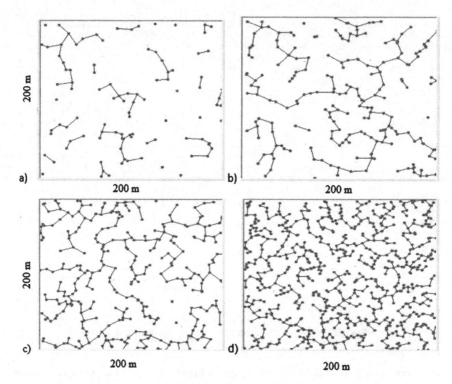

Fig. 6. Network connectivity for density: (a) 100, (b) 200, (c) 300, (d) 700

5 Conclusion and Future Work

Future traffic development brings new requirements and challenges to future mobile systems, so 5G networks will be a combination of different enabling technologies and D2D, as one of these technologies, will play mean role in future wireless networks. In this work, we analysed the potential scenarios of D2D communications in 5G network. Using the simulation model, we have been estimating the probability of D2D connectivity under different conditions. As first, these initial results confirm our theoretical presumption that the area with high density of terminals can provide very high connectivity.

Moreover, we also identified the potential issues of low-densified areas which can be solved by adding temporary nodes located in the area between groups by using of UAVs. Our approach can be also applied to investigate the D2D potential in specific use cases which is highly required by network operators to evaluate their intended investments into the D2D technologies.

Acknowledgement. Research described in this paper was financed by the National Sustainability Program under grant L01401 of Brno University of Technology. For the research, infrastructure of the SIX Center was used. The reported study was funded within the Agreement

№ 02.a03.21.0008 dated 24.11.2016 between the Ministry of Education and Science of the Russian Federation and RUDN University.

References

1. Ericsson: 5G RADIO ACCESS, Ericsson Paper White, June 2013
2. NSN: White Paper, Looking ahead to 5G, NSN White Paper, December 2013
3. Andreev, S., Pyattaev, A., Johnsson, K., Galinina, O., Koucheryavy, Y.: Cellular traffic offloading onto network-assisted device-to-device connections. IEEE Commun. Mag. 52(4), 20–31 (2014)
4. 3GPP: 3GPP Release 12 (2015). www.3gpp.org/specifications/releases/68-release-12. Accessed 8 Feb 2015
5. 3GPP: TR 23.703 v12.0.0, Study on architecture enhancements to support Proximity-based Services (ProSe) (Release 12), February 2014
6. Erdős, P., Rényi, A.: On random graphs I. Publ. Math. Debrecen 6, 290–297 (1959)
7. Erdős, P., Rényi, A.: On the evolution of random graphs. Publ. Math. Inst. Hungar. Acad. Sci. 5, 17–61 (1960)
8. Erdős, P., Rényi, A.: On the evolution of random graphs. Bull. Inst. Int. Statist. Tokyo 38, 343–347 (1961)
9. Masek, P., Muthanna, A., Hosek, J.: Suitability of MANET routing protocols for the next-generation national security and public safety systems. In: Balandin, S., Andreev, S., Koucheryavy, Y. (eds.) NEW2AN/ruSMART 2015. LNCS, vol. 9247, pp. 242–253. Springer, Heidelberg (2015)
10. Masek, P., Zeman, K., Hosek, J., Tinka, Z., Makhlouf, N., Muthanna, A., Novotny, V.: User Performance gains by data offloading of LTE mobile traffic onto unlicensed IEEE 802. 11 links. In: Proceedings of the 38th International Conference on Telecommunication and Signal Processing, TSP 2015, 1 Prague, Czech Republic, s. 1–5. Asszisztencia Szervezo Kft (2015). ISBN 978-1-4799-8497
11. 3GPP: TR 36.843, v12.0.1, Study on LTE Device to Device Proximity Services; Radio Aspects (Release 12), March 2014
12. Fodor, G., et al.: Design aspects of network assisted device-to-device communications. IEEE Commun. Mag. 50(3), 170–177 (2012)
13. Bettstetter, C.: On the connectivity of ad hoc networks. Comput. J. 47(4), 432–447 (2004). doi:10.1093/comjnl/47.4.432
14. Pitsiladis, G.T., Panagopoulos, A.D., Constantinou, P.: A spanning-tree-based connectivity model in finite wireless networks and performance under correlated shadowing. IEEE Commun. Lett. 16, 842–845 (2012)
15. Andreev, S., Moltchanov, D., Galinina, O., Pyattaev, A., Ometov, A., Koucheryavy, Y.: Network-assisted device-to-device connectivity: contemporary vision and open challenges. In: Proceedings of 21th European Wireless Conference European Wireless (2015)
16. Pyattaev, A., Hosek, J., Johnsson, K., Krkos, R., Gerasimenko, M., Masek, P., Ometov, A., Andreev, S., Sedy, J., Novotny, V., Koucheryavy, Y.: 3GPP LTE-assisted Wi-Fi-direct: trial implementation of Live D2D technology. ETRI J. 37(5), 877–887 (2015). ISSN 1225-6463

Queuing Model with Unreliable Servers for Limit Power Policy Within Licensed Shared Access Framework

Konstantin Samouylov[1], Irina Gudkova[1,2(✉)], Ekaterina Markova[1], and Natalia Yarkina[1]

[1] RUDN University, 6 Miklukho-Maklaya St., Moscow 117198, Russia
{samuylov_ke, gudkova_ia, markova_ev}@pfur.ru,
nat.yarkina@mail.ru
[2] Institute of Informatics Problems, Federal Research Center "Computer Science and Control" of Russian Academy of Sciences (IPI FRC CSC RAS), 44-2 Vavilova Str., Moscow 119333, Russia

Abstract. Shared access to spectrum by several parties seems to become one of the most promising approaches to solve the problem of radio spectrum shortage. The framework proposed by ETSI, licensed shared access (LSA), gives the owner absolute priority in spectrum access, to the detriment of the secondary user, LSA licensee. The latter can access the spectrum only if the owner's QoS is not violated. If the users of both parties need continuous service without interruptions, the rules of shared access should guarantee the possibility of simultaneous access. Balancing the radio resource occupation between parties could take quite a long time compared to the dynamics of the system due to the coordination process by the national regulation authority (NRA). We examine a scheme of the simultaneous access to spectrum by the owner and the LSA licensee that minimizes the coordination activities via NRA. According to this scheme, when the owner needs the spectrum, the power of the LSA licensee's eNB/UEs is limited. From the LSA licensee's perspective, the scheme is described in the form of a queuing system with reliable (single-tenant band) and unreliable (multi-tenant band) servers. We show that the infinitesimal generator of the system has a block tridiagonal form. The results are illustrated numerically by estimating the average bit rate of viral videos, which varies due to aeronautical telemetry corresponding to the owner's traffic.

Keywords: Licensed shared access · Limit power policy · Queuing system · Unreliable servers · Blocking probability · Average bit rate

The reported study was funded within the Agreement № 02.a03.21.0008 dated 24.11.2016 between the Ministry of Education and Science of the Russian Federation and RUDN University, by RFBR according to the research projects No. 15-07-03608 a and 16-37-00421 mol_a.

O. Galinina et al. (Eds.): NEW2AN/ruSMART 2016, LNCS 9870, pp. 404–413, 2016.
DOI: 10.1007/978-3-319-46301-8_34

1   Introduction

The demand for mobile broadband services as well as the volume of traffic increases every year [1, 2]. A considerable amount of frequency resources is needed to provide to users services with a required level of quality of service (QoS) [3, 4]. The problem of resource shortage can be solved by means of the shared access to spectrum by several entities, implemented, for instance, by using the licensed shared access (LSA) framework [5, 6]. LSA framework [7] can improve the efficiency of resource usage and ensure the access to a spectrum which otherwise would be underused. The spectrum is shared between the owners (incumbents) and a limited number of LSA licensees (e.g., mobile network operators). The LSA licensee has access to both bands – the single-tenant band assigned only to it and the multi-tenant band assigned also to the incumbent. The LSA implementation is required to guarantee the QoS for all users, the strictest requirement being not to interrupt users in service due to the incumbent accessing spectrum.

For the shared access, ETSI [8] proposes to use the spectrum allocated for aeronautical and terrestrial telemetry or specific applications including cordless cameras, portable video links, and mobile video links. Various policies of interference coordination between two entities could be considered. The authors of [3] propose three of them: the so-called ignore policy [8, 9], shutdown policy, and limit power policy. The latter implies managing the user equipment (UE) power in uplink and eNodeB (eNB) power in downlink.

In the paper, we consider the case described in [3, 9, 10]. The airport (incumbent) has a frequency band for telemetry with airplanes (air traffic control, ATC). The mobile operator (LSA licensee) also has access to it, thus having its own single-tenant band and the incumbent's multi-tenant band. We assume the users of the mobile operator to watch short videos (e.g., viral) [11] in high quality. At the time when the airplane is communicating with ATC, ATC asks the mobile operator to limit the interference around the airplane. The interference threshold is achieved by reducing the downlink power of the eNB creating interference with the airplane (limit power policy). This results in a bit rate decrease [12] on the multi-tenant band so that the users continue watching video (but in a lower quality). Note that the users will continue to get service at a degraded bit rate after the release of the multi-tenant band by the airport. This is due to the fact that any changes require additional signaling procedures and potential coordination with the national regulation authority (NRA), which could lead to intolerable delays. We also assume that, on the multi-tenant band, new requests are accepted at the maximum bit rate only when all users at the degraded bit rate have finished watching videos.

The paper is organized as follows. In Sect. 2, we propose a mathematical model of the LSA framework with the limit power policy. In Sect. 3, we analyze numerically the performance measures: the blocking probability, the average bit rate, and the utilization factor. Section 4 concludes the paper.

2 Mathematical Model

2.1 General Assumptions and Parameters

We consider a single cell of mobile network with an overlaid LSA framework and one service that generates streaming traffic. We suppose that the single-tenant band has the total capacity of C_1 bandwidth units (b.u.) whereas the multi-tenant band has the total capacity of C_2 b.u. Each request processed on the single-tenant band is served at the guaranteed bit rate (GBR) d_{max}. The number of resources allocated to the request on the multi-tenant band equals to d_{max} or d_{min} depending on the state of the multi-tenant band – operational or unavailable.

Let the arrival rate λ be Poisson distributed and let the service time be exponentially distributed with mean μ^{-1}. Then, we denote the corresponding offered load as $\rho = \lambda/\mu$. We assume that the multi-tenant band goes into unavailable mode with rate α and recovers into operational mode with rate β. Recovery and failure intervals follow the exponential distribution. Let us introduce the following notation:

- n_1 – the number of single-tenant band users;
- n_2^{max} – the number of multi-tenant band users when the multi-tenant band is operational;
- n_2^{min} – the number of multi-tenant band users when the multi-tenant band is unavailable;
- s – the state of the multi-tenant band, s equals to 1 if the band is operational and s equals to 0 if the band is unavailable;
- $N_1 = \left\lfloor \frac{C_1}{d_{max}} \right\rfloor$ – the maximum number of single-tenant band users;
- $N_2 = \left\lfloor \frac{C_2}{d_{max}} \right\rfloor$ – the maximum number of multi-tenant band users.

2.2 Limit Power Policy

Let us consider in more detail the policy of reducing the corresponding UE's uplink power by the eNB in order to meet the interference constraints. First of all, we determine the rules for accepting requests for service, provided that the UE's uplink power is not yet limited and is at its maximum.

Given the above considerations, when a new request arrives, four scenarios are possible:

- The request will be accepted for service on the single-tenant band, if the request finds the single-tenant band having not less than d_{max} free b.u.
- The request will be accepted for service on the multi-tenant band, if the request finds the single-tenant band having less than d_{max} b.u. free, the multi-tenant band is operational, i.e. $s = 1$, and having not less than d_{max} b.u. free.
- Otherwise, the request will be blocked without any after-effect on the corresponding Poisson process.

If the power is limited due to the incumbent's need for resources and the single-tenant band is totally occupied, then QoS on the multi-tenant band is degraded. In this case, the multi-tenant band goes into "unavailable" mode and the bit rates of all requests in service on the multi-tenant band switch from the maximum d_{max} to the minimum d_{min} value. When the multi-tenant band recovers, the bit rates are not switched back and all users that have been degraded continue to receive service at bit rate d_{min}. It should be noted that the multi-tenant band has the following property: requests can be served at the maximum bit rate, i.e. the multi-tenant band goes into operational mode, only when the service of all requests at the minimum bit rate is completed.

2.3 System of Equilibrium Equations

According to the above considerations, we can describe the LSA operation by a Markov process $\mathbf{X}(t) = \left\{ \left(N_1(t), N_2^{max}(t), N_2^{min}(t), S(t) \right), \ t \geq 0 \right\}$ on the state space

$$
\begin{aligned}
X = \Big\{ & n_1 = 0, \ldots, N_1, \ \ n_2^{max} = 0, \ldots, N_2, \ \ n_2^{min} = 0, \ \ s = 1 \\
& \vee \ n_1 = 0, \ldots, N_1, \ \ n_2^{max} = 0, \ \ n_2^{min} = 0, \ldots, N_2, \ \ s = 0 \Big\}.
\end{aligned}
\tag{1}
$$

State space (1) can be subdivided into two subspaces: $\{ n_1 = 0, \ldots, N_1, \ \ n_2^{max} = 0, \ldots, N_2, \ \ n_2^{min} = 0, \ s = 1 \}$ if the multi-tenant band is operational and requests can be served at the maximum bit rate, and $\{ n_1 = 0, \ldots, N_1, \ \ n_2^{max} = 0, \ \ n_2^{min} = 0, \ldots, N_1, \ \ s = 0 \}$ if the multi-tenant band is unavailable and requests continue their service at the minimum bit rate. Figure 1 shows the structure of the state space, considering the two subspaces.

The corresponding Markov process $\mathbf{X}(t)$, which representing the system's states, is described by the system of equilibrium equations

$$
\begin{aligned}
& p\left(n_1, n_2^{max}, n_2^{min}, s\right) \left[\lambda \cdot 1(n_1 < N_1) + \lambda \cdot 1\left(n_1 = N_1, \ n_2^{max} < N_2, \ s = 1\right) \right. \\
& + \beta \cdot 1\left(s = 0, \ n_2^{min} = 0\right) + n_2 \mu \cdot 1\left(n_2^{min} > 0\right) + n_1 \mu \cdot 1(n_1 > 0) + n_2^{max} \mu \cdot 1\left(n_2^{max} > 0\right) \\
& \left. + \alpha \cdot 1(s = 1) \right] = p\left(n_1 + 1, n_2^{max}, n_2^{min}, s\right)\left[(n_1 + 1)\mu \cdot 1(n_1 < N_1)\right] \\
& + p\left(n_1, n_2^{max} + 1, 0, 1\right)\left[(n_2^{max} + 1)\mu \cdot 1\left(n_1 = N_1, \ n_2^{max} < N_2, \ s = 1\right)\right] \\
& + p\left(n_1 - 1, n_2^{max}, n_2^{min}, s\right)\left[\lambda \cdot 1(n_1 > 0)\right] \\
& + p\left(n_1, n_2^{max} - 1, 0, 1\right)\left[\lambda \cdot 1\left(n_2^{max} > 0, n_1 = N_1\right)\right], \quad \left(n_1, n_2^{max}, n_2^{min}, s\right) \in X,
\end{aligned}
\tag{2}
$$

where $\left(p\left(n_1, n_2^{max}, n_2^{min}, s\right)\right)_{\left(n_1, n_2^{max}, n_2^{min}, s\right) \in X} = \mathbf{p}$ is the stationary probability distribution.

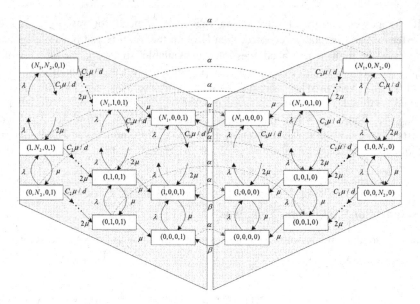

Fig. 1. The state space.

2.4 Infinitesimal Generator

The system probability distribution is numerically computed as the solution of the system of equilibrium equations $\mathbf{p} \cdot \mathbf{A} = \mathbf{0}$, $\mathbf{p} \cdot \mathbf{1}^T = 1$, where $\mathbf{A}$ is the infinitesimal generator of Markov process $\mathbf{X}(t)$. Let us denote $n = 0, \ldots, N_1 + N_2$ the number of users. If the lexicographical order on state space X is defined as

$$\left(n_1, n_2^{max}, n_2^{min}, s\right) < \left(n_1', n_2'^{max}, n_2'^{min}, s'\right) \text{ if and only if } n_1 + n_2^{max} + n_2^{min} < n_1' + n_2'^{max} + n_2'^{min}$$

or $n_1 + n_2 + m = n_1' + n_2'^{max} + n_2'^{min}$ and $\left(n_1 > n_1' \text{ and } n_1 = n_1' \text{ or } s > s'\right)$, then

(1) Infinitesimal generator $\mathbf{A}$ is a block tridiagonal matrix and has the form

$$\mathbf{A} = \begin{bmatrix} \mathbf{N}_0 & \mathbf{\Lambda}_0 & 0 & \cdots & & 0 \\ \mathbf{M}_1 & \mathbf{N}_1 & \ddots & \cdots & & \vdots \\ 0 & \ddots & \ddots & \ddots & & 0 \\ \vdots & \vdots & \ddots & & \ddots & \mathbf{\Lambda}_{N_1+N_2-1} \\ 0 & \cdots & 0 & & \mathbf{M}_{N_1+N_2} & \mathbf{N}_{N_1+N_2} \end{bmatrix}.$$

(2) Blocks $\mathbf{\Lambda}_n$, $n = 0, \ldots, N_1 + N_2 - 1$ of the upper diagonal have the sizes

$$\dim \mathbf{\Lambda}_n = \begin{cases} (2n+2) \times (2n+4), & n = 0, \ldots, N_2 - 1, \\ (2N_2+2) \times (2N_2+2), & n = N_2, \ldots, N_1 - 1, \\ (2(N_1+N_2-n)+2) \times (2(N_1+N_2-n)), & n = N_1, \ldots, N_1 + N_2 - 1, \end{cases}$$

and the following form:

$$\Lambda_n\left((n_1, n_2^{'max}, n_2^{'min}, s), \left(n_1', n_2^{'max}, n_2^{'min}, s'\right)\right)$$

$$= \begin{cases} \lambda, & n_1' = n_1 + 1, & n_2^{'max} = n_2^{max}, & n_2^{'min} = n_2^{min}, & s' = s & \text{or} \\ & n_1' = n_1 = N_1, & n_2^{'max} = n_2^{max} + 1, & n_2^{'min} = n_2^{min} = 0, & s' = s = 1, \\ 0, & \text{otherwise.} \end{cases}$$

(3) Blocks $\mathbf{M}_n$, $n = 1, \ldots, N_1 + N_2$ of the lower diagonal have the sizes

$$\dim \mathbf{M}_n = \begin{cases} (2n+2) \times 2n, & n = 1, \ldots, N_2, \\ (2N_2 + 2) \times (2N_2 + 2), & n = N_2 + 1, \ldots, N_1, \\ (2(N_1 + N_2 - n) + 2) \times (2(N_1 + N_2 - n) + 4), & n = N_1 + 1, \ldots, N_1 + N_2, \end{cases}$$

and the following form:

$$\mathbf{M}_n\left((n_1, n_2^{max}, n_2^{min}, s), \left(n_1', n_2^{'max}, n_2^{'min}, s'\right)\right)$$

$$= \begin{cases} n_1 \mu, & n_1' = n_1 - 1, & n_2^{'max} = n_2^{max}, & n_2^{'min} = n_2^{min}, & s' = s, \\ n_2^{max} \mu, & n_1' = n_1, & n_2^{'max} = n_2^{max} - 1, & n_2^{'min} = n_2^{min} = 0, & s' = s = 1, \\ n_2^{min} \mu, & n_1' = n_1, & n_2^{'max} = n_2^{max} = 0, & n_2^{'min} = n_2^{min} - 1, & s' = s = 0, \\ 0, & \text{otherwise.} \end{cases}$$

(4) Blocks $\mathbf{N}_n$, $n = 0, \ldots, N_1 + N_2$ of the main diagonal have the sizes:

$$\dim \mathbf{N}_n = \begin{cases} (2n+2) \times (2n+2), & n = 0, \ldots, N_2 - 1, \\ (2N_2 + 2) \times (2N_2 + 2), & n = N_2, \ldots, N_1, \\ (2(N_1 + N_2 - n) + 2) \times (2(N_1 + N_2 - n) + 2), & n = N_1 + 1, \ldots, N_1 + N_2. \end{cases}$$

and the following form:

$$\mathbf{N}_n\left((n_1, n_2^{max}, n_2^{min}, s), \left(n_1', n_2^{'max}, n_2^{'min}, s'\right)\right)$$

$$= \begin{cases} \alpha, & n_1' = n_1, & n_2^{'max} = n_2^{max} = 0, & n_2^{'min} = n_2^{min} = 0, & s' = s - 1 & \text{or} \\ & n_1' = n_1, & n_2^{'max} = 0, & n_2^{'min} = n_2^{max}, & s' = s - 1, \\ \beta, & n_1' = n_1, & n_2^{'max} = n_2^{max} = 0, & n_2^{'min} = n_2^{min}, & s' = s + 1, \\ *, & n_1' = n_1, & n_2^{'max} = n_2^{max}, & n_2^{'min} = n_2^{min}, & s' = s, \\ 0, & & \text{otherwise,} \end{cases}$$

where $* = -\left(\lambda \cdot 1\{n_1 < N_1\} + n_1\mu \cdot 1\{n_1 > 0\} + \lambda \cdot 1\left\{n_1 = N_1, \ n_2^{\max} < N_2, \ s = 1\right\} + \right.$
$\left. + n_2^{\max}\mu \cdot 1\left\{n_2^{\max} > 0\right\} + n_2^{\min}\mu \cdot 1\left\{n_2^{\min} > 0\right\} + s\alpha + (1-s)\beta\right).$

3 Numerical Analysis

3.1 Performance Measures

Having found the probability distribution $p\left(n_1, n_2^{\max}, n_2^{\min}, s\right)$, $\left(n_1, n_2^{\max}, n_2^{\min}, s\right) \in X$, one can compute the performance measures of the considered scheme: the probability B that a request is blocked, the average bit rate $\bar{d}$, the average bit rate $\bar{d}(C_2)$ on the multi-tenant band, and the utilization factor UTIL of the bands:

$$B = \sum_{i=0}^{N_2} p(N_1, 0, i, 0) + p(N_1, N_2, i, 0), \tag{3}$$

$$\bar{d} = \frac{\displaystyle\sum_{\left(n_1, n_2^{\max}, n_2^{\min}, s\right) \in X/(0,0,0,0),(0,0,0,1)} \frac{n_1 d_{\max} + n_2^{\max} d_{\max} + n_2^{\min} n_1 d_{\min}}{n_1 + n_2^{\max} + n_2^{\min}} \cdot p\left(n_1, n_2^{\max}, n_2^{\min}, s\right)}{\displaystyle\sum_{\left(n_1, n_2^{\max}, n_2^{\min}, s\right) \in X/(0,0,0,0),(0,0,0,1)} p\left(n_1, n_2^{\max}, n_2^{\min}, s\right)}, \tag{4}$$

$$\bar{d}(C_2) = \frac{\displaystyle\sum_{\left(n_1, n_2^{\max}, n_2^{\min}, s\right) \in X \neg n_2^{\max} \neq 0 \vee n_2^{\max} \neq 0} d_{\max} \cdot p\left(n_1, n_2^{\max}, 0, 1\right) + d_{\min} \cdot p\left(n_1, 0, n_2^{\min}, 1\right)}{\displaystyle\sum_{\left(n_1, n_2^{\max}, n_2^{\min}, s\right) \in X \neg n_2^{\max} \neq 0 \vee n_2^{\max} \neq 0} p\left(n_1, n_2^{\max}, n_2^{\min}, s\right)}, \tag{5}$$

$$\begin{aligned} \text{UTIL} \cdot C = & \sum_{\left(n_1, n_2^{\max}, n_2^{\min}, s\right) \in X: \ n_2^{\min} = 0, \, s = 1} \left(n_1 + n_2^{\max}\right) d^{\max} \cdot p\left(n_1, n_2^{\max}, 0, 1\right) + \\ & + \sum_{\left(n_1, n_2^{\max}, n_2^{\min}, s\right) \in X: \ n_2^{\max} = 0, \, s = 0} \left(n_1 d^{\max} + n_2^{\min} d^{\min}\right) \cdot p\left(n_1, 0, n_2^{\min}, 1\right). \end{aligned} \tag{6}$$

3.2 Numerical Example

Let us assume that users view short video clips, e.g. viral video, the length of which is about 20–30 s. The video is in high quality at bit rate $d_{\max} = 2$ Mbps. If a part of the frequency band has to be returned, the mobile operator reduces the corresponding eNB uplink power, whereby the bit rate decreases to $d_{\min} = 0.7$ Mbps. This bit rate $d_{\min}$ also allows users to browse video, but in lower quality. Finally, let us assume that the multi-tenant band goes into unavailable mode every hour (3600 s) or every four hours

(14400 s) on average and the recovery takes around one minute. Table 1 summarizes the initial data of the example. Note that 1 b.u. for the example under consideration equals to 1 Mbps.

Table 1. System parameters

Parameter description	Notation	Value
Peak bit rate for single-tenant band	C_1	20 Mbps [14]
Peak bit rate for multi-tenant band	C_2	20 Mbps [14]
Average service time of one user	μ^{-1}	30 s
Average time when multi-tenant band is available	α^{-1}	3540 s, 14340 s
Average time when multi-tenant band is unavailable	β^{-1}	60 s [3]
Maximum bit rate	d_{max}	2 Mbps [11]
Minimum bit rate	d_{min}	0.7 Mbps [11]
Offered load	ρ	$0 \div 30$

The figures below show the behavior of each performance measure under examination – blocking probability B (Fig. 2), average bit rates $\bar{d}$ and $\bar{d}(C_2)$ serving requests on both bands or multi-tenant band respectively (Fig. 3), and utilization factor UTIL (Fig. 4) – for different values of α^{-1} (the average time when the multi-tenant band is available). All three figures show that the less multi-tenant band goes into "unavailable" mode, the better the performance metrics that characterize the impact of LSA on the QoS, namely, the blocking probability is lower, whereas the average bit rate and the utilization factor are higher.

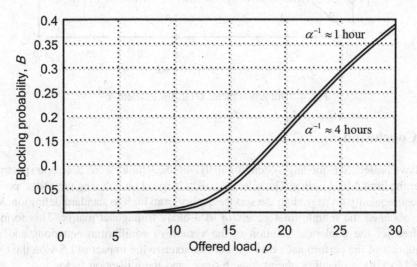

Fig. 2. Blocking probability B for different α^{-1}

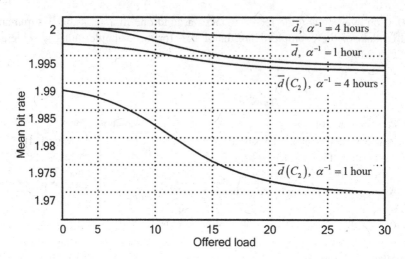

Fig. 3. Average bit rates $\overline{d}$ and $\overline{d}(C_2)$ for different α^{-1}

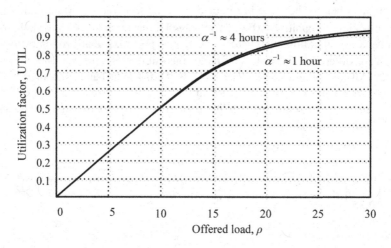

Fig. 4. Utilization factor UTIL for different α^{-1}

4 Conclusion

We have presented a queuing system for analyzing the simultaneous access to spectrum under the limit power policy. The selected policy is based on reducing the eNBs' power and consequently on degrading the service quality from high to standard definition. We have obtained the infinitesimal generator as a block tridiagonal matrix. This form is required for the numerical solution of the system of equilibrium equations and the calculation of the performance metrics that characterize the impact of LSA on the QoS – the blocking probability, the average bit rate, and the utilization factor.

Acknowledgment. The authors are grateful to the students of the Applied Probability and Informatics Department of RUDN University Anastasia Feduro and Dmitry Polouektov for computing the numerical example.

References

1. Cisco Visual Networking Index: Forecast and Methodology, 2015–2020 (2016)
2. Andrews, J., Buzzi, S., Choi, W., Hanly, S.V., Lozano, A., Soong, A.C.K., Zhang, J.C.: What will 5G be? IEEE J. Sel. Areas Commun. **32**, 1065–1082 (2014)
3. Ponomarenko-Timofeev, A., Pyattaev, A., Andreev, S., Koucheryavy, Y., Mueck, M., Karls, I.: Highly dynamic spectrum management within licensed shared access regulatory framework. IEEE Commun. Mag. **54**(3), 100–109 (2015)
4. Shorgin, S.Y., Samouylov, K.E., Gudkova, I.A., Galinina, O.S, Andreev, S.D.: On the benefits of 5G wireless technology for future mobile cloud computing. In: 1st International Science and Technology Conference Modern Networking Technologies (MoNeTec): SDN & NFV, pp. 151–154 (2014)
5. Buckwitz, K., Engelberg, J., Rausch, G.: Licensed shared access (LSA) – regulatory background and view of administrations. In: CROWNCOM (invited paper), pp. 413–416 (2014)
6. Ahokangas, P., Matinmikko, M., Yrjola, S., Mustonen, M., Luttinen, E., Kivimäki, A., Kemppainen, J.: Business models for mobile network operators in licensed shared access (LSA). In: DYSPAN, pp. 407–412 (2014)
7. Gomez-Miguelez, I., Avdic, E., Marchetti, N., Macaluso, I., Doyle, L.E.: Cloud-RAN platform for LSA in 5G networks – tradeoff within the infrastructure. In: Communications, Control and Signal Processing, pp. 522–525 (2014)
8. ETSI TS 103 113 Mobile broadband services in the 2300 MHz 2400 MHz band under Licensed Shared Access regime (2013)
9. Borodakiy, V.Y., Samouylov, K.E., Gudkova, I.A., Ostrikova, D.Y., Ponomarenko A.A., Turlikov, A.M., Andreev, S.D.: Modeling unreliable LSA operation in 3GPP LTE cellular networks. In: 6th International Congress on Ultra Modern Telecommunications and Control Systems ICUMT-2014, pp. 490–496 (2014)
10. Gudkova, I.A., Samouylov, K.E., Ostrikova, D.Y., Mokrov, E.V., Ponomarenko-Timofeev, A.A., Andreev, S.D., Koucheryavy, Y.A.: Service failure and interruption probability analysis for licensed shared access regulatory framework. In: 7th International Congress on Ultra Modern Telecommunications and Control Systems ICUMT-2015, pp. 123–131 (2015)
11. Live encoder settings, bitrates and resolutions. YouTube Help (2016). https://support. google.com/youtube/answer/2853702?hl=en
12. Borodakiy, V., Gudkova, I., Markova, E., Samouylov, K.: Modelling and performance analysis of pre-emption based radio admission control scheme for video conferencing over LTE. In: ITU Kaleidoscope Academic Conference, pp. 53–59 (2014)
13. Neuts, M.F.: Matrix-Geometric Solutions in Stochastic Models: An Algorithmic Approach. The John Hopkins University Press, Baltimore (1981). 332 p.
14. 3GPP TS 36.300 Evolved Universal Terrestrial Radio Access (E-UTRA) and Evolved Universal Terrestrial Radio Access Network (E-UTRAN); Overall description; Stage 2: Release 13 (2015)

Correlation Properties Comparative Analysis of Pseudorandom Binary Sequences and LTE Standard ZC Sequences

Vladimir Lavrukhin, Vitaly Lazarev$^{(\boxtimes)}$, and Alexander Ryjkov

The Bonch-Bruevich St. Petersburg State University of Telecommunications,
Saint-Petersburg, Russia
lavrukhin@sut.ru, laviol.94@gmail.com, aryjkov@mail.ru

Abstract. The paper describes correlation properties comparative analysis of pseudorandom binary sequences such as M-sequences, Gold codes, Hadamard ordered Golay sequences and LTE standard ZC sequences. In particular, the research considers primary synchronization signal (PSS) and random access preamble. Generation processes and the special features of ZC sequence based signals used in LTE standard are shown. Autocorrelation and cross correlation functions of appropriate signals are presented, correlation coefficients are evaluated.

Keywords: Pseudorandom binary sequence · ZC sequence · Correlation · LTE · Synchronization

1 Introduction

Scrambling sequences and reference signals are of great usage for downlink and uplink transmissions in the 21 century communication standards. In the UTRA standard Hadamard ordered binary orthogonal sequences for the channel separation and for channelization codes forming are applied. The primary synchronization channel in the UTRA standard is organized upon ordered by Hadamard generalized Golay sequences. Besides channel codes and primary synchronization channel, various Gold code scrambling sequences formed by modulo 2 addition of two different M-sequences are used in the UTRA standard.

Several types of M-sequences are used to form the secondary synchronization signal (SSS) in the LTE standard. All of these sequences are pseudorandom binary sequences (PRBS) and in perfect case their cross-correlation functions are close to zero. In addition to PRBS, non binary sequences – discrete complex ZC-sequences are widely used in LTE. ZC-sequences are used as reference demodulation signals in SC-FDMA based uplink channel. In a downlink channel ZC-sequences are used to form primary synchronization signal (PSS) [1, 3]. ZC-sequences are also used as sounding signals, random access preambles and for signal traffic in uplink transmissions [1, 2].

In this article we provide an overview of pseudorandom binary sequences and ZC-sequences correlation properties by calculating their autocorrelation and

© Springer International Publishing AG 2016
O. Galinina et al. (Eds.): NEW2AN/ruSMART 2016, LNCS 9870, pp. 414–425, 2016.
DOI: 10.1007/978-3-319-46301-8_35

cross-correlation functions. We also show examples of these sequences generation process and determine emission value limits of correlation functions.

2 Overview of Gold Codes and M-Sequences in LTE

To provide coherent detection of uOFDM signal in radio channel eNB in downlink and UE in uplink add CRS (Cell-Specific Reference Signal) to the data transmission.

CRS are predefined complex values used at the receiver end to make the required phase correction and amplitude scaling of the information signals. Downlink CRS are based on pseudorandom Gold codes sequences. Special Positioning Reference Signals (PRS) are aslo used in LTE for reliable measuring of subscriber's location. These PRS are also based on Gold sequences. In addition scrambling codes in uplink and downlink are also organized by Gold codes.

According to [9] pseudo-random sequences mentioned above are based on Gold sequences of length 31. The output sequence $c(n)$ of length M_{PN}, where $n = 0, 1, \ldots$ $M_{PN} - 1$, is defined as:

$$
\begin{aligned}
c(n) &= (x_1(n+N_C) + x_2(n+N_C)) \bmod 2 \\
x_1(n+31) &= (x_1(n+3) + x_1(n)) \bmod 2 \\
x_2(n+31) &= (x_2(n+3) + x_2(n+2) + x_2(n+1) + x_2(n)) \bmod 2
\end{aligned}
\tag{1}
$$

where $N_C = 1600$, and the first M-sequence defined as $x_1(0) = 1$, $x_1(n) = 0$, $n = 1,2,\ldots30$.

The second M-sequence is determined by the value of C_{init}:

$$
C_{init} = \sum_{i=0}^{30} x_2(i) \cdot 2^i
\tag{2}
$$

The value of C_{init} depends on the $c(n)$ sequence further usage.

As an example of this sequence let us take a PUCCH (Physical Uplink Control Channel) signal, where C_{init} is generally determined by cell identifier N_{ID}^{Cell}, which can take a value from 0 to 503. The examples of $c(n)$ sequence autocorrelation function for $N_{ID}^{Cell} = 1$ and the cross-correlation function for $N_{ID}^{Cell} = 0$ and 1 are shown in Figs. 1 and 2, respectively.

Downlink Secondary Synchronization Signals (SSS) based on M-sequences of length 31. SSS provides frame synchronization during signal receiving.

There are three different M-sequences used to generate $d(n)$ symbols of SSS: $s(n)$, $c(n)$ and $z(n)$ sequences of length 31:

$$
\begin{aligned}
\tilde{s}(n) &= \begin{matrix} 1111 - 111 - 11 - 1 - 111 - 1 - 1 - 1 - 1 \\ -1111 - 1 - 11 - 1 - 1 - 11 - 11 - 1 \end{matrix} \\
\tilde{c}(n) &= \begin{matrix} 1111 - 11 - 11 - 1 - 1 - 11 - 1 - 1111 \\ -1 - 1 - 1 - 1 - 111 - 1 - 11 - 111 - 1 \end{matrix} \\
\tilde{z}(n) &= \begin{matrix} 1111 - 1 - 1 - 111 - 1 - 11 - 1 - 1 - 1 - 1 - 1 \\ 1 - 1111 - 111 - 11 - 11 - 1 - 1 \end{matrix}
\end{aligned}
\tag{3}
$$

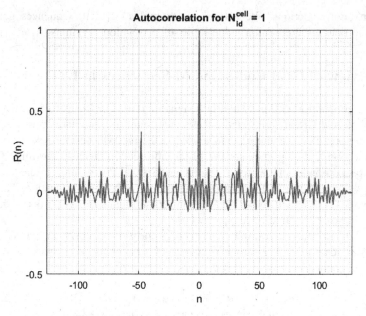

Fig. 1. The *c(n)* sequence autocorrelation function

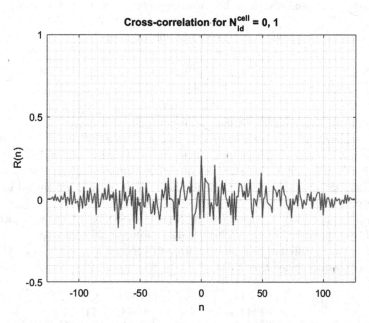

Fig. 2. The cross-correlation function of $c(n)$ sequences for values $N_{ID}^{Cell} = 0$ and $N_{ID}^{Cell} = 1$

416   V. Lavrukhin et al.

Wherein the basic M-sequence $s(n)$ is scrambled by M-sequence $c(n)$, and the odd symbols of SSS is secondary scrambled by M-sequence $z(n)$:

$$d(2n) = \begin{cases} s^{(m_0)}(n)c_0(n) & in \ subframe \ 0 \\ s^{(m_1)}(n)c_0(n) & in \ subframe \ 5 \end{cases}$$
$$d(2n+1) = \begin{cases} s^{(m_1)}(n)c_1(n)z^{(m_0)}(n) & in \ subframe \ 0 \\ s^{(m_0)}(n)c_1(n)z^{(m_1)}(n) & in \ subframe \ 5 \end{cases} \quad (4)$$

$c_0(n)$ and $c_1(n)$ sequences are differed by shift, that is defined by $N_{ID}^{(2)}$:

$$c_0(n) = \tilde{c}((n + N_{ID}^{(2)}) \bmod 31)$$
$$c_1(n) = \tilde{c}((n + N_{ID}^{(2)} + 3) \bmod 31) \quad (5)$$

In (4) all symbols of $d(n)$, $s(n)$, $c(n)$ and $z(n)$ sequences take values of $+1$ and -1. The difference between $s^{(m0)}$ and $s^{(m1)}$, as well as between $z^{(m0)}$ and $z^{(m1)}$, is in a cyclic shift of corresponding M-sequences by m_0 and m_1 elements. Autocorrelation functions and cross-correlation functions of these sequences are shown in Figs. 3 and 4, respectively.

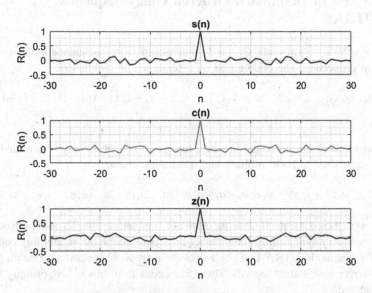

Fig. 3. The autocorrelation function of SSS signal sequences.

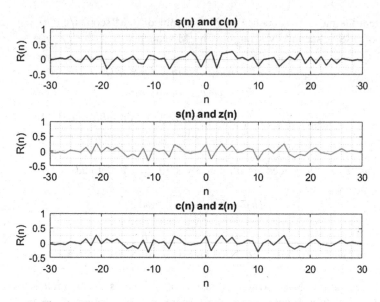

Fig. 4. The cross-correlation function of SSS signal sequences

3 Overview of Hadamard Ordered Golay Sequences in UTRA

Primary synchronization code C_{psc} is a generalized Golay sequence with high autocorrelation properties. The code based on a sequence of 16 bits [9]:

$$a = \ <x_1, x_2, x_3, \ldots, x_{16}> \ = \ <1, 1, 1, 1, 1, 1, -1, -1, 1, -1, 1, -1, 1, -1, -1, 1> \tag{6}$$

C_{psc} consists of 16 direct and inversed repeats of synchronously transmitted vector a (at the same time in both in-phase and quadrature channels):

$$C_{psc} = (1+j) \times \ <a, a, a, -a, -a, a, -a, -a, a, a, a, -a, a, -a, a, a> \tag{7}$$

After synchronizing itself with the network, i.e. setting the TS start, UE starts to analyze secondary synchronization channel S-SCH to define the scrambling codes that cover BTS signals. In UTRA-FDD networks there are 8192 specified scrambling codes that may cover base station signals. Thus, 512 codes from this set are primary and the rest are secondary.

In the S-SCH 16 different chip codes are used, where each chip code is a vector of 256 chips obtained by Hadamard scheme:

$$H_0 = (1)$$
$$H_k = \begin{pmatrix} H_{k-1} & H_{k-1} \\ H_{k-1} & -H_{k-1} \end{pmatrix}, k \geq 1 \tag{8}$$

Denote H_8 as a matrix containing 256 rows. Each row of the matrix consists of 256 elements of $h_m(i)$, $i = 0...255$, where m – the number of the row. Let the vector z contains 16 elements:

$$z = <b, b, b, -b, b, b, -b, -b, b, -b, b, -b, -b, -b, -b, -b> , \qquad (9)$$

where each symbol b is a vector, that composed from elements of the vector a with inverted signs in the second half of the vector a:

$$b = <x_1, x_2, x_3, x_4, x_5, x_6, x_7, x_8, -x_9, -x_{10}, -x_{11}, -x_{12}, -x_{13}, -x_{14}, -x_{15}, -x_{16}>$$

16 codes of S-SCH are generated by the following scheme:

$$C_{ssc,k} = (1+j) \times <h_m(0) \times z(0), h_m(1) \times z(1), ..., h_m(255) \times z(255)> , \qquad (10)$$
$$k = 1, 2, 3, ..., 16; m = 16 \cdot (k-1)$$

After consecutive reading of S-SCH codes in 15 consecutive TS, UE defines the start of the frame and the code group number. The examples of $C_{ssc,1}$ sequence autocorrelation function and $C_{ssc,1}$ and $C_{ssc,2}$ sequences cross-correlation function are shown in Figs. 5 and 6, respectively.

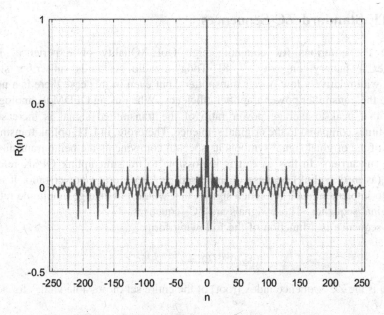

Fig. 5. The autocorrelation function of a $C_{ssc,1}$ sequence

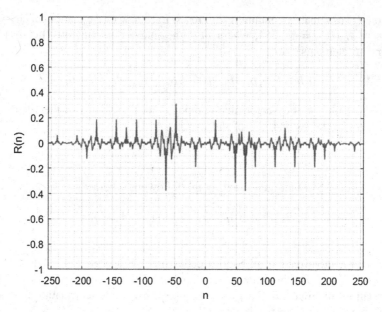

Fig. 6. The cross-correlation function of $C_{ssc,1}$ and $C_{ssc,2}$ sequences

4   LTE Standard ZC-Sequences

One of the criterions for obtaining high QoE (Quality of Experience) is the rechargeable battery longevity [7, 8]. Today's radio standards subscriber stations operate with relatively low-power transmitters, but even in this case there is a need to increase the transmitter power amplifier efficiency. When using OFDM technology it is known that peak-to-average power ratio of the transmitted signal is increased by several times compared to the original sequence. Therefore, in LTE uplink transmission instead of the original signal symbols its spectral components are being transmitted on OFDM subcarriers. In this case it is impossible to transmit uplink QPSK reference signals (as in downlink) because of increased peak-to-average power ratio. It is necessary to implement such reference signals that would not change amplitude ratios of the original sequence. Those signals are ZC-sequences.

ZC-sequence is a function of the following form:

$$(X_{ZC})^{(q)}(k) = e^{-j \cdot \pi \cdot q \frac{k(k+1)}{M_{ZC}}} \tag{11}$$

where q is the ZC-sequence index (root) of the entire set of possible values for a given length M_{ZC}.

ZC-sequences belong to a CAZAC (Constant-Amplitude Zero-Auto-Correlation) class sequences and have the following properties:

- a constant signal amplitude;
- a zero cross-correlation of the same root sequence with different cyclic shifts and low correlation of different sequences for certain values of q.

Signal phase shifts based on ZC-sequence depends on the parameters q and M_{ZC}, wherein M_{ZC} should be a prime number – only in this case such sequences have optimal correlation properties.

In downlink Primary Synchronization Signal (PSS) three ZC-sequences with fixed q indexes (25, 29 and 34) are used. For each of these sequences there is a physical layer identifier $N_{ID}^{(2)} = 0;\ 1;\ 2$. Thus root indexes $q = 25$, 29 and 34 correspond to the values of $N_{ID}^{(2)} = 0$, 1 and 2.

In practice it is useful when building a network based on a three-sector cell structures: there are three base stations in eNB, and each of them belongs to the same group identity, but has its own $N_{ID}^{(2)}$.

PSS consists of 62 symbols ($n = 0\ldots61$) in the following form:

$$d_q(n) = \begin{cases} e^{\frac{-j\pi q n(n+1)}{63}}, & \text{if } n = 0, 1, \ldots, 30 \\ e^{\frac{-j\pi q(n+1)(n+2)}{63}}, & \text{if } n = 31, 32, \ldots, 61 \end{cases} \tag{12}$$

Sequences used to generate PSS are shown in Fig. 7a.

PSS cross-correlation properties are investigated, and resulting cross-correlation functions are presented in Fig. 7b. The last figure confirms the low cross-correlation level of PSS from neighboring sectors.

As mentioned above ZC-sequences are used to generate reference uplink signals in PUCCH and PUSCH channels in contrast to the classic downlink digital reference signal.

The uplink reference signal is generated using the cyclic shift α of the base ZC-sequence $\bar{r}_{u,v}(n)$:

$$r_{u,v}^{(\alpha)}(n) = e^{j\alpha n}\bar{r}_{u,v}^{(\alpha)}(n) \tag{13}$$

In (13) n is the number of the subcarrier. The reference signal length $M = 12m$, where m is the number of allocated resource blocks in the physical channel.

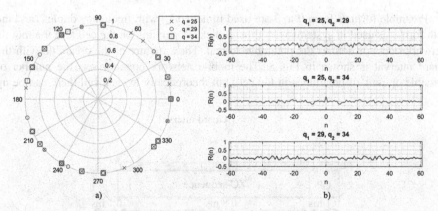

Fig. 7. ZC-sequences with different q indexes (a) and the Cross-correlation functions of PSS signals from neighboring sectors (b)

If the number of resource blocks is three or more, then the base sequence is the ZC-sequence of the following form:

$$r_{u,v}(n) = x_q(n \bmod M_{ZC}) \qquad 0 \le n \le 12m \tag{14}$$

where respective ZC-sequence:

$$X_q(k) = e^{\frac{-j\pi q k(k+1)}{M_{ZC}}} \qquad 0 \le k \le M_{ZC} - 1 \tag{15}$$

Root ZC-sequence is selected so that the length M_{ZC} is to be equal to the maximum prime number less than M. Thus, if three resource blocks are allocated (36 subcarriers), ZC-sequence of length 36 elements is required, while a nearest simple integer value is 31. Therefore in (14) and (15) $M_{ZC} = 31$. The cell identifier N_{ID}^{cell} and M_{ZC} value determine the choice of q in (15). It is possible to arrange 30 groups of base sequences $\bar{r}_{u,v}(n)$ of index u, whereby a mobile network operator can work with a fixed distribution of u between cells or use group jump of index u, varying group numbers from the one subframe to another. In this case v can be equal to 0 or 1, and that influences on the calculation of q [1].

Let us further look at ZC-sequences used to generate random access preambles. According to [10] there are five preamble formats. Their values shown in Table 1.

Table 1. Random access preamble formats

Preamble format	Preamble length, ms	Preamble length with guard interval, ms	Amount of subframes	Recommended cell radius, km
0	0.800	0.903	1	<14
1	0.800	1.484	2	~75
2	1.600	1.803	2	~28
3	1.600	2.284	3	~108
4	0.133	0.148		

Preamble formats from 0 to 3 are used in networks with frequency duplex, and the 4-th format is used in systems with time duplex. Depending on the cell size the mobile network operator defines the preamble format. The structure of the preamble with the guard interval is shown in Fig. 8. The mobile network operator sets the number of possible options of preambles in the cell, but theoretically in each cell there can be up

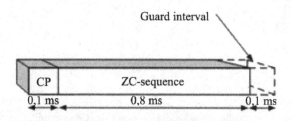

Fig. 8. Structure of preamble with guard interval

to 64 preambles. Random access from subscriber station in the LTE standard can be based on competition or without it. When competition based access is used, subscriber station selects preamble randomly from all of the available preambles. When subscriber station receives the transmit command from the network (e.g. handover, synchronization recovery, etc.) it gets the number of the preamble, that has to be used to access a network without the competition.

First the set of 64 preamble sequences for each cell is evaluated for all available cyclic shifts of root ZC-sequence (in ascending order of shifts) using a logical index RACH_ROOT_SEQUENCE, where RACH_ROOT_SEQUENCE transmitted as part of system information. If not all of 64 preambles can be generated from a single root ZC-sequence, additional preamble sequences are obtained from the root sequences of subsequent logical indexes until they receive all of 64 preambles. The logical root sequence is cyclic: the logical index 0 follows the index of 837. The relation between the index of the logical root sequence and the index q of physical root sequence is specified in [9] for all preamble formats. Thus, digits for physical root sequence are selected so that preamble cross-correlations is minimal.

ZC-sequence with the q-th root is defined as:

$$x_q(n) = e^{\frac{-j\pi q n(n+1)}{N_{ZC}}}, \qquad 0 \le n \le N_{ZC} - 1 \tag{16}$$

where length N_{ZC} for different preamble formats is given in Table 2.

Table 2. Preamble formats and N_{ZC} values

Preamble format	N_{ZC}
0–3	839
4	139

Preambles with regions of zero correlation of length $N_{CS} - 1$ are determined from ZC-sequences with q-th root and cyclic shifts by the following form:

$$x_{q,v}(n) = x_q((n + C_v) \bmod N_{ZC}) \tag{17}$$

where the cyclic shift is given as

$$C_v = \begin{cases} vN_{CS}, & v = 0, 1, \ldots, \left[\frac{N_{ZC}}{N_{CS}}\right] - 1, & N_{CS} \ne 0 \\ 0, & & N_{CS} = 0 \end{cases} \tag{18}$$

Equation (18) used for an unlimited variants of cyclic shifts. In [9] there is an option of cyclic shifts constraints that are used when subscriber station is moving with high speed to eliminate unwanted correlation effects caused by the Doppler frequency shift.

The N_{CS} value specified in Table 3, taken from [9], for types of preambles from 0 to 3, where zeroCorrelationZoneConfig parameter set by the operator.

Table 3. N_{CS} values for generating preambles of format from 0 to 3

zeroCorrelationZoneConfig	N_{CS}	zeroCorrelationZoneConfig	N_{CS}
0	0	8	46
1	13	9	59
2	15	10	76
3	18	11	93
4	22	12	119
5	26	13	167
6	32	14	279
7	38	15	419

N_{CS} value depends on cell radius and the higher it is, the greater must be the value of N_{CS} to eliminate possible cross-correlations of preambles sent by stations near the node or on the cell edge.

As mentioned above, the maximum number of preamble sequences within the cell is 64. Let us take an example of these sequences generation and estimate their correlation properties. Let's set the parameters:

zeroCorrelationZoneConfig = 5, which corresponds to $N_{CS} = 26$ (Table 3)

RACH_ROOT_SEQUENCE = 4 for base sequence that corresponds to the physical root sequence number $q = 120$ [4];

RACH_ROOT_SEQUENCE = 5 for neighboring base sequence that corresponds to the physical root sequence number $q = 719$ [8];

Preamble format 0, which corresponds to $N_{ZC} = 839$ [8];

Let us get 64 different preamble sequences in the following way:

Sequence (0) = base sequence without cyclic shift;

Sequence (1) = a cyclic shift of the base sequence by 1*26 samples;

Sequence (2) = a cyclic shift of the base sequence by 2*26 samples;

Sequence (3) = a cyclic shift of the base sequence by 3*26 samples;

…

Sequence (31) = a cyclic shift of the base sequence by 31*26 samples;

Sequence (32) = a cyclic shift of the base sequence to the neighboring sequence (substitution $q = 120$ to $q = 719$);

Sequence (33) = a cyclic shift of the new base sequence by 1*26 samples;

Sequence (34) = a cyclic shift of the new base sequence by 2*26 samples;

Sequence (35) = a cyclic shift of the new base sequence by 3*26 samples;

…

Sequence (63) = a cyclic shift of the new base sequence by 31*26 samples.

An example of ZC-sequence obtained by the expression (16) and used as the base sequence with the above given parameters, shown in Fig. 9a.

The resulting cross-correlation function shown in Fig. 9b confirms the low level of cross-correlation of preambles with neighboring numbers of physical root sequence ($q_4 = 120$, $q_5 = 719$). The simulations having been carried out has shown that emission value limits of preamble sequences normalized cross correlation functions are less than 0.05.

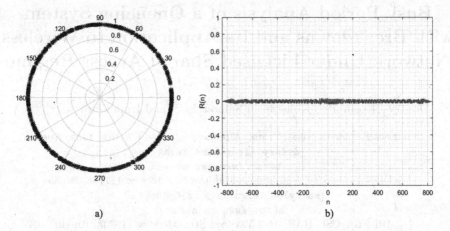

a)   b)

Fig. 9. ZC-sequence preamble with $q = 120$ (a) and the cross-correlation function of preamble sequences with $q = 120$ and $q = 719$ (b)

5   Conclusion

For the first time, in the 4-th generation mobile communication LTE standard in addition to previously used pseudorandom binary sequences not digital, but discrete analog sequences are introduced. Investigation has shown that ZC-sequences have necessary correlation properties and allow to minimize interference when transmitting synchronization signals, reference signals and preambles.

References

1. Ahmadi, S.: LTE-Advanced. A Practical Systems Approach to Understanding the 3GPP LTE Releases 10 and 11 Radio Access Technologies. Academic Press, Oxford (2014). 1116p.
2. Gelgor, A., Pavlenko, I., Fokin, G., Gorlov, A., Popov, E., Lavrukhin, V.: LTE base stations localization. In: Balandin, S., Andreev, S., Koucheryavy, Y. (eds.) NEW2AN/ruSMART 2014. LNCS, vol. 8638, pp. 191–204. Springer, Heidelberg (2014)
3. Myung, H.G., Goodman, D.J.: Single Carrier FDMA. A New Air Interface for Long Term Evolution. Wiley, New York (2008)
4. Gold, R.: Optimal binary sequences for spread spectrum multiplexing (Corresp.). IEEE Trans. Inform. Theor. **13**(4), 619–621 (1967)
5. Silva, C., Dolecek, G., Harris, F.: Cell search in long term evolution systems: primary and secondary synchronization. In: 2012 IEEE 3rd Latin American Symposium on Circuits and Systems (LASCAS) (2012)
6. Rahman, T., Sacchi, C.: A cooperative radio resource management strategy for mobile multimedia LTE uplink. In: 2014 IEEE Aerospace Conference (2014)
7. Yahui, H., Yinlong, L., Xu, Z., Zhen, X.: Towards QoE-based resource allocation schemes in SC-FDMA systems. J. China Univ. Posts Telecommun. **22**(5), 63–100 (2015)
8. E-UTRA Physical channels and modulation. http://www.3gpp.org/ftp/Specs/archive/36_series/36.211
9. UTRA Spreading and modulation (FDD). http://www.3gpp.org/DynaReport/25213.htm
10. RACH. http://www.sharetechnote.com/html/RACH_LTE.html

Busy Period Analysis of a Queueing System with Breakdowns and Its Application to Wireless Network Under Licensed Shared Access Regime

Dmitry Efrosinin[1,2], Konstantin Samouylov[2], and Irina Gudkova[2,3(✉)]

[1] Johannes Kepler University Linz, Altenbergerstrasse 69, 4040 Linz, Austria
dmitry.efrosinin@jku.at
http://www.jku.at
[2] RUDN University, 6 Miklukho-Maklaya St, Moscow 117198, Russia
{samuylov_ke,gudkova_ia}@pfur.ru
http://www.rudn.ru
[3] IPI FRC CSC RAS, 44-2 Vavilova Str., Moscow 119333, Russia

Abstract. Licensed shared access (LSA) framework is becoming one of the promising trends for future 5G wireless networks. Two main parties are involved in the process of sharing the frequency band – the primarily user (owner) and the secondary user (licensee). From the LSA licensee's perspective, who has access to the band when the owner does not need it, the band is unreliable and its customers (e.g. users of wireless network) suffer from possible service interruptions. This can only occur when there is at least one customer in service (i.e. in busy period). The aim of this paper is to estimate the impact of the LSA band unreliability to the LSA licensee within the period when some interruptions are possible. The metric is the relation between the number of service interruptions and the number of customers served during a busy period. We model the occupancy of the LSA band as a multi-server homogeneous queueing system with finite and infinite buffer size and deal with the busy period analysis. The system is analyzed in steady state by deriving expressions in terms of the Laplace-Stiltjes transforms for the continuous time distributions and in terms of the probability generating functions for the number of customers served during a busy period and number of failures which occur in a busy period. The non-monotonic nature of some probabilistic measures is identified. Some illustrative examples are added to the paper.

Keywords: Licensed shared access · Multi-server queueing system · Non-reliable servers · Busy period · Number of failures

The reported study was funded within the Agreement No. 02.a03.21.0008 dated 24.11.2016 between the Ministry of Education and Science of the Russian Federation and RUDN University.

© Springer International Publishing AG 2016
O. Galinina et al. (Eds.): NEW2AN/ruSMART 2016, LNCS 9870, pp. 426–439, 2016.
DOI: 10.1007/978-3-319-46301-8_36

1   Introduction

The main problem facing modern wireless networks is the exponential traffic growth due to the increased requirements for bit rates and bandwidth. One of the promising issue is a flexible spectrum usage technique based on the shared access to frequency band by several parties. ETSI proposes the regulatory framework named licensed shared access (LSA) [1–3] to coordinate it. The current owner of spectrum rights of use is called the incumbent; the party operating a wireless network, which holds individual rights of use to LSA spectrum, is the LSA licensee. Spectrum usage is binary by nature, i.e. it is used by either the incumbent or the LSA licensee, in such a way that the incumbent has the primarily access to frequency band. ETSI proposes to deploy the LSA framework in the 2 300–2 400 MHz band, which is now used for aeronautical and terrestrial telemetry as well as for ancillary services such as cordless cameras, portable and mobile video links.

The researches in LSA mainly focus on analyzing ETSI initiatives [4–6]. Only several papers propose simulations or mathematical models. The authors of [7] developed a simulation model of the scenario where the incumbent uses LSA frequency band for aeronautical telemetry and proposed three shared access algorithms. One of them is the use of LSA spectrum by either the incumbent or the LSA licensee. It was also modeled as unreliable queuing systems in papers [9] and [8] where the authors mainly analyzed performance measures from the wireless network customers' perspective, such as blocking probability and interruption probability. The analysis of more aggregated metric showing the availability of frequency band to the LSA licensee during the period where there are at least one customer (busy period) is the subject this paper. We model the occupancy of the LSA band as a multi-server homogeneous queueing system with finite and infinite buffer size.

It is clear that the queueing situation for the proposed system have the feature that the servers can fail and be repaired simultaneously. The non-reliable multi-server queueing systems have been intensively studied. Some results can be found in [10], for the queues with balking and reneging, in [13] for the queues with impatient customers. The multi-server queueing models with controllable vacation was studied in [11] and with synchronous vacations in [14]. Despite a sufficient wide spectrum of the proposed results in the area of multi-server queueing systems, the missing link to an applicability of such systems with servers subject to simultaneous breakdowns is a performance analysis of the system in a busy period. The busy period is probably the most important characteristic value describing the first-passage time to the empty state after an arrived customer finds the system empty. Performance characteristics of the busy period, the number of customers served and the number of failures are employed to construct the cost functional for controllable queueing models. Our analysis of the busy period includes the derivation of the Laplace-Stieltjes transforms and probability generating functions of the length of the busy period, number of customers served and failures in a busy period. The numerical inversion of the proposed transforms is realized to calculate the values of the corresponding probability density and distribution functions.

The paper is organized as follows. Sections 2 and 3 are devoted to the busy period analysis in a finite and infinite buffer case respectively. Some illustrative numerical examples related to the LSA framework in a wireless network are discussed in Sect. 4.

In further sections we will use the notations $\mathbf{e}, \mathbf{e}_j$ and I, respectively, for the column-vector of dimension 2 consisting of 1's, the column-vector of dimension 2 with 1 in the j-th position and 0 elsewhere, and an identity matrix of dimension 2×2.

2   Busy Period Analysis for a Finite Queue

2.1   Stationary Probabilities

The paper deals with a multi-server $M/M/c$ queueing model with c homogeneous servers subject to simultaneous failures which take place independently on the states of servers. The customers arrive to the system according to a Poisson process with intensity λ. The service time are exponentially distributed with intensity μ. The life time of the servers is exponentially distributed with intensity α. After the failure the repair process starts immediately and takes exponentially distributed time with intensity β. The inter-arrival, service, life and repair times are assumed to be mutually independent.

Let $N(t)$ and $D(t)$ denote, respectively, the number of customers in the system and the state of the servers at time t. The two-dimension process

$$\{X(t)\}_{t \geq 0} = \{N(t), D(t)\}_{t \geq 0} \tag{1}$$

is a continuous-time Markov chain with state space given by

$$E = \{x = (n, d) : 0 \leq n \leq r, d \in \{0, 1\}\},$$

where $d = 0$ or $d = 1$ means that the servers are in a failed state or in a operational mode and $r < \infty$. The states $x \in E$ are partitioned as follows:

$$\mathbf{n} = \{(n, 1), (n, 0)\}, \ 0 \leq n \leq r.$$

Then, the infinitesimal generator of the Markov chain $\{X(t)\}_{t \geq 0}$ has the form

$$\Lambda = \delta(Q_{1,0}, Q_{1,1}, \ldots, Q_{1,c}, \underbrace{Q_{1,c}, \ldots, Q_{1,c}}_{r-c}, Q_{1,c+1})$$

$$+ \delta^-(Q_{2,1}, Q_{2,2}, \ldots, Q_{2,c} \underbrace{Q_{2,c}, \ldots, Q_{2,c}}_{r-c-1}) + \delta^+(\underbrace{Q_{0,1}, \ldots, Q_{0,1}}_{r-1}), \tag{2}$$

δ, δ^+ and δ^- stands for the diagonal, upper diagonal and low diagonal matrix respectively. The coefficient matrices in (2) are given by

$$Q_{0,1} = \lambda E, \ Q_{2,k} = k \begin{pmatrix} \mu & 0 \\ 0 & 0 \end{pmatrix}, \ 1 \leq k \leq c, \ Q_{1,0} = \begin{pmatrix} -(\lambda + \alpha) & \alpha \\ \beta & -(\lambda + \beta) \end{pmatrix} \tag{3}$$

$$Q_{1,k} = Q_{1,0} - Q_{2,k}, \ 1 \leq k \leq c, \ Q_{1,c+1} = Q_{1,c} + Q_{0,1}.$$

Let the macro-vector $\boldsymbol{\pi}$, partitioned as $\boldsymbol{\pi} = (\boldsymbol{\pi}_0, \boldsymbol{\pi}_1, \ldots, \boldsymbol{\pi}_r)$, denote the stationary probability vector of matrix $\Lambda = [\lambda_{xy}]_{x,y \in E}$ with elements $\pi_{(n,d)}$, where

$$\boldsymbol{\pi}\Lambda = \mathbf{0}, \ \boldsymbol{\pi}\mathbf{e} = 1.$$

The computation of the stationary distribution is reduced to solving a finite or infinite block tridiagonal system. There are a number of methods of solution. We apply here a block forward-elimination-backward substitution approach.

Theorem 1. *The stationary probabilities can be calculated by*

$$\boldsymbol{\pi}_n = \boldsymbol{\pi}_r \prod_{j=1}^{r-n} M_{r-j}, \ 0 \leq n \leq r-1 \tag{4}$$

and $\boldsymbol{\pi}_r$ is a unique solution of the system of equations

$$\boldsymbol{\pi}_r \sum_{n=0}^{r} \prod_{j=1}^{r-n} M_{r-j}\mathbf{e} = 1, \tag{5}$$

$$\boldsymbol{\pi}_r[Q_{1,c+1} + M_{r-1}Q_{0,1}] = \mathbf{0},$$

where the matrices M_{r-j} are defined by

$$M_0 = -Q_{2,1}Q_{1,0}^{-1}, \tag{6}$$

$$M_k = -Q_{2,k+1}[Q_{1,k} + M_{k-1}Q_{0,1}]^{-1}, \ 1 \leq k \leq r-1.$$

Proof. The statement follows by application of the block forward-elimination-backward substitution method.

Once the vector $\boldsymbol{\pi}$ is computed, a variety of classical performance characteristics can be routinely evaluated. We are interested in calculation for the mean measures on the busy period. Among them are the expected amount of time $\mathbb{E}[T_0]$ in a regenerative cycle (mean iterarrival time λ^{-1} plus mean length of busy period $\bar{\Phi}$) during which the servers are in a failed state or the expected amount of time $\mathbb{E}[T_1]$ when all c servers are operational and busy,

$$\mathbb{E}[T_0] = \sum_{n=0}^{r-1} \boldsymbol{\pi}_n\mathbf{e}_2(\lambda^{-1} + \bar{\Phi}) = \frac{\sum_{n=0}^{r-1} \boldsymbol{\pi}_n\mathbf{e}_2}{\lambda\boldsymbol{\pi}_0\mathbf{e}}, \tag{7}$$

$$\mathbb{E}[T_1] = \sum_{n=c}^{r-1} \boldsymbol{\pi}_n\mathbf{e}_1(\lambda^{-1} + \bar{\Phi}) = \frac{\sum_{n=c}^{r-1} \boldsymbol{\pi}_n\mathbf{e}_1}{\lambda\boldsymbol{\pi}_0\mathbf{e}}.$$

In the next two sections we provide a busy period analysis in finite and infinite buffer case. In this regard we propose the following remarks.

Remark 1. The proposed relations include multiplication and inversion of matrices. It is clear that if r is large some computational difficulties may occur. Hence the approximation of the finite buffer system by infinite analogue seems to be a reasonable alternative to the proposed analysis. In this case the buffer size must be enough large to satisfy the following inequality for the estimated parameter r,

$$\varepsilon = \mathbb{P}[N(t) > r] \geq \sum_{k=r+1}^{\infty} \pi_n \mathbf{e} = \pi_c R^{r+1-c}(I - R)^{-1}\mathbf{e}, \, r > c. \tag{8}$$

where ε is a prespecified small value, vector π and the matrix R are evaluated in the next section for the infinite buffer queueing system.

Remark 2. In light traffic case, e.g. if $\rho \ll 1$, the probability $\mathbb{P}[N(t) > r = c+1]$ can take very small values. Hence the queueing system under study can be treated as a $M/M/\infty$ system with breakdowns. In this case it is not difficult to derive the probability generating functions

$$\tilde{P}_0(z) = \sum_{n=0}^{\infty} p_{(n,0)} z^n, \; \tilde{P}_1(z) = \sum_{n=0}^{\infty} p_{(n,1)} z^n, \, |z| \leq 1,$$

in form

$$\tilde{P}_0(z) = e^{\frac{\lambda}{\mu}(z-1)} \frac{\alpha}{\alpha + \beta} \left(\frac{\beta}{\beta + \lambda(1 - z)}\right)^{1+\frac{\alpha}{\mu}},$$

$$\tilde{P}_1(z) = e^{\frac{\lambda}{\mu}(z-1)} \frac{\beta}{\alpha + \beta} \left(\frac{\beta}{\beta + \lambda(1 - z)}\right)^{\frac{\alpha}{\mu}}. \tag{9}$$

Obviously this transform represents the sum of Poisson distributed random value with parameter λ/μ and weighted negative binomial with parameters $(\alpha/\mu + 1, \beta/(\lambda + \beta))$ and $(\alpha/\mu, \beta/(\lambda + \beta))$ distributed random values.

2.2   Duration of the Busy Period

Under the busy period of the $M/M/c$ queueing system we will understand the duration of the time interval starting when a new arrival finds the system empty (the group of states at level $\mathbf{0}$) and ends when the system becomes empty again at a service completion. Further we make some notations:

Φ – the duration of the busy period, Φ_x – the first-passage time to the level $\mathbf{0}$ give the initial state is $x \in E$, $f_{\Phi_x}(t) = \frac{1}{dt}\mathbb{P}[\Phi_x \in [t, t + dt)]$ – the distribution density function (PDF) of Φ_x, $\tilde{\varphi}_x(s) = \int_0^{\infty} e^{-st} f_{\Phi_x}(t)dt$, $\mathrm{Re}[s] \geq 0$ – the Laplace-Stiltjes transform (LST). The LSTs are partitioned according to the number of customers in the system,

$$\tilde{\varphi}_0(s) = (\tilde{\varphi}_{(0,1)}(s), \tilde{\varphi}_{(0,0)}(s))' = \mathbf{e},$$
$$\tilde{\varphi}_n(s) = (\tilde{\varphi}_{(n,1)}(s), \tilde{\varphi}_{(n,0)}(s))', \, 1 \leq n \leq r.$$

Theorem 2. *The LST $\tilde{\varphi}_1(s)$ is given by*

$$\tilde{\varphi}_1(s) = \sum_{i=1}^{r} \prod_{j=1}^{i-1} U_j(s) V_i(s), \tag{10}$$

where

$$U_1(s) = -(Q_{1,1} - sI)^{-1} Q_{0,1}, \; V_1(s) = -(Q_{1,1} - sI)^{-1} Q_{2,1} \mathbf{e}, \tag{11}$$
$$U_k(s) = -(Q_{1,k} - sI + Q_{2,\min\{k,c\}} U_{k-1}(s))^{-1} Q_{0,1}, \; 2 \le k \le c,$$
$$V_k(s) = -(Q_{1,k} - sI + Q_{2,\min\{k,c\}} U_{k-1}(s))^{-1} Q_{2,\min\{k,c\}} V_{k-1}(s), \; 2 \le k \le r-1,$$
$$V_r(s) = -(Q_{1,c+1} - sI + Q_{2,c} U_{r-1}(s))^{-1} Q_{2,c} V_{r-1}(s).$$

Proof. By virtue of the first-step analysis one gets

$$(s + \alpha + \lambda + \min\{c,n\}\mu)\tilde{\varphi}_{(n,1)}(s) = \alpha\tilde{\varphi}_{(n,0)}(s) + \lambda\tilde{\varphi}_{(n+1,1)}(s) \tag{12}$$
$$+ \min\{c,n\}\mu\tilde{\varphi}_{(n-1,1)}(s), \; 1 \le n \le r-1,$$
$$(s + \beta + \lambda)\tilde{\varphi}_{(n,0)}(s) = \beta\tilde{\varphi}_{(n,1)}(s) + \lambda\tilde{\varphi}_{(n+1,0)}(s), \; 1 \le n \le r-1,$$
$$(s + \alpha + c\mu)\tilde{\varphi}_{(r,1)}(s) = \alpha\tilde{\varphi}_{(r,0)}(s) + c\mu\tilde{\varphi}_{(r-1,1)}(s),$$
$$(s + \beta)\tilde{\varphi}_{(r,0)}(s) = \beta\tilde{\varphi}_{(r,1)}(s).$$

The last system is expressed in matrix form, namely

$$- (Q_{1,1} - sI)\tilde{\varphi}_1(s) = Q_{0,1}\tilde{\varphi}_2(s) + Q_{2,1}\mathbf{e},$$
$$- (Q_{1,\min\{n,c\}} - sI)\tilde{\varphi}_n(s) = Q_{0,1}\tilde{\varphi}_{n+1}(s) + Q_{2,\min\{n,c\}}\tilde{\varphi}_{n-1}(s), \; 2 \le n \le r-1,$$
$$- (Q_{1,c+1} - sI)\tilde{\varphi}_r(s) = Q_{2,c}\tilde{\varphi}_{r-1}(s).$$

By applying to the last system the forward elimination backward substitution method we can deduce that

$$\tilde{\varphi}_n(s) = U_n(s)\tilde{\varphi}_{n+1}(s) + V_n(s), \; 1 \le n \le r-1, \tag{13}$$
$$\tilde{\varphi}_r(s) = V_r(s),$$

where the matrices $U_n(s)$ and $V_n(s)$ satisfy the recursive relations (11). By backward substitution in (13) yields automatically (10) which completes the proof. Since the busy period starts by visiting a state of the level $\mathbf{0} = \{(0,1), (0,0)\}$, for the unconditional LST $\tilde{\varphi}(s) = \int_0^\infty e^{-st} f_\Phi(t) dt$ and LT $\tilde{\Phi}(s) = \int_0^\infty e^{-st} F_\Phi(t) dt$ we have

$$\tilde{\varphi}(s) = \frac{\pi_0}{\pi_0 \mathbf{e}} \tilde{\varphi}_1(s), \; \tilde{\Phi}(s) = \frac{1}{s}\tilde{\varphi}(s). \tag{14}$$

The value of the PDF $f_\Phi(t)$ at point $t = 0$ is defined as $f_\Phi(0) = \lim_{s \to \infty} s\tilde{\varphi}(s) = \mu$.

Remark 3. The moments $\mathbb{E}[\Phi^n]$ of the busy period can be obtained by recursive computation of the conditional busy period moments. The idea consists in differentiation of (12) n times at point $s = 0$. The first moment can be evaluated by

$$\bar{\Phi} = \mathbb{E}[\Phi] = \frac{1}{\lambda}\left(\frac{1}{\pi_0 e} - 1\right). \tag{15}$$

2.3   Number of Customers Served in a Busy Period

The next interesting descriptor is the number of customers served in a busy period. Its study complements the busy period analysis since it provides a discrete counterpart of the length of busy period. Define the following notations:

Ψ – the number of customers served in a busy period L, Ψ_x – the number of customers served in a time L_x, $f_{\Psi_x}(k) = \mathbb{P}[N_x = k]$ – PDF of N_x, $\tilde{\psi}_x(z) = \sum_{k=1}^{\infty} f_{\Psi_x}(k)z^k, |z| \leq 1$ – the probability generating function (PGF), $\bar{\Psi} = \mathbb{E}[\Psi]$ and $\bar{\Psi}_x = \mathbb{E}[\Psi_x]$ – the corresponding moments and

$$\tilde{\psi}_n(z) = (\tilde{\phi}_{(n,1)}(z), \tilde{\phi}_{(n,0)}(z))', 1 \leq n \leq r,$$
$$\bar{\Psi}_n(z) = (\bar{\Psi}_{(n,1)}, \bar{\Psi}_{(n,0)})', 1 \leq n \leq r.$$

For the conditional density $f_{\Psi_x}(k)$ via the law of the total probability we obtain,

$$f_{\Psi_x}(k) = \frac{\lambda_{xy'}}{\lambda_x}f_{\Psi_{y'}}(k-1) + \sum_{y \neq x, y'}\frac{\lambda_{xy}}{\lambda_x}\psi_y(k), \tag{16}$$

where $\lambda_x = \sum_{y \neq x}\lambda_{xy}$. The first term in the right hand side stands for the transition due to customer departure, whereas the second term includes other possible transitions. In terms of the PGF the latter equality can be expressed as

$$\tilde{\psi}_x(z) = \frac{z\lambda_{xy'}}{\lambda_x}\tilde{\psi}_{y'}(z) + \sum_{y \neq x, y'}\frac{\lambda_{xy}}{\lambda_x}\tilde{\psi}_y(z). \tag{17}$$

Theorem 3. *The PGF $\tilde{\psi}(z)$ is given by*

$$\tilde{\psi}(z) = \frac{\pi_0}{\pi_0 e}\sum_{i=1}^{r}\prod_{j=1}^{i-1}H_j(z)K_i(z), \tag{18}$$

where the matrices $H_k(z), 1 \leq k \leq r - 1$, and $K_k(z), 1 \leq k \leq r$, are obtained from (11) by substituting $s = 0$ and replacing the matrices $Q_{2,k}$ by $zQ_{2,k}$.

Proof. By means of the first-step analysis we get the system of equations governing the dynamic of the generating function $\tilde{\psi}_x(z)$. For the states of the level $\mathbf{0} = \{(0,1), (0,0)\}$ obviously $\tilde{\psi}_{(0,d)}(z) = 1, d \in \{0,1\}$. The resulting system is of the form (12) for $s = 0$, but the service rates μ are replaced by $z\mu$. By expressing these equations in matrix form we can get the result (18) for the unconditional PGF $\tilde{\psi}(z)$.

Remark 4. The moments $\mathbb{E}[\Psi^n]$ of the number of customers served in a busy period can be obtained by recursive computation of the conditional moments $\mathbb{E}[\Psi_x^n]$. The corresponding equations are obtained by differentiation of the system from the previous proof n times at point $z = 1$. The first moment can be evaluated by

$$\bar{\Psi} = \mu\bar{\Phi} = \frac{\mu}{\lambda}\Big(\frac{1}{\pi_0\mathbf{e}} - 1\Big). \tag{19}$$

2.4 Number of Failures in a Busy Period

Another characteristic measure which can be treated as a discrete counterpart of the length of busy period is the number of failures in this period. Denote by Θ the number of failures in a busy period with density $f_\Theta(t)$ and by $\tilde{\theta}(z)$ the corresponding PGF.

Theorem 4. *The PGF $\tilde{\theta}(z)$ is given by*

$$\tilde{\theta}(z) = \frac{\pi_0}{\pi_0\mathbf{e}} \sum_{i=1}^{r} \prod_{j=1}^{i-1} H_j(z)K_i(z), \tag{20}$$

where $H_k(z), 1 \leq k \leq r - 1$, and $K_k(z), 1 \leq k \leq r$, are obtained from (11) by substituting $s = 0$ and replacing $Q_{1,k}$ by $Q_{1,k}(z) = Q_{1,k} - \alpha(1 - z)\mathbf{e}_1 \otimes \mathbf{e}_2'$.

Proof. The system of equations for the conditional PGF $\tilde{\psi}_x(z)$ is obtained from (12) for $s = 0$, but the failure rates α from the right hand side are replaced by $z\alpha$. By simple algebraic manipulations and subsequent matrix form presentation we get (20).

Remark 5. The moments $\mathbb{E}[\Theta^n]$ of the number of customers served in a busy period can be obtained by recursive computation of the conditional moments $\mathbb{E}[\Theta_x^n]$. The corresponding equations are obtained by differentiation of the system from the previous proof n times at point $z = 1$. The first moment can be evaluated by

$$\bar{\Theta} = \alpha\bar{\Phi} = \frac{\alpha}{\lambda}\Big(\frac{1}{\pi_0\mathbf{e}} - 1\Big). \tag{21}$$

3 Busy Period Analysis for an Infinite Queue

3.1 Stationary Probabilities

To guarantee the existence of the stationary state probabilities it is needed to derive the corresponding stability condition. To accomplish this we define matrix $Q = Q_{0,1} + Q_{1,c} + Q_{2,c}$. According to the results for the QBD processes in [12], the necessary and sufficient condition for ergodicity of the process $\{X(t)\}_{t\geq 0}$ is of the form $\mathbf{p}Q_{2,c} > \mathbf{p}Q_{0,1}$, where the row-vector $\mathbf{p}$ is given by $\mathbf{p}Q = \mathbf{0}$ and

$\mathbf{pe} = 1$. After some simple routine manipulation, the ergodicity condition turns out to be

$$\rho = \lambda \mathbb{E}[B] = \frac{\lambda}{c\mu}\left(1 + \frac{\alpha}{\beta}\right) < 1, \tag{22}$$

where $\mathbb{E}[B]$ is a mean effective service time.

Theorem 5. *If condition (22) holds, the stationary probabilities can be calculated by*

$$\boldsymbol{\pi}_n = \boldsymbol{\pi}_c \prod_{j=1}^{c-n} M_{c-j}, \, 0 \le n \le c - 1, \tag{23}$$

$$\boldsymbol{\pi}_n = \boldsymbol{\pi}_c R^{n-c}, \, n \ge c,$$

and $\boldsymbol{\pi}_c$ is a unique solution of the system of equations

$$\left[\sum_{n=0}^{c-1}\prod_{j=1}^{c-n} M_{c-j} + (I - R)^{-1}\right]\mathbf{e} = 1, \tag{24}$$

$$\boldsymbol{\pi}_c[M_{c-1}Q_{0,1} + Q_{1,c} + RQ_{2,c}] = \mathbf{0}.$$

where M_{c-j} satisfies the recursive relations (6) and R is the minimal non-negative solution to the square matrix equation,

$$R^2 Q_{2,c} + RQ_{1,c} + Q_{0,1} = \mathbf{0} \tag{25}$$

and is of the form

$$R = \begin{pmatrix} \frac{\lambda}{c\mu} & \frac{\lambda\alpha}{c\mu(\lambda+\beta)} \\ \frac{\lambda}{c\mu} & \frac{\lambda}{\lambda+\beta}\left(\frac{\alpha}{c\mu}+1\right) \end{pmatrix}. \tag{26}$$

Proof. For the boundary states $(n, d), 0 \le n \le c - 1$, the probabilities can be evaluated in the same way as for the model with a finite buffer. If $n \ge c$, then due to [12], the solution is obtained in geometric form. An appeal to the matrix equation (25) and relation $RQ_{2,c}\mathbf{e} = Q_{0,1}\mathbf{e}$ yields (26).

3.2   Duration of the Busy Period

Theorem 6. *The LST $\tilde{\varphi}_1(s)$ is given by*

$$\tilde{\varphi}_1(s) = \sum_{i=1}^{c}\prod_{j=1}^{i-1} U_j(s)V_i(s), \tag{27}$$

where $U_k(s)$ and $V_k(s)$ for $1 \le k \le c - 1$ are defined as before, and

$$V_c(s) = -(Q_{1,c} - sI + Q_{2,c}U_{c-1}(s) + Q_{0,1}\Omega(s))^{-1}Q_{2,c}V_{c-1}(s). \tag{28}$$

The matrix $\tilde{\Omega}(s)$ is the minimal non-negative solution of the quadratic matrix equation

$$\tilde{\Omega}(s) = -(Q_{1,c} - sI)^{-1}Q_{2,c} - (Q_{1,c} - sI)^{-1}Q_{0,1}\tilde{\Omega}^2(s). \tag{29}$$

Proof. The analysis of the LST of the first-passage time to the level $\mathbf{0}$ in QBD processes is related to the study of the so-called fundamental period (see e.g. Neuts [12]). The basic idea consists in the following. Denote by $w_{dd'}^{(k)}(t)$ the conditional probability that starting in state $(c+k,d)$ the process $\{X(t)\}_{t\geq0}$ reaches the states at level $\mathbf{c} = \{(c,1),(c,0)\}$ for the first time no later than time t and does so by entering the state $(c,d'),d' \in \{0,1\}$. Denote by $\tilde{\omega}_{dd'}^{(k)}(s) = \int_0^{\infty} e^{-st}dw_{dd'}^{(k)}(t), \mathrm{Re}[s] \geq 0$, the corresponding LST and set the matrix $\tilde{\Omega}^{(k)}(s) = (\tilde{\Omega}_{dd'}^{(k)}(s))$, which is of the size 2×2. Then by employing a first-step conditional argument is can be shown that $\tilde{\Omega}^{(k)}(s) = (\tilde{\Omega}(s))^k$ and the matrix $\Omega(s)$ satisfies the quadratic matrix relation (29). Moreover, for every s, $\tilde{\Omega}(s)$ is the minimal non-negative solution of (29) and can be evaluated by the following iterative scheme,

$$\tilde{\Omega}_0(s) = \mathbf{0}, \tag{30}$$
$$\tilde{\Omega}_{n+1}(s) = -(Q_{1,c} - sI)^{-1}Q_{2,c} - (Q_{1,c} - sI)^{-1}Q_{0,1}\tilde{\Omega}_n^2(s), \quad n \geq 0.$$

Conditioning on the state of the first entrance to level $\mathbf{c}$ from states of level $\mathbf{c}+1$,

$$\tilde{\varphi}_{c+1}(s) = \tilde{\Omega}(s)\tilde{\varphi}_c(s). \tag{31}$$

For the level $\mathbf{c}$ the following relation holds,

$$-(Q_{1,c} - sI)\tilde{\varphi}_c(s) = Q_{0,1}\tilde{\varphi}_{c+1}(s) + Q_{2,c}\tilde{\varphi}_{c-1}(s). \tag{32}$$

Due to recursive relations (13) for $1 \leq n \leq c$, $\tilde{\varphi}_{c-1}(s) = U_{c-1}(s)\tilde{\varphi}_c(s) + V_{c-1}(s)$. The last equality together with (31) yields the result (27).

3.3   Number of Customers Served in a Busy Period

For the number of customers served and number of failures in a busy period the following two statements are obtained similarly to the previous section.

Theorem 7. *The PGF $\tilde{\psi}(z)$ is given by*

$$\tilde{\psi}(z) = \frac{\boldsymbol{\pi}_0}{\boldsymbol{\pi}_0 \mathbf{e}} \sum_{i=1}^{c} \prod_{j=1}^{i-1} H_{\bar{j}}(z)K_i(z), \tag{33}$$

where $H_k(z)$ and $K_k(z), 1 \leq k \leq c-1$, as before and

$$K_c(z) = -z(Q_{1,c} + zQ_{2,c}H_{c-1}(z) + Q_{0,1}\Omega(z))^{-1}Q_{2,c}K_{c-1}(z).$$

Here $\Omega(z)$ satisfies the matrix equation

$$\Omega(z) = -Q_{1,c}^{-1}(zQ_{2,c} + Q_{0,1}\Omega^2(z)). \tag{34}$$

3.4 Number of Failures in a Busy Period

Now we present the result for the number of failures in a busy period.

Theorem 8. *The PGF $\tilde{\theta}(z)$ is given by*

$$\tilde{\theta}(z) = \frac{\pi_0}{\pi_0 e} \sum_{i=1}^{c} \prod_{j=1}^{i-1} H_j(z) K_i(z), \tag{35}$$

where $H_k(z)$ and $K_k(z), 1 \leq k \leq c-1$, were obtained before and

$$K_c(z) = -(Q_{1,c}(z) + Q_{2,c}H_{r-1}(z) + Q_{0,1}\Omega(z))^{-1}Q_{2,c}K_{c-1}(z). \tag{36}$$

Here $\Omega(z)$ satisfies the matrix equation,

$$\Omega(z) = -Q_{1,c}^{-1}(z)(Q_{2,c} + Q_{0,1}\Omega^2(z)). \tag{37}$$

4 Application to Wireless Network Under LSA Regime and Numerical Example

Let us consider the incumbent holding spectrum rights of use and the mobile operator holding individual rights of use to LSA frequency band. The arrival rates of requests for access to band as well as average time of band occupancy are assumed to be equal to α, β^{-1} and λ, μ^{-1} for the incumbent and mobile operator correspondingly. From the incumbent's perspective, the considered system is always reliable due to the absolute priority access to LSA band. Whereas from the mobile operator's perspective, the system is unreliable when the incumbent uses LSA band. Thereby, we model LSA operation as a queueing system with breakdowns. We consider a scenario of aeronautical telemetry with airplane flies [7], which do not depend on time of the day – 6 flies per day ($\alpha^{-1} = 240\,\text{min}$), 12 flies per day ($\alpha^{-1} = 120\,\text{min}$), 24 flies per day ($\alpha^{-1} = 60\,\text{min}$), and 36 flies per day ($\alpha^{-1} = 40\,\text{min}$). The takeoff speed of all airplanes in the airport are the same and the average time during which an airplane is flying over one cell is equal to $\beta^{-1} = 1\,\text{min}$. We assume that one cell of wireless network has capacity of $c = 30$ users who perform calls during the day with average duration μ_i^{-1} of 3 min, 12 min, or 1.5 min.

In this section we present numerical results including the probabilistic characteristic $\mathbb{E}[T_0]$, the first moments and the numerical inversion of the unconditional transforms $\tilde{\phi}(s), \tilde{\psi}(s)$ and $\tilde{\theta}(s)$. The formulas for the infinite buffer case were implemented. In the following examples we fix $c = 30, \lambda = 1, \mu = 1/3, \alpha = 1/120$ and $\beta = 1$. In Fig. 1 we plot the value $\mathbb{E}[T_0]$ for varying λ and α. It should be noticed that the curves are not monoton. They start to decrease to a minimum and then these curves exhibit monoton increasing shapes by increasing λ. As expected, when α increases, the amount of time in a regenerative cycle when the servers are blocked is getting higher. In Fig. 2, we display four curves for the density function $f_\Phi(t)$. They are obtained by numerical inversion of expression

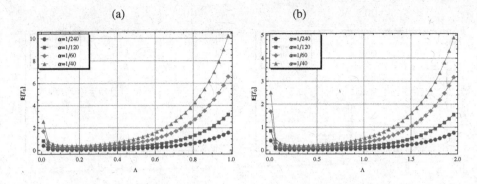

Fig. 1. The value $\mathbb{E}[T_0]$ versus λ and $\mu = 1/6$ (a) and $\mu = 1/3$ (b)

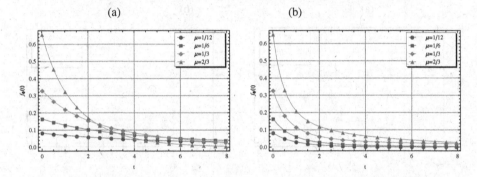

Fig. 2. The density $f_\Phi(t)$ versus μ and $\lambda = 1/10$ (a) and $\lambda = 1$ (b)

(14) for the infinite buffer case. It can be seen that $f_\Phi(0) = \mu$ in accordance with Tauberian result. The presented curves exhibit decreasing shapes with heavier tails for higher λ and smaller μ. The corresponding first moment take the following values,

$$\bar{\Phi} = \{23.544, 8.322, 3.541, 1.641\}, \text{ figure labeled by "(a)" and}$$
$$\bar{\Phi} = \{175159.166, 418.389, 19.521, 3.539\}, \text{ figure labeled by "(b)".}$$

In Fig. 3 we analyze the effect of arrival and service intensity on the number of service completions in a busy period. The lowest value at point $k = 1$ corresponds to the case $\lambda = 1$ and $\mu = 1/12$ and this curve exhibits the heaviest tail. The moments of the number of service completions Ψ are presented below,

$$\bar{\Psi} = \{1.962, 1.387, 1.180, 1.094\}, \text{ figure labeled by "(a)" and}$$
$$\bar{\Psi} = \{14596.597, 69.731, 6.507, 2.359\}, \text{ figure labeled by "(b)".}$$

In Fig. 4, we realize the numerical inversion algorithm to get the probability densities $f_\Theta(k)$ for the number of retrials. The value at point $k = 0$ corresponds to the probability that the service in a busy period will be without breakdowns.

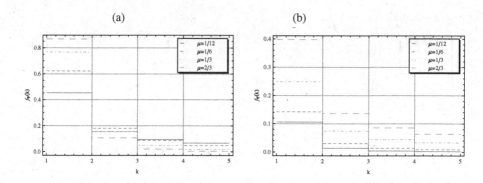

Fig. 3. The density $f_\Psi(k)$ versus μ and $\lambda = 1/10$ (a) and $\lambda = 1$ (b)

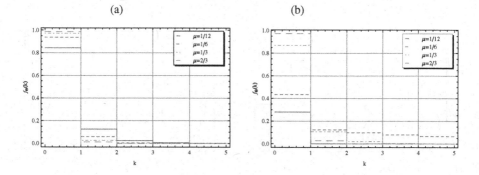

Fig. 4. The density $f_\Theta(k)$ versus μ and $\lambda = 1/10$ (a) and $\lambda = 1$ (b)

Obviously the highest value belongs to the case with lowest load factor ρ when $\lambda = 1/10$ and $\mu = 2/3$. Further we display the mean number of failures in a busy period,

$$\bar{\Theta} = \{0.196, 0.069, 0.029, 0.014\}, \text{ figure labeled by "(a)" and}$$
$$\bar{\Theta} = \{1456.667, 3.487, 0.161, 0.029\}, \text{ figure labeled by "(b)".}$$

The present paper is our first step to analyzing the busy period of wireless network under LSA regime as a queueing system with breakdowns. Our further research activities aim at other different performance measures like the number of interrupted customers, the time interval during which the service is interrupted.

References

1. ETSI TR 103 113: Electromagnetic compatibility and Radio spectrum Matters (ERM); SystemReference document (SRdoc); Mobile broadband services in the 2 300 MHz – 2 400MHz frequency band under Licensed Shared Access regime (2013)

2. ETSI TR 103 154: Reconfigurable Radio Systems (RRS); System requirements for operationof Mobile Broadband Systems in the 2 300 MHz - 2 400 MHz band under Licensed Shared Access (LSA) (2014)
3. ETSI TR 103 235: Reconfigurable Radio Systems (RRS); System architecture and high levelprocedures for operation of Licensed Shared Access (LSA) in the 2 300 MHz – 2 400 MHz band (2015)
4. Hasan, N.U., Ejaz, W., Ejaz, N., Kim, H.S., Anpalagan, A., Jo, M.: Network selection and channel allocation for spectrum sharing in 5G heterogeneous networks. IEEE Access **4**, 980–992 (2016)
5. Buckwitz, K., Engelberg, J., Rausch, G.: Licensed Shared Access (LSA) - regulatory background and view of administrations. In: 9th International Conference on Cognitive Radio Oriented Wireless Networks, CROWNCOM, pp. 413–416 (2014)
6. Gundlach, M., Hofmann, J., Markwart, C., Mohyeldin, E.: Recent advances on LSA in standardization, regulation, research and architecture design. In: 1st International Workshop on Cognitive Cellular Systems, CCS, pp. 1–5 (2014)
7. Ponomarenko-Timofeev, A., Pyattaev, A., Andreev, S., Koucheryavy, Y., Mueck, M., Karls, I.: Highly dynamic spectrum management within licensed shared access regulatory framework. IEEE Commun. Mag. **5**(3), 100–109 (2016)
8. Borodakiy, V.Y., Samouylov, K.E., Gudkova, I.A., Ostrikova, D.Y., Ponomarenko-Timofeev, A.A., Turlikov, A.M., Andreev, S.D.: Modeling unreliable LSA operation in 3GPP LTE cellular networks. In: 6th International Congress on Ultra Modern Telecommunications and Control Systems, ICUMT, pp. 390–396 (2014)
9. Gudkova, I.A., Samouylov, K.E., Ostrikova, D.Y., Mokrov, E.V., Ponomarenko-Timofeev, A.A., Andreev, S.D., Koucheryavy, Y.A.: Service failure and interruption probability analysis for Licensed Shared Access regulatory framework. In: 7th International Congress on Ultra Modern Telecommunications and Control Systems, ICUMT, pp. 123–131 (2015)
10. Al-Seedya, R.O., El-Sherbinya, A.A., El-Shehawyb, S.A., Ammar, S.I.: Transient solution of the $M/M/c$ queue with balking and reneging. Comput. Math. Appl. **57**(8), 1280–1285 (2009)
11. Kea, J.-C., Linb, C.-H., Yangb, J.-Y., Zhangc, Z.G.: Optimal (d, c) vacation policy for a finite buffer $M/M/c$ queue with unreliable servers and repairs. Appl. Math. Model. **33**(10), 3949–3962 (2009)
12. Neuts, M.F.: Matrix-Geometric Solutions in Stochastic Models. The John Hopkins University Press, Baltimore (1981)
13. Vinodhini, G.A.F.: Transient solution of a multi-server queue with catastrophes and impatient customers when system is down. Appl. Math. Sci. **92**(8), 4585–4592 (2014)
14. Yue, D., Yue, W.: Analysis of an $M/M/c/N$ queueing system with balking, reneging, and synchronous vacations. In: Yue, W., Takahashi, Y., Takagi, H. (eds.) Advances in Queueing Theory and Network Applications, pp. 165–180. Springer, New York (2009)

Evaluating a Case of Downlink Uplink Decoupling Using Queuing System with Random Requirements

Eduard Sopin[1(✉)], Konstantin Samouylov[1], Olga Vikhrova[1],
Roman Kovalchukov[1], Dmitri Moltchanov[1,2], and Andrey Samuylov[1,2]

[1] Department of Applied Probability and Informatics,
Peoples' Friendship University of Russia, Moscow, Russia
{esopin,ovikhrova,rkovalchukov}@sci.pfu.edu.ru
[2] Tampere University of Technology, Tampere, Finland
{andrey.samuylov,dmitri.moltchanov}@tut.fi

Abstract. The need for efficient resource utilization at the air interfaces in heterogeneous wireless systems has recently led to the concept
of downlink and uplink decoupling (DUDe). Several studies have already
reported the gains of using DUDe in static traffic conditions. In this paper
we investigate performance of DUDe with stochastic session arrivals patterns in LTE environment with macro and micro base stations. Particularly, we use a queuing systems with random resource requirements and
to calculate the session blocking probability and throughput of the system. Our results demonstrate that DUDe association approach allows to
significantly improve the metrics of interest compared to conventional
downlink-based association mechanism.

Keywords: DL and UL decoupling · RBs allocation · Signal-to-noise
ratio · LTE-advanced · Pathloss · Heterogeneous networks

1 Introduction

The predicted increase in the user traffic demands places extreme requirements
on the future evolution of mobile systems, often referred to as fifth generation (5G) networks [1,2]. In addition to physical layer improvements including
advanced modulation and coding, and MIMO techniques, over the last decade
the researchers proposed a number of network solutions providing performance
improvements including the use of small (micro/pico/femto) cells [3], client-
relays [4], direct in-band and out-of-band device-to-device communications [5].
All these concepts target aggressive spatial frequency reuse promising substantial
area capacity gains.

Heterogeneous deployment featuring multiple serving layers such as
macro/micro/pico/femto is seen as a pragmatic and cost-effective way to significantly enhance the capacity of modern LTE cellular networks. In homogeneous
networks, both uplink (UL) and downlink (DL) have been associated to the base
station according to its transmitting power in DL direction. In heterogeneous
networks (HetNet) this type of association is no longer effective.

© Springer International Publishing AG 2016
O. Galinina et al. (Eds.): NEW2AN/ruSMART 2016, LNCS 9870, pp. 440–450, 2016.
DOI: 10.1007/978-3-319-46301-8_37

In a typical HetNet scenario, the DL coverage area in a macro cell is much larger than that of a small cell because of large disparity in signal transmit power. At the same time, mobile devices have nearly the same transmit powers in the UL and thus nearly the same range. Consequently, there are some locations with higher received power from a macro cell, but better path loss to a small cell, resulting in increased efficiency of decoupled cell association. This concept, termed DL/UL Decoupling (DUDe), has already been supported by different analytical and simulation results, and shown significant gains in network performance and load balancing. It has been first introduced as a technique to improve performance of HetNets in [6]. The impact that decoupled association has on the average throughput has been analyzed in [7] by using the stochastic geometry framework to derive the association probability for UL and DL. Analytical results shown in [8] highlight that using different association strategies for uplink and downlink lead to significant improvement in joint uplink-downlink rate coverage over the coupled associations.

The existing studies on DUDe do not take into account the traffic dynamics of the system. In this paper, we fill this gap by proposing a queuing model that explicitly includes dynamics of sessions arrivals and reflects resource utilization in both micro and macro cells. The metrics of interest are the session loss probability and the system throughput. Using these metrics, we compare the conventional UL/DL association mechanism with DUDe scheme, and show that the latter provides substantial gains.

The rest of the paper is organized as follows. Section 2 describes the DUDe mechanism and provides system parameters mapping to the parameters of queuing systems with random requirements. The mathematical framework for the estimation of radio resource requirements distribution is introduced in Sect. 3. Numerical results are presented in Sect. 4. Section 5 concludes the paper.

2   Model Description

We consider a characteristic HetNet scenario with DUDe association as shown on Fig. 1. Without the loss of generality, we consider a single small cell eNB (SeNB) within the coverage of a single macro cell eNB (MeNB). According to DUDe, the DL and UL channels of a user can be associated to different eNBs: DL association is based on the maximum received power from eNB, and UL association is done according to the minimal path loss to minimize the energy consumption of a UE [8]. Thus, we have two borders, and when a user crosses one of them, the association of either the DL or UL channels will switch to another eNB.

1. If a UE is located inside a circle or radius D_{DL} centered at SeNB, then both the DL and UL channels will be associated to SeNB;
2. If a UE is further than D_{DL}, but is still inside the circle of radius D_{UL}, then the UE switches its DL association to a more powerful MeNB, while keeping its association with SeNB in UL;
3. Otherwise, UE has a coupled association to MeNB.

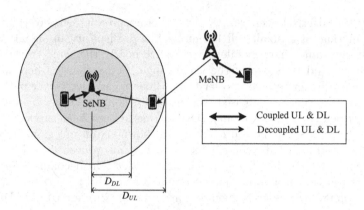

Fig. 1. Illustration of coupled SeNB association, DUDe and coupled MeNB association

Session dynamics in HetNets can be described in terms of queuing systems with limited resources and random requirements. Initialization of new user sessions in the scenario is modeled by Poisson process. Each user session has a specific duration (service time of a customer) and requirements for radio resources. The latter depend on the type of service, required bitrate, radio signal path loss and can be approximated by a specific probability distribution. Different types of services provided by the network can be described by several arrival flows in the queuing system.

Queuing systems with limited resources and random requirements under Poisson arrivals have been first investigated in [9], and the further analysis is provided in [10,11]. The main results that are used to evaluate the performance gains of the DUDe concept are described below.

Consider a queuing system with L incoming mutually independent Poison flows with rates $\lambda_1, \lambda_2, \ldots, \lambda_L$ and N servers, depicted on Fig. 2. We consider a separate customer arrival flow for each service type and cell association.

Each customer, in order to be served, demands some amount of resources of M types. Serving times have exponential distribution with rate μ_l, where l is the customer type. Let $\mathbf{R} = (R_1, \ldots, R_M)$ denote the vector of available resources. A new l type customer will occupy $\mathbf{r}_l \geq 0$ resources, where $\mathbf{r}_l = (r_{l1}, \ldots, r_{lM})$ is a multidimensional RV. A new customer will be dropped if there are not enough resources to meet its requirements or all servers are busy.

At some instant t system behavior can be described by a random process $X(t) = (n(t), \theta(t), \gamma(t))$, where $n(t)$ is the total number of customers in system, $\theta(t) = (\theta_1(t), \theta_2(t), \ldots, \theta_{n(t)}(t))$ is a vector of customer types ranked by remaining service time, $\gamma(t) = (\gamma_{1,i}, \ldots, \gamma_{M,i})_{i \leq n(t)}$ - is a matrix of occupied resources, and $\gamma_{m,\bullet}(t) = \sum_{i=1}^{n(t)} \gamma_{m,i}(t) \leq R_m$.

Denote $p_{l_k, \mathbf{r}_{l_k}}$ the probability that k-th customer of l-type will require $\mathbf{r}_{l_k}$ resources and $p_{l, \mathbf{r}_l}$ the probability that a customer of l-type will require $\mathbf{r}_l$ resources. Then $p_{l, \mathbf{r}_l}^{(k_l)}$ is the probability that k_l customers of l-type will require $\mathbf{r}_l$,

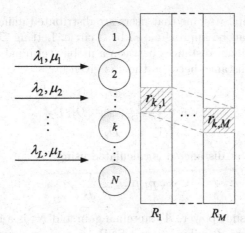

Fig. 2. Multiserver queueing system with L types of customer and M types of resources

where $p_{l,\mathbf{r}_l}^{(k_l)}$ is k_l-fold convolution of probabilities $p_{l,\mathbf{r}_l}$. According to [11] the stationary probabilities distribution of $X(t)$ is

$$q_{l_1,\ldots,l_k}^k(\mathbf{r}_{l_1},\ldots,\mathbf{r}_{l_k}) = q_0 p_{l_1,\mathbf{r}_{l_1}} \cdots p_{l_k,\mathbf{r}_{l_k}} \prod_{i=1}^k \frac{\lambda_{l_i}}{\sum_{j=1}^i \mu_{l_j}}, \tag{1}$$

$$q_0 = \left(1 + \sum_{k=1}^N \sum_{\mathbf{r}_{l_1}+\ldots+\mathbf{r}_{l_k}=\mathbf{R}} p_{l_1,\mathbf{r}_{l_1}} \cdots p_{l_k,\mathbf{r}_{l_k}} \prod_{i=1}^k \frac{\lambda_{l_i}}{\sum_{j=1}^i \mu_{l_j}}\right)^{-1} \tag{2}$$

If we group $p_{l_k,\mathbf{r}_{l_k}}$ over the types of customers then stationary probabilities (1) and (2) can be calculated as the analog of convolutional algorithm for closed networks:

$$q_{1,\ldots,L}^k(\mathbf{r}_1,\ldots,\mathbf{r}_L) = q_0 p_{1,\mathbf{r}_1}^{(k_1)} \cdots p_{L,\mathbf{r}_L}^{(k_L)} \frac{\rho_1^{k_1}}{k_1!} \cdots \frac{\rho_L^{k_L}}{k_L!} \tag{3}$$

$$q_0 = \left(1 + \sum_{n=1}^N \sum_{k_1+\ldots+k_L=n} \sum_{\mathbf{r}_1+\ldots+\mathbf{r}_L=\mathbf{R}} p_{1,\mathbf{r}_1}^{(k_1)} \cdots p_{L,\mathbf{r}_L}^{(k_L)} \frac{\rho_1^{k_1}}{k_1!} \cdots \frac{\rho_L^{k_L}}{k_L!}\right)^{-1} \tag{4}$$

where $p_{i,\mathbf{r}_i}^{(k_i)} = \sum_{0<\mathbf{j}<\mathbf{R}} p_{i,\mathbf{r}_i-\mathbf{j}}^{(k_i-1)} p_{l_k,\mathbf{j}}$ is a k_i-fold convolution of probabilities.

$p_{l_1,\mathbf{r}_{l_1}},\ldots,p_{l_k,\mathbf{r}_{l_k}}$ for all $l_k = i$ and $\sum_{k=1}^{k_i} \mathbf{r}_{l_k} = \mathbf{r}_i$. Stationary probabilities (3), (4) are independent from the order of customers in the system.

3   Resource Requirements Analysis

To apply the analytical results to multitier heterogeneous cellular network, we first need to estimate the probability mass function (PMF) of user session resource requirements. Assume that users are distributed uniformly on the coverage area, which can be approximated by a circle. Letting D_1 and D_2 be the minimum and maximum distances between a mobile terminal and eNB, respectively, the CDF of distance between them is given by

$$F_d(x) = \begin{cases} 0, & x < D_1 \\ \frac{x^2 - D_1^2}{D_2^2 - D_1^2}, & x \in [D_1; D_2] \\ 1, & x > D_2 \end{cases}.$$ (5)

The SNR values at distance d is calculated using

$$SNR(d) = \frac{PA}{N_p d^2}$$ (6)

where P is the transmit power, A - antenna gain and N_p is a noise power.

Using the distance distribution and SNR we could write the joint CDF of SNR in DL and UL $F_{SNR,i}(x,y) = P\{SNR_{DL,i} < x; SNR_{UL} < y\}$ for both coupled and decoupled association, where i takes values M (for MeNB) or S (for SeNB).

For the coupled association, the distances between user equipment and base station are equal to each other. Then, the CDF $F_{SNR,i}(x,y)$ can be calculated as:

$$F_{SNR,i}(x,y) = P\left\{\frac{P_i A_i}{N_p d^2} < x; \frac{P_U A_U}{N_p d^2} < y\right\} =$$
$$= P\left\{\frac{P_i A_i}{N_p d^2} < x; \frac{P_i A_i}{N d_M^2} < y \cdot \frac{P_i A_i}{P_U A_U}\right\} = P\left\{\frac{P_i A_i}{N_p d^2} < \min\left(x; y \cdot \frac{P_i A_i}{P_U A_U}\right)\right\} = \quad (7)$$
$$= 1 - F_d\left(\max\left(\sqrt{\frac{P_i A_i}{N_p x}}; \sqrt{\frac{P_U A_U}{N_p y}}\right)\right),$$

where P_U is transmit power of UE and A_U - its antenna gain.

For the decoupling scenario, assume that distances from UE and eNBs in the UL and DL are independent of each other. Thus, the joint CDF of SNR in DL and UL is

$$F_{SNR}(x,y) = P\left\{\frac{P_M A_M}{N_p d_M^2} < x; \frac{P_U A_U}{N_p d_S^2} < y\right\} =$$
$$= P\left\{\frac{P_M A_M}{N_p d_M^2} < x\right\} \cdot P\left\{\frac{P_U A_U}{N_p d_S^2} < y\right\} = \quad (8)$$
$$= \left(1 - F_d\left(\sqrt{\frac{P_M A_M}{N_p x}}\right)\right) \cdot \left(1 - F_d\left(\sqrt{\frac{P_U A_U}{N_p y}}\right)\right).$$

3GPP specifications [12] define 15 CQI (Channel Quality Indicator) indices and assign a MCS (Modulation and Coding Scheme) with specific spectral efficiency for each CQI index. In [14], SNR bounds were evaluated to achieve the BLER value to be less than 10 % for each MCS (Table 3).

LTE standard supports $K = 15$ values for MCS, so denote S_j a margin of SNR where $j = \overline{1, K}$ and assume $S_0 = 0$ and $S_{K+1} = \infty$. Let $\pi_{k,j}^i$ be the probability that a user session is assigned to CQI k in DL and CQI j in UL, where i takes values M (for association with MeNB), S (for association with SeNB) and D (for decoupled association). Then, the PMF of joint CQI distribution in UL and DL is

$$\pi_{k,j}^i = F_{SNR,i}(S_{k+1}, S_{j+1}) - F_{SNR,i}(S_k, S_j), \quad 0 \le k, j \le K.$$ (9)

According to 3GPP standards, each CQI level corresponds to specific spectral efficiency. Therefore, using the spectral bandwidth of a RB (resource block) one can estimate the maximum achievable bitrate on the RB with specific CQI level. We also assume that 3 out of 14 symbols are used for signaling during a subframe [14]. We denote the ratio of data symbols to the overall number of symbols per a subframe as payload coefficient ε. For clarity, we assume that RBs are allocated to a user session without fragmentation. This assumption can be related whenever needed.

Table 1 presents the notation for service requirements used for estimation of CDF of resource requirements $F(x)$.

Table 1. Notation for service requirements

Parameter	Description
U	Number of service types
$\omega,\ kHz$	RB spectral bandwidth
$e_j,\ bps/Hz$	Spectral efficiency for CQI j
$\varepsilon = 0,76$	Payload coefficient
$C_j = \alpha\omega\,e_j,\ bps$	Maximum achievable bitrate on one RB for CQI j
$V_u^{DL},\ bps$	Required rate for service type u in DL
$V_u^{UL},\ bps$	Required rate for service type u in UL
For coupled association: $r_{ij}^u = \left\lceil \frac{V_u^{DL}}{C_i} \right\rceil + \left\lceil \frac{V_u^{UL}}{C_j} \right\rceil$ For DUDe: $r_{ij}^u = \left\{ \left\lceil \frac{V_u^{DL}}{C_i} \right\rceil ; \left\lceil \frac{V_u^{UL}}{C_j} \right\rceil \right\}$	Required number of RBs for service u with CQI i in DL and CQI j in UL

The parameter r_{ij}^u denoting the number of RBs required by a user session type u, where i and j are the CQI levels in DL and UL accordingly, and the probability distribution of the received CQI levels $\pi_{k,j}^i$ unambiguously define the probability distribution of resource requirements that is used for performance analysis of queuing system with random requirements.

4   Numerical Results

In this section, we demonstrate the numerical comparison of traditional association and DUDe approach is presented. We consider two types of popular video services: video calls in high quality (HQ) (user session type 1) and group video conferences with three participants (user session type 2). We first analyze performance measures of coupled eNB association scenario and then association scenario with DUDe.

For the case of traditional coupled UL and DL we consider two incoming flows of service type 1 requests with rates λ_1 and λ_3, and two flows of service type 2 requests with rates λ_2 and λ_4 for MeNB and SeNB accordingly. In case of DUDe, we add two more flows of requests with rates λ_5 and λ_6 for each session type. Based on the average number of UEs per eNB from [13] and simulation deployment from [8], we take the following assumptions for the numerical analysis:

1. UEs are uniformly distributed within a macro cell (Mcell) and 13 small cells (Scells) within the Mcell border.
2. The number of user requests for video call in HQ (user session type 1) is twice bigger than the corresponding number of group video sessions (type 2 sessions).
3. Average session duration is 1 min (Table 2).

Table 2. System load parameters

Parameter	Value	
	For coupled DL and UL	For DUDe
λ_1(Mcell)	4,33	1,08
λ_2(Mcell)	2,17	0,54
λ_3(Scell)	0,45	0,45
λ_4(Scell)	0,23	0,23
λ_5(DUDe)	-	3,25
λ_6(DUDe)	-	1,625
μ	1	1
ρ	7,175	7,175

Both Mcell and Scell utilize different frequency bands, 10 MHz bandwidth each, which corresponds to 50 available RBs. Based on eNBs parameters, traffic rate recommendation for user sessions and spectral efficiency for each CQI we evaluate the PMF of user session resource requirements, Table 3.

For video calls in HQ Skype recommends the traffic rate in UL and DL of at least 500 Kbps. For group video conferences the requirement in DL is much bigger than in UL and considered to be 1.5 Mbps, while the traffic rate in UL is specified to be 512 Kbps, Table 4.

Signal transmit power, noise power and antenna gain is in accordance with [13]. The minimum distance D_1 between UE and SeNB is recommended to be at least 2 m while the same parameter for Mcell is considered to be 15 m. We assume that Mcell radius is 1 km and the maximum Scell radius is 75 m, see Table 5. Note that DUDe is feasible when the distance between UE and SeNB is more than 75 m.

Table 3. User session resource requirements distribution

CQI	MCS	Spectral efficiency	SNR (dB)	$\left\lceil\frac{V_1}{C_i}\right\rceil$	$\left\lceil\frac{V_2^{DL}}{C_i}\right\rceil$	$\left\lceil\frac{V_2^{UL}}{C_i}\right\rceil$
0	—	—	< -6.7536	—	—	—
1	QPSK, 78/1024	0.15237	$-6,7536 < -4,9620$	19	58	20
2	QPSK, 120/1024	0.2344	$-4,9620 < -2,9601$	13	38	13
3	QPSK, 193/1024	0.3770	$-2,9601 < -1,0135$	8	24	8
4	QPSK, 308/1024	0.6016	$-1,0135 < 0,9638$	5	15	5
5	QPSK, 449/1024	0.8770	$0,9638 < 2,8801$	4	10	4
6	QPSK, 602/1024	1.1758	$2,8801 < 4,9185$	3	8	3
7	16QAM, 378/1024	1.4766	$4,9185 < 6,7005$	2	6	2
8	16QAM, 490/1024	1.9141	$6,7005 < 8,7198$	2	5	2
9	16QAM, 616/1024	2.4063	$8,7198 < 10,515$	2	4	2
10	64QAM, 466/1024	2.7305	$10,515 < 12,45$	2	4	2
11	64QAM, 567/1024	3.3223	$12,45 < 14,348$	1	3	1
12	64QAM, 666/1024	3.9023	$14,348 < 16,074$	1	3	1
13	64QAM, 772/1024	4.5234	$16,074 < 17,877$	1	2	1
14	64QAM, 873/1024	5.1152	$17,877 < 19,968$	1	2	1
15	64QAM, 948/1024	5.5547	$>19{,}968$	1	2	1

Table 4. Service rate requirements

Skype service type	Required rate in DL	Required rate in UL
Video calling in HQ	500 kbps	500 kbps
Group video (3 person)	1,5 kbps	512 kbps

For the assumed parameters, we demonstrate examples of probability distribution of resource requirements. Tables 6 and 7 show the probability that a new user session will occupy r RBs. In Table 8, we follow the same idea but as far as we decouple UL and DL a new user session will require some RBs in both Scell and Mcell.

Table 5. Network parameters

	UL Scell	DL Scell	UL Mcell	DL Mcell	UL DUDe (Scell)	DL DUDe (Mcell)
Transmitting power, P	0,2 W	0,25 W	0,2 W	40 W	0,2 W	40 W
Antenna gain, A	0 dBi	2 dBi	0 dBi	15 dBi	0 dBi	2 dBi
Distance, $[D_1; D_2]$	[2–75] m	[2–75] m	[15–1000] m	[75–1000] m	[75–125] m	[100–1000] m
Noise, N_p	6 dB	9 dB	5 dB	9 dB	6 dB	9 dB

Table 6. Resource distribution for video group in Mcell

r	3	4	5	6	7	8	10	11	13
P_r^2	0.0369	0.0513	0.0292	0.0944	0.1269	0.1731	0.1653	0.1428	0.18

Table 7. Resource distribution for video calling in Scell

r	2	3	4	5	6	7
P_r^1	0.0469	0.1717	0.0944	0.1935	0.2866	0.2067

Table 8. Recourse distribution for video group for DUDe

r	{2,0}	{2,1}	{2,2}	{2,3}	{2,4}	{2,5}	{2,8}	{2,13}
P_r^2	0.403	0.000834	0.003953	0.002869	0.004253	0.006884	0.010638	0.004146
r	{3,0}	{3,1}	{3,2}	{3,3}	{3,4}	{3,5}	{3,8}	{3,13}
P_r^2	0.5198	0.001076	0.005092	0.003707	0.005492	0.008876	0.13723	0.005346

The results for blocking probability analysis show the significant gain of using DUDe approach, see Fig. 3. When UL traffic is offloaded to a SeNB, the probability that UE's association decreases from 15 % to 0.2 %. The similar performance improvement is observed for DL, where the blocking probability reduces by 23.2 %.

Another important network performance parameter is the average number of occupied resources. Figure 4 shows the benefit of DUDe in terms of the average allocated RBs in SeNB and MeNB. Thus, the DUDe scheme can balance load at the expense of underutilized small cells in modern HetNet.

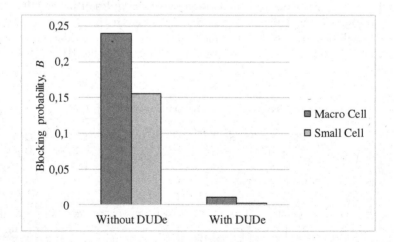

Fig. 3. Blocking probabilities in Scell and Mcell for two types of Skype service

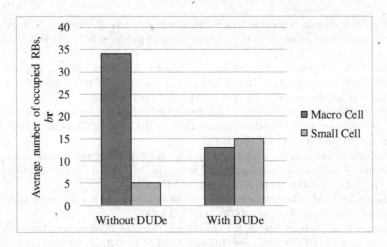

Fig. 4. Number of occupied RBs in Scell and Mcell for two types of Skype service

5 Conclusion

In this paper we analyzed the performance of wireless systems with conventional DL-based and DUDe user association algorithms. Based on required rate for each type of multimedia services, we determined the distribution of number of RBs needed for their transmission. The method can be easily adopted for a smaller resource unit, such as decile RB, in order to improve approximation. The analytical modelling showed that under user session dynamics the advantage of DUDe concept is still significant compared to static simulation results in recent papers.

Acknowledgments. This work was partially supported by RFBR, projects No. 15-07-03051, 15-07-03608, 16-37-60103, 16-07-00766. The reported study was funded within the Agreement 02.03.21.0008 dated 24.11.2016 between the Ministry of Education and Science of the Russian Federation and RUDN University.

References

1. Ericsson mobility report, Ericsson AB (2015). https://www.ericsson.com/mobility-report
2. Cisco Visual Networking Index: Global Mobile Data Traffic Forecast Update (2014–2019). http://www.cisco.com/c/en/us/solutions/collateral/service-provider/visual-networking-index-vni/mobile-white-paper-c11-520862.html
3. Andrews, J., Claussen, H., Dohler, M., Rangan, S.: Femtocells: past, present, and future. IEEE JSAC **30**(3), 497–508 (2012). IEEE
4. Lee, J., Wang, H., Andrews, J., Hong, D.: Outage probability of cognitive relay networks with interference constraints. IEEE Trans. Wirel. Commun. **10**(2), 390–395 (2011). IEEE

5. Fodor, G., Parkvall, S., Sorrentino, S., Wallentin, P., Lu, Q., Brahmi, N.: Device-to-device communications for national security and public safety. IEEE Access **2**(1), 1510–1520 (2014). IEEE

6. Boccardi, F., Andrews, J., Elshaer, H., Dohler, M., Parkvall, S., Popovski, P., Singh, S.: Why to decouple the uplink and downlink in cellular networks and how to do it. IEEE Commun. Mag. **54**(3), 110–117 (2016). IEEE

7. Singh, S., Zhang, X., Andrews, J.: Joint rate and SINR coverage analysis for decoupled up-link-downlink biased cell associations in HetNets. IEEE Trans. Wirel. Commun. **14**(10), 5360–5373 (2015). IEEE

8. Elshaer, H., Boccardi, F., Dohler, M., Irmer, R.: Downlink and uplink decoupling: a disruptive architectural design for 5G networks. In: Global Communications Conference (GLOBECOM), pp. 1798–1803. IEEE (2014)

9. Tikhonenko, O.M.: Generalized Erlang problem for service systems with finite total capacity. Probl. Inf. Transm. **41**(3), 243–253 (2005)

10. Naumov, V.A., Samouylov, K.E.: On the modeling of queuing systems with multiple re-sources. PFUR Bull. Ser. Inform. Math. Phys. **3**, 58–62 (2014)

11. Naumov, V., Samouylov, K., Sopin, E., Yarkina, N., Andreev, S., Samuylov, A.: LTE performance analysis using queuing systems with finite resources and random requirements. In: 2015 7th International Congress on Ultra Modern Telecommunications and Control Systems and Workshops (ICUMT), pp. 100–103. IEEE (2015)

12. ETSI TS 136.213. Evolved Universal Terrestrial Radio Access (E-UTRA). Physical layer procedures. Release 13 (2016)

13. ETSI TR 36.931. Evolved Universal Terrestrial Radio Access (E-UTRA). Radio Frequency (RF) requirements for LTE Pico Node B (2013)

14. Reyhanimasoleh, A.: Resource allocation in uplink long term evolution. Ph.D. thesis (2013)

NEW2AN: NoC and Positioning

Intra-CPU Traffic Estimation and Implications on Networks-on-Chip Research

Dmitri Moltchanov[1]([✉]), Arkady Kluchev[2], Pavel Kustarev[2],
Karolina Borunova[3], and Alexey Platunov[2]

[1] Department of Communications Engineering,
Tampere University of Technology, Tampere, Finland
dmitri.moltchanov@tut.fi
[2] Department of Embedded Systems, ITMO University, St.-Petersburg, Russia
{kluchev,kustarev,platunov}@lmt.ifmo.ru
[3] Faculty of Infocommunications Networks and System,
St.-Petersburg State University of Telecommunications, St.-Petersburg, Russia
karolina.borunova@sut.ru

Abstract. General purpose networks-on-chip (GP-NoC) are expected to feature tens or even hundreds of computational elements with complex communications infrastructure binding them into a connected network to achieve memory synchronization. The experience accumulated over the years in network design suggests that the knowledge of the traffic nature is mandatory for successful design of a networking technology. In this paper, based on the Intel CPU family, we describe traffic estimation techniques for modern multi-core GP-CPUs, discuss the traffic modeling procedure and highlight the implications of the traffic structure for GP-NoC research. The most important observation is that the traffic at internal interfaces appears to be random for external observer and has clearly identifiable batch structure.

Keywords: Networks on chip · Wireless network on chip · Intra-CPU communications · Traffic estimation

1 Introduction

In the beginning of 21st century the development of general-purpose central processing units (GP-CPUs) has reached the level, where it became more beneficial to scale the computational power horizontally by parallelizing the computations than to continue increasing the clock frequency. Addressing this issue, major CPU manufactures, Intel and AMD, presented their dual-core CPUs in 2005, spawning the era of multi-core CPUs. Starting from a simple integration of two computing nodes on a single chip and providing shared access to RAM, they have nowadays evolved to truly multi-core systems of 4, 8 and beyond computing nodes on a single ship with deep integration between the components and dynamic threads redistribution between the cores [1,2].

O. Galinina et al. (Eds.): NEW2AN/ruSMART 2016, LNCS 9870, pp. 453–464, 2016.
DOI: 10.1007/978-3-319-46301-8_38

Evolution of the technological process will allow to integrate more cores to the chip in the coming future. Providing effective memory synchronization between the chip components becomes more challenging with the number of cores growing. One of the typical solutions is based on the *Last Level Cache (LLC)*, often integrated on chip and serving as a gateway between the cores to maintain coherency of the shared data. The connectivity is provided by a bus, connecting all the cores to LLC, memory and input/output controllers. This puts restrictions on a multi-core CPU design as additional space is required for wired communications subsystem. Furthermore, the size of LLC grows with the number of cores taking more and more space inside the chip, thus, limiting the space left for the cores [3]. The proposed 3D chip design paradigm [4] can address some of these issues by spreading the cores and cache memory between several layers. However, the question of enabling efficient communications still remains open.

Over the last decade many solutions have been proposed for design of intercore communications in many core versions of GP-CPUs, often called generalpurpose networks-on-chips (GP-NoC), including both wired and wireless ones. For wired solutions waveguide-based mm-waves, common bus and optical communications have been suggested. Researchers also addressed different topology designs including ring, grid, mesh and various derivatives of these such as torus, tree, and segmented ring. For wireless designs, proposals advocating the use of mm-wave band, 10–300 GHz, and THz band, 0.1–3 THz [5,6] are popular. Wireless topologies range from star to routed grids, meshes and small-worlds.

Most of the studies on GP-NoC designs approach the problem from communications system perspective proposing solutions targeting various communications and networking mechanisms and paying less attention to the internal structure and the needs of a CPU including memory synchronization. The latter is, however, of paramount importance as it dictates the routes of traffic inside a CPU and may affect the resulting performance of a chip imposing strict requirements on the loss and latency of data delivery process. Thus, the proposed architectures should be supplied with detailed analysis of performance under realistic traffic conditions.

In this paper we present traffic estimation methodologies for modern GP-CPUs and demonstrate how they can be used jointly to identify an accurate intra-CPU traffic model. We also discuss implications of the traffic characteristics on GP-NoC research. The rest of the paper is structured as follows. Background information is provided in Sect. 2. Intra-CPU traffic estimation methodologies are introduced in Sect. 3. Selective benchmarking of internal interfaces is discussed in Sect. 4. Traffic modeling procedure and implications for GP-NoC research are presented in Sect. 5. Conclusions are drawn in the last section.

2   Modern Network-on-Chip

2.1   General-Purpose NoCs

NoCs have been a topic of intensive research and development over the last decade. A NoC can be defined as a complex communication system integrating multiple computing nodes within the system-on-chip. There are a number of examples of successful applications of NoC concept with graphical processing units (GPU) being possibly the most widely known to large audience. The reason for an extraordinary increase in performance of such systems is mainly due to the nature of tasks allowing for their perfect parallelization leading to simple processing elements and well-defined traffic patterns.

The application of the NoC concept to GP-CPUs is more complex as the tasks to be performed greatly vary in their specifics, the level of parallelization and, thus, may require intensive exchange of information between computational elements placing additional requirements on the design of the intra-CPU communications infrastructure.

2.2   Cache Subsystem and Associated Interfaces

The cache subsystem is an integral part of modern CPUs including general-purpose ones and those targeted to a certain application. The core idea of using caches is to bring data and instructions closer to the computational units such that they are quickly available upon request. The latter is achieved by using small and fast memory arrays, nowadays, up to few tens of megabytes (MB). To speed-up the access further caches are often physically located on chip [3]. Comprehensive predictive algorithms are used to cache data.

In multi-core GP-CPUs ensuring the same view of memory for all cores is one of the most complicated problems. The protocols performing this functionality are known as cache coherence protocols and their operation depends on the system architecture. Below, we concentrate on Intel GP-CPUs.

The cache subsystem of modern Intel GP-CPUs consists of multiple cache layers, denoted as Lx, see Fig. 1. Separate L1 and L2 caches are associated with each core. The size of L1 cache is typically in the range of few tens of KBs and is divided into data and instruction caches. L2 cache is exclusively used for data; its size is typically on the order of few hundreds of KBs. Finally, L3 cache is a shared cache used for synchronization of lower level caches and its size is typically several times more than their aggregated size. Particularly, for Intel Haswell architecture it varies between 2 and 20 MBs. To connect L2 caches to a single shared L3 cache local bus is currently used [7]. To ensure coherent view of memory for all the cores an inclusive cache maintenance protocol is used by Intel. Inclusiveness ensures that the data present at one level of hierarchy also exist at all other levels. This significantly reduces the complexity of the protocol at the expense of non-perfect memory utilization.

2.3   Extensions of the Modern Architecture

To the best of our knowledge, only few studies on NoCs design published over the last decade took into account details of prospective NoC applications. The research mostly concentrated on the new topologies, routed solutions and mechanisms for efficient communications between computational elements. However,

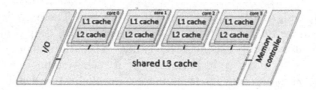

Fig. 1. A typical hierarchical cache architecture.

taking into account strict latency requirements on the information propagation in the cache subsystem of modern GP-CPUs [8], even the simplest possible routed solutions may not satisfy them. Many studies concentrated on the topology design presuming the perfect memory synchronization. Taking into account that most of the latency time is spent for cacheline search, not for communications itself, routed solutions may not be realistic for GP-NoCs.

CPU vendors continue to evolve the current cache architecture. There are various reasons behind this ranging from the miniaturization of the technological processes allowing to fit 8 cores and 20 MB of L3 cache on a chip using 22 nm technology [1,2] and promising extensions to 16 and 32 cores in the future to development of new concepts such as 3D stacked designs [4] allowing for efficient short-distance interconnects between them, etc. Preserving the classic hierarchical cache design and possibly adding additional layers whenever needed keeps the latency at satisfactory level. Backward compatibility with well tested cache coherence protocols is another reasons for conservative approach.

Even preserving the current cache architecture, the increase the number of cores on a chip to 16, 32 and/or extension to "vertical" 3D integration might lead to new bottlenecks in the communications infrastructure. Thus, understanding of intra-CPU traffic dynamics is not only crucial for design of future GP-NoCs but for smooth evolution of the current architecture.

3   Intra-CPU Traffic Estimation

There are three different approaches to study intra-CPU traffic. These are (i) logic analysis of cache coherence protocols, (ii) indirect measurements and (iii) system simulations. Below, we describe them specifying the type of knowledge about intra-CPU traffic they bring to the analyst.

3.1   Microarchitecture-Level Analysis

According to the first approach, one could try to reveal traffic characteristics analyzing the logic of various components of a CPU and its cache subsystem. The information covering the internal CPU organization can be retrieved from the open documentation. Even though it typically lacks precision and details, such analysis could provide a valuable starting point for more detailed investigations.

There are two fundamental problems preventing from getting detailed knowledge of intra-CPU traffic by performing logical system analysis. First, the information about specifics of the architecture is often not detailed enough. For instance, throughput and latency characteristics of the internal interconnects may vary depending on the dynamics of the computational process, while the documentation provides the "best-case" values only. Another example is that the knowledge of the cache coherence protocols for such complex systems does not fully define the way how the controllers work as implementations may differ depending on how the controllers interact to perform transactions. Design choices like snooping versus directory-based protocols or invalidate versus update protocol noticeably influence the traffic footprint on the intra-CPU interfaces. Secondly, static evaluation does not provide the information about traffic dynamics, that is, distribution of load in time. Overcoming these issues requires direct or indirect measurements or system level simulations.

3.2   System Measurements

In most modern general-purpose CPU there are no direct mechanisms to measure the traffic on an interface of interest. However, using Intel CPU one can still try to infer the amount of traffic indirectly relying on the so-called performance counters available starting from Sandy Bridge family of CPUs [9]. Among others, these counters provide the information about the accumulated number of cache misses to a certain cache layer in a time interval of fixed duration.

Consider, for example, how to obtain the amount of uplink traffic at L2–L3 interface. The L2–L3 interface is only used when there are no cachelines containing the addressed data in L1/L2 caches. In this case, the read request to L3 cache is sent. Processing this request, the cacheline is found in L3 or RAM and sent back to L2. The total amount of cachelines sent per time unit at L2–L3 interface equals to the value of L2 misses during this period. In 64-bit CPU architecture, the cache line size is 64 bytes, while the read request length is 8 bytes. Thus, the total traffic on L2–L3 interface is $T = L2_m(8 + 64)$, where $L2_m$ is the number of L2 misses per time unit.

To access the values of the performance counters one could rely on existing tools such as perf [10] or Intel Performance Counter Monitor (PCM, [11]). The latter is a certified tool developed by Intel that is capable to monitor the CPU activity and periodically report the values of counters. Note, these tools can only be run in full operating system (OS) environment, where the background processes bias the measured statistics. Thus, this type of measurements is useful when one is interested in realistic OS conditions. An example of data for different applications obtained using Intel PCM is shown in Fig. 2. The point of interest was the L2–L3 interface of Intel eight-core CPU i7-5960X Haswell-E architecture featuring 20 Mb of L3 cache, clock rate 3.0 GHz, and 16 Gb DDR4. The build of Linux Kubuntu 14.10 with kernel version 3.16 was used. The tests were ranging from background traffic to reading of each 64 byte of data (64B test, always ensures L2 misses) to sequential reading (1B test, always ensures L2 hits) to complex applications including Skype, game, AES encryption and video

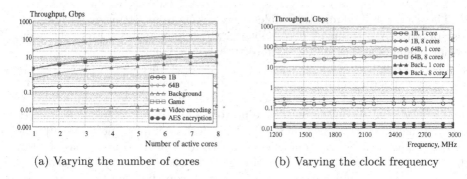

(a) Varying the number of cores   (b) Varying the clock frequency

Fig. 2. Traffic characteristics of intra-CPU interfaces.

playback. The number of active cores was set to the number of simultaneously run applications and each applications was assigned a core. As one may observe, the traffic load increases as the number of applications increases and reaches values of 110 Gbps for 64B test.

3.3   Cycle-Accurate CPU Simulations

The discussed methods are not capable to provide detailed traffic structure at the transactional level. The approach that potentially allow to get accurate resolution is cycle-accurate CPU simulations. Nowadays, there are a number of simulators supporting x86 architecture, MARSS [12], Gem5 [13], zSim [14], and SST [15]. All of them are very flexible allowing for detailed time-stamping of events making them suitable for our task. zSim and SST are tailored at system simulations of extremely large systems featuring hundreds of cores and, compared to MARSS and Gem5, lack detailed control functionality. The comparison of projects is shown in Table 1.

Table 1. Comparison of cycle-accurate simulation environments.

Simulator	Libraries and instruments	Citation count	Downloads	Last update
SST	OpenMPI, Boost, DRAMSim2	27	558	August 2015
Gem5	SWIG, protobuf	213	5302	August 2015
ZSim	PIN	10	551	July 2015
MARSSx86	QEMU, PTL-Sim, SDL	41	473	April 2014

We implemented a typical Intel x86 architecture in Gem5 including all major features and components of Intel architecture. The chosen cache size and latency parameters are typical for modern general-purpose CPUs, however, slight deviations from the real values of particular systems should not change the resulting time series qualitatively. The cache subsystem was assumed to be inclusive with 64 KB/2 ns, 2 MB/12 ns and 16 MB/30 ns size/latency at L1, L2 and L3 caches. The model explicitly takes into account delays associated with information retrieval and emulates the pipelining capability. Systems with 1, 2, 3, 4, 8 and 16 cores have been simulated. The clock frequency was set to 3.0 GHz.

To emulate a typical load at intra-CPU interfaces we selected a number of tests covering various aspects of program code including reading, writing and sorting routines, more comprehensive recursive factorial estimation and Euclid's greatest common divisor algorithms involving divisions and multiplications, to complex ones including AES encryption/decryption and compression using zlib. In multi-core configurations the number of simultaneously run tests were set equal to the number of operational cores. In overall, 70 tests have been performed. The output of the simulation is stored in well-known ASCII-based value change dump (vcd) format. To obtain time-series data the selected objects have been saved in "timing analyzer" (tim) format and then parsed using a specifically written C program.

Figure 3 illustrates the time series of the traffic at the L2–L3 and L3-DRAM interfaces for two tests, *write* and *evklid*, for different number of cores by showing busy interface indicators $I_A + b$, where A is the event of busy interface, b is the constant added to distinguish between traces for different number of cores. We can make two qualitative conclusions: (i) the traffic at both interfaces has a

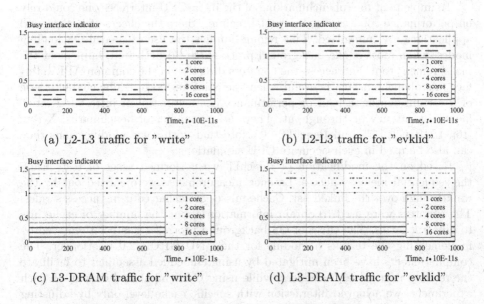

(a) L2-L3 traffic for "write"   (b) L2-L3 traffic for "evklid"

(c) L3-DRAM traffic for "write"   (d) L3-DRAM traffic for "evklid"

Fig. 3. Traffic structure at L2–L3 and L3-DRAM interfaces.

stochastic structure and (ii) the traffic has clearly identifiable batches and gaps between them.

The above mentioned observations have a number of important consequences for GP-NoC research. First, working with CPU traffic researchers expect a certain degree of deterministic behavior. However, due to a number of advanced mechanisms used in modern CPUs including pipelines, cache coherence protocols and unknown apriori flow of instructions, even for a single core it has clear stochastic component at both illustrated interfaces. The most simple stochastic traffic model is a homogeneous Poisson process. The second observation clearly rejects this hypothesis for L2–L3 interface implying that more complex models have to be used. The data suggests that batch processes might be a good choice for intra-CPU traffic while for L3-DRAM interface Poisson process can be tried first.

4   Benchmarking of Interfaces

An in-depth look into CPU behavior is often performed by selective measurements targeted on a certain CPU subsystem. Nowadays, there are a number of frameworks, often called microbenchmarks, allowing for analysis of microarchitecture-level details including CPU communications subsystem. Due to their widespread use for software optimization and hidden hardware bottlenecks identification, microbenchmarking has received significant attention in the past. Benchmarking can be used to obtain detailed information about intra-CPU interfaces including the throughput and latency needed to construct realistic simulation models.

To implement microbenchmarking of the intra-CPU interfaces one could rely on performance counters in the special regime, where the effects of background applications are minimized. This regime implies the use of special OS loading modes, when most of the background processes are deactivated, implement the measurement tools and access the counters directly via the common API. Interface microbenchmarking is useful when one is interested in traffic characteristic of a certain synthetic load or performance characteristics of intra-CPU interfaces, e.g. latency or throughput. There are a number of benchmarks, X-Ray [16], LMBench [17], and BenchIT [8]. Note that with some modifications they can also be used in cycle-accurate CPU simulations.

Based on the modification of BenchIT we performed a set of tests using the sequential memory access. Latency measurements were carried out by the walkthrough over the linked list. Cache misses are achieved using access strides. The counters were used to obtain information about the number of cache hits to different levels. The effect of OS background processes has been removed by implementing the test as a module for the GNU GRUB 2.0 bootloader. The compiler effects have been mitigated by using embedded assembler to facilitate the reading procedure procedures, while using C for input and output. In each experiment, we achieved interaction with specific cache level only by adjusting the access strides.

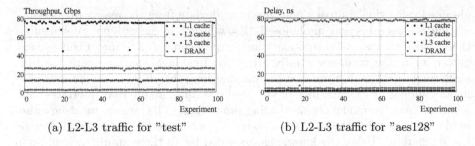

(a) L2-L3 traffic for "test"   (b) L2-L3 traffic for "aes128"

Fig. 4. Measurements of throughput and latency of intra-CPU interfaces.

The values obtained in one hundred successive experiments are visualized in Fig. 4 while the throughput and delay metrics are shown in Table 2. Note that these characteristics represent throughput and latency up to a certain cache level and thus include overhead related to search procedures. The obtained results deviates from the claimed values by 25 %.

Table 2. Throughput and latency of intra-CPU interfaces

Metric	L1 cache	L2 cache	L3 cache	DRAM
Throughput, Gbps	9.33	3.25	1.63	0.37
Latency, ns	1.50	4.18	12.72	77.32

5 Traffic Modeling and Implications for GP-NoC

5.1 Identification of a Model

Contrarily to the network environment the closed nature of GP-CPU development does not allow to completely rely on a single approach for identifying intra-CPU traffic properties and formulating accurate traffic models. All considered approaches contributes to the overall understanding we can build upon to come up with detailed description of traffic properties.

The microarchitecture-level analysis using publicly available documentation provides the first step towards a traffic model. Fixing the set of algorithms and architectural decisions helps to specify the tools needed at later stages. First, the analysis of the functionality of a CPU and, particularly, of the cache coherence protocol, allows to understand the effect of different subsystems and make decision about the level of detail for simulations models. The cycle-accurate system level simulations, when performed correctly by taking into account all major mechanisms implemented in modern GP-CPU, allows to understand the nature of the traffic at different intra-CPU interfaces and many particular details pertaining to its small-scale behavior. The absolute values of the traffic patterns obtained using this approach may, however, deviate from the reality due to simulation abstractions and undisclosed "know-hows" in algorithms' implementations. These effects, are of secondary importance compared to major conclusions pertaining to the small-scale traffic structure. Real measurements performed by perf, Intel PCM or a similar tool available for other GP-CPUs, adds to the understanding of exact values of the traffic volume at internal interfaces and allows to tune the model obtained using simulations. The measurements are also needed to determine the delays between different cache levels and parameterize the simulations. Using the knowledge provided by all three approaches one can come up with a detailed traffic model for modern GP-CPUs.

The knowledge of the system architecture and the cache coherence protocols also allows to extend the traffic model to the case of more cores or even different internal communications subsystems including those to be used in future GP-NoCs. The former can be done using extrapolation techniques. The latter is much more difficult and may require extensive system level simulations as measurements may not be feasible.

5.2   Implications for GP-NoC Research

The conducted investigation reveals that the traffic at internal interfaces of modern multi-core CPUs is stochastic. Moreover, the traffic process has a clearly identifiable batch structure. There are two important factors contributing to this effect. First, the structure of modern GP-CPUs and associated caching, instruction prediction and execution algorithms is extremely complex with multiple mechanisms such as pipelining, replication and branch prediction causing similar operands to be handled differently, depending on the context. The second effect is attributed to the software being executed at a certain moment of time manifesting itself in an a-priori unpredictable flow of instructions.

It is important to note that the intra-CPU traffic is not inherently random but appears to be random for an *external observer*. In other words, fixing a certain type of a CPU and the set of processes currently running, the evolution of the traffic can be, in principle, obtained deterministically. However, similarly to the network traffic, the dimension of the problem is extremely large and highly sensible to small changes in the input data preventing from efficient deterministic handling methods and forcing the stochastic view of intra-CPU traffic.

The results revealed by the reported study reflects the way how prospective GP-NoCs has to be researched and developed. First, the specification of the NoC communications infrastructure cannot be done without the detailed analysis of the feasible cache coherence protocols and/or development of new protocols. Even the best possible topology optimizing the delay of data transfer between two arbitrary nodes in a core network may not be the best possible solution due to the nature of the traffic flows, dictated by the needs of the computational process and the cache coherence protocol. As experience accumulated in the networking research tells us, the structure of the traffic may get more complicated when GP-CPUs are extended to tens or hundreds of cores and more complex cache coherence protocols will be used.

6   Conclusions

We discussed traffic estimation methodologies for modern GP-CPUs and analyzed the implications of the revealed traffic properties for GP-NoC research. Intra-CPU traffic modeling is a complex task as traffic volume and properties depend on system architecture, implementation and behavior of software in runtime. The complexity of modern CPUs and the lack of detailed microarchitecture-level documentation for consumer products adds to those issues. Our main conclusions are that (i) the intra-CPU traffic nature in modern GP-CPUs is random (ii) the traffic process at L2–L3 interfaces has more complex structure than Poisson and has clearly identifiable batch behavior. These two observations has to be taken into account when estimating performance of intra-CPU communications subsystems for prospective NoCs.

The identification of the accurate traffic model for intra-CPU traffic requires the use of all three considered approaches, microarchitecture-level analysis, field measurements and cycle-accurate CPU simulations. Taking into account complex structure of the traffic even in modern simple bus topology of the cache subsystem, future investigation of the optimal intra-CPU communications, in addition to the structure of the network itself, shall consider the effects of the cache coherence protocols and instructions handling procedures.

Acknowledgement. The authors are grateful to Alexander Antonov from ITMO University and Vitaly Petrov from Tampere University of Technology for insightful comments that allowed to improve this paper.

References

1. Intel: Intel Core i7-5960x processor extreme edition. Technical specifications (2015). http://ark.intel.com/products/82930/Intel-Core-i7-5960X-Processor-Extreme-Edition-20M-Cache-up-to-3_50-GHz . Accessed 07 June 2015
2. AMD: AMD FX Series Processors. Technical specifications (2015). http://www.amd.com/en-us/products/processors/desktop/fx. Accessed 07 June 2015
3. Mittal, S., Vetter, J., Li, D.: A survey of architectural approaches for managing embedded DRAM and non-volatile on-chip caches. IEEE Paral. Distrib. Comput. Syst. **26**, 1524–1537 (2015)
4. Davis, W., Wilson, J., Mick, S., Xu, J., Hua, H., Mineo, C., Sule, A., Steer, M., Franzon, P.: Demystifying 3D ICs: the pros and cons of going vertical. IEEE Des. Test Comput. **22**, 498–510 (2014)
5. Petrov, V., Moltchanov, D., Koucheryavy, Y.: Interference and SINR in dense terahertz networks. In: IEEE 82nd Vehicular Technology Conference (VTC Fall), pp. 1–5, September 2015
6. Petrov, V., Moltchanov, D., Koucheryavy, Y.: On the efficiency of spatial channel reuse in ultra-dense THz networks. In: 2015 IEEE Global Communications Conference (GLOBECOM), pp. 1–7, December 2015
7. Hammalund, P.: Haswell: the fourth-generation intel core processor. IEEE Micro **2**, 6–20 (2014)
8. Molka, D.: Memory performance and cache coherency effects on an Intel nehalem multiprocessor system. In: Proceedings of IEEE PACT, pp. 78–86 (2009)

9. Intel 64 and IA-32 architectures software developers manual combined volumes. Technical specifications, Intel Corporation (2014)
10. Perf: Technical report. https://perf.wiki.kernel.org. Accessed 02 Sept 2015
11. Intel Performance Counter Monitor: Software tool, Intel Corporation. http://www.intel.com/software/pcm. Accessed 02 Sept 2015
12. MARSSx86 - micro-architectural and system simulator for x86-based systems. Software package, GNU License. http://marss86.org/marss86/. Accessed 02 Sept 2015
13. Binkert, N.: The Gem5 simulator. ACM SIGARCH Comput. Archit. News **39**, 1–7 (2011)
14. Sanches, D., Kozyrakis, C.: ZSim: fast and accurate microarchitectural simulation of thousand-core systems. In: Proceedings of ACM ISCA, pp. 475–486 (2013)
15. The structural simulation kit (SST). In: Proceedings of ACM ISCA, Tutorial. http://www.ece.cmu.edu/calcm/isca2015. Accessed 02 Sept 2015
16. Yotov, K., Pingali, K., Stodghill, P.: Automatic measurement of memory hierarchy parameters. In: Proceedings of ACM SIGMETRICS, p. 181 (2005)
17. Staelin, C.: LMBench: an extensible micro-benchmark suite. Softw. Pract. Exp. **35**(11), 1079–1105 (2011)

Indoor Positioning in WiFi and NanoLOC Networks

Mstislav Sivers[1], Grigoriy Fokin[2(✉)], Pavel Dmitriev[2],
Artem Kireev[2], Dmitry Volgushev[2],
and Al-odhari Abdulwahab Hussein Ali[2]

[1] St. Petersburg State Polytechnical University, St. Petersburg, Russia
m.sivers@mail.ru
[2] The Bonch-Bruevich St. Petersburg State, University of Telecommunications,
St. Petersburg, Russia
grihafokin@gmail.com, {p-dmitriev,kireyev}@list.ru,
{d.volgushev,abdwru2011}@yandex.ru

Abstract. In this paper we compare the indoor positioning techniques of RToF in nanoLOC and RSSI fingerprinting in WiFi networks experimentally and highlight the impact of orientation during primary measurement acquisition for increasing location accuracy in the case of NLOS and multipath signal propagation conditions. Resulting accuracy estimates confirm known results and reveal that radiomap construction with primary RSSI measurements in four, angular directions can improve positioning accuracy by 0.5 m in comparison with traditional fingerprinting in deployed WiFi and location dedicated nano-LOC networks.

Keywords: Wifi · NanoLOC · RSSI · Round-trip time of flight · Radiomap · Positioning accuracy

1 Introduction

Accurate positioning of mobile user equipment and automated guided vehicles is an important trend in next-generation mobile communications for delivering location based services (LBS), such as tracking transport vehicles and even a strong demand in such applications as emergency services. Traditional Global Navigation Satellite Systems (GNSS) such as Global Positioning System (GPS) and GLObal NAvigation Satellite System (GLONASS) provide rather accurate position estimates outdoor, however for indoor environment this is not the case because of non-line of sight (NLOS) conditions as a result of reduced satellite visibility and its weak signal inside buildings.

Available solutions for indoor location include special wireless technologies such as Redpin [1], Subpos [2] and NanoLOC [3] which are dedicated for positioning and give accurate estimates, for example, NanoLOC networks use Symmetrical Double-Sided Two Way Ranging (SDS-TWR) based on Round-trip Time of Flight (RToF) and declares ranging accuracy up to 1 m [4]. However deploying dedicated networks for location based services is not so economical as utilizing widely deployed

© Springer International Publishing AG 2016
O. Galinina et al. (Eds.): NEW2AN/ruSMART 2016, LNCS 9870, pp. 465–476, 2016.
DOI: 10.1007/978-3-319-46301-8_39

existing WiFi networks. That's why several approaches are proposed for positioning in wireless local area networks based on WiFi [5] among which fingerprinting is considered to be most accurate.

Wireless networks employ several positioning techniques which can be classified by the primary measurements: Received Signal Strength Indication (RSSI), Angle of Arrival (AoA), Time of Arrival (TOA), Time Difference of Arrival (TDOA) [6].

These techniques have both drawbacks and benefits. TOA and TDOA have highest accuracy however require precise synchronization of measurement units [7]. AoA methods are not widely used due to complexity and cost of transceivers and antenna modules [8]. RSSI are most attractive approach in existing WiFi networks since mobile devices receive its primary signal strength measurements during standard operation [9].

The aim of this paper is to compare the indoor positioning techniques of RToF in dedicated NanoLOC and RSSI fingerprinting in deployed WiFi networks experimentally and highlight the impact of orientation during primary measurement acquisition for increasing location accuracy in the case of NLOS and multipath conditions.

The material in the paper organized as follows. Indoor positioning techniques experimentally implemented in WiFi and NanoLOC networks with algorithms, protocols and related works is presented in the second part. Experimental analysis including scenarios, conditions, implementation features and obtained results are presented in the third part. Finally, we draw the conclusions in the fourth part.

2   Indoor Positioning Techniques

Indoor positioning techniques experimentally validated during investigation include RSSI based weighted centroid and fingerprinting in WiFi and SDS-TWR based on RToF in NanoLOC networks.

2.1   Weighted Centroid Algorithm

Weighted centroid [10] is an improved technique of classic centroid [9], where RSSI measurements are used as weighting coefficients for range measurements rRSSIi, collected by WiFi access points (AP) with predefined location as illustrated in Fig. 1.

To improve accuracy of location estimation primary RSSI measurements are transformed to mW using the following expression:

$$RSSIP_n = (P \cdot 10^{\frac{RSSI_n}{20}})^g, \tag{1}$$

where P = 1 mW, n = 1...N, N - number of AP and g is a static degree to determine reasonable difference between lower and higher RSSI measurements; the higher g, the less is the impact of weak signals on positioning results. Such arrangement by received signal power helps to improve accuracy of location estimation, so that lower values give less influence. Optimal selection of g depends on area configuration and amount of AP [11]. Resulting formula for location estimation using weighted centroid:

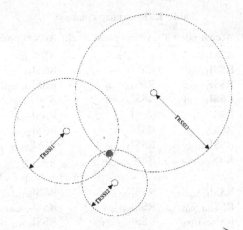

Fig. 1. Weighted centroid algorithm for WiFi networks

$$Pos(x_{est}, y_{est}) = \left(\sum_{n=1}^{N} W_n x_n, \sum_{n=1}^{N} W_n y_n \right), \tag{2}$$

where x_{est} and y_{est} – estimated position coordinates, x_n and y_n, - RSSI measurements from AP, W_n weighting coefficient:

$$W_n = \frac{RSSIP_n}{\sum\limits_{n=1}^{N} RSSIP_n}. \tag{3}$$

2.2 Fingerprinting Method

Fingerprinting method is based on measuring the RSSI from all accessible AP in a certain number of points, called "reference" which results in database called "radiomap" [12]. Radiomap contains RSSI measurements at the indicated reference point (RP). The location of the target object is calculated based on comparing the current RSSI measurements of target object with the data stored in radiomap. The algorithm for calculating the location of the target object consists of two phases: the offline acquisition and the online positioning [13].

Offline phase is the training phase. A location fingerprint database must be constructed. Algorithm is the following: first, we divide the targeted positioning area into grids, usually the grid is 1 m*1 m; the coordinate of each grid is regarded as RP. Then, at each RP RSSI is collected. Each RP corresponds to a vector R. Typical R = (r_1, r_2, ..., r_N) consists of RSSI values from N APs.

It is a known fact [12] that orientation could improve precision of radiomap construction, that's why let's define $RSSI_{mn}$ in four angular directions, where m - index of reference point, m = 1...M. The resulting radiomap database is presented in Table 1.

Table 1. Radiomap database

№ RP	Access point 1	Access point 2	...	Access point N
1	$RSSI_{11\text{-}0°}$	$RSSI_{12\text{-}0°}$	...	$RSSI_{1N\text{-}0°}$
	$RSSI_{11\text{-}90°}$	$RSSI_{12\text{-}90°}$		$RSSI_{1N\text{-}90°}$
	$RSSI_{11\text{-}180°}$	$RSSI_{12\text{-}180°}$		$RSSI_{1N\text{-}180°}$
	$RSSI_{11\text{-}270°}$	$RSSI_{12\text{-}270°}$		$RSSI_{1N\text{-}270}$
2	$RSSI_{21\text{-}0°}$	$RSSI_{22\text{-}0°}$	...	$RSSI_{2N\text{-}0°}$
	$RSSI_{21\text{-}90°}$	$RSSI_{22\text{-}90°}$		$RSSI_{2N\text{-}90°}$
	$RSSI_{21\text{-}180°}$	$RSSI_{22\text{-}180°}$		$RSSI_{2N\text{-}180°}$
	$RSSI_{21\text{-}270°}$	$RSSI_{22\text{-}270°}$		$RSSI_{2N\text{-}270}$
...	...	...	...	...
M	$RSSI_{M1\text{-}0°}$	$RSSI_{M2\text{-}0°}$	...	$RSSI_{MN\text{-}0°}$
	$RSSI_{M1\text{-}90°}$	$RSSI_{M2\text{-}90°}$		$RSSI_{MN\text{-}90°}$
	$RSSI_{M1\text{-}180°}$	$RSSI_{M2\text{-}180°}$		$RSSI_{MN\text{-}180°}$
	$RSSI_{M1\text{-}270°}$	$RSSI_{M2\text{-}270°}$		$RSSI_{MN\text{-}270}$

Online phase is the location determination phase. The basic idea is to determine the measure of similarity of current vector measurement with each of the values existing in radiomap. In this phase, target object observes a sample RSSI vector S ($RSSI_1$, $RSSI_2$, ..., $RSSI_N$), which is then to be matched in database. The location with best match of RSSI in radiomap database should be the estimated location.

To find best match of RSSI Holder's [13] condition can be used:

$$\|x_p\| = \left(\sum |x_i|^p\right)^{1/p},$$ (4)

where $p \geq 1$ and is generally natural value.

Another conditions which could be used include Manhattan distance ($p = 1$) and Euclidean distance ($p = 2$).

Typical algorithms based on the above mentioned conditions include nearest neighbor in signal space algorithm (NN) [14] and K-nearest neighbor algorithm (KNN) [15].

NN - is the main and easiest algorithm to determine the location. The idea is to compute the Euclidean distance (ED) between vector S from user and the recorded vector R = ($RSSI_1$, $RSSI_2$,..., $RSSI_M$) in database, using formula:

$$ED = \min \text{dis}(S,R), \quad \text{dis}(S,R) = \sqrt{\sum_{n=1}^{N}(RSSI_N - RSSI_M)^2}.$$ (5)

The smallest Euclidean distance between R and S mean the smallest distance between user current position and reference point recorded in database.

In general, fingerprinting method operates according to the diagram in the Fig. 2, which includes two phases: (A) offline and (B) online. Offline phase consists of collecting primary RSSI measurements (1) and creating database and forming radiomap

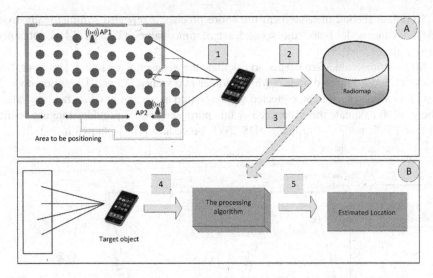

Fig. 2. Fingerprinting algorithm for WiFi networks

for algorithm processing (2,3). Online phase includes gathering RSSI measurements by user equipment (4) and then processing it by algorithm (5) to estimate location.

One of the main disadvantages of fingerprinting technique is radiomap data construction, support and relevance especially if the number of RP is large. For example, if the surrounding space has changed due to furniture location or AP were added/removed, radiomap database need to be actualized which is rather difficult and time consuming.

2.3 NanoLOC Technology

NanoLOC technology employs SDS-TWR based on RToF and measures a ranging signal sent by an anchor (measurement unit) and an acknowledgement sent back from the tag (target object) which allows to cancel out the requirements for clock synchronization and eliminate the effect of clock drift/offset of ranging measurements taken by both the tag and the anchor and results in a reasonably accurate measurement even in the most challenging of environments like NLOS [3].

During the SDS-TWR measurements a signal propagates from one node to a second node and back to the original node (Round Tripping - or Two Way Ranging). The double time a signal propagates from node 1 (Tag) to node 2 (Anchor) is measured by node 1. This can be done by following methodology. First, node 1 (tag) requests ranging and sends a Data packet to node 2 (Anchor) which automatically returns an Ack packet. Then, node 1 generates T1 using the time when the Data packet was sent and the time when the Ack packet was received. This value T1 is stored in memory on node 1 and used along with known value of the signal propagation speed to calculate the distance between two devices. Finally, T2 is the measured value of processing a Data packet received from node 1. This value T2 is stored in memory on node 2.

To obtain second measurement the entire process is repeated symmetrically from node 2 to the node 1 and the second set of time values (T3 and T4) is obtained accordingly.

All four time values are needed to calculate the distance were found. However, two values, T1 and T4, are on node 1 while another two values, T2 and T3, are on node 2. These four values need to be collected on one station as a set of pramary ranging values to be used to calculate the distance. For this purposes node 2 provides ranging results (T2 and T3) to node 1. Complete SDS-TWR procedure [16] is shown in Fig. 3.

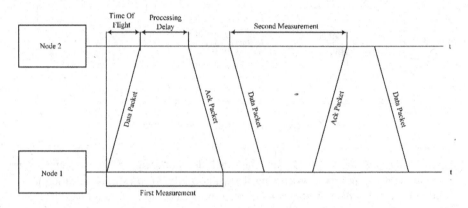

Fig. 3. Symmetrical double-sided two way ranging

With all four time values available, the distance (r) between Node 1 and Node 2 can be generated using the following formula:

$$r = \frac{(T1 - T2) + (T3 - T4)}{4},$$ (6)

where (T1-T2) – first measurement; T1 - propagation delay time of a round trip between node 1 and node 2; T2 – processing delay in node 2; (T3-T4) – second measurement; T3 - propagation delay time of a round trip between node 2 and node 1; T4 – processing delay in node 1.

The nanoLOC localization system consists of three main devices: anchor, tag, base station. Anchors are used as reference points with predefined coordinates. A mobile tag determines its location by estimate distance between itself and each of the reference points. The measurement result is transmitted to the base station. The tag position can be calculated by trilateration. Base station is responsible for initializing and configuring the nanoLOC network and for sending start and stop ranging commands. After sending ranging request it receives back raw ranging data (measured distances) and then passes this data on to the PC with following processing in nanoLOC location GUI software.

During initialization, the nanoLOC location software sends a request to the base station to start searching for all alive tags. When a tag is found (but not located), then the MAC address of this tag is added to the list of "alive" tags. The MAC address of anchors is set in location software manually and is sended to tags before ranging starts. After initialization, location software transmits ranging start command to all alive tags via the base station. Tag estimates the distance to all of the anchors and sends measurement results to the base station, which transmits them to the location software to calculate and display tag coordinates.

3 Experimental Analysis

In this section we describe experimental analysis including scenarios, conditions, implementation features and obtained results. Experiment was conducted in St. Petersburg State University of Telecommunications in the classroom of size $7,9 \times 8,2$ m with six WiFi APs and four nanoLOC anchors. Primary RSSI measurements were collected by laptop with software package WirelessMon [17].

Positioning accuracy was evaluated in terms of Root Mean Square Error (RMSE) position estimate in meters with respect to true position in Cartesian coordinates:

$$\Delta = \frac{1}{P} \left(\sum_{p=1}^{P} \sqrt{(x_{est} - x_a)^2 + (y_{est} - y_a)^2} \right), \qquad (7)$$

where P – number of measurements for one estimated location, x_a and y_a – actual position coordinates. To evaluate and compare the performance of indoor positioning techniques in WiFi and NanoLOC networks through primary measurements gathering, processing and visualization we developed Matlab [18] software.

3.1 WiFi Positioning

To evaluate weighted centroid and fingerprinting techniques in WiFi networks we arranged twenty-eight RPs according to Fig. 4.

Classical Nearest Neighbor (NN-1) algorithm selects one value from the radiomap, closest relative to the current RSSI measurement. NN with four angular directions (NN-4) algorithm selects several values from the radiomap closest to the current RSSI measurement with respect to its orientation and takes its average. Visualization of actual and estimated locations for weighted centroid and fingerprint positioning in WiFi networks during one measurement trial is illustrated in Figs. 5 and 6.

Results of ten measurement trials for weighted centroid and fingerprint positioning in WiFi networks are illustrated in Table 2.

From Table 2 we can conclude that weighted centroid gives poor accuracy in small indoor area while it could be better in bigger distances [9]. Nearest Neighbor is more accurate: average RMSE $\Delta = 2,63$ m for NN-1 and $\Delta = 2,12$ m for NN-4.

To evaluate SDS-TWR in nanoLOC networks we arranged four anchors according to Fig. 7.

Visualization of actual and estimated locations for SDS-TWR positioning in nanoLOC networks during one measurement trial is illustrated in Fig. 5 and 6.

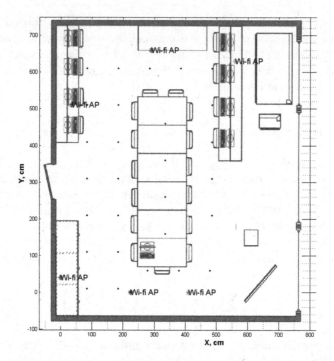

Fig. 4. WiFi APs and RPs arrangement for radiomap construction

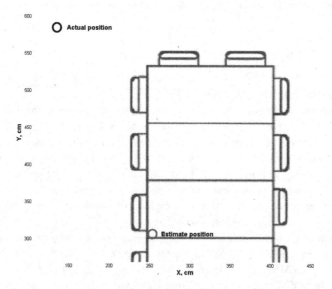

Fig. 5. Weighted centroid positioning results in WiFi network

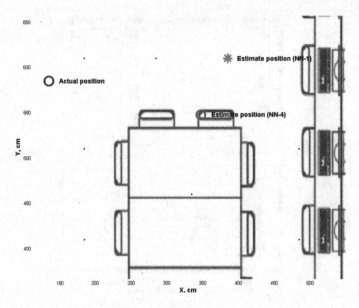

Fig. 6. Fingerprint positioning results in WiFi network

Table 2. Weighted centroid and fingerprint positioning results in WiFi networks

Actual coordinates		Weighted centroid RMSE	Nearest neighbor (NN-1) RMSE	Nearest neighbor (NN-4) RMSE
x_a	y_a	Δ, m	Δ, m	Δ, m
0,85	4,6	19,90	0,5	0,35
1,35	0,85	22,35	0,90	1,42
1,35	5,85	18,69	2,80	1,85
1,85	0,6	22,25	5,60	2,73
1,85	2,6	20,68	3,55	2,66
1,85	4,6	19,20	1,50	1,88
3,35	6,1	17,07	2,70	1,82
4,35	3,6	18,35	1,41	1,62
5,7	3,1	18,00	2,66	2,21
6	0,9	19,77	4,70	4,66
Average Δ		19,63	2,63	2,12

Results of ten measurement trials for SDS-TWR positioning in nanoLOC networks are illustrated in Table 3 (Fig. 8).

From Table 3 we can conclude that SDS-TWR gives average RMSE $\Delta = 2,63$ m which is greater than declared ranging accuracy up to 1 m and is caused by NLOS in some points.

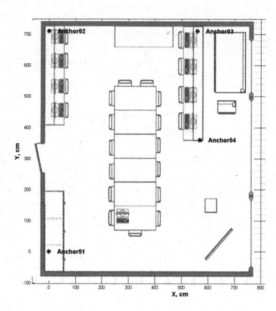

Fig. 7. NanoLOC configuration

Table 3. SDS-TWR positioning results in nanoLOC networks

Actual coordinates		NanoLoc RMSE
x_a	y_a	Δ, m
0,85	4,6	3,58
1,35	0,85	5,52
1,35	5,85	0,69
1,85	0,6	2,01
1,85	2,6	1,76
1,85	4,6	1,77
3,35	6,1	1,01
4,35	3,6	0,84
5,7	3,1	0,55
6	0,9	3,06
Average Δ		2,63

4   Conclusion

In this paper we evaluated the positioning accuracy of RToF location in nanoLOC and RSSI fingerprinting location in WiFi networks experimentally and received RMSE $\Delta = 2,63$ m for SDS-TWR, $\Delta = 2,12$ m for NN with four angular directions (NN-4) and $\Delta = 2,63$ m for NN-1. Performed experiment confirm known results and reveal that radiomap construction with primary RSSI measurements in four angular directions can

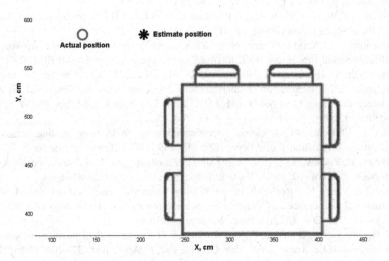

Fig. 8. SDS-TWR positioning in nanoLOC networks

improve positioning accuracy by 0.5 m in comparison with traditional fingerprinting in deployed WiFi and location dedicated nanoLOC networks in the case of NLOS and multipath signal propagation conditions.

References

1. Bolliger, P.: Redpin-adaptive, zero-configuration indoor localization through user collaboration. In: Proceedings of the First ACM International Workshop on Mobile Entity Localization and Tracking in GPS-Less Environments, Association of Computing Machinery, pp. 55–60 (2008)
2. SubPos a «Dataless» Open-Source WiFi positioning system. http://www.subpos.org/
3. Real Time Location Systems. A White Paper from Nanotron Technologies GmbH, Version 1.02. http://nanotron.com/EN/pdf/WP_RTLS.pdf
4. Röhrig, C., Müller, M.: Localization of sensor nodes in a wireless sensor network using the nanoLOC TRX transceiver. In: Vehicular Technology Conference, pp. 1–5. IEEE (2009)
5. Yim, J.: Comparison between RSSI-based and TOF-based indoor positioning methods. Int. J. Multimedia Ubiquit. Eng. **2**, 221–234 (2012)
6. Galov, A., Moschevikin, A., Voronov, R.: Combination of RSS localization and ToF ranging for increasing positioning accuracy indoors. In: 11th International Conference on ITS Telecommunications (ITST), pp. 299–304. IEEE (2011)
7. Sivers, M., Fokin, G.: LTE positioning accuracy performance evaluation. In: Balandin, S., Andreev, S., Koucheryavy, Y. (eds.) NEW2AN/ruSMART 2015. LNCS, vol. 9247, pp. 393–406. Springer, Heidelberg (2015)

8. Kireev, A., Fokin, G.: Radio direction finding of LTE emissions using mobile spectrum monitoring station with circular antenna array. Trudy Nauchno-issledovatel'skogo instituta radio, vol. 2, pp. 68–71 (2015). ISSN:0134-5583

9. Kolodziej, K.W., Hjelm, J.: Local Positioning Systems: LBS Applications and Services. CRC Press, Boca Raton (2006). pp. 102, 145, 151

10. Blumenthal, J: Weighted centroid localization in zigbee-based sensor networks. In: Intelligent Signal Processing IEEE International Symposium, pp. 1–6. IEEE (2007)

11. Schuhmann, S., Herrmann, K., Rothermel, K., Blumenthal, J., Timmermann, D.: Improved weighted centroid localization in smart ubiquitous environments. In: Sandnes, F.E., Zhang, Y., Rong, C., Yang, L.T., Ma, J. (eds.) UIC 2008. LNCS, vol. 5061, pp. 20–34. Springer, Heidelberg (2008)

12. Moka, E., Retscher, G.: Location determination using WiFi fingerprinting versus WiFi trilateration. J. Location Based Serv. **1**(2), 145–159 (2007). Taylor & Francis

13. Dmitriev, P., Pisarev, S., Sivers, M.: Analysis of methods and algorithms of positioning. In: Wifi Networks, vol. 10, p. 44. Vestnik Sviazy (2015). ISSN:0320–8141

14. Feng, J., Wang, Q.: Research of positioning technique based on wireless LAN. In: International Conference on Computer Science and Information Technology (ICCSIT 2011), vol. 51, pp. 202–207. IACSIT Press, Singapore (2012)

15. Liu, H., Darabi, H., Banerjee, P., Liu, J.: Survey of wireless indoor positioning techniques and systems. IEEE Trans. Syst. Man Cybern. Part C Appl. Rev. **37**, 1067–1080 (2007). IEEE

16. nanoPAN 5375 Development Kit User Guide Version 1.1 (2007). http://nanotron.com/EN/pdf/Factsheet_nanoPAN_5375_Dev_Kit.pdf

17. WirelessMon software. http://www.passmark.com/products/wirelessmonitor.htm

18. The MathWorks, Inc. http://www.mathworks.com

NEW2AN: ITS

Pilot Zone of Urban Intelligent Transportation System Based on Heterogeneous Wireless Communication Network

Vladimir Grigoryev, Igor Khvorov, Yury Raspaev, Vladimir Aksenov, and Anna Shchesniak[✉]

Saint Petersburg National Research University of Information Technologies, Mechanics and Optics (ITMO University), St. Petersburg, Russia
vgrig@rdnet.ru, {khvorov,vladaksi}@labics.ru, raspaev@mail.ru, anna.schesnyak@scaegroup.com

Abstract. The primary aim of Intelligent Transportation Systems (ITS) is to increase road safety as well as traffic efficiency. In order to meet strict and varying with ITS applications requirements, we propose using a combination of cellular, short range and Ultra High Frequency (UHF) communication technologies. Our approach was inspired by Communication access for Land Mobile (CALM). In this paper, we introduce, describe and discuss a distributed heterogeneous pilot ITS network. The technologies in focus are UHF communications for high reliability and robustness, Dedicated Short Range Communications (DSRC) for direct vehicular communications, WiMAX and LTE for broadband services. We describe the services that were chosen and how they map to the technologies. The characteristics of the technologies are translated into frequency-territorial planning, which derives the number of base stations and their locations. Finally, we show how the capabilities of the pilot network are demonstrated. This pilot network is a first step towards investigation of heterogeneous ITS in real urban communication surroundings.

1 Introduction

Modern public transport can already be equipped with on-board devices to implement a number of applications: route monitoring, fare collection, video surveillance of interior and road conditions, emergency alarm transmission [1]. Data transmission from these applications can be realized by the means of different communication technologies, e.g., Dedicated Short Range Communications (DSRC). When a vehicle is connected to a network through one or more communication technologies it becomes a part of Intelligent Transport Systems (ITS) [2].

Conceptually, ITS are not bound to vehicular communications [3]. However, in this paper, we refer to ITS in terms of connected road transport. The main purpose of ITS is to enhance traffic safety through information exchange between vehicles (vehicle-to-vehicle, V2V) as well as between vehicles and infrastructure (vehicle-to-infrastructure, V2I) [1]. The next important ITS application is traffic efficiency, e.g., reducing traffic jams due to vehicular traffic flow control or tolling

© Springer International Publishing AG 2016
O. Galinina et al. (Eds.): NEW2AN/ruSMART 2016, LNCS 9870, pp. 479–491, 2016.
DOI: 10.1007/978-3-319-46301-8_40

over the air. Finally, vehicular communications may be utilized for infotainment. For example, infotainment services could be music streaming, online gaming and contextual information. All three ITS applications have in common that they require ubiquitous network connectivity in a moving vehicle.

Except for ubiquitous connectivity requirement, safety, traffic efficiency and infotainment have intrinsically contradicting requirements to enabling communications technology [1]. For the safety applications, the most important parameters are reliability, robustness and low delay. The requirements in terms of available transmission rates are low. Entertainment, however, requires high transmission rates, but is more tolerant to delay and packet loss. This is why choosing a single communication technology is a challenging (when not an impossible) task.

Currently, European Telecommunications Standards Institute (ETSI) sees Dedicated Short Range Communications (DSRC), or 802.11p, as a primary communication technology for ITS implementation [2]. There is a number of pilot studies solely dedicated to DSRC for ITS investigation, e.g., Connected Vehicle Safety Pilot in the USA [4], Cooperative Intelligent Transport system Initiative (CITI) Project in Australia [5], or a research project by Araniti et al. [6]. Despite research interest and its maturity, DSRC is not widespread due to unbounded delays and non-deterministic Quality of Service (QoS) [7]. Furthermore, in order to achieve a proper performance, a large fleet of cars is required to be equipped with an extra and rather expensive DSRC modem. At the same time, cellular modems are either already inbuilt in most vehicles, or are widely available and inexpensive. Moreover, the cellular infrastructure is either already in place with almost ubiquitous coverage. Park et al. in [7] propose starting the distribution of vehicular safety systems with a special smartphone application. The advantages of such an approach are as following: no significant investments into communication architecture and independency from a car manufacturer. However, a smartphone-based solution would share the communication resources with the rest of the network users, setting in question the overall performance. Moreover, the phone can be stolen, lost or simply left somewhere. Generally, there are a number of projects that study specific technologies and an influence of ITS realization on different aspects of city life, e.g., CO2 emission. More details on European ITS projects can be found in [8]. This paper does not aim at exhaustive description of all the ITS projects.

Despite considerable amount of research on ITS, to the best knowledge of the authors, this is a first implemented heterogeneous ITS pilot network with the following communication technologies: Ultra High Frequency (UHF) communications of standard Citran [9]; Dedicated Short Range Communications (DSRC) [10]; Worldwide Interoperability for Microwave Access (WiMAX) [11,12] and Long-Term Evolution (LTE) [13].

The goal of this paper is to present a real-life implemented and demonstrated pilot network, so that it could serve as a descriptive basis for further research work. Future research work includes verifications, measurements, and experiments that are to be based on the pilot network. The description here concentrates on the heterogeneous pilot network architecture and its realization.

The manuscript is organized as following. First, the main services and prospective technologies are determined and described in Subsect. 2.1. Second, the Frequency-Territorial Planning (FTP) was carried out and presented in Subsect. 2.2. Here, FTP determines how many Base Stations (BSs) are needed for the ubiquitous coverage, where to place them and which frequencies to assign. Section 2 is concluded with the resulting logical architecture of the pilot network and its measurement capabilities. Then, in Sect. 3, three pilot network demonstration scenarios are described. The last section concludes this work.

2   Pilot Public ITS Access Network

Pilot networks allow not only testing of communication technologies, but development and test of the network roll-out process. This is especially important for such large-scale projects as a dedicated city-wide access network for ITS implementation [14]. With this motivation Saint-Petersburg National Research University of Information Technologies, Mechanics and Optics with support of the Transport Committee of St. Petersburg, Russian Federation formed a working group to build a heterogeneous pilot network for ITS. The project started with the technology choice, included pre-project investigations, e.g. [15], and network planning, as well as construction, commissioning and demonstration.

This section describes our approach to the pilot network realization in the logical sequence of project work. First, in Subsect. 2.1 we describe the services and technologies. It is a first step as the set of services forms the set of possible realization technologies and in the end the entire architecture. Second, in Subsect. 2.2 we present the methodology of Frequency-Territorial Planning (FTP) as it determines not only the number of Base Stations (BSs), but also the cost of network implementation [9]. Finally, Subsect. 2.3 presents resulting pilot network logical architecture in the current state.

2.1   Services and Technologies

A clear set of services must be chosen in order to choose properly the technologies during the phase of a (pilot) network planning. Depending on the country, the road situation and level of technological development, the set of services and, thus, technologies could vary greatly. In our pilot network we focus on the services that are relevant for St. Petersburg:

1. Safety:
   – Dispatching communication with the driver;
   – Real-time video surveillance of interior and traffic situation, face recognition, event triggered monitoring;
   – Control of priority lanes, safety control, intruder detection;
   – Emergency communications and alarms;
   – Public emergency alerts.
2. Traffic efficiency:

- Transport traffic monitoring, route control, verifying that public transport traffic complies with its regulations and timetables;
- Automated traffic management, real-time travel and passenger traffic monitoring;
- Information boards at bus stops.

3. Commercial applications (infotainment in public transport):
- Wi-Fi in the cabin and on stations, contextual advertising.

For the pilot network, the following technologies were selected according to the corresponding services: UHF communications of standard "Citran" (later referred to as "UHF") [9], DSRC [10], WiMAX [11] and LTE [13]. As the communication equipment is not designed to the last letter of standard specifications, we refer to the specific equipment manufacturers.

UHF was chosen for implementation of emergency and voice communications, navigation, periodical transmission of telematics information as vehicle location position or speed. UHF is considered as an alternative to General Packet Radio Service (GPRS) and a back-up network due to its high robustness and reliability. The drawback of UHF is low communication rates (range of kbps), so the range of services to be implemented with it is limited. The equipment utilized for the pilot network was purchased from "Communication Technology".

DSRC provides only meshed V2V communications for the pilot network. The main purpose of DSRC use is ensuring a minimal delay for the safety services. It was not chosen for V2I services as [9] showed prohibitive costs. The equipment manufacturuours for this network is from MikroTik, Latvia.

WiMAX and LTE support high-speed services such as video streaming, real-time and other infotainment. WiMAX equipment was purchased from Alvarion, Israel. LTE module is a standard Mobile Network Operator (MNO) modem.

Table 1 shows the main parameters of the equipment utilized in the pilot zone. Note, that the access to the LTE network was provided by one of the MNO, and the network was viewed as a black box. Therefore, it was not possible to influence the network. All the other networks were dedicated.

2.2   Frequency-Territorial Planning

One of the core access network design problems is Frequency-Territorial Planning (FTP). FTP defines access network structure, locations of BSs sites, antenna types, height and orientation. FTP aims to provide required coverage area with predetermined communication quality. Moreover, during FTP a frequency allocation plan for BSs is developed and Electro-Magnetic Compatibility (EMC) of the network under design with existing networks is investigated.

FTP for our case study was performed in an operator-like way: using professional software package and appropriate input planning parameters. The utilized software package is ICS Telecom from ATDI, France. Usage of a specialized program instead of manual planning provides the following benefits: reliable, repeatable results and automation of the process. The main task here is to gather the correct input parameters and then interpret the results from the program.

Table 1. Equipment parameters used in the pilot network

-	Working frequencies, MHz	Channel width	Multiple access	Duplex
UHF	385–388, 442–450	25 kHz	TDMA	FDD
WiMAX	3400–3600	5, 7, 10 MHz	OFDMA	TDD
LTE	2600	20 MHz	OFDMA	TDD
DSRC	5855–5925	10 MHz	CSMA/CA	TDD

The wave propagation model that was used for calculations was 525/526 ICS Designer-a, inbuilt in the software and based on ITU-R recommendations [16,17]. Technical parameters of the equipment were taken from equipment documentation. The height of the hanger was taken based on the average height of the buildings, where the equipment could be installed. For UHF, WiMAX and LTE it was 56 m, e.g., top of a building, and for DSRC it was the average height of a traffic light pole −7m. These parameters are an input to the actual planning.

The outcome of planning process (calculated by the program) is presented in Fig. 1. Here, the main performance parameter is Received Signal Strength Indication (RSSI) as it defines the resulting coverage (together with receiver sensitivity). Figure 1(a–c) show the RSSI for various technologies: UHF, WiMAX

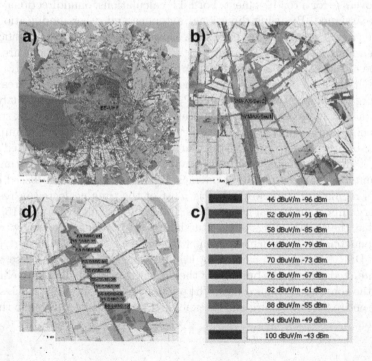

Fig. 1. Examples of frequency-territorial planning calculation results.

and DSRC respectively. Figure 1(d) shows a color palette that reflects which RSSI corresponds to which color, using two units: dBuV/m and dBm.

Figure 1(a) shows the resulting range of a single UHF BS. It can be seen that UHF allows a sustainable uniform coverage with a radius up to 6 km. However, a rather strong signal can reach areas far away from the required coverage area. In this case, if the BSs are not divided in frequency, it may cause severe interference between them, degrading the performance of the entire network. So, in order to cover the area of St. Petersburg without suburbs (about 600 km^2) only nine UHF BSs are required. These BS shall have four sectors with two nominal frequencies per sector.

Figure 1(b) illustrates the result of FTP for WiMAX. The figure shows that WiMAX functions well in tunnels of streets, which is beneficial for urban ITS implementation. FTP results show that WiMAX allows reliable communications with mobile units (vehicles) with a radius of two km. For covering the area of St. Petersburg 70 BSs are needed. Detailed investigations on what the number of BSs means for the cost of the network are presented in [9]. In the pilot network, LTE has not been deployed yet, rather taken as a "black box" from an MNO. Thus we do not consider LTE in this papers analysis even though the FTP for it was carried out.

For DSRC with the FTP resulted in not less than 1500 Road Side Units (RSUs) for St. Petersburg. This is due to the line-of-sight requirements. Figure 1(c) shows how densely these RSUs must be located to provide a full DSRC coverage for a road segment. For FTP calculations, omnidirectional antennas were assumed. By using directional antennas and other modifications, the communications range can be increased. However, generally any modifications lead to an increase in the equipment costs. Thus, these modifications are out of the scope of the paper. We considered only the off-the-shelf equipment.

Here, it shall be noted that FTP was carried out for the entire area of St. Petersburg (600 km^2). Based on the FTP results, a techno-economic analysis [9] was conducted. It was shown that the overall cost of the access network strongly depends on country-specific factors and cannot be generalized. For example, for St. Petersburg's case study it turned out that a DSRC-based dedicated access network is more expensive than a dedicated LTE-based. The dominating costs were country specific certification payments. These payments are based on the country's regulatory framework that is most of the time neglected in research, but is very important in real life projects. In another regulative study [15] it was shown that the shortcomings of the Russian regulatory framework may severely limit commercial ITS implementation, when based on DSRC. In short, at the moment DSRC equipment cannot be legally utilized in Russia. These studies were conducted prior to the building of the pilot network and allowed to eliminate some of the considered technologies. The pilot network is, of course, much smaller than the entire St. Petersburg and its exact architecture is introduced in the next subsection.

2.3   Pilot Network Architecture

Geographically, the pilot network consists of three road segments with a total length of about 5 km, shown on a map in Fig. 2. This geographical area is covered by an MNO LTE network, and is partly covered by WiMAX pilot network. The transport network provides 70 Mbps dedicated link with round trip delay of 10 ms from an MNO "RadioNet". The location of the closest WiMAX BS is also shown in Fig. 3, where the green lines show the directions of the two BS sectors. In the areas that are outside the WiMAX coverage, it is possible to test direct vehicular communications with DSRC. The pilot network includes three buses (as users) that were enhanced with communication equipment [9]. The scheme of the pilot network and the equipment inside the buses are shown in Fig. 3.

The logical elements in Fig. 3 follow a color map. Green indicates that an element can communicate through public cellular network (LTE). Blue indicates that the communications are executed through WiMAX, and purple shows the run through two or more networks. Mesh here shows direct V2V communication without any additional infrastructure.

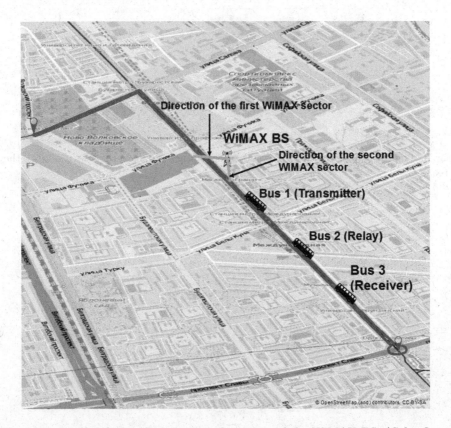

Fig. 2. Geographical map: bus routes and position of the WiMAX BS. (Color figure online)

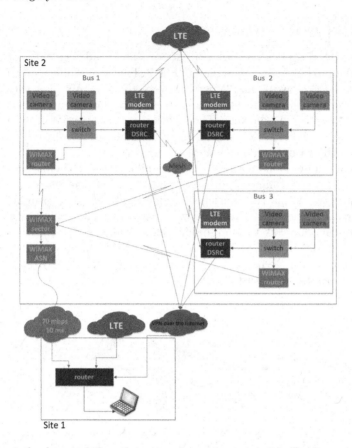

Fig. 3. Logical scheme of the pilot network: equipment inside the buses and at the sites.

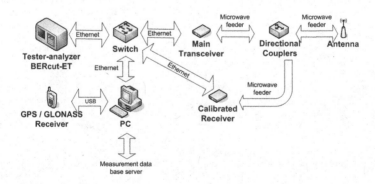

Fig. 4. Transceiver scheme with measurement equipment.

In our pilot network the LTE network is publicly available as well. Thus, the network itself is depicted as a black box, as the MNO provided only the access to the network. By this means, all the network settings and configuration remained solely under MNO control and were not disclosed. In this case the network serves not only the buses, but also regular users. Virtual Private Network (VPN) was used to isolate our traffic from the regular users. Section 3 presents already demonstrated scenarios based on this logical scheme.

The UHF is not yet a part of the deployed network and thus it is not presented in the logical scheme. UHF was shown to be a cheap technology that, due to high robustness, can be a good back-up dedicated network [9]. UHF equipment is already purchased and is planned to be deployed.

Apart from the equipment under investigation, the metrological one was also included. Figure 4 depicts the measurement scheme. This scheme is the same for buses and BSs. All the measurements equipment could be easily transported due to compact size. The central measurement node is a tester-analyzer "BERcut-ET". It allows assessing actual network coverage and measuring the following parameters: Transmission rate, Number of errors, Bit Error Rate (BER), Delay and Jitter, etc. GPS/GLONASS receiver allows accurate positioning for the moving vehicles (buses). Calibrated receiver is needed to measure the signal strength (RSSI and Signal to Noise and Interference Ratio, SNIR) at the maim transceiver's input/output. The signal is supplied through a directional coupler. These functionalities make our pilot network a powerful tool for technology assessment.

3   Demonstration Scenarios

Three demonstration scenarios were used to verify pilot network operation and to demonstrate its capabilities. These scenarios represent the services that are already available and show them over different technologies. The buses were moving along the route depicted in Fig. 2 with the allowed speed. All three scenarios include video transmission. Video is an important part of safety applications, e.g., video surveillance [18], and infotainment, e.g., video streaming [1]. In the future we plan to expand the list of implemented scenarios to cover a greater number of services and options.

3.1   Scenario I: WiMAX vs. LTE for Video Transmission in V2I

In the first scenario, two video streams are transmitted from three buses over a dedicated WiMAX access network. For comparison, we utilize transmission over a public LTE network. Figure 5 shows a logical diagram of this scenario. Three buses are equipped with two mounted surveillance cameras and WiMAX and LTE modems. The buses are connected in parallel to the respective networks. Video is transmitted to the Control and Monitoring Center (CaMC). The uplink video streams had the following parameters: H.264 and 512 kbps per stream.

After the implementation of the scenario the following answers were obtained. The use of existing networks is possible. In this case for video streaming can be used only 3G or 4G networks. The main conditions for using these networks are the complete coverage of selected area and a low load by the network customers. If the conditions are not satisfied, the existing network cannot meet the requirements of the selected services. The use of dedicated communication network allows to get the following technical and economic benefits: provide the required QoS for all selected services, make a flexible change of network parameters when adding new or deleting existing services.

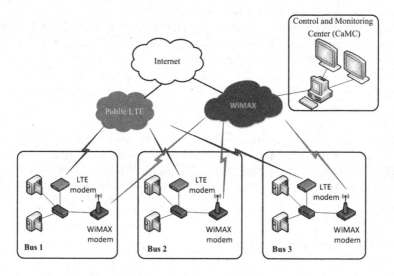

Fig. 5. Network configuration for demonstration Scenarios I and II.

3.2 Scenario II: LTE and WiMAX Multiservice Transmission in V2I

The second scenario aims at evaluating the quantitative and qualitative network performance characteristics for multiservice transmission, i.e., simultaneous transmission of voice, video and data under different priorities and over different networks.

During the demo, the downlink data rate (for TCP and UDP) was 2–5 Mbps from the CaMC/Internet to each bus. The rest of the services were transmitted in both directions. Voice was transmitted with a high priority (simulating an emergency call) at 8 kbps (codec G.729) or at 64 kbps (codec G.711), a single voice stream per bus. Video stream characteristics were the same as in the first scenario.

As in the first scenario, two video streams were transmitted simultaneously over LTE and WiMAX networks from each bus to CaMC, see Fig. 5. Here, only

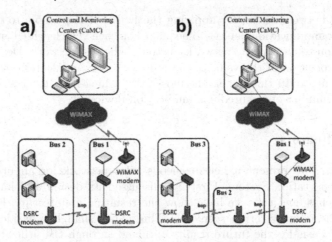

Fig. 6. Network configuration for demonstration Scenario III.

the set of services was changed in comparison to the first scenario. After the implementation of the scenario the following answers were obtained. Differences in the implementation of multi-service between existing networks and dedicated networks begin to appear as the utilization of existing networks by their subscribers was growing. Existing and dedicated communications network show similar response to changes in the composition of multi-services. Increasing demands from the services leads to higher degrees of network resources utilization. When approaching the degree of utilization to one flow control mechanisms provided in both networks, produce the same results.

3.3   Scenario III: Video Transmission over DSRC in V2V

In the third scenario three buses communicate in V2V fashion, i.e., directly. DSRC equipment forms an infrastructureless mesh-network to demonstrate and investigate video transmission over DSRC, see Fig. 2. The receiving bus further transmits video to the CaMC over cellular network, as it was demonstrated for CaMC.

Figure 6(a) shows the direct transfer scenario: two video streams from Bus 2 were sent directly over DSRC to Bus 1 which sends the data further over WiMAX network to the CaMC. Figure 6(b) represents transmission with an intermediate relay (DSRC-capable vehicle) in between: receiver and transmitter on the data link layer of OSI model "do not see" the relay and "think" transmission being direct.

The following answers were obtained after the implementation of the scenario. A transmission of video streams from one vehicle to another, either directly or by multi-hop, is possible. At the same time, depending on the relative position of the vehicles the number of video streams and their parameters are varying a lot. The closer the vehicles to one another, the more video streams may be transmitted between them. By applying the DSRC it is possible to reduce the number of communication devices to be placed on each vehicle which are required for connection to an existing networks or to a dedicated network. Depending on the relative position of the vehicles the number of communication devices may be reduced up to 10 times, i.e., the closer the vehicles are to each other, the smaller the number of required communication devices is.

4   Conclusions

In this paper, we present a heterogeneous pilot network for an urban Intelligent Transportation System. First, we discussed its design and implementation approaches, and then we have show the resulting architecture. Finally, we described the demonstration scenarios. This paper shall be considered as an explanatory basis for the future results obtained through the utilization of the pilot network. Through the deployment of the pilot network the following goals have been already achieved:

- Techno-economic evaluation of UHF, Wi-Fi, DSRC and LTE (and WiMAX) for roll-out of a dedicated ITS access network [9]. It was shown that the network cost strongly depends on the country (city) specific parameters and cannot be generalized.
- Network planning and demonstration scenarios are formalized in the current paper. It is shown, what is our approach, which steps are the most important and the first demonstration scenarios.

In the future work, we plan to move from network planning and demonstration towards technology assessment. First, the pilot network is planned to be enhanced with UHF equipment. Then, more scenarios are planned to be deployed. Finally, future work includes measurements and theory verifications on both individual technologies and on the heterogeneous system.

References

1. Dar, K., Bakhouya, M., Gaber, J., Wack, M., Lorenz, P.: Wireless communication technologies for ITS applications [topics in automotive networking]. IEEE Commun. Mag. **48**(5), 156–162 (2010)
2. ETSI: Intelligent Transport Systems. http://www.etsi.org/technologies-clusters/technologies/intelligenttransport?highlight=YToxOntpOjA7czozOiJpdHMiO30=
3. Himayat, N., Yeh, S.-P., Panah, A.Y., Talwar, S., Gerasimenko, M., Andreev, S., Koucheryavy, Y.: Multi-radio heterogeneous networks: architectures and performance. In: Proceedings of International Conference on Computing, Networking and Communications (ICNC), pp. 252–258. IEEE (2014)
4. United States Department of Transportation: Connected Vehicle Safety Pilot. http://www.its.dot.gov/safety_pilot/
5. The Intelligent Transport Systems Portal: An insight into C-ITS in Australia through the NSW CITI Project. http://erticonetwork.com/an-insight-into-c-its-in-australia-through-thensw-citi-project/

6. Araniti, G., Campolo, C., Condoluci, M., Iera, A., Molinaro, A.: LTE for vehicular networking: a survey. IEEE Commun. Mag. **51**(5), 148–157 (2013)
7. Park, Y., Ha, J., Kuk, S., Kim, H., Liang, C.-J.M., Ko, J.: A feasibility study and development framework design for realizing smartphone-based vehicular networking systems. IEEE Trans. Mob. Comput. **13**(11), 2431–2444 (2014)
8. Rafiq, G., Talha, B., Patzold, M., Gato Luis, J., Ripa, G., Carreras, I., Coviello, C., Marzorati, S., Perez Rodriguez, G., Herrero, G., et al.: What's new in intelligent transportation systems: an overview of European projects and initiatives. IEEE Veh. Technol. Mag. **8**(4), 45–69 (2013)
9. Grigoryev, V., Khvorov, I., Raspaev, Y., Grigoreva, E.: Intelligent transportation systems: techno-economic comparison of dedicated UHF, DSRC, Wi-Fi and LTE access networks: case study of St. Petersburg, Russia. In: Conference of Telecommunication, Media and Internet Techno-Economics (CTTE), pp. 1–8. IEEE (2015)
10. Jiang, D., Delgrossi, L.: IEEE 802.11 p: towards an international standard for wireless access in vehicular environments. In: Proceedings of IEEE Vehicular Technology Conference, pp. 2036–2040. IEEE (2008)
11. Teo, K.H., Tao, Z., Zhang, J.: The mobile broadband WiMAX standard [standards in a nutshell]. IEEE Signal Process. Mag. **24**(5), 144–148 (2007)
12. Ometov, A., Andreev, S., Turlikov, A., Koucheryavy, Y.: Characterizing the effect of packet losses in current WLAN Deployments. In: Proceedings of 13th International Conference on ITS Telecommunications (ITST), pp. 331–336. IEEE (2013)
13. Sesia, S., Toufik, I., Baker, M.: LTE: The UMTS Long Term Evolution. Wiley Online Library, New York (2009)
14. Kuznetsov, V., Raspaev, Y., Tarakanov, S., Khvorov, I.: ITS elements' design within realization of a pilot zone. ELECTROSVYAZ **10**, 24–27 (2013)
15. Grigoryev, V., Khvorov, I., Vlasov, V., Grigoreva, E.: Legislative particularities in the Russian Federation of the radio spectrum allocation for radio electronic equipment built in vehicles. In: Proceedings of IEEE 80th Vehicular Technology Conference, pp. 1–6. IEEE (2014)
16. ITU-R Recommendation P.525-2: Calculation of free-space attenuation (1994)
17. ITU-R Recommendation P.525-2: Propagation by diffraction (2009)
18. Belyaev, E., Vinel, A.V., Jonsson, M., Sjöberg, K.: Live video streaming in IEEE 802.11 p vehicular networks: demonstration of an automotive surveillance application. In: Proceedings of INFOCOM Workshops, pp. 131–132 (2014)

Connectivity of VANET Segments Using UAVs

Pavel Shilin, Ruslan Kirichek[(✉)], Alexander Paramonov,
and Andrey Koucheryavy

State University of Telecommunication, St. Petersburg, Russia
beerkol00500@gmail.com, kirichek@sut.ru,
alex-in-spb@ya.ru, akouch@mail.ru

Abstract. The main topic of the researches is VANET (Vehicle Ad-hoc Network). VANET is a peer-to-peer network based on IEEE 802.11p standards and
group standards IEEE 1609 Wireless Access in Vehicular Environments
(WAVE). Another current line of research is the UAVs. In 2015, the scientific
works, oriented to research of a possibility of UAVs use for the VANET networks, began to appear. The scientific works present issues concerning connection of separately located network nodes by means of UAVs.

In this paper, we suggest evaluation of the possibility of creating flying
VANET nodes. We will consider the model of the communication network of
several isolated vehicles' segments using UAVs. We will carry out the modelling and calculations in order to determine the maximum number of segments
that can service the node based on UAVs for several types of call flows and
describe circuit of preparation and statistical data production in the context of
real network segment.

Keywords: WAVE · DSRC · IEEE 802.11p · IEEE 1609 · VANET · UAV

1 Introduction

VANET [1] is gradually becoming more recognized and interesting topic of research.
Many countries have accepted radio-frequency agreements that permit use of working
range 5.85–5.925 GHz for DSRC [2] applications. The IEEE organization has also
established a working group IEEE 1609 [3], which develops family standards for
WAVE and IEEE 802.11p [4] and defines network architecture, set of services and
interfaces, collectively, designed to create a wide range of solutions, including
improving safety of road traffic, navigation, automated movement of vehicles one after
another. The node is a vehicle in this network, which exchanges data with other
vehicles through wireless radio. The nodes of such network are equal in rank. But if we
check the architecture, there are two main types of connections V2V (Vehicle to
Vehicle) and V2I (Vehicle to Infrastructure) [5, 6, 23], and initially there are no any
connections with flying nodes, but many authors of researches use fly nodes based on
copters for improve connectivity [7–9].

Another, slightly similar class networks are Flying Ubiquitous Sensor Networks,
which were designed for applications of monitoring and data collection and management, creation of temporary communication nodes using UAVs. Having set of

© Springer International Publishing AG 2016
O. Galinina et al. (Eds.): NEW2AN/ruSMART 2016, LNCS 9870, pp. 492–500, 2016.
DOI: 10.1007/978-3-319-46301-8_41

programs and algorithms on the board, UAVs is able with high effectivity to serve terrestrial nodes, to exchange information with each other, to coordinate its actions in the air, ensuring a high coverage of radio accessibility unlike terrestrial wireless networks [10, 11]. The objective compound of disparate groups of vehicles can be addressed both by a single UAV, and by a swarm of UAVs [12].

All these opportunities are one of the ways to improve connectivity and VANET network efficiency. If we combine the analyzed types of networks, we can create a new network node that combines the capacities of these networks. It is expected that the use of UAVs will improve the distance, connectivity, routing [13] of the discrete segments of the network, unable to join the network directly with each other because of the conditions. General representation of the network, considered in the article, is shown in Fig. 1.

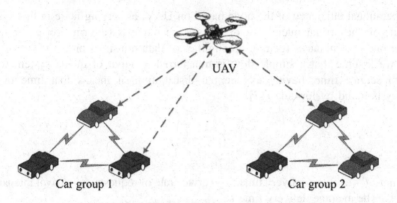

Fig. 1. Idea concept

Concept network was tested on the test bed [14] on the model network [15] of lab of Internet of Things. We will solve this problem with the help of submission of the UAV in the form of queuing system. Such method of the decision is applied in other papers [16] where submission of the UAV in the form of queuing system is required.

2 The Network Model

The considered network model is presented in Figs. 2 and 3, consists of G_n groups of vehicles, UAV, which is static and at some point on h height. For example, on Figs. 2 and 3 we used three car groups. Vehicle groups interact with UAV via radio channel. R-value describes the radius of the possible interaction between UAV and nodes, which depends on the height h of UAV flight. This model does not take into account the terrain and weather conditions.

As the model uses the resources of many available transportation vehicles, connected in a single data transmission network via UAV, it is a queue system, shown schematically in Fig. 2.

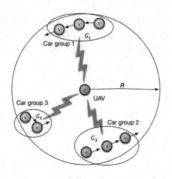

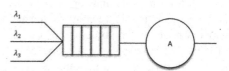

Fig. 2. Network model   **Fig. 3.** Queue model

Operational efficiency of the node based on UAV, as serving node in this system, depends on the arrival intensity λ of the service requests (information packets) and service rate μ of requests (getting possibilities of data transmission).

If we assume that a simple flow is transferred to input of queue system with a random service time, having exponential distribution, it means that time of data delivery is found by formula [17]:

$$T = \frac{\frac{1}{\mu}}{1 - \frac{\lambda}{\mu}} \tag{1}$$

Where T – average delivery time, μ – service rate of requests, λ – arrival intensity of requests. The average delivery time is:

$$T = W + t_{cp} \tag{2}$$

Where W –queue time,
t_{cp}- Average service time, which is defined as:

$$t_{cp} = \frac{1}{\mu} \tag{3}$$

In order to show the arrival intensity of requests λ, it is necessary to simplify the formula:

$$\frac{1}{T} = \frac{1 - \frac{\lambda}{\mu}}{\frac{1}{\mu}} \tag{4}$$

$$\lambda = \mu - \frac{1}{T} \tag{5}$$

The probability of requests' receipt from each group of vehicles within the service radius in relation to QS is calculated by the formula of Poisson distribution:

$$P_m(t) = \frac{(\lambda S(t))^m}{m!} e^{-\lambda S(t)} \qquad (6)$$

Where λ – requests in time,
$S(t) = 2R$ – Square of service sector.
The time required for service of single request is calculated as:

$$\tilde{t} = \frac{t(l)}{\beta} + t_{pre} \qquad (7)$$

Where $t(l)$ – time required for transmission, depending on the long packet.
t_{pre} - time required for preparation of data transmission is the amount:

$$t_{pre} = t_{AIFC\lceil AC \rceil} + t_{backoff} + t_{phy} \qquad (8)$$

$$AIFSN_{ac} = SIFS + AIFS * SlotTime \qquad (9)$$

Where SIFS - Short Inter Frame Space 32 µs, AIFSN [AC], AIFS time from package of access modes (AC).

$$SIFS(32\mu s) + AIFSN_{ac} * SlotTime = 32 + 2 \times 13 = 58\mu s$$
$$BackOff\ Interval = \lceil 0...CWmin\lceil AC \rceil \rceil * aSlotTime \approx 2 * 16 = 32\mu s$$

Where [0...CWmin[AC]] – sampled value of number, t_{phy} – 53.3 ~ 55.7 µs.

It is interesting to note that not all requests can be processed for lack of node accessibility due to interferences, fading effects and other features of signal propagation.

Limit of vehicles' number can be calculated as:

$$V_{lim} = \frac{T_{cch}}{T_{app}} \qquad (10)$$

Where T_{cch} – channel rate, T_{app} – time required for data transmission.

3   Calculations and Modeling

To evaluate the possibility and operating efficiency of the system, band of software packages OMNETPP [18], MIXIM [19], Veins [20] was used for forming a software simulation environment that includes environment model of radio wave propagation, physical and link layer, protocol stack.

According to the specifications for the adequacy of the modeling results, it is necessary to use several communication channels: control channel CCH, and service channel SCH. In its turn, the occurrence intervals of one or another message types for CCH depend on the message type, and vary from 10 to 50 Hz. Consequently, some message types can be transmitted. The switching interval is T_{cch} equal to 1000 ms.

In the process of flows modelling, messages flow was generated by each group of vehicles according to the specification of IEEE 802.11p WAVE by WSMP Protocol [21] (Fig. 4).

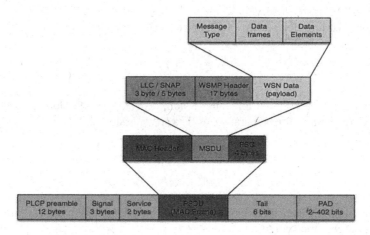

Fig. 4. Implementation message in WSMP protocol

To calculate the maximum limit of nodes we have formula below. Let us assume that all vehicles start transmission, making sure that the channel is clear, in strict sequence one after another. The transmission time for each node is calculated as:

$$t_0 = t_{tx} \tag{11}$$

$$t_{1-n} = t_{n-1} + t_{tx}$$

Where t_{tx}– total time required to transmit data from one node.

According to the WAVE specification, the size of the transmitted data depends on the type of the transmitted message (PSDU) and ranges within 1016–1376 bits. We used the value of 1200 bits (150 bytes) in the modelling (Table 1).

Table 1. The simulation parameters

Standard	IEEE 802.11p
Frequency of control channel	5890 MHz
Frequency of service channel	5870 MHz
Channel width	10 MHz
Transmitted power	100 mW
Receiver sensitivity	−94 dBm
Connection speed	12 Mbit
Minimum interval between calls (int)	1 s
Simulation time	60 s

When using the model of random calls flow, moment of data transmission t_n for different nodes is determined by the formula:

$$t_n = t_0 + t_{off} \qquad (12)$$

$$t_{off} = f[0; 1 * int]$$

where int – current time.

4   Output Evaluation

We used metrics as the analyzed parameters: the number of successfully accepted packets, the residence time of the channel in busy state, and the total number of packets lost. The total number of lost packets is composed of the number of packets, lost due to low differential level of signal/noise and failure of packet's acceptance due to transmit mode (Fig. 5).

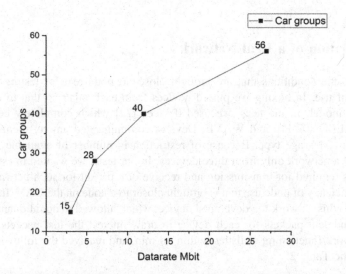

Fig. 5. Results simulation for determined calls model

Modelling outputs in time of determined calls flow show the number of nodes that can operate without collisions. According to the researches [21], which studied only separate nodes based on vehicles, we got much the same limits. This proves that the node based on UAV is no different from any other node of VANET network.

The second stage was simulation of random calls flow model. In practice, this model is more common and more accurately presents calls distribution. This model demonstrates an increase of the total losses, depending on the number of groups of vehicles in a geographical area (Fig. 6).

The percentage of losses relative to the total number of transmitted packets: 2 groups – 1 %, 5 groups - 1.5 %, 10 groups - 6.7 %, 15 groups - 19.7 %.

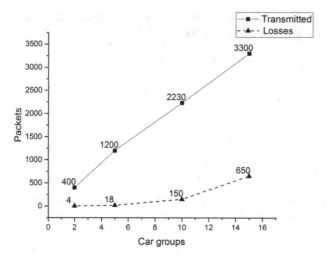

Fig. 6. Results simulation for random calls model

5   Simulation of a Real Network

We created the conditions that are brought closer to node real for testing UAVs of efficiency of use. In boxing we placed devices from each other so that to create the researched model. In the tests, we used devices [22] which completely conform to standards IEEE 802.11p and WAVE. Devices communicated among themselves the Basic Safety Message type. Because of restriction in number of available to us, we could build a network only from three devices. In our tests, we wanted to estimate the time delays required for transmission and receive of a packet for such network and to estimate efficiency of node use that is brought closer to a node on the UAV. In addition, we added to this network the developed device, which allowed to count quantity of the accepted and sent packets for each device to draw interest the lost packets. We collected network functioning statistics within 30 min, and received the following results shown in the Table 2.

Table 2. Results real simulation

Metrics/Devices	OBU-Car1 (Locomate OBU)	OBU-Car2 (Locomate OBU)	OBU-UAV (Locomate mini)
Packet received	39279	41048	32238
Rx latency per packet (usec)	54664	52040	69852
Packet sent	21448	21482	21430
Tx latency per packet (usec)	100119	100021	100020
Packets sent for device	42912	42878	42930
Percent packet lost	8.47	4.27	24.91

6 Conclusion

In the study, the following results were obtained:

1. The node based on the UAV can be the same node of the VANET network as well as the car.
2. The limit of car groups, which are possible for servicing generally, depends on the radius of service UAV, data rate of the channel, length data and messages. This value can be counted mathematically, but it is more convenient to simulate.
3. When calculating model of random time of calls, it is defined that, increase in car groups affects transmission quality of data. Percentage ratios are calculated from what it is possible to define types of traffic, or services, which can function on such network.
4. According to the total of the received preliminary results of real simulation, we could wait time on transmission and reception, each packet from a network based on three devices.
5. According to the total of the received preliminary results, the UAV node shows great importance of losses of packets. Possibly all the reason of such behavior is cut in mobility of a node and diminutiveness of the equipment, which perhaps possesses several worst characteristics.

7 Future Work

In the subsequent operations, we plan to organize experiments with UAV use on a flight polygon to estimate efficiency of use of such network.

Acknowledgment. The reported study was supported by RFBR, research project No. 15 07-09431a "Development of the principles of construction and methods of self-organization for Flying Ubiquitous Sensor Networks".

References

1. Hartenstein, H., Laberteaux, K.: VANET Vehicular Applications and Inter-Networking Technologies (2009)
2. Vehicle Safety Communications Project Final Report, U. S. Dept. Trans., Nat. Highway Traffic Safety Admin., Rep. DOT HS 810 591 (2006)
3. IEEE P1609.4-2010 - IEEE Standard for Wireless Access in Vehicular Environments (WAVE) 2010
4. IEEE 802.11p-2012 - IEEE Standard for Information technology– Local and metropolitan area networks– Specific requirements– Part 11: Wireless LAN Medium Access Control (MAC) and Physical Layer (PHY) Specifications Amendment 6: Wireless Access in Vehicular Environments (2012)
5. Car 2 Car Communication Consortium Manifesto. Overview of the C2C–CC System

6. ETSI TS 102 636–3 V1.1.1 (2010–03): Intelligent Transport Systems (ITS); Vehicular Communications; GeoNetworking; Part 3: Network architecture. — European Telecommunications Standards Institute (2010)

7. Mase, K.: Wide-area disaster surveillance using electric vehicles and helicopters. In: 2013 IEEE 24th Annual International Symposium on Personal, Indoor, and Mobile Radio Communications (PIMRC), pp. 3466–3471 (2013)

8. Dorrell, D., Vinel, A., Cao, D.: Connected vehicles - advancements in vehicular technologies and informatics. IEEE Trans. Ind. Electron. **62**(12), 7824–7826 (2015)

9. Koucheryavy, A., Vladyko, A., Kirichek, R.: State of the art and research challenges for public flying ubiquitous sensor networks. In: Balandin, S., Andreev, S., Koucheryavy, Y. (eds.) NEW2AN/ruSMART 2015. LNCS, vol. 9247, pp. 299–308. Springer, Heidelberg (2015)

10. Sahingoz, O.K.: Mobile networking with UAVs: opportunities and challenges. In: 2013 International Conference on Unmanned Aircraft Systems (ICUAS), pp. 933–941 (2013)

11. Kirichek, R., Paramonov, A., Vareldzhyan, K.: Optimization of the UAV-P's motion trajectory in public flying ubiquitous sensor networks (FUSN-P). In: Balandin, S., Andreev, S., Koucheryavy, Y. (eds.) NEW2AN/ruSMART 2015. LNCS, vol. 9247, pp. 352–366. Springer, Heidelberg (2015)

12. Kirichek, R., Paramonov, A., Koucheryavy, A.: Swarm of public unmanned aerial vehicles as a queuing network. In: Vishnevsky, V., Kozyrev, D. (eds.) DCCN 2015. CCIS, vol. 601, pp. 111–120. Springer, Heidelberg (2016). doi:10.1007/978-3-319-30843-2_12

13. Oubbati, O.S., Lakas, A., Lagraa, N., Yagoubi, M.B.: CRUV: connectivity-based traffic density aware routing using UAVs for VANets. In: 2015 International Conference on Connected Vehicles and Expo (ICCVE), pp. 68–73 (2015)

14. Kirichek, R., Koucheryavy, A.: Internet of things laboratory test bed. In: Zeng, Q.-A. (ed.) WCNA 2014. LNEE, vol. 348, pp. 485–494. Springer, Heidelberg (2016)

15. Kirichek, R., Vladyko, A., Zakharov M., Koucheryavy, A.: Model networks for internet of things and SDN. In: Proceedings of the 18th International Conference on Advanced Communication Technology (ICACT), pp. 76–79 (2016)

16. Kirichek, R., Paramonov, A., Koucheryavy, A.: Flying ubiquitous sensor networks as a queueing system. In: Proceedings of the 17th International Conference on Advanced Communication Technology (ICACT), pp. 127–132 (2015)

17. Allen, A.O.: Probability. Statistics and Queueing Theory. PHI Learning, New Delhi (2009)

18. OMNeT++ Network Simulation Framework. http://www.omnetpp.org/

19. MiXiM (mixed simulator). http://mixim.sourceforge.net/

20. Veins - open source vehicular network simulation framework. http://veins.car2x.org/

21. Lee, J.-M., Woo, M.-S., Min, S.-G.: Performance analysis of WAVE control channels for public safety services in VANETs. Int. J. Comput. Commun. Eng. **2**(5), 563–570 (2013)

22. Arada systems web site. http://www.aradasystems.com/products/

23. Vinel, A., Vishnevsky, V., Koucheryavy, Y.: A simple analytical model for the periodic broadcasting in vehicular ad-hoc networks. In: IEEE Globecom Workshops, GLOBECOM 2008

NEW2AN: Network Issues

The Analysis of Abnormal Behavior of the System Local Segment on the Basis of Statistical Data Obtained from the Network Infrastructure Monitoring

Ilya Lebedev, Irina Krivtsova, Viktoria Korzhuk$^{(\boxtimes)}$,
Nurzhan Bazhayev, Mikhail Sukhoparov, Sergey Pecherkin,
and Kseniya Salakhutdinova

ITMO University, Lomonosova str. 9, Saint Petersburg, Russia
{lebedev,ikr,vika}@cit.ifmo.ru,
nurzhan_nfs@hotmail.com, mikhailsukhoparov@yandex.ru,
pecherkin.sa@gmail.com, kainagr@mail.ru
https://cit.ifmo.ru

Abstract. The wireless network of low-power devices of Smart home and Internet of Things is considered. The signs of unauthorized access are defined. The analysis of the characteristics of systems based on wireless technologies obtained from the experiment results of passive monitoring and active polling of device forming the network infrastructure is conducted. The state-analyzing model based on the identity, quantity, frequency and temporal characteristics is presented. Evaluation of the information security state is focused on analyzing of the system normal functioning profile excluding the search of signatures and characteristics of anomalies under different kinds of attacks. The accumulation of data for decision-making is carried out by comparison of the statistical information of service message from the terminal nodes in passive and active modes. The proposed model can be used to determine the technical characteristics of the devices of wireless ad-hoc networks and to make recommendations concerning the information security state analysis.

Keywords: Information security · Wireless networks of soft spaces · Vulnerability · Availability of devices · The model of information security

1 Introduction

By reason of the emergence of a large number of mobile devices connected to the Internet, the implementation of processes of reception, transmission, processing of the incoming information outside of the controlled area leads to the need for information security.

Wireless technologies used in automatic control systems, Smart Cities, Smart Homes, the Internet of Things, Soft Spaces are especially vulnerable. And, considering the first two areas have typical adapted protection means, in others due to lack of standardization today the developers pay little attention to solutions in the field of security [1–3].

© Springer International Publishing AG 2016
O. Galinina et al. (Eds.): NEW2AN/ruSMART 2016, LNCS 9870, pp. 503–511, 2016.
DOI: 10.1007/978-3-319-46301-8_42

Usual household "intelligent" items ("smart" microwaves, coffee makers, washing machines), which make up the network segment of the Internet of things or Smart Home, provide an opportunity to detect and identify itself by generating messages and can be affected by the outside attacker or malicious user. Unification of means of interaction ensuring by individual manufacturers of "smart" appliances and configuration processes given to regular users who do not have relevant qualifications, create preconditions for carrying out of various attempts to control such devices from outside [4, 5].

There is the number of signs determining that someone is trying to access the system from the outside. Detection of abnormal network traffic, incorrect or inappropriate command in specified situation, identification of the large number of repeated events might be harbingers of unauthorized access attempts [6, 7].

Thus, the protection of information flows for the purpose of ensuring the integrity of transmitted data is one of the urgent tasks for various low-power systems.

In this regard, the number of tasks aimed at implementation of external monitoring of events of information security of "intelligent" devices occurs.

2 Statement of Problem

The common solution for the organization of interaction between devices is the wireless network consisting of set of nodes. The network infrastructure is the combination of physical and logical components that provide connectivity, security, routing, management, access and other required properties of the network [8].

The considered devices providing coupling of various devices and devices of "Soft spaces", in the majority are low powered and allow the receiving, processing and transmitting the limited type of messages.

To detect anomalous behavior it is possible to use data that reflects the state of the system, which can be applied in the statistical analysis [9, 10]. At the initial stage of operation after deployment it is possible to assess the different characteristics of the intensity of informational and service messages, the response times for service requests, the frequency of unrecognized and missed messages. Such characteristics can be obtained in the result of passive monitoring and active polling of devices.

Thus, it is necessary to determine characteristics of many elements of the controlled system enabling to identify its anomalous state relative to the "normal" functioning with the given probability.

3 Description of the Approach

Traditionally, the detection of the actions of the attacker or malicious user involves the identification of unexpected parameters of the network packets (false addresses, the flags of the messages and requests of the connections raised at the same time, traffic analysis network). However, the analysis of such events at low levels of network interaction in order to identify information security incidents is a challenging task even for professionals. It requires knowledge of networking protocols and narrowly specialized for devices by specific manufacturers.

One of the possible ways of system state analysis can be carried out on the basis of statistical data of the protocols of application level of nodes communication of the low-power devices using passive and active monitoring. In the first case the network device listening and statistical analysis of events as sending and receiving different types of messages are implemented. In the second case the following actions are performed: the request from the monitoring system in the form of various commands, analysis of time delays and changes of download of individual computing resources, the collation of identification information and settings.

Therefore, the most attractive to the average user are the detection methods that can be implemented in external independent devices.

Tracking the series of events is related to the increase of number of the following occasions in the network:

- Emergence of the unrecognized messages;
- Emergence of repetitive messages;
- Increase in the number of reported faults and failures;
- Increase in the number of broadcast and service messages;
- Occurrence of delays entailing the change in the statistical traffic of informational and service messages;
- Change of the delays of device responses to broadcast requests for various modes of operation;
- Increase in the number of lost messages;
- Change of frequency of informational and service messages.

It allows considering the model of the information security state analysis in the form of identification (I), quantitative (N), frequency (f) and time (T) characteristics. The profile of the system functioning is determined by the tuple of signs:

$$F = <I, N, f, T > \qquad (1)$$

Each of the signs represents a vector of values that is changeable during the time. Depending on the mode of operation, the change in the statistical portrait of network and devices functioning is observed.

Taking into account the low power of the devices providing the network infrastructure, evaluating the state of information security is easier to carry out using the profiles of normal functioning of the system.

4 The Experiment

In different modes of system operation we can experience anomalies that require more detailed study on the possibility of unauthorized access. Determination of quantitative and frequency data indicating unrecognized and missed messages, obtaining of information about the final state of the nodes on the basis of statistical data of application-level of interaction between low-power device nodes protocols by passive and active monitoring allows the construction of the classifier.

At the same time, these signs can be not the result of abuse only, but also the result of random errors and failures of equipment.

Figures 1 and 2 show the analyzed system. Informational and service messages circulate between nodes A and B, the device C is designed for collecting information. In the first case, the device C listens on the network and generates the statistical data sample. In the second case, it additionally sends requests to the devices and measures the variety of characteristics (Fig. 2A).

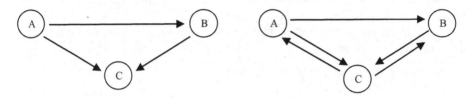

Fig. 1. The system scheme

In the experiment naive Bayes classifier was implemented. Its advantage is the small amount of training data needed to estimate the parameters required for classification.

$$C = argmax_{h \in H} \frac{p(X/h)p(h)}{p(X)} \tag{2}$$

where h, X are the predicted and preceding events and function of P is the probabilities of these events and its consequences (P = m/n where m is the number of occurred events, n is the number of all events).

In order to form the decision rule we use the data obtained in the active and passive monitoring modes:

- Relative frequencies of system state classes;
- Total number of sign characteristics in the designated grades for analysis;
- Relative frequency of signs within each class;
- Number of sample signs.

In the considered experiment the attack of malicious users on the application level stands in changing the intensity of the arriving messages (interception and playback of informational messages, service massage flooding, which causes the processes of association, authentication, dissociation, deauthentication and connection requests). The sequence of commands sent to the functioning device was selected randomly.

The data is accumulated by means of comparing of statistical information of service message of the terminal nodes in passive and active modes.

Figures 2, 3 and 4 shows the range of analyzed system states considered in the experiment where the results of the functioning are combined in the statistical data according to the types of messages (Figs. 3A and 4A).

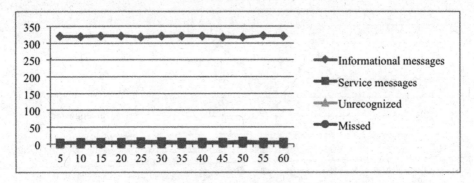

Fig. 2. The system in the normal state

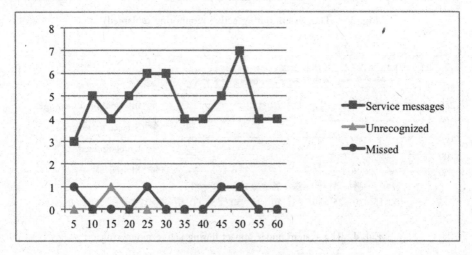

Fig. 2A. The system in the normal state (enlarged)

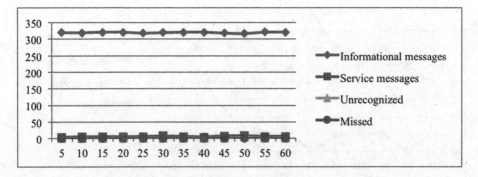

Fig. 3. The system during active monitoring

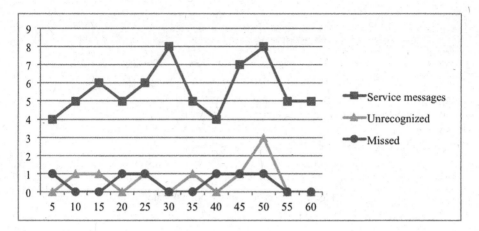

Fig. 3A. The system during active monitoring (enlarged)

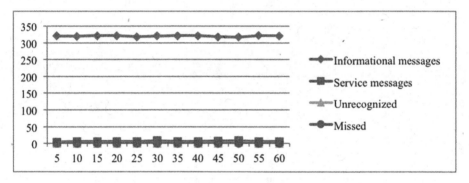

Fig. 4. The system under impact during active monitoring

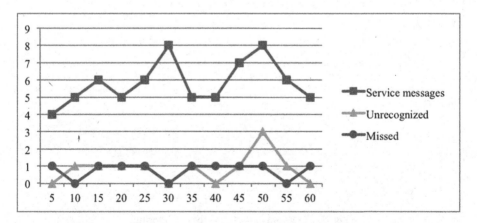

Fig. 4A. The system under impact during active monitoring (enlarged)

The standard tuple shown in Fig. 5 is based on statistical information obtained at certain time intervals related to the device functioning (Table 1).

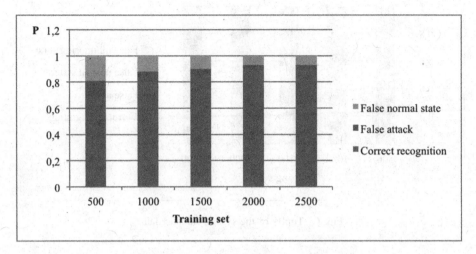

Fig. 5. Tuples of the system statistic data

Table 1. Tuples of system statistic data

Time	5 s	10 s	...	N s
Informational messages	320	319	...	321
Service message	3	5	...	5
Unrecognized	0	0	...	0
Missed	1	0	...	0
...				
Delay time per message	0.001	0.001	...	0.002

The number of action was carried out while receiving and analyzing of quantitative indicators of system service informational messages for different modes.

1. Switching the system to the required mode of operation for training;
2. Analysis of the system state characteristics;
3. Formation of the analyzed characteristics tuple;
4. Receiving messages from devices, accumulation of statistics and creation of the studied indices database;
5. Processing of accumulated statistical data, comparison with the received data and assessment of the current state.

Figures 5 and 6 present the results of the classifier for different system states.

To assess the qualitative characteristics it is necessary to choose different indicators and its groups. However, even the statistics obtained on the basis of the conducted experiment shows different kind of response of the analyzed system. Thus, it is sufficient for the probabilistic state determination.

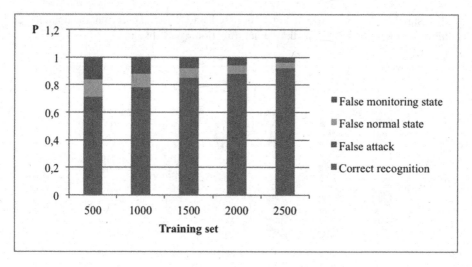

Fig. 6. Tuples of the system statistic data

5 Conclusion

The method of information security state monitoring of the of low-power device network segment on the basis of statistical data of the interaction of functioning control transceivers of the application-level of local information system is proposed. That allows obtaining probabilistic characteristics of the information security state. The sign of proposed approach is the ability to adapt quickly to the local networks of low-power devices produced by different manufacturers of Internet Of Things and Smart Home solutions.

To implement this type of monitoring it is not necessary to develop complex system applications.

References

1. Kumar, P., Ylianttila, M., Gurtov, A., Lee, S.-G., Lee, H.-J.: An efficient and adaptive mutual authentication framework for heterogeneous wireless sensor networks-based applications. MDPI Sens. **14**(2), 2732–2755 (2014)
2. Sridhar, P., Sheikh-Bahaei, S., Xia, S., Jamshidi, M.: Multi agent simulation using discrete event and soft-computing methodologies. Proc. IEEE Int. Conf. Syst. Man Cybern. **2**, 1711–1716 (2003)
3. Page, J., Zaslavsky, A., Indrawan, M.: Countering security vulnerabilities using a shared security buddy model schema in mobile agent communities. In: Proceedings of the First International Workshop on Safety and Security in Multi-agent Systems (SASEMAS 2004), pp. 85–101 (2004)
4. Zikratov, I.A., Lebedev, I.S., Gurtov, A.V.: Trust and reputation mechanisms for multi-agent robotic systems. In: Balandin, S., Andreev, S., Koucheryavy, Y. (eds.) NEW2AN/ruSMART 2014. LNCS, vol. 8638, pp. 106–120. Springer, Heidelberg (2014)

5. Wyglinski, A.M., Huang, X., Padir, T., Lai, L., Eisenbarth, T.R., Venkatasubramanian, K.: Security of autonomous systems employing embedded computing and sensors. IEEE Micro **33**(1), 80–86 (2013). Art. no. 6504448

6. Lebedev, I., Korzhuk, V.: The monitoring of information security of remote devices of wireless networks. In: Balandin, S., Andreev, S., Koucheryavy, Y. (eds.) NEW2AN/ ruSMART 2015. LNCS, vol. 9247, pp. 3–10. Springer, Heidelberg (2015)

7. Prabhakar, M., Singh, J.N., Mahadevan, G.: Nash equilibrium and Marcov chains to enhance game theoretic approach for VANET security. In: Kumar, M.A., Selvarani, R., Kumar, T.V. S. (eds.) Proceedings of ICAdC. AISC, vol. 174, pp. 191–199. Springer, Heidelberg (2013)

8. Bazhayev, N., Lebedev, I., Korzhuk, V., Zikratov, I.: Monitoring of the information security of wireless remote devices. In: 9th International Conference on Application of Information and Communication Technologies, AICT 2015 - Proceedings 7338553, pp. 233–236 (2015)

9. Nikolaevskiy, I., Lukyanenko, A., Polishchuk, T., Polishchuk, V.M., Gurtov, A.V.: isBF: scalable in-packet bloom filter based multicast. Comput. Commun. **70**, 79–85 (2015)

10. Al-Naggar, Y., Koucheryavy, A.: Fuzzy logic and Voronoi diagram using for cluster head selection in ubiquitous sensor networks. In: Balandin, S., Andreev, S., Koucheryavy, Y. (eds.) NEW2AN/ruSMART 2014. LNCS, vol. 8638, pp. 319–330. Springer, Heidelberg (2014)

11. Chehri, A., Hussein, T.: Moutah survivable and scalable wireless solution for e-health and emergency applications. In: Proceedings of the 1st International Workshop on Engineering Interactive Computing Systems for Medicine and Health Care, EICS4MED 2011, Pisa, Italy, pp. 25–29 (2011)

Improving the Efficiency of Architectural Solutions Based on Cloud Services Integration

Vladimir V. Glukhov, Igor V. Ilin[✉], and Oxana Ju. Iliashenko

Peter the Great St. Petersburg Polytechnic University, St. Petersburg, Russia
{vicerector.me, ilyashenko_oyu}@spbstu.ru,
ilyin.igor@eei.spbstu.ru

Abstract. Recently, many Russian companies are trying to transform their businesses through new Internet services to increase productivity, reduce costs, minimize risks and extend the capabilities of the enterprise. Increasingly such problems are solved with the help of media access to a variety of cloud-based services. The key features of such services are self-service on-demand, broadband network access, resource pooling in the pools, instant flexibility, measurability of the services provided. The acquisition of cloud solutions for enterprise becomes profitable alternative to their own IT infrastructure as cloud can save money for the hardware purchase and support. Cloud solutions help to adjust the amount of resources used in real-time, paying only for actual power consumed. In addition, this approach allows you to streamline business processes within the company, since most of the issues relating to IT, outsourced to an external provider. The paper investigates the possibilities to increase the efficiency of architectural solutions of IP-telephony company sphere by improving the quality of services sold. This paper analyzes the existing methods of valuation services. Method of assessing the call-centers efficiency has been selected. This method is based on the approved KPI business processes. The possibilities of the call-center optimization and improvements of KPI business processes by increasing the level of Software as a Services (SaaS). The main criteria of successful work of call-centers have been described in the paper, to explore the market of existing cloud-based solutions, to assess the possibility of introducing cloud services in architectural solution and to estimate the obtained figures.

Keywords: Cloud computing · Services · Architectural solution · The KPI business process · Call-centers

1 Introduction

Nowadays information and communication technologies and services are the key factor in the development of all spheres of public life. Most major companies and government agencies to handle the flow of calls use IP-telephony phone system – call-centers based on a computer-telephony integration.

Information and communication services provided by call-centers, provide support, promotion and expansion of businesses in different spheres of activity. In call-centers they use software and hardware system for the mass processing of incoming calls and outgoing calls from clients. Modern call centers are treated not only phone calls, but

© Springer International Publishing AG 2016
O. Galinina et al. (Eds.): NEW2AN/ruSMART 2016, LNCS 9870, pp. 512–524, 2016.
DOI: 10.1007/978-3-319-46301-8_43

also for the treatment of electronic and paper mail, fax, working with references in the online live chat. One of the main indicators of call-centers efficiency is to manage the flow of customer complaints, which directly determines the quality of customer service.

Call-center design concept joins Customer Relationship Management technology (CRM) and technology of Computer Telephony Integration (CTI). It is aimed at improving the efficiency of the organization by providing personalized customer service at the highest level, regardless of where and how a contact occurs. In terms of Information Technology Call Center is a telephone call center, the principle of construction is based on routing according to certain rules agent calls that are being developed in the company and allow efficiently and effectively serve customers, as well as maintaining a queue of calls, allowing customers not to hear busy signal".

The functionality of the call-center implemented on the basis of specialized application servers, the core of the system is a software solution. It is a communication platform with all the features of classical automatic telephone exchange. It provides extensive functionality to manage the flow of customer requests (calls), such as queue management, routing calls, interactive voice response, etc. The system also includes database servers [1]. Thus, the provision of call-center and transfer content are the result of the integration of hardware and software, which from the point of view of the enterprise architecture is the interaction process layer and the application layer through the infrastructure services [2]. Quality of service implemented directly affects the efficiency of data processing at the application layer, and thus the performance metrics for call-center as a whole, in particular, to call processing KPI business processes and the level of services provided to the customer. The key business process of call-centers is handling calls from customers. Thus, to improve the efficiency of key business processes you need to change the existing architectural solution by improving infrastructure services. One of the possible ways to improve services is transfer to a cloud application. The article deals with the architectural solution for call-center, and studied the effect of infrastructure services on the performance indicators [3].

2 Analysis of the Existing Architectural Solutions Call-Center

The service-oriented architecture plays an important role to analyze the architecture, which allows to evaluate the effectiveness of aligned layers, as well as the quality of the services that these layers are aligned [4].

Given the scale and complexity of enterprise architecture, such an analysis requires the selection of assessment methods. At the heart of each method is based on a specific algorithm that helps not only visually assess the built architecture, but also to make a comparison of alternative solutions and make informed management decisions. There are a number of methods and approaches that help architects, taking into account the views of stakeholders, to compare alternative architectures and therefore carry out an informed choice of design solutions (options) in the search for compromises between such aspects as cost, quality, and performance. In addition to these aspects, decision-makers need to understand the impact of changes in various architectural solutions.

For the analysis of the enterprise architecture are 4 methods: quantitative, functional, analytical and the method of modeling (simulation) [2]. The choice of method causes the approach to the analysis of architecture conducted to assess the effectiveness of the business operation as a whole and find bottlenecks [5]. Due to the nature of the selected companies – call-center – and a set of criteria that a company must meet, it was chosen as the standard valuation method for call-centers based on the approved key performance indicators of business processes. In this approach, the company holds monthly analysis of key performance indicators. Standard KPIs for the business processes are as follows:

1. SL (Service Level) – indicator to show the percentage of customers who have received the answer of the operator of the time, less than the set (the number of calls to the total number of calls);
2. LCR (Lost Call Rate) – an indicator indicating the percentage of customers who did not wait for an answer operator and dropped his challenge (the ratio of the number of lost calls, distributed in all the total number of calls);
3. FCR (First Call Resolution) – percentage of client's complaints who needed only one call for the provision of services [6].

After the analysis of business processes work within a certain time period, in 1 month for instance, we make a decision on carrying out changes to the existing architecture.

The first stage of the work was to analyze the work of call-center and the study of main business processes. For a survey of the business processes of the company must be:

1. Identify the main business process;
2. Identify and analyze key performance indicators of the business process;
3. Identify the "weak link" of architecture, to find a missing or unused items;
4. Describe the action to solve the problem and the expected results;
5. Implement the changes and to analyze the results.

Consider the existing algorithm of the call-center and call processing script in the call-center (Fig. 1). After receiving a call comes in caller ID. Then, the system determines the number of agents and a queue length of calls formed. If the queue is longer than the set value (in the call-center value is $n*2+1$, where n-number of operators), the call is cleared, thereby completing the call occurs. If the queue is less than the numerical value, the call is placed in the waiting queue and processed by the operator after the previous processing applications.

The main business processes of the call center are a call processing from the client and business service – automatic processing of the application (client request). Application service presented automatic processing of calls [7], which is being implemented Asterisk application component, which is a free decision of the CTI c open source software for creating communication applications [8]. Application component uses two service infrastructure: the definition of the number of operators and determining the length of the queue, which in turn are implemented by means of the appropriate hardware and software:

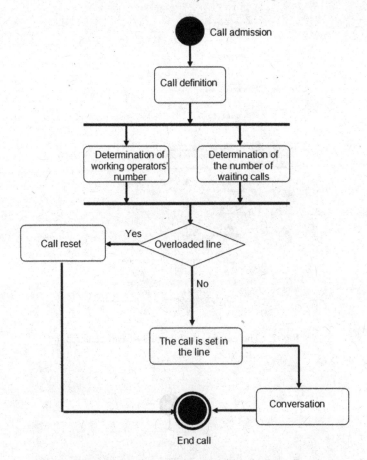

Fig. 1. The initial call processing algorithm in the call-center.

- Cluster infrastructure services with Asterisk interface;
- The two databases – cdr and queue_log;
- Devices – server installed a firewall and Internet connection.

The existing call center architecture is shown in Fig. 2.

Under the key business processes in the enterprise architecture we mean a set of interrelated activities or tasks designed to create a certain product or service to consumers [2]. The interaction of suppliers and consumers of services is carried out through business services. Business services, in turn, can be realized through the application services (services, showing the automated behavior) [9], which provides the application layer, and infrastructure services provided externally visible functionality and units related to the process layer.

In the basic architecture shown in Fig. 3, the main business process is processing a call that implements the business service "Automatic processing of applications" and applications used by service "Automatic calls processing". Automated call handling is implemented application components (modular, deployable piece of software systems)

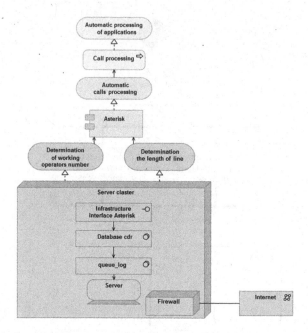

Fig. 2. The existing architecture of call-center

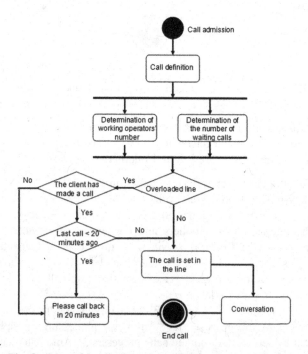

Fig. 3. Proposed call processing algorithm in the call-center.

Asterisk, which in turn uses two infrastructure services, "Determination of the working operators' number" and "Determination of the length of line". The implementation of these services performed by a node "Cluster Server", represented by the computing resource, combining software and hardware systems. Node "server cluster" consists of infrastructure interface (access point) Asterisk, used by system software, database and cdr queue_log, which are in turn used by the device "Server", which is a physical resource with data processing capabilities. Also the node element is a firewall that is associated with a network (communication environment) the Internet.

Analysis of the above performance indicators: SL (Service Level), LCR (Lost Call Rate) and FCR (First Call Resolution) showed that the productivity and efficiency indicators were low, therefore, requires optimization and reorganization of call-center's work. For it is proposed to change the existing application processing algorithm by modifying the services and transfer them to the cloud solution. In general, service modification may be accomplished in three different ways:

1. Changes in existing services and their characteristics;
2. The addition of new services;
3. Removal of services.

Application processing algorithm change cannot be based on the modification or removal of existing services, as they are the basic set of services in the architecture of the call-centers. To implement changes in the existing processing algorithm is proposed to integrate into a service-oriented architecture of the new service.

According to the criteria, the increased load on the channels often is due to subscribers, making many to call the call-center for a short period of time due to lack of information on the state of the line and the required waiting time. Now client, calling in less than 20 min after the previous call, will play the IVR (interactive voice response – a system of pre-recorded voice message): "All operators are busy. Please call no earlier than 20 min". Thus, we propose the following implementation of the algorithm: after receiving a call comes in caller ID. Then, the system determines the number of agents and a queue length of calls formed. If the queue is less than the numerical value, the call is placed in the waiting queue and processed by the operator after the previous processing applications. In a situation where all more than the set value, the call to the database, where re-checked subscriber appeal. If there is one, and the challenge was accomplished in less than 20 min ago, the subscriber is offered to call back in 20 min, and the call is terminated. In the case when the subscriber re-accesses, and after more than 20 min, it is placed in a call waiting queue operator response. When the operator answers the call completion occurs. Figure 3 shows the modified call processing script in the call-center.

The proposed algorithm can be implemented using the possibility of modifying the application services built on the traditional model, through the revision of the programming code. Another way is to convey the implementation of all services to cloud-based solution. There is a cloud computing platform adaptation expertise in the existing service-oriented architecture (SOA) [10–12].

3   Possibilities of Cloud Implementation of Services in the Solutions

Consider the feasibility of services at various levels with the use of cloud computing – the approach to the construction of information systems, in which the consumer uses required IT resources in the form of web services from external providers.

According to Genesys Laboratories [13] one of the main directions of development of the call-centers customer service industry is to move to the cloud: companies, that were first to apply the cloud, began the transition in 2012 – two years earlier than predicted. Now, a business of all sizes in all branches takes the cloud as a standard.

There are four main categories of cloud services that can be provided to the customer (Fig. 4): IaaS (rent of computing resources of data centers and storage systems), PaaS (rental environment for creating applications for software developers), SaaS (delivery of ready client software via a web browser) and BPaaS (provide business processes as cloud services). Cloud Solutions "substitute" for users their own information infrastructure, specific software and hardware platform, software or business services for the client. The first three listed areas are experiencing rapid growth. The greatest demand is seen on SaaS products, for which $ 39.8 billion was spent in 2014, and IDC analysts believe that by 2018 these costs will rise to $ 82.7 billion [14]. Volume of IaaS market will grow from $ 8.7 billion in 2014 to $ 24.6 billion by 2018, the corresponding indicators are submitted for PaaS – increase from $ 8.1 billion to $ 20.3 billion. BPaaS segment has appeared relatively recently, but according to Forrester Research forecasts, by 2020 it will exceed the traditional segment of the infrastructure cloud service IaaS and approach to the segment of PaaS.

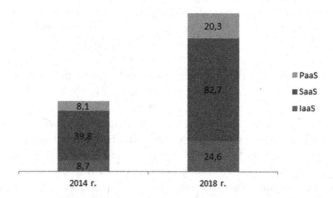

Fig. 4. Cloud market in 2014–2018 years, $billion *(Source: IDC, 2014)*

Nowadays, cloud service has three main characteristics that distinguish it from the normal service:

- Regime "on-demand resources";
- Elasticity;
- Resources association;
- Independence from the infrastructure controls.

4 Integration of Cloud Solutions in a Service-Oriented Call-Center Architecture

In the situation of migration of all the architectural layers in the cloud, target architecture would look like as follows (Fig. 5). However, there is a need to consider the feasibility of the transition to the use of cloud-based solutions in all layers of the architecture. In the framework of this architectural solution, there already exists IT-infrastructure, which has been building up over several years. In this connection, it is not always possible or appropriate to completely abandon the existing resources and move them to the cloud. Transference of services to the cloud allows the use of online cloud (SaaS) services for the organization of call-center activity without the need to purchase equipment and control systems. In the basic architecture of call-center under consideration, there is a standard component of Asterisk application implemented into IT infrastructure. Most cloud solutions realize the mechanism that is used to connect databases (MySQL, PostgreSQL) through an ODBC driver to the Asterisk server for the organization of incoming and outgoing telephone.

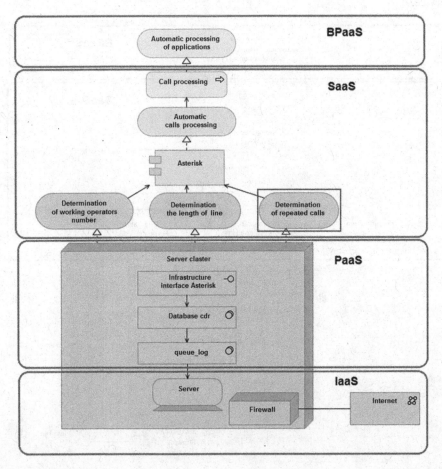

Fig. 5. Target architecture of call-center, which realizes all of the levels in cloud computing.

In addition, there are certain requirements for the control of call-center data. In such cases, it is advisable to talk about a hybrid cloud, when for solving specific tasks it is more efficient to use public clouds. For example, when it comes to new systems in the field of mobility, e-commerce, IP-telephony, analytics, or solving the problems of development and testing. If it is necessary to increase the efficiency of use of available IT-infrastructure, it is recommended to implement a private cloud. This will allow to reduce the cost of its maintenance and development, increase its flexibility and to reduce the terms of IT-projects. In general, the relation between the solutions, using both private and public clouds in the telecommunications sector, is approximately the same.

Based on the conducted analysis, it its proposed to implement a modification of service structure in terms of the inclusion of the new service "Definition of callbacks" and transfer the entire set of services ("Definition of callbacks", "Determination of the number of operators", "Determination of the queue length", "Automatic processing of calls", "Automatic processing of the application") to the cloud application.

Figure 6 shows the target architecture corresponding to the use of cloud-based solutions in terms of the implementation of business processes and software as a service.

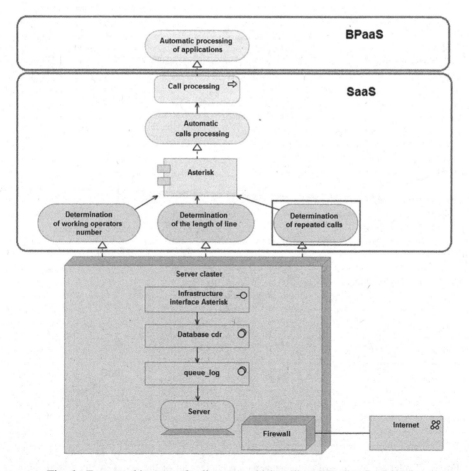

Fig. 6. Target architecture of call-center, which realizes BPaaS и SaaS levels.

Addition of a new service and distribution of services in the cloud application "Call Center" that provides call processing, lets you stream processing of incoming calls and outbound marketing campaigns. Call Center significantly expands the possibilities of virtual PBX, adding functions related to the monitoring of processing of customer services, operating personnel, service performance measurement. The call comes into the view of Call Center at the stage when the caller is listening to the voice menu (IVR), or inclusion in the compounds waiting queue with the employee. Besides, with the help of Caller ID and contact cards, the system instantly recognizes already known customers or partners, and displays information about them instead of phone numbers.

The proposed architecture can be implemented using a variety of solutions offered on the market of cloud computing [15, 16]: Mango Office, Naumen Contact Center, Terrasoft Contact Center, etc. All of them are compatible with the existing IT infrastructure of call-center and ensure the implementation of the required services. Implementation of the proposed architecture was realized using cloud solutions "Call Center" of the company Mango Office.

5 Results and Discussions

As a result of the implementation of services viewed above into cloud solutions, index (P) of dropped and not answered calls has decreased (Fig. 7), and the speed of serving the customers (less than 20 s) has increased (Fig. 8). Thus, a key indicator of the effectiveness of Service Level has increased, and the indicator of Lost Call Rate has decreased (Fig. 9).

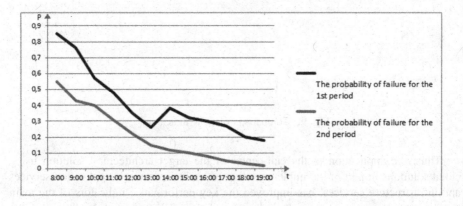

Fig. 7. Comparison of probabilities of failure.

Also during the second time period, key performance indicator "First Call Resolution" has increased (indicator of clients, who made the call for the first time) (Fig. 10).

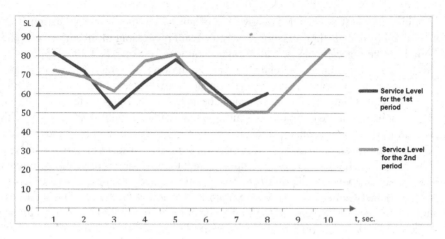

Fig. 8. Comparison of SL indicators.

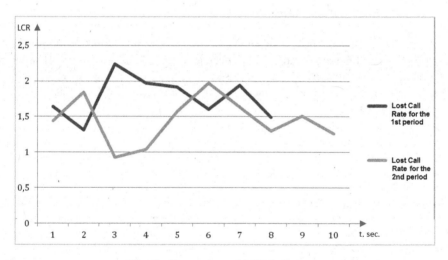

Fig. 9. Comparison of LCR indicators.

Thus, the application in the call-center of the target architectural solution using cloud solutions in part of the implementation of business services, application services and infrastructure services, has improved the key performance indicators of the main business process and, as a result, indicators of the efficiency of the call-center as a whole.

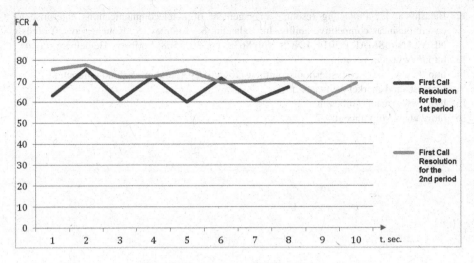

Fig. 10. Comparison of FCR indicators.

References

1. Organization of the call center: [electronic resource]. Encyclopedia KPI. http://dgalkin.com/material/kpi-service/. Accessed 30 Mar 2016
2. Lankhorst, M.: Enterprise Architecture at Work. Modelling, Communication, Analysis. Springer, Heidelberg (2013). 338 p.
3. Kapilevich, O.L., Markov, N.G.: Information business performance management system based on KPI. Bull. Tomsk Polytech. Univ. **317**(5) (2010)
4. Ilyin, I.V., Levina, A.I.: The integration of the project management approach into the business architecture model of the company. Sci. Tech. Vedomosti St. Petersburg State Polytech. Univ. Econ. **185**(6–2), 74–82 (2013). C-15
5. Shirokova, S.V.: Project Management. Project Management Implementation of Information Systems for the Enterprise: Textbook. Benefit/S.V. Shirokova – SPb: Publishing House of the Polytechnic University Press, 56 p. (2012)
6. KPI Performance CALL-centers: [electronic resource]. Official site of the Company "NORBIT". http://www.norbit.ru/press/articles/1869.html. Accessed 30 Mar 2016
7. Goldstein, B.S., Pinchuk, A.V., Suhovitsky, A.L.: IP-telephony. In: Goldstein, B.S., Pinchuk, A.V., Suhovitsky, A.L.-M. (eds.) Radio and Communications, 336 p. (2001)
8. Bembieva, A., Namiot D.: Voice dialogs for Asterisk. Int. J. Open Inf. Technol. **1**(7) (2013)
9. Ilyin, I.V., Ilyashenko, O.J., Levina, A.I., Shirokova, S.V., Dubgorn, A.S.: Formation project for the integration of large data processing technologies in enterprise architecture. Science Week SPbPU Materials Science Forum with International Participation. Interdisciplinary Sections and Plenary Meetings of the Institutions. St. Petersburg, pp. 92–102 (2015)
10. Zhang, L.J., Zhou, Q.: CCOA: cloud computing open architecture. In: IEEE International Conference on Web Services, ICWS 2009, pp. 607–616 (2009)
11. Gluhov, V.V., Ilin, I.V.: Project portfolio structure in a telecommunications company. In: Balandin, S., Andreev, S., Koucheryavy, Y. (eds.) NEW2AN/ruSMART 2014. LNCS, vol. 8638, pp. 509–518. Springer, Heidelberg (2014)

12. Balashova, E.: Projecting resource management of a telecommunications enterprise to ensure business competitive ability. In: Balandin, S., Andreev, S., Koucheryavy, Y. (eds.) NEW2AN/ruSMART 2014. LNCS, vol. 8638, pp. 502–508. Springer, Heidelberg (2014)
13. http://www.genesys.com
14. http://www.forbes.com/sites/louiscolumbus/2015/01/24/roundup-of-cloud-computing-forecasts-and-market-estimates-2015
15. http://www.mango-office.ru
16. http://www.voipoffice.ru

Optimization of Selection Strategies for P2P Streaming Network Based on Daily Users' Behavior and Users' Distribution over Time Zones

Yuliya Gaidamaka and Ivan Vasiliev[(✉)]

Department of Applied Probability and Informatics,
RUDN University, Moscow, Russia
{ygaidamaka,ivasilyev}@sci.pfu.edu.ru

Abstract. In this paper, an optimization problem of selection strategies for peer-to-peer (P2P) live streaming network is discussed. To solve the problem, the simulation model of P2P live streaming network is developed. The model considers daily peers behavior, their distribution over time zones, collisions, time lags between the server and a peer, lags between peers, and three types of selection strategies: neighbor selection strategy, peer selection strategy, and chunk selection strategy. Daily peers' behavior is defined as the distribution of the number of online users by the time of day. Initial data for the peers distribution over time zones and their daily behavior are taken from the known Internet sources. The aim of the research is to find an appropriate solution of the proposed optimization problem and to show how the choice of a certain set of selection strategies affects the key characteristics of P2P streaming networks. The results of the conducted numerical analysis show the increase of the network performance up to 16,25 %.

Keywords: Peer-to-peer · P2P · Live streaming network · Playback continuity · Daily users' behavior · Users' distribution over time zones · Optimization problem · Selection strategy

1 Introduction

Major service providers in the online TV market, such as Zattoo, Pulse, iPlayer and others, make use of P2P technology to provide their top-level services with minimum expenses [13–15]. A fairly complete overview on P2P technical aspects with an extensive list of references was done by Yue et al. [3]. One of the most remarkable advantages of P2P networks is high performance-cost ratio. It allows for-profit companies to minimize expenses of technical equipment. More importantly, P2P technology not only enables efficient network resources use, but also

The reported study was partially supported by RFBR, research project No. 14-07-00090.

O. Galinina et al. (Eds.): NEW2AN/ruSMART 2016, LNCS 9870, pp. 525–535, 2016.
DOI: 10.1007/978-3-319-46301-8_44

reduces the load on the server that is the source of the video data. Other advantages of P2P technology are specified in [1,2,14,16].

Per contra, P2P networks also have disadvantages. The main weaknesses include relatively long start-up latency, a rather big data transmission delay, insufficient level of security, inter-peer playback lag, and playback discontinuity. While information security is not required, the goal is to minimize the transmission delay, and hence lack of data in the network [3,4,9,10,17]. To solve the optimization problem for these performance measures, mathematical models and simulators should be developed. Known mathematical and simulation models for streaming P2P-network pay much attention to investigation of the buffering mechanism [13–16] and usually take into account lags [5,6]. A lag means the delay of data transmission from the server to the peer, as well as the data transmission delays between peers (inter-peer playback lag). These models allow to carry out a qualitative analysis of the key performance measures of video streaming P2P networks, i.e. the probability of playback continuity, that is the probability of watching video with no pauses.

In order to get closer to reality and assess the adequacy of the individual models, to give at least the recommendations against some of the problems before they occur in a real P2P networks, we propose a new model that can help to avoid troubleshooting. The presented model takes into account the geographical location of each peer, remoteness of peers from the server and from each other and their daily behavior. In contrast to recent results (see, i.g. [13]), in this paper, the problem is solved taking into account the distribution of users across time zones.

Previously, we have studied various models of streaming P2P networks that have focused on the study of the optimal downloading strategy with the criterion of maximizing the probability of playback continuity of the video stream [13–16]. Then, the model was modified to take into account the users' behavior by introducing the probability of arrival of new users, as well as the probability of their leaving the network. The negative effect of peers churns on playback continuity was observed and investigated [4]. The analytical model [14,16] also gives the correct understanding of lags' impact on the network performance.

Previously, in [13], we showed the impact of daily peers behavior and their distribution over time zones on the key performances of P2P streaming networks. The difference in results between known basic model of P2P streaming network and a new model with peers distribution over time zones is up to 30 %. Thus, we have shown the need for additional analysis, taking into account a variety of new network parameters and, above all, the choice of optimal selection strategies: neighbor selection strategy, peer selection strategy, and chunk selection strategy. To do this, we organize the article as follows. In Sect. 2, a simulation model of a P2P live streaming network is discussed. The model takes into account video data buffering mechanism, daily peers' behavior and peers distribution over time zones. Also, a detailed algorithm of chunk exchange between peers in P2P live streaming networks is described and the main performance measures are defined. In Sect. 3, the motivation of selection strategies optimization is submitted. Here, the description of selection strategies is presented, including neighbor selection strategy, peer selection strategy, and chunk selection strategy. Also, an optimization problems description is formulated. Here, a mission statement is described, and all the ins and outs of solution research are delineated. In Sect. 4, an optimization approach is proposed. In Sect. 5, the numerical analysis and case study is performed. The conclusion of this paper is presented in Sect. 6.

2   Model Description

In this section, the simulation model of video data distribution in a P2P live streaming network with buffering mechanism is summarized. Previously developed model [13] was improved by taking in consideration buffer selection strategies, peers' geolocation and behavior [12,13]. In contrast to the previous model, in this paper, besides chunk selection strategy, two more strategies were introduced - neighbor selection strategy and peer selection strategy. The choice of strategies has a significant effect on the P2P-network performance measures, including the probability of playback continuity, the probability of chunk availability, and the probability of chunk selection [7,16].

We consider the basic model of a P2P network with N peers and a single server, transmitting only one video stream, which we developed in [16]. The process of video stream playback is divided into time slots, the length of each time slot corresponding to the playback time of one chunk. Each peer has a buffer designed to accommodate $M + 1$ chunks, where the buffer positions are numbered from 0 to M: 0-position is to store the freshest chunk just received from the server, the other M-positions, $1..M$, are to store chunks, already received during the past time slots or that will be downloaded in the coming time slots. The buffer M-position is to store the oldest chunk that will be moved out from the buffer for playback during the next time slot. Than, a state of n-th peer is represented in the form of a vector $\mathbf{x}(n) = (x(n,0), x(n,1),.., x(n,M))$, where $x(n,m) = 1$ if the n-th peer has a chunk at the buffers position m, and $x(n,m) = 0$ otherwise [1,15,16].

We consider a predetermined set of original parameters for each of N peers. They are upload u and download d rates, value lag of a lag, and a set $\mathbf{B}$ of neighbors [8,11,13]. A neighbour is a peers from which one can download data. A lag is the number of time slots between sending and receiving entities, thus, lag reflects the quantitative characteristics of a chunk delay. The algorithm works in such a way that, within a group, the peer selects a neighbor to download data from using the criterion of minimum lag between the neighbors, regardless of their time zones distribution. As an example, suppose that $Peer1$ is located in Poland and its neighbor, $Peer2$, is in Moscow, i.e. they are from different time zones. Suppose that $Peer3$ is located in Angola, in the same time zone as $Peer1$, and they are also neighbors. In this example, $Peer1$ selects $Peer2$ because they have the smallest lag, although $Peer2$ is located in a different time zone. The set of neighbors for each peer is formed according to the neighbor selection strategy depending on upload and download rates and lags. The neighbor selection is one of the target function parameters for optimization problems.

There are three types of chunk selection strategies: Latest First (LF) according to which a peer selects the newest chunk in the network; Greedy (Gr) is a strategy where the closest chunk to playback is selected; Mixture (**Mi**) that is a combination of the previous ones. The algorithm of peers actions at each time slot is described below according to the protocol of data distribution in P2P live streaming networks [15].

1. At the beginning of each time slot, the chunk at the M-position of the buffer is going to be played if it is present. Video data in the buffers shifts one position towards the end of the buffer. 0-position is nulled.
2. The server randomly chooses a peer and loads the newest chunk to 0-position of its buffer.
3. Each peer that was not chosen by the server selects a target peer from the set of neighbors to download a chunk during the current time slot. Target peers selection is carried out in accordance to the peer selection strategy.
4. If collision takes place, the peer gets nothing during the current time slot. Otherwise, it selects a chunk to download according to the chunk selection strategy. If there is an available chunk to download, the loading starts. Otherwise, the peer gets no chunks during the current time slot.

It should be noticed, that in a real network each peer is able to join the video stream and to disjoin it at any time slot and at any step of the algorithm. We say that a peer joins or disjoins immediately after the first step of the algorithm. The first difference from the basic model [13] is that in the model three strategies were considered: neighbor selection strategy, peer selection strategy, and chunk selection strategy. The second difference is that the presented model takes into account peers' geolocation and twenty-four hours peers' activity [12,13].

The number of peers in the network is not constant. Every peer stays online a random amount of time each day, with the average value of $0 < HO < 86400 = 24\,h \times 60\,min \times 60\,s$: $HoursOnline(n) \sim P(HO)$, $n = 1, ..., N$. Here, $86\,400$ is the number of time slots when modelling one astronomical day with one second as one time slot: $24\,h \times 60\,min \times 60\,s = 86400$ time slots.

Let us introduce the parameter of users' activity, which reflects the behavior of peers in the streaming network: $UserActivity(n) \sim RAND(1..UA), n = 1, ..., N; UA > 1$, where UA is the maximal number of peer's joining the network within a day.

This parameter shows how often a peer joins the network, disjoins it and switches channels. Here, $UserActivity(n) = 1$ means that within a day n-th peer once came to a network and was online during the random time $HoursOnline(n)$ without switching to other channels. $UserActivity(n) = i$, $1 \leq i \leq UA$, means that the n-th peer joined the network i times per day including switching channels, and each session lasted exactly $HoursOnline(n)/i$ time slots. Thus, e.g. if $UserActivity(n) = 100$ and $HoursOnline(n) = 15\,000$ the n-th peer per day during 15 000 time slots (seconds) carries out 100 sessions of 150 time slots each [13].

Figure 1 presents the distribution of the number of online peers versus the time of the day [13]. To simplify the modeling in this research, we investigate the influence of TV watchers behavior only. The graph shows that the peak of the online users is between 6 p.m. and 12 a.m. While the minimum number of TV watchers is from 2 a.m. to 7 a.m.

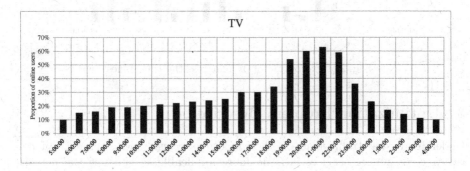

Fig. 1. Distribution of users in the network

In accordance with the distribution in Fig. 1, parameters $UserActivity = (UserActivity(n))$, $n = 1, ..., N$, and $HoursOnline = (HoursOnline(n))$, $n = 1, ..., N$, correspond to randomly generated intervals when peers are online. Let $TimeOnline = (TimeOnline(n, t))$, $n = 1, ..., N$; $t = 1, ..., T$, be a binary matrix of the $N \times T$ size, where T is the number of time slots in the simulation. The matrix indicates time slots when peers are online: $TimeOnline(n, t) = 1$ if the n-th peer at the t-th time slot is online, and $TimeOnline(n, t) = 0$ otherwise. Thus, if $TimeOnline(n, t) = 1$ and $TimeOnline(n, t + 1) = 0$, then the n-th peer left the network at the $(t + 1)$-th time slot, and $TimeOnline(n, t) = 0$ and $TimeOnline(n, t + 1) = 1$ say that the n-th peer joined the network at the $(t + 1)$-th time slot.

Peers churns significantly influence the key performance measures. So, when a new peer joins the network it still has no data for exchange with other peers, but it uses other peers resources to download content. Similarly, when a peer disjoins the network, it stops to participate in distribution of already downloaded video chunks. For a proper peers churns simulation, it is important to take into account the distribution of users over time zones. In this paper, the total of N peers in the network are divided in a random way across time zones according to the distribution shown in Fig. 2 [13]. One can see that the majority of the users is located in -5, $+1$, and $+8$ time zones, which include the USA, Canada, Europe, and China that are the most populated and technologically developed regions. In the model, splitting peers across time zones allows to reduce the probability of the global splashes corresponding to mass connections and disconnections of users.

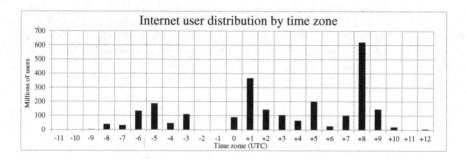

Fig. 2. Distribution of users across the time zones

3   Optimization Problem Description

Under increasing traffic load on the modern networks and emergence of new resource-consuming services it is no more applicable to enhance the network and computing capacities by means of new equipment procurement only. It is crucial to take all possible advantages of alternative ways to cope with the ongoing increase of gross adds. Below, we propose one of approaches to handle the problem.

Since the number of peers cannot be limited, while the size of peers buffer is fixed, and the upload and download speeds and lags are random values and cannot be managed on the application layer of the OSI/ISO model, the service providers can variate some other parameters of P2P streaming network.

We propose to explore the impact of selection strategies and to solve the optimization problem if it makes financial sense in terms of CAPEX/OPEX ratio. It is known that there are three selection strategies in P2P streaming networks: neighbor selection strategy, peer selection strategy, and chunk selection strategy. In this paper we analyze the influence of neighbor selection strategy and chunk selection strategy to reach the desired goals while leaving the peer selection strategy for the further studies. Further, we provide the description of the optimization problem, while in Sect. 4, an approach of seeking the suboptimal strategy is proposed and, for the sake of brevity, not described in details, but illustrated only. In fact, this approach is an expert decision, and the numerical results show its reasonability.

Let us set up the optimization problem. For the fixed input data, it is necessary to increase the probability of playback continuity PV by manipulating the parameters of neighbor selection strategy and chunk selection strategy.

It is to be recalled that $\mathbf{CSS} \in \mathbf{Mi}(CSS_1, CSS_2, M^*), \mathbf{Mi}(CSS_2, CSS_1, M^*)$, where $CSS_i \in \{LF, Gr\}, i = 1, 2$. The demarcation point M^* is the index of the buffer position. We split the buffer with M^* and apply alternately first CSS_1 and second CSS_2 strategies on corresponding sides of the buffer. The index M^* varies within the range $1 \le M^* \le M$. In case $M^* = 1$ or $M^* = M$, the **Mi** chunk selection strategy becomes one of the typical $LF = \mathbf{Mi}(LF_1, Gr_2, 40) = \mathbf{Mi}(Gr_2, LF_1, 1)$ or $Gr = \mathbf{Mi}(Gr_1, LF_2, 40) = \mathbf{Mi}(LF_2, Gr_1, 1)$ strategy depending on the major chunk selection strategy. LF is the Latest First chunk selection strategy, when a chosen chunk to download is the rarest chunk in the network, while the Gr (Greedy) is the chunk selection strategy, according to whitch a chosen chunk to download is the most popular chunk. Sinse $1 \leq M^* \leq M$, the total amount of possible chunk selection strategies is equal to $4 \cdot M$.

The neighbor selection strategy **NSS** is a rule in accordance to which peers from different groups are included into peers neighbor lists,

$$
\mathbf{NSS} = \begin{pmatrix} NSS\,(1,1) \ldots NSS\,(1,I) \\ \vdots \quad \ddots \quad \vdots \\ NSS\,(I,1) \ldots NSS\,(I.I) \end{pmatrix},
$$

where $NSS(i,j)$ is the number of peers from j-group that are included into the neighbor list of a peer from i-group, $1 \leq i, j \leq I$, where I is the number of peers groups that differ in the lag value. Note that the sum of the row is equal to B

$$
\sum_{i=1}^{I} NSS\,(i,j) = B, i = 1, ..., I,
$$

and the elements of matrix **NSS** satisfy the inequality

$$
0 \leq NSS\,(i,j) \leq B, i, j = 1, ..., I.
$$

Thus, for example, for $I = 3$ groups, $NSS(1,1)$ peers from the 1st group, $NSS(1,2)$ peers from the 2nd group, and $NSS(1,3)$ peers from the 3rd group are included into the neighbor list of each peer from the 1st group.

Let us find the number of possible variants to compose a neighbor list. For each i-th group, $1 \leq i \leq I$, it is acceptable to interpret the problem of neighbor list generation as a combination problem: "How many ways are there to distribute B neighbors among I groups". It is easy to see that the number of the variants equals to $\frac{(B+I-1)!}{(I-1)!B!}$. Since the generation of a neighbor list for each group of peers is independent, the total amount of variants to compose a neighbor list equals to $\left(\frac{(B+I-1)!}{(I-1)!B!} \right)^I$.

If to join the neighbor selection strategy and chunk selection strategy, then the total amount of different combinations equals to $4\,(M+1) \left(\frac{(B+I-1)!}{(I-1)!B!} \right)^I$. For example, it is $\frac{(M+1)(B+1)^3(B+2)^3}{2} \approx 1,1 \cdot 10^{11}$ for $I = 3$, buffers size $M = 40$, and the number of neighbors $B = 60$.

Since we define the probability of playback continuity PV as a key performance measure, we formulate the optimization problem

$$
F = PV\,(\mathbf{NSS}, \mathbf{CSS}) \to \max
$$

where PV is a probability of playback continuity. The optimization problem is a nondeterministic polynomial time complete problem. In next section an approach of finding a suboptimal strategy **NSS** and optimal strategy **CSS** is described.

4   Optimization Approach

Neighbor selection strategy and chunk selection strategy are mutually independent. It means that the optimization problem may be considered as composition $F = F_1 \circ F_2$, where $F_1 = PV\,(\mathbf{NSS}) \to \max$ and $F_2 = PV\,(\mathbf{CSS}) \to \max$.

Subproblem $F_2 = PV\,(\mathbf{CSS}) \to \max$ can be solved by searching the values of the dependent **CSS** parameter. To find the suboptimal solution of the subproblem $F = F_1 \circ F_2$, where $F_1 = PV\,(\mathbf{NSS}) \to \max$, we fix the number of peers groups, taking $I = 3$. The peers lags in the groups equal to $lag(n) \in \{0, 10, 20\}, n = 1, ..., N$, respectively. For the neighbor selection strategy $\mathbf{NSS} = \begin{pmatrix} B/3\ B/3\ B/3 \\ B/3\ B/3\ B/3 \\ B/3\ B/3\ B/3 \end{pmatrix}, B = 60$, it is known [15] that the peers from the 1st group with $lag(n) = 0$ have the smallest probability of playback continuity PV. For this reason we keep the peers distribution in the neighbor list for the 2nd and 3rd groups of peers $NSS\,(i, j) = B/3, i, j = 2, 3$.

The proportion of peers from all the groups in the neighbor list for the 1st group of peers is varied within the range from 0 to B with the step 4: $0 \leq NSS\,(1, j) \leq B, j = 1, ..., 3$ as set forth below and illustrated in the corresponding diagrams.

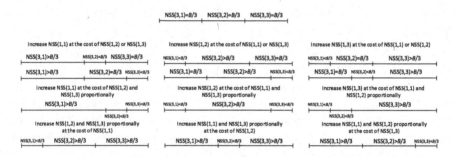

Fig. 3. Illustration of approach for finding suboptimal solution

Having found the optimal proportion of peers from different groups in the neighbor list of the 1st group, keeping the found result, we carry out the same numerical experiment for peers from the 2nd group in the same way. As we obtained the proportion of neighbors for the 2nd group of peers we carry out the same numerical experiment for the 3rd group. Thus, we have formulated a simple and obvious way to find suboptimal solutions of the optimization problem, and now we can proceed to its numerical analysis.

5   Numerical Analysis

The aim of numerical experiment is to find the optimal chunk selection strategy and a suboptimal neighbor selection strategy. Earlier, the results of P2P

streaming network simulations with ordinary set of strategies have been analyzed in [14,15] and discussed in [13,16]. The following inputs have been chosen
for modelling: the number of peers is $N = 300$; the size of peers buffer is $M = 40$;
the upload rate of every peer is $u = N - 1$; the download rate of every peer is
$d = 1$; the number of time slots is $T = 10^6$. As 1 time slot equals 1 sec, T
corresponds to about 12 astronomical days; the values of peers' lags are 0, 10,
and 20. As for the peer selection strategy, choice of target peer is random and
equiprobable. First of all, let us find optimal chunk selection strategy. As shown
in Fig. 4 the optimal chunk selection strategy is $\mathbf{Mi}(LF_1, Gr_2, 8)$ which deliv-
ers the highest value of PV. Likewise implementation of the chunk selection
strategies $\mathbf{Mi}(LF_1, Gr_2, M^*)$ with $M^* = 6, ..., 14$ shows quite good results. The
increase of the probability of playback continuity is up to 3 % against LF chunk
selection strategy, and up to 10 % against Gr strategy.

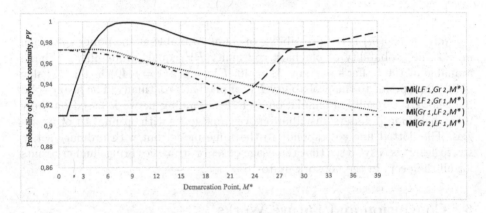

Fig. 4. Probability of playback continuity

In the next step, we numerically show the existence of suboptimal neighbor
selection strategy. As it is described in Sect. 5, we found suboptimal neighbor
selection strategy step by step by means of simulation:

$$\mathbf{NSS}: \begin{pmatrix} B/3 & B/3 & B/3 \\ B/3 & B/3 & B/3 \\ B/3 & B/3 & B/3 \end{pmatrix} \rightarrow \begin{pmatrix} B & 0 & 0 \\ B/3 & B/3 & B/3 \\ B/3 & B/3 & B/3 \end{pmatrix} \rightarrow$$

$$\rightarrow \begin{pmatrix} B & 0 & 0 \\ B/2 & B/2 & 0 \\ B/3 & B/3 & B/3 \end{pmatrix} \rightarrow \begin{pmatrix} B & 0 & 0 \\ B/2 & B/2 & 0 \\ B/3 & 0 & 2B/3 \end{pmatrix}.$$

In Fig. 5, probability of chunk availability increases up to 16 % due to implementations of optimal chunk selection strategy and suboptimal neighbor selection
strategy against the basic model with basic input data.

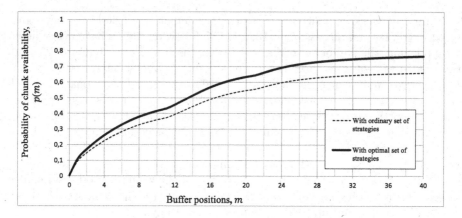

Fig. 5. Probability of chunk availability

Note that, $p(m)$ is the probability of chunk available at the buffer's position m. So, the probability of playback continuity PV equals probability of chunk available at the buffer's position $M = 40$: $PV = p(M) = p(40)$ in Fig. 5, value $p(40)$ corresponds to the probability PV of playback continuity. The heavy solid line shows the results of simulation of P2P streaming network based on daily users' behavior and users' distribution over time zones with optimal set of strategies. The dotted line corresponds to the same model but with ordinary set of strategies. It is easy to see that the usage of set of strategies brings higher results for all the peers in the network up to 16 %.

6   Conclusion and Future Works

We try to construct a model of a P2P streaming network, which is the most approximate to the reality. Preliminary numerical analysis shows that more research is necessary to find the optimal strategies, buyout will improve the performance of the network. To do this, it is necessary to formulate the appropriate optimization problems, to find ways to solve them, even if numerical, and conduct computer experiments using the simulator described above. It is already clear that it is necessary to modify the strategies used in P2P streaming network.

Acknowledgment. The reported study was partially supported by the RFBR, research project No. 14-07-00090. The authors gratefully thank Prof. Konstantin Samouylov for initiating this research and very valuable advice on research design.

References

1. Zhao, Y., Shen, H.: A simple analysis on P2P streaming with peer playback lags. In: Proceedings of the 3rd International Conference on Communication Software and Networks, Xian, China, pp. 396–400 (2011)

2. Gu, Y., Zong, N., Ed., Zhang Y., Piccolo F., Duan S.: Survey of P2P streaming applications. In: IETF, pp. 1–22 (2015)
3. Yue, G., Wei, N., Liu, J., Xiong, X., Xie, L.: Survey on scheduling technologies of P2P media streaming. J. Netw. **6**(8), 1129–1136 (2011)
4. Hei, X., Liu, Y., Ross, K.W.: IPTV over P2P streaming networks: the mesh-pull approach. IEEE Commun. Mag. **46**(2), 86–92 (2008)
5. Payberah, A.H.: Live streaming in P2P and hybrid P2P-cloud environments for the open internet. Doctoral thesis in Information and Communication Technology Stockholm, Sweden, pp. 1–119 (2013)
6. Li, X., Loguinov, D.: Stochastic models of pull-based data replication in P2P systems. In: 14-th IEEE International Conference on Peer-to-Peer Computing, pp. 1–10 (2014)
7. Abinaya, R., Ramachandran, G.: Efficient p2p video sharing scheme in online social network. Int. J. Eng. Sci. **3**(5), 23–27 (2014)
8. Bradai, A., Ahmed, T., Boutaba, R., Ahmed, R.: Efficient content delivery scheme for layered video streaming in large-scale networks. J. Netw. Comput. Appl. **45**, 1–14 (2014)
9. Medjiah, S., Ahmed, T., Boutaba, R.: Avoiding quality bottlenecks in P2P adaptive streaming. IEEE J. Sel. Areas Commun. **32**(4), 734–745 (2014). Institute of Electrical and Electronics Engineers
10. Alghazawy, B.A., Fujita, S.: A scheme for maximal resource utilization in peer-to-peer live streaming. Int. J. Comput. Netw. Commun. **7**(5), 13–28 (2015)
11. Li, S., Ya, L., Wang, A.: A social network based bandwidth sharing model for P2P streaming service. Int. J. u- e- Serv. Sci. Technol. **8**(2), 171–180 (2015)
12. Li, Z., Kaafar, M.A., Salamatian, K., Xie, G.: User behavior characterization of a large-scale mobile live streaming system. In: International World Wide Web Conference Committee, Italy, pp. 307–313 (2015)
13. Gaidamaka, Y., Vasiliev, I., Samouylov, K., Shorgin, S.: An approach to the modeling of the P2P streaming network based on peers' geolocation and activity. In: Proceedings of the 10th International Conference on Digital Society and eGovernments, Venice, Italy, pp. 13–17 (2016)
14. Gaidamaka, Y., Samouylov, K., Shorgin, S., Medvedeva, E., Vasiliev, I.: Optimizing performance measures by peer selection strategy in P2P streaming network. In: Proceedings of the 27th European Conference On Operational Research, Glasgow, Great Britain, p. 96 (2015)
15. Gaidamaka, Y., Samuylov, A.K., Medvedeva, E.G., Vasiliev, I., Abaev, P.O., Ya, S.S.: Design and software architecture of buffering mechanism for peer-to-peer streaming network simulation. In: Proceedings of the 29th European Conference on Modelling and Simulation, Germany, Digitaldruck Pirrot GmbH, pp. 682–688 (2015)
16. Gaidamaka, Y., Vasiliev, I., Samuylov, A., Samouylov, K., Shorgin, S.: Simulation of buffering mechanism for peer-to-peer live streaming network with collisions and playback lags. In: Proceedings of the 13th International Conference on Networks, Nice, France, pp. 86–91 (2014)
17. Ometov, A., Masek, P., Urama, J., Hosek, J., Andreev, S., Koucheryavy, Y.: Implementing secure network-assisted D2D framework in live 3GPP LTE deployment. In: 2016 IEEE International Conference on Communication Workshops (ICC), Kuala Lumpur, Malaysia, pp. 749–754 (2016)

NEW2AN: SDN

Comprehensive SDN Testing Based on Model Network

Andrei Vladyko, Ammar Muthanna$^{(\boxtimes)}$, and Ruslan Kirichek

State University of Telecommunication, St. Petersburg, Russia
{vladyko,kirichek}@sut.ru, ammarexpress@gmail.com

Abstract. Software-defined networking (SDN) can significantly automate and facilitate network management by allowing their programming. In this paper we show the SDN concept, method of testing, and the main characteristics of the SDN controller. The study based on the model network in the laboratory high-speed backbone networks DWDM and programmable networks SUT. The paper presents results of a collaborative functioning of SDN and DWDM-network. In this paper we created graphical interface over Cbench and added several advanced features in C++, also conducted stress testing SDN controllers.

Keywords: SDN · Network modeling · Testing

1 Introduction

The concept of software-defined networking (SDN) [1, 2] suggests the separation of data transfer and data management; logically centralized data management level; virtualization of physical network resources; as well as a single, unified interface (Open Flow) between the control plane and data plane (not depend on the vendors) [3, 4]. The following levels are distinguished in the SDN network structure [5]. The first one is the level of networking applications, at this level are implemented various network management functions like management of data flows in the network, security management, traffic monitoring, QoS management, policy management, and so on. The control Level maintains network topology, and also provides unified interface to application level. Network infrastructure level includes network device of the SDN network (Open Flow switches) and data channels. The controller is a key element of the SDN network, it acts as the brain of the entire network. Performance and network capabilities are directly related to the controller characteristics. The controller itself is a network operating system installed on a dedicated physical server. The main controller functions include management of network devices; topology management (construction of network topology, add/delete new network elements), application management and control of available server resources (threads, cores). The approach to SDN allows to automate and facilitate network management by allowing their 'programming', to build a flexible, scalable networks that can easily adapt to the changing conditions of functioning and the needs of users. Implementation of this approach, in particular, should have a great impact on the management of the network infrastructure in the data center (DC), corporate networks, WAN, home networks and mobile cellular networks

© Springer International Publishing AG 2016
O. Galinina et al. (Eds.): NEW2AN/ruSMART 2016, LNCS 9870, pp. 539–549, 2016.
DOI: 10.1007/978-3-319-46301-8_45

5G, for the efficient allocation of radio resources by centralizing, and seamless mobility using different technologies and common control plane.

Model network is a prototype of the current and projected network, which is built on the appropriate equipment with a given level of abstraction [6]. Usage of such a network can carry out comprehensive testing of equipment, both in normal functioning and under load with stress regimes [7–9]. Model networks can be used for testing the hardware and software products and applications that are parts of existing and future networks [10–12]. The rest of paper is organized as follows. Section 2 analyzes controllers involved in the study and their characteristics. Section 3 shows the experimental investigation and model network. Section 4 presents testing results. Section 5 describes the methodology allowed to evaluate the basic characteristics of SDN network controller. Last Sect. 6 is conclusion, which summarized our work.

2   Controllers Involved in the Study and Their Characteristics

Floodlight Controller open source, supported by an open community of developers. It's developed on the basis of Beacon Controller platform in Java, has the Apache license which can be used for any purpose. Open Daylight is an open design with a modular and flexible platform. This controller is implemented in software, within its own virtual java-machine (jvm). Thus, it can be deployed on any operating system and hardware platform that supports Java. Mul is designed in C, has a multithreaded kernel-level infrastructure. It supports multi-level interface for network applications. The main objective in the development of this controller was to ensure the performance and reliability you need when deploying heavy-loaded networks. Beacon Cross-platform, modular Open Flow controller in Java. Beacon is used in many research projects and test deployments. Beacon is used in an experimental data center at the Stanford, where he runs 100 virtual and physical switches 20. It works on multiple platforms, ranging from high-performance multi-core Linux-based servers to Android smart phones. Maestro network operating system developed in Rice University. Maestro provides interfaces to implement modular network management applications to access and modify the network status, as well as coordination of their interaction. Despite the fact that this project is aimed at creating Open Flow controller, Maestro is not limited to Open Flow-networks. Maestro is developed on Java (the platform itself and its components) and it's universal for different operating systems and architectures. Ryu implemented in Python. Provides API, which is easy to create new network management applications. It supports various network device management protocols such as Open Flow, NETCONF, OF-config. Fully supports OF v. 1 0, 1 2, 1 3, 1 4, 1 5.

The main characteristics of the controller are:

- Performance: Speed of processing flows - the number of requests from the switches to be processed by the controller in a second – (flows/second) and delay - the time it takes the controller to process one request – (milliseconds).
- Scalability: change performance metrics while increasing the number of connections to the switches, change performance metrics while increasing the number of

leaf nodes in the network; and changes in performance metrics while increasing the number of processor cores.
- Resource capacity: CPU cores; Usage physical memory.
- Reliability: the number of failures during the test; and uptime for a given load profile.
- Rationale for the test criteria: the selection criteria for assessing the performance of the controller is based on the previously developed network devices testing methods: RFC 2889 - the testing methodology LAN switching network devices; and RFC 2544 - the testing methodology for connecting network devices.

3   The Experimental Investigation and Model Network

In the laboratory 5-segments model network is launched that enables to carry out comprehensive testing of both real and virtual settings of the Internet of Things [5]. In addition, the laboratory has a test bench for capture and subsequential analysis of the network traffic on any part of the network model. The block diagram of the model network is shown in Fig. 1.

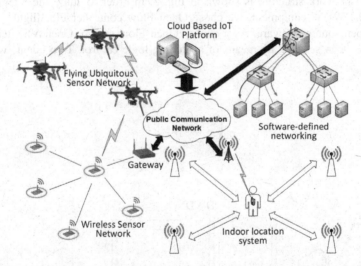

Fig. 1. Block diagram of a model network internet of things laboratory [5]

The study was conducted based on the model network of high-speed backbone systems DWDM and programmable networks laboratory in SUT. The modern transport network is becoming increasingly urgent problem with the growth of traffic and volume of transmitted data, to solve this problem, it is necessary to increase the capacity. Bandwidth fiber networks can be increased by using two methods: increasing the level of STM-signal or introduce the technology of dense wavelength division multiplexing (Dense Wavelength Division Multiplexing - DWDM). This technology involves the separation of the fiber spectral bandwidth into multiple optical channels. That is, one pair of parallel optical fiber carries multiple independent channels (each at its wavelength), which improves the system transmission bandwidth. In the laboratory we Used DWDM equipment companies T8 based on multi-platform 'Volga' in the laboratory. Platform 'Volga' allows you to organize up to 96 channels at 100 Gbit/s per channel. 'Volga' - the only Russian DWDM-platform that supports flows up to 100 Gbit/s and has the official status of the ministry's 'equipment of Russian origin.' In the experiment we used transponders dual 10 Gigabit channel (Ethernet, SDH, Fiber Chanel, OTN) «Desna' aggregators and channels of various formats (SDH, Ethernet, Fiber Chanel, OTN) to 2.5 Gbit per second. DWDM system was designed to enhance the capacity of each channel up to 100 gigabits per second. The system cost 40 - channel multiplexers with active channel and adjusting the power in the system. Intercom system to the point - the point is organized on a separate channel width of the GbE 1, that allows not only to confidently and quickly transmit/receive official information, but also to the need to pass it extra data. Control units have RSTP function for the organization MESH - links and switching equipment in the ring. The Developers of optical transport networks strive to create the most automated network, and to manage them based on programming, in order to minimize operating costs and to provide new services and applications faster and more efficient way.

Model network structure is shown in Fig. 2. In order to study the interaction of SDN and DWDM equipment, we chose Open Flow controller Floodlight, company Nicira, open source software. As a quality Open Flow switch, Open VSwitch virtual switch has been selected by means of switching logic control. Two clients were also

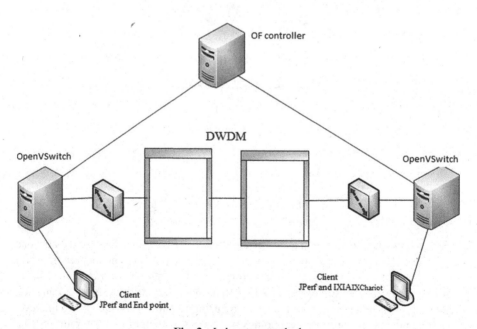

Fig. 2. Laboratory testbed

used as PC with installed software on them JPerf and IXIA IXChariot. One of the PC acting as a client and the other as a server. The physical architecture of the experimental test bed consist of server hardware component with processor: Intel Xeon E3-1220 V2 3.10 GHz (4 cores, 4 threads); memory: 3 board Foxline FL1600D3U11-4G DDR3 4 GB DIMM; and hard disk: Hitachi HDS721010CLA332 1000 GB. And server hardware component with tool for testing cbench; also controllers like OpenDaylight - Lithium SR3, Floodlight - v 1.0, Mul - v 3.2.7, Beacon - v 1.0.2, Maestro - v 0.2.0, and Ryu - v 4.1.

For the Controller performance we chose the relation between the performance and number of connected switches:

- The dependence of performance on the number of connected switches (1, 5, 10, 20, 50, 100), for a fixed number of hosts (10^4);
- The dependence of performance on the number of end nodes (10^3), (10^4), (10^5), (10^6), (10^7), with a fixed number of switches (20);
- The dependence of performance on the number of CPU cores;

For Controller Latency, the dependence of delay on the number of connected hosts (10^3), (10^4), (10^5), (10^6), (10^7), for a fixed number of switches (1).

In this paper was created GUI over Cbench and several additional functions in C++, with the introduction of the initial values for testing and graphic output bandwidth or delay, depending on the selected test mode. Also we added the ability to change the size of outgoing packets up to 2048 bytes, then we added displaying the average value of the controller to process a single package on the results of each test. Also, the function has been added to set the number of packets being sent and set the interval between them. Utility benchmarking Cbench controllers includes operating modes as the speed measurement mode processing streams and delay measurement mode. The algorithm works like Cbench simulates N switches Open Flow; and creates N of Open Flow sessions to the controller. The work results are shown as one of the test: the number of threads, mounted in a second controller (flow/s) (summed over all the switches) and Cbench: minimum, maximum and average value of the RPS (requests/sec) in all tests. In Fig. 3 we can see an example of using the utility.

```
root@sdn:/home/sdn/oflops/cbench# cbench -c 192.168.0.1 -p 6653 -m 10000 -l 2 -s 1 -M 10000 -t
cbench: controller benchmarking tool
   running in mode 'throughput'
   connecting to controller at 192.168.0.1:6653
   faking 1 switches offset 1 :: 2 tests each; 10000 ms per test
   with 10000 unique source MACs per switch
   learning destination mac addresses before the test
   starting test with 0 ms delay after features_reply
   ignoring first 1 "warmup" and last 0 "cooldown" loops
   connection delay of 0ms per 1 switch(es)
   debugging info is off
13:50:57.311 1   switches: flows/sec:  758115   total = 75.770910 per ms
13:51:07.412 1   switches: flows/sec:  257941   total = 25.792235 per ms
RESULT: 1 switches 1 tests min/max/avg/stdev = 25792.24/25792.24/25792.24/0.00 responses/s
```

Fig. 3. Example of the used Cbench utility

In Fig. 4 showed created GUI over Cbench.

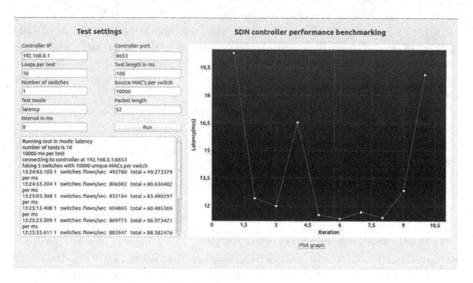

Fig. 4. GUI over Cbench

4   Testing Results

The studies were conducted using IXIA IxChariot program that was used to generate traffic and monitor the quality of the service parameters. For testing, the following types of traffics were chosen: FPS, HTTP, Bit Torrent, IPTV, VoIP. In order to study the interaction of DWDM and network equipment SDN, we conducted tests on two circuits network using DWDM equipment and network segment SDN and using the network segment SDN. To solve similar problems, we planned to develop an integrated methodology for testing and localization of bottlenecks in the structure of SDN. To this end, it is planned to conduct a series of natural experiments, by results of which will be proposed optimal modes of interaction DWDM and SDN network equipment (Table 1)

Test results showed that DWDM equipment does not introduce significant delays in the transmission of different types of traffic, such as: FPS, BitTorrent, VoIP, IPTV of SDN segment. During the experiment, it was found that HTTP traffic is transmitted via a DWDM equipment bandwidth reduction of 32 %, and increase in response time on the client side of 47 %. This HTTP traffic dependence on DWDM equipment is caused by the specifics of the equipment used and its configuration. In particular, a more detailed study revealed errors in the 1000Base-T network adapter driver. To solve similar problems, there is a plan to develop a comprehensive methodology for testing and localization of bottlenecks in the SDN structure. To this end, there is a plan to conduct a series of field experiments, the results of which will offer optimal modes of interaction of DWDM SDN network equipment.

For high-quality results display, the controllers have been divided into high pro-ductive (Maestro, Mul), average productive (Floodlight, OpenDaylight) and

Table 1. The experimental results

	FPS		BitTorrent		HTTP		VoIP		IPTV
	Throughput, Mbps	Resp Time, seconds	Throughput, Mbps	Resp Time, seconds	Throughput, Mbps	Resp Time, seconds	Throughput, Mbps	Resp Time, seconds	Throughput, Mbps
SDN	0,0151	16,233	5,306	2,472	0,181	0,057	3,929	0,001	1,452
SDN+ DWDMM	0,0153	16,299	5,489	2,391	0,124	0,084	3,575	0,001	1,452

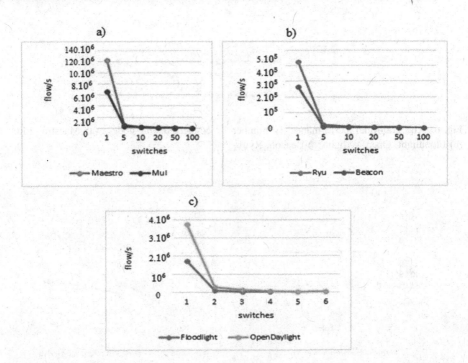

Fig. 5. The controller performance vs number of switches: (a) Maestro, Mul; (b) Ryu, Beacon; (c) Floodlight, OpenDaylight

unproductive (Ryu, Beacon). The dependence on the number of controller's performance switches shown in Fig. 5.

Increasing the number of connected switches, the number of processing flows received from each switch decreased. Controller performance depending on the number of connected end nodes shown in Fig. 6.

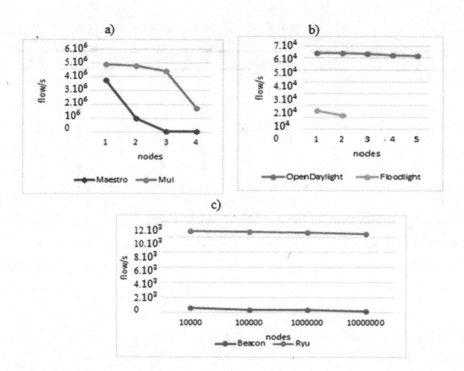

Fig. 6. The controller performance vs number of connected end nodes: (a) Maestro, Mul; (b) Floodlight, OpenDaylight (c) Beacon, Ryu

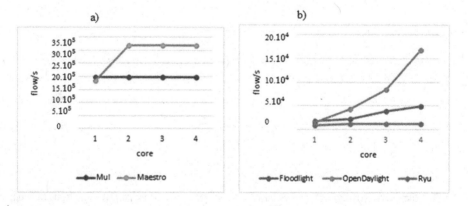

Fig. 7. The performance of the controller vs number of processor core (flows): (a) Maestro, Mul; (b) Floodlight, OpenDaylight, Ryu

With increasing the number of connected of end nodes was decreased in the number of flows processed controller. Controller performance dependence on the number of core (flow) of the processor shown in Fig. 7.

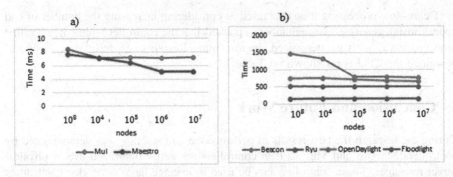

Fig. 8. Controller delay vs number of connected end nodes: (a) Maestro, Mul (b) Floodlight, Open Daylight, Ryu, Beacon

In case of increasing the number used by processor cores, the amount of increased flows processed by the controller. The dependence of controller delay on the number of connected of end nodes is shown in Fig. 8.

By increasing the amount of end nodes, considering a fixed number of switches, the processing flow controller decreased.

5 The Methodology Allowed to Evaluate the Basic Characteristics of SDN Network Controller

Performance and scalability:

- An increase in the number of connected switches, the number of processing threads coming from each switch was reduced;
- Increasing the number of end nodes, the number of reduced flows processed by the controller, but only slightly;
- Increasing the number of used network operating system of processor cores, the number of processed increased flow controller.

Table 2. Resource use controllers and CPU cores

Controller	OpenDaylight	Floodlight	Mul	Maestro	Ryu
The amount of server memory (GB)	2.1	7.3	0.8	1.2	4.1
Loading processor cores (%)	three core 96-99%, 85-87% single core	four cores 97-100%	4 cores 90-98%	4 cores 70-98%	3 7-18% kernel, kernel 1 99%

Delay-flow processing time decreased as considering increasing the number of end nodes (hosts), generate a large number of packets that the controller within the specified time processed with greater speed, in less time. Resource capacity - memory and download the CPU cores shown on Table 2.

6 Conclusion and Future Work

During the research the best results in performance and stability was demonstrated by controllers Maestro and Mul, as data controllers are better optimized use of physical server resources. These controllers can be used to manage large networks; Controllers Open Daylight, Floodlight and Ryu showed low-average performance. These controllers can be used for smaller networks; Beacon controller becomes unstable, which is unacceptable in the implementation of the SDN. Further research is expected to conduct a comprehensive testing of proprietary SDN controller, the reliability of their operation at high load for a long time, as well as the reaction to the ill-formed message.

On the basis of the available software solutions for network benchmarking SDN controller to develop its own testing tool with enhanced functionality and intuitive graphical interface. Testing the interaction of DWDM and SDN network equipment proved that DWDM equipment does not significantly affect the transfer of these types of traffic like a FPS, Bit Torrent, VoIP, IPTV segment SDN network. On the HTTP traffic is affected by factors beyond the control of DWDM equipment, such as the specificity used in SDN network equipment, its configuration, as well as failures in the 1000Base-T.

Further research is planned to test Open Flow traffic passing through the public communication network and to make appropriate comparisons.

Acknowledgment. The reported study was supported by RFBR, research project No. 15 07-09431a "Development of the principles of construction and methods of self-organization for Flying Ubiquitous Sensor Networks".

References

1. Kreutz, D., Ramos, F., Verissimo, P., Rothenberg, C., Azodolmolky, S., Uhlig, S.: Software-defined networking: a comprehensive survey. Proc. IEEE **103**(1), 14–76 (2015)
2. Xia, W., Wen, Y., Foh, C., Niyato, D., Xie, H.: A survey on software-defined networking. IEEE Commun. Surv. Tutor. **17**(1), 27–51 (2015)
3. OpenFlow. https://www.opennetworking.org/sdn-resources/openflow
4. Araniti, G., Cosmas, J., Iera, A., Molinaro, A., Morabito, R., Orsino, A.: OpenFlow over wireless networks: performance analysis. In: 2014 IEEE International Symposium on Broadband Multimedia Systems and Broadcasting, pp. 1–5. IEEE (2014)
5. Software-Defined Networking (SDN) Definition. https://www.opennetworking.org/sdn-resources/sdn-definition

6. Kirichek, R., Vladyko, A., Zakharov, M., Koucheryavy, A.: Model networks for internet of things and SDN. In: 2016 18th International Conference on Advanced Communication Technology (ICACT), pp. 76–79. IEEE (2016)
7. Kirichek, R., Koucheryavy, A.: Internet of things laboratory test bed. In: Zeng, Q.A. (ed.) Wireless Communications, Networking and Applications. LNEE, vol. 348, pp. 485–494. Springer, Heidelberg (2016)
8. Koucheriavy, A., Vasiliev, A., Lee, K.: Methods of testing the NGN technical facilities. In: The 7th International Conference on Advanced Communication Technology, ICACT 2005, vol. 1, pp. 317–319. IEEE (2005)
9. Andreev, D., Savin, K., Shalaginov, V., Zharikova, V., Ilin, S.: Limited values of network performance and network productivity estimation approach for services with required QoS. Service benchmarking. In: Balandin, S., Koucheryavy, Y., Hu, H. (eds.) NEW2AN 2011 and ruSMART 2011. LNCS, vol. 6869, pp. 463–474. Springer, Heidelberg (2011)
10. Recommendation ITU-T Q.3900 Methods of testing and model network architecture for NGN technical means testing as applied to public telecommunication networks. ITU, September 2006
11. Recommendation ITU-T Q.3950 Testing and model network architecture for tag-based identification systems and functions. ITU, November 2011
12. Choumas, K., Makris, N., Korakis, T., Tassiulas, L., Ott, M.: Testbed innovations for experimenting with wired and wireless software defined networks. In: 2015 IEEE 35th International Conference on Distributed Computing Systems Workshops, pp. 87–94. IEEE (2015)

Increasing the Efficiency of IPTV by Using Software-Defined Networks

Yuri Ushakov[1], Petr Polezhaev[1(✉)], Leonid Legashev[1],
Irina Bolodurina[1], Alexander Shukhman[1], and Nadezhda Bakhareva[2]

[1] Department of Mathematics and Computer Science,
Orenburg State University, Orenburg, Russian Federation
{unpk,newblackpit}@mail.ru, silentgir@gmail.com,
shukhman@gmail.com, ipbolodurina@yandex.ru
[2] Department of Computer Science and Engineering,
Volga Region State University of Telecommunications and Informatics,
Samara, Russian Federation
nadin1956_04@inbox.ru

Abstract. The problem of control and routing of broadband multimedia traffic has been considered in this paper. We suggest SDN-based network infrastructure for IPTV. Within the infrastructure IGMP and PIM are replaced by our OpenFlow-based protocol. To control multicast quality, we also propose real-time selective marking of IP packets by means of unique tags on input device, so monitoring respective stream quality and correcting it if necessary. A simulation model of broadband multimedia multicasting infrastructure based on SDN has been developed. The infrastructure performance has been studied on OMNET++ simulator. The experiment shows a growth of switch performance up to twice as much when the OpenFlow is used for processing multicast streams.

Keywords: IPTV · SDN · OpenFlow · Multicasting · Simulation

1 Introduction

At present, there is a tendency of online video spiking in the whole Internet traffic. According to Cisco within the following 5 years online video consumption will grow up to 85 % in the USA entire Internet traffic. Globally by 2019 video online IP traffic is going to grow 3 times as much as now. It is likely to be 2 zettabytes per year [1]. Within mobile traffic audio and video streams hold up to 55 % and this figure is growing [2].

Around the world multimedia transferring technologies (TV, radio, video on demand) are getting gradually digitalized. The most progressive and cheapest way of arranging a multicast network is to use data transfer on the existing networks with IP stack protocols. IPTV is the most widely used service in this field. Unlike stream multicasting with HTTP such as videos from web sites, IPTV is based on multicasting, the content distributed on several client devices. Currently the demand is growing for IPTV services but there are some obstacles to its wider use.

© Springer International Publishing AG 2016
O. Galinina et al. (Eds.): NEW2AN/ruSMART 2016, LNCS 9870, pp. 550–560, 2016.
DOI: 10.1007/978-3-319-46301-8_46

The new data transfer technologies for provider internal networks and wide use of digital TV devices lead to growing demand for IPTV and VoD services. Most providers base their IPTV implementation on channel multicast. Each client has to have a device or a program working through IGMP to receive UDP streams. All the intermediate devices within provider network also have to support IGMP and multicast.

The first problem is control and routing of broadband multimedia traffic. Simultaneous multicasting and transcoding on all the channels is resource-consuming, so multicasting starts on demand only. In this case the client sends a notice to be connected to a certain multicasting group (e.g. chooses the IPTV channel). The demand reaches a device making the decision on multicasting or otherwise redirecting the demand to another network with a multicasting device. The multicasting started, the IGMP snooping sends the stream only onto the demanded client devices. There are different IGMP versions and modes of multicasting and routing (dense mode, sparse mode, sparse-dense mode). Each IPTV provider uses different IGMP versions for different channels, which requires corresponding streams transcoding on streaming servers.

For distributed networks with routers different protocols are needed such as PIM or Multicast OSPF. They follow the IGMP routing, and neighbor routers are connected by IP tunnels, a multicasting tree is built for each multicasting. In "Dense mode" the tree covers all the routes like in case of multicasting within one network. In other modes the tree is built starting from the clients, the IGMPv3 setting the routes based on Source-Based Trees, Implicit Join, or Shared Trees.

However, most routers support only PIM. While multicasting PIM and IGMP interact, each of them routes the traffic in its own way: IGMP uses the preset commutation channels, PIM uses multicasting routing protocols such as MOSPF or DVMRP. Figure 1 shows the process of IGMP request, and Fig. 2 – the process of multicasting.

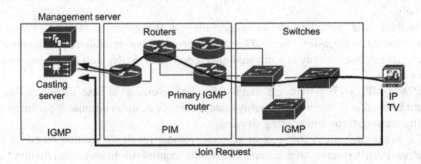

Fig. 1. Process of IGMP request

As a rule, IGMP uses the same route for the request and response, due to the absence of parallel routes or loops on L2 within one VLAN. PIM is based on routing tables with different possible routes. Coordination of routers for optimal load on channels is complicated, the settings have to be performed by qualified personnel manually.

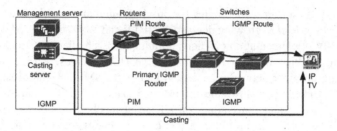

Fig. 2. Process of multicasting

IGMP low adaptivity on the one hand, and the complicated PIM setting (over 10 auxiliary protocols may be involved) on the other hand, lead to expensive IPTV deployment and low quality services. There are hardly any technologies for easy setting of multicasting in heterogeneous networks, the existing provider equipment rarely supports all the technologies. Many home client switches have very low multicasting performance, some cheaper devices loose the quality even for 5 streams.

Several services such as "Video on demand" require simultaneous unicasting and multicasting for the same stream, which requires additional equipment. Therefore, the main IPTV drawbacks arise due to inflexible and obsolete technologies.

Another important problem in multimedia multicasting is the quality decrease because of the wrong frame order, skipped frames, artefacts, significant delays, higher jitter. The existing solutions are expensive, they require complicated settings, and cannot correct the visual defects in real time.

2 Related Work

Rebuilding of multicasting trees in case of network failures has been considered in [3–8].

Medard, Xue and others [3, 4] suggest the use of preplanned reserving based on reservation of multicasting trees. The algorithms involve centralized computations, which can be effective only in software-defined networks (SDN) with easily determined client states.

P2MP MPLS [5] provides a method for fast rerouting in case of failure, but it cannot be used in networks with highly changeable client states because it performs the routing starting from multicasting devices.

Li, Wang and others [6] suggest an improvement for this algorithm based on the effective P2MP reserve tree computations, the control of network bandwidth and reserve routing choice. The P2MP MPLS allows to recover fast after failures, but still it cannot be used in networks with high client state changeability.

Medard, Finn, and others [3] have suggested the algorithms for redundant tree computation in vertex-redundant or edge-redundant graphs. These algorithms can be used for preplanned reservation of multicast routes. Mochizuki, Shimizu, and others [7] present a similar algorithm, which also minimizes the number of edges in the trees. The authors state the necessity of centralized computations for the above algorithms, but do not use the SDN technologies.

Kotani, Suzuki, and others [8] consider SDN to solve the problem of fast tree rebuilding in multicasting with minimum frame losses in case of network failure. The authors suggest two trees for each group: the main one, and the reserve one having no or minimum common edges.

Kim, Liu, and others [9] give a detail research for reducing zapping time such as including/excluding a client in/from the multicasting group. The problem is solved in [10–12].

Keshav, Paul [10] suggest precalculated routing with Prim's algorithm, the distance from source node assumed as the weight of each link. This algorithm also requires centralized computations, but it is not efficient without SDN technologies. Ratnasamy, Ermolinskiy and others [11] suggest the use of unicast routes for multicasting forming an overlay network. MPLS using for multicasting traffic control is described in [12], but this approach lacks scalability and flexibility compared to SDN technologies.

3   IPTV Implementation Based on Software-Defined Networks

We suggest a network infrastructure based on software-defined networks (Polezhaev et al. [13–15]) for IPTV (Fig. 3).

It consists of Internet access system, SDN routers on the distribution layer, SDN switches on the access layer, and client's networks.

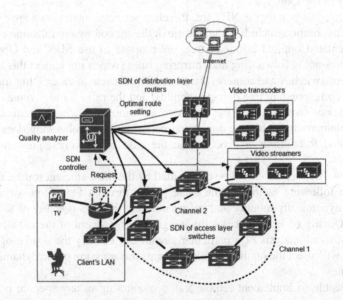

Fig. 3. Multicasting in SDN network

Video streams are received from Internet or satellites. Video transcoders convert the streams into one format. Video streamers are stream sources in the network infrastructure.

The SDN controller should contain modules, which implement the broadband multimedia multicast routing algorithm based on SDN, the algorithm for collecting information about the topology and network state, the broadband multimedia traffic quality analyzer based on SDN.

A client network can include different equipment such as routers, access points, TV devices.

The SDN approach allows to intercept a multicast connection request on the first controlled switch directly and to set an optimal route for each stream in a dynamic mode considering the channel and network equipment load. It is possible to use an arbitrary number of parallel routes, and there is no need to use and configure the PIM.

The route, once installed, does not require any controller interference, the controller is used only for IGMP emulation works to the client side on the final switch. If a client has SDN equipment, there is no need to emulate the IGMP.

The set of all data routes from the video streamer to clients forms a multicast tree. Construction of optimal multicast tree can be formalized as solution of the Steiner tree optimization problem for the directed graph of network infrastructure. The vertices of this graph are network nodes (servers, switches, routers, client devices), the directed edges are network links. The weights of the arcs are residual bandwidths and current delays on the ports. In addition, there is the source of multicast stream and the client vertices. It is necessary to construct a directed tree that each client vertex will accessible from the source, while optimizing some function that reflects the satisfaction of the stream QoS to the QoE.

In general, this problem is NP-hard, therefore artificial intelligence approaches can be used for its solution, including the genetic or the ant colony optimization algorithms.

For calculated optimal multicast tree, we propose to use SDN and OpenFlow to install corresponding forwarding and mirroring rules (which implement this tree) to the flow tables of switches and routers. If a new user connects to an existing multicasting stream, the corresponding tree quickly rebuilds, and the rules for new routes are added to flow tables of switches, then the old rules are removed. Switches connected to the client's equipment drop possible duplicated packets, so tree rebuilding does not affect the end users. Rebuilding also occurs, in the case of network equipment or links failures.

Each new IPTV channel will be transmitted by the most efficient routes, taking into account the following factors: if the route passes through OpenFlow switches, it is possible to dynamically monitor and ensure not only the QoS (Quality of Service), but also QoE (Quality of Experience) – the integrated assessment of the streaming media quality. Statistical analysis of a particular stream can identify the source of problems before they will affect the quality of video streams transmission or the dynamic changes of their routes.

It is possible to implement remote traffic monitoring in any specific point of the network by resending selected traffic from it to quality analyzer (Fig. 4).

We propose to selectively mark packets with unique tags at the entrance to the network using the OpenFlow header rewriting mechanism. Then, when the packet stays

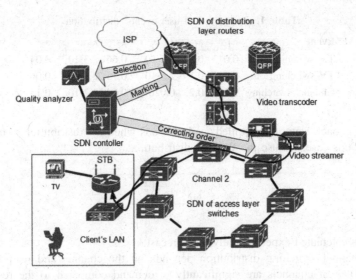

Fig. 4. Correcting the order of packets in the flow sequence

in the queue of video streamer or converter, quality analyzer verifies the need for corrections in the video stream (i.e., reordering packets, removing of corrupted packets), and, if necessary, sends a correction command to the converter queue (the delay is not more than 1 s) or to the switch immediately after the streamer. If necessary, the analyzer can also modify parameters of video streaming.

4 Simulation Model of Multimedia Broadband Traffic Infrastructure for SDN

At present, a provider cannot control multicast trees on access layer and intermediate L2 switches. Main links can be overload at 200–300 IPTV channels, if the number of HD channels grows the overload grows too.

The article [16] shows the statistical regularities and probability distributions for channel switching within a large IPTV network. Starting and finishing of video channel watching have different distributions and intensities at different times. There are peaks with the normal distribution and high intensity from 7.00 to 7.30 am, around 9 am or 12. In addition, there are minimums between the peaks with other parameters, usually with the exponential distribution. Absolute maximum's time is in the evening when a network is sure to be overload.

Based on research data from [16] we assume $f(x) = \sum_{i=1}^{n} a_i\lambda_i e^{-\lambda_i x}$ for the probability density function of client's events, where λ_i is event intensity, a_i is weighting coefficient, $\sum a_i = 1$.

The experimental parameters for $n = 3$ (minimal possible number of events) are presented in Table 1.

Table 1. Parameters of user events' distribution

Event	λ_1	a_1	λ_2	a_2	λ_3	a_3
TV switching on	0,013	0,3	$3,3 \cdot 10^{-3}$	0,66	$2,3 \cdot 10^{-4}$	0,04
TV switching off	0,032	0,19	$2,5 \cdot 10^{-3}$	0,75	$2,4 \cdot 10^{-4}$	0,06
Channel switching	2,1	0,23	0,026	0,64	$3,2 \cdot 10^{-3}$	0,13

This approach cannot be applied for low loads when the distribution is not exponential. In this case, we use the Weibull distribution:

$$f(x) = \frac{k_i x^{k-1} e^{-\left(\frac{x}{\mu_i}\right)^k}}{\mu_i^k}, \tag{1}$$

where k_i is calculated experimentally, by forecast $k_i < 1$.

The channel switching distribution depends on the channel and its popularity. Around 10 % of channels are significantly in demand compared to the rest of the channels, and the switching intensity inside this group is exponential.

For multicast traffic study a simple model has been created based on OMNET++ simulator (Fig. 5) and IGMP/PIM example. The network consists of the main ring 10 Gb/s, one video streamer (server) and clients with IPTV devices. The clients request new channels through IGMP according to the exponential distribution of channel switching time. In addition, clients receive other traffic in «light browsing» mode (2 requests per second, 100–500 Kb answers on the uniform distribution).

Switches with IGMP support [16] transfer the requests to the video streamer and connect the clients to a new multicasting group, excluding them from previous one. IGMP is simulated by PIM-SM to follow the events. For STP simulation OSPF is used with predefined metrics. Channels are AVC/H264 4 Mb/s UDP streams with frame size from 1300 to 1450 bytes. The video streamer uses UDPBasicApp, the client uses UDPSink.

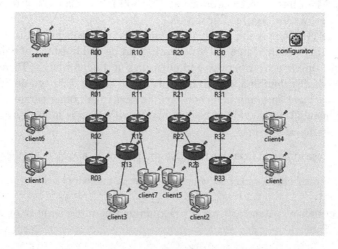

Fig. 5. Simulation model for multicast IPTV

a)

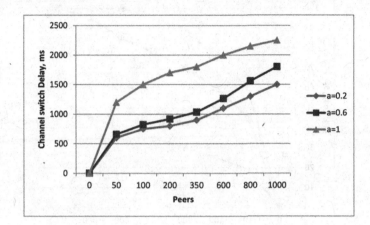

b)

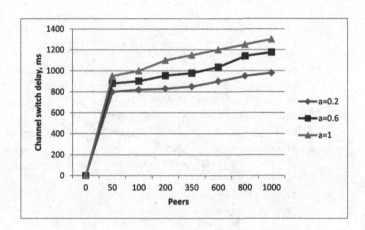

Fig. 6. Channel switching delays (a) IGMP; (b) OpenFlow.

See the experiment parameters in Table 1.

With the number of requests exceeding 50 at one device the average 500–700 ms delay grows for channel switching. With the number of requests exceeding 600, the packet loss is above 0.05 % and delay grows above 1 s (IPTV buffer overflow) and consequently video artefacts are possible.

IPTV traffic is generated at the same time with 1 Mb/s browsing traffic per client.

Within the experiment the simulation model was to be used for IGMP tree and then for OpenFlow tree.

We have used the range from minimum to maximum load (f_{light}, f_{hard}) to calculate the function values:

a)

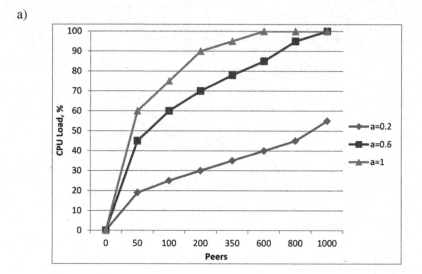

b)

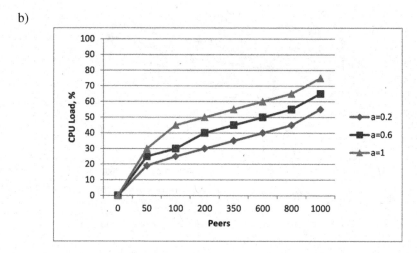

Fig. 7. Switch CPU load (a) IGMP; (b) OpenFlow.

$$\text{Delay}(x,a) = a \cdot \text{Delay}_{\text{light}}(x) + (1 - a) \cdot \text{Delay}_{\text{hard}}(x), \qquad (2)$$

$$\text{Load}(x,a) = a \cdot \text{Load}_{\text{light}}(x) + (1 - a) \cdot \text{Load}_{\text{hard}}(x). \qquad (3)$$

Channel switching delays are presented in Fig. 6.

In addition, the switch CPU load was determined. The experimental results are presented in Fig. 7.

Packet losses are also found in case of time out at switches and in case of IPTV packet life time out.

The figures show that the use of OpenFlow for multicast processing is more efficient both for delay and performance. The use of OpenFlow makes the load capacity of switches up to twice as big in the multicast mode.

5 Conclusion

The problem of control and routing broadband of multimedia traffic has been considered as well as the problem of low multicasting quality. The analysis of existing publications shows no complete solution of the both.

This paper suggests SDN-based network infrastructure for IPTV. Within the infrastructure IGMP and PIM are replaced by our OpenFlow-based protocol, which also intercepts IGMP requests from client equipment not supporting OpenFlow. The multicast tree is rebuilt in case of network failures or it is necessary to optimize multicasting for new clients switching in. To control multicast quality, we suggest real-time selective marking of IP packets by means of unique tags on input device, so monitoring respective stream quality and correcting it if necessary.

The simulation model of broadband multimedia multicasting infrastructure based on SDN has been developed.

The infrastructure performance has been studied on OMNET++ simulator. The experiment shows a growth of switch performance up to twice as much when the OpenFlow is used for processing multicast streams.

In the future, we plan to implement the proposed multimedia broadband traffic infrastructure for IPTV on a real software-defined network and to study its performance, scalability and flexibility.

Acknowledgments. The research has been supported by the Ministry of Education of Orenburg region (project no. 30 on 30 June 2016), the Russian Foundation for Basic Research (projects 15-07-06071, 16-07-01004, 16-47-560335, and 16-29-09639), the Government of Orenburg region (16-47-560335) and the President of the Russian Federation within the grant for young scientists and PhD students (SP-2179.2015.5).

References

1. Cisco Visual Networking Index: Forecast and Methodology, 2014–2019 White Paper, Cisco VNI (2015). http://www.cisco.com/c/en/us/solutions/collateral/service-provider/ip-ngn-ip-next-generation-network/white_paper_c11-481360.html
2. Cisco Visual Networking Index: Global Mobile Data Traffic Forecast Update 2014–2019 White Paper. Cisco VNI (2015). http://www.cisco.com/c/en/us/solutions/collateral/service-provider/visual-networking-index-vni/white_paper_c11-520862.html
3. Medard, M., Finn, S.G., Barry, R.A.: Redundant trees for preplanned recovery in arbitrary vertex-redundant or edge-redundant graphs. IEEE/ACM Trans. Netw. **7**, 641–652 (1999)
4. Xue, G., Chen, L., Thulasiraman, K.: Quality-of-service and quality-of-protection issues in preplanned recovery schemes using redundant trees. IEEE J. Sel. Areas Commun. **21**(8), 1332–1345 (2003)

5. Yasukawa, S.: Signaling requirements for point-to-multipoint traffic-engineered MPLS label switched paths (LSPs). RFC 4461 (Informational), Internet Engineering Task Force. http://www.ietf.org/rfc/rfc4461.txt

6. Li, G., Wang, D., Doverspike, R.: Efficient distributed MPLS P2MP fast reroute. In: 25th IEEE International Conference on Computer Communications, pp. 1–11 (2006)

7. Mochizuki, K., Shimizu, M., Yasukawa, S.: CAM05-3: Multicast tree algorithm minimizing the number of fast reroute protection links for P2MP-TE networks. In: Global Telecommunications Conference, pp. 1–5 (2006)

8. Kotani, D., Suzuki, K., Shimonishi, H.: A design and implementation of OpenFlow controller handling IP multicast with fast tree switching. In: 2012 IEEE/IPSJ 12th International Symposium on Applications and the Internet (SAINT), pp. 60–67. IEEE (2012)

9. Kim, E., Liu, J., Rhee, B., Cho, S., Kim, H., Han, S.: Design and implementation for reducing zapping time of IPTV over overlay network. In: International Conference on Mobile Technology, Application and Systems, New York, pp. 1–7 (2009)

10. Keshav, S., Paul, S.: Centralized multicast. In: Proceedings of the Seventh Annual International Conference on Network Protocols, p. 59. IEEE Computer Society, Washington (1999)

11. Ratnasamy, S., Ermolinskiy, A., Shenker, S.: Revisiting IP multicast. SIGCOMM Comput. Commun. Rev. **36**, 15–26 (2006)

12. Martinez-Yelmo, I., Larrabeiti, D., Soto, I., Pacyna, P.: Multicast traffic aggregation in MPLS-based VPN networks. IEEE Commun. Mag. **45**(10), 78–85 (2007)

13. Konnov, A.L., Legashev, L.V., Polezhaev, P.N., Shukhman, A.E.: Concept of cloud educational resource datacenters for remote access to software. In: Proceedings of 11th International Conference on Remote Engineering and Virtual Instrumentation (REV), Polytechnic of Porto (ISEP) in Porto, Portugal, 26–28 February 2014, pp. 246–247 (2014)

14. Polezhaev, P., Shukhman, A., Ushakov, Y.: Network resource control system for HPC based on SDN. In: Balandin, S., Andreev, S., Koucheryavy, Y. (eds.) NEW2AN/ruSMART 2014. LNCS, vol. 8638, pp. 219–230. Springer, Heidelberg (2014)

15. Polezhaev, P., Shukhman, A., Konnov, A.: Development of educational resource datacenters based on software defined networks. In: Proceedings of 2014 International Science and Technology Conference "Modern Networking Technologies (MoNeTec)", Moscow, Russia, pp. 133–139 (2014)

16. Qiu, T., Ge, Z., Lee, S., Wang, J., Xu, J., Zhao, Q.: Modeling user activities in a large IPTV system. In: Proceedings of the 9th ACM SIGCOMM Conference on Internet Measurement Conference, pp. 430–441 (2009)

Fuzzy Model of Dynamic Traffic Management in Software-Defined Mobile Networks

Andrei Vladyko[1], Ivan Letenko[1(✉)], Anton Lezhepekov[1],
and Mikhail Buinevich[2]

[1] The Bonch-Bruevich Saint-Petersburg State University of
Telecommunications, Saint-Petersburg, Russia
vladyko@sut.ru, letenko@gmail.com,
anton.lezhepekov@mail.ru
[2] Saint-Petersburg University of State Fire Service of EMERCON of Russia,
Saint-Petersburg, Russia
bmv1958@yandex.ru

Abstract. Nowadays, in mobile networks, latency-sensitive services may compete for the bandwidth of other services, thus degrading overall performance. Software-defined mobile networks opening many new possibilities for dynamic traffic management. This gives chances to ensure strict requirements of the service quality in changing conditions. We considered two requirements for the quality of service: bandwidth and latency. Providing the required bandwidth is relatively simple compared to the end-to-end delay, as its guarantee requires a complex model that takes into account the mutual influence of flows throughout the entire path. Our model includes the following metrics: maximum channel utilization, traffic priority, the number of "hops". Fuzzy balancer module has been developed for the Floodlight controller in Java. This module calculates the weights of the links in accordance with the proposed method. Simulation network was held in Mininet environment. During the experimental implementation, it has been shown that simple algorithm based on mathematical apparatus of fuzzy logic allows dynamically adapting the network to the change of the traffic volume, as well as its structure.

Keywords: SDN · Software-Defined Mobile Networks · Fuzzy model · Dynamic traffic management · Floodlight controller

1 Introduction

Software-Defined Mobile Networks (SDMN) is a perspective approach of mobile networks construction, which is based on the concepts and technologies of Software-Defined Networking (SDN) and Network Function Virtualization (NFV) [1, 2]. The development of these technologies nowadays outpaces their use among mobile operators in LTE[1] networks. Their implementation opens up the capabilities of building new solutions to traffic stream optimization in mobile networks by using modern techniques of separation of control plane (control-plane) and packet forwarding

---

[1] LTE – Long-Term Evolution.

© Springer International Publishing AG 2016
O. Galinina et al. (Eds.): NEW2AN/ruSMART 2016, LNCS 9870, pp. 561–570, 2016.
DOI: 10.1007/978-3-319-46301-8_47

plane (data-plane). Experience and current trends in the use of these technologies will help to increase the network utility, reduce energy costs and the number of network failures on the side of the operator, as well as the overall increase in the quality of services provided on the network subscriber side [3].

In this article, we will research the possibility of traffic path multi-objective optimization based on the weight coefficients of networks segments. The problem of the existing traffic priority distribution mechanism in mobile networks is the lack of workload accounting at a particular route segment in the through-channel provided to the network subscriber.

The aim of the work is to provide a routing mechanism that ensures strict requirements fulfillment regarding the service quality for the highest possible number of streams, and to ensure the maximum network utility.

2 Background

Presently, the separation of the data-plane and control-plane is one of the most effective approaches to solving existing problems of mobile operators regarding the network resources optimization in order to increase the services quality.

The ability to provide reliable data transfer with guaranteed low latency is critical, e.g. for the transmission of voice data by the Voice over LTE technology, as voice services are currently sharing channel with data transfer services.

Mechanism of ensuring the quality of service QoS[2] in LTE marks the traffic streams with the class identifier QCI (QoS Class Identifier). Each class has the corresponding QoS parameters for the given traffic type, namely priority, acceptable latency and the number of lost packets, guaranteed data transfer rate (GBR).

Thus, the network must provide the transfer of isolated traffic streams with different demand of network resources (quality of service).

In this paper, we consider two requirements for the quality of service: bandwidth and latency. Providing the required bandwidth is relatively simple compared to the end-to-end latency. To provide bandwidth it is enough to control the residual connection throughput, and to use speed limiters for the streams at the input routers for restricting traffic entering the network from the stream source. Dynamic allocation of traffic streams allows distributing uniformly the load across the network devices and reducing delays in stream processing queues.

The purpose of the proposed model is to fulfill the QoS requirements for the new stream without violating the requirements for the existing streams.

Despite the fact that there are technologies that allow to calculate the bandwidth as a function of the required latency and other QoS requirements, this paper focuses on the pass-through latency, because providing it requires a complex model that takes into account the mutual influence of streams across the entire route.

---

[2] QoS – Quality of Service.

3 Load Balancing Model

With traffic optimization there is always a problem of finding a compromise between the load balancing and minimizing the route length [4]. While minimizing the route may cause the overload of individual network segments, load balancing tends to use more connections than is necessary for load distribution, which leads to the waste of network resources and increases latency. The proposed routing model solves this problem by multi-objective optimization of the route «cost». The model includes the following metrics: maximum channel utilization, traffic priority, the number of «hops». In contrast to the other algorithms (e.g. WSP), which optimize the load only at the «bottleneck», the proposed algorithm takes into account the load on all the connections at the route.

Achieving optimal load balancing requires satisfaction of the several criteria.

First: Maximize route throughput, i.e. maximize the residual throughput at the connection, which is the «bottleneck».

Second: Minimize the number of priority streams at the route (not to create «bottlenecks» of priority streams), and thereby fulfill QoS requirements for the maximum number of streams.

Third: Minimize the number of «hops». Since the route with a large number of «hops» will likely overlap (interfere) with other routes at one of the connections.

Let us consider a detailed model of latency formation. Transmission latency is an additive metric consisting of the propagation delay, buffering and processing delays. From the traffic management standpoint, the buffering delay is the key one, which is determined by the current queue length and the type of queue processing algorithm.

In the queue constructed according to the priorities, the latency value will depend firstly on the speed with which packets are received in the streams having a higher priority, and secondly, on how quickly the connection can transmit packets. Thus, when a new stream is added to the queue, it's latency will be affected by the streams having a higher priority, while it, in turn, will increase the latency of streams with low priority.

Buffering delay D_f of the stream f, introduced by the switch, is described by the following function:

$$D_f = f(q, R_{1:q}, C), \tag{1}$$

where q - stream priority in the queue, $R_{1:q}$ – sum of the required stream throughputs having the equal or higher priorities, and C - line throughput.

Function for the buffering delay calculation is defined by the type of queue processing algorithm, e.g. FIFO[3], WFQ[4], HTB[5], WRR[6] etc.

---

[3] FIFO – First In First Out.
[4] WFQ – Weighted Fair Queueing.
[5] HTB – Hierarchical Token Bucket.
[6] WRR – Weighted Round Robin.

$$\sum D_f < D_s \qquad\qquad (2)$$

Sum of delays at all switches of the route must be less than D_s – minimum required pass-through latency.

The controller must choose a position in the queue along the whole route in order to satisfy the requirements for all the streams, e.g. if the stream at one of the switch ports received low priority, then to ensure total pass-through latency at the next switch the controller must assign that stream a higher priority.

For this, as suggested in [5], at each connection (switch port) maximum possible position in the queue is calculated for the stream, in which it does not affect the current streams, and the minimum possible position, in which its latency requirements are fulfilled, thus we get a range of possible priorities (positions in the queue). The resulting range can be used as metric in edge weight calculation.

3.1   Mathematical Apparatus

It is known that the problem of route multi-objective optimization with multiple route cost metrics is NP-hard[7]. The problem is further complicated due to conflicting optimization criteria.

Fuzzy logic is an effective tool for solving multi-objective optimization problems with potentially conflicting criteria. Fuzzy logic allows representing the values of different criteria as linguistic terms, which represent the level of belonging to a particular term in the form of a number in the range $\{0, 1\}$.

Let us introduce the network in the form of a directed graph G (N, L), where N - set of vertices (hosts), and L – set of edges (connections).

Consider the membership functions for the optimization criteria suggested above.

To evaluate the edge load based on throughput we use the membership function depicted in Fig. 1, where BW_{min}[8] and BW_{max}[9] are the minimum and maximum available residual throughput of network edges, respectively. Thus, p_{xy} takes the minimum value at the edge, which is the «bottleneck» in the network.

To evaluate the edge load based on the number of high-priority streams l_{xy}, we use the same membership function (Fig. 1), but this time BW_{min} and BW_{max} are the minimum and maximum position in the queue. In the case of SSF algorithm proposed in [5], for membership function value assignment it is possible to use the range width of the available priorities (queues).

Membership function of the third criterion h_{sd} is calculated not for the individual edges, but for whole route. In Fig. 2, the maximum and minimum possible route length is denoted as H_{min} and H_{max}, respectively, m value is chosen to be 0.75.

---

[7] NP-hard – Non-deterministic polynomial-time hard.

[8] BW_{min} – Minimal bandwidth.

[9] BW_{max} – Maximal bandwidth.

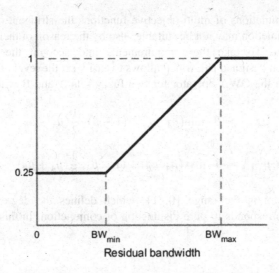

Fig. 1. Membership function of the edge load based on throughput

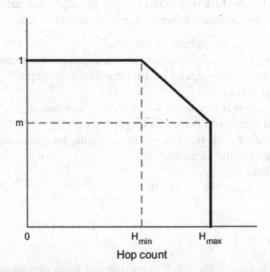

Fig. 2. Membership function of the path preference based on number hops.

We use fuzzy s- and t-norm-operators. Generally, s-norm and t-norm operators are implemented using *max* and *min* functions:

$$\mu_{A \cap B}(x) = \min(\mu_A(x), \mu_B(x)) \tag{3}$$

$$\mu_{A \cup B}(x) = \max(\mu_A(x), \mu_B(x)) \tag{4}$$

For some formulations of multi-objective functions the stringent operators of disjunction and conjunction may not be suitable, also by the reason of their indifference to individual criteria. To take these requirements into account the Jager's OWA-operator[10] has been designed [6], which allows to easily set the level of disjunction and conjunction. Then the OWA-operator for two fuzzy sets A and B will be as follows:

$$\mu_{A \cap B} = \beta \cdot \min(\mu_A, \mu_B) + (1 - \beta) \cdot \frac{1}{2}(\mu_A + \mu_B) \tag{5}$$

$$\mu_{A \cup B} = \beta \cdot \max(\mu_A, \mu_B) + (1 - \beta) \cdot \frac{1}{2}(\mu_A + \mu_B) \tag{6}$$

where β - constant in the range $\{0, 1\}$, which defines the degree to which the OWA-operator corresponds to pure disjunction or conjunction. In this paper $\beta = 0.8$.

3.2   Algorithm

The proposed algorithm starts with calculating the edge weights of the traffic with requested priority by the first two criteria. Then the algorithm calculates k-possible routes, chooses the best one according to the fuzzy inference rule.

1. Calculate the residual throughput of the edges.
2. Calculate the range of available queues on the edges for the requested traffic class.
3. Calculate weights of the edges based on residual throughput p_{xy} and number of high-priority streams l_{xy}.
4. Calculate the k - best routes using weights from previous step.
5. Choose the best route considering its length h_{sd}.
6. If there is a conflict with the low-priority streams, then rearrange routes for the respective streams in the same way.
7. Set rules to routers.
8. Collect statistics, adjust the constants.

3.3   Experiment

Development and testing of the model experimental implementation was carried out with Floodlight controller in the Eclipse development environment. Floodlight controller module has been developed in Java language, which calculates the edge weights according to the method proposed above. The module proposed in [7] was used for the implementation of QoS features. Network simulation was conducted in Mininet environment [8], which includes the virtual switch Open vSwitch. For implementation of the QoS mechanism in this environment the standard queues of the Traffic Control subsystem of Linux operating system are used. For queue construction the HTB

---

[10] OWA-operator – Ordered Weighted Averaging aggregation Operator.

method was used (Hierarchical Token Buffer). Package planner based on HTB allows to share the bandwidth among multiple traffic classes and provides a mechanism for bandwidth borrowing.

Standard topology module of Floodlight controller calculates the best route with the minimum number of «hops», using the same weight for all connections. In this paper, the standard module has been modified for the calculation of several best routes based on the metrics described above. Similar to work [9], for the implementation of the decision making algorithms we used jFuzzyLogic library written in Java.

For experimental implementation, we used the value of the required bandwidth for the traffic with the same or higher priority as the l_{xy} metric.

The metrics were calculated based on statistics obtained by the standard Floodlight REST API. To obtain more accurate data, in the future we plan to use external solutions of data transfer level, such as sFlow.

In Mininet environment the network has been modeled, which configuration is shown in Fig. 3.

The network consists of four switches and four hosts. The traffic was being generated between the hosts H1 and H4, as well as H3 and H2, respectively. For broadcasting, receiving and saving video to file VLC[11] media player was used iperf utility was used to generate the background TCP[12] traffic.

HTTP[13] and UDP[14] protocols were used as transport for video streams. Queues configuration on the switches was configured in such a way that the video traffic had priority over the iperf traffic.

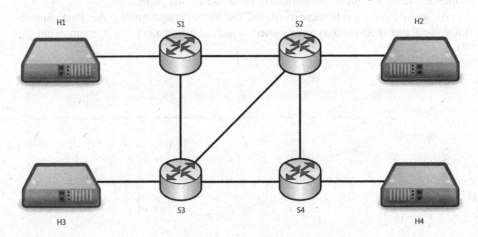

Fig. 3. Network scheme

---

[11] VLC – VideoLAN Client.

[12] TCP – Transmission Control Protocol.

[13] HTTP – HyperText Transfer Protocol.

[14] UDP – User Datagram Protocol.

We have conducted three sets of experiments. In the first set of experiments, we have used a standard implementation of the Floodlight controller topology module, in the second set the topology module has been changed, and the residual bandwidth has been used as the edge weight, in the third set of experiments the route has been chosen according to the proposed algorithm.

Testing was conducted according to the following scenario:

1. Form queues with different priorities at the switches.
2. Start video stream transmission between the hosts H1 and H4 via HTTP.
3. Start video stream transmission between the hosts H3 and H2 via UDP.
4. 60 s later using iperf we have simulated constant traffic between the hosts H1 and H4, as well as H3 and H2.
5. Add another video stream between the hosts H1 and H4 via HTTP.

In the first case, for all the streams the shortest route was selected, thus the data traffic interfered with the video traffic. In the second case, the data traffic controller selected an alternative route (e.g., route S1-S3-S4 has been selected between the hosts H1 and H4), which helped avoid the overloading of the shortest route, however, both video streams between the hosts H1 and H4 were on the same route. In the third case, by using the proposed algorithm the controller has selected different routes for video, data traffic has been transferred through the same routes, but with lower priority. Thus, it became possible to transmit the maximum amount of streams and achieve greater utilization of the network.

Figure 4 shows the plots of iperf streams throughput for the two cases, the standard implementation and the implementation proposed in this paper.

As an objective evaluation metric of the video image quality the Peak Signal-to-Noise Ratio (PSNR) is typically used, which can be calculated by comparing the

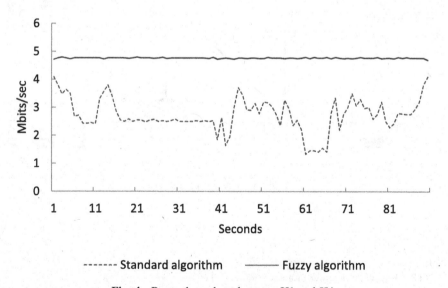

Fig. 4. Route throughput between H1 and H4

original and transmitted video. Figure 5a depicts the PSNR values for the video transmitted over UDP between the hosts H3 and H2 with the standard controller implementation, whereas Fig. 5b - with the implementation proposed in this paper.

The graph in Fig. 5a clearly illustrates the decline in the average PSNR level and the reduction of the frames number due to interference of the video stream and the data stream.

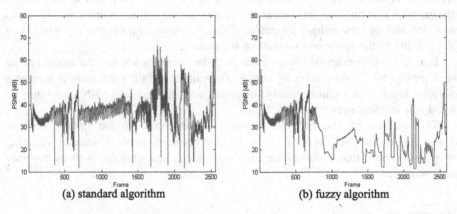

(a) standard algorithm   (b) fuzzy algorithm

Fig. 5. PSNR values for video between the hosts H3 and H2.

For demonstration of the interference of the data traffic and HTTP-video Fig. 6 shows the graphs of congestion window values of TCP protocol.

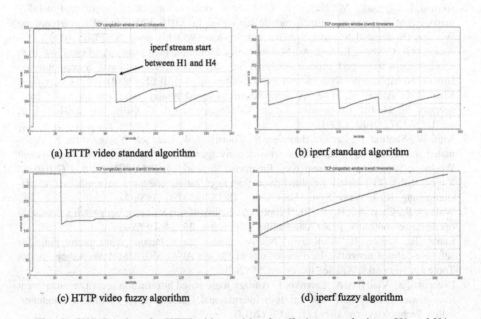

(a) HTTP video standard algorithm   (b) iperf standard algorithm

(c) HTTP video fuzzy algorithm   (d) iperf fuzzy algorithm

Fig. 6. CWND values for HTTP video and iperf traffic between the hosts H1 and H4

4 Conclusion

The results of the study confirm the importance of applying SDN to separate the data stream plane from the control plane in mobile networks.

In this paper, we propose the model of dynamic load balancing for software-defined networks that takes into account the mutual influence of traffic flows with different priorities. Traffic control algorithm, designed based on the proposed model, allows dynamically adapting the network to the change of the traffic volume, as well as its structure, and ensures optimal allocation of network resources, thereby providing their accessibility to the maximum number of streams.

During the experimental implementation, it has been shown that the model can be implemented by a computationally simple algorithm based on mathematical apparatus of fuzzy logic, which can be easily integrated with the existing SDN controllers of wired and wireless networks.

This method can be used for resources optimization of the mobile network depending on the load, as well as for service chaining, network features mapping (mapping), optimizing traffic of various applications and QoS, as well as mobility management.

References

1. Chen, T., Matinmikko, M., Chen, X., Zhou, X., Ahokangas, P.: Software defined mobile networks: concept, survey, and research directions. IEEE Commun. Mag. 53(11), 126–133 (2015)
2. Ahmad, I., Liyanage, M., Namal, S., et al.: New concepts for traffic, resource and mobility management in software-defined mobile networks. In: 2016 12th Annual Conference on Wireless On-demand Network Systems and Services (WONS), pp. 1–8. IEEE (2016)
3. Araniti, G., Cosmas, J., Iera, A., Molinaro, A., Morabito, R., Orsino, A.: OpenFlow over wireless networks: performance analysis. In: 2014 IEEE International Symposium on Broadband Multimedia Systems and Broadcasting, pp. 1–5. IEEE (2014)
4. Khan, J.A., Alnuweiri, H.M.: A fuzzy constraint-based routing algorithm for traffic engineering. In: Global Telecommunications Conference, 2004, GLOBECOM 2004, vol. 3, pp. 1366–1372. IEEE (2004)
5. Kim, W., Sharma, P., Lee, J., Banerjee, S., Tourrilhes, J., Lee, S.J., Yalagandula, P.: Automated and scalable QoS control for network convergence. In: Internet Network Management Workshop/Workshop on Research on Enterprise Networking (INM/WREN), p. 1 (2010)
6. Yager, R.R.: On ordered weighted averaging aggregation operators in multicriteria decisionmaking. IEEE Trans. Syst. Man Cybern. 18(1), 183–190 (1988)
7. Wallner, R., Cannistra, R.: An SDN approach: quality of service using big switch's floodlight open-source controller. Proc. Asia-Pacific Adv. Netw. 35, 14–19 (2013)
8. Lantz, B., Heller, B., McKeown, N.: A network in a laptop: rapid prototyping for software-defined networks. In: Proceedings of the 9th ACM SIGCOMM Workshop on Hot Topics in Networks, Article No. 19. ACM (2010)
9. Dotcenko, S., Vladyko, A., Letenko, I.: A fuzzy logic-based information security management for software-defined networks. In: 16th International Conference on Advanced Communication Technology, pp. 167–171. IEEE (2014)

NEW2AN: Satellite Communications

GNSS Attitude Determination Based on Antenna Array Space-Time Signal Processing

Igor Tsikin and Elizaveta Shcherbinina[✉]

Institute of Physics, Nanotechnology and Telecommunications,
Peter the Great St. Petersburg Polytechnic University, Saint-Petersburg, Russia
tsikin@mail.spbstu.ru, lizspbstyle@gmail.com

Abstract. This paper presents Global Navigation Satellite Systems (GNSS) attitude determination approach based on solution of direction-finding (DF) problem. The carrier phase GNSS concept is used to achieve precise attitude parameters (pitch, roll and yaw angles). This approach consists in maximization of resulting functions constructed on different combinations of R-functions. Effective attitude parameters estimations are maximum likelihood estimations. Well known algorithms (least-square, QUEST, etc.) of likelihood function maximization are computationally complicated. Besides these algorithms, reference phase differences can be used to maximize likelihood function. Reference phase differences are easily implemented and have advantages in computing costs. Angles accuracy is higher, when presented approach compared to ML, when reference phase differences are used in both cases. DF concept for attitude determination is discussed in this paper. Two different modifications of resulting function using for attitude determination are considered. Efficiency of described methods is expressed in terms of angles accuracy and abnormal error rate.

Keywords: Attitude determination · Direction-finding problem · Space-time processing · Carrier phase measurements

1 Introduction

Interferometric methods of attitude determination based on global navigation satellite systems are widely distributed nowadays [1–5]. Maximum likelihood (ML) method provides effective attitude parameters (angles yaw α, pitch β, roll γ) estimations [5, 6]. Different ways to maximize likelihood function are assumed: based on solving of Wahba's problem (for example algorithm QUEST and its different modifications) [6–9], conventional nonlinear least – square algorithm [4, 10], etc. These algorithms are computationally complicated [6].

In this context, maximization method based on reference phase differences (RPD) has certain advantages [11]. Angles accuracy depends on quantity of RPD values graduations. Along with ML method, other approach in attitude determination applications can be used [12, 13]. This approach is based on direction-finding (DF) concept. It consists in obtaining angles estimations by different combinations of

O. Galinina et al. (Eds.): NEW2AN/ruSMART 2016, LNCS 9870, pp. 573–583, 2016.
DOI: 10.1007/978-3-319-46301-8_48

R-functions maximization. Each R-function is constructed on transformation of spatial power distribution [14]. The advantages of such method in comparison with ML are showed in [13], when in both cases RPD are used. Nevertheless, in mentioned papers proposed results are concerned the heuristic combinations of the signals from several satellites. In particular, approach based on R-functions summation was considered. Angles are estimated by searching the maximum of resulting function, which is the sum of R-functions. Each R-function is calculated separately for each satellite. In the same time, it is interesting to compare this combination R-functions to another one, which based on multiplication of these functions for separate satellites. It can be supposed, that due to global maximum of resulting function narrowing angles accuracy should increase. Also, it should be pointed out, that in [12] R-function was obtained only for special case of antenna array (AA) with three elements.

In this paper angles estimation method based on multiplication of R-functions is analyzed. Antenna arrays with random number of elements are discussed.

2   Attitude Determination Based on R-Function

2.1   Direction-Finding Approach in Attitude Determination

Differences between carrier phase measurements of signals, which are received by array elements, (AE) are the basis of interferometric methods. The information about angles between direction to the satellites and vectors, which are formed by AE pairs, is contained in these differences. Problem of attitude determination is solved by carrier phase differences, when separate satellites coordinates, location of AE on navigation object (NO) and location of NO as material point are known.

Pay attention to the fact, that attitude determination problem might be considered as inverse to direction finding problem. Principal difference this problem from DF problem is consisted in known NO location, including attitude parameters in last case. In the same time, emitter (in this case satellite) coordinates are only known in attitude determination. Values of α, β, γ should be measured. Therefore, look first at DF problem.

In common case, DF problem can be decided as problem of searching spatial power maximum. In particular case of one emitter, which is considered as satellite, spatial power estimation for antenna array with M elements can be expressed in follow form, where noises on AE are statistically independent with equal variances and zero means [12]:

$$\hat{P}(\mathbf{w}) = \mathbf{w}^H \left(\frac{1}{N} \sum_{n=1}^{N} \left(\mathbf{F}(t_n)\mathbf{F}^H(t_n) \right) + \sigma^2 \mathbf{I} \right) \mathbf{w}, \tag{1}$$

where $\mathbf{w} = [|w_1| \cdot e^{j\theta_1}, \ldots, |w_M| \cdot e^{j\theta_M}]^T$ - vector-column of weight coefficients; $\mathbf{F}(t) = [F_1(t), \ldots, F_M(t)]^T$ - vector-column of signal complex envelopes on AE, $F_m(t) = |a_m| e^{j\varphi_m} S(t)$, φ_m – signal phase, which is received on m-antenna element from emitter; $S(t)$ – signal complex envelop of emitter; a_m – complex propagation coefficient between emitter and m-AE, $\mathbf{N}(t) = [N_1(t), \ldots, N_M(t)]^T$ – vector-column of additive noise complex envelopes on AE и $\mathbf{I}$ – identity matrix. Signs T and H denote transpose and Hermitian conjugation operations. Main information about spatial orientation of emitter is contained in phase differences values. So amplitudes of all weight coefficients are equal 1: $|w_1| = |w_2| = \ldots = |w_M| = 1$. The following equation is obtained:

$$\hat{P}(\mathbf{w}) = \left| e^{-j\theta_1} \quad \cdots \quad e^{-j\theta_M} \right| \cdot \begin{vmatrix} \frac{1}{N}\sum\limits_{n=1}^{N} |F_1(t_n)|^2 + \sigma^2 & \cdots & \frac{1}{N}\sum\limits_{n=1}^{N} F_1(t_n)F_M^*(t_n) \\ \vdots & \vdots & \vdots \\ \frac{1}{N}\sum\limits_{n=1}^{N} F_M(t_n)F_1^*(t_n) & \cdots & \frac{1}{N}\sum\limits_{n=1}^{N} |F_M(t_n)|^2 + \sigma^2 \end{vmatrix}$$

$$\cdot \begin{vmatrix} e^{j\theta_1} \\ \cdots \\ e^{j\theta_M} \end{vmatrix}. \tag{2}$$

Path differences of signals, incoming the AE, do not differ much in comparison with distance to the emitter. So the amplitudes of complex propagation coefficients are equal:

$$|a_1| = \ldots = |a_M| = |a|. \tag{3}$$

Mean power estimation of receiving signal is following:

$$\hat{P}_{Snp} = \frac{1}{N}|a|^2 \sum_{n=1}^{N} |S(t_n)|^2. \tag{4}$$

Considering (3) and (4) Eq. (2) can be converted in the following way:

$$\hat{P}(\mathbf{w}) = \hat{P}_{Snp}\left(M + 2 \cdot \sum_{k=1}^{K} \cos(\Delta\varphi_k - \Delta\theta_k)\right) + M\sigma^2, \tag{5}$$

where K is the number of bases in AA, which can be found as $M = C_N^2 = N!/((N-2)! \cdot 2!)$, $\Delta\varphi_k = \varphi_i - \varphi_j$ and $\Delta\theta_k = \theta_i - \theta_j$ for $i \neq j, i = 1\ldots M, j = 1\ldots M$. To maximize the function $\hat{P}(\mathbf{w})$ is sufficient to find the maximum of the following equation:

$$\hat{R} = \sum_{k=1}^{K} \cos(\Delta\varphi_k - \Delta\theta_k). \tag{6}$$

Function $\hat{R}$ reaches the maximum in conditions: $\Delta\theta_k = \Delta\varphi_k$, where $\Delta\varphi_k$- true values of signals phase differences. Actually, values $\Delta\varphi_k$ are unknown and in practice they can be obtained only as estimations of measured phase differences $\Delta\tilde{\varphi}_k$. Then maximization of following function will be implemented as solution of DF problem:

$$R = \sum_{k=1}^{K} \cos(\Delta\tilde{\varphi}_k - \Delta\theta_k). \tag{7}$$

Values of R-function can be obtained for emitter directions by substituting different values of parameters $\Delta\theta_k$, which are defined different emitter directions. Position of maximum R-function value corresponds to emitter direction estimation.

Initial data for maximization $\hat{P}(\mathbf{w})$ are values of signals complex envelopes on array elements, and for R-function these values are estimations of measured phase differences $\Delta\tilde{\varphi}_k$. Estimation of phase differences can be accomplished after optimal correlation processing unit, when signals structure is known. In general, sum of signals from all visible GNSS satellites is received on AE, when the user knows code- (GPS) or frequency- (GLONASS) GNSS signals division mechanism. Angles α, β, γ are assumed to be known, when satellite direction finding problem is being solved. Respectively, satellites directions are determined by the values $\Delta\theta_k$ and known angles.

Then go back to the attitude determination problem. Satellites directions are known, when attitude determination problem is being solved. Specifically, azimuth and elevation angles are being calculated, using coordinates of satellite and navigation object in geocentric coordinate system. In this case, values α, β, γ are unknown. So values of R-function can be obtained for different angles α, β, γ by substituting different values of parameters $\Delta\theta_k$, which now are defined by different attitude parameters.

Minimum two satellites [15] are required for unique solution of attitude determination problem. Thus, joint use of several R-functions for different satellites is necessary. Various methods of R-functions joint use are possible. In particular, in R-functions combination, based on summation, is discussed:

$$R_\Sigma = \sum_{l=1}^{L} R_l, \tag{8}$$

where $R_l = \sum_{k=1}^{K} \cos(\Delta\varphi_k^{(l)} - \Delta\theta_k^{(l)})$ denotes R-function for l-satellite.

Other resulting function modification, which affords angles estimations, is now considered. This modification is based on multiplication of functions R_l for different satellites:

$$R_\Pi = \prod_{l=1}^{L} R_l. \tag{9}$$

Angles accuracy is expected to increase because the main maximum width of the resulting R_Π-function is reducing in comparison with R_Σ-function. Comparative analysis of two mentioned approaches for constructing resulting functions is of interest to carry out. It is also important to investigate as normal and abnormal measurements errors.

2.2 Resulting Functions Maximization Based on Reference Phase Differences

In paper reference phase differences are proposed to consider as maximization method of resulting functions (8) and (9):

$$\Delta\psi_m^{(l)} = \frac{2\pi}{\lambda} \left[(x_m \sin \alpha_s^{(l)} + y_m \cos \alpha_s^{(l)}) \cdot \cos \beta_s^{(l)} + z_m \sin \beta_s^{(l)} \right], \qquad (10)$$

where $\alpha_s^{(l)}$, $\beta_s^{(l)}$ - l-satellite's azimuth and elevation. Coordinates (x_m, y_m, z_m) of baseline m-vectors are obtained in accordance with the equation:

$$x_m = x_{AEk} - x_{AEn}, y_m = y_{AEk} - y_{En}, z_m = z_{AEk} - z_{AEn}, \qquad (11)$$

where $(x_{AEk}, y_{AEk}, z_{AEk})$ - coordinates of k-AE in local coordinate system, which is associated with NO.

Reference phase differences for each specific satellite are phase differences values of signals, which are received on AE. These values are being calculated in accordance with Eq. (10) before or during measurements. Matching all possible angle combinations to the coordinates of baseline m-vectors is obtained using transformation matrix from the fixed coordinate system to the local coordinate system. Hence, RPD values corresponding to different α, β, γ are derived.

So, some table of RPD values is obtained. In this table each angles' combination is matched KL set of RPD values. In the end, values of R_Π, R_Σ functions can be obtained for all possible angles α, β, γ by substitution these sets in these resulting functions. Position of maximum resulting functions values defines attitude parameters estimations $\hat{\alpha}, \hat{\beta}, \hat{\gamma}$.

3 Simulation and Simulation Results

3.1 Data

Comparative analysis mentioned below methods of constructing resulting functions was accomplished for antenna array with minimum number of elements ($M = 3$), which is necessary for attitude determination. These AA are used in small unmanned aerial vehicles (UAV). AE location on navigation object, which is given as UAV, is illustrated on Fig. 1. Values of baselines between AE are equal 2 m and $\sqrt{2}$ m respectively.

GPS constellation was used for attitude determination. It was fixed in specific location of Saint Petersburg on 20 of February 2016. In Table 1 azimuths and elevations of satellites are showed.

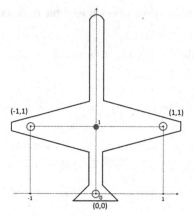

Fig. 1. AE location on UAV

Table 1.

$\alpha_s^{(l)}$	$\beta_s^{(l)}$
193.89	60.40
262.08	60.56
104.82	42.74
192.77	34.33
108.15	34.05
294.98	32.07
56.72	27.59
318.59	24.66
217.21	16.15

3.2 Simulation

Gaussian approximation of phase difference measurements [16] distribution is used in simulation for large values of signal to noise ratio.

$$W(\Delta\varphi_{kn}) = = \frac{1}{\sqrt{2\pi}\sigma_{\Delta\varphi}} \exp\left(-\frac{1}{2}\frac{(\Delta\varphi_{kn} - \Delta\hat{\varphi}_{kn})^2}{\sigma_{\Delta\varphi}^2}\right), \tag{12}$$

where $\left(\sigma_{\Delta\varphi}\right)^2$, $\Delta\hat{\varphi}_{kn}$ denote variances and means values $\Delta\varphi_{kn}$ respectively. Mean values of these phase differences are equal to the RPD which have been calculated for true spatial position (α_0, β_0, γ_0). Standard deviations (STD) of these phase differences $\sigma_{\Delta\varphi}$ are determined by (SNR, q) on array elements [16]:

$$\left(\sigma_{\Delta\varphi}^{(l)}\right)^2 = \frac{1}{q^{(l)}}. \tag{13}$$

In this paper case of equality SNR for different satellites is considered. In simulation various realizations of phase differences measurements were obtained for chosen true attitude parameters α_0, β_0, γ_0.

3.3 Results and Analysis

Minimum values of normal and abnormal measurement errors were considered as selection criterion between two resulting functions R_Π, R_Σ. On Fig. 2 STD values $\sigma_{\hat{\alpha}}, \sigma_{\hat{\beta}}, \sigma_{\hat{\gamma}}$ for different values of SNR are showed. It should be pointed out that view presented on Fig. 3 dependencies is the same for other possible true attitude parameters.

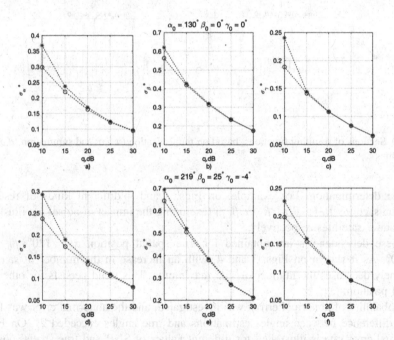

Fig. 2. Standard deviations of yaw (a,d), pitch (b,e) and roll (c,f) angles for two true spatial positions

Derived dependencies confirm assumption about advantage of the method, based on multiplication in the normal errors area.

Interferometric methods are characterized by measurement ambiguity when baseline between AE is of the order of the satellite signal's wavelength or larger. There are many different methods of ambiguity resolution [1, 3, 8, 17, 18]. On the other hand, when presented in this paper approach of attitude determination is used, this problem is almost solved when a large number of satellites. The R_Π, R_Σ functions have a lot of side maximums, when only two satellites are used. This fact leads to abnormal errors in the

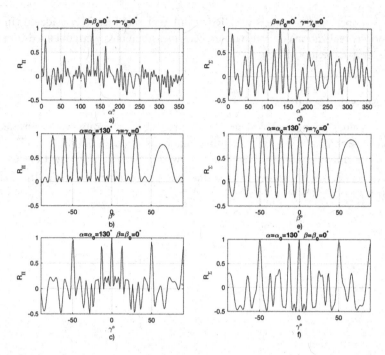

Fig. 3. Sections of resulting functions, based on multiplication (a,b,c) and summation (d,e,f) for 2 satellites

attitude determination. For example, on Figs. 3 and 4 different kinds of resulting function sections $R_\Pi(\alpha, \beta, \gamma), R_\Sigma(\alpha, \beta, \gamma)$ (excluding the impact of noise) are illustrated for 2 and 7 satellites respectively.

These dependencies are obtained for true spatial position $\alpha_0 = 130°$, $\beta_0 = 0°$, $\gamma_0 = 0°$. As it shown on Figs. 3 and 4, with an increase in the number of satellites extremely decreases the number of side maximums. This trend persists for other true spatial position.

Probability of abnormal errors was investigated, and the abnormal error was fixed, when difference between angles estimations and true angles exceeded 2°. On Fig. 5 abnormal error rate is illustrated for different values of SNR and true spatial position mentioned above.

In case of abnormal error analysis, method of combination R-functions, based on multiplication, leads to a larger abnormal error rate than the summation method.

So, as it follows from simulation results, method of R-functions combination, based on multiplication, does provide the lowest variance of the measurement estimation in normal errors area. However, this method is inferior to one, based on R-functions summation, in abnormal errors area. Angles estimations STD derived by these two methods are similar for SNR values larger than 15 dB. Because the typical SNR values for GNSS are between 15 and 20 dB, preference is given to resulting function R_Σ.

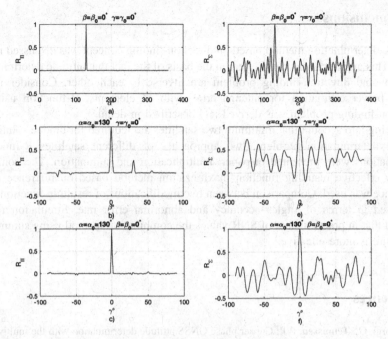

Fig. 4. Sections of resulting functions, based on multiplication (a,b,c) and summation (d,e,f) for 7 satellites

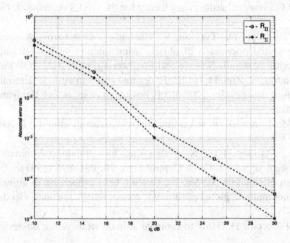

Fig. 5. Abnormal error rate for different values of signal-to-noise ratio and two resulting functions

4   Conclusions

Attitude determination method based on direction-finding concept was discussed in this paper. This approach was considered on the basis of the fact that attitude determination problem and direction-finding problem are inverse to each other. Considering the spatial power estimation for antenna array with M elements, R-function using in direction-finding problem was derived and described in detail.

Taking into account that minimum two satellites are required for unique solution of attitude determination problem, two approaches of different satellites R-functions combinations were presented in paper: multiplication and summation. The computationally effective resulting function maximization method based on reference phase difference was used. Comparison between two modifications of resulting functions was expressed in terms of angles accuracy and abnormal error rate. Simulation results showed that in practical area of SNR values the combination method using summation procedure is more effective.

References

1. Giorgi, G., Teunissen, P.J.: Carrier phase GNSS attitude determination with the multivariate constrained LAMBDA method. In: 2010 IEEE Aerospace Conference, pp. 1–12. IEEE (2010)
2. Giorgi, G., Teunissen, P.J.: GNSS carrier phase-based attitude determination (2012)
3. Teunissen, P.J.G.: Integer least-squares theory for the GNSS compass. J. Geodesy **84**, 433–447 (2010)
4. Fateev, Y.L., Dmitriev, D., Tyapkin, V., Kartsan, I., Goncharov, A.: Phase methods for measuring the spatial orientation of objects using satellite navigation equipment. In: IOP Conference Series: Materials Science and Engineering, p. 012022. IOP Publishing (2015)
5. Markel, M., Sutton, E., Zmuda, H.: IEEE: optimal anti-jam attitude determination using the global positioning system. In: 2002 IEEE Position Location and Navigation Symposium, pp. 12–19 (2002)
6. Shuster, M.: Maximum likelihood estimation of spacecraft attitude. J. Astronaut. Sci. **37**, 79–88 (1989)
7. Crassidis, J.L., Markley, F.L.: New algorithm for attitude determination using global positioning system signals. J. Guid. Control Dyn. **20**, 891–896 (1997)
8. Bing, W., Lifen, S., Guorui, X., Yu, D., Guobin, Q.: Comparison of attitude determination approaches using multiple global positioning system (GPS) antennas. Geodesy and Geodyn. **4**, 16–22 (2013)
9. Markley, F.L., Mortari, D.: How to estimate attitude from vector observations, pp. 1979–1996 (2000)
10. Chmykh, M.K., Fateev, Y.L.: Algorithms of object three-dimensional orientation determination based on global satellite navigation systems. In: 2nd International Conference on Satellite Communications - Proceedings of ICSC 1996, vol. 1–4, pp. 227–229 (1996)
11. Davydenko, A.S., Makarov, S.B.: Application of a method of a reference phases difference for definition of spatial object orientation. St. Petersburg State Polytech. Univ. J. Comput. Sci. Telecommun. Control Syst., 39–46 (2013)

12. Venediktov, V.T., Tsikin, I.A., Shcherbinina, E.A.: Satellite navigation signals processing in the framework of space. St. Petersburg State Polytech. Univ. J. Comput. Sci. Telecommun. Control Syst., 29–38 (2013)
13. Tsikin, I.A., Shcherbinina, E.A.: Comparative analysis of attitude determination algorithms based on reference phase differences using global navigation satellite system signals. In: The 18-th International Conference Digital Signal Processing and Its Applications, vol. 2, pp. 572–577, Moscow, Russia (2016)
14. Krim, H., Viberg, M.: Two decades of array signal processing research: the parametric approach. Sig. Process. Mag. IEEE **13**, 67–94 (1996)
15. Attitude Determination. Global Positioning System: Theory and Applications, vol. II, pp. 519–538. American Institute of Aeronautics and Astronautics (1996)
16. Tsikin, I.A., Shcherbinina, E.A.: Interferometric attitude estimation by maximum likelihood using reference phase difference. Electromag. Waves Electron. Syst. **19**, 30–37 (2014)
17. Wang, B., Miao, L.J., Wang, S.T., Shen, J.: A constrained LAMBDA method for GPS attitude determination. GPS Solut. **13**, 97–107 (2009)
18. Fateev, Y.L., Dmitriev, D.D., Tyapkin, V.N., Ishchuk, I.N., Kabulova, E.G.: The phase ambiguity resolution by the exhaustion method in a single-base interferometer. ARPN J. Eng. Appl. Sci. **10**, 8264–8270 (2015)

Angle-of-Arrival GPS Integrity Monitoring Insensitive to Satellite Constellation Geometry

Igor Tsikin and Antonina Melikhova[⊠]

St. Petersburg Polytechnic University, 29 Politechnicheskaya St.,
St. Petersburg 195251, Russia
tsikin@mail.spbstu.ru, antonina_92@list.ru

Abstract. Signals in global navigation satellite systems (GNSS) due to weak power are vulnerable to structural interferences which can lead to a significant deviation of the position solution from its true value. In aviation, UAV-controlling or some another life critical applications misleading coordinate information is a great threat so that procedures to detect such GNSS integrity failure are under a big concern. This paper is focused on decision-making algorithm for the failure detection applied to Angle-of-Arrival (A-o-A) integrity monitoring method in a case when the fixed decision threshold is preset in accordance with false alarm probability restricted for all possible observing satellite constellations. Decision threshold value was obtained for a different number of satellites by statistical simulations for quite a number of randomly generated satellite constellations with suitable geometric dilution of precision (GDOP) level. Minimum number of navigation signals was estimated for the situation when the simplest three elements antenna array implemented on compact UAV is used for a direction-finding procedure. As a result A-o-A integrity monitoring efficiency was estimated for real GPS satellite constellation under conditions when decision threshold was fixed as insensitive to satellite constellation geometry.

Keywords: Global navigation satellite systems · Interference mitigation · Spoofing detection · Antenna array

1 Introduction

Users of global navigation satellite systems (GPS, Galileo, GLONASS, etc.) estimate their coordinates by the analysis of satellite signals radio-navigation parameters (doppler frequency shifting and transmitting time delay) [1]. Open service signals [2], commonly used in avionic [3], UAV-control systems [4], etc. [5], are vulnerable to spoofing threats, so that obtained coordinates can be completely mislead by impact of structural interference signal sources [6]. User navigation equipment (point A on Fig. 1), located in the area of such false navigation signal source (FNSS on Fig. 1) coverage, will detect false signals instead of legal, and as a result it will lead to incorrect information about user's position (point A' on Fig. 1).

The situation, when the value of the position error exceeds an admissible limit, is referred to the integrity failure [7]. The procedure to detect such a failure is known as integrity monitoring [8] which provide the user with a warning that the satellite

© Springer International Publishing AG 2016
O. Galinina et al. (Eds.): NEW2AN/ruSMART 2016, LNCS 9870, pp. 584–592, 2016.
DOI: 10.1007/978-3-319-46301-8_49

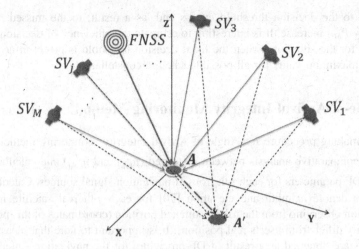

Fig. 1. GNSS integrity failure

navigation system cannot be used for positioning measurements. Well known civilian integrity monitoring methods [9] suffer from some shortcomings such as weight and overall dimensions [10], low noise immunity of additional communication channel between static reference station and user [11], etc.

On the other hand, in the situations, when all spoofing signals are transmitted from a single source, their Angles-of-Arrival (A-o-A) [12, 13] can be used to detect such a failure [12]. Based on this principle A-o-A integrity monitoring method involves comparative analysis of carrier phase [12, 14, 15] or direction-finding (DF) [12, 16] measurements for all navigation signals. Such measurements can be obtained during spatial signal processing using antenna arrays with a small number (2…7) of elements [17]. Arrays with three elements are widespread on drones and preferable for integrity monitoring system implementation. This kind of antenna arrays is analyzed in this work.

Obviously, the situation of integrity failure, when all signals are transmitted by a common source, is characterized by equal values for all measured DF parameters (azimuth μ and elevation η) [16]. But under fault-free GNSS conditions these measured parameters have different values for all navigation signals because they are transmitted by separate sources located on GNSS constellation satellites.

Of course, if an attacker uses multiple sources of fake signals, measured parameters can have different values for the navigation signals. To detect such a kind of integrity failures some another distinctive feature of differences between directions to real satellite vehicles (these directions are known values) and to interference signal sources should be used.

Earlier decision-making algorithms [16, 18, 19] for A-o-A integrity monitoring method were constructed with optimal decision threshold estimation for each observing satellite constellation. Practical implementation of such algorithms is not an easy way.

More simple approach can be based on fixed decision threshold for all possible satellite constellations. During the optimization procedure this fixed threshold can be preset in accordance with limitations on false alarm P_{FA} probability which must be averaged or maximized for all possible constellations. First variant leads to the undesirable situations of P_{FA} probability increase for some constellations. The second one leads to the decision threshold excess and, as a result, to the missed detection probability P_{MD} increase. It is interesting to estimate the efficiency of decision-making algorithm for the situations when the fixed decision threshold is preset in accordance with P_{FA} maximum value for all possible satellite constellations.

2 Angle-of-Arrival Integrity Monitoring Method

Decision-making procedure for Angle-of-Arrival integrity monitoring method can be based on comparative analysis between measured ($\mu_{msd}^{(j)}$ and $\eta_{msd}^{(j)}$) and calculated ($\mu_{clc}^{(j)}$ and $\eta_{clc}^{(j)}$) DF parameters for each j-th from M navigation signal sources. Calculated DF parameters denote azimuth and elevation [20] for each j-th real satellite in GNSS constellation observing from the users estimated position (coordinates of this point may significantly differ from user's real position). It is important to note that measured DF parameters are obtained as a result of DF-procedure for the navigation signal source/ sources at the users real position [21].

In case, when distances between elements in a small antenna array are less than half of the GNSS signals carrier wavelength, measured DF parameters are normally distributed, and their standard deviations are not greater than 10 degrees [22]. For example, in the practical interest area (10...20 dB [23]) of $h^2 = E_0/N_0$ ratio (where E_0 is a received GNSS signal energy and the $N_0/2$ is a mean power spectral density of white Gaussian noise (WGN) at the input of the optimal signal processing block), DF-parameters $\mu_{msd}^{(j)}$ and $\eta_{msd}^{(j)}$ are normally distributed and have standard deviations $\sigma_{\mu,j}^{msd}$ and $\sigma_{\eta,j}^{msd}$ in the range 1...10 degrees [22] which are much greater than respective values $\sigma_{\mu,j}^{clc}$ and $\sigma_{\eta,j}^{clc}$ for calculated DF-parameters $\mu_{clc}^{(j)}$ and $\eta_{clc}^{(j)}$, which are also normally distributed [22].

Arithmetical transformations for likelihood ratio like [24] lead to final form of constructed decision-making algorithm:

"GNSS integrity failure is detected if

$$\sum_{j=1}^{M} \frac{\left(\mu_{clc}^{(j)} - \mu_{msd}^{(j)}\right)^2}{\left(\sigma_{\mu,j}^{clc}\right)^2 + \left(\sigma_{\mu,j}^{msd}\right)^2} + \sum_{j=1}^{M} \frac{\left(\eta_{clc}^{(j)} - \eta_{msd}^{(j)}\right)^2}{\left(\sigma_{\eta,j}^{clc}\right)^2 + \left(\sigma_{\eta,j}^{msd}\right)^2} + \sum_{j=1}^{M} \left\{ \frac{2}{M} \mu_{msd}^{(j)} \sum_{k=1}^{M} \mu_{msd}^{(k)} - \left(\mu_{msd}^{(j)}\right)^2 \right\} \Big/ \left(\sigma_{\mu,j}^{msd}\right)^2 -$$

$$- \sum_{j=1}^{M} \frac{1}{M^2} \left(\sum_{j=1}^{M} \mu_{msd}^{(j)} \right)^2 \Big/ \left(\sigma_{\mu,j}^{msd}\right)^2 + \sum_{j=1}^{M} \left\{ \frac{2}{M} \eta_{msd}^{(j)} \sum_{k=1}^{M} \eta_{msd}^{(k)} - \left(\eta_{msd}^{(j)}\right)^2 \right\} \Big/ \left(\sigma_{\eta,j}^{msd}\right)^2 -$$

$$- \sum_{j=1}^{M} \frac{1}{M^2} \left(\sum_{j=1}^{M} \eta_{msd}^{(j)} \right)^2 \Big/ \left(\sigma_{\eta,j}^{msd}\right)^2 > \ln \Lambda_0, \qquad (1)$$

otherwise GNSS fault-free conditions are detected".

3   Decision Threshold Estimation

Decision-making threshold estimation procedure can be processed on basis of Newman-Pierson strategy, which assumes P_{MD} minimization when P_{FA} value is fixed. It is clear that fixed decision threshold value provides different P_{FA} and P_{MD} values depending on observing satellite constellation geometry and antenna array structure that characterize distribution parameters for the random values in (1). So the fixed decision threshold $\ln \Lambda_0^*$ providing P_{FA} probability no greater than some predetermined value must be estimated on basis of the probability characteristics for different observing constellations.

Observing satellite constellations in modern GNSS are characterized by geometric dilution of precision γ [25] which value for near Earth surface located users is no greater than 5 [26]. Due to that fact this paper is focused on constellations providing $\gamma < 5$. Typical GPS constellation observed from the point (60°00'30.8"N 30°22'28.8"E) at the time 09:10 GMT on 03.03.2016 is presented on Fig. 2a. This constellation provides geometric dilution of precision value close to 2. Satellites numbers on this figure correspond to the system numbers in GPS.

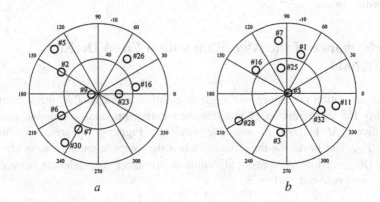

Fig. 2. Real (a) and randomly generated (b) satellite constellations with $\gamma = 1.8$

Monte-Carlo statistical simulations were made for a quite a number G (no more than 200, because further G increment will not lead to the notable results deviation) of randomly generated satellite constellations which were selected in accordance with criterion $\gamma < 5$. One of such randomly generated constellations is presented on Fig. 2b. It can be noticed that real and randomly generated constellations with equal γ differ little from each other.

False alarm probability $P_{FA}^{(i)}$ for situation, when three elements array with distances between elements is near half of a wavelength is used for DF-procedure, were obtained depending on decision threshold $\ln \Lambda_0$ for each i-th of such constellations for different M satellites number in use ($M \le 9$) when $h^2 = 10\,\mathrm{dB}$ (lower border of practical interest area [23]). As a result $P_{FA}^{(i)}$ upper bound $\max_i \left\{ P_{FA}^{(i)} \right\}_M$ was achieved for the situation

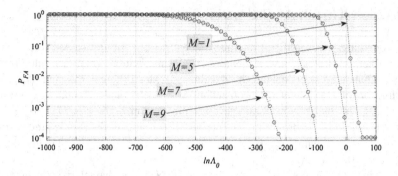

Fig. 3. $\max_i \left\{ P_{FA}^{(i)} \right\}_M$ dependences on $\ln \Lambda_0$ for $L = 3$ and $h^2 = 10\,\text{dB}$

when antenna array with three elements is used for DF-procedure. This simulation results for different M number of currently available satellites are presented on Fig. 3.

Presented results were obtained for algorithm (1) implementation and enable to estimate decision threshold value providing P_{FA} for all observing satellite constellations in admissible limits.

4   Performance of the Algorithm with a Fixed Decision Threshold

The next simulation procedure was made to estimate averaged $\bar{P}_{MD}$ missed detection probability for all satellite constellations used during previous simulation procedure with different M number of available satellites. Figure 4 illustrates dependences between $\bar{P}_{MD}$ and $\ln \Lambda_0$ for the situation when the simplest antenna array ($L = 3$) is used for DF procedure. As a result algorithm performance (dependences between $\bar{P}_{MD}$ and P_{FA}) are presented on Fig. 5.

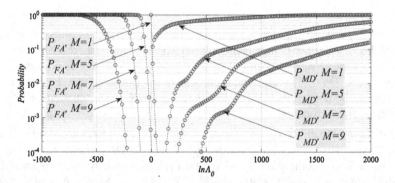

Fig. 4. Dependences between $\bar{P}_{MD}$ and $\ln \Lambda_0$ for $L = 3$ and $h^2 = 10\,\text{dB}$

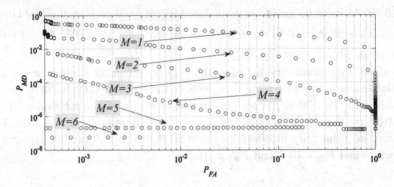

Fig. 5. Dependences between $\bar{P}_{MD}$ and $\max_i\left\{P_{FA}^{(i)}\right\}_M$ for $L = 3$ and $h^2 = 10\,\text{dB}$

Presented on Fig. 5 algorithm (1) performance analysis enables us to conclude that fixed decision threshold presetting allows to achieve suitable probability based characteristics ($\max_i\left\{P_{FA}^{(i)}\right\}_M < 10^{-5}$ and $\bar{P}_{MD} < 10^{-5}$) even for a small (no greater than 5) visible satellites number when simplest antenna array ($L = 3$) is used for DF-procedure. Required visible satellites number can be provided by single GNSS constellation (GPS only as example [27]). Under difficult reception conditions such as urban or indoor areas required performances can be provided by multiple GNSS constellations (GPS, ГЛОНАСС, Galileo [28]).

Finally algorithm (1) performance was tested for fixed decision threshold selected in accordance with the recent simulation results (Fig. 3) when the real satellite constellation presented on Fig. 1a is used for decision-making procedure. Figure 6 shows dependences between P_{MD} and $\ln \Lambda_0$, P_{FA} and $\ln \Lambda_0$ obtained for mentioned constellation. Dotted and dash-dot lines on this figure illustrate dependences between $\max_i\left\{P_{FA}^{(i)}\right\}_M$ and $\ln \Lambda_0$, dependences $\bar{P}_{MD}$ and $\ln \Lambda_0$, respectively for $M = 9$ and

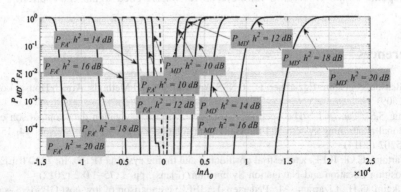

Fig. 6. Dependences between P_{FA} and $\ln \Lambda_0$, between P_{MD} and $\ln \Lambda_0$ for satellite constellation presented on Fig. 2b and $L = 3$

$h^2 = 10$ dB. As it can be seen from these dependences fixed decision threshold must be preset on value $\ln \Lambda_0^* \approx 0$ in the restrictions on $\max_i \left\{ P_{FA}^{(i)} \right\}_M < 10^{-5}$ for all possible constellations with 9 visible satellites.

Required P_{FA} probability for algorithm (1) implemented for this real satellite constellation can be provided by its own decision threshold which value in the practical interest h^2 area (10...20 dB) is not greater than -500 (solid line for $P_{FA} = 10^{-5}$ and $h^2 = 10$ dB on Fig. 6). It is clear that despite such threshold overstatement (-500 comparing to 0) integrity monitoring system all the same will provide required performance so that $P_{FA} < 10^{-5}$ and $P_{MD} < 10^{-5}$.

5   Conclusions

Decision-making algorithm to detect GNSS integrity failure is analyzed in this paper. Its performance in accordance with the antenna array structure and visible satellites numbers is achieved for typical satellite constellations. Unlike another algorithms which assume various optimal decision threshold for each of visible satellite constellations, presented novel algorithm uses fixed decision threshold for all possible of them. The fixed decision threshold is selected to provide restrictions on $\max_i \left\{ P_{FA}^{(i)} \right\}_M$ for all possible satellite constellations. Simulation results for such situation showed that despite inevitable losses caused by threshold fixing integrity monitoring system anyway will provide users with required performance ($P_{FA} < 10^{-5}$, $P_{MD} < 10^{-5}$) even in the situations when only 5...7 satellites are visible. Simulation results were confirmed by considering a situation when the real GPS satellite constellation was used for decision-making procedure. It was shown that in real situation the algorithm with fixed decision threshold would provide required integrity monitoring efficiency.

It is possible to use the simplest antenna array with 3 elements to detect integrity failures with required performance which is actual for example for compact UAVs. Also, in case if the receiver detects the situation that GNSS signals are spoofed, antenna array spatial signal processing can be used for interference signal cancelation.

References

1. Gleason, S., Gebre-Egziabher, D.: GNSS Applications and Methods. Artech House, London (2009)
2. Jan, S.S., Tao, A.L.: The open service signal in space navigation data comparison of the global positioning system and the BeiDou navigation satellite system. Sensors **14**, 15182–15202 (2014)
3. Bartone, C.G.: IEEE: a terrestrial positioning and timing system (TPTS). In: 2012 IEEE/ION Position Location and Navigation Symposium (Plans), pp. 1175–1182 (2012)
4. Elkaim, G.H., Lizarraga, M., Pedersen, L.: IEEE: comparison of low-cost GPS/INS sensors for autonomous vehicle applications. In: 2008 IEEE/ION Position, Location and Navigation Symposium, vol. 1–3, pp. 285–296 (2008)

5. Wang, J.L., Liao, C.S., Inst, N.: Time and frequency dissemination system for synchronization applications at TL. In: Proceedings of the ION 2015 Pacific PNT Meeting, pp. 985–992 (2015)
6. Stubberud, S.C., Kramer, K.A.: IEEE: threat assessment for GPS navigation. In: 2014 IEEE International Symposium on Innovations in Intelligent Systems and Applications (INISTA 2014), pp. 287–292 (2014)
7. Azoulai, L.: ION: GNSS threats and aviation, mitigation techniques, alternatives and regulation. In: Proceedings of the 24th International Technical Meeting of the Satellite Division of the Institute of Navigation (ION GNSS 2011), pp. 1897–1906 (2011)
8. Wang, E.S., Yue, X.D., Pang, T., Zhang, Z.X.: Research on GPS receiver autonomous integrity monitoring algorithm in the occurrence of two-satellite faults. In: International Conference on Electronic, Information and Computer Engineering (ICEICE) (Year)
9. Tao, H.Q., Li, H., Lu, M.Q.: A GNSS anti-spoofing method based on the cooperation of multiple techniques. In: China Satellite Navigation Conference (CSNC) 2015 Proceedings, vol I 340, pp. 205–215 (2015)
10. Khanafseh, S., Roshan, N., Langel, S., Chan, F.C., Joerger, M., Pervan, B.: IEEE: GPS spoofing detection using RAIM with INS coupling. In: 2014 IEEE/ION Position, Location and Navigation Symposium - Plans 2014, pp. 1232–1239 (2014)
11. Dvorska, J., Podivin, L., Musil, M., Zaviralova, L., Kren, M., Inst, N.: GBAS CAT II/III business aircraft flight trials and validation - phase 1. In: Proceedings of the 27th International Technical Meeting of the Satellite Division of the Institute of Navigation (ION GNSS 2014), pp. 822–834 (2014)
12. Bitner, T., Preston, S., Bevly, D., Inst, N.: Multipath and spoofing detection using angle of arrival in a multi-antenna system. In: Proceedings of the 2015 International Technical Meeting of the Institute of Navigation, pp. 822–832 (2015)
13. Xu, K.J., Nie, W.K., Feng, D.Z., Chen, X.J., Fang, D.Y.: A multi-direction virtual array transformation algorithm for 2D DOA estimation. Sig. Process. 125, 122–133 (2016)
14. Psiaki, M.L., Powell, S.P., O'hanlon, B.W.: GNSS spoofing detection using high-frequency antenna motion and carrier-phase data, pp. 2949–2991, Nashville (2013)
15. Magiera, J., Katulski, R.: Accuracy of differential phase delay estimation for GPS spoofing detection. In: 2013 36th International Conference on Telecommunications and Signal Processing (TSP), pp. 695–699 (2013)
16. Magiera, J., Katulski, R.: Detection and mitigation of GPS spoofing based on antenna array processing. J. Appl. Res. Technol. 13, 45–57 (2015)
17. Melikhova, A., Tsikin, I.: Antenna array with a small number of elements for angle-of-arriving GNSS integrity monitoring. In: 2016 39th International Conference on Telecommunications and Signal Processing (TSP). IEEE, pp. 190–193 (2016)
18. Daneshmand, S., Jafarnia-Jahromi, A., Broumandan, A., Lachapelle, G.: IEEE: a GNSS structural interference mitigation technique using antenna array processing. In: 8th IEEE Sensor Array and Multichannel Signal Processing Workshop (SAM), pp. 109–112 (Year)
19. Tsikin, I.A., Melikhova, A.P.: Optimization of angle-of-arrival GPS integrity monitoring. In: Balandin, S., Andreev, S., Koucheryavy, Y. (eds.) NEW2AN/ruSMART 2015. LNCS, vol. 9247, pp. 722–728. Springer, Heidelberg (2015)
20. Montenbruck, O., Schmid, R., Mercier, F., Steigenberger, P., Noll, C., Fatkulin, R., Kogure, S., Ganeshan, A.S.: GNSS satellite geometry and attitude models. Adv. Space Res. 56, 1015–1029 (2015)
21. Wang, X.R., Aboutanios, E., Amin, M., Pui, C.Y., Inst, N.: Off-grid high resolution DOA estimation for GNSS circular array receivers. In: Proceedings of the 27th International Technical Meeting of the Satellite Division of the Institute of Navigation (ION GNSS 2014), pp. 2260–2267 (2014)

22. Melikhova, A., Tsikin, I.: Angle of arrival method for global navigation satellite systems integrity monitoring. St. Petersburg State Polytech. Univ. J. Comput. Sci. Telecommun. Control Syst. **212**(1), 37–49 (2015)
23. Kaplan, E., Hegarty, C.: Understanding GPS: Principles and Applications. Artech House, London (2005)
24. Tsikin, I.A., Melikhova, A.P.: The angle-of-arrival integrity monitoring efficiency under multiple observations. Radioengineering **9**, 69–77 (2015)
25. Chen, C.S.: Weighted geometric dilution of precision calculations with matrix multiplication. Sensors **15**, 803–817 (2015)
26. Teng, Y.L., Wang, J.L.: New characteristics of geometric dilution of precision (GDOP) for multi-GNSS constellations. J. Navig. **67**, 1018–1028 (2014)
27. Guochang, X.: GPS: Theory, Algorithms and Applications. Springer Science & Business Media, Heidelberg (2007)
28. Petrovski, I.G.: GPS, GLONASS, Galileo, and BeiDou for Mobile Devices: From Instant to Precise Positioning. Cambridge University Press, Cambridge (2014)

NEW2AN: Signals and Circuits

Investigation of Questions of Non-harmonic Signal Scattering on Impedance Structures

Alexander F. Kryachko[1]([✉]), Mikhail A. Kryachko[2],
Kirill V. Antonov[2], Yakov Y. Levin[2], and Igor E. Tyurin[2]

[1] Peter the Great St. Petersburg Polytechnic University, St. Petersburg, Russia
alex_k34.ru@mail.ru
[2] Saint-Petersburg State University of Aerospace Instrumentation,
St. Petersburg, Russia
kartovan@gmail.com

Abstract. On the basis of diffraction problem known solutions on a wedge in the case of harmonic effects using frequency method was analyzed the UWB pulse scattering on the impedance wedge. The features of the diffraction different kinds of pulses were revealed. We were defined the influence of the wedge's electrical and geometrical parameters, sensing conditions and monitoring, as well as probing signal spectrum shape and the diffraction pulse energy. In study of the time dependence influence the type of probe pulse on the scattering field was tested a time-frequency method.

Keywords: Impedance wedge · UWB impulse · Diffraction · Frequency method

1 Introduction

It is known that in ultra-wideband (UWB) radar [1] the scattering of the probe pulse at specific local centers largely determines the nature of the total field of diffraction. Under certain conditions the main contribution to the total effective scattering cross section (SCS) of the object with complex geometric shapes makes the scattering at its edges and kinks. Therefore, practically important task is the study of the diffraction of a UWB pulse for the impedance wedge, which in some cases can be used as an electro-dynamic model in the study of scattering on the edges of the real object. Currently well enough developed methods for solving diffraction problems for the electromagnetic field in the regime of steady harmonic oscillations. It is much harder to solve the diffraction problems in cases where the primary source varies in time non-sinusoidally. To solve this unsteady problem also different methods have been developed [2]. Most of them are based on the use of the superposition principle, which is used for the solution of any linear differential equations. In some cases, it seems appropriate to use frequency method [3], the main advantage of which is the ability to apply for known solutions found for the case of diffraction of the waves monochromatically, in the study of non-stationary scattering. The resulting field is defined as a superposition of responses to the elementary impact of the spectral components of UWB pulses.

© Springer International Publishing AG 2016
O. Galinina et al. (Eds.): NEW2AN/ruSMART 2016, LNCS 9870, pp. 595–603, 2016.
DOI: 10.1007/978-3-319-46301-8_50

Under the UWB signal we mean a signal with a large relative width of the spectrum [4]. The width of the spectrum $\Delta f = f_B - f_H$, where f_B, f_H are the upper and lower frequencies in the signal spectrum, and $f_0 = (f_B + f_H)/2$ – average frequency.

An indicator of the high bandwidth signal is defined by the expression:

$$\mu = \Delta f / f_0 \tag{1}$$

Signals, for which $\mu \geq 1/2$, are considered of UWB. The interest in nonstationary diffraction is caused by the scattering of pulses with the rate of the bandwidth $\mu \geq 1/2$ greatly differs from phenomena occurring during diffraction of the harmonic signal and requires separate consideration.

Let the impedance wedge with the opening angle F (Fig. 1), formed by two semi-infinite edges, which in the case of H-polarization are characterized by the normalized impedances $Z_{\pm}/Z_0$ (Z_0 - the impedance of free space), diffracted ultra-wideband pulse at an angle φ_0.

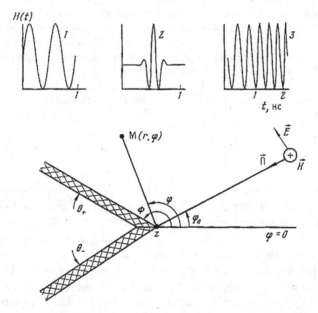

Fig. 1. Impedance wedge and the kinds of probing UWB pulses: 1-rectangular pulse; 2 - perfect sounding pulse; 3 - linear frequency modulation pulse.

Under H-polarization is understood to be the case when the incident field vector H parallel to the edge of the wedge. The z-axis of a cylindrical coordinate system aligned with the edge of the wedge.

2 Main Part

To be determined the time dependence of the diffracted field in the observation point $M(r, \varphi)$. For this the frequency method is used [3]. The exact solution for the case of harmonic vibrations has the form:

$$\dot{H}_z = -\frac{\dot{A}_{0i}}{4\Phi} \int_\gamma \exp[-ikr\cos\alpha] \frac{\psi(\alpha + \varphi)\cos(\pi\varphi_0/2\Phi)d\alpha}{\psi(\varphi_0)[\sin(\pi(\alpha + \varphi)/2\Phi) - \sin(\pi\varphi_0/2\Phi)]} \quad (2)$$

Part of subintegral expression of the function $\psi(\delta)$ is expressed through the function of Maluzinetz ψ_m [5]. In the case of diffraction of a plane IBM at impedance for the wedge Eq. (2) gives the value of the field in any point of space outside the wedge. However, its direct use in numerical simulations is difficult because the integral is not expressed through known functions. Away from the edges (i.e. when the condition $kr \gg 1$) it can be calculated by an approximate asymptotic method [5].

Full field in this case is represented as the following sum of asymptotic series:

$$\dot{H}_z^{\text{ПЛ}}(r,\varphi) \cong \dot{H}_{z\,\text{ДИФ}}^{\text{ЦИЛ}} + \dot{H}_{z\,\text{ПАД}}^{\text{ПЛ}} + \dot{H}_{z\,\text{ОТР+}}^{\text{ПЛ}} + \dot{H}_{z\,\text{ОТР-}}^{\text{ПЛ}} + \dot{H}_{z\,\text{ПОВ±}} \quad (3)$$

where

$$\dot{H}_{z\,\text{ДИФ}}^{\text{ЦИЛ}} = A_0\pi\cos\left[\pi\varphi_0/2\Phi\right]\left[\sqrt{2\pi kr}\times\left(2\Phi\psi(\varphi_0)\right)\right]^{-1}\times$$

$$\times\left[\left[\psi(\varphi-\pi)\right]\sin\left[\pi(\varphi-\pi)/(2\Phi)\right]-\sin\left[\pi\varphi_0/2\Phi\right]\right]^{-1}-$$

$$-\left[\left[\psi(\varphi+\pi)\right]\sin\left[\pi(\varphi+\pi)/(2\Phi)\right]-\sin\left[\pi\varphi_0/2\Phi\right]\right]^{-1}\exp\left[i(kr+\pi/4)\right]$$

characterizes cylindrical IBM, scattered in space edge wedge; $H_{z\,\text{ПАД}}^{\text{ПЛ}}$ and $H_{z\,\text{ОТР±}}^{\text{ПЛ}}$ respectively, and z-components of magnetic field intensity vector of incident and reflected from the faces of the wedge of electromagnetic waves. Component $\dot{H}_{z\,\text{ПОВ±}}$ determines the impedance at the excited faces of the wedge surface electromagnetic waves, which in the case of inductive impedance faces, and in the absence of losses is sustained and propagated along the edges of the wedge to infinity.

In practice there is no need to compute all partial components. It is enough to analyze the behavior of those who make the main contribution to the Effective Surface Scattering (ESS).

According to the frequency method, the approximate solution for diffraction at the wedge impedance UWB can be found as a result of the inverse Fourier transform of the spectral density of the probing signal and function reflecting the dependence of field strength with frequency in the case of steady-state harmonic oscillations. It can be written as:

$$\dot{H}(t) = \frac{1}{2\pi} \int\limits_{\infty}^{\infty} \dot{F}(i\omega)\dot{H}_z(\omega, \varphi, r) \exp[i\omega t]d\omega \qquad (4)$$

where $\dot{F}(i\omega) = \int\limits_{\infty}^{\infty} S(t)\exp[-i\omega t]dt$ - is the spectral density of the probing signal, and $\dot{H}_z(\omega, \varphi, r)$ — the solution to this problem for the case of harmonic oscillations (3).

Bulky the algorithm for computing special functions of Maluzintza $\psi_m(z)$ can be replaced by more simple approximation of its values [5]:

$$\psi_m(z) = \exp\left\{\frac{1}{2}(U + iV)\right\} \qquad (5)$$

where

$$U = -0.3 \sum_{n=1}^{5} \frac{\text{ch}[(0.3n - 0.15)x] \cos[(0.3n - 0.15)y] - 1}{(0.3n - 0, 15)\text{ch}[\pi(0.3n - 0.15)/2]\text{sh}[2\Phi(0.3n - 0.15)]};$$

$$V = 0.3 \sum_{n=1}^{5} \frac{\text{sh}[(0.3n - 0.15)x] \sin[(0.3n - 0.15)y]}{(0.3n - 0.15)\text{ch}[\pi(0.3n - 0.15)/2]\text{sh}[2\Phi(0.3n - 0.15)]}.$$

In the study of the influence of the form of the time dependence of the incident field at the diffraction of a pulse with fixed terms of sensing and monitoring the electrical and geometrical characteristics of impedance wedge is advisable to use a time-frequency method.

First you need to define the functional dependence of field strength $\dot{H}_z(\omega, \varphi, r)$ on frequency in case of harmonic oscillations. Then the result is inverse Fourier transform of this function finds the spatial impulse response of the wedge:

$$h(t) = \frac{1}{2\pi} \int\limits_{\infty}^{\infty} \dot{H}_z(\omega, \varphi, r) \exp[i\omega t]d\omega \qquad (6)$$

The time dependence of the scattered field can now be defined as the result of the convolution of the spatial impulse response signal and the time dependence of the field strength of the probing signal:

$$\dot{H}(\tau) = \int\limits_{\infty}^{\infty} \dot{H}_{\text{зонд}}(\tau)h(t - \tau)d\tau \qquad (7)$$

While of numerical calculations, we have analyzed the diffraction of a UWB pulse of several types (Fig. 1). Survey have shown that the form of the time dependence of diffraction of the pulse (DP) at the point of observation, its magnitude, duration, and energy largely depend on the geometrical and electro-physical parameters of impedance wedge, and environment sensing and monitoring.

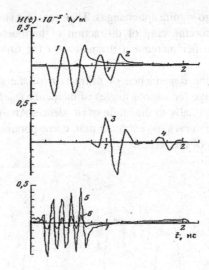

Fig. 2. Time dependence of the diffraction pulses: Im $(\theta_{\pm}) = 0{,}01$ (1); 0,5 (2); F = 100° (3); 140° (4); $\varphi = 40°$ (5, 6)

Figure 2 shows the time dependence of diffracted pulses of different types: 1, 2 - rectangular; 3, 4 - perfect probe pulse (impulse type 2) (Fig. 1); 5, 6 - signal with linear-frequency modulation (chirp signal).

The increase in the reactive component of the impedance edges of the wedge (Fig. 2, curves 1, 2) leads to a decrease of the amplitude of DI. This is due to the intensive excitation of surface waves on the edges of the wedge. Dependency analysis 3 and 4 in Fig. 2 allows us to conclude that the decrease in the opening angle of the wedge (the increasing angle F) essentially changes the shape of the scattered pulse. The position of the point of observation (curves 5 and b in Fig. 2) significantly affects the amplitude and duration of the scattered pulse. Type of the time dependence of the probe pulse and its spectrum under other equal conditions has a significant impact on the process of scattering at the wedge. One reason for this is the frequency dependence of the impedance faces of the wedge [5]. Compare for example a rectangular pulse and an ideal sounding pulse (Fig. 1, curves 1, 2). The proportion of the shape of the spectrum of a rectangular pulse and spatial phase characteristics of the wedge is such that with the diffraction of this pulse on the wedge increases the level of low-frequency components compared with the components of the middle part of the spectrum. Strongly reduced level of high-frequency spectrum components still makes a negligible contribution to the pulse energy. The above reason leads to a decrease in the energy of the signal delays of the rise and fall, increasing the pulse duration (Fig. 2, curves 1 and 2).

The shape of the amplitude and phase spectrum of the ideal pulse is such that its spectrum undergoes even greater changes than in the case described above. There is a shift of average frequency and spectrum shift in the lower frequency area. Change energy and type of time dependence DI (Fig. 2, curves 3, 4) are bigger than in the first case. The nature of the changes of the diffraction of the pulse depends on the bandwidth of the signal μ. With increasing μ to a value approximately equal to 0.5, the shape of the pulse does not undergo significant changes. Therefore, for such signals it is possible to use the ratio found for the case of diffraction of harmonic waves (with certain amendments). With a further increase μ the nature of the time dependence of DI is changed substantially.

In Fig. 3 is shown the dependence of diffraction of the pulse energy from the opening angle of the wedge for various angles of incidence φ_0. Moreover, the angle of observation φ is chosen equally to the angle of incidence. Analysis of energy changes of DI showed that all the curves have a maximum corresponding to the case of mirror reflection from the faces of the wedge.

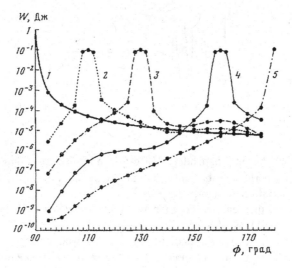

Fig. 3. Dependence of the energy DI from the opening angle of the wedge: $Im(\theta_\pm) = -0.5$; $F = 0°(1)$; $20°(2)$; $40°(3)$; $70°(4)$; $90°(5)$

In case of equality to zero of the values of the angle of incidence and the angle of observation (curve 1), diffraction of the pulse energy takes the highest value of all studied cases (curves 2–5) when the opening angle of the wedge F, equal to the value 0.5 $\square$, i.e. when the wedge degenerates into an infinite impedance plane. This is explained by the fact that in the impedance plane in the absence of inhomogeneity there are no excited surface waves.

Equality:

$$\lim_{\theta_\pm \to 0} \left| \dot{H}_{z\,\text{ПОВ}\pm} \right| = 0;$$

can be considered as a proof of asymptotic stability of solutions to change the opening angle of the wedge.

At other angles of incidence of the pulse, the position of the maximum is also determined by the ratio $F = \varphi_0 + 0.5\pi$ (to $\varphi_0 = \varphi = 20°$, $\Phi_m = 110°$, to $\varphi_0 = \varphi = 40°$, $\Phi_m = 130°$, etc.). The value of the opening angle of the wedge corresponding to the maximum, shifts towards higher values. Such character of change of energy DI is determined by the dependence of the energy of surface waves from the opening angle of the wedge.

The scatter plots (Fig. 4) have a clearly cut diffraction. In their structure two characteristic petals are distinguished, the position of which does not depend on the magnitude of the impedance and is determined only by the values of the angles φ_0 and F, i.e. the geometry of the problem. In case of equality of the impedance faces of the petals are arranged symmetrically relative to the axis of the wedge.

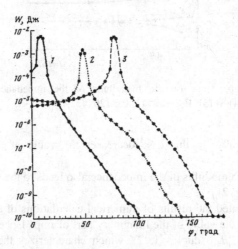

Fig. 4. Scatter plot of $Im(\theta_\pm) = -0.5$; F = 100° (1); 140° (2); 170° (3).

Therefore, Fig. 4 shows only one petal. The direction of one of them coincides with a boundary region of a mirror reflection of the illuminated face $\varphi' = 2\Phi - \pi + \varphi_0$. The energy of surface waves in this case is minimal, and the main contribution to the scattered field introduces diffraction component. When you remove the angle from the maximum observation, energy of the diffraction field made by surface wave increases, and the energy of the diffraction component is reduced.

Reducing the opening angle of the wedge (Fig. 4, curves 2, 3) change the conditions of excitation of surface waves on the sides and causes a shift in the position and magnitude of the maximum petal of the scattering diagram.

The position of the second petal depends on the condition $\varphi_0 < \pi - \Phi$. If this condition is fulfilled, then the corresponding equal angle $\varphi'' = -2\Phi + \pi - \varphi_0$. Otherwise this angle is $\varphi'' = \varphi_0 - \pi$. When reducing the opening angle of F, petals will approach and in case of equality of angle of aperture value 0.5π will merge into one. This phenomenon can be given a physical explanation. Back scattering patterns also have two petals. However, the main petals correspond to the angles of mirror reflection from the faces ($\varphi' = \Phi - \pi, \varphi'' = -\Phi + \pi$).

The increase of the module $|\theta_\pm|$ results in the scattering of H-polarized wave to the fact that a significant part of the energy of the incident pulse is scattered along the edges

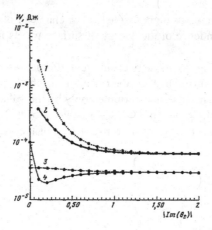

Fig. 5. DI Energy Dependence on the magnitude of the impedance: $\varphi = \varphi_0 = 40°$; Re $(\theta_\pm) = 0.1$ (1); 0.5 (2, 4); 0 (3); F = 100° (1, 2); 120° (3, 4)

of the wedge. Naturally, in this case decreases the value of the $|\dot{H}_{z\,\Pi OB\pm}|$ sector $|\varphi| < 60°$ (Fig. 5, a).

The increase in the modulus of the impedance also leads to the change of DI type in the time domain (Fig. 2).

In Fig. 5 is presented the results of numerical calculations of energy values of the diffraction of a pulse the value of the imaginary part of impedance for different values of the parameter Re $|\theta_\pm|$ (curves 1, 2), which characterizes the heat losses at the wedge's edge. With increasing real part of $\theta_\pm$, the energy of the surface waves decreases, and as faster as the more imaginary part $\theta_\pm$ is. The total field energy also decreases. Increase of the impedance value leads to an increase in the proportion of surface waves energy. The total field energy by increasing the reactive part of the impedance from 0 to 1 (curves 1, 2) decreases by 2 orders of magnitude. When reducing the opening angle of the wedge (curves 3, 4) the influence of impedance in comparison with the described case is not essential.

3   Conclusions

Summarizing the results of the work, it can be noted that the process of diffraction of UWB pulses at the impedance wedge is significantly different from the case of diffraction of harmonic waves. Type of diffraction of the pulse, its energy depend on the angle of signal arrival, the position of the observation point, geometrical and electro-physical parameters of the wedge, the type of the probe pulse. These characteristics determine the conditions of excitation of surface waves, redistribution of energy between the partial components of the scattered field. The method of calculation of the diffraction field change in time during irradiation of impedance wedge ultra-wideband pulse has a simplicity and physical clarity, saves computational resources (since it contains only one operation of integration).

References

1. Astanin, P.Y., Kostylev, A.A.: Ultrawideband radar meters. Publishing House of the USSR Ministry of Defense (1983)
2. Ankudinov, V.E., Romanov, A.E.: Foreign radioelectronics, no. 41, p. 6 (1991)
3. Zernov, N.V.: Reports as USSR, T. 80, no. 91, p. 33 (1951)
4. Astanin, L.Y., Kostylev, A.A.: Fundamentals of ultra-wideband radar measurements. Radio and communication (1989)
5. Kryachko, A.F., Likhachev, V.M., Smirnov, S.N., Stashkevich, A.I.: Theory of scattering of electromagnetic waves in angular structures of Saint Petersburg, Nauka (2009)
6. Nertao, J., Volakis, I., Senior, T., et al.: IEEE Trans. V, AR-N 9, R. 1083 (1987)
7. Markov, G.T., Chaplin, A.F.: Excitation of electromagnetic waves. Radio and communication (1983)

Joint Use of SEFDM-Signals and FEC Schemes

Dmitry Vasilyev, Andrey Rashich[(⊠)], and Dmitrii Fadeev

Radio and Telecommunication Systems Department,
Peter the Great St. Petersburg Polytechnic University, St. Petersburg, Russia
rashich@cee.spbstu.ru

Abstract. The combination of multicarrier signals with nonorthogonal frequency spacing (spectrally efficient frequency division multiplexing, SEFDM) and forward error correction (FEC) schemes is analyzed for LTE convolutional and turbo codes. BER performance of coded and uncoded OFDM and SEFDM is considered while keeping the spectral efficiency constant. Also puncturing patterns are selected for providing different code rates of the LTE turbo-code encoder with very small step size. It is shown that compared with uncoded OFDM the best value of SEFDM compression factor providing better BER performance can be found. But for coded OFDM coded SEFDM always performs worse due to waterfall behaviour of BER curves, thus restricting the application borders of SEFDM.

Keywords: OFDM · NOFDM · SEFDM · Multicarrier FTN · LTE turbo code

1 Introduction

In the fifth-generation (5G) wireless networks there is a gross need for the data rates increase. There are two obvious ways to do that: increase the signal bandwidth or use high order constellations. Frequency domain resource is extremely expensive in wireless systems. At the same time, big sizes of constellations require transponder's amplifiers with the corresponding dynamic range. This attracted attention to SEFDM signals which are considered as a candidate to be used in 5G wireless network PHY.

It is shown [1–3] that transition from OFDM signals to SEFDM can increase spectral efficiency in 2...3 times. Besides, SEFDM signals possess high stability in channels with frequency-selective fading, like OFDM does.

In this paper we compare BER performance of coded or uncoded OFDM-signals and coded SEFDM-signals provided all of them have the same spectral efficiency. We want to find out if there is any optimal combination of code rate and bandwidth compression factor for SEFDM-signals that provide better BER performance than OFDM-signals.

In this research two popular types of FEC encoders are used: convolutional code ([171 133], 7) encoder (CC) with base rate 1/2 and LTE turbo-code (4, [13 15], 13) encoder with base rate 1/3.

The rest of the paper is organized as follows: Sect. 2 outlines the SEFDM system model. Section 3 introduces the FEC schemes and puncturing patterns. Simulation results are provided in Sect. 4. Section 5 summarizes the paper.

O. Galinina et al. (Eds.): NEW2AN/ruSMART 2016, LNCS 9870, pp. 604–611, 2016.
DOI: 10.1007/978-3-319-46301-8_51

2 System Model

Complex envelope of the considered SEFDM-signals on the symbol duration T can be expressed as following:

$$s(t) = \sum_{k=-N/2}^{N/2-1} C_k e^{j2\pi k \Delta f t}, \ t \in [0; \ T].\tag{1}$$

The following notations are used in (1): N – number of subcarriers, Δf – frequency spacing between adjacent subcarriers, C_k – complex modulation symbol of the k-th subcarrier, $\Delta f = \alpha/T$, $0 < \alpha \leq 1$. For OFDM $\alpha = 1$, while for SEFDM: $\alpha < 1$. We also use guard intervals in the frequency domain to reduce aliasing effects:

$$C_k = 0, k = -N/2, \ldots, -N/2 + N_{GI\_left} - 1, N/2 - N_{GI\_right} - 1.\tag{2}$$

In (1) N_{GI_left} – number of subcarriers that determines left guard interval (area of low frequencies), N_{GI_right} – number of subcarriers that determines right guard interval (area of high frequencies).

In general, the subcarriers in SEFDM-signals are not orthogonal to each other and the ISI is presented in SEFDM. This leads to BER performance degradation of SEFDM compared to OFDM. The main idea, we consider in this paper, is that the bandwidth compression (values of α) of SEFDM can be exchanged for lower FEC coder rates to overcompensate ISI in SEFDM and improve overall BER performance. In other words given the same spectral efficiency for coded OFDM and coded SEFDM the combination of α and SEFDM code rate providing the best BER performance is to be found.

For SEFDM code rate R_{SEFDM} and OFDM code rate R_{OFDM} the condition for equal spectral efficiency (the same bandwidth and modulation for both OFDM and SEFDM) is the following:

$$R_{SEFDM} = \alpha R_{OFDM}.\tag{3}$$

Several possible combinations of α and R_{SEFDM} for OFDM with $R_{OFDM} = 2/3$, 3/4 and 5/6 are shown in Table 1.

Table 1. Coded OFDM- and coded SEFDM-signals comparison

OFDM	SEFDM	
R_{OFDM}	α	R_{SEFDM}
2/3	9/16	3/8
	5/8	5/12
	3/4	1/2
	7/8	7/12
	15/16	5/8
3/4	3/4	9/16
	1/2	3/8
5/6	3/4	5/8
	1/2	5/12

The considered system model is presented on Fig. 1. The transmitter includes FEC encoder (CC or turbo-code), puncturing scheme, bit interleaver for 100 SEFDM symbols and SEFDM-modulator. SEFDM-modulator operates in frequency domain using IFFT. We use the first algorithm proposed in [4] to generate time domain samples of SEFDM-symbol. Only the QPSK modulation is considered. After the SEFDM-modulator time domain samples of SEFDM-signal enter the AWGN channel.

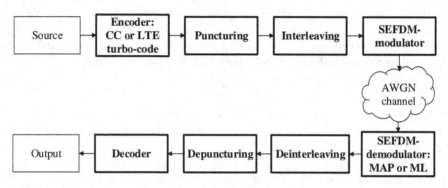

Fig. 1. System model

The receiver performs demodulation (maximum likelihood (ML) for small number of subcarriers [4] and MAP approximation with parameter $K = 7$ [5] for big number of subcarriers in OFDM- and SEFDM-signals), bit deinterleaving, depuncturing and decoding (soft input Viterbi for CC and max-log-map iterative decoder with 8 iterations for LTE turbo code).

The following two cases are considered in the rest of the paper: uncoded OFDM vs coded SEFDM and coded OFDM vs coded SEFDM. In the second case the same FEC schemes are used both for OFDM and SEFDM, but code rates are different. Also SEFDM-signals with low (10, IFFT size 16) and big (1200, IFFT size 2048) number of subcarriers are explored.

3   FEC Schemes and Puncturing Patterns

We consider two popular FEC schemes for joint use with SEFDM and OFDM signals: CC and LTE turbo-code. The CC is of type ([171 133], 7) with base code rate 1/2. The LTE turbo-code with two component CCs (4, [13 15], 13) has base code rate 1/3. Such choice is motivated by the fact that CC ([171 133], 7) is one of the most widely used encoder in wireless systems and low code rate (1/3) of the LTE turbo-code allow to obtain large set of higher code rates via puncturing. This turbo-code is also used in the prototypes of 5G network PHY layer standards.

Puncturing patterns for CC ([171 133], 7) are widely known. They are represented in Table 2.

Table 2. Puncturing patterns for the CC ([171 133], 7) with base code rate 1/2

Code rate	Puncturing pattern	Code rate	Puncturing pattern
1/2	$\begin{pmatrix} 1 \\ 1 \end{pmatrix}$	5/6	$\begin{pmatrix} 1 & 0 & 1 & 0 & 1 \\ 1 & 1 & 0 & 1 & 0 \end{pmatrix}$
2/3	$\begin{pmatrix} 1 & 0 \\ 1 & 1 \end{pmatrix}$	7/8	$\begin{pmatrix} 1 & 0 & 0 & 0 & 1 & 0 & 1 \\ 1 & 1 & 1 & 1 & 0 & 1 & 0 \end{pmatrix}$
3/4	$\begin{pmatrix} 1 & 0 & 1 \\ 1 & 1 & 0 \end{pmatrix}$		

Puncturing patterns for LTE turbo-code are well known only for several popular code rates. That is why we analyzed various patterns by BER performance simulations to provide different code rates with low step size. The results are presented on Fig. 2 and in Table 3. Simulation parameters were the following: signal type – OFDM; subcarrier modulation – QPSK; number of subcarriers – 2048; number of informational (used) subcarriers – 1200; codeblock size – 6000...10000 bit.

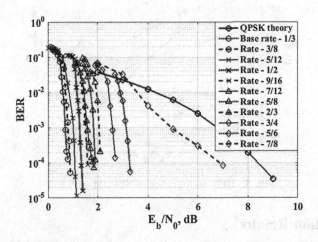

Fig. 2. BER performance of LTE turbo-code with various puncturing patterns

Table 3. Puncturing patterns for the LTE turbo-code (4, [13 15], 13) with base code rate 1/3

Code rate	Puncturing pattern	Code rate	Puncturing pattern
3/8	$\begin{pmatrix} 1 & 1 & 1 \\ 1 & 1 & 1 \\ 1 & 0 & 1 \end{pmatrix}$	7/12	$\begin{pmatrix} 1 & 1 & 1 & 1 & 1 & 1 & 1 \\ 0 & 1 & 0 & 1 & 0 & 1 & 0 \\ 0 & 0 & 1 & 0 & 1 & 0 & 0 \end{pmatrix}$
5/12	$\begin{pmatrix} 1 & 1 & 1 & 1 & 1 \\ 0 & 1 & 1 & 1 & 0 \\ 1 & 1 & 0 & 1 & 1 \end{pmatrix}$	5/8	$\begin{pmatrix} 1 & 1 & 1 & 1 & 1 \\ 0 & 0 & 1 & 0 & 0 \\ 0 & 1 & 0 & 1 & 0 \end{pmatrix}$
1/2	$\begin{pmatrix} 1 & 1 \\ 1 & 0 \\ 0 & 1 \end{pmatrix}$	2/3	$\begin{pmatrix} 1 & 1 & 1 & 1 \\ 0 & 1 & 0 & 0 \\ 0 & 0 & 1 & 0 \end{pmatrix}$
9/16	$\begin{pmatrix} 1 & 1 & 1 & 1 & 1 & 1 & 1 & 1 & 1 \\ 0 & 1 & 0 & 1 & 0 & 1 & 0 & 1 & 0 \\ 0 & 0 & 1 & 0 & 1 & 0 & 1 & 0 & 0 \end{pmatrix}$	3/4	$\begin{pmatrix} 1 & 1 & 1 & 1 & 1 & 1 & 1 & 1 & 1 \\ 1 & 0 & 1 & 0 & 0 & 0 & 0 & 0 & 0 \\ 0 & 1 & 0 & 0 & 0 & 0 & 0 & 0 & 0 \end{pmatrix}$
5/6	$\begin{pmatrix} 1 & 1 & 1 & 1 & 1 & 1 & 1 & 1 & 1 & 1 \\ 0 & 1 & 0 & 0 & 0 & 0 & 0 & 0 & 0 & 0 \\ 1 & 0 & 0 & 0 & 0 & 0 & 0 & 0 & 0 & 0 \end{pmatrix}$		
7/8	$\begin{pmatrix} 1 & 1 & 0 & 1 & 1 & 1 & 1 & 1 & 1 & 1 & 1 & 1 & 1 & 1 \\ 1 & 0 & 1 & 0 & 0 & 0 & 0 & 0 & 0 & 0 & 0 & 0 & 0 & 0 \\ 0 & 1 & 0 & 0 & 0 & 0 & 0 & 0 & 0 & 0 & 0 & 0 & 0 & 0 \end{pmatrix}$		

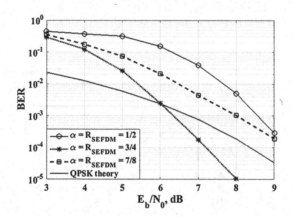

Fig. 3. BER performance of SEFDM with CC

4 Simulation Results

We start from the first case of research: compare uncoded OFDM to coded SEFDM. BER performance simulations results for SEFDM with CC and LTE turbo-code for various code rates and bandwidth compression factors are presented on Figs. 3 and 4. For the case of CC the simulation was performed for low number of subcarriers (10 used subcarriers and IFFT size is 16); ML demodulator (exhaustive search in time

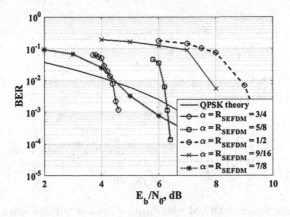

Fig. 4. BER performance of SEFDM with LTE turbo-code

domain) was implemented. For the case of LTE turbo code SEFDM-signal with 1200 used subcarriers and IFFT size of 2048 was considered. For such a big number of subcarriers the MAP approximation [5] was used.

From Figs. 3 and 4 we can see that for SEFDM the best values of bandwidth compression and code rate exist providing better BER performance than uncoded OFDM. Despite the type of demodulator and encoder, for the both experiments this value is 3/4.

Thus, SEFDM signals usage approach based on equality of the bandwidth compression and code rate provides up to 3 dB energy gain from OFDM without encoder.

Now we turn to the second case of research: compare coded OFDM with coded SEFDM. The corresponding simulation results are presented on Figs. 5, 6 and 7. The simulation parameters were the following: number of used subcarriers 1200, IFFT size 2048. On Figs. 5, 6 and 7 BER performance of coded SEFDM with various code rates and bandwidth compression factors are compared to coded OFDM with code rates of 2/3, 3/4 and 5/6.

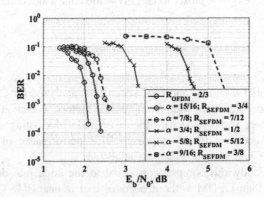

Fig. 5. BER performance of OFDM with $R_{OFDM} = 2/3$ and SEFDM with various R_{SEFDM}

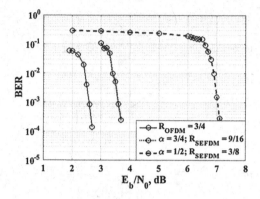

Fig. 6. BER performance of OFDM with R_{OFDM} = 3/4 and SEFDM with various R_{SEFDM}

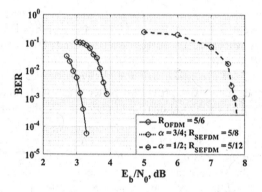

Fig. 7. BER performance of OFDM with R_{OFDM} = 5/6 and SEFDM with various R_{SEFDM}

Coded SEFDM demonstrates worse results in comparison with coded OFDM. For all OFDM code rate values SEFDM BER performance curves are to the right of the OFDM. SEFDM curves come closer to OFDM while bandwidth compression increases and aspires to 1.

5   Conclusions

The joint use of SEFDM-signals and FEC schemes is analyzed for CC and LTE turbo code for various code rates, bandwidth compression factors. The main advantage of this approach is that there is no need to difficult upgrade of existing devices for using SEFDM signals. According to this approach BER performance of SEFDM against OFDM signals with and without FEC was analyzed.

SEFDM with bandwidth compression equal to the code rate demonstrate better BER performance than OFDM without encoder if it is near 3/4. This allows to get about 3 dB energy gain by using SEFDM with proposed parameters compared to OFDM.

Despite that, there is only energy loss while using SEFDM (with corresponding parameters) compared to coded OFDM. Thus, limits of this approach applicability were denoted. SEFDM signals with FEC can be used in telecommunication systems working in good channel conditions as the alternative of OFDM PHY layer technology.

References

1. Yang, X., Ai, W., Shuai, T., Li, D.: A fast decoding algorithm for non-orthogonal frequency division multiplexing signals. In: International Conference on Communications and Networking in China (CHINACOM), pp. 595–598, August 2007
2. Kanaras, I., Chorti, A., Rodrigues, M., Darwazeh, I.: An overview of optimal and sub-optimal detection techniques for a non orthogonal spectrally efficient FDM. In: International Symposium on Communication and Information Technologies, pp. 460–465, September 2009
3. Bharadwaj, S., Krishna, N., Sudheesh, P., Jayakumar, M.: Low complexity detection scheme for NOFDM systems based on ML detection over hyperspheres. In: International Conference on Devices and Communications (ICDeCom), pp. 1–5, February 2011
4. Kislitsyn, A.B., Rashich, A.V., Tan, N.N.: Generation of SEFDM-signals using FFT/IFFT. In: Balandin, S., Andreev, S., Koucheryavy, Y. (eds.) NEW2AN/ruSMART 2014. LNCS, vol. 8638, pp. 488–501. Springer, Heidelberg (2014)
5. Rashich, A., Kislitsyn, A., Dmitrii, F., Tan, N.: FFT-based trellis receiver for SEFDM signals. In: GLOBECOM 2016 (2016, accepted for publication)

Possibilities of "Nyquist Barrier" Breaking by Optimal Signal Selection

Sergey V. Zavjalov[1], Sergey B. Makarov[1], Sergey V. Volvenko[1(✉)], and Shen De Yuan[2]

[1] Peter the Great St. Petersburg Polytechnic University, St. Petersburg, Russia
zavyalov_sv@spbstu.ru, {makarov,volk}@cee.spbstu.ru
[2] Jiangsu Normal University, Suychzhou, China
vanli@inbox.ru

Abstract. A possibility of overcoming the "Nyquist barrier" by finding the optimal waveform for a binary signal is investigated. The same BER performance as of BPSK signals is required. This problem can be viewed as an optimization problem. Parameters to be optimized are the rate of decay with frequency of out-of-band emissions, duration of signals and BER performance. BER performance is determined by a cross-correlation coefficient. Solutions to the optimization problem are obtained numerically. These solutions have the form of the coefficients of the truncated Fourier series of the waveforms obtained under different restrictions. Corresponding power spectra are analyzed. It is shown that the doubling of data rate leads to 30 % increase in bandwidth. Spectral efficiency can be increased by the use of longer signals. But the increase in signals duration leads to the increase of peak-to-average ratio of random sequence of signals. At the same time BER performance degrade insignificantly. Additional energy losses are no more than 0.5 dB.

Keywords: Optimal waveform · "Nyquist barrier" · BER performance · Data rate · Bandwidth · Cross-correlation coefficient · Optimization problem

1 Introduction

Overcoming the "Nyquist barrier" [1–3] leads to significant energy losses. These losses can be reduced by using optimal signals [4–8] with waveforms derived under constraints imposed on the values of bandwidth F, out-of-band emissions (OOBE), signal energy E and duration T_c, as well as on the cross-correlation coefficient K_0 [8]. The K_0 determines the energy losses related to significant intersymbol interference (ISI). Such interference appears with the increase of binary data transmission rate $R = 1/T$ where T is the duration of data bit exceeding the Nyquist rate $R_N = 1/(\alpha T)$ for $\alpha < 1$ [2, 3].

Comparison of solutions to the optimization problem, even in the case of the binary channel alphabet, results in appearance of ambiguity when comparing the spectral efficiency of the optimal signals with Nyquist signals. It is linked on the one hand to the correct comparison of bandwidth F of finite length signals, and on the other hand, with the influence of peak-to-average power ratio $PAPR$ of the random sequence of signals on the change of average power P_{av} of transmitted signals.

© Springer International Publishing AG 2016
O. Galinina et al. (Eds.): NEW2AN/ruSMART 2016, LNCS 9870, pp. 612–619, 2016.
DOI: 10.1007/978-3-319-46301-8_52

The communication channel is modeled as AWGN channel with the bandwidth F, and the shape of the frequency response repeating the shape of the power spectrum of a random signal sequence.

The goal of this work is to find the possibilities of increasing the spectral efficiency R/F of signals under the conditions of overcoming the Nyquist barrier by finding the optimal waveform taking into account different constraints.

2 Signal Model and Optimization Functional

Let us consider spectrally efficient signals with QPSK modulation [9] as the signal model. In generalized form the truncated r-th realization of the random sequence consisting of L signals of duration T_c with the carrier frequency ω_0 can be written as follows:

$$
\begin{aligned}
\zeta^{(r)}(t) = A_0 \sum_{k=0}^{L} a(t - k/(\alpha T)) d_i^{(k)} \cos(\omega_0 t) \\
+ A_0 \sum_{k=0}^{L} a(t - k/(\alpha T)) d_q^{(k)} \sin(\omega_0 t),
\end{aligned}
\tag{1}
$$

where A_0 is the amplitude; $a(t)$ is the function that defines the envelope shape; $d_i^{(k)}$, $d_q^{(k)}$ are the binary symbols of the respective quadrature components; $d_i^{(k)} = 1$ at $i = 1$ and $d_i^{(k)} = -1$ at $i = 0$.

Note that in Eq. (1) the single signal duration is assumed to be $T_c > T$. The power spectrum $|S_a(f)|^2$ of a random sequence of such signals is determined by the spectrum of real envelope $a(t)$ of the quadrature component and is equal to:

$$
|S_a(f)|^2 = A_0^2 \left| \int_{-\infty}^{+\infty} a(t) e^{-j2\pi f t} dt \right|^2 .
\tag{2}
$$

One of the criteria, which determines the behavior of $|S_a(\omega)|^2$ within the frequency band F and outside of this band is the required rate of decrease of OOBE with frequency [7, 8] determined by the weighting function $g(f) = 1/f^{2n}$ $(n = 1, 2, \ldots)$. Solution to the optimization problem using this criterion is linked to the minimization of a functional given by [4, 5, 7, 8]:

$$
J = A_0^2 \frac{1}{2\pi} \int_{-\infty}^{+\infty} g(f) |S_a(f - f_0)|^2 df.
\tag{3}
$$

If the solution to (3) is to be found numerically, it is necessary to define the coefficients of the truncated Fourier series expansion of $a(t)$:

$$a(t) = \frac{a_0}{2} + \sum_{k=1}^{m} a_k \cos\left(\frac{2\pi}{T}kt\right). \tag{4}$$

Let us suppose that $a(t)$ is even in the $[-T_c/2; T_c/2]$ interval. When using (4), the initial functional (3) may be converted into a function of several variables in the following way [8]:

$$J\left(\{a_k\}_{k=1}^{m}\right) = \frac{T_c}{2} \sum_{k=1}^{m} \left(\frac{2\pi}{T_c}k\right)^{2n} a_k^2. \tag{5}$$

In this case, the optimization problem is converted into a problem of minimizing the function of several variables (5). The optimization constraints are:

- fixed signal energy E;
- fixed signal duration T_c;
- restriction on the value of K_0 [8], which determines the ISI level and BER performance

$$\max_{k=1\dots(L-1)} \left\{ \int_{0}^{(L-k)T} a(t)a(t-kT)dt \right\} < K_0 \text{ (for } T_c = LT). \tag{6}$$

Using this approach, the problem of finding the value of peak-to-average power ratio of random sequence of signals with duration T_r is reduced to calculation of *PAPR* after minimization of function (5). Let us present the formula for calculation of *PAPR* in the following form:

$$PAPR = \max_{r}\{P^{(r)}(t)\} \left/ \frac{1}{D}\sum_{r=1}^{D} P_{av}(r), \right. \tag{7}$$

where $P^{(r)}(t)$ is the power of the r-th realization of the random sequence (1); D is the number of possible realizations of the random sequence (1); $P_{av}(r)$ is the average power of the r-th realization of the random sequence (1):

$$P_{av}(r) = \frac{1}{T_r} \int_{0}^{T_r} P^{(r)}(t)dt. \tag{8}$$

3   The Optimization Results

Let us review the results of optimization of (5) with constraints considered earlier. Here, we assume that the value of K_0 in (6) is equal to 10^{-2} that provides the BER performance of BPSK signals that is close to the potential limit. The rate of OOBE decay with frequency is equal to $1/f^4$ ($n = 2$), Fig. 1 shows the shapes of the real envelope $a(t)$ (Fig. 1a, c and e) and plots of normalized spectra $|S_a(f)|^2/|S_a(0)|^2$ (Fig. 1 b, d and h) for the optimal signal durations where $T_c = 2T, 4T, 6T$.

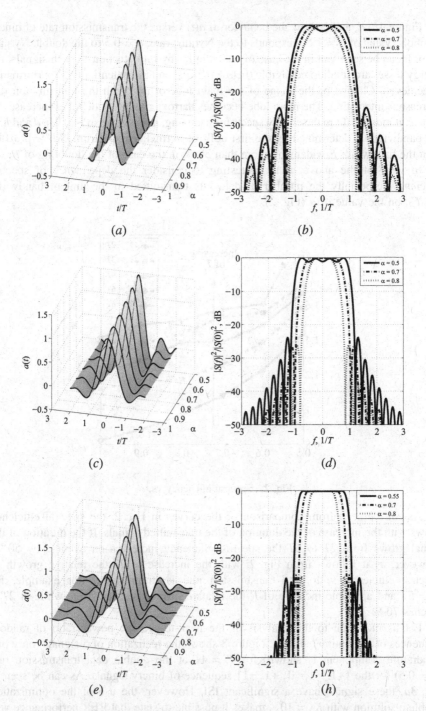

Fig. 1. Real envelopes of optimal signals and respective energy spectra

Figure 1*b*, *d*, and *h* show the behavior of $a(t)$ versus the transmission rate of binary symbols (value of $\alpha = 1$ corresponds to the Nyquist rate, $\alpha = 0.5$ to the double Nyquist rate). It can be seen that in the case of coherent bit-by-bit detection of such signals, the energy losses are small to negligible (below 0.5 dB) and the potential BER performance is achieved. Change of the shape of the main lobe of $a(t)$ with the transmission rate increase is interesting. The main lobe becomes narrower and its side lobes increase. Of course, it must result in the energy spectrum widening, as is shown in Fig. 1*b*, *d* and *h*. If the bandwidth is determined by the first nulls of normalized spectrum $|S_a(f)|2/|S_a(0)|$ 2 then the bandwidth F widens by more than 30 % in the case of the doubling of R.

In view of the above, it is interesting to consider the dependence of spectral efficiency, especially the number of data bits transmitted in the unit of bandwidth $1/(\alpha F)$, on the value of α (Fig. 2).

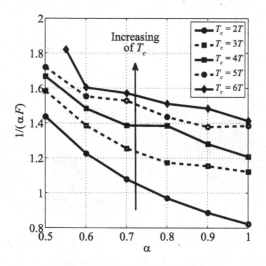

Fig. 2. Spectral efficiency vs. α

As can be seen from comparison of the curves in Fig. 2, the spectral efficiency grows with the increase of the duration of the transmitted signals. If the duration of the signal grows from $2T$ to $6T$ the spectral efficiency increases on average by 50 %. However, as it follows from Fig. 2, with the increase of T_c the relative growth of spectral efficiency with data transmission rate is diminishing. For example, for $T_c = 5T$ and a double increase of R the relative growth reaches 24 % while at $2T$ it reaches 70 %.

Let us move on to the analysis of the peak-to-average power ratio of random sequences of the optimal signals. Figure 3 shows the realization of 5 signals (only one quadrature component is shown) for $T_c = 4T$ at the double data transmission rate ($\alpha = 0.5$) for the $\{+1, -1, -1, +1, -1\}$ sequence of binary signals. As can be seen in Fig. 3*a*, these signals have a significant ISI. However, the use of the optimization problem solution with $K_0 = 10^{-2}$ makes it possible to state that BER performance will not be reduced by more than 0.5 dB. Figure 3*b* provides an example of the normalized instantaneous power of a random sequence of these signals with a carrier frequency of 20 MHz.

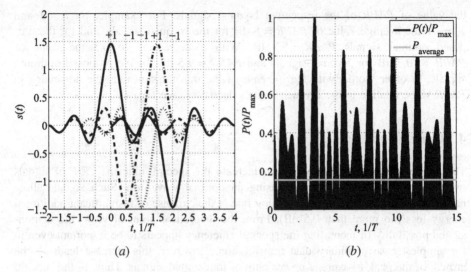

Fig. 3. Signal sequence (*a*) and normalized instantaneous power (*b*) level

It is obvious that there is a great difference between the peak and average power, which determines the value of *PAPR*. Even in a short sequence of signals there are significant oscillations of the instantaneous power that indicate a high value of the peak-to-average power ratio.

The results of investigation of dependence of the *PAPR* on α are shown in Fig. 4. The increase of T_c leads to the growth of *PAPR*. Yet it should be noted that for each value of T_c and α the value of *PAPR* of the proposed optimal signals will be lower than

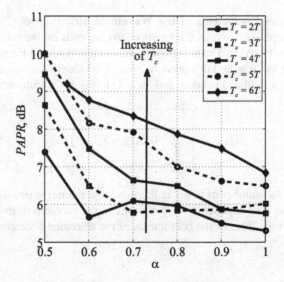

Fig. 4. Peak-to-average power ratio vs. α

the value of *PAPR* of the respective Nyquist signals. For example, for $\alpha = 1$ and $T_c = 4T$ the average value of *PAPR* = 8 dB for the Nyquist signals, and for the proposed optimal signals *PAPR* = 5.7 dB. With α decreasing to 0.5 the value of *PAPR* = 10.3 dB for Nyquist signals and *PAPR* = 9.5 dB for the proposed optimum signals. In other words, with the increase of the transmission rate, the advantage in *PAPR* value relative to the Nyquist signals decreases (see Fig. 4).

4 Conclusions

This work shows that it is possible to increase the spectral efficiency *R/F* of signals overcoming the "Nyquist barrier" using an optimal waveform. Such signals allow nearly doubling of the data transmission rate at the expense of insignificant additional energy loss (no more than 0.5 dB). From the theoretical point of view, the demonstrated possibility of increasing the spectral efficiency appears to be important even in the simplest case of binary data transmission. However, this increase leads to the growth of the peak-to-average power ratio of transmitted signals. Thus, as the spectral efficiency increases by 50 %, the peak-to-average power ratio increases on average by 37 %. Nevertheless, the peak-to-average power ratio of the proposed optimal signals is lower than the peak-to-average power ratio of the Nyquist signals for the same transmission parameters. It is interesting to perform similar studies for the channel alphabet of more than two symbols and for different types of modulation, for example, OFDM.

Appendix A

The Fourier Series Coefficients of the Waveform $a(t)$. It should be noted that the accuracy of the representation of the function $a(t)$ depends on the number of terms of the truncated Fourier series. The number of terms in the truncated Fourier series can be determined using the standard deviation between the values of the functions $a_m(t)$ and $a_{m-1}(t)$, which are calculated using m and $(m - 1)$ terms of the truncated Fourier series:

$$\varepsilon(m) = \sqrt{\int_{-T/2}^{T/2} \left(a_m(t) - a_{m-1}(t)\right)^2 dt}.$$

If we restrict the number of terms in the truncated Fourier series to $m = 7$ then the standard deviation ε is less than 10^{-2}. Solutions to the problem of finding an optimal waveform $a(t)$, in the form of the coefficients of the truncated Fourier series, are listed in Table A.1.

Table A.1. The Fourier series coefficients of the waveform $a(t)$ for $\alpha = 0.7$, $K_0 = 10^{-2}$, $n = 2$ and different values of T_c.

T_c	2	4	6
a_0	0.845590	0.422444	0.279605
a_1	0.753031	0.418061	0.280992
a_2	0.271849	0.415527	0.278880
a_3	−0.037072	0.250595	0.283111
a_4	0.011446	0.018148	0.231501
a_5	−0.004471	−0.012500	0.058885
a_6	0.002148	0.005848	−0.014814
a_7	−0.001153	−0.003375	0.007840
a_8	0.000671	0.002036	−0.004541

References

1. Gattami, A., Ringh, E., Karlsson, J.: Time localization and capacity of faster-than-Nyquist signaling. In: 2015 IEEE Global Communications Conference (GLOBECOM), pp. 1–7 (2015)
2. Mazo, J.E.: Faster-than-Nyquist signaling. Bell Syst. Tech. J. **54**, 1451–1462 (1975)
3. Zhou, J., Li, D., Wang, X.: Generalized faster-than-Nyquist signaling. In: 2012 IEEE International Symposium on Information Theory Proceedings (ISIT), pp. 1478–1482 (2012)
4. Xue, W., Ma, W.-Q., Chen, B.-C.: Research on a realization method of the optimized efficient spectrum signals using legendre series. In: 2010 IEEE International Conference on Wireless Communications, Networking and Information Security (WCNIS), vol. 1, pp. 155–159 (2010)
5. Xue, W., Ma, W., Chen, B.: A realization method of the optimized efficient spectrum signals using fourier series. In: International Conference on Wireless Communications Networking (2010)
6. Yoo, Y.G., Cho, J.H.: Asymptotic optimality of binary faster-than-Nyquist signaling. IEEE Commun. Lett. **14**, 788–790 (2010)
7. Zavjalov, S.V., Makarov, S.B., Volvenko, S.V.: Application of optimal spectrally efficient signals in systems with frequency division multiplexing. In: Balandin, S., Andreev, S., Koucheryavy, Y. (eds.) NEW2AN/ruSMART 2014. LNCS, vol. 8638, pp. 676–685. Springer, Heidelberg (2014)
8. Zavjalov, S.V., Makarov, S.B., Volvenko, S.V., Xue, W.: Waveform optimization of SEFDM signals with constraints on bandwidth and an out-of-band emission level. In: Balandin, S., Andreev, S., Koucheryavy, Y. (eds.) NEW2AN/ruSMART 2015. LNCS, vol. 9247, pp. 636–646. Springer, Heidelberg (2015)
9. Zavjalov, S.V., Makarov, S.B., Volvenko, S.V.: Nonlinear coherent detection algorithms of nonorthogonal multifrequency signals. In: Balandin, S., Andreev, S., Koucheryavy, Y. (eds.) NEW2AN/ruSMART 2014. LNCS, vol. 8638, pp. 703–713. Springer, Heidelberg (2014)

Reduction of Energy Losses Under Conditions of Overcoming "Nyquist Barrier" by Optimal Signal Selection

Sergey V. Zavjalov, Sergey B. Makarov, and Sergey V. Volvenko[✉]

Peter the Great St. Petersburg Polytechnic University, St. Petersburg, Russia
zavyalov_sv@spbstu.ru, {makarov,volk}@cee.spbstu.ru

Abstract. A problem of overcoming "Nyquist barrier" for single-frequency signals minimum energy losses is discussed. Main constrains for this optimization problem are the rate of decay of out-of-band emissions and BER performance. BER performance is determined by the cross-correlation coefficient. The optimization problem is solved numerically. The solutions of optimization problem with different constraints give the envelopes of optimal signals. The simulation showed that energy losses can be significantly reduced by using the optimal signals. Additional energy losses for the doubled data rate are no more than 0.5 dB. In contrast the use of the Nyquist signals in the same conditions leads to BER performance degradation up to 8 dB for 20 % data rate increase.

Keywords: "Nyquist barrier" · BER performance · Bandwidth · Cross-correlation coefficient · Optimization problem

1 Introduction

Let us consider the classic formulation of the problem of signal reception under the conditions of overcoming the "Nyquist barrier". It is necessary to recall [1, 2] the formula of the channel capacity C of a baseband communication channel:

$$C = F \cdot \log_2(1 + P_s/P_n), \tag{1}$$

where F is the bandwidth of the communication channel; P_s is the signal power; P_n is the power of additive white Gaussian noise (AWGN) in the channel.

The closer the data rate R to C the more complex the signal-code construction and reception algorithm are. Let us assume that the complexity is acceptable when the transmission rate $R = 0.75C$. In this case, for relatively simple systems, we have:

$$R \leq 0.75 \cdot F \cdot \log_2(1 + P_s/P_n) = 0.375 \cdot R_N \cdot \log_2(1 + P_s/P_n) \tag{2}$$

where $R_N = 2F$ is the Nyquist speed.

Therefore, it is easy to derive the following estimation of the required increase of signal-noise ratio (SNR):

O. Galinina et al. (Eds.): NEW2AN/ruSMART 2016, LNCS 9870, pp. 620–627, 2016.
DOI: 10.1007/978-3-319-46301-8_53

$$P_S \big/ P_n > 2^{(1/0.375)R/R_N} - 1. \tag{3}$$

Exceeding of the "Nyquist barrier", for example, by two times, requires increasing signal power about 150 times compared to transmission speed $R = R_N/2.66$, which provides the maximum use of the communication channel's capacity.

The attempts of overcoming the "Nyquist barrier" with minimum energy losses have been undertaken in numerous studies [1–3]. Main results were obtained by solving the optimization problem of finding the Nyquist signal waveform that provides maximum concentration of energy in a frequency band [4, 5]. The duration of signals $T_s > T$ was fixed during optimization.

When evaluating the frequency and energy efficiency, it is necessary to consider the following two factors. First, the reception of signals under such conditions occurs with significant intersignal interference (ISI). Appearance of ISI is related to the increased duration of signals $T_s > T$. The ISI level is related to the value of cross-correlation coefficient K_0 [6]. The illustration of this for $R = 2R_N$ is given in Fig. 1a that shows a random sequence of the truncated Nyquist signals of the $\sin(x)/x$ type for the sequence of binary symbols $d_i(k)$.

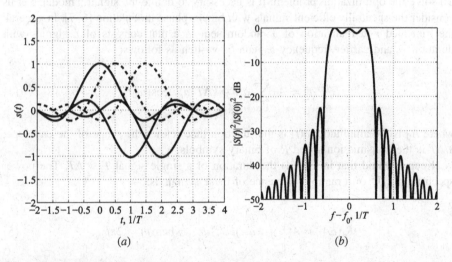

Fig. 1. Truncated Nyquist signals with duration $T_s = 6T$ for $R = 2R_N$ (a) and corresponding energy spectrum (b).

In this notation the index $k = 1 \ldots L$ gives the symbol place in the sequence and index i determines the binary symbol value. Index i has a value of either 1 or 0 and $d_1(k) = 1$, and $d_0(k) = -1$. Figure 1a shows the signals consisting of a sequence of symbols $\{1;1;-1;-1;1\}$. The signal duration is $T_s = 6T$. As is clear from Fig. 1a there is significant ISI that reduces BER performance. Also there is an ambiguity in determining the bandwidth F (Fig. 1b). Indeed, for the Nyquist signals of infinite duration the communication channel bandwidth is equal to the signal bandwidth and is $F = 1/T$. For signals of limited duration, F may be determined by different ways. For example, by selecting a certain level of the normalized power spectrum $|S(f)|^2/|S(0)|^2$ (Fig. 1b). For example, for the level of -30 dB the bandwidth is $F = 1.7/T$; for -40 dB the value is $F = 2.6/T$. Selection of the criterion of determination of F influences the value of spectral efficiency (number of data bits transmitted per unit of channel's bandwidth).

As has been previously noted, the truncated Nyquist signals are transmitted with considerable ISI. It can be shown that when $R = 2R_N$ the energy losses is above 15 dB irrespective of T_s. The reduction of the energy losses can be achieved by switching from the "classic" Nyquist time limited signals to the signals found as a result of the solution of an optimization problem.

The Work Objective. Reduction of energy requirements under the conditions of overcoming the "Nyquist barrier" by optimizing the signal waveform with constraints on the rate of decay of out-of-band emissions (OOBE) and on the correlation coefficient, determined by intersymbol interference.

2   Model of Signals

To solve the optimization problem, it is necessary to define the signal's model. Let us consider the spectrally efficient signals with binary phase modulation [6, 7]. In general, the truncated r-th realization of a random sequence that consists of L signals with duration T_s and carrier frequency ω_0 can be written as follows:

$$\zeta^{(r)}(t) = A_0 \sum_{k=0}^{L} a(t - k\Delta T)d_i^{(k)} \cos(\omega_0 t) \qquad (4)$$

where A_0 is the amplitude; $a(t)$ is the function that defines the shape of the envelope; $1/\Delta T$ is the transmission speed R of binary symbols.

Keep in mind that in Eq. (4) the duration of a single signal $T_s > \Delta T$. The power spectrum $|S_a(f)|^2$ of a random sequence of these signals is:

$$|S_a(\omega)|^2 = A_0^2 \left| \int_{-\infty}^{+\infty} a(t)e^{-j\omega t}dt \right|^2 \text{, where } \omega = 2\pi f. \qquad (5)$$

3   Optimization Problem

When formulating the optimization problem, it is first of all necessary to choose the criterion that determines the behavior of the energy spectrum of the synthesized signals within and outside the band F. This is required for the unambiguous comparison of bandwidths F of the truncated Nyquist signals (Fig. 1b) and optimal signals. The most general criterion is that of the specified rate of decay with frequency of OOBE combined with the minimum distortion of the spectrum $|S_a(f)|^2$ within the band. As additional constraints we introduce the requirements of constant energy E and duration

$T_s > \Delta T$ of the signal, the minimal ISI level that provides maximum reception noise immunity. Solution to the optimization problem satisfying the criterion of the specified rate of decay of the OOBE level can be found by minimizing the following equation [7–10]:

$$J = \frac{1}{2\pi} \int\limits_{-\infty}^{+\infty} g(f)|S(f)|^2 df = A_0^2 \frac{1}{2\pi} \int\limits_{-\infty}^{+\infty} g(f)|S_a(f - f_0)|^2 df \qquad (6)$$

where $g(f) = 1/f^{2n}$ is the weighting function, and the parameter n sets the rate of decay of the OOBE level with frequency.

The solution can be found in the form of coefficients of expansion of $a(t)$ into the truncated Fourier series:

$$a(t) = \frac{a_0}{2} + \sum_{k=1}^{m} a_k \cos\left(\frac{2\pi}{T} kt\right). \qquad (7)$$

Let us assume that the function $a(t)$ is symmetric within the interval $[-T_s/2; T_s/2]$. Using (7) Eq. (6) can be transformed into a function of several variables [7]:

$$J(\{a_k\}_{k=1}^{m}) = \frac{T_c}{2} \sum_{k=1}^{m} \left(\frac{2\pi}{T_c} k\right)^{2n} a_k^2. \qquad (8)$$

Thus the initial optimization problem is transformed into the problem of finding the minimum of the function of several variables (8). It important to note that the restrictions on signal energy, conditions that determine the rate of decay of the OOBE level, and constraint on the cross-correlation coefficient K_0 [7] must all be satisfied:

$$\max_{k=1\ldots(L-1)} \left\{ \int\limits_{0}^{(L-k)T} a(t)a(t - kT)dt \right\} < K_0 \text{ (for } T_s = LT). \qquad (9)$$

Results of Optimization. The results of solving the optimization problem are analyzed for the rate of decay of the OOBE level of $1/f^4$ ($n = 2$) which is the same as for the truncated Nyquist signals (Fig. 1b). The real envelopes of the optimal signals are shown in Fig. 2a for a sequence of symbols $\{1;1;-1;-1;1\}$. The duration of signals is $T_s = 6T$ and the cross-correlation coefficient is selected as $K_0 = 5 \cdot 10^{-2}$. The transmission speed is $R = 2R_N$. As is shown in Fig. 2a these signals also have significant ISI. As it will be shown below, this ISI does not significantly reduce BER performance. Figure 2b shows the normalized power spectrum $|S_a(f)|^2/|S_a(0)|^2$. For the level of $|S_a(f)|^2/|S_a(0)|^2 = -30$ dB the value $F = 2.6/T$; for -40 dB the value $F = 3.1/T$. These values of the bandwidth are wider than of the truncated Nyquist signals by 40 % and 16 % respectively. It can be seen that the optimal waveform with the selected rate of decay of the OOBE level has lower spectral efficiency.

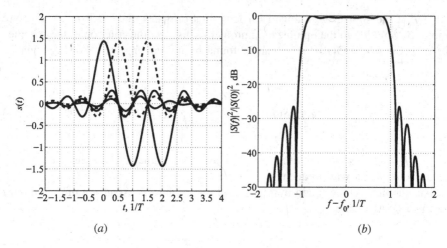

Fig. 2. Sequence of envelopes of optimal signals for the case of $R = 2R_N$ (a) and corresponding normalized power spectrum (b).

Let us consider changes in the shape of the real signal envelope $a(t)$ for various optimization problem parameters. Figure 3 shows the envelopes $a(t)$ of optimal signals having duration $T_s = 6T$ at various data rates: from $R = 1.1$ $R_N = 1.1/T$ to $R = 1.8$ $R_N = 1.8/T$ (Fig. 3a). As is clear from the figure, the maximum value of $a(t)$ increases and the duration of the main lobe decreases with the increase in R. This causes changes in the peak-to-average power ratio (PAPR) of the random signal sequence. Let us also remark here that the real envelope is bipolar in the region of $|t| \in [0.5T, 3T]$. The bipolarity of the real envelope is explained by the value of $K_0 < 0.1$. If no constraints on K_0 is set then the real envelop will be unipolar.

Figure 3b shows the shapes of the real envelopes $a(t)$ of optimal signals for various values of K_0. It is obvious that the influence of this parameter on the shape of $a(t)$ is less than in the previous case (Fig. 3a). We can make an assumption that PAPR of a

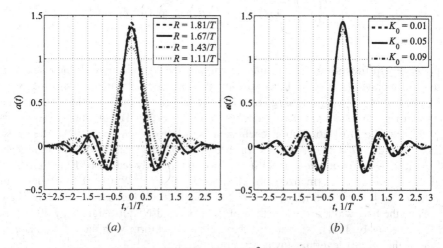

Fig. 3. Signal envelopes for $K_0 = 10^{-2}$ (a) and $R = 1.81/T$ (b).

random sequence of such signals with the fixed value of data rate R will be practically independent of the value of K_0.

To analyze BER performance of optimal signals a computer simulation of the transmission system was carried out for a channel with constant parameters, additive white Gaussian noise (AWGN) and the bandwidth F.

4   Simulation

Figure 4 shows the flow diagram of a simulation process. The simulation was performed using Matlab. The objective of the simulation modeling was to evaluate BER performance of optimal signals at different data rates above the "Nyquist barrier".

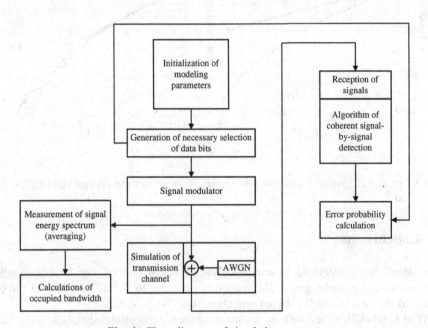

Fig. 4. Flow diagram of simulation process.

The input parameters of the model: T_s, R, envelope $a(t)$, SNR, $f_0 = 10$ MHz. Generation of the data sequence is performed after initialization of the parameters. The sequence of 10^7 bit was used to evaluate the error probability for each value of the signal-to-noise ratio.

The generated symbols arrive at the input of the "Signal modulator" unit. From the output of this unit the signal goes to the input of the communication channel simulation unit where AWGN is added to the signal at a specified signal-to-noise ratio.

The mixture of the signal and noise is processed by the "Reception of signals" unit. The coherent bit-by-bit detection algorithm [11, 12] is used as a reception algorithm. The decisions about the received symbols are used for calculation of error probability in the respective unit.

Figure 5 shows the results of analysis of BER performance. Figure 5*a* shows how error probability depends on signal-to-noise ratio E_b/N_0 in case of reception of binary signals with the Nyquist envelope of duration $T_s = 6T$ (see Fig. 1*a*). It follows from the analysis of this graph that even 20 % increase of R above R_N leads to energy losses of at least 8 dB for error probability of $4 \cdot 10^{-2}$. For the optimal signals of the same duration BER performance is close to the potentially achievable. Energy losses are less than 0.5 dB (Fig. 5*b*).

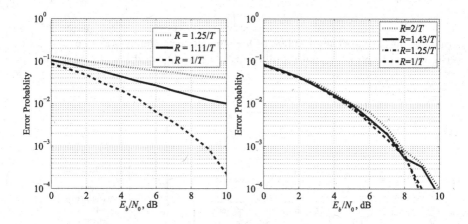

Fig. 5. BER performance for the Nyquist signals ($T_s = 6T$) (*a*) and optimal signals ($T_s = 6T$, $K_0 = 5 \cdot 10^{-2}$) (*b*).

5   Conclusions

The signal energy required to overcome the "Nyquist barrier" can be significantly reduced using optimal signals. The reduction of losses to 0.5 dB is achieved for the two-fold increase of the bit stream rate above the "Nyquist barrier".

The bandwidth F occupied by a random sequence of optimal signals is wider than of the truncated Nyquist signals. The widening of F depends on the selection of the criterion of bandwidth determination and ranges from 40 % to 16 %.

References

1. Gattami, A., Ringh, E., Karlsson, J.: Time localization and capacity of faster-than-Nyquist signaling. In: 2015 IEEE Global Communications Conference (GLOBECOM), pp. 1–7 (2015)
2. Zhou, J., Li, D., Wang, X.: Generalized faster-than-Nyquist signaling. In: 2012 IEEE International Symposium on Information Theory Proceedings (ISIT), pp. 1478–1482 (2012)
3. Mazo, J.E.: Faster-than-Nyquist signaling. Bell Syst. Tech. J. **54**, 1451–1462 (1975)

4. Halpern, P.: Optimum finite duration Nyquist signals. IEEE Trans. Commun. **27**, 884–888 (1979)
5. Hua, Y., Sarkar, T.K.: Design of optimum discrete finite duration orthogonal Nyquist signals. IEEE Trans. Acoust. Speech Signal Process. **36**, 606–608 (1988)
6. Zavjalov, S.V., Makarov, S.B., Volvenko, S.V., Xue, W.: Waveform optimization of SEFDM signals with constraints on bandwidth and an out-of-band emission level. In: Balandin, S., Andreev, S., Koucheryavy, Y. (eds.) NEW2AN/ruSMART 2015. LNCS, vol. 9247, pp. 636–646. Springer, Heidelberg (2015)
7. Yoo, Y.G., Cho, J.H.: Asymptotic optimality of binary faster-than-Nyquist signaling. IEEE Commun. Lett. **14**, 788–790 (2010)
8. Xue, W., Ma, W., Chen, B.: A realization method of the optimized efficient spectrum signals using Fourier series. In: International Conference on Wireless Communications Networking and Mobile Computing (2010)
9. Xue, W., Ma, W.-q., Chen, B.-c.: Research on a realization method of the optimized efficient spectrum signals using legendre series. In: 2010 IEEE International Conference on Wireless Communications, Networking and Information Security (WCNIS), vol. 1, pp. 155–159 (2010)
10. Zavjalov, S.V., Makarov, S.B., Volvenko, S.V.: Application of optimal spectrally efficient signals in systems with frequency division multiplexing. In: Balandin, S., Andreev, S., Koucheryavy, Y. (eds.) NEW2AN/ruSMART 2014. LNCS, vol. 8638, pp. 676–685. Springer, Heidelberg (2014)
11. Zavjalov, S.V., Makarov, S.B., Volvenko, S.V.: Nonlinear coherent detection algorithms of nonorthogonal multifrequency signals. In: Balandin, S., Andreev, S., Koucheryavy, Y. (eds.) NEW2AN/ruSMART 2014. LNCS, vol. 8638, pp. 703–713. Springer, Heidelberg (2014)
12. Zavjalov, S.V., Makarov, S.B., Volvenko, S.V., Balashova, A.A.: Efficiency of coherent detection algorithms nonorthogonal multifrequency signals based on modified decision diagram. In: Balandin, S., Andreev, S., Koucheryavy, Y. (eds.) NEW2AN/ruSMART 2015. LNCS, vol. 9247, pp. 599–604. Springer, Heidelberg (2015)

Root-Raised Cosine versus Optimal Finite Pulses for Faster-than-Nyquist Generation

Anton Gorlov[✉], Aleksandr Gelgor, and Van Phe Nguyen

Peter the Great St. Petersburg Polytechnic University, St. Petersburg, Russia
anton.gorlov@yandex.ru, a_gelgor@mail.ru,
nvphe1905@gmail.com

Abstract. In the paper we compare two approaches of intentional introduction of inter-symbol interference (ISI) into single-carrier signal to improve its bandwidth efficiency. The first approach is "faster-than-Nyquist" signaling (FTN) with infinite rrc-pulses. The second approach is generation of multi-component signals (MCS) utilizing optimal finite pulses. In the optimization problem, we used a criterion of minimal bandwidth comprising 99 % of signal power. The maximal level of ISI and peak-to-average power ratio were used as additional constraints. The comparison of optimal MCS and FTN is given in the bandwidth and energy consumptions plane for the fixed computational complexity of the receiver. In addition, we considered modified energy consumptions taking values of PAPR into account. It is shown that optimal MCS provide lower consumptions with respect to FTN with rrc-pulses.

Keywords: Faster-than Nyquist · Optimal finite pulses · Multicomponent signals · Bandwidth efficiency · Energy efficiency

1 Introduction

A trend of data transmission rate increase can be explicitly observed in the changing of generations of digital communication. In simplified terms, the second generation (2G) has tried to provide a bitrate up to 1 Mbits/s for a single user, in the third generation (3G) an upper bound has been increased to 10 Mbits/s, and it has been wished for the fourth generation (4G) standard to provide a 100 Mbits/s bitrate. It is also known that the changing of generations of telecommunication systems imply an improvement of other network parameters such as latency, a throughput, a flexibility of time-frequency resource allocation, and a reduction of physical size of devices.

In the evolution from the 2G to the 4G the rise of user bitrate is reached by following means: a signal constellation size increase, new efficient forward error correction schemes, wider occupied bandwidth, and efficient methods for inter-symbol-interference (ISI) cancellation. Almost all of these systems utilize orthogonal signals: there is no intentional ISI in 3G and 4G systems.

Similar changes took place in other telecommunication standards. For instance, European standards for satellite digital video broadcasting DVB also have two generations: DVB-S1 and DVB-S2. The changes of physical layer are utilization of larger signal constellations, and more efficient forward error correction schemes. These

© Springer International Publishing AG 2016
O. Galinina et al. (Eds.): NEW2AN/ruSMART 2016, LNCS 9870, pp. 628–640, 2016.
DOI: 10.1007/978-3-319-46301-8_54

standards do not define any means for ISI cancellation since, with a good approximation, the channel can be considered with a single-path propagation. The DVB standards also define orthogonal signaling.

It was shown by Mazo in 1975, that a transition from orthogonal signals to signals with controlled ISI is an efficient way to improve spectral efficiency [1]. The results were obtained for orthogonal signals with linear modulation:

$$y(t) = \sum_k C_k a(t - kT), \tag{1}$$

where C_k are symbols uniformly chosen from a M-symbols constellation, T is a symbol transmission period, and $a(t)$ is a pulse which meets the orthogonality condition:

$$\int_{-\infty}^{\infty} a(t)a(t - kT)dt = 0, \, k = \pm 1, \, \pm 2, \ldots . \tag{2}$$

Signals based on pulses which meet the condition (2) are called full response signals (FRS), otherwise the term partial response signals (PRS) is used.

For signals (1) Mazo proposed to increase symbol transmission rate making it higher than in the case of FRS. These signals can be written as:

$$y(t) = \sum_k C_k a(t - k\tau T), \tag{3}$$

where $\tau = 1$ for original signals without ISI (FRS), and $\tau < 1$ for signals with controlled ISI (PRS). Mazo showed that in this approach, for BPSK constellation and sinc-pulse used in (3), a 25 % gain of spectral efficiency is possible with zero losses of energy efficiency with respect to orthogonal signals; it corresponds to a transition from $\tau = 1$ to $\tau = 0.8$.

The power spectrum of sinc-pulse and, hence, power spectrum of random sequence (3) has rectangular shape. Thus, in the $1/T$ bandwidth, it is possible to transmit and detect symbols with the $1.25/T$ symbol rate that more than the theoretical Nyquist limit of $1/T$ symbols per second. According to this fact, Mazo proposed the term "Faster-than-Nyquist" signaling (FTN).

Obviously, a transition to PRS leads to a necessity of utilization of more computationally complex detection algorithms. For instance, for signals with linear modulation the Viterbi algorithm can be employed instead of a conventional matched filter receiver.

Despite the Mazo results, the FTN approach has not been implemented in telecommunication systems. A possible reason is the high computational complexity of receivers which must perform real time signal processing. Nevertheless, there are some publications which claim that the prospective systems (such as systems of fifth generation (5G)) will utilize signals with controlled ISI [2].

Apparently, a development of signal processing units has made it possible to implement complex algorithms for PRS detection. In particular, a candidate for physical layer of 5G systems is SE-FDM technique [3]. The SE-FDM signals can be derived from OFDM signals by allocation of subcarrier signals closer to each other, with a frequency spacing lower than in the orthogonal mode. Obviously, each subcarrier spectrum has shape of sinc function, and such signals are frequency-domain analog of Mazo signals.

Since the first FTN publication, some new papers developing this approach have appeared. In particular, Liveris and Georghiades [4] have proposed utilization of rrc-pulses

$$a(t) = \frac{1}{T^{1/2}} \frac{\sin\{(\pi t/T)(1-\alpha)\} + (4\alpha t/T)\cos\{(\pi t/T)(1+\alpha)\}}{(\pi t/T)\{1 - (4\alpha t/T)^2\}}, \ 0 \le \alpha \le 1, \quad (4)$$

(the sinc pulse is a particular case of (4) with $\alpha = 0$), which are widely used in many communications standards for definition of orthogonal signals. It has been shown in [4], that introduction of ISI into signals with rrc-pulses similarly to (3) can also improve spectral efficiency with zero energy efficiency losses.

We should notice that for any non-zero value of roll-off α, the rrc-pulse spectrum is not rectangular. Therefore, the signal bandwidth computed with any criterion can be larger than $1/T$. It means that utilization of $\tau < 1$ does not necessarily lead to exceeding the Nyquist limit, and, to be meticulous, we can deal with "PRS without FTN" or with "PRS with FTN". Nevertheless, we will use the term "FTN with rrc-pulses" to refer the signal generation technique (3) with rrc-pulses. To highlight the case $\tau = 1$ we will use the term "FRS with rrc-pulses".

In [5] it has been suggested to generate PRS with optimal pulses instead of the generation technique (3) with conventional rrc- or sinc-pulses. The optimal pulses provided maximal free Euclidean distance (D_{free}) for signal sequences based on a chosen signal constellation, and for a chosen ISI window length and for a bandwidth $W_{99\%}$, comprising 99% of signal power. The free Euclidean distance is similar to free distance of convolutional codes but the former is computed with Euclidean metric instead of Hamming metric. This term is correct since a PRS generator can be considered as a finite state machine and described by a trellis. As a result, authors obtained characteristics of PRS outperforming characteristics of signals from [1, 4].

All of noted works consider infinite pulses. On the one hand, it increases gains in spectral and energy efficiencies, but on the other hand, the problem of practical implementation of shaping filters remains relevant. It also remains unknown how spectral characteristics vary with pulses truncation which takes place in practice.

In [6, 7], to overcome negative effects of pulse truncation we proposed formulation and numerical solving of optimization problem for finite pulses for linear modulation (1):

$$a(t) = 0, \ |t| > LT/2, \quad (5)$$

where L is a natural number showing pulse duration as a number of symbol periods T. This approach allows to obtain shaping filter impulse response with no additional transformations of infinite pulses. We utilized a criterion of minimal bandwidth $W_{99\%}$.

As an additional constraint, we introduced a signal peak-to-average power ratio (PAPR) as a ratio of the maximal signal power $P_{\max}$ to the average power P_{av}:

$$\mathrm{PAPR} = P_{\max}/P_{\mathrm{av}}. \tag{6}$$

The value of PAPR is especially critical in a design of satellite telecommunication standards. Finally, we introduced the ISI measure called maximal group correlation MGC. The value of MGC gives the maximal normalized power of ISI noise at a matched filter output at reference time instants:

$$MGC = ACF(0)^{-1} \sum_{k=-L+1 k \neq 0}^{L-1} |ACF(k)|, \tag{7}$$

$$ACF(k) = \int\limits_{-LT/2}^{LT/2} a(t)a(t - kT)dt.$$

The introduction of this ISI measure instead of the conventional free Euclidean distance allowed significant reduction of a number of problem constraints.

The utilization of finite pulses in a linear modulation makes it possible to write a convenient decomposition of signal into separate components:

$$y(t) = \sum_{p=0}^{L-1} y_L^{(p)}(t) = \sum_{p=0}^{L-1} \sum_k C_k^{(p)} a_L^{(p)}(t, k) = \sum_{p=0}^{L-1} \sum_k C_k^{(p)} a(t - pT - kLT). \tag{8}$$

A peculiarity of each component is in non-overlapping neighboring pulses. Thus, there is no ISI in a single component, but signals of different components interfere. To refer such cases of linear modulation with finite pulses we introduced the term "multicomponent signals" (MCS). In cases of optimal pulses utilization, we will call our signals as optimal multicomponent signals.

For estimation of optimal MCS gain in [6, 7] we compared bandwidth and energy consumptions of these signals with consumptions of conventional signals with orthogonal rrc-pulses. Since the optimal signals provided minimal values of bandwidth $W_{99\%}$, we computed the bandwidth consumptions as:

$$\beta_F = W_{99\%}/R, \tag{9}$$

where R is a transmission bitrate (bits/s). The energy consumptions were computed as a signal-to-noise ratio required for the fixed value of bit error rate (the value BER $= 10^{-4}$ was considered in [6, 7]):

$$\beta_E = h^2 = E_{\mathrm{bit}}/N_0. \tag{10}$$

Since the optimization problem included additional constraints of signal PAPR, we also considered the modified energy consumptions:

$$\beta_E^* = \text{PAPR} \cdot \beta_E. \tag{11}$$

A comparison of β_E values is a conventional approach, and, for instance, it allows an estimation of gain provided by ISI introduction with respect to increase of signal constellation size. A comparison of β_E^* values allows a gain estimation with signal PAPR variation taken into account, that is important for a design of power amplifiers.

We have shown in [6, 7] that the optimal 8-component signals based on QPSK constellation provide a significant gain with respect to orthogonal signals based on QPSK, 16-QAM and 64-QAM constellations. Thus, we have confirmed Mazo results and proved that the introduction of ISI is more effective than the increase of signal constellation size. The penalty for bandwidth and energy efficiency gain is a significant increase of receiver's computational complexity.

The objective of this paper is a comparison of bandwidth and energy consumptions for FTN with rrc-pulses and for optimal MCS with a fixed receiver's computational complexity. In other words, our goal is the comparison of two approaches for PRS generation: the utilization of optimal finite pulses (MCS) and non-optimal infinite rrc-pulses (FTN). The retention of receiver's computational complexity, at first, is required for the correct comparison of these two approaches. At second, the results will be closer to practical implementation, whereas some other works represent asymptotic characteristics of signals.

If the optimal MCS outperform FTN, we will show a way to improve SE-FDM signals characteristics, which is in the transition from sinc-pulses to optimal finite pulses. If successful, the optimal MCS will be a candidate for prospective communications standards including 5G systems.

2 Simulation Details

For comparison of optimal MCS with FTN with rrc-pulses in terms of energy consumptions, we implemented a simulator, which performed a generation, a passing through a channel with additive white Gaussian noise, and a detection of these signals. In all modes we considered QPSK constellation, and needed values of bandwidth consumptions were set by handling of ISI. For optimal MCS, stronger ISI was set by increase of MGC from the minimal value MGC = 0, or by reduction of D_{free} from the maximum $D_{\text{free}} = \sqrt{2}$. For FTN a stronger ISI corresponds to a reduction of τ from the maximal value $\tau = 1$.

2.1 MCS Parameters

We considered five sets of pulses for optimal MCS. In all sets, the optimization was done with the criterion of minimal bandwidth $W_{99\%}$, comprising 99% of signal power. A numbers of components and, hence, pulses' durations were equal in all sets: $L = 8$. These sets were obtained under different types of additional constraint. The values of MGC, D_{free}, and signal PAPR were bounded, and pulse's symmetry was or was not required (Table 1).

Table 1. Additional constraints for five sets of optimal pulses

Set number	Constraint type	Pulse's type	Notation
1	MGC	Symmetric	SMGC-pulses
2	MGC	Asymmetric	AMGC-pulses
3	Dfree	Symmetric	SDfree-pulses
4	Dfree	Asymmetric	ADfree-pulses
5	PAPR	Symmetric	PAPR-pulses

For the PAPR constraint, we considered only symmetric pulses because for this type of constraint a utilization of asymmetric pulses do not lead to any different signal characteristics in comparison with the case of symmetric pulses [7]. Each set contained optimal pulses obtained for several values of particular constraint from an allowed range. The detailed technique of pulse optimization and some particular pulse shapes are represented in [6, 7].

2.2   FTN Parameters

For FTN with rrc-pulses we considered 11 values of roll-off factor α: from 0 to 1 with a 0.1 step. For each value of roll-off we considered 11 values of τ: from 1 down to 0.5 with a 0.1 step, and from 0.5 down to 0.25 with a 0.05 step (1, 0.9, 0.8, 0.7, 0.6, 0.5, 0.45, 0.4, 0.35, 0.3, 0.25).

It is known, that rrc-pulses are infinite, and a shape of rrc-pulse in time domain is utilized as shaping filter impulse response in both of FRS and FTN modes. In order to design a shaping FIR-filter for these schemes an original pulse is multiplied by a finite window function. Additionally, this truncation procedure provides a transformation from FRS and FTN with infinite rrc-pulses to MCS since pulses become finite.

We performed the truncation of pulse with a rectangular window with a duration equal to integer number of symbol periods T. A particular window durations were chosen by a criterion of comprising no less than $E_{PART} = 99.9\%$ of energy of original infinite pulse in its truncated version. To clarify this criterion, we should remind that the signal efficiency was estimated in (β_F, β_E) and (β_F, β_E^*) planes, and we tried to make the truncation which does not deteriorate these consumptions values with respect to the case of infinite pulses. It was required, that the values of $W_{99\%}$, PAPR and h^2 for the chosen BER do not change after the truncation. As we can see from Table 2, the values of $W_{99\%}$ do not change after $E_{PART} = 99\%$, the limit values of h^2 can be reached with $E_{PART} = 99.9\%$ (there is only a slight 0.1 dB loss for $\alpha = 0, 0.1$). The PAPR value is increased with an increase of E_{PART}, but the difference is small for all values of roll-off $\alpha > 0$ at the transition from $E_{PART} = 99\%$ to $E_{PART} = 99.9\%$. Thus, the truncation with $E_{PART} = 99.9\%$ provides limit values of $W_{99\%}$ and h^2; and, for this truncation mode, the values of PAPR are reduced especially for the roll-off $\alpha = 0$. This reduction of PAPR can be considered as a some gain from pulses' truncation.

Notice that non-rectangular window functions can also be utilized for the truncation. On the one hand, these window functions can reduce a level of spectrum side lobs, that has small effect on the occupied bandwidth $W_{99\%}$. On the other hand, a choice of

Table 2. A dependence of $W_{99\%}$, h^2, and PAPR from the truncation parameter E_{PART}

E_{PART}	α										
	0	0,1	0,2	0,3	0,4	0,5	0,6	0,7	0,8	0,9	1
$W_{99\%}$											
90%	1,22	1,29	1,26	1,26	1,23	1,28	1,34	1,40	1,48	1,55	1,63
95%	1,08	1,06	1,08	1,13	1,20	1,26	1,34	1,41	1,48	1,56	1,63
99%	0,99	1,02	1,07	1,13	1,20	1,27	1,34	1,41	1,48	1,56	1,63
99,9%	0,99	1,02	1,07	1,13	1,20	1,27	1,34	1,41	1,48	1,56	1,63
99,99%	0,99	1,02	1,07	1,13	1,20	1,27	1,34	1,41	1,48	1,56	1,63
h^2 (dB) for BER = 10^{-4}											
90%	14,4	13,6	12,8	11,9	11,1	8,4	8,4	8,4	8,4	8,4	8,4
95%	10,3	9,8	9,2	11,9	11,1	10,4	9,7	9,2	8,4	8,4	8,4
99%	8,9	8,9	8,7	9,1	8,6	8,4	8,4	9,2	8,8	8,5	8,4
99,9%	8,5	8,5	8,4	8,4	8,4	8,4	8,4	8,4	8,4	8,4	8,4
99,99%	8,4	8,4	8,4	8,4	8,4	8,4	8,4	8,4	8,4	8,4	8,4
PAPR (dB)											
90%	2,54	3,60	3,35	3,02	3,11	2,87	3,16	3,10	3,31	3,51	3,56
95%	5,41	5,68	5,27	4,25	3,43	3,29	3,41	3,34	3,46	3,57	3,70
99%	8,82	7,38	5,81	4,62	3,67	3,44	3,53	3,44	3,53	3,63	3,77
99,9%	12,45	7,68	5,95	4,72	3,74	3,48	3,57	3,47	3,55	3,65	3,78
99,99%	15,05	7,71	5,97	4,73	3,75	3,49	3,57	3,48	3,56	3,65	3,79

particular window function can be considered as a kind of pulse optimization thus representing an independent problem. However, in this work, we intend to compare optimal MCS with conventional FTN with rrc-pulses.

Let us consider a utilization of truncated rrc-pulses for FTN generation. For instance, if $\tau = 1$ and some pulse of duration LT is utilized, then modulation symbols go to shaping filter with a $1/T$ rate, and output signal contains L components. Now, if we increase symbol rate and send them to the shaping filter $1/\tau$ times faster ($\tau < 1$), it means a reduction of symbol period: $T' = \tau T < T$. As a result, a new pulse duration is increased in terms of number of symbol periods: $L' = L/\tau > L$. Therefore, the transition from FRS with rrc-pulses (or, equally, from MCS with truncated rrc-pulses of duration LT) to FTN with a factor of bandwidth consumptions reduction τ is equal to the increase of number of components from L to L/τ. In general, to obtain a new integer number of components with no losses of pulse's definition accuracy the value $L' =$ ceil (L/τ) should be utilized, where ceil() is a rounding to the right nearest integer. The particular L values depending from τ and α are represented in Table 3 for the truncation with $E_{PART} = 99.9\%$.

To sum up, our simulation for FTN was done with finite pulses. Thus, in both cases of optimal MCS and FTN a shaping filter had finite impulse response.

Table 3. A duration of rrc-pulse as a number of symbol periods L depending on roll-off factor α and factor of bandwidth consumptions reduction τ for the truncation with $E_{PART} = 99.9\%$

τ	α										
	0	0.1	0.2	0.3	0.4	0.5	0.6	0.7	0.8	0.9	1
1	203	15	9	7	5	5	4	4	4	3	3
0.9	226	17	10	8	6	6	5	5	5	4	4
0.8	254	19	12	9	7	7	5	5	5	4	4
0.7	290	22	13	10	8	8	6	6	6	5	5
0.6	339	25	15	12	9	9	7	7	7	5	5
0.5	406	30	18	14	10	10	8	8	8	6	6
0.45	452	34	20	16	12	12	9	9	9	7	7
0.4	508	38	23	18	13	13	10	10	10	8	8
0.35	580	43	26	20	15	15	12	12	12	9	9
0.3	677	50	30	24	17	17	14	14	14	10	10
0.25	812	60	36	28	20	20	16	16	16	12	12

2.3   Detection Algorithm Details

A generation of L-component signal is similar to an encoding by convolutional code with a code constraint length equal to $L - 1$ [6, 7]. During the MCS generation, fragments of pulse modulated by information symbols are sent to a "shift register" instead of information bits in the encoding. The signal is generated by summation of all modulated fragments stored in the register. According to this analogy between linear modulation and convolutional encoding, for MCS detection we implemented the Viterbi algorithm which is usually utilized for decoding of convolutional codes. This algorithm computed Euclidean metrics for each branch in a trellis. These metrics are Euclidean distances between the received and ideal sequences of samples.

The MCS signal was generated in frames of N modulation symbols. During the demodulation by the Viterbi algorithm, the initial and the final states of shift register were known. Decisions about transmitted sequences were made after a processing of entire frames, in other words, the traceback depth was set to TBLen = $N + L - 1$. It was done in order to obtain the best performance because in this mode the Viterbi algorithm actually performs a computationally efficient exhaustive search detection. The initialization and the termination of shift register can be implemented by insertion of pilot symbols or guarding intervals between neighboring frames. For analogy, for signals generated in frequency domain (SE-FDM signals), these procedures are provided by guarding frequency intervals.

Taking into account the chosen TBLen value and utilization of QPSK constellation in all modes, the computational complexity of the receiver was $O((N + L - 1) \cdot 2^{(L - 1)})$. As we can see from the Table 3, the complexity of the Viterbi algorithm is extremely high for some combinations of α and τ. Because of this fact, we decided to estimate energy consumptions of MCS and FTN with the fixed complexity of the receiver. Since the optimal pulses for MCS were obtained for the $L = 8$ duration, it was decided to implement the Viterbi algorithm with $2^{8 - 1}$ states no matter which pulse duration had been utilized during the generation. This reduced-complexity detection can be implemented by consideration of 8-component signal instead of L_{TX}-component signal, where $8 \leq L_{TX}$. In other words, we truncated the pulse at the detection stage whereas the finite pulse with larger duration was utilized in the transmitter. In this technique, in the receiver it is reasonable to consider a central part of pulse since rrc-pulses contain stronger oscillations with higher amplitudes in the center, and this central part causes stronger ISI with respect to pulse's edges.

3 Simulation Results

During the simulation, we obtained a lot of numerical results for different modes of FTN and optimal MCS. A direct comparison of the results in form of BER curves is rather complicated. In this section, we represent the results on the (β_F, β_E) and the (β_F, β_E^*) planes for the fixed BER = 10^{-4} and for the frame length $N = 1000$ unless other values of N are defined.

At first, we consider the results for FTN with rrc-pulses. As we can see from Fig. 1, the asymptotic characteristics from [1] have not been reached for the $\alpha = 0$: the energy loss for $\tau = 0.8$ is about $(13.5 - 8.4) = 5.1$ dB. For $\tau = 1$, a utilization of the Viterbi algorithm instead of the matched filter receiver has led to a 0.4 dB energy loss. These energy losses can be explained by small ISI window length used in the receiver and by the fact that we analyzed energy consumptions corresponding to the BER = 10^{-4}. In addition, by the transition in the receiver from $L = 8$ to $L = 10$ and 12 the energy losses were reduced to 3.4 and 2.5 dB respectively for $\tau = 0.8$. Nevertheless, resultant values of energy consumptions were far from asymptotic characteristics [1]. With $L = 8$

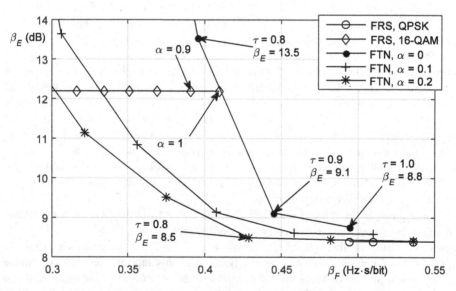

Fig. 1. Bandwidth and energy consumptions curves for FRS with rrc-pulses for a matched filter receiver (FRS, QPSK and FRS, 16-QAM), and for FTN with rrc-pulses for the Viterbi receiver with $L = 8$

a transition from BER = 10^{-4} to BER = 10^{-5} led to a 10 dB increase of energy loss, and the BER-curve did not reach BER = 10^{-6}.

Another way to reduce energy losses is a reduction of the frame size N to values near or less than L. For the reduced frame size, the ISI caused by non-orthogonal pulse can not be "accumulated" to the maximal possible level. By initialization and termination of the trellis, an additional energy gain can be obtained. For instance, for the frame size $N = 10$ and $\alpha = 0$ roll-off, the energy losses were 0.8 and 0.3 dB for $\tau = 0.8$ and 1 respectively. A reduction of N is more reasonable for frequency-domain ISI (in SE-FDM signals) than for time-domain because in time-domain it leads to a significant increase of percentage of pilot symbols or guard intervals.

In addition, we can see from Fig. 1 that better results can be reached with increase of α. For instance, for $\alpha = 0.2$ and $\tau = 0.8$, energy losses are about 0.1 dB with respect to signals with zero ISI.

For further comparison, we chose the combinations of α and τ that provide minimal consumptions for FTN with rrc-pulses. These minimal consumptions are shown in Fig. 2. Notice, that most of the "optimal" combinations contain values of roll-off near 0.5.

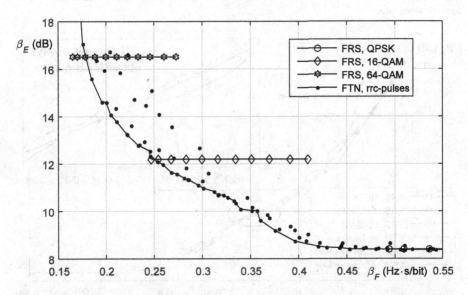

Fig. 2. An illustration of choice of "optimal" α and τ combinations for FTN with rrc-pulses

Let us compare FTN with rrc-pulses and optimal MCS. It follows from Fig. 3, that the choice of MGC constraint is not reasonable because FTN signals outperform optimal MCS in the range $\beta_F > 0.35$. Optimal MCS outperform FTN only in the range of strong ISI ($\beta_F < 0.2$), and in the range $0.2 < \beta_F < 0.35$ there is no apparent advantage of optimal MCS over FTN.

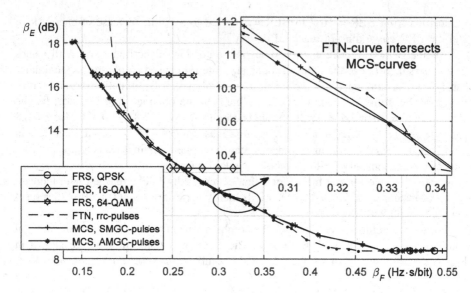

Fig. 3. A comparison of FTN with rrc-pulses and MCS with SMGC and AMGC-pulses in the consumptions plane

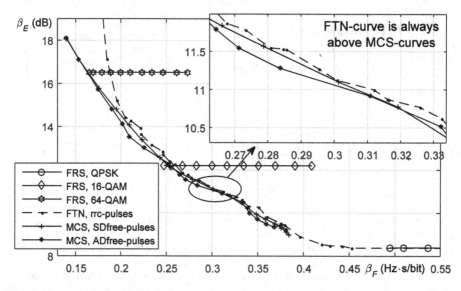

Fig. 4. A comparison of FTN with rrc-pulses and MCS with SDfree and ADfree-pulses in the consumptions plane

We can see from Fig. 4, that the optimal pulses obtained with the D_{free} constraint provide the best results in the (β_F, β_E) plane.

In some cases, FTN and optimal MCS with SDfree-pulses have equal consumptions, but FTN never outperform optimal MCS. Thus, consumptions of MCS with SDfree-pulses correspond to ultimate consumptions of FTN with rrc-pulses. A transition to ADfree-pulses for MCS leads to an additional energy gain. We can see, that optimal MCS with D_{free} constraint, as well as MCS with MGC constraint, provide a significant energy gain with respect to FTN in the range of strong ISI ($\beta_F < 0.2$). In general, the energy gain of optimal MCS varies from 0.2 to 2 dB.

We should notice, that for the constraint $D_{\text{free}} = \sqrt{2}$ the potential energy characteristics of orthogonal pulses have not been reached by MCS with SDfree as well as with ADfree-pulses. It can be explained by the fact that the D_{free} give an asymptotic BER performance, and the values of energy consumptions will be equal for FRS and PRS in the range of very low values of BER.

It follows from Fig. 5, that consumptions of MCS with PAPR-pulses correspond to ultimate consumptions of FTN in the (β_F, β_E^*) plane. This observation is similar to the case of SDfree-pulses in the (β_F, β_E) plane. In other words, in the (β_F, β_E^*) plane, the consumptions of MCS with PAPR-pulses are lower than for FTN with rrc-pulses. The energy gain of optimal MCS varies in different ranges of β_F. The maximal values of energy gain can be observed in $\beta_F < 0.2$ and $\beta_F > 0.45$ ranges, they are about 2 and 0.9 dB respectively.

From Fig. 5 we can also see that, in the (β_F, β_E^*) plane, ISI introduction is more efficient than constellation order increase. However, from Figs. 3 and 4 it follows that, in the (β_F, β_E) plane, ISI introduction provides gain only with respect to FRS with QPSK.

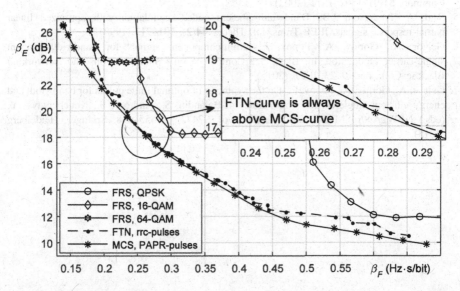

Fig. 5. A comparison of FTN with rrc-pulses and MCS with PAPR-pulses in the modified consumptions plane

4 Conclusions

A utilization of optimal MCS provides lower values of bandwidth and energy consumptions with respect to FTN with rrc-pulses for the fixed computational complexity of the receiver. There are optimal combination of α and τ for FTN with rrc-pulses, but these combinations can be found by an exhaustive search procedure. The advantage of MCS is that the rigorous constraints can be included in the optimization problem, and combinations of different types of constraints are also allowed. Finally, the optimal pulses for MCS are finite that is important for practical implementation.

References

1. Mazo, J.E.: Faster-than-Nyquist signaling. Bell Syst. Tech. J. **54**(8), 1451–1462 (1975)
2. Dai, L., Wang, B., Yuan, Y., Han, S., Chih-Lin, I., Wang, Z.: Non-orthogonal multiple access for 5G: solutions, challenges, opportunities, and future research trends. IEEE Commun. Mag. **53**(9), 74–81 (2015)
3. Kanaras, I., Chorti, A., Rodrigues, M.R.D., Darwazeh, I.: Spectrally efficient FDM signals: bandwidth gain at the expense of receiver complexity. In: International Conference on Communications ICC, pp. 1–6. IEEE (2009)
4. Liveris, A.D., Georghiades, C.N.: Exploiting faster-than-Nyquist signaling. IEEE Trans. Commun. **51**(9), 1502–1511 (2003)
5. Said, A., Anderson, J.B.: Bandwidth-efficient coded modulation with optimized linear partial-response signals. IEEE Trans. Inf. Theory **44**(2), 701–713 (1998)
6. Gelgor, A., Gorlov, A., Popov, E.: Multicomponent signals for bandwidth-efficient single-carrier modulation. In: Black Sea Conference on Communications and Networking BlackSeaCom, pp. 19–23. IEEE (2015)
7. Gelgor, A., Gorlov, A., Popov, E.: On the synthesis of optimal finite pulses for bandwidth and energy efficient single-carrier modulation. In: Balandin, S., Andreev, S., Koucheryavy, Y. (eds.) NEW2AN/ruSMART 2015. LNCS, vol. 9247, pp. 655–668. Springer, Heidelberg (2015)

Optimal Input Power Backoff of a Nonlinear Power Amplifier for FFT-Based Trellis Receiver for SEFDM Signals

Andrey Rashich and Dmitrii Fadeev[✉]

Peter the Great St. Petersburg Polytechnical University, St. Petersburg, Russia
andrey.rashich@gmail.com, fadeev_dk@spbstu.ru

Abstract. In this paper BER performance of FFT-based trellis receiver for multicarrier signals with spectrally efficient frequency division multiplexing (SEFDM) is evaluated taking into account the impact of nonlinear distortions caused by nonlinear power amplifier (PA). Values of input power back-off (IBO) corresponding to the best BER performance are obtained. The energy efficiency is quantified with energy loss with respect to the system with ideal linear PA. The energy efficiency of considered system is compared with the one of SEFDM system with detection algorithm based on decision diagram and no coding.

Keywords: OFDM · SEFDM · BER performance · Input back-off · IBO · Iterative receiver · MAP algorithm · 5G

1 Introduction

Orthogonal Frequency Division Multiplexing (OFDM) has been widely used in wireless communications. Such signals have high spectral efficiency and good BER performance in channels with inter-symbol interference (ISI).

The extension of OFDM signals is spectrally efficient frequency division multiplexing (SEFDM) signals. This kind of signals has higher spectral efficiency due to smaller subcarrier spacing. On the other hand, nonorthogonal frequency spacing between subcarriers causes energy loss in comparison with OFDM. SEFDM is the possible technique to be used in the 5th generation wireless systems (5G) [1].

The majority of papers devoted to SEFDM BER performance does not take into account the effect of transmit and receive signal path. SEFDM signals have high peak-to-average power ratio (PAPR) which may cause significant distortions in the power amplifier (PA) output signal.

It was shown in previous work that the optimal value of input power back-off corresponding to the best BER performance for SEFDM signals are different from the one for OFDM signals [2]. For significant decrease of frequency spacing even small distortions can affect BER performance. It should be noted that the optimal value of IBO could vary for different demodulation algorithms too.

SEFDM demodulation algorithms continue to be improved. Particularly, a new FFT-based trellis receiver for SEFDM signals is proposed [3]. A FEC scheme is used in

© Springer International Publishing AG 2016
O. Galinina et al. (Eds.): NEW2AN/ruSMART 2016, LNCS 9870, pp. 641–647, 2016.
DOI: 10.1007/978-3-319-46301-8_55

concatenation with SEFDM modulation. Iterative SEFDM demodulator with the log-MAP algorithm for SEFDM-symbols demodulation/demapping is proposed.

The aim of this work is to determine the optimal value of IBO for a PA to obtain the best BER performance of FFT-based trellis receiver for SEFDM signals.

2 Model Description

The SEFDM signal is a multicarrier signal that consist of a number of symbols following one after another. One SEFDM symbol of duration T can be written as:

$$x(t) = \frac{1}{\sqrt{T}} \sum_{n=0}^{N-1} s_n e^{j2\pi n \Delta f t} \tag{1}$$

where N is the number of subcarriers, s_n is the complex manipulation function for n-th subcarrier, $\Delta f = \alpha/T$ is the frequency separation between adjacent subcarriers, α is the bandwidth compression factor.

In this work solid-state power amplifier (SSPA) model is used. AM/AM and AM/PM conversions are expressed as [4]:

$$G[|x(t)|] = \frac{g_0 |x(t)|}{\left[1 + \left(\frac{|x(t)|}{x_{sat}}\right)^{2p}\right]^{\frac{1}{2p}}} \tag{2}$$

$$\Phi[|x(t)|] = 0$$

where g_0 is the amplifier gain, x_{sat} – is the saturation level, p is the parameter to control the AM/AM sharpness of the saturation region and equals 2.

Typical AM/AM conversion curve in dB scale is shown at Fig. 1. The value of the input power 0 dB corresponds to saturation level x_{sat}, the value of the output power 0 dB corresponds maximum output power $g_0 \cdot x_{sat}$.

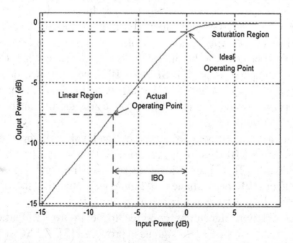

Fig. 1. AM/AM conversion curve for SSPA model with $p = 2$ in dB scale

Maximum PA power efficiency is obtained when operating point is at saturation level. As mentioned above, SEFDM signals exhibit high PAPR, therefore input power back-off (IBO) is required to shift the operating point to ensure that the PA operates within linear region, as shown in Fig. 1.

IBO shows the value of the average input power decrease compared with saturation level, in dB, and can be written as

$$IBO = 10 \log_{10}\left(P_{sat}/P_{avg}\right) \qquad (3)$$

On the one hand, higher IBO corresponds to less output power and less BER performance. On the other hand, higher IBO corresponds to lower distortions caused by the PA.

For performance estimation, Total Degradation (TD) metric is used [5]. Actually, TD is the energy loss with respect to the system with ideal linear PA. TD can be written as:

$$TD = SNR_{PA}(IBO) - SNR_{AWGN} + IBO \quad [in \; dB] \qquad (4)$$

where SNR_{AWGN} is the SNR which is required to achieve a target BER in AWGN channel and $SNR_{PA}(IBO)$ is the SNR which is required to achieve a target BER taking into account distortions caused by the PA at a given IBO.

SEFDM signals show energy loss in comparison with OFDM which inversely depends on α due to inter-carrier interference (ICI). Let us introduce modified metric TD^* taking into account energy losses caused by both ICI and PA nonlinearity:

$$TD^* = SNR_{PA}(IBO) - SNR_{BEST} + IBO \quad [in \; dB] \qquad (5)$$

where SNR_{BEST} is the SNR which is required to achieve a target BER in AWGN channel for OFDM signals with the same FEC code as used in SEFDM demodulator.

The MatLab simulation model was developed to evaluate BER performance, TD and TD* metrics (Fig. 2).

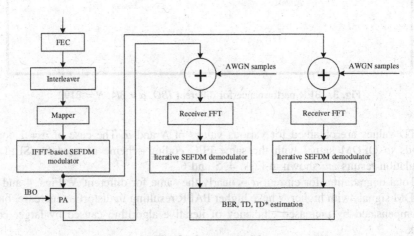

Fig. 2. System model

The model for SNR_{AWGN} estimation includes generation of SEFDM symbols, AWGN channel and the receiver of SEFDM-signals. SSPA model unit is added to the model to obtain SNR_{PA} for the set of IBO values.

The algorithm for SEFDM generation based on Inverse Fast Fourier Transform (IFFT) [6] is used. Informational bits are previously coded by (3, [5, 7]) convolutional code of rate = 1/2. Modulation scheme is QPSK.

FFT-based trellis receiver for SEFDM is used [3]. It employs iterative SEFDM demodulator with the log-MAP algorithm for SEFDM-symbols demodulation/ demapping. Its structure is quite similar to turbo equalizer schemes. In this model the number of iterations is three.

3 Simulation Results

BER curves are obtained for various values of N and α. Figure 3 shows BER curves for $N = 8192$, $\alpha = 3/4$ for the set of IBO values. There is no energy loss when IBO = 10 dB. For another considered values of IBO there is energy loss, which is higher for less IBO. TD metric helps us compare energy losses corresponding to different N and α.

Fig. 3. BER performance for different *IBO*, $\alpha = 3/4$, $N = 8192$

TD values are obtained for various values of N and α. The case of $\alpha = 1$ corresponds to OFDM signal with the same FEC coding scheme as used for SEFDM. Simulation results are shown at Figs. 4, 5 and 6.

Total degradation for current α is nearly the same for different N (Figs. 4 and 5). SEFDM signals with higher N have higher PAPR resulting in distortion increase, but it is compensated by increased efficiency of iterative algorithm caused by larger code block.

Fig. 4. TD vs IBO for different N, $\alpha = 1/2$

Fig. 5. TD vs IBO for different N, $\alpha = 3/4$

TD versus IBO curves for different α are shown at Fig. 6. Minimum value of TD is larger for lower α.

In previous work TD metrics has been obtained for SEFDM system with detection algorithm based on decision diagram and no coding [2]. The energy efficiency of FFT-based trellis receiver is higher for all corresponding values of α. Total degradation for SEFDM with $\alpha = 3/4$ is nearly the same as for OFDM signals and no coding.

TD* metric is evaluated to show energy losses caused by both ICI and PA non-linearity in comparison with coded OFDM signals (Fig. 7).

Fig. 6. TD vs IBO for different α, $N = 8192$

Fig. 7. TD* vs IBO for different α, $N = 8192$

For IBO > 6 the key impact to the difference between curves is caused by ICI. Minimum value of TD* is 2.4, 4.3 and 7.1 dB for α = 3/4, 5/8 and 1/2, respectively. Corresponding values of IBO are 0, 2 and 4 dB.

Out of band emission level is the same as one considered in previous work [2].

4 Conclusion

In this paper, BER performance of FFT-based trellis receiver for SEFDM is evaluated taking into account the impact of nonlinear distortions caused by nonlinear power amplifier.

The MatLab simulation model was developed to evaluate BER performance. Values of IBO corresponding to the best BER performance are obtained. The energy efficiency is quantified with energy loss TD with respect to the system with ideal linear PA. The energy efficiency of considered system is higher in comparison with the SEFDM one with detection algorithm based on decision diagram and no coding.

The second metric TD^* is introduced to show energy losses caused by both ICI and PA nonlinearity in comparison with coded OFDM signals. Minimum value of TD^* is 2.4, 4.3 and 7.1 dB for $\alpha = 3/4$, 5/8 and 1/2, respectively. Corresponding values of IBO are 0, 2 and 4 dB.

References

1. Xu, T., Darwazeh, I.: Spectrally efficient FDM: spectrum saving technique for 5G? In: 2014 First International Conference on 5G for Ubiquitous Connectivity (5GU), pp. 273 – 278, 26–28 November 2014
2. Fadeev, D.K., Rashich, A.V.: Optimal input power backoff of a nonlinear power amplifier for SEFDM system. In: Balandin, S., Andreev, S., Koucheryavy, Y. (eds.) NEW2AN/ruSMART 2015. LNCS, vol. 9247, pp. 669–678. Springer, Heidelberg (2015)
3. Rashich, A., Kislitsyn, A., Dmitrii, F., Tan, N.: FFT-based trellis receiver for SEFDM signals. In: GLOBECOM 2016, under consideration
4. Thompson, S.C., Proakis, J.G., Zeidler, J.R.: The effectiveness of signal clipping for PAPR and total degradation reduction in OFDM systems. In: Global Telecommunications Conference, 2005, GLOBECOM 2005, vol. 5, p. 2811. IEEE
5. D'Andrea, A.N., et al.: RF power amplifier linearization through amplitude and phase distortion. IEEE Trans. Commun. **44**(11), 1477–1484 (1996)
6. Kislitsyn, A.B., Rashich, A.V., Tan, N.N.: Generation of SEFDM-signals using FFT/IFFT. In: Balandin, S., Andreev, S., Koucheryavy, Y. (eds.) NEW2AN/ruSMART 2014. LNCS, vol. 8638, pp. 488–501. Springer, Heidelberg (2014)

The Allan Variance Usage for Stability Characterization of Weak Signal Receivers

Yuriy V. Vekshin[1] and Alexander P. Lavrov[2(✉)]

[1] Institute of Applied Astronomy of Russian Academy of Sciences,
St-Petersburg, Russian Federation
yuryvekshin@yandex.ru
[2] Peter the Great St. Petersburg Polytechnic University,
St-Petersburg, Russian Federation
lavrov@ice.spbstu.ru

Abstract. The paper is devoted to the Allan variance usage for output signal fluctuation analysis of weak signal receivers. The tri-band microwave receiving system for the new radio telescope RT-13 of Institute of Applied Astronomy of Russian Academy of Sciences is considered. The Allan variance correction for signals with «dead time» in data acquisition is developed. Output stability of the receiver is investigated in terms of noise fluctuation type – white, flicker, drift, etc., and their time stability intervals calculation. Investigations are performed in *S*-, *X*-, and *Ka*- receiver bands. The influence of the input cryogenic stage temperature on output receiver signal is considered also.

Keywords: Receiver stability · Allan variance · Cryogenic receiver · Noise signal analysis · Flicker-noise · Drift · Measurement dead time · Corrected Allan variance

1 Introduction

The sensitivity of weak signal receivers and accuracy of obtained data are limited by instability of their parameters. Receiver output signal fluctuations have components with different power spectral density – white noise, flicker noise, drift, and other spectral components, which limits fluctuations decrease with increasing the averaging time. The Allan variance (AV) analysis allows to identify noise types and their levels, and also to determine the time stability interval of receiver – optimal averaging time, at which the best sensitivity is achieved. Initially, AV was used to analyze the frequency standards stability [1], but in recent times, it has been used to receiver stability research [2–4]. In this paper, AV is applied for the stability analysis of the tri-band receiving system for the new radio telescope RT-13 of Institute of Applied Astronomy RAS.

2 Allan Variance: Calculation and Correction

The Allan variance $\sigma_A^2(\tau)$ – is the variance of the averaged over the time interval τ first differences of the processed data y values, sampled with a period $T, \tau = n \cdot T$, $n = 1, 2, \ldots$:

© Springer International Publishing AG 2016
O. Galinina et al. (Eds.): NEW2AN/ruSMART 2016, LNCS 9870, pp. 648–657, 2016.
DOI: 10.1007/978-3-319-46301-8_56

$$\sigma_A^2(\tau) = \frac{1}{2(N-1)} \cdot \sum_{j=1}^{N-1} (\bar{y}_{j+1}(\tau) - \bar{y}_j(\tau))^2. \tag{1}$$

The Allan variance $\sigma_A^2(\tau)$ is associated with the noise power spectral density (PSD) $S(f) = h_\alpha/f^\alpha$: $\sigma_A^2(\tau) = K_\alpha \tau^{\alpha-1}$, where $K_0 = h_0/2\tau, K_1 = h_1 2 \cdot \ln 2, K_2 = h_2 \pi^2 \cdot 2\tau/3$ [1].

We have made modeling of noises with different PSD (and its combination) in LabVIEW. AV calculations for these noises are presented in log-log scale at Fig. 1, where: 1 – is the white noise, decrease of $\sigma_A^2(\tau)$ is 10 dB/decade; 2 – is the $1/f$ noise, $\sigma_A^2(\tau)$ – const; 3 – is the $1/f^2$ noise, increase of $\sigma_A^2(\tau)$ is 10 dB/decade; 4 – is the sum of the white and the $1/f$ noise, $\sigma_A^2(\tau) = K_0/\tau + K_1$; 5 – is the sum of the white and the $1/f^2$ noise, $\sigma_A^2(\tau) = K_0/\tau + K_2 \cdot \tau$, 6 – is the linear drift, increase of $\sigma_A^2(\tau)$ is 20 dB/decade. There is a minimum on the $\sigma_A^2(\tau)$ plot, when the analyzed signal is the sum of several noise types. This minimum gives an optimal averaging time τ_{OPT} of noise signal y, at which the considered system is stable and averaging of signal y gives decrease of its variance.

In practical measurements sample registration is often characterized by the presence of the dead time DT between samples, when the averaging time τ is less than the sample period T, $DT = T - \tau$. In this case, the Allan variance correction, calculated by (1), is required, since $DT = 0$ is supposed in (1) [4, 5]. We have developed the Allan variance correction technique.

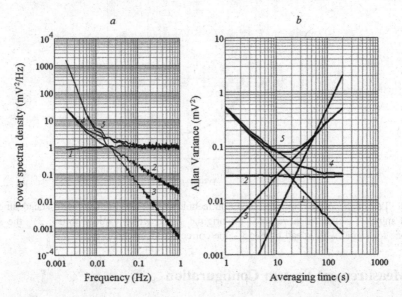

Fig. 1. Relationship between the Allan variance (*a*) and the power spectral density (*b*) for different noise types

The Allan Variance plot for measurements with «dead time» – $\sigma^2_{ADT}(\tau)$ is calculated on multiple to period T intervals. $\sigma^2_{ADT}(\tau)$ plot is approximated by linear combination of power functions, that are supposed to noise basic components: white noise, $1/f$ noise, $1/f^2$ noise, linear drift:

$$\sigma^2_{ADT}(\tau) = K_0/\tau + K_1 + K_2\tau + K_3\tau^2, \qquad (2)$$

weights K_i is determined by approximation. Then K_i are corrected by multipliers B_i:

$$\sigma^2_{AC}(\tau) = B_C\sigma^2_{ADT}(\tau) = B_0K_0/\tau + B_1K_1 + B_2K_2\tau + B_3K_3\tau^2, \qquad (3)$$

where $B_0 = \tau/T$, $B_1 = B_2 = B_3 = 1$.

The correction technique is tested in numerical experiments with models of noise signals. The result for the sums of noises: white noise, $1/f$ noise, $1/f^2$ noise, linear drift, is presented at Fig. 2: 1 – AV $\sigma^2(\tau)$ for original signal without «dead time», $\tau = T = 0.1$s; 2 – AV$\sigma^2_{ADT}(\tau)$ for signal with «dead time», obtained by decimation the original signal, $\tau = 0.1$ s, $T = 1$ s («dead time» $DT = 0.9$ s); 3 – AV$\sigma^2_{AC}(\tau)$ after correction accordingly (3). The corrected AV $\sigma^2_{AC}(\tau)$ coincides with the AV for original signal without «dead time». Standard deviation of the relative difference of these plots does not exceed 5 % when averaged over all values of τ. This AV correction gives good coincidence with original AV, when τ/T not very small ($\tau/T > 0.01$).

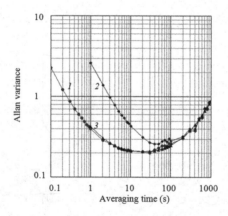

Fig. 2. Test of the Allan variance correction technique for measurements with «dead time» on model signal: 1 – the Allan variance for original signal without «dead time»; 2 – the Allan variance for signal with «dead time»; 3 – the corrected Allan variance

3 Measurement System Configuration

Stability investigations are conducted for the tri-band receiving system of the radio telescope RT-13. Radio telescope RT-13 with a mirror diameter of 13.2-m is a part of the two-element radio interferometer of new generation, created in the Institute of

Applied Astronomy of Russian Academy of Sciences at the stations of "Quasar" VLBI network [6]. The radio interferometer is designed for the well-timed determination of Universal Time (UT1).

The radio telescope receiving system (Fig. 3) operates at two circular polarizations simultaneously in the following bands: S-band (2.2–2.6 GHz), X-band (7.0–9.5 GHz), Ka-band (28–34 GHz) [7]. The tri-band feed and input low-noise amplifiers with microwave isolators are placed in cryostat (the cryogenic unit) and cooled using microcooler by the closed-cycle cryogenic system to the temperature below 20 K, that significantly reduces the receiver noise temperature. The frequency converters split up, filter and convert amplified by cryogenic unit input microwave signals to intermediate frequency band 1–2 GHz – the operating band of the digital acquisition system. The commutator provides additional gain and subchannels to record selection. Calibration unit contains adjustable noise generators for each band. The noise temperature Tr (referenced to input), total gain Gtotal of the receiving system and gain of the receiving system units G are presented at Table 1.

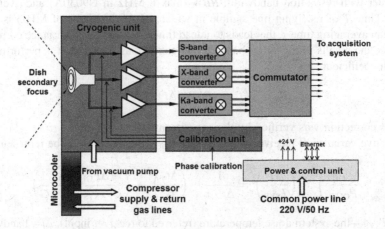

Fig. 3. Block diagram of the radio telescope tri-band receiving system

Table 1. Main parameters of the receiving system

Parameter	S-band	X-band	Ka-band
Tr, K	20	15	50
Gtotal, dB	103	99	95
Gcryogenic unit, dB	30	30	30
Gconverter, dB	48	44	40
Gcommutator, dB	25	25	25

Long-term (few hours) recording of the receiving system output signals P_{out} was performed for stability measurements (measurement scheme at Fig. 4). The special noise load (Low temperature Noise Generator), cooled by a liquid nitrogen (made for calibrating tasks) was installed on receiving system input. For stability measurements of single units in receiver chain their inputs were match loaded.

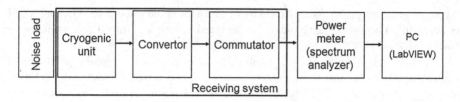

Fig. 4. Measurement scheme

Signals at commutator outputs were registered by power meter Agilent N1914A with the 8487D power sensors (in X- and Ka-bands) and by spectrum analyzer Agilent N9030A (in S-band), managed by LabVIEW.

The spectrum analyzer was used as a spectral-selecting registration system due to the presence of RF interference (mainly communication networks 3G and Wi-Fi). Signal analyzer sweeps frequency band Δf over sweep time ST using relatively narrow-band filter with resolution bandwidth RBW (max 8 MHz in N9030A) and averaging time τ. Time T of receiving one sample in wideband Δf (hundreds of MHz) is much more than averaging time τ, that leads to «dead time». So, the Allan variance correction to reduce the results to the entire band Δf and continuous registration is performed by (3) with coefficient

$$B_0 = \tau/T = (RBW/\Delta f) \cdot (ST/T). \qquad (4)$$

This correction was verified by the experiments.

Relative variance of receiver output power P_{OUT} fluctuations can be represented as

$$\left(\frac{\Delta P_{OUT}}{P_{OUT}}\right)^2 = \left(\frac{\Delta T}{T_{SYS}}\right)^2 = \frac{1}{\Delta f \cdot \tau} + \left(\frac{\Delta G}{G}\right)^2 + \left(\frac{\Delta P_{OUT\ REG}}{k \cdot T_{SYS} \cdot \Delta f \cdot G}\right)^2, \qquad (5)$$

where T_{SYS} – the system noise temperature (referred to receiver input), Δf – bandwidth, $\Delta G/G$ – relative total gain fluctuation, $\Delta P_{OUT\ REG}$ – data acquisition system noise, k – Boltzmann constant. This representation is convenient when single part contribution in total variance is analyzed.

The relative total gain fluctuation is the sum of the relative gain fluctuations of the receiver single stages (G_1 and G_2):

$$\left(\frac{\Delta G}{G}\right) = \left(\frac{\Delta G_1}{G_1}\right) + \left(\frac{\Delta G_2}{G_2}\right). \qquad (6)$$

If the relative gain fluctuation of each stage is independent, the total variance σ is the sum of variances of single stage gain fluctuations

$$\sigma^2 = \sigma_1^2 + \sigma_2^2. \qquad (7)$$

So, we can try to predict fluctuations of all system, if single stage fluctuations are known, or, contrariwise, to distinguish gain fluctuation of single stage if fluctuations of all system and fluctuations of other stages are known.

If the Allan variance plot for relative output power fluctuations of one cascade with gain G_1 is known, the addition in the Allan variance figures of relative output power fluctuations of combination of cascades with gain G_1 and G_2 can be related with gain fluctuation G_2, assuming that noise temperature T_{SYS} is stable and data acquisition system noise is negligible:

$$\left(\frac{\Delta P_{OUT}}{P_{OUT}}\right)^2 = \left(\frac{\Delta T}{T_{SYS}}\right)^2 = \frac{1}{\Delta f \cdot \tau} + \left(\frac{\Delta G_1(\tau)}{G_1}\right)^2 + \left(\frac{\Delta G_2(\tau)}{G_2}\right)^2. \tag{8}$$

The first term in (8) is due to the so-called "radiometric noise" at radiometer output with no gain fluctuations. Radiometric noise is sensitivity for ideal radiometer [2].

4   Investigation Results

Stability investigations of the receiving system output signals are performed in «warm» mode (physical temperature inside cryogenic unit at the second stage of microcooler $T_{2st} = 300$ K) and «cold» mode (physical temperature inside cryogenic unit at the second stage of microcooler $T_{2st} = 10$ K). Results of calculating of the relative Allan *deviation* $\sigma_{A\,REL}(\tau)$ for the receiving system output signals in X-band (7–8 GHz) are presented at ·Fig. 5. The relative Allan *deviation* is the Allan deviation for relative fluctuations $\Delta P/P$. At Fig. 5 plot *1* is $\sigma_{A\,REL}(\tau)$ for output signal of the receiving system in «warm» mode, *2* – the same $\sigma_{A\,REL}(\tau)$ plot for «cold» mode, *3* – theoretical

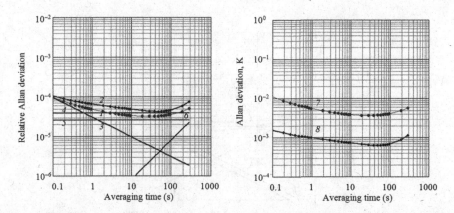

Fig. 5. X-band receiver stability: *1* – $\sigma_{A\,REL}(\tau)$ plot for the receiving system output signal in «warm» mode; *2* – $\sigma_{A\,REL}(\tau)$ plot for the receiving system output signal in «cold» mode; *3* – theory – white noise for 1 GHz bandwidth; *4* – 1/f noise component of *2*; *5* – 1/f noise component of *1*; *6* – drift component of *1* or *2*; *7* – absolute $\sigma_A(\tau)$ plot for the receiving system output signal in «warm» mode; *8* – absolute $\sigma_A(\tau)$ plot for the receiving system output signal in «cold» mode

$\sigma_{A\ REL}(\tau)$ plot, corresponding to case of white noise for 1 GHz bandwidth. Plot 1 is approximated by sum of white noise, $1/f$ noise 5 and drift 6. Plot 2 is approximated by sum of white noise, $1/f$ noise 4 and drift 6. It is obvious that $1/f$ noise component in «warm» mode is under than in «cold» mode in terms of relative Allan deviation. The «warm» noise load (with temperature $T_L = 300$ K) was installed on the receiving system input during these experiments. To evaluate absolute sensitivity of the receiving system, output signal was calibrated in K degrees, and temperature of «warm» noise load T_L was substracted, the results in terms of absolute Allan deviation $\sigma_A (\tau)$ in «warm» mode ($Tr = 115$ K) is presented at plot 7, in «cold» mode ($Tr = 15$ K) is presented at plot 8. The stability interval of the receiving system in «cold» mode (plot 8) is about 30 s (the Allan deviation decreases on this interval). Receiver signal chain without cryogenic unit is also investigated, the $\sigma_{A\ REL}(\tau)$ plot is coincide with the plot 1, that indicates, that in «warm» mode instability of subsequent stages (converter and commutator) is dominant. When considering the entire receiving system in «cold» mode, $1/f$ noise of cryogenic unit is dominant in the range of averaging time 1–100 s. At averaging time greater than 100 s main source of instability is drift of subsequent stages.

Results of the stability investigations of the receiving system output signals in Ka-band (28–29 GHz) are shown at Fig. 6. Plot 1 indicates $\sigma_{A\ REL}(\tau)$ of the entire receiver in «cold» mode, plot 2 corresponds $\sigma_{A\ REL}(\tau)$ for receiver signal chain without cryogenic unit, plot 3 – theoretical $\sigma_{A\ REL}(\tau)$ plot, corresponding to case of white noise for 1 GHz bandwidth. Plot 1 is approximated by sum of white noise, $1/f$ noise 4 and $1/f^2$ noise 6. Plot 2 is approximated by sum of white noise 3, $1/f$ noise 5 and $1/f^2$ noise 7. It is obvious that $1/f$ noise component of cryogenic unit is above than for receiver chain without cryogenic unit in terms of relative Allan deviation and plot 1 lies above plot 2. It indicates that dominant source of instability is cryogenic unit. In the range of output signal averaging time 1–100 s fluctuation of Ka-band receiver is flicker-noise (the

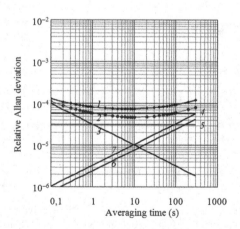

Fig. 6. Ka-band receiver stability: 1 – for the entire receiver chain in «cold» mode; 2 – for receiver chain without cryogenic unit; 3 – theory – white noise for 1 GHz bandwidth; 4 – $1/f$ noise component of 1; 5 – $1/f$ noise component of 2; 6 – $1/f^2$ noise component of 1, 7 – $1/f^2$ noise component of 2

Allan deviation is constant), after 100 s averaging time $1/f^2$ noise appears. Time stability interval is about 1 s.

Stability investigations of the receiving system output signals in S-band are performed in 2290–2360 MHz band (this band is free from the RF interference) using swept spectrum analyzer. Results are presented at Fig. 7; plot 1 is $\sigma_{A\ REL}(\tau)$ plot for the receiving system output signal in «cold» mode, registered by swept spectrum analyzer, so it is signal with «dead time». So, the Allan deviation correction (see (3) and (4)) is applied, the result of correction is plot 2. Corrected $\sigma_{A\ REL}(\tau)$ is approximated by sum of white noise 3 (70 MHz bandwidth), $1/f$ noise (plot 4) and drift (plot 5). Flicker-noise is dominant at 3–100 s averaging times, at longer intervals drift appears. Time stability interval is about 3 s.

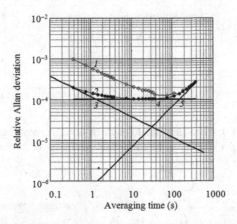

Fig. 7. S-band receiver stability: 1 – not corrected $\sigma_{A\ REL}(\tau)$, with «dead time»; 2 – corrected $\sigma_{A\ REL}(\tau)$, without «dead time»; 3 – theory – white noise for 70 MHz bandwidth; 4 – $1/f$ noise component of 2; 5 – drift component of 2

The influence of the input cryogenic stage temperature on the output receiver signal is investigated. Figure 8 shows the feed temperature T_{FEED} variation (a) and the simultaneous relative variation of X-band output signal (b) 100 min after the change of the "cold" load to the "warm" load on the receiving system input. The correlation coefficient (c) for these signals (a, b) is calculated after deduction of linear and quadratic drifts. Signals correlation is visible, the correlation coefficient $K_{COR1} = -0.4$, indicating that the feed temperature effects on the cryogenic unit transfer coefficient. The influence coefficient of the T_{FEED} on the $\Delta P/P$ variation $K_{INF1} = -0.047$ 1/K, it is estimated as

$$K_{INF1} = K_{COR1} \cdot RMS(\Delta P/P)/RMS(\Delta T). \qquad (9)$$

The influence of the second stage temperature T_{2ST} of microcooler in cryogenic unit is illustrated at Fig. 9. Variations with 1 Hz frequency (microcooler operating frequency) are observed (Fig. 9a). This spectral component is slightly visible in the amplitude spectrum of X-band output signal $S_{\Delta P/P}$ also (Fig. 9b). The influence coefficient of T_{2ST}

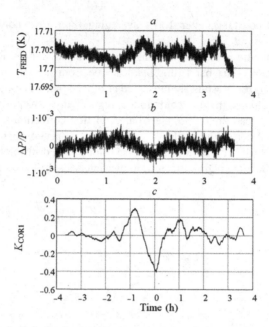

Fig. 8. The influence of the feed temperature on the X-band receiver output signal: a – the feed temperature; b – the output signal; c – the correlation coefficient of the feed temperature and the output signal

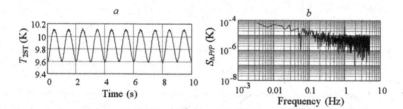

Fig. 9. The influence of the second stage temperature of microcooler in cryogenic unit on the X-band receiver output signal: a – the second stage temperature; b – amplitude spectrum of the output signal

ripples on the $\Delta P/P$ variation is calculated by the ratio of 1 Hz harmonics in its amplitude spectra: $K_{INF2} = 1.7 \cdot 10^{-4}$ 1/K. Calculation of correlation coefficient K_{COR2} for signals T_{2ST} and $\Delta P/P$ shows 1 Hz ripple also, and $K_{COR2} \approx 0.02$.

5 Conclusions

The Allan variance is useful to characterize and compare stability of different type's receivers when steady state condition at its input is realized. It allows distinguishing the contribution of the different noise components (white noise, $1/f$ noise, $1/f^2$ noise, drift) in analyzing data at specific time intervals. The Allan variance can be applied to noise and stability measurements of receiver as whole device, so also to separate cascades in receiver chain, in last case one can analyze cascades' gain fluctuations and look for its main contributions. For case of receiver output signal data acquisition with «dead time» the Allan variance correction technique is developed and it is applied for wideband noise power measurements using swept spectrum analyzer. The Allan variance was used for investigations of output signal stability of the modern very sensitive radio telescope receiving system and its chain stages in S-, X-, and Ka-bands. Dominant noise components and the receive time stability intervals are determined. Time stability intervals – the optimal averaging time of the output signal, at which minimum of the noise variance is achieved, for the receiver in «cold» operation mode are: 3 s for S-band, 30 s for X-band, and 1 s for Ka-band. Flicker-noise is dominant component for averaging time in range from few seconds to 100 s. At longer averaging times drift appears in S-, X- bands and $1/f^2$ noise appears in Ka-band. Stability of the receiving system output signals in «warm» and «cold» receiver modes is compared for X-band. Absolute Allan deviation values in «cold» mode are below than in «warm» mode, but $1/f$ noise component is above than in «warm» mode. Comparing relative fluctuations of the receiver chain stages allows to determ main sources of instability. Fluctuations of the first cryogenic stage are dominant in S-, X-, and Ka-bands: they exceed fluctuations of the subsequent stages (frequency converters and commutators). The influence of the input cryogenic stage temperature variation on the receiver output signal is obtained also using correlation and spectrum analysis.

We consider the presented results show that the Allan variance usage is productive and informative approach and it can be applied to study instabilities of various types' multi-stage receivers.

References

1. Rutman, J.: Characterization of phase and frequency instabilities in precision frequency sources: fifteen years of progress. Proc. IEEE **66**, 1048–1075 (1978)
2. Land, D.V., Levick, A.P., Hand, J.W.: The use of the Allan deviation for the measurement of the noise and drift performance of microwave radiometers. Meas. Sci. Technol. **18**, 1917–1928 (2007)
3. Gonneau, E., Escotte, L.: Low-frequency noise sources and gain stability in microwave amplifiers for radiometry. IEEE Trans. Microw. Theor. Tech. **60**(8), 2616–2621 (2012)
4. Shan, W., Zhenqiang, L., Shengca, S., Yang, J.: Gain stability analysis of a millimeter wave superconducting heterodyne receiver for radio astronomy. In: Proceedings of Asia-Pacific Microwave Conference, pp. 477–480 (2010)
5. Barnes, J.A., Allan, D.W.: Variances based on data with dead time between the measurements. NIST Technical Note 1318 (1990)
6. Ipatov, A.V.: A new-generation interferometer for fundamental and applied research. Phys.-Usp. **56**(7), 729–737 (2013)
7. Ipatov, A., Ipatova, I., Mardyshkin, V., Evstigneev, A., Khvostov, E., Chernov, V.: Tri-band system for the Russian interferometer. In: IVS 2014 General Meeting Proceedings of the VGOS: The New VLBI Network, pp. 118–12. Science Press, Beijing (2014)

An Extremely Flexible, Energy, and Spectral Effective Green PHY-MAC for Profitable Ubiquitous Rural and Remote 5G/B5GIoT/M2M Communications

Alexander Markhasin[(⊠)]

Siberian State University of Telecommunications, Novosibirsk, Russia
almar@risp.ru

Abstract. In this paper, the fundamental PHY-MAC throughput limits and extremum of the energy, power, spectral efficiency invariant criteria are proved. The invariant criteria are constructed relying on Shannon's m-ary digital channel capacity which a rich palette of the technically interpreted PHY-MACs parameters consider. Therefore, the invariant criteria as very suitable for research and design of an 5G extremely performance problems are found. The PHY-MACs smart distributed control techniques which able implements "on-the-fly" the limits close and invariant criterion optimization or trade-off is proposed. Such PHY-MAC's smart control techniques represent a key disruptive technologies meet the 5G/B5G network challenges.

Keywords: 5G/B5G · Rural · PHY-MAC · Green · Profitable

1 Introduction

Unacceptably high investments are required into deployment of the optic core infrastructure for ubiquitous wide covering of sparsely populated rural, remote, and difficult for access (RRD) areas using the recent (4G) and also forthcoming (5G) broadband radio access (RAN) centralized techniques, characterized by short cells ranges, because their profitability boundary exceeds a several hundred residents per square kilometer. Furthermore, the unprecedented requirements and new features of the forthcoming Internet of Things (IoT), machine-to-machine (M2M), smart city, and also many other machine type IT-systems lead to a breakthrough in designing extremely intensive technologies for future 5G/B5G wireless systems which will be able to reach in real time the performance extremums, trade-off optimums and fundamental limits [1–3]. Recently, a number of 5G extremely intensive solutions were proposed which are suitable mainly for well-urbanized areas: ultra-dense networks [4], massive MIMO [5] and M2M for smart city [6], disruptive 5G PHY technology [2].

For weakly urbanized areas, we offer extremely effective green [8] techniques as an approach for ubiquitous profitable covering by 5G/B5G IoT/M2M/H2H multifunctional communications of the RRD territories [8, 9]. Practically, the necessary and sufficient conditions for RRD areas profitability border overcoming envisages three extremal RRDs networking performances': (i) the hyper long range hypercells' radically

© Springer International Publishing AG 2016
O. Galinina et al. (Eds.): NEW2AN/ruSMART 2016, LNCS 9870, pp. 658–669, 2016.
DOI: 10.1007/978-3-319-46301-8_57

distributed cost-effective Multifunctional Hyperbus Architecture (MFHBA) [8, 9] and MFHBA mission-critical convergent techniques – (ii) the extremely energy-effective green PHY [11], and (iii) the supreme throughput capacity multifunctional MAC [11]. Convergent PHY-MACs shall close the Shannon's fundamental limits or extremums using the multifunctional optimal "on-the-fly" control techniques [8, 12].

Recently [3, 7], the spectral (SE) and energy (EE) efficiency criteria are expressed usually through the Shannon's capacity of the continuous channels with additive white Gaussian noise (AWGN) [13] which, in principle, allow to study only the PHY potential efficiency values depending directly on three spectral-power basic parameters, i.e., bandwidth ΔF_s, signal power P_s and AWGN noise power P_n. In [10, 14] so called invariant criteria of spectral (ICSE), power (ICPE) and energy ICEE) efficiency were first introduced for orthogonal spread spectrum m-ary signals. The invariant criteria are constructed relying on Shannon's m-ary digital channel capacity which considers a rich palette of the technically interpreted PHY-MAC parameters. Therefore, the invariant criteria were found very suitable for research and design of an 5G extremely performance problems.

In this paper, we generalize and develop the results of our above cited researches of the RRD radically distributed multifunctional device-centric MFHBA architecture, fundamental RRD PHY and information-theoretic RRD MAC limits and extremums focused on the 5G/B5G extremely performance issues and also PHY-MAC multifunctional optimal control techniques which meet green profitable ubiquitous rural and remote 5G/B5G IoT/M2M/H2H communications. The conceptual vision of the green profitable ubiquitous RRD 5G/B5G IoT/M2M/H2H architecture is also presented.

2   Extremely Green and Cost-Effective Ubiquitous RRD 5G PHY

2.1   Vision of Extremely Green and Effective 5G PHY for RRD Areas

As stated above, the widespread cell range of the recent (4G) and forthcoming (5G) generations of radio access technologies (RAN) does not exceed few kilometers. The sparsely populated RRD areas differ by low density up to a few tens residents. Hence, the really indispensable approach for overcoming the 5G RRDs economical barrier lead to extremely increasing of the number of the effectual subscribers, i.e., to increasing of the air interface range of broadband hypercells by several ten times through approaching the fundamental Shannon limits of spectral (SE) and power (PE) efficiency. Usually [3, 7], the SE and PE efficiency criteria are expressed through the Shannon's capacity of the continuous channels with additive AWGN noise [13]

$$C = \Delta F_s \log_2(1 + P_s/P_n), \tag{1}$$

where ΔF_s is bandwidth, P_s – signal power, P_n – noise power, $P_n = \Delta F_s N_0$, N_0 – signal-sided spectral power noise density, in Watt-per-Hertz. In the channel output, or receiver input, power characteristic P_s/P_n is called the signal-to-noise-ratio (SNR).

However, the continuous channel throughput capacity (1) allow to study only the potential efficiency PHY values depending directly on three spectral-energy basic parameters. So called invariant criteria of spectral, power and energy efficiency [10] allow to solve an optimization or trade-off problem depending on the set of real conditions and parameters of the radio channel, methods of signal coding, formation, modulation, transmitting, receiving, processing, decoding, etc. Two invariant efficiency criteria were first introduced for the wireless physical layer with orthogonal spread spectrum m-ary signals in [14] basing on Shannon's m-ary digital channel capacity. As in [10], let us introduce an invariant efficiency criterion for modern 5G PHY relying on SINR [15] approaches. The invariant criterion for spectral efficiency (ICSE) was introduced as the digital channel Shannon capacity per Hertz ((bit/sec)/Hz):

$$c_F(m, g, B_s) = C_m(g, B_s)/(B_s/2) \tag{2}$$

where g is channel-side, or receiver input, mean square signal power invariant variable expressed via signal-to-interference plus noise ratio (SINR) [15],

$$g^2 = P_s/(P_i + P_n) \tag{3}$$

P_s, P_i, P_n are, respectively, signal, interference, and noise powers, B_s is frequency-time invariant variable named as signal's base, $B_s = 2\Delta F_s T_s$, T_s is m-ary signal duration. Further, $C_m(g, B_s)$ is m-ary digital channel Shannon capacity in bit-per-symbol [10],

$$\begin{aligned} C_m(g, B_s) = \log_2 m + [1 - p_m(g, B_s)] \log_2[1 - p_m(g, B_s)] \\ + p_m(g, B_s) \log_2[p_m(g, B_s)/(m - 1)], \end{aligned} \tag{4}$$

where $p_m(g, B_s)$ is m-ary symbol's error probability (SER) [15] defined through invariant variable $h(g, B_s) = g\sqrt{B_s/2}$ expressed, in turn, through receiver's output ratio signal energy per symbol to signal-sided spectral power additional Gaussian interference plus noise density $N_{0_{in}} = N_{0_i} + N_{0_n}$, i.e., energies SINR, or ESINR [10]:

$$h^2 = E_s/N_{0_{in}} = P_s B_s/[2(P_i + P_n)] = g^2 B_s/2. \tag{5}$$

An invariant criterion for power efficiency (ICPE) was introduced as the signal-to-interference plus noise ratio (SINR) per m-ary digital channel Shannon capacity per Hertz: SINR/[(bit/sec)/Hz] [10]:

$$w(m, g, B_s) = g^2/c_F(m, g, B_s). \tag{6}$$

One can express a power efficiency criterion (6) through various measure units: dBm per (bit/sec)/Hz, Watt per (bit/sec)/Hz, and also convert it to energy efficiency invariant criterion (ICEE) in Joule per (bit/sec)/Hz. Moreover, through invariant criterion for power efficiency (6) one can express the invariant criteria for cover efficiency (ICCE) in Watt/(bit/sec)/Hz/square km, i.e., ICPE per area covering by cell radius $R_c(m, g, B_s)$ and also invariant criterion for investment (cost) efficiency (ICIE) through CAPEX calculated as some invariant function $F_I[w(m, g, B_s)]$ divided into area covering $\pi R_c^2(m, g, B_s)$.

Based on the introduced invariant criteria, we can formulate the following RRD-aimed breakthrough qualities and techniques capable to implement the perfect green 5G PHY for hyperrange space/wireless mediums corresponding to rural ubiquitous IoT/M2M/H2H 5G communications:

- design an advanced set of the orthogonal broadband m-ary OFDM-CDMA like waveforms corresponding to a perfect green 5G PHY for hyperrange space/wireless mediums which are well adapted to cognitive interference-robust "on-the-fly" control and approaching the trade-off extremums or fundamental limits of the spectral/power/energy/economics efficiency criteria [10];
- refine the green 5G PHY disruptive approaches for the potentially reachable energy-saving techniques of hyperrange rural area cost-effective covering;
- increase the channel-side ratio SINR (3) in pure ecological way of improvement both the denominator (reduce an interference [18]), and the numerator (smarter increase a beamforming and antenna gain, as Friis models), close to the fundamental limits without the rise of transmitter power;
- reaching continuously the fundamental minimum [10] power consumption criterion ICPE representing an imperative law for smart green PHY optimization and trade-off problems;
- as in [3], developing the profitability-power-efficiency-aimed fundamental trade-offs for rural green 5G networks in practical invariant variables notions.

2.2   Fundamental Limits and Extremums of 5G PHY

The fact that the value of invariant function $F(x_1, x_2...)$ does not change by substitution instead of every x_i argument's his $x_i^*(x_1, x_2, ...)$ invariant maps may be suitable for universal appropriateness research all measures: the information, the power, the covering, and the investment (i.e., profitability) measures. Let us denote by U the set of possible values of the invariant parameters (m, g, B_s). In a specific optimization problem some invariant variables are free and other parameters are fixed. We denote the set of possible values of the free variables by V, $V \in U$. Next we can formulate a set of general optimization problems [10].

Power Efficiency Optimization Problem. For ICPE (6), we can formulate the general optimization problem

$$w(m, g, B_s) \rightarrow \min, \qquad (7)$$

where a free variable belongs to V. It is necessary to bind the problem (7) with the constraint on the least permissible value $[c_F]_{\min}$ of ICSE

$$c_F(m, g, B_s) \geq [c_F]_{\min} \qquad (8)$$

and, possibly, the constraints on the permissible values of cover efficiency ICCE and investment efficiency ICIE, i.e. profitability. The example of the numerical analysis [10] of ICPE optimization problem is shown in Fig. 1.

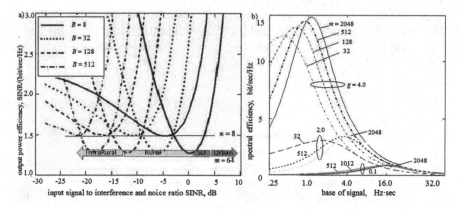

Fig. 1. The graphs of general optimization problems of green invariant efficiency criterion: (a) close fundamental minimum limit (infimum) of power efficiency ICPE [10]; (b) close upper limits of spectral efficiency ICSE (calculation: T. Pereverzina, D. Shatsky).

Studying the Fig. 1a and [14], we can formulate a fundamental power-consumption

Statement 1: The minimal specific power consumption (6) $w_{\min}(m, g, B_s)$ per (bit/sec)/ Hz for fixed alphabet size m, and free g and B_s for both Gaussian noise and interference is a universal power constant which depends neither on the signal base B_s nor on SINR (3).

Let $w_{\min}^*(m, g^*, B_s^*)$ be some minimum point on graph of Fig. 1a which was expressed in SINR-per-(bit/Hz)/sec. We can express this minimum value in Joule-per-(bit/Hz)/sec using the invariant relationship $w_{Jc}^*(m, g^*, B_s^*) = w_{\min}^*(m, g^*, B_s^*) \times N_{0_{in}}^* B_s^*/2$, where $N_{0_{in}}$ is the value realized in minimum point of both Gaussian interference and signal-sided noise spectral power density as in (5), $N_{0_{in}} = N_{0_i} + N_{0_n}$, in Watt-per-Hertz. Moreover, we can express this minimum value in Joule-per-bit $w_{Jb}^*(m, g^*, B_s^*) = w_{Jc}^*(m, g^*, B_s^*) \times B_s^*/2$.

Spectral Efficiency Optimization Problem. For ICSE (2), we can formulate the general optimization problem [10]:

$$c_F(m, g, B_s) \rightarrow \max \tag{9}$$

with respect to free variables belonging to the subset $V, V \in U$. The problem (9) expediently be bound with constraints on the infimum value of ICPE criterion

$$w(m, g, B_s) = \text{const}(m) \equiv w_{\inf}(m, g^*, B_s^*) + o(w), \tag{10}$$

where $o(w)$ is Landau Small Symbol. The Eqs. (9) and (10) determine the fundamental extremum of the invariant power efficiency ICPE (6) as in

Statement 2: The fundamental local maximum, or conditional supremum, of the invariant spectral efficiency (2) under the condition of minimal power consumption, or conditional infimum (10), equals the solution of the problem (9).

Figure 1b shows three subsets of ICSE spectral efficiency optimization graphs accordingly to three fixed values of SINR invariant variable ($g = 0.5/1.0/2.0$) and different signal alphabet sizes m with dependency from signal base B_s variations. In fact, the given series of SINR express the changes of channel quality from very poor up to average. Observing the numerical optimization graphs according to the given SINR series and correlating the signal complexity with signal base B_s values presented on Fig. 1b graphs and [15], we can state the following fundamental ICSE.

Statement 3: Optimal signals according to the spectral efficiency criterion (2) should be more complicated as the quality of SINR of the channel is worse, and, on the contrary, these signals should be easier when the quality of the SINR is better.

The above formulated statements lead to extremely green strategies of minimal power consumption and energy saving for both the ultra-dense urban and also the ultra-covering or extremely cost-effective rural optimization problems.

2.3   Fundamental Limits of m-ary Orthogonal Signal Interference

As shown in [16], the errors of inaccurate fulfillment of conditions of mutual signals orthogonality inevitably generate the intra-cell and inter-cell interference that determines the available value of the SINR ratio. The SINR value, in turn, limits the capacity of cellular cell. In [16], an advanced calculation method of the CDMA network capacity is offered, which allows to consider the dependences "SINR versus not strict orthogonality errors" directly through the orthogonal signals autocorrelation and mutual correlation functions. It is shown, that it is possible to raise many times the SINR or network capacity by reduction of the signal orthogonality errors. The statements concerning fundamental limits for interference power are proved [16]:

Statement 4: If the errors E caused of the not-strictly orthogonality of the intra-cell m-ary orthogonal signals ensembles can be reduced as wished, then the power P(E) of intracell interference can be asymptotically decreased up to as much as small values:

$$\lim_{E \to 0} P_{\text{intra-cell}}(E) = \lim_{E \to 0} \sum_{j=0, j \neq i}^{n-1} \frac{1}{T} \sqrt{M[K_{ji}^2(t, E_j)]} = 0, \tag{11}$$

where $K_{ji}^2(t, E_j)$ is the intra-cell mutual correlation function.

Figure 2 explains the impact of the reduction of the signal orthogonality errors on the raise many times of the SINR or the network capacity.

Statement 5: If the errors E caused by the not-strict orthogonality of the inter-cell m-ary orthogonal signals ensembles can be reduced as wished, then the power P(E) of inter-cell interference can be asymptotically decreased to small values of an order of Landau Big Symbol $0(M[a], 1/\sqrt{n})$:

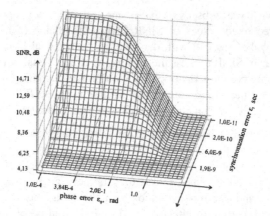

Fig. 2. 3D graphs SINR versus standard deviations of the synchronization ε_t and phase error ε_ϕ by thermal noise -113.101 dB [18].

$$\lim_{E \to 0} P_{\text{inter-cell}} = \lim_{E \to 0} \sum_{J \in G \setminus I} \sum_{J \in G \setminus I} \sum_{jj}^{n-1} M[a_{jj}]/T \times$$

$$\sqrt{M[K_{j,i}^2(t, E_{jj})|_{jj \in J, i \in I}]} = 0(M[a], 1\sqrt{n}), \tag{12}$$

where $K_{j,i}^2(t, E_{jj})$ is the inter-cell mutual correlation function, $M[a]$ is the weighted average of the space path loss indexes a_{jj}, n – the degree of the generating M-sequence polynomial.

3   Extremely Flexible and QoS-Guaranteed Distributed Multifunctional RRD 5G MAC

3.1   Vision of the Distributed Multifunctional Perfect 5G MAC

Assume that at some time t some quantity N_t of machine type network's i^{th} devices'/ users' which defined by k^{th} data service classes, G_{ikt} input traffics intensities, S_{ikt} output traffic intensities, total traffic $G_t = \sum_{i,k} G_{ikt} \leq C_{MAC}$, where $C_{MAC} = \max_{\{G_t, t\}} \sum_{i,k} S_{ikt}(G_t)$ is MAC useful throughput, is active. Let we denote further by X_{ikt} the really values of service parameters by $[X_{ikt}]$ – their required values, and by Y_{it} – i^{th} device's bandwidth resource. In our vision, the perfect machine type (MTC) rural 5G MAC protocol represents a MTC-enhanced flexible multifunctional distributed long-delay medium access control technology (MFMAC [8]), including also the functions of guaranteed dynamical control up to real time ("on the fly") of the bandwidth resources $\{Y_{it}\}$ allocation, guaranteed dynamical control accordingly to k^{th} data service classes of the traffic parameters $\{S_{ikt}\}$ and soft/different QoS parameters $\{X_{ikt}\}$, i.e., personally guaranteed Quality of Experience (QoE) for any user/device.

The required qualities of a perfect rural 5G MAC protocol may be implemented as MTC-aimed enhancements of the multifunctional distributed long-delay medium access control techniques [8, 12] exactly:

- high efficiency, tolerance, and lower latency [12], higher throughput and minimal overheads both come nearing fundamental limits [11] for a distributed multiple access control to long-delay wireless/space mediums;
- high controllability, reliability, stability, flexibility, and guarantee of distributed dynamical ("on the fly") control of broadband RAN technologies [8, 9, 12];
- multifunctional and universality abilities that rely on the dynamically controlled and adaptive ATM-like smart unified protocol MAC, i.e., MFMAC [8, 12], through the entire wireless networking hierarchy – core, backbone, and access networks;
- fully mesh all-device-centric radio access architecture all_device-to-all_device (DmD, m>>2) relies on the multipoint-to-multipoint (MPMP) [9] Virtual Space/ Wireless ATM Hyperbus topology with fully distributed QoS-guaranteed multi-functional long-delay MAC [8];
- cost-effective completely distributed (grid-like) all-IP/MPLS over ATM-MFMAC Hyperbus (MFHBA) that implements the data packet selecting technique rather than packet switching technique [8, 9].

3.2   Fundamental Limits of the Distributed MAC

As showed [11], the real reachable throughput for various MAC protocols depends on ensuring the "MAC collective intellect" that contains a plenitude of information about the real-time state of the multiple access processes in geographically distributed queues, and also on the normalized overhead for provisioning QoS. What is the minimum reachable, or infimum, MAC overhead? And what is the potential reachable maximum throughput, or fundamental limit of potential capacity, of the ideal MAC protocol? It is reasonable to find the MAC overhead infimum as the Shannon entropy of the distributed multiple access processes based on the Markov models of distributed queues, and to find the potential capacity of MAC protocols as a function of the overhead infimum.

Let we define the real throughput capacity for real MAC protocol specified by real structural specifications and system parameters Γ, and by real medium conditions Ψ including presence of errors as

$$C_{\Gamma,\Psi} = \max_{\{G \in F_G\}} S_{\Gamma,\Psi}(G), \tag{13}$$

where F_G is the field of the possible values of input traffic intensity G. As in [11], we define the MAC throughput fundamental limit as supremum of the real throughput (13) on the set F_Γ of MAC protocol's possible structural specifications and system parameters by given medium conditions Ψ, i.e., as potential capacity,

$$C_\Psi^{\text{sup}} = \sup_{\{G \in F_G, \Gamma \in F_\Gamma\}} S_{\Gamma,\Psi}(G) = M[\tau]/(M[\tau] + \delta_\Psi^{\text{inf}}) = 1/(1 + v_\Psi^{\text{inf}}), \tag{14}$$

where δ_Ψ^{inf} is the potential reachable minimum, or infimum, of the time resource overhead for medium access control per data unit/packet by duration $M[\tau]$,

$$\delta_\Psi^{\text{inf}} = \inf_{\{\Gamma \in F_\Gamma\}} \delta_{\Gamma,\Psi}, \tag{15}$$

v_Ψ^{inf} is the normalized value of the infimum of overhead (15) according $M[\tau]$. The MACs overhead and throughput fundamental limits for widespread queueing models of distributed multiple access systems TDMA determine the statements [11]:

Theorem 1: If the TDMA system is described by an infinite model of equivalent centralized M/M/1 queue Γ_0 by Ψ_0 zero errors channel conditions, then the value of minimum reachable overhead on distributed MAC control is equal to

$$\inf_{\{\Gamma \in F_\Gamma\}} v_{\Gamma,\Psi_0} = v_{\Gamma_0,\Psi_0}^{\text{inf}} = [2 + H(\tau)]/BM[\tau], \tag{16}$$

and the potential throughput capacity of the ideal MAC is equal to

$$\sup_{\{\Gamma \in F_\Gamma; G \in F_G\}} S_{\Psi_0,\Gamma}(G) = C_{\Psi_0}^{\text{sup}} = 1/(1 + (2 + H(\tau))/BM[\tau]), \tag{17}$$

where B is the bit rate, $M[\tau]$ is the mean duration of traffic packets, $H(\tau)$ is the entropy of the packets duration distribution given by the geometric law [11].

Theorem 2: If the TDMA system is described by the infinite model of equivalent centralized M/D/1 queue Γ_0 under conditions described in Theorem 1, then the value of minimum reachable expenses on distributed MAC control is equal to

$$\inf_{\{\Gamma \in F_\Gamma\}} v_{\Gamma,\Psi_0} = v_{\Gamma_0,\Psi_0}^{\text{inf}} = 1,854/BM[\tau] \tag{18}$$

and the potential throughput capacity of the ideal MAC protocol is equal to

$$\sup_{\{\Gamma \in F_\Gamma; G \in F_G\}} S_{\Psi_0,\Gamma}(G) = C_{\Psi_0}^{\text{sup}} = (1/(1 + 1,854/BM[\tau]). \tag{19}$$

We observe in Fig. 3, that the MAC's total entropy, i.e., overhead infimums (16) and (19), depends mainly from the data slots duration law indeterminacy. The M/M/1/* systems family must be characterized by greatest entropy in accordance with its exponential law's greatest indeterminacy. Opposite them, the M/D/1/* systems which are described by deterministic duration law ensure the least entropy, therefore – the greatest MAC potential throughput capacity (17) and (20). As proved in [11], the adaptive controlled multiple access MAC protocols with deterministic packet size and, hence – the least overheads, allow to reach to a fundamental limit of a MACs throughput capacity which, in turn, as much close to 1,0 as it's wished. The fully distributed ATM-like multifunctional MAC technology (MFMAC) [8, 12] meets the above breakthrough qualifications.

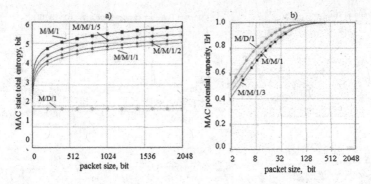

Fig. 3. MACs state entropy, or overhead infimum (a), and MACs throughput supreme (b) versus packets duration laws, SER = 1.0E-3.

The disruptive MFMAC technology uses the recurrent M-sequences (RS) MAC addressing opportunities [8, 14] in order to organize a RS-token tools "all-in-one" for high effective multiple access to long-delay space medium, soft QoS provision and distributed dynamical control of traffic parameters and bandwidth resources [8, 12, 14] approaching the sublimit of throughput capacity. The M-subsequences $A_j = a_{j-(n-1)}$, $a_{j-(n-2)}, \ldots, a_j$ serve as RS-identifiers of the unique MAC addresses and other protocol subjects. Some subset of ith "personal" identifiers $\{A^i_{jkt} | k = 1, 2, \ldots, m_{it}\} = B_{it}$ are dynamically assigned to each ith station on a decentralized basis by Shannon-Fano method for passing of the user's request in proportion to the required bandwidth resource $[Y_{it}]$.

4 Concept of Ubiquitous IoT/M2/H2H Green RRD 5G System

The RRDs extremely green device-centric Hypercelle is explained in Fig. 4. A conceptual look of the IP over DVB-2S multifunctional satellite-based fully distributed hybrid 5G networking technology RCS-MFMAC for RRD areas is explained in Fig. 5.

The hybrid architecture relies on implementation of the QoS-guaranteed multifunctional 5G machine type MAC perfect rural PHY-MAC techniques basing on the developing of the advanced delay-tolerant 5G ATM-like MPMP MFMAC technologies [8, 12] which in turn should be adapted to conditions of the satellite platforms' DVB-2S-RCS [10], VSAT, etc. The main breakthrough drivers for RRD-oriented 5G communications include also a push MFMAC-based next generations of wireless asynchronous transfer mode (ATM/MFMAC), of multi-protocol label switching (MPLS/MFMAC), and also of IP over DVD-S/MFMAC integrated networking technologies [9].

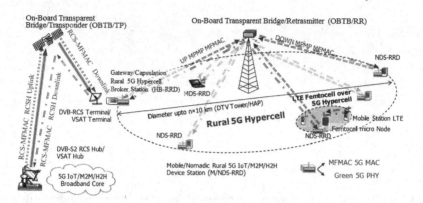

Fig. 4. Rural extremely green 5G Hypercelle.

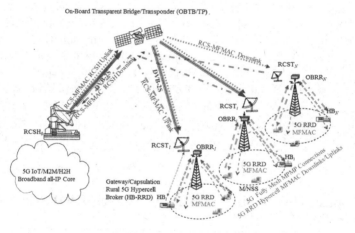

Fig. 5. Ubiquitous Green Rural 5G Hybrid Architecture.

5   Conclusion

In this paper, the green, ecological and cost-effective advanced approach to creation of the flexible QoS-guaranteed ubiquitous 5G IoT/M2M/H2H multifunctional RRD communications has been designed. Offered approach rely on implementation of the extremely flexible, energy and spectral effective 5G PHY-MAC techniques: (i) smarter increase of the SINR through beamforming/antenna/orthogonality gain, without rise of the transmitter power; (ii) closing "on-the-fly" the fundamental minimum of power consumption ICPE; (iii) providing "on-the-fly" the profitability/power efficiency aimed fundamental trade-offs for rural green 5G networks in practical invariant variables notions. It should be noted the key mission critical opportunities of a perfect rural 5G MAC: (j) the reachable low overhead close to fundamental infimum; (jj) the flexible 5G scheduler adapt "on-the-fly" the superframe formats and optimally allocate the massive machine type and also multiservice traffic by equal ATM-like minimal bandwidth block per second, without superframe overflow or redundancy.

References

1. Cicconetti, C., de la Oliva, A., Chieng, D., Zúñiga, J.C.: Extremely dense wireless networks. IEEE Commun. Mag. **53**(1), 88–89 (2015)
2. Boccardi, F., Heath Jr., R.W., Lozano, A., Marzetta, T.L., Popovski, P.: Five Disruptive Technology Directions for 5G. IEEE Com. Mag. **52**(2), 74–80 (2014)
3. Chen, Y., Zhang, S., Xu, S., Li, G.Y.: Fundamental trade-offs on green wireless networks. IEEE Commun. Mag. **49**(6), 30–37 (2011)
4. Cao, A., Gao, Y., Xiao, P., Tafazolli, R.: Performance analysis of an ultra-dense network with and without cell cooperation. In: International Symposium on Wireless Communication Systems (ISWCS), pp. 51–55 (2015)
5. Larsson, E.G., Edfors, O., Tufvesson, F., Marzetta, T.L.: Massive MIMO for next generation wireless systems. IEEE Commun. Mag. **52**(2), 186–195 (2014)
6. Solanas, A., Patsakis, C., Conti, M., Vlachos, I.S., Ramos, V., Falcone, F., Postolache, O., Pérez-Martínez, P.A., Di Pietro, R.: Smart health: a context-aware health paradigm within smart cities. IEEE Commun. Mag. **52**(8), 74–81 (2014)
7. Chih-Lin, I., Rowell, C., Han, S., Xu, Z., Li, G., Pan, Z.: Toward green and soft: a 5G perspective. IEEE Commun. Mag. **52**(2), 66–73 (2014)
8. Markhasin, A.: Conception of satellite-based ubiquitous & multifunctional mobile and wireless all-IP environment 4G with fully distributed, mesh, and scalable architecture for RRD areas. In: The 18th Annual IEEE International Symposium on Personal, Indoor and Mobile Radio Communications (PIMRC), Athens, Greece, #275, pp. 1–5 (2007)
9. Markhasin, A.: Satellite-based fully distributed mesh hybrid networking technology DVB-S2/RCS-WiMAX for RRD areas. In: Proceedings of the 5th Advanced Satellite Multimedia Systems Conference (ASMS/SPCS), Cagliari, Italy, pp. 294–300 (2010)
10. Markhasin, A., Kolpakov, A., Drozdova, V.: Optimization of the spectral and power efficiency of m-ary channels in wireless and mobile systems. In: Proceedings of the Third IEEE International Conference in Central Asia on Internet (ICI), Tashkent, Uzbekistan, #177, pp. 1–5 (2007)
11. Markhasin, A.: Shannon bounds for large-scale wireless MAC's potential capacity in presence of errors. In: Proceedings of the Eleventh ACM International Conference on Modeling, Analysis, and Simulation of Wireless and Mobile Systems (MSWiM), Vancouver, Canada, pp. 169–176 (2008)
12. Markhasin, A.: QoS-oriented medium access control fundamentals for future all-MPLS/ATM satellite multimedia personal communications 4G. In: Proceedings of the IEEE International Conference on Communications (ICC), Paris, France, pp. 3963–3968 (2004)
13. Shannon, C.: A mathematical theory of communication. Bell Syst. Tech. J. **27**, 379–423, 623–656 (July, October 1948)
14. Markhasin, A.: Architecture of Packet Radio Networks. Science Publishing, Novosibirsk (1984). 144 p. (in Russian)
15. Gu, Q.: RF System Design of Transceivers for Wireless Communications. Springer, New York (2004). 604 p.
16. Markhasin, A., Svinarev, I., Belenky, V.: Impact analysis of orthogonal signal errors on network capacity with code division. Vestnik SibSUTIS **4**, 98–109 (2014). (in Russian)

Application of Microwave Photonics Components for Ultrawideband Antenna Array Beamforming

S.I. Ivanov, A.P. Lavrov$^{(\boxtimes)}$, and I.I. Saenko

Peter the Great St. Petersburg Polytechnic University, St-Petersburg, Russia
{s.ivanov,lavrov,saenko}@ice.spbstu.ru

Abstract. We consider beamforming arrangement for ultrawideband antenna array that can currently provide the required true time delay capabilities by using the units and elements available at the market of modern components of fiber-optical telecommunication systems. The essential parameters of accessible analog microwave photonic link's main components as well as performance characteristics of the complete photonic link assembly have been measured. The beamformer scheme based on true-time-delay technique, DWDM technology and fiber chromatic dispersion has been designed. The developed and tentatively investigated beamformer model is mainly composed of commercial microwave photonic components. The results of beamformer model experimental testing jointly with the wideband linear antenna array in 6–15 GHz frequency range show no squint effect while steering the antenna pattern and good accordance with the calculation estimates.

Keywords: Phased array antenna · True time delay · Photonic beamforming · Fiber optic link · Fiber dispersion

1 Introduction

For different applications, particularly in electronic warfare, phased array antenna systems, microwave antenna signal distribution and so on microwave photonics potentially offer the well known advantages of large instantaneous bandwidth, low loss in signal transmitting, small size and weight, high resistance to electromagnetic interference and flexible designing [1–3]. Especially ultrawideband large antenna arrays could benefit from development of photonic beamformers based on true-time-delay (TTD) technique. The application of microwave photonic components and links in such beamformers could yield significant improvements to their performance parameters. So, in the last two decades intensive researches have been carried for phased array antenna (PAA) control using both of TTD and microwave photonic techniques. It has been shown that applying the microwave photonic devices for PAA beamforming could afford the means to overcome conventional radio electronic beamformer limitations [4, 5]. Therefore, many beamforming architectures based on microwave photonics have been proposed and evaluated, both in transmitting and reception modes. Most of the proposed architectures using fiber optic components to

© Springer International Publishing AG 2016
O. Galinina et al. (Eds.): NEW2AN/ruSMART 2016, LNCS 9870, pp. 670–679, 2016.
DOI: 10.1007/978-3-319-46301-8_58

provide control of one- or multibeam antenna array pattern require specially developed units, such as low-noise, fast tunable lasers with narrow spectral linewidths, chirped fiber Bragg gratings, reflection fiber segments with precise lengths, optical filter arrays and others. Numerous research over the past ten years have yielded significant improvements in the key performance parameters and have advanced the state-of-the-art for those special microwave-photonic components [6–8], meanwhile, implementation of widely developed analog fiber-optic link components to compose the antenna array beamformer (BF) arrangement has been extensively investigated [9–11]. Some of proposed architectures were found to be the most suitable for this way of BF prototype designing; however there are special requirements (namely, phase-frequency dependence, intrinsic time delays in photonic transmitter and receiver modules) which the BF photonic components have to meet anyway while the fiber-optic communication links manufacturers do not take them into account. Examined here photonic beamformer model for PAA in receive mode is based on dense wavelength division multiplexing (DWDM) components and chromatic dispersion of a single mode optical fiber and implemented with units and components available at the market of modern fiber-optic communication systems [10, 11]. We consider the results of a kind of BF experimental model assembling and adjustment and its initial performance investigation including the microwave receiving linear (1-D) PAA far-field pattern measurement in the 6–15 GHz frequency band.

2 Photonic Beamforming System Configuration

The developed beamformer architecture based on the components implementing a DWDM technology approach and optical fiber chromatic dispersion is shown in Fig. 1 [9, 11, 12]. This scheme is designed for the N element's linear PAA and uses optical comb – a set of N lasers with different but uniformly spaced (step $\Delta\lambda$) wavelengths, the total wavelength band is $(N - 1) \cdot \Delta\lambda$. The RF signals from the antenna elements A_1–A_N modulate a set of laser diodes. Electrooptic conversion is achieved either by direct modulation of lasers or by using an external modulators as represented in Fig. 1 (scheme elements M). Further, the intensity-modulated optical carriers are combined into a single fiber by a multiplexer (MUX N x 1 unit) and fed into a time delay unit (TDU). As all the optical carriers share with the same light paths, the time delay differences between adjacent channels are produced by the fiber chromatic dispersion D measured in ps/(nm·km). Time delay $\tau = \Delta\lambda \cdot L \cdot D$ introduced between adjacent channels results in the respective tilting of PAA beam. Photodiode (PD) at TDU output converts sum of delayed intensity-modulated optical carriers back to microwave domain. In Fig. 1 one can see 2 fiber segments (single mode fibers) with lengths L_1 and L_2 successively inserted between the multiplexer and the photodiode thus giving (together with "0" - length position) 3 interchannel delay values: 0, τ_1 and τ_2 corresponding to 0, θ_1 and θ_2 beam pointing angles. The "Correction Delays" unit contains fiber segments with strictly sized up lengths equalizing the interchannel delays from RF sources (scheme elements A_1–A_N) to the MUX output. Hence, it corresponds to a flat phase law tilted 0^0 to PAA base line – the PAA beam directed normal to PAA base. The RF signals modulating the comb of optical wavelengths are coherently summed at the photodetector output while optical carriers are summed incoherently. So due to chosen fiber segment lengths and thus specified interchannel delays RF signals received from the corresponding direction are combined in phase.

In the considered architecture of 1-D beamformer the number of switching beams can be increased that only requires increased number of dispersive fiber segments. Also a multibeam mode could be realized (for example as in [9]) by using of a higher splitting ratio after MUX and a set of photodiodes.

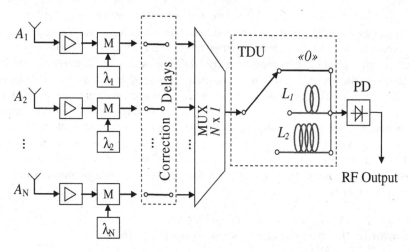

Fig. 1. Architecture of photonic beamformer based on DWDM and fiber chromatic dispersion interchannel delays

3 Beamformer Model Arrangement Design

The beamformer model developed for initial demonstration of linear PAA beam steering is based mainly on microwave photonic components commercially available on the market of fiber optic telecommunication links (radio over fiber link - RFoFL). Among the photonic key components depicted in Fig. 1 the units converting radio signal into optical one (laser jointly with modulator) and backward (photoreceiver) are crucial for BF correct performance. In our BF model arrangement the transmitters and receiver modules from analog fiber-optic links OTS-2-18 from Emcore [13] were used which can operate at 0.1...18 GHz bandwidth. The transmitter block diagram is depicted in Fig. 2. In-built microprocessor-based transmitter control for laser bias and temperature as well as Mach-Zehnder modulator bias provides better consistent performance operation and allows for appreciably reduce the measurement results variations. Transmitter unit includes three separate optically combined elements: DFB laser with wavelength corresponding to ITU grid with 100 GHz step, LiNbO$_3$ Mach-Zander modulator (MZM) with half-wave voltage approx. 5.8 V, optical 10 dB directional coupler and photodiode for MZM quadrature working point stabilization. Transmitter optical power in a fiber is about 10 dBm. The receiver module has an additional RF

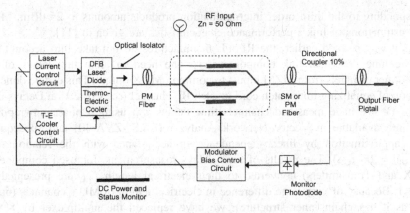

Fig. 2. Block diagram of OTS-2-18 link (Emcore) transmitter

amplifier with 15 dB gain after photodetector. Shown in Fig. 3 frequency responses of two links for frequency band up to 20 GHz are $| S21(f) |$ for link 'Ch 28' (line '1') and for 'Ch 32' (line '2'), where channel numbers correspond to ITU grid numbering. Other links have parallel $| S21(f) |$ frequency behavior. The traceable curve ripples we expect to follow from the features of receiver module optical entrance.

The measured link gain ranges from −3 dB at 1 GHz to −7 dB at 18 GHz giving the respective links gain about −6 dB for frequency 10 GHz. The 1 dB compression point $P_{IN\ 1dB} = 16.5$ dBm, and the minimum input level is $P_{IN\ MIN} = -144$ dBm/Hz for the receiver output noise $P_N = -155$ dBm/Hz, averaged through the total frequency bandwidth 18 GHz. We have measured also transfer function P_{OUT} (P_{IN}) for one of OTS-2-18 links for two input RF signals with frequencies near 4 GHz and equaled powers. As the result, the 1 dB compression dynamic range comes to 160.5 dB/Hz and the spurious free dynamic range is about 113 dB/Hz$^{2/3}$. The input signal level IP$_3$

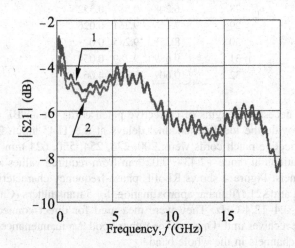

Fig. 3. Frequency responses for channel links 28 and 32

corresponding to the third order intermodulation products amounts to 25 dBm. More detail estimations of link's performance characteristics are given in [11].

As it was pointed earlier, the RFoFL manufactures do not take into account the intrinsic time delays in link components, so one need to align the delays of all beamformer channels from MZM inputs to common MUX output. This can be done by insertion of equalizing fiber patch cord in each BF channel (see Correction Delays unit in Fig. 1). We have measured channels time delays τ_{CH} using a phase unwrapping procedure available in Vector Network Analyzer (R&S, ZVA 40) and subsequent phase approximation by linear dependence in accordance with the relation: arg $[S21\ (f)] = 2\pi f \cdot \tau_{CH})$. The results of time delays measurements for used components (MUX and Transmitters) converted to their electrical length L_{EL} are presented in Table 1. Because of very large difference in electrical length of MUX channels (up to 6 m, as it has chain inner structure), we have replaced the multiplexer by 8 × 1 splitter/combiner (LiNbO$_3$ Planar Light Circuit, PLC). In view of the fact of Ch 32 transmitter large electrical length, we exclude it from lengths' alignment procedure, limiting BF model actual number of channels up to 5. The values ΔL_{EL} in column 4 are given relative to electrical length L_{EL} of Ch 30 and they have to be corrected by insertion of additional fiber patch cords. Optical losses in 8 × 1 combiner channels were 11.5 ± 0.35 dB (including FC/APC connectors), channels electrical lengths were in range 1558.1 ± 1.2 mm. Pointed deviation of the lengths is considerably less than that of the multiplexer. One have in mind that in fiber-optic link 1 dB change in optical power leads to 2 dB change in RF output power and we do not use any alignment additional attenuators.

Table 1. Electrical lengths of BF components

ITU Ch #	MUX	Transmitters	
	L_{EL}, m	L_{EL}, m	ΔL_{EL}, m
27	4.975	9.638	0.403
28	6.208	9.587	0.352
29	7.156	9.261	0.026
30	8.359	9.235	0.0
31	9.689	9.289	0.054
32	10.607	13.295	4.06

Further, the necessary lengths of corrective patch cords for all BF channels were calculated to provide the identical channel delays at the TDU input. The calculated lengths of 5 corrective patch cords were: 200, 226, 254, 550, 624 mm and they were made with deviations in range +2.43--−0.02 mm from required values and inserted in BF model channels. Figure 4 shows RFoFL phase-frequency characteristics, as deviation $\Delta\varphi(f)$ from arg[S21 (f)] linear approximation, for 5 transmitters (Ch 28–Ch 31) in frequency band 0.4–12.4 GHz. They were measured for direct connection of transmitters with one receiver unit. One can see rather good $\Delta\varphi$ maintenance in near +/−8° limits for these channels in the whole band.

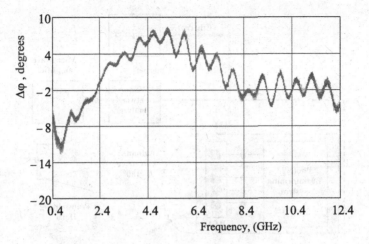

Fig. 4. Phase-frequency characteristics of five channels

TDU in our BF model include consistently connected segments of SM fibers with lengths labeled 0, L_1, L_2 corresponding to PAA beam pointing angles θ_0, θ_1, θ_2 in accordance with relation $L = d \cdot \sin(\theta)/(c \cdot D \cdot \Delta\lambda)$, where d is the distance between neighboring antenna elements and c is the speed of light in a vacuum. Figure 5 shows the measured channel dispersion delays resulted from connection L_1 or L_2 fiber segment compared with the calculated ones on the assumption of dispersion index being equal to 17 ps/nm·km at $\lambda = 1550$ nm for standard SMF 28 type fiber. It can be seen good accordance the expected and measured values.

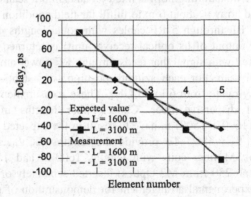

Fig. 5. Calculated and measured channels delays for two beam point angles.

4 BF with PAA Testing and Results

Steering performance of BF model assembled 1-D antenna array is characterized through far-field pattern measurements in anechoic chamber. Figure 6 schematically depicts experimental setup for these measurements. Actually available receiving PAA

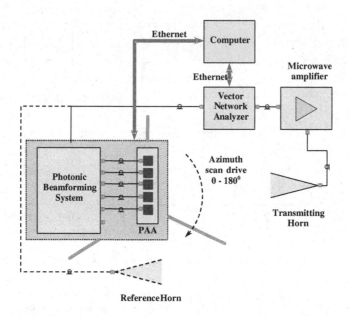

Fig. 6. Diagram of arrangement for demonstration of photonic BF beam steering performance

has 5 radiating elements and some inactive elements around them, radiating elements being Vivaldi patch antennas elements spacing $d = 20$ mm, working frequency range 6–15 GHz. The wideband transmitting horn fed by a vector network analyzer (with additional RF power amplifier) radiates a microwave signal. The PAA under test located in front of the transmitting horn is fixed on an azimuth scan drive. The distance between the horn and array is about 6 m to fulfill far-field condition requirement. PAA was connected with BF through 5 RF cables with equal lengths 40 cm. Microwave signal from BF (RF output of its optical receiver unit) is returned to vector network analyzer and compared with signal that feeds through RF power amplifier transmitting horn during azimuth scan. For each azimuth position – 1° step – network analyzer scans wide frequency range from 6 to 15 GHz. These measurements were done for 3 different lengths of dispersive fiber. Chosen fiber lengths in TDU $L_0 = 0$ m, $L_1 = 1600$ m, $L_2 = (1600 + 1500)$ m have to result in the expected beam point angles $\theta_0 = 0°$, $\theta_1 = 18.9°$, $\theta_2 = 38.7°$ To provide the lengths L_0, L_1, L_2 we used short patch-cord and two SM fiber coils with lengths 1600 m and 1500 m (2 coils in Rack-mount Fiber Lab 250 from M2 Optics) inserted separately or one after another. Photograph of the experimental arrangement for demonstration of photonic BF beam steering performance is shown in Fig. 7.

Figure 8 represents the far-field patterns measured with photonic BF for 3 optical fiber lengths at frequencies 6 (red), 9 (blue), and 12 GHz (green), respectively. Insertion of 0, 1600 and 3100 m optical fibers shifts the antenna pattern to $\theta_0 = 1.3°$ (red), $\theta_1 = 18.8°$ (blue), and $\theta_2 = 36.8°$ (green), respectively. The measured shifts are close to the calculated ones when time delays between two neighboring radiating elements are 0, 21.57, 43.14 ps, see Fig. 5. Small "0"- beam deviation from the normal

Fig. 7. The PAA with BF located on an azimuth scan drive.

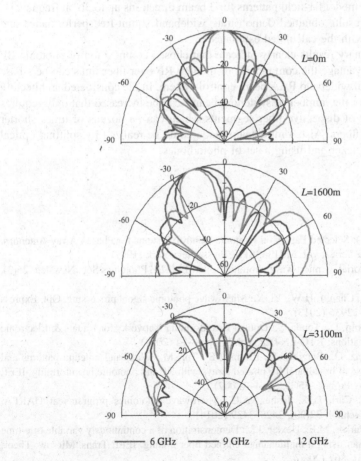

Fig. 8. Control of beam position of the 5-element PAA at 6, 9, and 12 GHz (Color figure online)

position (0°) was generated by residual differences about a few tenth of a millimeter in total channel electrical lengths (including RF cables between PAA and the BF model inputs, with no RF phase fine adjustment used).

The sidelobe level for $\theta_0 = 1.3°$ is approx. −13 dB in accordance with the uniform amplitude distribution. One can see no squint effect in wide frequency range for PAA with designed photonic BF model.

5 Conclusions

The photonic beamformer model to form and steer a beam of microwave linear phase antenna array in the receive mode has been developed and demonstrated. We have investigated the essential characteristics of commercially available new analog fiber optic links main components included in BF model based on DWDM technique and fiber dispersion. Experimental demonstration of the designed model beamforming features includes actual measurement of 5-element microwave linear array antenna with the developed BF model far-field patterns for 3 beam directions up to 36° at frequencies 6..12 GHz. The results obtained demonstrate wideband squint-free performance and good agreement with the calculated estimates.

These preliminary results clearly indicate that chosen composition of photonic BF model gaining advantage the components of modern RF over fiber links can be a base for development of wideband PAA beam control system. In the considered architecture of 1-D beamformer the number of switching beams can be increased that only requires increased number of dispersive fiber segments which can be dozens of times shorter when using DCF fibers. Also a multibeam mode could be realized by splitting optical signal after multiplexer and using a set of photodiodes.

References

1. Riza, N.A. (ed.): Selected Papers on Photonic Control Systems for Phased Array Antennas. SPIE Milestone Series, vol. MS136. SPIE Press, New York (1997)
2. Yao, J.P.: A tutorial on microwave photonics – part II. IEEE Photonics Soc. Newslett. **26**(3), 5–12 (2012)
3. Minasian, R.A., Chan, E.H.W., Yi, X.: Microwave photonic signal processing. Opt. Express **21**(18), 22918–22936 (2013)
4. Chazelas, J., Dolfi, D., Tonda, S.: Optical Beamforming Networks for Radars & Electronic Warfare Applications. RTO-EN-028, vol. 9, pp. 1–14 (2003)
5. Rotman, R., Raz, O., Barzilay, S., Rotman, S., Tur, M.: Wideband antenna patterns and impulse response of broadband RF phased arrays with RF and photonic beamforming. IEEE Trans. Antennas Propag. **55**(1), 36–44 (2007)
6. Richard, W.R., Carl, L.D., Joshua, A.C.: Microwave photonics programs at DARPA. J. Lightwave Technol. **32**(20), 3428–3439 (2014)
7. Hunter, D.B., Parker, M.E., Dexter, J.L.: Demonstration of a continuously variable true-time delay beamformer using a multichannel chirped fiber grating. IEEE Trans. Microw. Theor. Tech. **54**(2), 861–867 (2006)

8. Meijerink, A., Roeloffzen, C.G.H., Meijerink, R., et al.: Novel ring resonator-based integrated photonic beamformer for broadband phased array receive antennas – part I: design and performance analysis. J. Lightwave Technol. **28**(1), 3–18 (2010)

9. Blanc, S., Alouini, M., Garenaux, R., Queguiner, M., Merlet, T.: Optical multibeamforming network based on WDM and dispersion fiber in receive mode. IEEE Microw. Theor. Tech. Trans. **54**(1), 402–411 (2006)

10. Lavrov, A.P., Ivanov, S.I., Saenko, I.I.: Investigation of analog photonics based broadband beamforming system for receiving antenna array. In: Balandin, S., Andreev, S., Koucheryavy, Y. (eds.) NEW2AN/ruSMART 2014. LNCS, vol. 8638, pp. 647–655. Springer, Heidelberg (2014)

11. Ivanov, S.I., Lavrov, A.P., Saenko, I.I.: Optical-fiber system for forming the directional diagram of a broad-band phased-array receiving antenna, using wave-multiplexing technology and the chromatic dispersion of the fiber. J. Opt. Technol. **82**(3), 139–146 (2015)

12. Yang, Y., Dong, Y., Liu, D., He, H., Jin, Y., Hu, W.: A 7-bit photonic True-time-delay system based on an 8 × 8 MOEMS optical switch. Chin. Opt. Lett. **7**, 118–120 (2009)

13. Optiva OTS-2 18 GHz Amplified Microwave Band Fiber Optic Links. http://products. emcore.com/avcat/images/documents/dataSheet/Optiva-OTS-2-18GHz-Amplified2.pdf

Study of Specific Features of Laser Radiation Scattering by Aggregates of Nanoparticles in Ferrofluids Used for Optoelectronic Communication Systems

Andrey Prokofiev[1], Elina Nepomnyashchaya[1], Ivan Pleshakov[1,2],
Yurii Kuzmin[1,2], Elena Velichko[1(✉)], and Evgenii Aksenov[1]

[1] Peter the Great St. Petersburg Polytechnic University, St. Petersburg, Russia
apr@inbox.com, elina.nep@gmail.com,
et.aksenov@gmail.com, {ivanple,iourk,
velichko-spbstu}@yandex.ru
[2] Ioffe Institute, St. Petersburg, Russia

Abstract. Ferromagnetic fluids are considered to be advanced materials both for the fundamental research and for possible applications, among which some integrated optic devices with the elements containing ferrofluids and controlled by an external magnetic field have recently been discussed. This work is devoted to the experimental study of the factors affecting the intensity and spatial distribution of the laser radiation scattered by the particle structures in ferrofluids in a zero magnetic field and in the presence of magnetic field with H = 1000 Oe. The samples of nanodispersed magnetite (Fe_3O_4) suspended in kerosene and in water were studied. Certain trends determining the scattering patterns were observed.

Keywords: Ferrofluids for optoelectronic systems · Light scattering · Z-scan · Laser correlation spectroscopy · Particles aggregation

1 Introduction

Ferromagnetic fluids or ferrofluids (FF) are colloids comprising a solid phase in the form of magnetic nanoparticles dissolved in liquids, such as kerosene, oil, and others. FF are usually produced by a chain of chemical reactions and are comprised of three major ingredients. The first is magnetically ordered nanoparticles with the sizes normally ranging from 1 to 30 nm. Another component of FF is a surfactant (oleic acid) which prevents the particles from aggregation. The two components mentioned above are suspended in a liquid carrier.

Ferromagnetic fluids are attractive as elements of optoelectronic systems because they can be built in systems and sensors that use optical fibers (such as modulating equipment, sensors, reflectors, interferometers, etc.) [1–3].

An exponentially growing interest of researchers to these systems stems from their remarkable properties. A comprehensive review of the problem can be found in [4, 5]. Though FFs were first produced a few decades ago, a lot of questions on their

O. Galinina et al. (Eds.): NEW2AN/ruSMART 2016, LNCS 9870, pp. 680–689, 2016.
DOI: 10.1007/978-3-319-46301-8_59

behaviour remain to be unanswered. Currently, rheological properties of these materials are mostly exploited. Computer hard drives, where magnetic fluids serve as bearings, may be considered as the most representative example.

FFs also possess peculiar nonlinear optical properties which are poorly studied, but which may provide a direct light control in modern photonic devices [6, 7].

When laser radiation is transmitted through a colloidal medium with magnetic nanoparticles, some important effects can be observed. First, magnetic nanoparticles can interact with the incident light, and magnetic particles conglomerate under the action of radiation field [8]. In addition, an applied magnetic field can dramatically change a FF sample optical properties. The commonly accepted point of view is that ferromagnetic particles form elongated clusters or chains under the influence of external field [4, 9]. Magnetic fluid as a poorly light transmitting substance can absorb a large part of incident radiation. This leads to substantial heating of FF samples. Therefore, when the light intensity is high enough, a number of effects related to the heat arise. These are thermomagnetic convection [10], Soret effect [11], and thermal lens [12]. The first one depends on the geometry of the sample [13], the two others depend on the chemical composition and applied magnetic field [14].

Light scattering appears to be the most important measurement technique for FF studies because it provides information on the mechanism of propagating light control. In this work the results of study of peculiarities of laser radiation scattering by aggregates of nanoparticles in ferrofluids are discussed.

2 Experimental Techniques

In order to understand the nanoparticle behavior, a number of experimental techniques are usually employed. For example, microphotography can be used to visualize the particle structures at the micro level [15]. However, the majority of standard techniques, such as scanning and probe microscopy, are insufficient for studying the nanosize magnetic particles directly in the colloidal solutions. Some noninvasive techniques, such as optical methods, should be used. They allow observation of changes in the agglomerate sizes without a strong influence on the characteristics of the surrounding medium and evaluation of their forms and microscopic structure. Besides, the optical methods allow one to analyze the system in real time. Therefore, the current study was devoted to the exploration of some indirect techniques, which, especially being used in combination, can render a valuable opportunity to investigate physical processes in FFs.

2.1 Samples

In this work the samples with suspended magnetite Fe_3O_4 were studied. The average particles diameter was about 10 nm and the typical concentrations were 0.02–0.2 vol. % (In order to achieve an acceptable transparency, the studied samples were diluted to a concentration of 0.02 vol. % of Fe_3O_4.). Another component of FF was a surfactant – oleic acid – which prevented the particles from aggregation. Kerosene and water served as solvents.

2.2   Laser Correlation Spectroscopy

The first method which was used to study the scattering properties of FFs was laser correlation spectroscopy. Among the optical approaches the laser correlation spectroscopy has many advantages in measuring the sizes and investigating the cluster formation in solutions [16]. The principle of the method involves measurements of an autocorrelation function of the light scattered by the sample. Correlation analysis of this signal provides information about translational and rotary Brownian diffusion. The autocorrelation function g(τ) is commonly characterized by an exponential behavior with the power depending on the size of particles (Fig. 1).

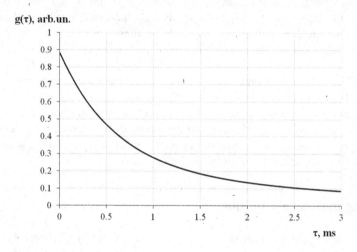

Fig. 1. Typical autocorrelation function of the light scattered by the particles' Brownian motion (measured on the model object)

The characteristic time is defined by the wave vector of scattered light and coefficient of translational diffusion [17]. For polydisperse solutions the autocorrelation function can be approximated as

$$|g(\tau)| = \int_0^\infty F(\Gamma)e^{-\Gamma\tau}d\Gamma, \tag{1}$$

where Γ is the diffusion broadening, and $F(\Gamma)$ is the contribution of the radiation component scattered by the particles of one size to the total intensity.

The diffusion broadening is related to the diffusion coefficient D as

$$\Gamma = Dq^2, \tag{2}$$

where $q = (4\pi n/\lambda)sin(\theta/2)$ is the scattering vector, n is the refractive index of the medium, λ is the wavelength, and θ is the angle of scattering.

By using the Stokes-Einstein formula

$$D = k_b T / 6\pi\eta R \qquad (3)$$

one can calculate the radius of the particles under study. Here, η is the viscosity of the medium, k_b is the Boltzmann constant, T is the temperature, and R is the particles' radius. In our study it was assumed that the temperature was constant within the laser spot.

Equation (1) belongs to the class of the so-called ill-posed problems. To solve this problem, a special algorithm based on regularization methods was used [18].

Since an applied magnetic field results in nanoparticle aggregation, it was expected that the laser correlation spectroscopy would yield a significantly greater measured particle diameter. In order to check this experimentally, a standard laboratory correlation spectrometer was supplemented with an electromagnet. The experimental setup is shown in Fig. 2.

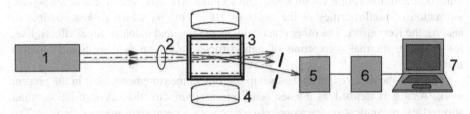

Fig. 2. Experimental setup for scattered radiation recording. 1 – laser radiation source; 2 – converging lens; 3 – sample; 4 – magnet; 5 – photomultiplier; 6 – oscilloscope; 7 – computer

A coherent light beam from a laser was transmitted through a converging lens and focused on the FF sample. The scattered light at an angle $\theta = 15$ degrees passed through a diaphragm and was detected by a photomultiplier. The signal from the photomultiplier was registered by an oscilloscope and passed to a computer for the processing. The magnetic field was generated by the DC coils and magnetic field strength was measured by the Hall probe.

2.3 Direct Measurement of Laser Radiation Scattering by Ferrofluid

One more method which was used to study the scattering properties of FFs refers to a direct measurement of the laser radiation intensity as a function of the scattering angle. The experimental setup is shown in Fig. 3.

In the experiment a translation stand was moved perpendicularly to the laser beam propagation axis. The maximum observed angle α was 52 degrees. A permanent magnet with the FF sample in-between its poles was located at the focal plane of a converging lens with focal length of 6 cm. In order to minimize absorption of the radiation and thus reduce the thermal effects, a He-Ne laser with the power in the range 0.7 to 4 mW was used as a light source ($\lambda = 632.8$ nm).

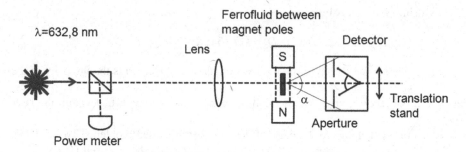

Fig. 3. Experimental setup for measuring scattered radiation from a ferrofluid sample

2.4 Z-Scan Technique

Z-scan technique is a comparatively recent but already well accepted method for the study of nonlinear properties of optical media [19]. It is often used to obtain nonlinear refraction and absorption coefficients. The technique is very sensitive and allows the estimation of nonlinearities of the order of 10^{-14} cm^2/W, which makes possible to analyze the Kerr effect. The other practically interesting and much stronger effects, like, for example, thermal convection of nanoparticles, also can be investigated in this manner.

Figure 4 shows an experimental setup for Z-scan measurements used in the present work. Axis z is defined as a laser beam propagation direction. A translation stand allowed the movement of the sample together with a permanent magnet along z. The magnetic field strength of the latter was around 1000 Oe. A cell with the FF having an optical path of 0.1 mm was placed in-between the magnetic poles. A converging lens with focal length f = 12 cm was utilized to form the beam with the radius depending on z. Thus, the displacement of the translation stand produced the laser spot with a variable light energy density on the sample. In this experiment the so-called closed aperture variant of the z-scan technique was applied, when all the radiation was collected at the photodetector. As a light source, a Nd:YAG laser operating at the second harmonic ($\lambda = 532$ nm) was employed.

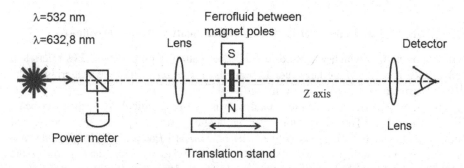

Fig. 4. A schematic illustration of a z-scan experiment

3 Results and Discussion

Before carrying out the principal experiments, the transmission bandwidths of the FFs were obtained - this was necessary for the choice of a light source with an appropriate wavelength. The transmission curve $T(\lambda)$ of the solution of 2.1 vol. % of magnetite in water is shown in Fig. 5 as an example. From these data a semiconductor laser KLM-G650-13-5 with a wavelength of 650 nm, high stability and a narrow spectral line was selected for further measurements.

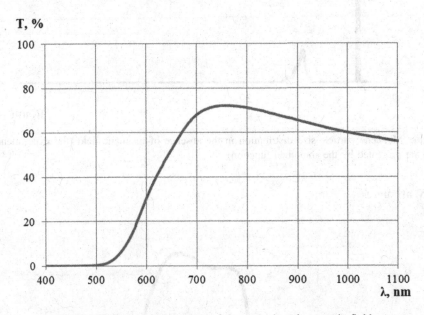

Fig. 5. The transmission curve of the water based magnetic fluid

The measurements by laser correlation spectroscopy were completed for the fields ranging from 0 to 10 mT. The experimental results are presented in Figs. 6 and 7.

The size of magnetic particles in the same FF was also investigated by means of electron microscopy [20]. It was found that the radii of individual particles did not exceed 7 nm. One can notice, however, that, as it can be seen from Fig. 6, even in the absence of magnetic field a certain percentage of particles appears to be aggregated. This is reflected by the presence of an additional peak in the size distribution at about 13 nm. It can imply an interaction of the incident optical radiation with the nanoparticles, resulting in cluster formation. The mechanism of this interplay is not well understood now. According to [21], photoinduced polarizability of magnetite may be altered in the presence of radiation, which can provoke instability of a colloidal system [22]. Another possibility is a partial aggregation of particles over the long storage time (it is an important issue to consider for the industrial use of FFs).

N, arb.units

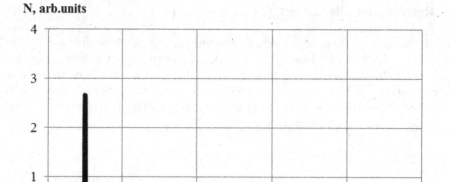

Fig. 6. Magnetic particle size distribution in the absence of magnetic field (the experimental data are presented by the smoothed function)

N, arb.units

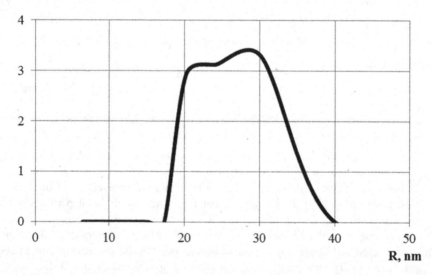

Fig. 7. Magnetic particle size distribution under the influence of magnetic field of 1 mT (the experimental data are presented by the smoothed function)

Under the action of magnetic field (Fig. 5) the size of the scattering objects in the sample significantly increases, up to 20–35 nm. Apparently in this case magnetic particles are built in conglomerates.

Thus, it may be concluded that the laser correlation spectroscopy is an efficient tool for studying the structures of magnetic particles in FF under the influence of an external magnetic field as well as in the presence of optical radiation.

Figure 8 presents a typical result of the experiments on scattering of the laser radiation. It is clear that the lower the incident radiation power, the greater the scattering. This can be explained by the fact that the absorbed light causes an increase in thermal motion of the aggregated particles, which in turn results in a partial destruction of the clusters. Therefore, a higher laser power results in a lower scattering from the FF sample.

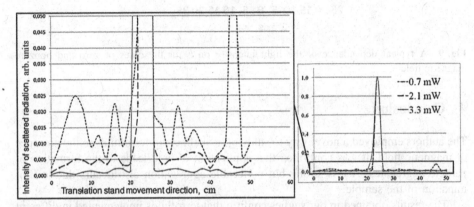

Fig. 8. Magnetic particles size distribution under the influence of magnetic field of 1 mT (the experimental data presented by the smoothed function)

A representative result of the experiments by Z-scan measurements is shown in Fig. 9. When the magnetic field is not applied, a slight valley is seen in the area of the beam waist ($z = 0$ cm), reflecting the fact that here only a small portion of the light is scattered or absorbed by the particles. It is obvious from the curve that the transmission depends on the energy density of the incident beam. This provides a strong support for the hypothesis of particle aggregation under the influence of light. The magnetic field application leads to a dramatic change in the dip in the transmitted light, which can be attributed to formation of large clusters and therefore increased light extinction.

In summary, the authors employed a number of techniques to study nonlinear optical effects in ferrofluids in the presence of magnetic field. It was confirmed that application of field leads to the formation of particle aggregates. A less strong but noticeable aggregation also occurs due to illumination of the sample. This photo-induced effect can be investigated by laser correlation spectroscopy.

Particle clusters cause a substantial part of the light scattering. When the light power is high enough, thermal effects strongly influence the scattering pattern. The thermal effects, though generally are able to destroy clusters by the convective motion of the fluid, result in some interesting effects appearing as a specific behavior of scattering which strongly depends on the applied magnetic field.

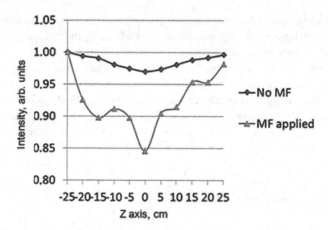

Fig. 9. A typical dependence of the light intensity on z for the cases of zero and nonzero magnetic field

4 Conclusion

The authors employed a novel way to study optical effects in ferrofluids in the presence of magnetic field. It was confirmed that application of field leads to the formation of particle aggregates. A less strong but noticeable aggregation also occurs due to illumination of the sample.

The results obtained in our studies confirm that ferrofluids implemented in different optical fiber systems may be used as elements which are operated by light or magnetic field. In particular, ferrofluid infiltrated microstructured optical fibers are promising materials for the use as controlled elements in modern optoelectronic communication devices, such as modulators, interferometers, sensors, etc.

Acknowledgments. The authors are grateful to E.E. Bibik for providing ferrofluid samples, T.V. Bocharova and A.V. Varlamov for the help in experiments.

References

1. Candiani, A., Margulis, W., Sterner, C., Konstantaki, M., Pissadakis, S.: Phase-shifted Bragg microstructured optical fiber gratings utilizing infiltrated ferrofluids. Opt. Lett. **36**, 2548–2550 (2011)
2. Candiani, A., Argyros, A., Leon-Saval, S.G., Lwin, R., Selleri, S., Pissadakis, S.: A loss-based, magnetic field sensor implemented in a ferrofluid infiltrated microstructured polymer optical fiber. Appl. Phys. Lett. **104**, 111106 (2014)
3. Deng, M., Huang, C., Liu, D., Jin, W., Zhu, T.: All fiber magnetic field sensor with ferrofluid-filled tapered microstructured optical fiber interferometer. Opt. Exp. **23**, 20668–20674 (2015)

4. Schere, C., Neto, A.: Ferrofluids: properties and applications. Braz. J. Phys. **35**, 718–727 (2005)
5. Taylor, R., Coulombe, S., Otanicar, T., et al.: Small particles, big impacts: a review of the diverse applications of nanofluids. J. Appl. Phys. **113**, 011301 (2013)
6. Agruzov, P.M., Pleshakov, I.V., Bibik, E.E., Shamray, A.V.: Magneto-optic effects in silica core microstructured fibers with a ferrofluidic cladding. Appl. Phys. Lett. **104**, 071108 (2014)
7. Zhao, Y., Lv, R., Zhang, Y., Wang, Q.: Novel optical devices based on the transmission properties of magnetic fluid and their characteristics. Opt. Lasers Eng. **50**, 1177–1184 (2012)
8. Hoffmann, B., Köhler, W.: Reversible light-induced cluster formation of magnetic colloids. J. Magn. Magn. Mater. **262**, 289–293 (2003)
9. Sawada, T., Hiroshiga, K., Matsuzaki, M., et al.: Visualization of clustering on nonmagnetic and ferromagnetic particles in magnetic fluids. In: SPIE Conference on Optical Diagnostics for Fluids/Heat/Combustion and Photomechanics for Solids, vol. 3783, p. 389 (1999)
10. Finlayson, B.A.: Convective instability of ferromagnetic fluids. J. Fluid Mech. **40**, 753–767 (1970)
11. Platten, J.K.: The Soret effect: a review of recent experimental results. J. Appl. Mech. **73**, 5–15 (2006)
12. Du, T., Yuan, S., Luo, W.: Thermal lens coupled magneto-optical effect in a ferrofluid. Appl. Phys. Lett. **65**, 1844–1847 (1994)
13. Suslov, S.A., Bozhko, A.A., Sidorov, A.S., Putin, G.F.: Thermomagnetic convective flows in a vertical layer of ferrocolloid: perturbation energy analysis and experimental study. Phys. Rev. E **86**, 016301 (2012)
14. Volker, Th., Blums, E., Odenbach, S.: Heat and mass transfer phenomena in magnetic fluids. GAMM-Mitt. **30**, 185–194 (2007)
15. Meng, Z.M., Liu, H.Y., Zhao, W.R., et al.: Effects of optical forces on the transmission of magnetic fluids investigated by Z-scan technique. J. Appl. Phys. **106**, 044905 (2009)
16. Broillet, S., Szlag, D., Bouwens, A., et al.: Visible light optical coherence correlation spectroscopy. Opt. Exp. **22**, 21944–21957 (2014)
17. Nepomnyashchaya, E., Velichko, E., Aksenov, E., Bogomaz, T.: Optoelectronic method for analysis of biomolecular interaction dynamics. IOP. J. Phys: Conf. Ser. **541**, 012039 (2014)
18. Nepomniashchaia, E.K., Velichko, E.N., Aksenov, E.T.: Solution of the laser correlation spectroscopy inverse problem by the regularization method. Univ. Res. J. **15**, 13–21 (2015)
19. Chapple, P.B., Staromlynska, J., Hermann, J.A., McKay, T.J., McDuff, R.G.: Single-beam Z-scan: measurement techniques and analysis. J. Nonlinear Opt. Phys. Mater. **6**, 251–293 (1997)
20. Bibik, E.E., Matygullin, B.Y., Raikher, Y.L., Shliomis, M.I.: Magnetostatic properties of magnetite colloids. Magnetohydrodynamics **9**, 58–62 (1973)
21. Milichko, V.A., Nechaev, A.I., Valtsifer, V.A., et al.: Photo-induced electric polarizability of Fe_3O_4 nanoparticles in weak optical fields. Nanoscale Res. Lett. **8**, 317–324 (2013)
22. Taboada-Serrano, P., Chin, C.J., Yiacoumi, S., et al.: Modeling aggregation of colloidal particles. Curr. Opin. Colloid Interface Sci. **10**, 123–132 (2005)

Acousto-Optic Switch Based on Scanned Acoustic Field

Alina Galichina, Elena Velichko$^{(\boxtimes)}$, and Evgeni Aksenov

Peter the Great Saint-Petersburg Polytechnic University, St.-Petersburg, Russia
aandreevna93@gmail.com, et.aksenov@gmail.com,
velichko-spbstu@yandex.ru

Abstract. A laboratory model of an acousto-optic switch in a combined implementation was created and investigated. The output optical elements and input fibers were implemented by using the space-wired interconnection, and the acousto-optic interaction took place in a planar waveguide. The operating abilities of the model and the possibility to build real devices are demonstrated.

Keywords: Acousto-optic switch · Bragg diffraction

1 Introduction

Considerable attention is given at present to development of all-optical communication systems. However, in spite of this, the circuit and packet switching is still performed by electronic switching devices (routers and cross-switches). For this reason, the development of optical channel switching devices is of vital importance.

The consideration of the applicability of acousto-optic control devices (AOCDs) in modern fiber-optic information systems shows that the use of AOCDs in the development of a number of subsystems can prove efficient [1]. Advantages of AOCDs are a high efficiency, information capacity, response speed and also a very simple technical implementation.

Alternatively to the space-wired interconnection (assembly of discrete elements), the device can be implemented in a combined or full optical integrated circuit [2, 3]. The major merits of the integrated implementation is a small size and a high vibration resistance, a high speed of operation due to a higher speed of the surface acoustic wave (SAW), and a low power consumption. The geometry of the acousto-optic interaction is shown in Fig. 1.

When a light wave is incident at the Bragg angle on an acoustic wave area, the electric field of the diffracted light wave is

$$E_{III} = \sum_{m=1}^{\infty} \sum_{i=-\infty}^{\infty} T_{mi}(x) \exp(-j\,\beta_{ymi}^{III}\,y) \exp(-j\,\beta_{zmi}^{III}(z-L)) \tag{1}$$

where L is the acousto-optic interaction length m, i is the diffraction order, m is the mode number, and βvmiIII is the projection of the wave number of the light flux in zone III (see Fig. 1) on the OY axis or the projection of the OZ mode of the i-th

O. Galinina et al. (Eds.): NEW2AN/ruSMART 2016, LNCS 9870, pp. 690–696, 2016.
DOI: 10.1007/978-3-319-46301-8_60

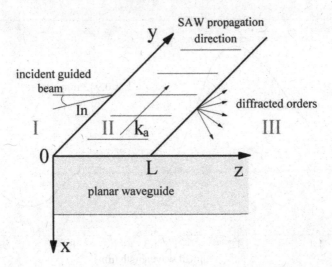

Fig. 1. Geometry of acousto-optic interaction, I - zone of the input light beam incident at the Bragg angle (θin), II - zone of propagation of surface acoustic wave with wave vector Ka, III - zone of diffraction orders

diffraction order. Equation (1) describes the electric field distribution for any form of diffraction for any electromagnetic wave [4].

From the fact that the electric field at the interface must be continuous, we obtain the following relation for the wave numbers

$$\beta_{ymi}^{III} = \beta_{vmi}^{II} = \beta_{vm0}^{III} + iK_a \tag{2}$$

Here the wave number is $\beta_m^2 = \beta_{zmi}^{III2} + \beta_{ymi}^{III2}$.

The most important parameter of an integrated optical switch is diffraction efficiency. Figures 2 and 3 show the +1 order diffraction efficiency for the TE0 and TM0 modes as functions of the wavelength (acousto-optic tunable filter mode). It can be seen from Figs. 2 and 3 that the diffraction efficiency strongly depends on the wave polarization [5].

In addition to the application in telecommunication, combined and integrated-optical AOCDs can be used in optical information processing systems. The advantage of the integrated-optical approach is a possibility to locate several devices on the same substrate. Figure 4 shows a diagram of an acousto-optic digital processor which consists of several switches. The input optical signal has frequency v_0 in all channels, its amplitudes at each moment of time are «an optical vector» a_k. In each channel the surface acoustic waves (SAW) have their own frequencies which obey the harmonic law $f_i = f_0 q^i$ ($i = 1, 2...N$). The SAW amplitude in each channel can be written as $\bar{b}_k = b_k(f_1, f_2...f_N)$. Here $a_k, b_k \in (0, 1)$. A vector of convolutions $\bar{c}$, where each of its components can be written as

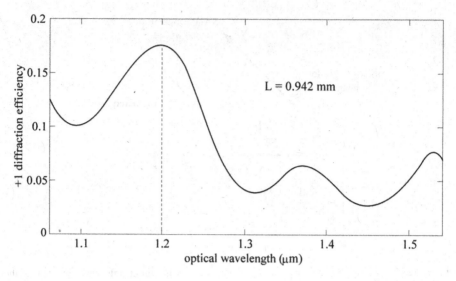

Fig. 2. Efficiency of the +1 diffraction order for mode TM0 as a function of light wavelength

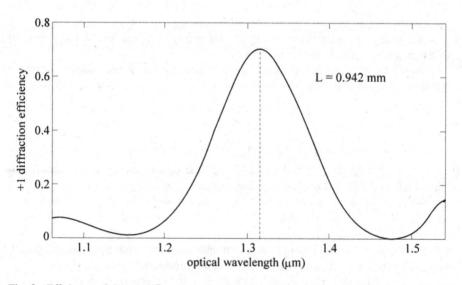

Fig. 3. Efficiency of the +1 diffraction order for mode TE0 as a function of light wavelength

$$ci = \sum_{i=1}^{k} b_{ij}a_i \tag{3}$$

is formed at the processor output. This algorithm is referred to as «digital multiplication via analog convolution» (DMAC algorithm)

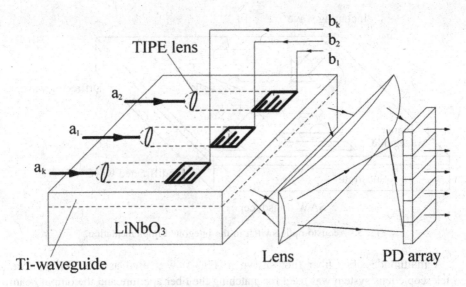

Fig. 4. Acousto-optic processor based on DMAC algorithm

The implementation of a multi-channel device that uses several frequencies of light waves $v_0 \ldots v_N$ allows one to realize a multi-channel calculation of convolutions [6].

The optical channel separation is performed by an acousto-optic tunable filter. The major difficulties encountered in the use of fully integrated optical technologies are related to the technological processes of growing layers with the desired electro-optic properties.

The application of the integrated-optical technologies is limited by two important factors, i.e., sizes and a high scattering in the waveguide plane. A reduction of the size to less than 1 micron is impossible because of the wavelengths used in the devices and final interaction lengths. So, size of the optical switch, even in the optical integrated circuit, makes this device inferior to the electronic counterparts [7].

2   Experimental

A combined acousto-optic 1×10-channel switch was implemented in the studies. An acousto-optic deflector which serves for the light beam deflection was built in the integrated optical implementation. The acousto-optic interaction took place in a thin-film glass waveguide DC 7059.

The output optics and receiving fibers were in the space-wired interconnection (separately from the substrate with the deflector). Therefore, this implementation can be regarded as a combined one [8]. A block diagram of the acousto-optic deflector is presented in Fig. 5.

The interaction length L was 3 cm. The optical beam wavelength was 635 nm. In the experiment, the operating frequency bandwidth and the central frequency of the switch $f_0 = 190$ MHz, $d_f = 20$ MHz were measured. Hence, the device can have 48 output channels.

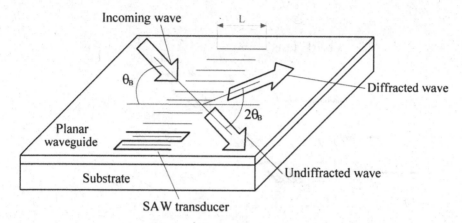

Fig. 5. Acousto-optic switch in the integrated implementation.

A multimode silica fiber (not shown in Fig. 4) was used as a receiving fiber. A telescopic lens system was used for matching the fiber aperture and the output beam (not shown in Fig. 4). It is necessary to consider the efficiency of excitation of the mode by the light beam incident at an angle on the fiber end when the light is launched into the optical fiber. Figure 6 shows the Gaussian beam which is incident on the end of the fiber with a gradient refractive index profile.

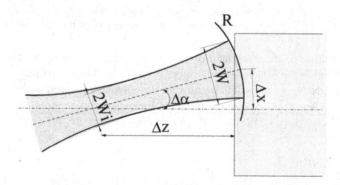

Fig. 6. Gaussian beam incident on the end of fiber

The beam axis is deflected at angle $\Delta\alpha$ with respect to the fiber axis and is shifted by distance Δh. The minimum spot radius w_i is at distance Δz from the fiber end, so the phase front of the beam at the fiber end has a curvature with radius R

$$R = \Delta z + \frac{w_i^4 k_m^2}{4\Delta z}, \tag{4}$$

where w_0 is the spot radius of the fundamental fiber mode, and k_m is the wave number of the matching medium between the beam and the fiber end. The dependence of the efficiency of excitation of the fundamental mode p_{00} on the spot radius of the

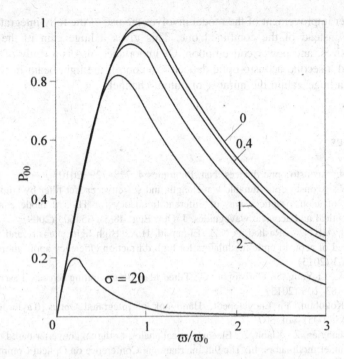

Fig. 7. Efficiency of excitation of fundamental mode of the fiber with a gradient refractive index profile as a function of relative size of the spot radius

fundamental fiber mode w_0 for this case is shown in Fig. 7. In the graph parameter σ is equal to $k_m w_0^2/R$. At a 100-% excitation efficiency, the minimum beam cross-section must be equal to the area of the input fiber end (the spot radius w_i was taken to be equal to the radius w_0 of the fundamental fiber mode) [9]. Any deviation from these conditions reduces the efficiency p_{00} because the excitation of higher order modes absorbs a part of power occurs.

The measured crosstalk of the output channels did not exceed the ambient illumination.

3   Conclusions

A combined acousto-optic switch for the potential use in fiber-optic networks as a traffic control device has been implemented. The results obtained in our studies suggest that it is possible to build a real device on the basis of the model we have developed.

The following results were obtained in our studies:

1. A laboratory model of an acousto-optic switch with 1×10 channels was built;
2. Characteristics of the basic switch elements and device as a whole were measured. The results confirmed the operating ability of the model;
3. Ways of optimization of the model parameters were found.

A further improvement of the model involves the use of the fully integrated-optical technology instead of the combined one. This gives a huge gain in the speed of operation, sizes, and power consumption. Of importance also is a study of the possibility to add a second acousto-optic deflector to control the light beam in two planes. This gives a huge gainin the number of output channels.

References

1. Savage, N.: Acousto-optic devices. Nat. Photonics **4**, 728–729 (2010)
2. Binh, L.N.: Acousto-optic tunable wavelength and selective packet filter by simultaneously launching of acousto-optic waves of different frequency for TE–TM mode conversion in diffused optical and acoustic waveguides. J. Opt. Eng. **48**(5), 054603 (2009)
3. Mohamed, A.E.N.A., Rashed, A.N.Z., El-hamid, H.A.: High light intensity and fast modulation speed of acousto optic modulators for high diffraction efficiency applications. IJRECE **1**(3), 52–63 (2013)
4. Jacques, A., Li-Yang, S., Christophe, C.: Tilted fiber Bragg grating sensors. Laser Photonics Rev. **7**(1), 83–108 (2013)
5. Iga, K., Kokubun, Y.: Encyclopedic Handbook of Integrated Optics. Taylor & Francis, New York (2005). vol. 507
6. Lei, X., Shangjan, Z., Xiaoli, Z.: Electro-optical analog-to-digital converter based on LiNbO$_3$ Mach-Zehnder modulators. In: The 9th International Conference on Optical Communications and Network, pp. 343–346 (2010)
7. Binh, L.N.: Photonic Signal Processing: Techniques and Applications. CRC Press, Boca Raton (2008). vol. 382
8. Peled, I., Kaminsky, R., Kolter, Z.: Acousto-optics bandwidth broadening in a Bragg cell based on arbitrary synthesized signal methods. J. Appl. Opt. **54**, 5065–5073 (2015)
9. Kozmaa, P., Kehlb, F., Ehrentreich-Förstera, E., Stammc, Ch., Biera, F.F.: Integrated planar optical waveguide interferometer biosensors: a comparative review. J. Biosens. Bioelectron. **58**, 287–307 (2014)

NEW2AN: Advanced Materials and Their Properties

Quantum Field Theoretical Approach to the Electrical Conductivity of Graphene

Galina L. Klimchitskaya[1,2], Vladimir M. Mostepanenko[1,2],
and Viktor M. Petrov[2(✉)]

[1] Central Astronomical Observatory at Pulkovo of the Russian Academy of Sciences,
St. Petersburg 196140, Russia
[2] Institute of Physics, Nanotechnology and Telecommunications,
Peter the Great Saint Petersburg Polytechnic University,
St. Petersburg 195251, Russia
vikpetroff@mail.ru

Abstract. The longitudinal and transverse electrical conductivities of graphene are calculated at both zero and nonzero temperature starting from the first principles of thermal quantum field theory using the polarization tensor in (2+1)-dimensional space-time. An expression for the universal conductivity of graphene found previously using different phenomenological approaches is confirmed. Both exact and approximate asymptotic expressions for the real and imaginary parts of the conductivity of graphene are derived and compared with the results of numerical computations. The obtained results can be used in numerous applications of graphene, such as in optical detectors, transparent electrodes and nanocommunications.

Keywords: Graphene · Electrical conductivity · Polarization tensor · Nanocommunications

1 Introduction

Among several novel materials of high promise for applications in nanotechnology, the two-dimensional sheet of carbon atoms called graphene occupies an outstanding position due to its unusual electrical, mechanical and optical properties [1,2]. These properties originate from the fact that at energies below a few eV the quasiparticles in graphene are massless Dirac fermions possessing the linear dispersion relation, but moving with the Fermi velocity $v_F \approx c/300$ rather than with the speed of light c. Thermal quantum field theory in two spatial dimensions provides the well developed formalism for theoretical description of such kind systems. Technological applications of graphene are numerous and diverse. Graphene coatings are used in optical detectors to increase an efficiency of light absorption on metallic surfaces [3]. Graphene-coated substrates are applied as transparent electrodes [4]. Deposition of graphene on silicon plates provides an excellent antireflection and is used in solar cells [5]. The electrical properties of

© Springer International Publishing AG 2016
O. Galinina et al. (Eds.): NEW2AN/ruSMART 2016, LNCS 9870, pp. 699–707, 2016.
DOI: 10.1007/978-3-319-46301-8_61

graphene make it prospective material for applications in electromagnetic communications, specifically in nanoantennas.

The main electronic property of graphene is its electrical conductivity, which was investigated in many papers both theoretically [6–18] and experimentally [19–23]. The most of theoretical results were obtained using the current-current correlation function, the Kubo formalism and Boltzmann's transport theory. It was found that even at zero temperature graphene possesses the universal finite frequency-independent conductivity σ_0 expressed in terms of fundamental constants e and $\hbar$. In so doing, the specific values of σ_0 obtained by different authors vary depending on the order of limiting transitions used in different formalisms [8].

In this paper, we investigate the electrical conductivity of pure graphene do not using any phenomenological approach, but starting from the first principles of thermal quantum field theory. This is achieved by using the polarization tensor of graphene, which was previously applied to study the Casimir effect in graphene systems [24–34] and the reflectivity properties of graphene and graphene-coated substrates [35–37].

The polarization tensor of graphene at $T = 0\,\mathrm{K}$ was calculated in Ref. [24]. The thermal correction to the polarization tensor was found at pure imaginary Matsubara frequencies in Ref. [25]. In Ref. [36] the thermal correction to the polarization tensor was determined along the real frequency axis. At zero temperature, we find the universal real conductivity of graphene obtained previously using different methods. At nonzero temperature, we obtain simple asymptotic expressions for both the real and imaginary parts of the conductivity of graphene. Specifically, it is shown that the imaginary part of conductivity changes its sign as a function of frequency and temperature.

2   Conductivity of Graphene at Zero Temperature

We consider the pure (pristine) graphene in the application region of the Dirac model. Our formalism can be generalized for the case of graphene with nonzero mass-gap parameter and chemical potential.

At both zero and nonzero temperature, the longitudinal (along the surface of graphene) and transverse (perpendicular to it) electrical conductivities of graphene can be expressed via the components of the polarization tensor of graphene as [38,39]

$$\sigma_\parallel(\omega, k, T) = -i\frac{\omega}{4\pi\hbar k^2}\,\Pi_{00}(\omega, k, T),$$

$$\sigma_\perp(\omega, k, T) = i\frac{c^2}{4\pi\hbar k^2\omega}\,\Pi(\omega, k, T). \tag{1}$$

Here, the polarization tensor is notated as $\Pi_{\alpha\beta}$ with $\alpha, \beta = 0, 1, 2$, ω is the frequency, k is the magnitude of the wave vector component in the plane of graphene, and the quantity Π is defined as

$$\Pi(\omega, k, T) \equiv k^2\Pi_{\mathrm{tr}}(\omega, k, T) + \left(\frac{\omega^2}{c^2} - k^2\right)\Pi_{00}(\omega, k, T), \tag{2}$$

where $\Pi_{\mathrm{tr}} \equiv \Pi_\alpha^\alpha$.

At $T = 0\,\mathrm{K}$ the 00-component of the polarization tensor and the quantity Π are given by [24, 36]

$$\Pi_{00}^{(0)}(\omega, k) = ie^2 \pi \frac{k^2}{\omega \eta(\omega, k)}, \qquad \Pi^{(0)}(\omega, k) = -ie^2 \pi \frac{\omega}{c^2} k^2 \eta(\omega, k), \qquad (3)$$

where

$$\eta \equiv \eta(\omega, k) = \sqrt{1 - \kappa^2(\omega, k)}, \qquad \kappa \equiv \kappa(\omega, k) = \tilde{v}_F \frac{ck}{\omega} \qquad (4)$$

and $\tilde{v}_F \equiv v_F/c$. For real photons on a mass-shell $k \leq \omega/c$ and, thus, $\kappa \leq \tilde{v}_F \ll 1$.

Substituting Eq. (3) in Eq. (1), one finds that both components of the conductivity of graphene at $T = 0\,\mathrm{K}$ are real and given by

$$\sigma_\parallel(\omega, k, 0) = \frac{e^2}{4\hbar \eta(\omega, k)} = \frac{\sigma_0}{\sqrt{1 - \kappa^2(\omega, k)}},$$

$$\sigma_\perp(\omega, k, 0) = \frac{e^2}{4\hbar} \eta(\omega, k) = \sigma_0 \sqrt{1 - \kappa^2(\omega, k)}, \qquad (5)$$

where the universal conductivity of graphene is defined as

$$\sigma_0 \equiv \frac{e^2}{4\hbar}. \qquad (6)$$

The same result was obtained by many authors using the Kubo formula [8–10, 12, 18]. In the local limit (the nonlocal corrections are of the order of 10^{-5}) one arrives at $\sigma_\parallel = \sigma_\perp = \sigma_0$.

3   Real Part of the Conductivity of Graphene at Nonzero Temperature

At $T \neq 0\,\mathrm{K}$, the 00-component of the polarization tensor and the quantity Π are represented as

$$\Pi_{00}(\omega, k, T) = \Pi_{00}^{(0)}(\omega, k) + \Delta_T \Pi_{00}(\omega, k, T),$$

$$\Pi(\omega, k, T) = \Pi^{(0)}(\omega, k) + \Delta_T \Pi(\omega, k, T), \qquad (7)$$

where $\Delta_T \Pi_{00}$ and $\Delta_T \Pi$ are the thermal corrections to the zero-temperature parts (3).

According to Eq. (1), the real part of conductivity of graphene is determined by the imaginary part of the polarization tensor. Using the results of Refs. [36, 37], the imaginary parts of the thermal corrections are given by

$$\mathrm{Im}\,\Delta_T \Pi_{00}(\omega, k, T) = \frac{4e^2 \omega}{\tilde{v}_F^2 c^2 \eta} Y_{00}(\omega, k, T),$$

$$\mathrm{Im}\,\Delta_T \Pi(\omega, k, T) = -\frac{4e^2 \omega^3 \eta}{\tilde{v}_F^2 c^4} Y(\omega, k, T), \qquad (8)$$

where

$$Y_{00}(\omega, k, T) = -\kappa^2 \int_{-1}^{1} \frac{dt}{e^{\beta}e^{-\beta\kappa t} + 1} \sqrt{1 - t^2},$$

$$Y(\omega, k, T) = -\kappa^2 \int_{-1}^{1} \frac{dt}{e^{\beta}e^{-\beta\kappa t} + 1} \frac{t^2}{\sqrt{1 - t^2}}, \qquad (9)$$

and $\beta \equiv \beta(\omega, T) = \omega/(2\omega_T) \equiv \hbar\omega/(2k_B T)$.

Using Eqs. (1), (3), (7), and (8), for the real part of conductivity one obtains

$$\mathrm{Re}\sigma_{\parallel}(\omega, k, T) = \frac{\sigma_0}{\eta} \left(1 + \frac{4}{\pi\kappa^2} Y_{00}\right),$$

$$(10)$$

$$\mathrm{Re}\sigma_{\perp}(\omega, k, T) = \sigma_0\eta \left(1 + \frac{4}{\pi\kappa^2} Y\right).$$

It is easily seen that the integrals (9) can differ from zero only under a condition $\beta\kappa \ll 1$. Expanding in (9) in powers of $\beta\kappa$, calculating the integrals and using Eq. (10), we arrive at

$$\mathrm{Re}\sigma_{\parallel(\perp)}(\omega, k, T) = \sigma_0 \left[1 - \frac{2}{e^{\beta} + 1} - \kappa^2 C_{\parallel(\perp)}(\beta) + O(\kappa^4)\right], \qquad (11)$$

where the functions $C_{\parallel}$ and $C_{\perp}$ are given by

$$C_{\parallel}(\beta) = \frac{\beta^2 e^{-\beta}(1 - e^{-\beta}) + 2(1 + e^{-\beta})^2(1 + 3e^{-\beta})}{4(1 + e^{-\beta})^3},$$

$$C_{\perp}(\beta) = \frac{3\beta^2 e^{-\beta}(1 - e^{-\beta}) - 2(1 + e^{-\beta})^2(1 + 3e^{-\beta})}{4(1 + e^{-\beta})^3}. \qquad (12)$$

Note that the relative contribution of the second order terms in Eq. (11) does not exceed 10^{-4}. By omitting these terms, we return to the previously known result [40]

$$\mathrm{Re}\sigma_{\parallel(\perp)}(\omega, T) \approx \sigma_0 \left[1 - \frac{2}{e^{\hbar\omega/(2k_B T)} + 1}\right] = \sigma_0 \tanh \frac{\hbar\omega}{4k_B T}. \qquad (13)$$

From Eq. (13) it is seen that at fixed temperature $\mathrm{Re}\sigma_{\parallel(\perp)}$ goes to σ_0 at high frequencies $\omega \gg 2\omega_T$ and to zero at low frequencies $\omega \ll 2\omega_T$. However, if the frequency is fixed, $\mathrm{Re}\sigma_{\parallel(\perp)}$ goes to σ_0 with vanishing temperature. This means that there is a discontinuity of the real part of conductivity of graphene as a function of ω and T at the point $(0,0)$.

As an example, in Fig. 1 we present the computational results for $\mathrm{Re}\sigma_{\parallel(\perp)}$ as a function of T by the lines 1, 2, 3, and 4 computed at $\omega = 10^{-1}$, 10^{-2}, 10^{-3}, and 10^{-4} eV, respectively. Computations are done using the exact Eq. (10). The computational results for $\mathrm{Re}\sigma_{\parallel}$ and $\mathrm{Re}\sigma_{\perp}$ are indistinguishable in the used scales.

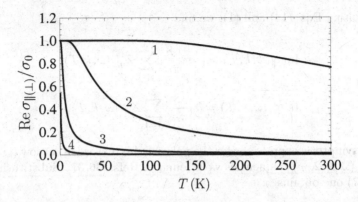

Fig. 1. The real part of the electrical conductivity of pure graphene as a function of T is shown by the lines 1, 2, 3, and 4 computed at $\omega = 10^{-1}$, 10^{-2}, 10^{-3}, and 10^{-4} eV, respectively.

4   Imaginary Part of the Conductivity of Graphene at Nonzero Temperature

According to Eq. (1), the imaginary part of the conductivity of graphene is determined by the real part of the polarization tensor. Taking into account that at $T = 0$ K the polarization tensor (3) is pure imaginary, $\mathrm{Im}\sigma_{\parallel(\perp)}$ is determined exclusively by the real part of the thermal correction to the polarization tensor.

According to the results of Refs. [36,37],

$$\mathrm{Re}\Delta_T \Pi_{00}(\omega, k, T) = \frac{8e^2\omega}{\tilde{v}_F^2 c^2} \sum_{j=1}^{3} Z_{00}^{(j)}(\omega, k, T),$$

$$\mathrm{Re}\Delta_T \Pi(\omega, k, T) = \frac{8e^2\omega^3}{\tilde{v}_F^2 c^4} \sum_{j=1}^{3} Z^{(j)}(\omega, k, T), \tag{14}$$

where the integrals $Z_{00}^{(j)}$ are defined as

$$Z_{00}^{(1)}(\omega, k, T) = \int_0^{1-\kappa} \frac{dy}{e^{\beta y}+1} \left\{ 1 - \frac{1}{2\eta} \left[\sqrt{(y+1)^2 - \kappa^2} + \sqrt{(y-1)^2 - \kappa^2} \right] \right\},$$

$$Z_{00}^{(2)}(\omega, k, T) = \int_{1-\kappa}^{1+\kappa} \frac{dy}{e^{\beta y}+1} \left[1 - \frac{1}{2\eta} \sqrt{(y+1)^2 - \kappa^2} \right], \tag{15}$$

$$Z_{00}^{(3)}(\omega, k, T) = \int_{1+\kappa}^{\infty} \frac{dy}{e^{\beta y}+1} \left\{ 1 - \frac{1}{2\eta} \left[\sqrt{(y+1)^2 - \kappa^2} - \sqrt{(y-1)^2 - \kappa^2} \right] \right\}.$$

The integrals $Z^{(j)}$ are obtained from $Z_{00}^{(j)}$ by the substitution

$$\sqrt{(y \pm 1)^2 - \kappa^2} \rightarrow \frac{(y \pm 1)^2}{\sqrt{(y \pm 1)^2 - \kappa^2}}. \tag{16}$$

Now, using Eqs. (14) and (1), we find

$$\text{Im}\sigma_\|(\omega, k, T) = -\sigma_0 \frac{8}{\pi \kappa^2} \sum_{j=1}^{3} Z_{00}^{(j)}(\omega, k, T),$$

$$\text{Im}\sigma_\perp(\omega, k, T) = \sigma_0 \frac{8}{\pi \kappa^2} \sum_{j=1}^{3} Z^{(j)}(\omega, k, T), \tag{17}$$

The asymptotic expressions for the polarization tensor (14) at low ($\omega \ll 2\omega_T$) and high ($\omega \gg 2\omega_T$) frequencies were found in Refs. [36,37]. Substituting them in Eq. (17) one obtains

$$\text{Im}\sigma_{\|(\perp)}(\omega, k, T) \approx \sigma_0 \frac{8 \ln 2}{\pi} \frac{k_B T}{\hbar \omega} \tag{18}$$

at low ω and

$$\text{Im}\sigma_{\|(\perp)}(\omega, k, T) \approx -\sigma_0 \frac{48\zeta(3)}{\pi} \left(\frac{k_B T}{\hbar \omega} \right)^3 \tag{19}$$

at high ω, where $\zeta(z)$ is the Riemann zeta function. Equations (18) and (19) lead to less than 1 % error under the conditions $\hbar \omega \lesssim 0.2 k_B T$ and $\hbar \omega > 70 k_B T$, respectively.

As an example, the exact Eq. (17) is used to compute $\text{Im}\sigma_{\|(\perp)}$ as a function of T. The computational results are presented in Fig. 2 by the lines 1, 2, and 3 computed at $\omega = 10^{-1}$, 10^{-2}, and 10^{-3} eV, respectively. It is seen that at high and low T (low and high ω) the asymptotic expressions of Eqs. (18) and (19), respectively, are well applicable.

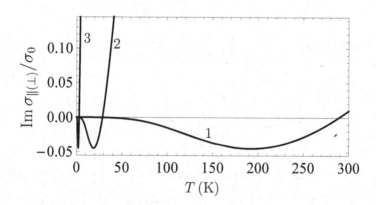

Fig. 2. The imaginary part of the electrical conductivity of pure graphene as a function of T is shown by the lines 1, 2, and 3 computed at $\omega = 10^{-1}$, 10^{-2}, and 10^{-3} eV, respectively.

5   Conclusions and Discussion

In the foregoing, we have calculated both the real and imaginary parts of the electrical conductivity of pure graphene at both zero and nonzero temperature. Our calculations are done starting from the first principles of thermal quantum field theory using the polarization tensor of graphene obtained earlier for some other purposes. In this way, we have confirmed an expression (1) for the universal conductivity of graphene σ_0 found previously using several phenomenological approaches and derived both exact and approximate asymptotic expressions for the longitudinal and transverse conductivities. The approximate expressions are found in a very good agreement with the results of numerical computations.

The obtained results can be used in numerous technological applications of graphene in optical detectors, transparent electrodes, antireflection coatings, and in nanocommunications. They can be generalized for graphene samples with nonzero mass-gap parameter and chemical potential using the more general expressions for the polarization tensor found recently in Refs. [36,41].

References

1. Katsnelson, M.I.: Graphene: Carbon in Two Dimensions. Cambridge University Press, Cambridge (2012)
2. Neto, A.H.C., Guinea, F., Peres, N.M.R., Novoselov, K.S., Geim, A.K.: The electronic properties of graphene. Rev. Mod. Phys. **81**, 109–162 (2009)
3. Jiang, X., et al.: Anti-reflection graphene coating on metal surface. Surf. Coat. Technol. **261**, 327–330 (2015)
4. Watcharotone, S., et al.: Graphene-silica composite thin films as transparent conductors. Nano Lett. **7**, 1888–1892 (2007)
5. Vajtai, R. (ed.): Springer Handbook of Nanomaterials. Springer, Berlin (2013)
6. Gusynin, V.P., Sharapov, S.G.: Transport of Dirac quasiparticles in graphene: hall and optical conductivities. Phys. Rev. B **73**, 245411-1–245411-18 (2006)
7. Katsnelson, M.I.: Zitterbewegung, chirality, and minimal conductivity of graphene. Eur. Phys. J. B **51**, 157–160 (2006)
8. Ziegler, K.: Robust transport properties of graphene. Phys. Rev. Lett. **97**, 266802-1–266802-4 (2006)
9. Falkovsky, L.A., Varlamov, A.A.: Space-time dispersion of graphene conductivity. Eur. Phys. J. B **56**, 281–284 (2007)
10. Stauber, T., Peres, N.M.R., Geim, A.K.: Optical conductivity of graphene in the visible region of the spectrum. Phys. Rev. B **78**, 085432-1–085432-8 (2008)
11. Pedersen, T.G.: Optical response and exitons in gaped graphene. Phys. Rev. B **79**, 113406-1–113406-4 (2009)
12. Lewkowicz, M., Rosenstein, B.: Dynamics of particle-hole pair creation in graphene. Phys. Rev. Lett. **102**, 106802-1–106802-4 (2009)
13. Palacios, J.J.: Origin of quasiuniversality of the minimal conductivity of graphene. Phys. Rev. B **82**, 165439-1–165439-6 (2010)
14. Moriconi, L., Niemeyer, D.: Graphene conductivity near the charge neutral point. Phys. Rev. B **84**, 193401-1–193401-5 (2011)
15. Buividovich, P.V., et al.: Numerical study of the conductivity of graphene monolayer within the effective field theory approach. Phys. Rev. B **86**, 045107-1–045107-8 (2012)

16. Dartora, C.A., Cabrera, G.G.: $U(1) \times SU(2)$ gauge invariance leading to charge and spin conductivity of Dirac fermions in graphene. Phys. Rev. B **87**, 165416-1–165416-5 (2013)

17. Louvet, T., Delplace, P., Fedorenko, A.A., Carpentier, D.: On the origin of minimal conductivity at a band crossing. Phys. Rev. B **92**, 155116-1–155116-11 (2015)

18. Merano, M.: Fresnel coefficients of a two-dimensional atomic crystals. Phys. Rev. A **93**, 013832-1–013832-5 (2016)

19. Tan, Y.-W., et al.: Measurement of scattering rate and minimum conductivity in graphene. Phys. Rev. Lett. **99**, 246803-1–246803-4 (2007)

20. Nair, R.R., et al.: Fine structure constant defines visual transparency of graphene. Science **320**, 1308 (2008)

21. Li, Z., et al.: Dirac charge dynamics in graphene by infrared spectroscopy. Nat. Phys. **4**, 532–535 (2008)

22. Mak, K.F., et al.: Measurement of optical conductivity of graphene. Phys. Rev. Lett. **101**, 196405-1–196405-4 (2008)

23. Horng, J., et al.: Drude conductivity of Dirac fermions in graphene. Phys. Rev. B **83**, 165113-1–165113-5 (2011)

24. Bordag, M., Fialkovsky, I.V., Gitman, D.M., Vassilevich, D.V.: Casimir interaction between a perfect conductor and graphene described by the Dirac model. Phys. Rev. B **80**, 245406-1–245406-5 (2009)

25. Fialkovsky, I.V., Marachevsky, V.N., Vassilevich, D.V.: Finite-temperature Casimir effect for graphene. Phys. Rev. B **84**, 035446-1–035446-10 (2011)

26. Bordag, M., Klimchitskaya, G.L., Mostepanenko, V.M.: Thermal Casimir effect in the interaction of graphene with dielectrics and metals. Phys. Rev. B **86**, 165429-1–165429-14 (2012)

27. Chaichian, M., Klimchitskaya, G.L., Mostepanenko, V.M., Tureany, A.: Thermal Casimir-Polder interaction of different atoms with graphene. Phys. Rev. A **86**, 012515-1–012515-9 (2012)

28. Klimchitskaya, G.L., Mostepanenko, V.M.: Van der Waals and Casimir interactions between two graphene sheets. Phys. Rev. B **87**, 075439-1–075439-7 (2013)

29. Klimchitskaya, G.L., Mohideen, U., Mostepanenko, V.M.: Theory of Casimir interaction from graphene-coated substrates using the polarization tensor and comparison with experiment. Phys. Rev. B **89**, 115419-1–115419-8 (2014)

30. Klimchitskaya, G.L., Mostepanenko, V.M.: Observability of thermal effects in the Casimir interaction from graphene-coated substrates. Phys. Rev. A **89**, 052512-1–052512-7 (2014)

31. Klimchitskaya, G.L., Mostepanenko, V.M., Sernelius, B.E.: Two approaches for describing the Casimir interaction in graphene: density-density correlation function versus polarization tensor. Phys. Rev. B **89**, 125407-1–125407-9 (2014)

32. Klimchitskaya, G.L., Mostepanenko, V.M.: Comparison of hydrodynamic model of graphene with recent experiment on measuring the Casimir interaction. Phys. Rev. B **91**, 045412-1–045412-5 (2015)

33. Klimchitskaya, G.L., Mostepanenko, V.M.: Origin of large thermal effect in the Casimir interaction between two graphene sheets. Phys. Rev. B **91**, 174501-1–174501-10 (2015)

34. Klimchitskaya, G.L.: Quantum field theory of the Casimir force for graphene. Int. J. Mod. Phys. A **31**, 1641026-1–1641026-13 (2016)

35. Klimchitskaya, G.L., Mostepanenko, V.M., Petrov, V.M.: Reflectivity properties of graphene and graphene-coated substrates. In: Balandin, S., Andreev, S., Koucheryavy, Y. (eds.) Internet of Things, Smart Spaces, and Next Generation Networks and Systems, pp. 451–458. Springer, Cham (2014)

36. Bordag, M., Klimchitskaya, G.L., Mostepanenko, V.M., Petrov, V.M.: Quantum field theoretical description for the reflectivity of graphene. Phys. Rev. D **91**, 045037-1–045037-19 (2015)

37. Klimchitskaya, G.L., Korikov, C.C., Petrov, V.M.: Theory of reflectivity properties of graphene-coated material plates. Phys. Rev. B **92**, 125419-1–125419-9 (2015)

38. Sernelius, B.E.: Retarded interactions in graphene systems. Phys. Rev. B **85**, 195427-1–195427-10 (2012)

39. Sernelius, B.E.: Casimir effect in systems containing 2D layers, like graphene and 2D electron gases. J. Phys.: Condens. Matter **27**, 214017-1–214017-25 (2015)

40. Falkovsky, L.A., Pershoguba, S.S.: Optical far-infrared properties of a graphene monolayer and multilayer. Phys. Rev. B **76**, 153410-1–153410-4 (2007)

41. Bordag, M., Fialkovsky, I.V., Vassilevich, D.V.: Enhanced Casimir effect for doped graphene. Phys. Rev. B **93**, 075414-1–075414-5 (2016)

Writing Ferroelectric Nanodomains in PZT Thin Film at Low Temperatures

Alexandr Vakulenko[1], Natalia Andreeva[1], Sergej Vakhrushev[1,2,3],
Alexandr Fotiadi[1], and Alexey Filimonov[1(✉)]

[1] Peter the Great St. Petersburg Polytechnic University,
195251 Polytechnicheskaya 29, St.-Petersburg, Russia
vakulenko705@gmail.com, nvandr@gmail.com,
s.vakhrushev@mail.ioffe.ru, fotiadi@rphf.spbstu.ru,
filalex@inbox.ru
[2] Ioffe Institute, 194021 Polytechnicheskaya 26, St.-Petersburg, Russia
[3] Saint-Petersburg State University,
199504 Ulyanovskaya str., 1, Petrodvorets, Russia

Abstract. Thin ferroelectric films are prospective materials for applications in the area of tunable microwave electronics as a base for varactors, phase shifters, delay lines, tunable filters and antennas. The most important technological aspect of using thin polar films in electronics is a possibility of miniaturization. By means of piezoresponse force microscopy technique, it is possible to create nanometer-sized areas (or ferroelectric domains) in thin films with preferable direction of polarization. Besides the fact that these domains could be used as a bit for mass storage application, it was found, that domain walls have their own properties, moreover, they are mobile. This circumstance could give rise to a new type of technology where mobile domain walls will be the "active ingredient" of the device.

In this work, we use a scanning piezoresponse force microscopy to investigate the process of writing and growth of ferroelectric domains in thin $PbZr_{0.3}Ti_{0.7}O_3$ film in a broad temperature range. It was found that even at 4.2 K nanoscale ferroelectric domains could be writing by application of short voltage pulses between the tip of atomic force microscope and extended bottom electrode. Based on the obtained experimental results the mechanism driven the ferroelectric domain dynamics in thin films at low temperatures was determined.

Keywords: Ferroelectric thin films · Piezoresponse force microscopy · Ferroelectric domain dynamics · Low temperatures

1 Introduction

One of the simplest and most effective ways of the energy independent information storage is to use of so-called ferroics. The term ferroics was first introduced by Aizu [1]. He determine ferroics in the following way: "A crystal is provisionally referred to as being "ferroic" when it has two or more orientation states in the absence of magnetic field, electric field, and mechanical stress and can shift from one to another of these states by means of a magnetic field, an electric field, a mechanical stress, or a

© Springer International Publishing AG 2016
O. Galinina et al. (Eds.): NEW2AN/ruSMART 2016, LNCS 9870, pp. 708–716, 2016.
DOI: 10.1007/978-3-319-46301-8_62

combination of these". The switchable parameter can be referred as the "order parameter" This class of materials includes ferroelectrics, ferromagnetics, ferroelastics and ferrotoroics. Small regions in ferroics with different order parameter directions are called "domains", the boundaries of these regions are called "domain walls". The domain walls are movable and often have physical properties, different from the bulk of the domain. This fact gave rise to a new type of devices, where domain walls play role of an "active element". The concept of domain wall electronics is most fully realized in magnetic wall electronics. A working principle of these devices is based on the high domain wall mobilities, velocities of magnetic walls can be supersonic. In case of ferroelectrics, domain walls are slower, but can interesting functional properties, for example, they could be conductive in a bulk insulator or semiconductor surrounding. Moreover, ferroelectric domains have smaller sizes compare to the magnetic ones which is an important technological aspect for device miniaturization. Despite the fact that ferroelectrics became an essential component for many electronic devices since early 1960s, the concept of using a single ferroelectric domain wall as an "active ingredient" in nanoelectronic devices is still under development and demands intensive researches directed to deeper understanding physical properties of ferroelectric materials.

An interest to study domain formation and domain wall motion in ferroelectric (FE) thin films is caused by the technological aspects of thin film application: the possibility of device miniaturization and their integration onto one substrate. Thus, starting from 2000s, thin ferroelectric films are widely applied and developed for memories, microwave electronic components and microdevices with pyroelectric and piezoelectric microsensors/actuators [2–9].

In this work, we used a scanning piezoresponse force microscopy (PFM) to study the process of FE nanodomains writing and domain wall motion in thin $PbZr_{0.3}Ti_{0.7}O_3$ film at low temperatures. This method allows imaging of FE domains with resolution up to 10 nm and is widely applied in ferroelectric researches for more than 10 years already. Regarding thin FE films, the domain wall motion has already been well studied with this technique at room temperatures [10–12]. The absence of well-developed techniques providing PFM measurements at low (cryogenic) temperatures results in a single works on FE domain dynamics in this temperature range. Thus, there is still not clear which physical mechanisms drive the domain wall motion at low temperatures.

2 Samples and Experiment

All results were obtained with the high-quality epitaxial $PbZr_{0.3}Ti_{0.7}O_3/LaSr_{0.7}Mn_{0.3}O_3$ bilayers, grew on (100)-oriented single-crystalline $SrTiO_3$ substrates by pulsed laser deposition. The PZT film has a nominal thickness of about 60 nm. Measurements were performed with a cryogenic atomic force microscope AttoAFM I (Attocube Systems, Germany) equipped with an external lock-in amplifier SR844 (Stanford Research Systems, CA) and a functional generator FC120 (Yokogawa Electric Corporation, Japan). The processes of the domain nucleation and growth in thin ferroelectric film were followed in the temperature range starting from 4.2 K to room temperature. We employed the gold coated silicon cantilevers NSG03/Au (NT-MDT, Russia) with the tip radius of curvature $r_{tip} \approx 35$ nm at the apex, resonant frequency $f_R \approx 100$ kHz and the force constant $k \approx 1$ N/m. The FE domain writing and reading with this technique was achieved by bringing a sharp conductive probe into a contact with a FE film and applying an alternating voltage bias to the tip in order to excite deformation of the sample surface through the converse piezoelectric effect. The resulting deflection of the cantilever is demodulated with a lock-in amplifier.

To study domain wall motion, nanoscale domain writing was done by applying dc voltage pulses to the bottom electrode while keeping the grounded PFM tip in contact with the film surface at a fixed point [13]. In order to minimize the influence of native polarization distribution on the domain formation, the film was initially poled in the upward direction by scanning the surface with the positive dc voltage $V = 5$–10 V applied between the PFM tip and the bottom electrode. Then we used negative writing voltages to create domains with a downward polarization. The duration of applied writing voltage pulses was varied in the range from 1 ms up to 100 s. The effective domain diameter has been calculated from the reversed domain area. More thorough results were obtained and analyzed for two temperature points $T = 120$ K and $T = 300$ K.

3 Results and Discussion

3.1 Theoretical Approach to the Domain-Wall Motion Description

The driving mechanism of domain-wall motion in thin FE films at the room temperature (RT) has already been studied [10–12]. In homogeneous defect-free single crystals, domain-wall motion is determined by the interaction of the domain wall with the crystal lattice. Nucleation and growth process is divided into two phases [12]: the formed embryonic domain (semi-ellipsoidal region of reversed polarization) grows rapidly along the film's depth until it crosses the whole film and transforms into a cylindrical 180° domain; and its slower expansion in the film plane. Wall velocity v exponentially depends on the field strength E according to the relation [11, 12]:

$$v(r) = v_\infty exp\left[-\frac{U_a}{k_B \cdot T} \cdot \left(\frac{V_c}{V(r)}\right)^{\mu}\right], \tag{1}$$

where U_a – typical activation energy; k_B – Boltzmann's constant; T – temperature; $V(r)$ – electric potential on the film surface at the distance r from the PFM tip; V_c represents some characteristic voltage; μ depends on the dimensionality of the wall and on the nature of the pinning potential; $t_\infty = h/v_\infty$. Electric field can be replaced with electric potential on the film surface divided by the film thickness $E = V/h$. Relation (1) could be written in the form:

$$v = v_\infty exp[-\frac{A}{T \cdot V(r)^{\mu}}], \tag{2}$$

where $A = U_a \cdot (V_c)^{\mu}/k_B$.

To fit experimental dependences of the domain diameter on the duration of dc voltage pulse we found $V(r)$ from the numerical solution of the Laplace's equation using finite difference method for the axially symmetric system. Geometry of the model is shown in Fig. 1. Permittivity of PZT film was assumed to be 500 [14, 15].

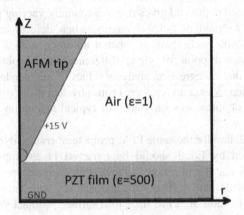

Fig. 1. Axially symmetric model of "PFM tip – PZT surface" using for $V(r)$ calculation.

Time t required for the domain growths to the radius R derives from integrating the $dt = dr/v(r)$, where for $v(r)$ Eq. 2 is used. Embryonic domain appears after very small $(10^{-7}–10^{-9}\mathrm{s})$ writing time and after that sidewise growth starts. Thus, integrating starts from the embryonic domain radius R_{min}. Radii-time dependence is given by the inverse function of $t(R)$:

$$t(R) = \int_{R_{min}}^{R} \frac{dr}{v(r)} = \int_{R_{min}}^{R} \frac{1}{v_\infty} \cdot e^{\frac{A}{T \cdot V(r)^\mu}} dr, \qquad (3)$$

Experimental data points were fitted by the Eq. 3 with a least square method, parameters v_∞, A and μ were taken as an independent variables in regression analysis.

3.2 Influence of the Poling Procedure on the Radii-Time Dependences of Nanodomains in Thin PZT Film

At first, to check the reproducibility of the obtained dependences of the domain radius on the pulse duration we conducted several series of measurements writing domains by different PFM probes at room temperature. All cantilevers were gold coated silicon cantilevers from NSG03/Au series with the same specification. The possible difference between them was due to the dispersion in their sizes provided by a manufacturer. It was found that radii-time dependences are strongly influenced by the cantilever (Fig. 2). We suppose, that the difference between these experimental dependencies occurs due to the dispersion in tip shapes and cantilever sizes, which vary from one probe to another and conditioned by the technological process of their manufacturing.

In our case this difference affects only on the value of domain radius, but not the shape of radii-time dependence – it is always logarithmic dependence with $\mu = 1$. This is an important result, because μ determines the mechanism driving the domain wall motion. If μ equals to 0.5–0.6 for two-dimensional domain walls, then the mechanism of wall motion called "random bond". In this case, defects in thin FE film locally modify the ferroelectric double-well depth and gives rise to a spatially varying pinning potential. If $\mu = 1$ then there is a "random field" scenario, when defects induce a local field, asymmetrizing the double well, there are spatial inhomogeneities in the electric field [11]. From mathematical point of view, difference in the obtained experimental dependencies means that in regression analysis values of independent variables v_{∞} and A change from one dependence to another. From physical point of view, it means that PFM tip determines or influences on values of typical activation energy U_a; characteristic voltage V_c or v_{∞}.

It should be noted, that for the same PFM probe temperature dynamics of radii-time dependences predicted by Eq. 3 should be retrieved in assumption that values of independent variables v_{∞}, A and μ don't change with temperature. Otherwise, it could mean an alteration of domain wall driving mechanism with temperature (in case of changing μ), or modification of "PFM tip – film surface" contact properties (in case of changing v_{∞} and A), as it was obtained for measurements radii-time dependences at room temperatures with different PFM probes (Fig. 2).

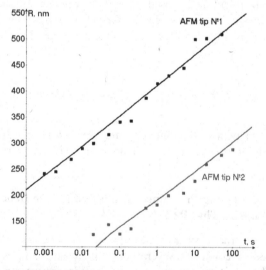

Fig. 2. Two experimental radii-time dependences obtained at room temperature with two different PFM probes of NSG03/Au series.

3.3   Experimental Results

The results of writing nanodomains by voltage pulses of different durations for several temperatures points are shown in Fig. 3. Domains in a bottom row were written by an

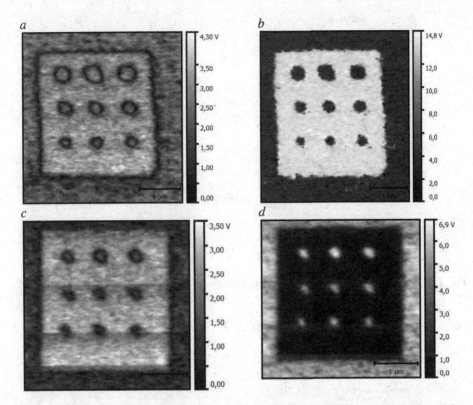

Fig. 3. Written domains in thin $PbZ_{0.3}Ti_{0.7}O_3$ film visualized by PFM: a – amplitude and b – phase of measured PFM signal at T = 300 K; c – amplitude and d – phase of measured PFM signal at T = 120 K.

application of dc voltage pulses with the duration of 1 s, in the middle - by 5 s dc pulses and in the upper row - by 20 s dc pulses. It could be seen, that domain diameter increases with increasing dc pulse duration at both temperatures.

3.4   Theoretical Analysis of the Experimental Results on Temperature Dynamics of Nanodomains in Thin PZT Film

Theoretically calculated with Eq. 3 radii-time dependences for temperatures 300, 200, 100 and 50 K in case of fixed values of v_∞, A and μ in all temperature range is given in Fig. 4.

Experimental results of domain size dependence on the dc pulse duration in the range of 1 ms–100 s for temperatures 120 K and 300 K are presented in Fig. 4a. As it could be seen, experimental data on temperature dynamics of radii-time dependences doesn't correspond to the theoretically predicted one (Fig. 5).

We consider several reasons, which could lead to the discrepancy between expected and observed temperature dynamics of radii-time dependences. First, we take into account changing the dielectric properties of a thin PZT film with temperature.

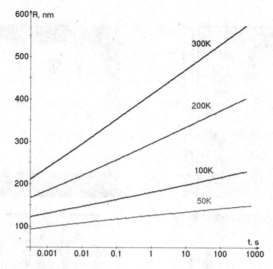

Fig. 4. Theoretically predicted radii-time dependences for temperatures 300, 200, 100 and 50 K.

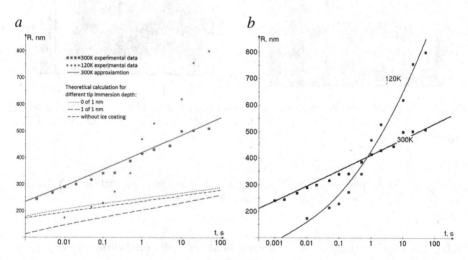

Fig. 5. *a* – the experimental radii-time dependences (rhombus and squares respectively) at 300 and 120 K and its approximation curves with Eq. 3 in assumption of fixed values of v_∞, A and μ; *b* - the experimental radii-time dependences (red and green respectively) at 300 and 120 K and its approximation curves with changing values of v_∞, A and μ. (Color figure online)

This should change an electric potential on a film surface, what, in turn, will influence on the approximation of experimental results with Eq. 3. Nevertheless, modeling this situation didn't improve the situation. Taking into account changing of the PZT film dielectric properties led to the proportional lift of the approximation curves and doesn't change the character of the radii-time dependence. Another possible reason of disagreement between theoretical and experimental results could follow from the presence of an additional layer on the sample surface. It could be a layer of ice formed due to the adsorbed water layer on the sample surface at room temperatures. All measurements were done in a helium atmosphere, not in ultrahigh vacuum, so, we can't except the presence of water layer on the surface. Dielectric properties of an ice layer change [16] with temperature causing the transformation of the "PFM tip – PZT surface" contact properties. We took into account this circumstance in our model (Fig. 1), adding a layer on the film surface with dielectric permittivity different from the film one. Dependence of the dielectric permittivity of this additional layer on temperature was chosen according to the temperature dynamics of dielectric properties of ice [16]. Thickness of the layer was set to 1 nm. Several cases were analyzed: PFM tip is on the layer, PFM tip penetrates on the entire depth of the ice layer. For both situations electric potential distribution on the sample surface was reconstructed and used in the regression analysis. Results of modeling is shown in Fig. 5a. It could be seen, that taking into consideration an additional surface layer on the FE thin film, leads to changing the slope of the approximation curves, but still don't explain the observed experimental results. Thus, we couldn't find a physical reason for the difference in the predictable by theory and experimentally obtained temperature dependence of radii-time dependences.

From mathematical point of view, it was possible to approximate experimental data on temperature behavior of FE domain sizes on the duration of applied dc voltage pulses only in the assumption that all independent variables change with temperature (Fig. 5b). In comparison with the results of the approximation at room temperature, where using different PFM tips led to different radii-time dependences, but for that approximation changing only two of independent variables v_∞ and A was required. For taking into account temperature dynamics of radii-time dependences, all three independent variables have to be altered for satisfactory approximation. Importantly, from the regression analysis of experimental results it is follow that the value of μ parameter changes from 1 at room temperature to 0.5 at T = 120 K. This fact indicates on switching the mechanism which drive the domain wall motion at low temperatures from random-field (at room temperatures) to random-bond scenario. Experimental results of nanodomain writing at 4.2 K give an additional evidence to this conclusion. We found that even at this temperature it was possible to write domain by applying dc voltage pulses between the PFM tip and the bottom electrode of the PZT film. But in this case, experimental radii-time dependences are also not in agree with theoretically predictable ones.

4   Conclusions

In the present work we demonstrate the possibility to write and visualize FE nanodomains in thin $PbZr_{0.3}Ti_{0.7}O_3$ film with a sharp PFM tip in the temperature range starting from 4.2 K and up to room temperature. It was demonstrated that experimental dependences of domain sizes on the duration of applied dc voltage pulse even at room temperatures are strongly influenced by the PFM probe parameters. Due to the dispersion in the PFM cantilever sizes, what is considered to be acceptable for most application of PFM technique, the results of nanodomain writing depend on the PFM tip parameters and are not reproducible. Temperature dynamics of experimental radii-time dependences doesn't follow the theoretically predicted one. We consider this fact as an evidence of changing the mechanism responsible for the domain wall motion at low temperatures.

Acknowledgement. This work of Andreeva N.V. and Vakulenko A.F. was supported by Russian President Grants for young scientists MK-7005.2016.8. The work of Filimonov A.V. was performed under the government order of the Ministry of Education and Science of RF.

References

1. Aizu, K.: Possible species of ferromagnetic, ferroelectric, and ferroelastic crystals. Phys. Rev. B. **2**, 754 (1970)
2. Dubois, M.A., Muralt, P.: PZT thin film actuated elastic fin micromotor. Ferroelectrics Freq. Control **45**(5), 1169–1177 (1998)
3. Luginbuhl, P., et al.: Microfabricated lamb wave device based on PZT sol-gel thin film for mechanical transport of solid particles and liquids. Microelectromech. Syst. **6**(4), 337–346 (1997)
4. Nemirovsky, Y., Nemirovsky, A., Muralt, P., Setter, N.: Design of novel thin-film piezoelectric accelerometer. Sens. Actuators A: Phys. **56**(3), 239–249 (1996)
5. Ledermann, N., et al.: Piezoelectric cantilever microphone for photoacoustic GAS Detector. Integr. Ferroelectrics **35**, 177–184 (2001)
6. Fujii, T., Watanabe, S.: Feedback positioning cantilever using lead zirconatetitanate thin film for force microscopy observation of micropattern. Appl. Phys. Lett. **68**, 467–468 (1996)
7. Park, J.Y., Yee, Y.J., Nam, H.J., Bu, J.U.: Micromachined RF MEMS tunable capacitors using piezoelectric actuators. In: IEEE Microwave Symposium Digest, vol. 3, pp. 2111–2114 (2001)
8. Baborowski, J., Ledermann, N., Muralt, P., Schmitt, D.: Simulation and characterization of piezoelectric micromachined ultrasonic transducers (pMUTs) based on PZT/SOI membranes. Int. J. Comput. Eng. Sci. **4**(3), 471–475 (2003)
9. Bernstein, J.J., et al.: Micromachined high frequency ferroelectric sonar transducers. Ferroelectrics Freq. Control **44**(5), 960–969 (1997)
10. Ganpule, C.S., et al.: Polarization relaxation kinetics and 180° domain wall dynamics in ferroelectric thin films. Phys. Rev. B **65**, 014101 (2001)
11. Tybell, T., et al.: Domain wall creep in epitaxial ferroelectric Pb(Zr0.2Ti0.8)O3 thin films. Phys. Rev. Lett. **89**, 097601 (2002)
12. Pertsev, N.A., et al.: Dynamics of ferroelectric nanodomains in BaTiO3 epitaxial thin films via piezoresponse force microscopy. Nanotechnology **19**, 375703 (2008)
13. Andreeva, N.V., Vakulenko, A.F., Petraru, A., et al.: Low-temperature dynamics of ferroelectric domains in $PbZr_{0.3}Ti_{0.7}O_3$ epitaxial thin films studied by piezoresponse force microscopy. Appl. Phys. Lett. **107**, 152904 (2015)
14. Yan, C., Minglei, Y., Qunying, Z., et al.: Properties of RF-sputtered PZT thin films with Ti/Pt electrodes. Int. J. Polym. Sci. **2014** (2014). Article ID 574684
15. Basu, T., Sen, S., Seal, A., Sen, A.: Temperature dependent electrical properties of PZT wafer. J. Electron. Mater. **45**, 2252–2257 (2016)
16. Kawada, S., Niinuma, J.: Curie-Weiss behavior of the static dielectric constant of the debye component in Ice Ih. J. Phys. Soc. Jpn. **43**, 715 (1977)

Principles of Constructive Synthesis of Electromagnetic Wave Radiators

Roman U. Borodulin[1], Boris V. Sosunov[1], and Sergey B. Makarov[2(✉)]

[1] Military Academy of Telecommunications named after Marshal of the SU
S.M. Budenny, St. Petersburg, Russia
borodulroman@yandex.ru, bsosunov@gmail.com
[2] Peter the Great St. Petersburg Polytechnic University, St. Petersburg, Russia
makarov@cee.spbstu.ru

Abstract. This paper presents the results of the generalization of the problems of the constructive synthesis (CS) and finding of factors influencing the bandwidth of synthesized small electrical size radiators. Definitions of various types of synthesis problems are collected, the definition of the CS is given and tasks of the CS of wideband radiators are defined. Thec paper includes a thorough analysis of small size wideband antennas of various types, discusses the causes of the bandwidth increase that are mainly due to the use of special techniques influencing the bandwidth. Potential capabilities of minimization of dimensions of various types of radiators are shown. Various methods of synthesis are compared using real world problems as examples. Main differences of the CS from parametric and structural-parametric synthesis are given. Strong and weak points of methods proposed by various authors are shown.

Keywords: Constructive synthesis · Structural-parametric synthesis · Geometrical shape of a radiator · Optimization methods · Target function

1 Introduction

Antenna synthesis is an inverse problem of electrodynamics, consisting of finding a system of sources creating electromagnetic field with the required structure [1]. Classical applied problem of radiating system synthesis is to find the field distribution producing the radiation pattern that satisfies the given requirements [2–7].

Therefore, the classical definition of the antenna synthesis problem is – to find the amplitude and phase distributions of current over the aperture of an antenna (internal problem) that creates the desired (given) radiation pattern (external problem).

Constructive antenna synthesis, in contrast with conventional approaches to synthesis, is a pure internal electrodynamics problem – finding of the geometric shape of a radiator that is smaller than one wavelength in size and has the electrical parameters that lies within predefined limits. The limitation on the size is set because large antennas can be classified as continuous or discrete antenna arrays for which the synthesis methods are already well developed.

O. Galinina et al. (Eds.): NEW2AN/ruSMART 2016, LNCS 9870, pp. 717–730, 2016.
DOI: 10.1007/978-3-319-46301-8_63

Currently the constructive synthesis is mainly used to find the current distribution over the surfaces of electrically small wideband radiators for which the natural impedance matching (without the use of special matching circuits) over the very wide operating frequency band (up to a decade) is the most important characteristic.

While many variants (types) of prototype radiators exist, not all of them are suitable for the application of the CS because some of them may have inherently narrow band features in their construction. Also most variants of simple antenna optimization (including geometrical shape optimization) cannot be considered the CS if some specific features of the CS are not present. Therefore, it is important to consider what differentiates the constructive synthesis from other methods.

2 Preliminary Definitions

Currently there are several standard terms describing the synthesis problems.

Parametric synthesis – the process of selecting the parameters of the elements of a synthesized object that satisfy a technical specification [8].

Structural-parametric synthesis – the process of finding the structure of a system and selecting some parameters of the elements of the structure that satisfy a technical specification [9].

Synthesis of radiating systems – the process resulting in finding the physically realizable field distribution over the aperture plane of an antenna of known geometry and type that produces a specified radiation pattern [2].

Synthesis of radiating systems (antenna arrays) includes the solution of several problems.

1. Antenna array synthesis that includes finding the size and shape of an antenna array and the amplitudes and phases of currents in the array elements starting from the specified radiation pattern [4].
2. Constructive antenna array synthesis that includes finding all dimensions of an antenna array that influence the electrical performance and produce a draft or drawing of the array. The term constructive synthesis of radiators implying the use of different mathematical apparatus was introduced to show the difference between the constructive synthesis of antenna arrays and small electrical size radiators.

None of the above definitions is suitable to describe the synthesis of electrically small antennas. Structural-parametric synthesis has a goal of finding the structure (construction) of an object and values of the elements of this structure and seems similar to CS.

Synthesis of radiating systems including CS is represented only by the antenna arrays. To avoid confusion, it is possible to formulate the problem of CS of wideband radiators in the similar way.

The task of constructive synthesis of wideband radiators can be formulated as follows. The type of a radiator is known, the requirements to its size, bandwidth and SWR are given. The task is to find the design of an antenna with such distribution of current that achieves the required SWR for a given impedance of a feeder.

Unlike the classical theory of radiating system synthesis in case of constructive synthesis of wideband radiators there are limitations on (radiation pattern) because of fundamental limitations existing for all electrically small radiators. These antennas may only have low directivity. Impedance matching with a feeder is important for electrically small antennas.

Currently constructive synthesis is used to modify some structural regions of an antenna of a given type that are mainly responsible for its natural matching with a feeding transmission line.

3 Design Features of Wideband Antennas Used in Constructive Synthesis

The task of increasing the bandwidth of antennas is very important. Usually at the same time it is desirable to decrease the size of an antenna in at least one of its spatial dimensions. This is achieved using various optimization algorithms or by simple variation (often based on intuition) of one or several structural parameters and measuring the influence of these changes on the reflection coefficient and trying to achieve good natural matching. It is interesting to examine wideband antenna designs in search of hidden potential capabilities helping to reduce their size if necessary.

For example [10], describes an antenna consisting of a monopole surrounded by dielectric rectangular bars forming a bird's nest like structure. The antenna is naturally matched over 4.2...6.7 GHz frequency band. Its dimensions are about 0.25*0.8 of the average wavelength. It has a conical radiation pattern with relatively constant shape at all operating frequencies and circular polarization.

The antenna is excited by a 13.5 mm high monopole. Since the average wavelength is 55 mm the monopole is about one fourth wavelength high. It is likely that the widening of the frequency band is achieved by placing dielectric slabs in the near field region of a monopole. Secondary field created by a dielectric spiral has such a strong influence on the antenna's input that the bandwidth of the whole system is significantly increased, the polarization becomes circular and the radiation pattern becomes similar to the one of a spiral antenna.

Here we can see the application of one of the fundamental properties of wideband antennas – the independence of angular shape from linear scaling which is used in spiral antennas. This feature is described in [11]. From the point of view of CS, it is also evident that the excitation by a narrow band monopole is limiting the potentially possible bandwidth and thus it is unsuitable as a CS prototype. Thus this peculiarity of design results in a narrowband effect limiting the bandwidth to an unimpressive figure.

Another approach to bandwidth increase is to intuitively add various distributed capacitive (or inductive) loads to an antenna. For example [12], presents a slot antenna etched in a conductive plate operating over 1.16...2.09 GHz frequency band.

Probably the bandwidth increasing effect in the initially narrowband (slot) antenna is achieved by adding a reactive load on the top. The influence of the load provides twofold increase in bandwidth. Here we see that the slot antenna itself is limiting the bandwidth but the effect of the top load is very instructive.

Often reactive loads are combined with the ultra-wideband parts of a system thus intuitively increasing the electrical length of an already wideband antenna. An example of this approach can be found in [13] where an antenna consisting of a conductive body of revolution and a parasitic ring shunted to a plane of the finite size is presented. The 2.15…14 GHz bandwidth is impressive, the size is relatively small (the height is 10 mm), the radiation pattern is similar to a monopole's.

Wideband effect is present from the very beginning because of use of an ultra-wideband element working together with the large size (compared to the wavelength) top capacitive. However, this is not enough and additionally the excitation region is separated by a circular slot reactively influencing the field. Finally, the addition of a 3-dimensional wideband monopole in the center amplifies the wideband effect which is improved by the top load. The 'lengthening' effect of the top load results in lowering of a low frequency limit, the cone shape of the central element results in the increase of a high frequency limit. The lowering of the input impedance is achieved by the widening the monopole, additional shunt 'inductances' parallel to the input of the antenna decrease the imaginary part of the input impedance. Here the three design features are mutually increasing the wideband effect. These features are a 3-D monopole (most important), top reactive load and shunt loads. However, in this case the CS also most likely will not be able to help to improve the design since the presence of a ring around the monopole is a narrowband factor. It is likely that the authors have reached the wideband limits of such a structure.

Some publications present good solutions that can be achieved relatively quickly by applying the CS. One example is [14] where three low profile unidirectional antennas are investigated. They have 2…15 GHz bandwidth and reflection coefficient of −10 dB. The length of all antennas is 140 mm, the sizes of the top side of the triangle are 58, 100 and 166 mm. The angles near the feed point are 60, 90 and 120°.

This is a curved planar triangular antenna with remote ground plane (the electrical distance to the shield is practically constant at any frequency) giving stable unidirectivity. This is a typical example of the requirements to the CS: to increase the bandwidth keeping the radiation pattern constant. The authors achieved optimal design by varying two parameters – the angle near the feed point and the radius of curvature of the loop formed by the ground plane and the body of the radiator – which can be done very fast if the CS is used. Unlike the previous example the use of a ground plane here is justified since it is necessary to achieve unidirectivity.

Three main design features helping to naturally match an antenna with a transmission line can be found from the above examples: the presence of a wideband element (equiangular, self-complementary), top capacitive load lowering the frequency and reactive elements (shunts, rings, slots) placed close to the radiator. Bad design features include the presence of narrowband elements connected to a feed line, fixed shields, etc.

It is reasonable to ask how the CS can be differentiated from parametric and structural-parametric synthesis. Can any optimization be called the CS? This question will be discussed in the next section.

4 Examples of Automated Design of Wideband Antennas that are Not Constructive Synthesis

Optimization of two-arm Archimedean spiral antenna in [15] is an example of the parametric synthesis. The optimization goal is to achieve distortionless radiation of an ultra-wideband pulse at any angle. The variable is: the angle between the radius of a spiral and the tangent to its surface. The structure in this case is a spiral, the parameter is the angle, and the result is the radiation pattern.

A good example of the structural-parametric synthesis is described in [16]. It is possible to see the main difference of this synthesis from CS.

The goal is to use the method of moments (method of analysis) and the genetic algorithm (method of synthesis) to get the output current in the form of an ultra-wideband pulse proportional to the incident field and at the same time achieve the low sidelobe level of the radiation pattern.

Here the structural-parametric synthesis of an ultra-wideband antenna with resistive insets is performed. The variable parameters are the values of the resistors. The structural parameter is the distance between the legs of the antenna. Without the resistors that would have been the constructive synthesis.

One of the first articles showing the transition from parametric to constructive synthesis was written by Johnson and Samii [17] and discusses the combination of the method of moments and the binary genetic algorithm applied to the search for the optimal design of a microstrip antenna (MSA). In the paper the design of a prototype antenna – a solid rectangular plate above the ground plane and naturally matched over a very low frequency band – is optimized to achieve double band matching at two frequencies selected by the authors. Modification of the MSA consisted of creating holes in it based on the results of work of the genetic algorithm. The objective function included several frequency points near the desired frequencies in which the minimization of the reflection coefficient was performed. In this case it is possible to find some distributed "RLC-parameters" that can be varied. It may sound strange but they are actually the holes of a fixed size made in random places. If the position of even one hole is changed the structure (shape) of the whole antenna is changed and its properties are significantly altered. If the mutual position of several holes is changed then it is hard to predict what shape of the prototype will give the desired result and the synthesis can be called construction synthesis.

The parametric synthesis of antennas is the simplest type of synthesis producing poor results. The influence of the structural-parametric synthesis of antennas is more complex. It can produce good results, but requires more time and resources since the number of controlled variables is small. Constructive synthesis is the most effective. It assumes that the structure of the antenna which does not have any lumped parameters will be changed. Since it has fundamental influence on all the distributed parameters of an antenna it produces good results in shorter time provided that the right prototype is selected.

Therefore, the main differences of CS from other types of synthesis are: the absence of lumped parameters, optimization algorithm changes many distributed parameters described by independent variables. These parameters can influence the target characteristic of an antenna (for example its matching with a feeding line over some frequency band). If the size of an antenna is increased to more than one wavelength CS transforms into constructive synthesis of continuous antenna arrays.

5   Examples of the Constructive Synthesis Problems

The term constructive synthesis allows differentiating between the synthesis of an electrically small antenna and the synthesis of a large radiating aperture.

The paper [18] presents the constructive synthesis of the geometrical shape of an asymmetric antenna. The shape is defined by several points connected by a spline. The shape of the rectangle playing the role of the ground plane is also varied. Authors call their method the synthesis of geometry.

1. There is a prototype antenna (a monopole with the top load);
2. There is a combination of analysis and optimization methods;
3. There are many distributed parameters defining by the antenna's shape;
4. There is an objective function – the absolute value of the reflection coefficient;
5. There are additional indirect characteristics influencing the main one – transfer coefficient of the communication channel and group delay;
6. The size of the antenna is limited to keep it electrically small.

The authors formulated the objective function containing the equations for the target characteristic of the antenna and additional characteristics describing the communication channel. All this and the limitations on the bandwidth and control variables' values reduced the search space significantly. They analyzed the dependence of 'quality indexes' on the number of control variables and found the 'optimal' number of control points for this geometrical shape (the number is 7).

From our point of view, the problems of this work leading to a poor result (the large size of the synthesized antenna – the height is 0.7λ average, the narrow frequency band (the frequency ratio $f_{max} / f_{min} = 2.25$)) are:

1. Bad selection of the prototype in which the features potentially widening the frequency band have little influence (top load, capacitance of the low part of the antenna). The spline used by the authors cannot produce for example a shunt, also it is impossible to achieve self-complementarity and current cut-off features;
2. Bad selection of the optimization method that works poorly in case of a large number of controlled variables;
3. Bad idea to formulate a part of the objective function in the form of an integral. The precision of numerical integration depends on the number of points used and since the frequency band is wide (2…9 GHz) the number of points may be large.

Slightly other goals were set in the paper [19] where the CS of an antenna of a simple geometrical shape – an isosceles triangle is presented (Fig. 1). Here some features of the CS are present, including the following:

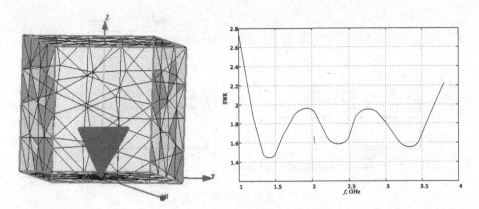

Fig. 1. The model of a triangular antenna inside the solution box and its bandwidth.

1. There is a prototype;
2. There is a combination of the finite element method (method of analysis) and Quasi-Newton method of optimization;
3. There are two independent control variables defining the width and height of the triangle;
4. There is an objective function – the limited SWR;
5. The size of the antenna is limited to keep it electrically small.

The result of the synthesis is the ratio of high to low frequency of three with the minimization of the triangular monopole height without any limitations on its width. The positive effect of using a small number of parameters for a triangular flat antenna of a limited size was shown.

The results proved the wideband efficiency of the prototype that has specific constructive features potentially influencing the bandwidth (Fig. 2). The sum of these features led to the desired geometrical shape, confirming in practice the independence of the angular parameter for this type of an antenna since the optimization algorithm set the angle at the feed point close to 90°. The current cut-off effect was also demonstrated (the limitation on height resulted in the limitation on the lowest operating frequency). The large size of the final structure (height 0.9λaverage) showed low efficiency of using a small number of control variables.

The benefit of the presented approach is the possibility of using fast optimization methods to synthesize structures described by a low number of independent control variables.

The same results could have been achieved by the structural-parametric synthesis by taking the coordinates of the corner point at the top of the radiator as a variable parameter. In our case the problem suitable for structural-parametric synthesis was solved by the CS which shows backward compatibility of the methods.

These results were taken into account in the paper [20] in which a planar antenna with a complex shape consisting of a triangular base and two trapezoidal segments was investigated (Fig. 2).

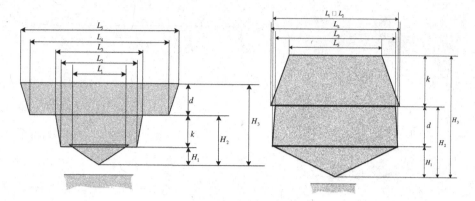

Fig. 2. The prototype and the final shape of the planar antenna.

Eight independent control variables were selected. The shape of the trapezoids allows achieving current cut-off producing overlapping sub-bands. The goal of the synthesis was to achieve the minimum height of the planar antenna having the highest to lowest frequency ratio of three with geometrical shape defined by a large number of control variables. All the features of the constructive synthesis listed above are present. Additionally, the indirect characteristic, influencing the main characteristic (limited SWR over a frequency band), that is the antenna's height was taken into account. After 287 iterations of the genetic algorithm sufficiently good results were obtained (Fig. 3). By the way several good shapes were obtained but the most practical was selected (Fig. 4). The height of the synthesized antenna is $0.375\lambda_{average}$, the maximum width – $0.4\lambda_{average}$. The radiation pattern in the high frequency part of the frequency band was checked after the synthesis: it was close to a toroidal shape that is characteristic for monopole antennas which was to be expected for a small size antenna.

The drawback of the method is again the form of the objective function, requiring many frequency points and thus influencing the duration of the optimization.

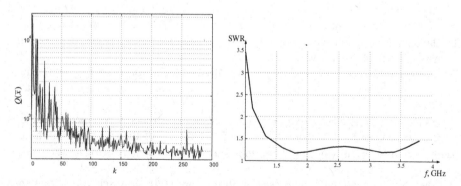

Fig. 3. The objective function vs. the number of iterations of the genetic algorithm (left) and the SWR over a frequency range.

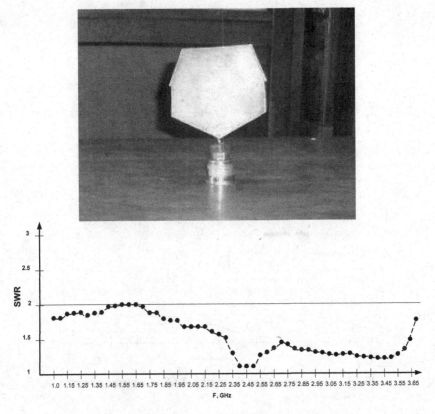

Fig. 4. The synthesized antenna and its SWR over a frequency range.

It may seem that the CS is good only for the synthesis of flat structures but this is not so. The CS of a more complex antenna was performed in [21] using the finite element method and genetic algorithm (Fig. 5).

The goal was to increase the bandwidth of a unidirectional antenna intended for use in the antenna array of a base station and consisting of several flat panels. The base station serves an area where customers may be at any position. The area is formed by the radiation patterns of all the panels. Therefore, the size of array element in this case is very important (see Table 1).

The task is typical for the CS but there is one obstacle – the presence of the ground plane. However, it is impossible to get rid of it in a unidirectional antenna. The positive factors are the presence of the distributed reactive elements and a large number of control variables. The principles of self-complementarity, equiangularity and current cut-off were not used and the bandwidth increase was small (Fig. 5). During the optimization the size of the antenna was reduced (Table 2) but the large number of frequency points used in the analysis increased the use of resources.

The construction synthesis is not limited to the presented examples because it can also be used to take into account the influence of a large object on which the electrically

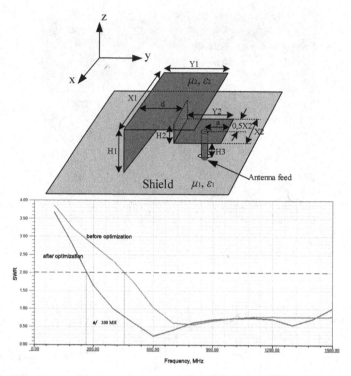

Fig. 5. The prototype and the results of CS of a planar antenna.

Table 1. The ranges of change of controlled variables.

X vector component	Geometric parameter of synthesizable construction	Interval and step of synthesis
h_1	H1	Const
h_2	H2	20–45 mm with step 7.5 mm
h_3	H3	5–20 mm with step 5 mm
x_1	X1	60–180 mm with step 15 mm
x_2	X2	30–75 mm with step 10 mm
y_1	Y1	40–90 mm with step 15 mm
y_2	Y2	30–75 mm with step 10 mm
d	d	30–70 mm with step 10 mm

small antenna is installed on the antenna performance. For example [22], examines a radiator used on a railway carriage.

The task was to place a wideband omnidirectional antenna under the roof of a railway carriage and that put significant limitations on the antenna's height. The CS method used was based on the combination of the time domain finite difference (FDTD) method and gradient Quasi-Newton optimization algorithm. The problem was to achieve an omnidirectional radiation pattern of an antenna installed on an object

Table 2. The synthesis result.

Geometric parameter synthesizable construction	Size before optimization	Size after optimization
H1	Const (65 mm)	Const (65 mm)
H2	40 mm	27.8 mm
H3	15 mm	9 mm
X1	115 mm	108.5 mm
X2	50 mm	31.5 mm
Y1	75 mm	54.7 mm
Y2	50 mm	32 mm
d	60 mm	42.5 mm

influencing its radiation pattern. At the same time the antenna must have the wide frequency band. A conical antenna with shunts was taken as a prototype (Fig. 6). All features of the CS are present.

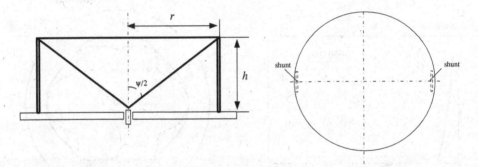

Fig. 6. A conical antenna with shunts placed under the roof of the railway carriage and its distributed parameters that are subject to optimization.

The required shape of the antenna was determined by varying the parameter ψ – the angle near the feed point and by minimization of the parameter h – the height of the antenna (Fig. 7). The shunts here are adding mechanical strength and slightly flatten the radiation pattern compensating for the influence of a long carriage.

The optimization was repeated producing the wideband antenna with improved radiation pattern (Fig. 8). Here is an example of excellent effect of 'helping' factors in case of limited search space and limited CPU power. This fact shows the effectiveness of the proposed method of the CS.

All the above examples were tested experimentally using real antennas and the results were better than expected. In the area of ultra-wideband (UWB) antennas good results were also achieved (ten times ratio of the highest to the lowest frequency) that cannot be presented here due to size limitations. One thing is for sure: the application of the formulated principles of the CS can significantly speed up the effective search for optimal design of an electrically small antenna. However, one significant drawback is present: the need to take into account the large number of frequency points in an objective function during the CS process.

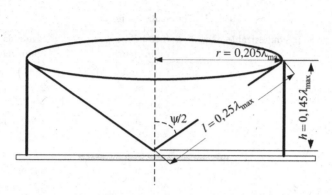

Fig. 7. Transformation of the synthesized conical antenna

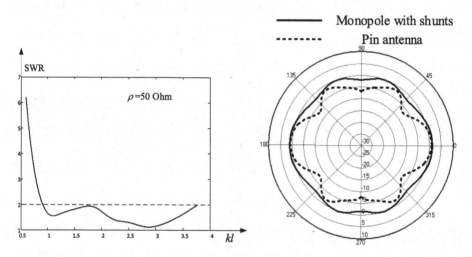

Fig. 8. The SWR over a frequency range and the radiation pattern of a monopole with shunts placed under the roof of the railway carriage at $f = 30$ MHz

6 Conclusion

Definitions of various types of synthesis techniques that find use in antenna synthesis are given in the paper. The definition of the constructive synthesis is given and the task of constructive synthesis of wideband radiators is formulated. The main difference from constructive synthesis of antenna arrays is shown. The difference is in the small electrical size that leads to the change in mathematical tools.

Various wideband antennas were analyzed in order to find main factors that can be used in synthesis to produce quantitatively better result in less time and using less computational resources. The factors include the independence of angular parameters of linear size, the principle of self-complementarity, the principle of current cut-off and the influence of reactive impedances.

The differences among the parametric, structural-parametric and constructive synthesis of wideband antennas are shown.

Advantages and drawbacks of the constructive synthesis of antennas of various types and frequency bands are shown as well as strong and weak points of used algorithms and approaches. The possibility of the reduction of the dimension of a search space using the main factors to which the synthesis algorithm is applied is shown.

Based on the above analysis the main principles that can be used to discriminate the constructive synthesis from other synthesis methods are formulated.

The presented methodology can be of significant help to developers of electrically small wideband antennas and, therefore, it has significant practical value.

References

1. Stutzman, W.L., Thiele, G.A.: Antenna Theory and Design. Wiley, Hoboken (2012)
2. Bakhrakh, L.D., Krementskiy, S.D.: Synthesis of Radiating Structures (Theory and Methods). Soviet Radio, Moscow (1974). (in Russian)
3. Bakhrakh, L.D., Litvinov, O.S.: The place of aperture synthesis in general antenna theory (review). Radiophys. Quant. Electron. **26**, 953–962 (1983)
4. Chaplin, A.F.: Analysis and Synthesis of Antenna Arrays. IRE AN SSSR, Moscow (1984). (in Russian)
5. Michelson, R.A., Schomer, J.W.: A three parameter antenna pattern synthesis technique. Microw. J. **8**, 88–94 (1965)
6. Safaaijazi, A.: A new formulation for the design of Chebyshev arrays. IEEE Trans. Antennas Propag. **42**, 439–443 (1994)
7. Smith, W.T., Stutzman, W.L.: A pattern synthesis technique for array feeds to improve radiation performance of large distorted reflector antennas. IEEE Trans. Antennas Propag. **40**, 57–62 (1992)
8. Abramov, O.V.: Parametric Synthesis of Stochastic Systems with Reliability Considerations. Nauka, Moscow (1992). (in Russian)
9. Akimov, S.V.: Investigation and development of methods of structural-parametric synthesis of microwave linear transistor amplifiers. Ph.D. SPbGUT (2002). (in Russian)
10. Leung, K.W.: Wideband circularly polarized dielectric bird-nest antenna with conical radiation pattern. IEEE Trans. Antennas Propag. **61**, 8 p. (2013)
11. Rumsey, V.: Frequency independent antennas. 1958 IRE Int. Convention Rec. **5**, 114–118 (1957)
12. Yongho, K., Morishita, H., Horiuchi, S., Atsumi, Y., Ido, Y.: Wideband planar folded dipole antenna with self-balanced impedance property. IEEE Trans. Antennas Propag. **56**, 6 p. (2008)
13. Iwaoka, H., Morishita, K., Yamauchi, J.: A wideband low-profile antenna composed of a conducting body of revolution and a shorted parasitic ring. IEEE Trans. Antennas Propag. **56**, 6 p. (2008)
14. Kishk, A.: A novel low-profile compact directional ultra-wideband antenna: the self-grounded bow-tie antenna. IEEE Trans. Antennas Propag. **60**, 7 p. (2012)
15. Radway, M.J., Filipovic, D.S.: Frequency-and time-domain performance of four-arm mode-2 spiral antennas. IEEE Trans. Antennas Propag. **60**, 8 p. (2012)

16. Chang-Hoi, A., Carin, L.: Nonuniform frequency sampling with active learning: application to wide-band frequency-domain modeling and design. IEEE Trans. Antennas Propag. **53**, 9 p. (2005)
17. Johnson, J.M., Rahmat-Samii, Y.: Genetic algorithms and method of moments (GA/MOM) for the design of integrated antennas. IEEE Trans. Antennas Propag. **47**, 1606–1614 (1999)
18. Lizzi, L., Viani, F., Azaro, R., Massa, A.: A PSO-driven spline-based shaping approach for ultrawideband (UWB) antenna synthesis. IEEE Trans. Antennas Propag. **56**, 2613–2621 (2008)
19. Borodulin, R.Y., Sosunov, B.V.: Constructive synthesis of phased array elements. Radiotechnics, antennas, microwave devices. Scientific-Technology Proceedings of SPbGPU **2**(169), pp. 47–53 (2013). (in Russian)
20. Borodulin, R.Y., Lukyanov, N.O., Sosunov, B.V.: Constructive synthesis of a planar radiator by genetic algorithm. Inf. Space Mag. **4**, 4–8 (2014). (in Russian)
21. Sosunov, B.V., Sheyanov, D.Y.: Constructive synthesis of planar radiators. Modern Theoretical and Applied Research Directions, pp. 44–46 (2012). (in Russian)
22. Borodulin, R.Y., Lukyanov, N.O., Sosunov, B.V.: Constructive synthesis of a low profile VHF band radiator for railway transport. Inf. Space Mag. **1**, 4–8 (2015). (in Russian)

Investigation of Electro-Physical and Transient Parameters of Energy Accumulating Capacitors Applied in Nanosecond and Sub-nanosecond High-Current Avalanche Switches

V.E. Zemlyakov[1], V.I. Egorkin[1], S.N. Vainshtein[2], A.V. Maslevtsov[3], and Alexey Filimonov[3(✉)]

[1] National Research University of Electronic Technology-MIET,
Bld. 1, Shokin Square, Zelenograd, Moscow 124498, Russia
vzml@rambler.ru, egorkin@qdn.miee.ru
[2] Electronics Laboratory, Department of Electrical Engineering,
University of Oulu, P.O. Box 4500, 90014 Oulu, Finland
vais@ee.oulu.fi
[3] Peter the Great St. Petersburg Polytechnic University,
Polytechnicheskaya 29, St.-Petersburg 195251, Russia
avm@spbstu.ru, filimonov@rphf.spbstu.ru

Abstract. Peak power of nanosecond and sub-nanosecond high-power pulsed optical transmitters was found to be drastically affected not only by the speed of the avalanche switch, but also by structure and geometry of the capacitive energy accumulator, and assembly construction. Together with trivial effect of parasitic inductance of the entire switching loop, it was found that some capacitors, even those designed for microwave use, are unable to release the charge in sub-nanosecond time scale when large signal operation (high current) is required. The problem was successfully solved using specially developed surface-mounted capacitors utilizing high-quality silicon nitride using plasma-deposition method. Finally miniature capacitors permitting direct assembling with avalanche transistor chips and laser diode with minimum possible (~ 1 nH) parasitic inductance withstand up to 600 V and allow nanosecond 20–30 A current pulse generation and optical pulses exceeding 30 W.

Keywords: Capacitors · Nanosecond switching transient · Electro-physical parameters · Plasma deposition · Miniature assembly

1 Introduction

There is a need to generate current pulses of a few nanoseconds in length with an amplitude of ~ 10–10^2 A across a low-ohmic load for a number of commercial applications. This concerns particularly the pumping of high-power broad-stripe laser diodes for laser radars and 3-D imaging utilizing time-of-flight (TOF) technique [1, 2]. One of the simplest, cheapest and most effective ways is to make use of high-voltage

O. Galinina et al. (Eds.): NEW2AN/ruSMART 2016, LNCS 9870, pp. 731–737, 2016.
DOI: 10.1007/978-3-319-46301-8_64

(~ 300 V) avalanche transistors [3, 4], which is the state-of-the art in optical radars (lidars) utilizing 3–10 ns in duration optical pulses from the laser diodes in 10–100 W power range. This duration of the optical pulse becomes a bottleneck of the lidars aiming at longest possible ranging range (preferably several kilometres) with high (around a decimetre) ranging precision. Indeed, the bandwidth of receiving channels based on avalanche detectors exceeds nowadays 300 MHz, and thus even 1 ns in duration optical pulses can be easily detected without noticeable reduction in the detector sensitivity and critical growth in the noise level. Peak power of nanosecond and sub-nanosecond high-power pulsed optical transmitters was found to be drastically affected not only by the speed of the avalanche switch, but also by structure and geometry of the capacitive energy accumulator, and assembly construction. Together with trivial effect of parasitic inductance of the entire switching loop, it was found that some capacitors, even those designed for microwave use, are unable to release the charge in sub-nanosecond time scale when large signal operation (high current) is required. This problem is typical of the capacitors utilizing dielectrics with high permittivity, while using "good" dielectric free of relaxation processes in nanosecond range creates the problems associated with relatively high parasitic inductance intrinsic of the capacitors of large area. Additionally a requirement of relatively high biasing voltage diminishes specific capacitance per unite area. Resolving this trade-off is a challenging task, which requires experimental and technological research using high-speed, high-current switches and various capacitors, rather than a routine selection of capacitors from different manufacturers trusting datasheets on frequency-dependent small-signal dielectric losses. An example illustrating the trade-off discussed above is shown in Fig. 1. The size of surface-mounted 1.25 mm $\times$ 1.25 mm capacitor chips is smaller than the size of multilayer NP0 (2 mm in length) capacitors, and accordingly parasitic inductance of the circuit loop including NP0 (curves 1a and 2a), found to be about 5.4 nH, is larger than that of 4 nH for COMPEX (curves 1b and 2b). Accordingly one could expect larger in amplitude and shorter in duration pulses for COMPEX capacitors (1b and 2b). Just the opposite is the case, however: the amplitude of the waveform 1a is larger than that of 1b for 270 pF capacitors, and for 120 pF pulse amplitude for COMPEX (2b) is not larger than for NP0 (2a). Furthermore, smaller pulse duration expected for smaller inductance has not been observed, and even relaxation oscillations well pronounced for NP0 (1a and 2a) are not visible in waveforms 1b and 2b: a "tail" in the current manifests itself instead meaning that COMPEX capacitors are unable releasing the accumulated charge in nanosecond scale, at least in large-signal mode.

The stated problem has been solved in this work using the capacitors manufactured by planar plasma-chemical deposition (PECVD) technology. Chemical reaction $SiH_4 + NH_3 + N_2 = Si_xN_yH_z + H_2$ under condition of radio-frequency plasma discharge at 200–400 °C is typically used for manufacturing high-quality dielectric films combining high specific capacitance with high breakdown voltage. Previous experience shows significant effect on the film parameters of the deposition conditions, such as substrate temperature, feed rate of reactants into the chamber and the ratio of reactants. Experimental selection of plasma chemical deposition parameters allows solving the problem of masking [5], passivation [6] and permits controlling the optical, chemical and electrical properties of the dielectric. In this study, we investigated the possibility

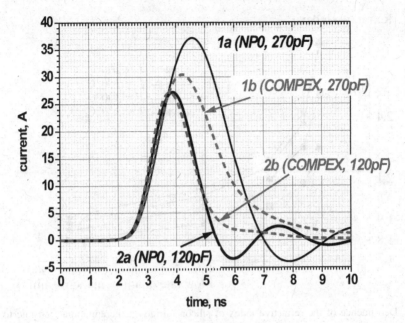

Fig. 1. Current pulses measured in a low-inductance loop involving 1 ohm load, an avalanche transistor and two different capacitors: multilayer ceramic capacitors of 2 mm in length with NP0 ceramic type, and surface-mounted COMPEX (types C-100 and C-120, 1.25 × 1.25 mm) capacitors.

of such a plasma chemical adjustment of the breakdown voltage and the specific capacitance to build on the basis of the PECVD method capacitors to be used in miniature pulsed nanosecond optical transmitters.

2 Samples Preparation and Measurement Procedure

Commonly, silicon-nitride films are deposited by the plasma-chemical method via the gas-phase reaction yielding silicon nitride ($SiH_4 + NH_3 + N_2 = Si_xN_yH_z + H_2$), which occurs under plasma-discharge conditions at substrate temperatures of 200–400 °C. It should be noted that silicon nitride obtained under the deposition conditions recommended by the equipment manufacturer demonstrated an unexpectedly high etching rate in the buffer etchant, despite a good refractive index (~ 1.95). The good optical properties of the insulator do not guarantee its quality as regards the stoichiometry and presence of impurities responsible for the high etching rate.

In [7], the composition of plasma-chemical silicon nitride was discussed and the possibility that it contains hydrogen (up to 40 at.%) chemically bonded to silicon and nitrogen atoms in various proportions was noted. To verify this hypothesis, we fabricated a set of samples with varied technological-process parameters. Silicon-nitride films with a thickness of 0.2 μm were deposited onto silicon wafers. The deposition temperature (200–320 °C) and the flow-rate ratio of monosilane and ammonia

Refractive index

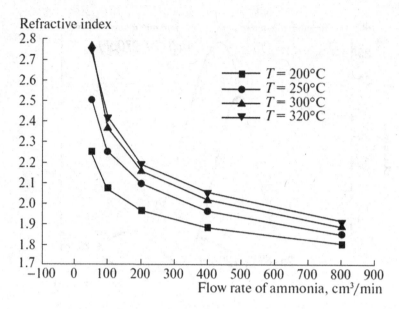

Fig. 2. Dependence of the refractive index of silicon nitride on the ammonia feed rate to the reactor.

(0.14–1.3) were varied. The discharge power and the working pressure in the reactor were maintained constant. Figure 2 shows how the refractive index of a silicon- nitride film depends on the flow rate of ammonia at a constant flow rate of monosilane at various temperatures. It can be seen that the optical constants of the insulator greatly depend on the technological process parameters. Next, we investigated the dependence

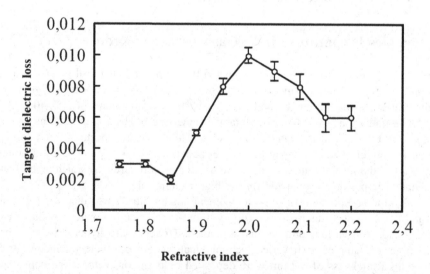

Fig. 3. The dependence of the dielectric loss tangent of the refractive index of the silicon nitride.

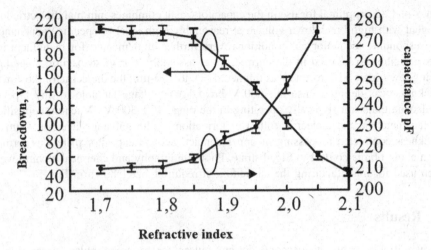

Fig. 4. The dependence of specific capacitance corresponding to sample area of 1 mm^2 (right) and the breakdown voltage (left) for the dielectric film thickness of 0.2 mkm.

of the breakdown voltage, the specific capacitance and dielectric loss tangent of the refractive index shown in Figs. 3 and 4.

Fairly complicated relation between breakdown voltage and specific capacitance is apparently associated with a configuration of hydrogen bonds in the molecule of silicon nitride [7]. It is obvious from the dependences shown in Figs. 3 and 4 that refractive

Fig. 5. Photo capacitor chip

index of 1.85 is optimal for use in the capacitors, as it combines minimal dielectric loss tangent with high breakdown voltage at reasonably high level of specific capacitance. Corresponding technological conditions were further implemented for manufacturing the capacitors to be used in the current driver assembly. For the capacitors manufacturing low-ohmic Si substrate was thinned down to 100 μm, the dielectric of 0.6 μm in thickness was deposed ensuring 600 V breakdown voltage suitable to any type of avalanche transistor (typically operating in the range 100–300 V). Vacuum deposition of the metal on the dielectric followed formation of the galvanic layer of 3 μm in thickness completed processing of upper contact. Second capacitor plate was formed by making ohmic contact to Si substrate. Photolithography and chemical etching were then used for manufacturing the capacitors of required size (shown in Fig. 5).

3 Results

Shown in Fig. 6 are nanosecond current pulses for pumping pulsed laser diodes achieved using the capacitors developed here. The results obtained using other commercially available capacitors are shown for sake of comparison. One can see that specially developed capacitor allows unique current pulse above 60 A in amplitude to be achieved at very moderate pulse duration of 2 ns, which is significantly higher than that achievable from NP0 due to large parasitic inductance intrinsic of technology of this capacitor. Unlike that COMPEX surface-mounted chip capacitor allows as low parasitic inductance as 1.7 nH to be achieved in the assembly. Used there dielectric quality does not suite the task, however, thus broadening the pulse up to 2 ns even at as low capacitor value as 290 pF and inductance of 1.7 nH. Also the current pulse amplitude remains fairly low due to dielectric properties.

Chip of 300 V avalanche transistor was used in an assembly with surface-mounted load resistor and a capacitor. The capacitor values for each capacitor type were selected

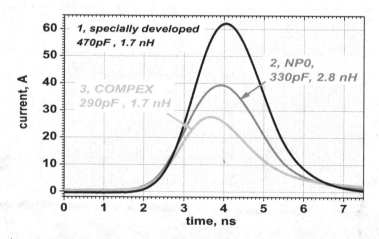

Fig. 6. Current waveforms recovered numerically from the voltage measured across 1 Ω load resistor on account of its parasitic 0.8 nH inductance.

for obtaining the same duration of the current pulse of 2 ns. Specially developed here surface mounted capacitor (curve 1) as well as COMPEX (curve 3), allows the parasitic inductance of the circuit loop to be diminished down to ~1.7 nH. Multilayer ceramic capacitor NP0 increases the total parasitic inductance to 2.8 nH due to its geometry and large size.

4 Conclusions

We thus conclude that no of commercially available capacitors can be recommended for nanosecond, high-power laser transmitter due to insufficient dielectric quality or large parasitic inductance due to large size and non-optimized construction. Specially developed here capacitor fits the task best and can be recommended for use in high-power transmitter when the pulse duration below 3 ns is required, which seems to be optimal for long-distance decimeter-precision radars.

This work was performed under the government order of the Ministry of Education and Science of RF.

References

1. Biernat, A., Kompa, G.: Powerful picosecond laser pulses enabling high-resolution pulsed laser radar. J. Opt. **29**, 225–228 (1998)
2. Kilpela, A., Kostamovaara, J.: Laser pulser for a time-of-flight laser radar. Rev. Sci. Instr. **68**, 2253–2258 (1997)
3. Herden, W.B.: Application of avalanche transistors to circuits with a long mean time to failure. IEEE Trans. Instrum. Meas. **IM-25**, 152–160 (1976)
4. Streetman, B.Q.: Solid State Electronic Devices, pp. 344–346. Englewood Cliffs, Inc., New York (1972)
5. Garmash, V.I., Egorkin, V.I., Zemlyakov, V.E., Kovalchuk, A.V., Shapoval, S.Y.: Semiconductors **49**(13), 1727–1730 (2015)
6. Yermolayev, D.M., Zemlyakov, V.E., Shapoval, S.Y., Marem'Yanin, K.M., Morozov, S.V., Gavrilenko, V.I., Fateev, D.V., Popov, V.V., Maleev, N.A., Sizov, F.F.: Solid-State Electron. **86**, 64–67 (2013)
7. Kovalchuk, A., Beshkov, G., Shapoval, S.: J. Res. Phys. **31**, 37 (2007)

NEW2AN: Economics and Business

Evaluating the Efficiency of Investments in Mobile Telecommunication Systems Development

Tatyana Nekrasova, Valery Leventsov[(⊠)], and Ekaterina Axionova

Peter the Great St. Petersburg Polytechnic University, Saint Petersburg, Russia
dean@fem.spbstu.ru, vleventsov@spbstu.ru,
director@eei.spbstu.ru

Abstract. The paper gives insight into the importance of economic evaluation of investments in Russian telecommunication companies. Qualitative case studies are based on scenario approach with risk analysis. The results include: risk classification of telecommunication companies in Russia, quantitative risk assessment; non-systematic risk premium assessment for mobile telecommunication services in Russia; quantitative evaluation of the major indicators of investment efficiency for telecommunication services.

Keywords: Telecommunication companies · Investment costs · Investment efficiency evaluation · Risk assessment · Simulation modeling

1 Introduction

At the moment there is serious progress in development of the Russian telecommunication network. Moreover, mobile communication systems experience an outstripping rate of growth. Appearance of mass market, which uses services of telecommunication companies, does not only require supporting the old ones, but also applying new marketing and technology approaches. Hence, research into investment activities aimed at development of the telecommunication infrastructure is becoming really important [3, 7, 9, 10].

The following ones are used as indicators of telecommunication investment efficiency: discounted payback period (PB), internal rate of return (IRR), net present value (NPV), profitability index (PI) [2, 4–6, 8].

None of these indicators individually is sufficient to determine efficiency of a project, so a decision about investment in the project must be taken given all the indicators.

When defining the efficiency of investment in mobile telecommunication systems, on-coming costs and performance are estimated within the limits of the calculation period (time horizon), which is 8 years in length [1, 7, 11]. The selected length of the calculation period correlates with the life of the core process equipment of the mobile network system and adequately reflects the changing trends in investment activities of mobile telecommunication companies, allowing estimating the projected values of the investment efficiency indicators.

© Springer International Publishing AG 2016
O. Galinina et al. (Eds.): NEW2AN/ruSMART 2016, LNCS 9870, pp. 741–751, 2016.
DOI: 10.1007/978-3-319-46301-8_65

When defining commercial effectiveness, real cash flow (balance of operating, investment and financial activities) acts as an effect at every stage of calculation. It is defined in order to estimate if the cash is sufficient for funding all the expenses needed for investment, production and financial activities of a telecommunication company. Positive balance of real money is a necessary criterion of investment project efficiency [1, 5, 7].

2 Main Points

The paper sees into 4 scenarios of a company's investment activity in the telecommunication market (all calculations are limited with the market size of 9000 people):

- scenario 1 – monopolistic limited market (6000 people) without any capability to expand. The size of profits increases (due to a growing number of subscribers up to a possible maximum) in case the prices of "phone conversations" remain unchanged and after that it becomes constant;
- scenario 2 – differently from option 1, a competitive market is reviewed. So, in order to hold the market whose size is 6000 people, every year 0.5 % fall in prices is provided for;
- scenario 3 – differently from option 1, the market can be expanded up to 9000 people;
- scenario 4 – differently from scenario 3, a competitive market is reviewed. To hold the achieved size of market, a fall in prices is provided for, similarly to scenario 2.

When redistributing the investment needed for servicing the market whose size is 9000 people, the following things are taken into account: the geographical zone of mobile network location; construction of cell sites, a commutation center and information security services.

Figures 1 and 2 show changes in real cash flow as running total and in profit margin for all the scenarios of investment activities depending on the competitive situation in the telecommunication market. These indicators have been chosen due to their highest importance for evaluating the financial standing of a mobile network operating company.

Figures 1 and 2 give evidence that:

- profit per unit has a constant value for scenarios 1, 3 and goes down for scenarios 2, 4;
- decrease in profit per unit (scenarios 2, 4), caused by decrease in prices of mobile telecommunication services due to the existing competition, does not result in a fall in real cash flow;
- despite the fact that the size of the market, profits and costs are invariable, the potential of the telecommunication company is growing by total amount of money resources (scenario 1);
- expansion of the monopolistic and competitive market (scenarios 3,4) by 1.5 times results in real cash flow growth by 45 % and 57 % in relation to the indicators of scenario 1 and scenario 2 respectively.

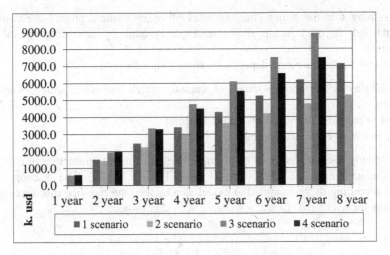

Fig. 1. Change in real cash flow as running total for scenarios 1–4

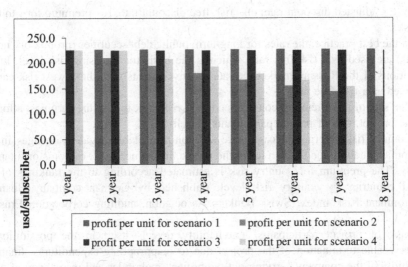

Fig. 2. Change in profit margin for scenarios 1–4

An investment development project of a mobile telecommunication company aimed at expansion of the size of mobile services is conceived based on quite definite presuppositions about investment and current costs, sales volumes of mobile telecommunication services and their prices, time frames of the project. Independently on the quality and validity of these presuppositions, future course of the events related to carrying out the project is always ambiguous. Therefore, it is necessary to review uncertainty and risk aspects.

Assuming that the major characteristics of an investment project are cash flow elements and discount factor, risk assessment is done by amending one of these parameters.

The size of discount rate depends on three major factors:

- availability of different sources of capital, which require a different level of payment;
- need to consider the value of money in time;
- risk factor of estimated earnings.

The following ones are methods of discount rate adjustment: production of a discount rate cumulative model, CAPM (Capital Asset Pricing Model). The discount rate is defined according to the cumulative model as a total of the risk-free discount rate and premiums for every risk factor, associated with this project.

$$e = ef + \sum_{i=1}^{I} r_i \tag{1}$$

where e – adjusted discount rate; ef – risk-free discount rate; r_i – premium for i-th type of risk.

In the best practice, the rates for long-term public debt securities (promissory notes, bonds) are used as a risk-free rate, which is the basis for assessing risk level. In the conditions of the Russian market the rate for investments with the lowest risk can be accepted as a risk-free rate.

The size of risk premium considers country risk, risk to lose the earnings, afforded by the project, risk of project participants' unreliability.

Country risk emerges in case there are non-projectable negative changes in the economic situation due to changes in the public, investment, tax, customs or financial policy. The premium for country risk is estimated according to the rankings of the world countries by country risk level, published by German company Business Environment Risk Index, Swiss Bankers Association, auditing corporation Ernst & Yung.

Risk of project participants' unreliability reveals itself in the possibility of unforeseeable termination of the project due to inappropriate expenditures, financial instability of the company carrying out the project, mala fides and insolvency of other participants of the project.

Non-systematic risk is caused by engineering, technological, organizational, financial and economic solutions, irregular fluctuations of production volumes and prices of services and resources, social changes (Fig. 3).

To a degree, any type of risk can occur during operation of a telecommunication company or system. In the assessment of non-systematic risks, occurring in the activities of a mobile telecommunication company, special attention is paid to the appraisal of novelty of the used technology, machinery and level to which the information transfer processes are known, need for R&D, quality risk, risk of dependence on foreign suppliers, currency risk, risk of poorly secured information transfer, which if increased can result in ineffective work of the company. Qualitative assessment of

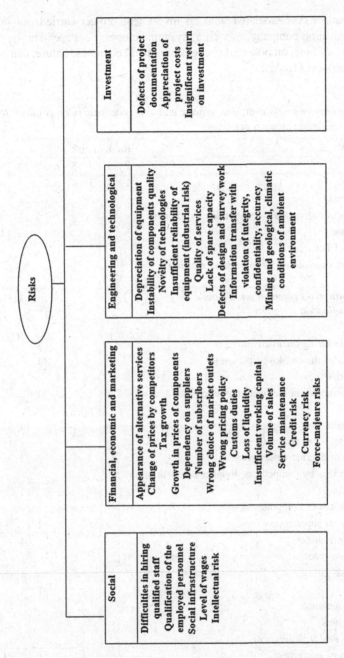

Fig. 3. Non-systematic risks common for the activities of a mobile telecommunication company, by groups

non-systematic risks associated with an investment project carried out in a mobile telecommunication company, as well as systematic ones, i.e. dependent on the market condition as a whole, on potential changes of general economic nature, can be made by a ranking method (Table 1).

Table 1. Qualitative assessment of systematic and non-systematic risks, common for operation of a mobile telecommunication company

Type of risk	Rank of risk									
1	*2*	*3*	*4*	*5*	*6*	*7*	*8*	*9*	*10*	*11*
	1	2	3	4	5	6	7	8	9	10
Systematic risks										
Aggravation of external socioeconomic environment						1				
Customs duties								1		
Possession of licenses									1	
Development capabilities for telecommunication market									1	
Non-systematic risks common for a mobile telecommunication company							1			
Social										
Difficulties in hiring qualified staff						1				
Qualification of the employed personnel						1				
Intellectual risk					1					
Social infrastructure in the telecommunication company						1				
Level of wages							1			
Financial and economic										
Appearance of alternative services						1				
Change of prices by competitors, tougher competition										1
Tax growth							1			
Growth in prices of components										
Dependency on suppliers						1				
Number of subscribers							1			
Force-majeure risks							1			
Currency risk							1			
Credit risk						1				
Service maintenance								1		
Volume of sales									1	
Engineering and technological										
Depreciation of equipment							1			
Mining and geological, climatic conditions of ambient environment							1			
Information transfer with violation of integrity, confidentiality, accuracy										1
Defects of design and survey work							1			

(*Continued*)

Table 1. (*Continued*)

Type of risk	Rank of risk									
Lack of capacity reserve									1	
Insufficient reliability of equipment (industrial risk)				1						
Novelty of technologies							1			
Instability of components' quality								1		
Quality of services									1	
Investment										
Insufficient return on investment							1			
Appreciation of project costs										1
Defects of project documentation										1
Quality of observations	0	0	0	1	1	8	8	5	6	5
Weighted total	0	0	0	4	5	48	56	40	54	50
Number of factors	34									
Weighted average value of risk	7.6									

An alternative method of non-systematic risks assessment is a technique based on the degree to which the equipment and technologies are used and the processes are known. In this case, risk premium can be defined by factors (Table 2).

Table 2. Assessment of non-systematic risk premium for mobile telecommunication services

Factors	Risk premium increment, %
1. Need for doing R&D by efforts of a research and development or/and design organization for over a year	8
2. Application of a new technology requiring using resources available in the free market	3
3. Uncertainty of demand for the provided mobile telecommunication services	0
4. Uncertainty of prices of the provided mobile telecommunication services	5
5. Instability of demand for the provided mobile telecommunication services	0.5
6. Uncertainty of the ambient environment during the projects (mining and geological, climatic conditions, aggression of the ambient environment)	5
7. Uncertainty in the process of developing the equipment and technology being used	0

To assess riskiness of investment in development of a telecommunication company listed in the stock market, it is possible to apply a CAPM, according to which the discount rate which considers risks has the following form:

$$e = e_f + \beta^*(e_m - e_f) \tag{2}$$

where e – adjusted discount rate; e_f – risk-free discount rate; β – beta-coefficient; e_m – yield of a mid-market block of securities.

β coefficient, which assesses investment risks, is determined on the basis of the stock market statistics information review and is defined as:

$$\beta = V_{pr.}\left(V_{an.pr.}\right)V_{mar.} \tag{3}$$

where $V_{pr.}$ – range of fluctuations in the previous period from the average rate of the company's stocks in %; $V_{an.\,pr.}$ – range of fluctuations in the market value of stocks for an analogues company; $V_{mar.}$ – range of fluctuations in the rate of stocks for the companies in all industries for the period reviewed.

Values of β coefficient are published by consulting agency AK&M. In case there is need for detailed analysis, the obtained value of risk premium can be divided between various types of simple risks.

In case the investment capital is set up from the company's own and borrowed funds, the discount rate is defined based on the weighted average capital cost model (WACC).

$$WACC = \sum_{i-1}^{I} d_i * x_i \tag{4}$$

where d_i – share of capital of i type; x_i – cost of capital of i type.

Thus, NPV is adjusted according to the following stages, Fig. 4.

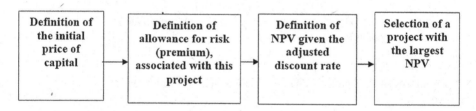

Fig. 4. Stages for defining NPV given risk

In case there has been no selection between alternative projects, adjusted discount rate allows assessing the level of planned cash receipts more realistically.

The research which has been carried out shows that investment in mobile connection networks is a high-yielding line of business and for the reviewed options it is $1.03–2.62 million of net present value to the initial level of investment of $515 k per market size of 9 k people. Herein, the payback period is 6 month whereas the profitability index is within a range of 4.4 ÷ 8.1.

The markets with the largest number of subscribers have an advantage (scenarios 3, 4) while the best indicators of investment efficiency correspond to option 3. According to the calculations, definition of an average subscription fee allows obtaining high efficiency results and, at the same time, offers an opportunity not to adjust price plans very often.

Risk assessment for an investment project through NPV size adjustment.

According to the global practice, there are a number of risk assessment models, which allow adjusting NPV size. The most common are: a simulation model (scenario method), a decision tree, sensitivity analysis, Monte Carlo method (simulation modeling), cash flow probability distribution analysis.

Sensitivity analysis implies examining the change of NPV size or any other indicator, which has been chosen as a key one (for example, IRR) in case the selected factors are changing, including such ones as: costs of services as a whole and their individual components, taxes, investment costs, volume of sold telecommunication services, prices of telecommunication services. In the structure of costs, changes in wages, costs of servicing the subscribers, costs of equipment can have the main impact on NPV size.

Thus, $NPV = f(x1,x2,xn)$, where xn- the n view factor.

A decision tree is an expanded model of sensitivity analysis, which requires comprehensive information.

The scenario method has the following stages, which are completed for each project:

1. Pessimistic forecast. Optimistic forecast. Most probable forecast;
2. Determination of probability for each of the possible scenarios to develop.
3. Base value modeling.
4. Determination of NPV for each forecast.
5. Determination of the range of variation and mean square deviation.

When using the scenario method to assess the degree of risk, basic elements of probability theory are used: mathematical expectation, dispersion, mean square deviation, variation coefficient.

Analysis of probabilistic distribution of cash flow allows evaluating the projected sizes of NPV and net earnings and review their probabilistic distribution. However, this method requires knowing the exact values of probabilities for all the options of cash receipts, which is quite complicated in reality.

Monte Carlo simulation modeling method creates an additional opportunity when assessing risk due to a capability to assess risk in an integrated manner on the basis of multiple simulations of the conditions for formation of the project efficiency indicators and their deviation from the estimated or mean value. This method helps to reveal a correlative connection between the base values, unlike sensitivity analysis, which studies, in an isolated manner, the effects of each factor on the final indicator. In risk analysis uses a wide range of information is used, which can be in the form of objective data or expert assessments. The result of risk analysis is expressed not as a single NPV, but as a probabilistic distribution of all possible values of this indicator.

1 – Change of NPV indicator due to a change in the costs of investments.
2 – Change of NPV indicator due to a change in the volume of services provided.
3 – Change of NPV indicator due to a change in the price of services.

Review of the obtained results shows that change in the price of provided services has the biggest impact on the change of the NPV indicator and proves that efficiency indicators for option 3 are the best (Fig. 5).

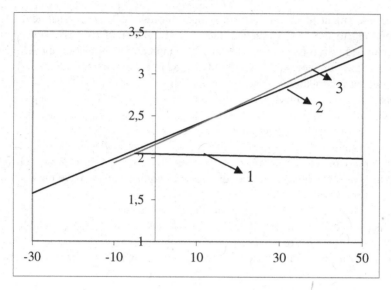

Fig. 5. Change of NPV indicator

In case of a dramatic change of the conditions in the telecommunication market, for example, if the prices of telephone conversations halve, the projected total profit for the calculation period of 8 years will also halve. Moreover, efficiency indicators for option 3, with the use of average subscription fee in a competitive market, will change as follows: NPV will fall to $1.05 million, PI – to 3.1. An investment project aimed at development of the mobile communication network becomes inefficient by NPV indicator with the size of the market equal to 9000 people in case the prices fall by 10 times.

However, with such a fall in prices, it is possible to project emergence of additional demand, which causes increase in the size of market and, consequently, growth of NPV.

3 Conclusion

As a result of the research, real cash flow has been formed with the use of an estimated level of prices, sales receipts, costs of production and sales of telecommunication services and investments. Based on it, the major indicators of investment efficiency have been defined – net present value, internal rate of return, payback period, profitability index – considering the risks in the discount rate built up according to the cumulative model and so the option with the best efficiency indicators has been selected. Sensitivity analysis of the NPV indicator has been carried out, which proves the advantage of the selected option. A threshold level of price change has been determined, after which an investment project becomes inefficient.

References

1. Adjemov, A.S.: Telecommunications, Infocommunications – What's Next?. PH Media Publisher, Moscow (2011)
2. Blank, I.A.: Fundamentals of Investment Management, vol. 1. Elga, Nika-Center, Kiev (2004)
3. Glukhov, V.V.: Appraisal of Scientific Research Results. SPbSTU, St. Petersburg (2002)
4. Glukhov, V.V.: Economics and Management in Infocommunications. Piter, St. Petersburg (2012)
5. Kovalev, V.V.: Investment Project Evaluation Methods. Finances and Statistics, Moscow (2000)
6. Kalinina, O.V.: Universal approach to building the progressive scale for income taxation. Actual Probl. Econ. **176**(2), 387–400 (2016)
7. Makarov, S.B.: Telecommunication Technologies: Introduction to GSM Technologies. Publishing center "Akademia", Moscow (2006)
8. Margolin, A.M.: Economic Appraisal of Investments, Moscow (2001)
9. Silkina, G.Y.: Innovation Processes in Knowledge Economy. Analysis and Modeling. Publishing House of the Polytechnic University, St. Petersburg (2014)
10. Berner, G.: Management in 20XX, Siemens (2004)
11. Trends in Telecommunication Reform 2009. Summary, ITU (2010)

Development of Project Risk Rating for Telecommunication Company

Sergei Grishunin and Svetlana Suloeva[(⊠)]

St. Petersburg State Polytechnical University, St. Petersburg, Russia
sergei.v.grishunin@gmail.com, emm@spbstu.ru

Abstract. We developed the project risk rating (PRR) for telecommunication companies. It provides qualitative risk scores assessment of capital expenditures (capex) projects to rank them by severity of exposures, to check their fit into the company's risk profile and, ultimately, to combine projects into the efficient project portfolio with the lowest risk given return. We discussed the definition, functions and advantages of investment controlling and presented the reference model of its main subsystem – project portfolio controlling responsible for building the efficient capex project portfolio. Then, we developed the model of PRR; worked out the example of PRR's scorecard and discussed the advantages of the PRR over the existing risk assessment tools in project portfolio management.

Keywords: Project portfolio management · Risk management · Controlling · Risk scoring models

1 Introduction

Telecommunication companies (telecoms) operate in a very complex and turbulent business environment which has recently becomes even more competitive and unpredictable. Capital expenditures (capex) projects of the telecoms are exposed to the growing number of exposures: from engineering, technological and operational to market, financial and regulatory [5]. The late discovery and evaluation of potential threats as well as their untimely and inefficient remediation may prevent capex projects from achieving their goals set in telecoms' strategic plans. According to the recent poll performed by the consulting company 'Ernst and Young' global telecom leaders named lack of necessary return of investments owning to the risks in the projects as one of the most important industry threats [3].

Our analysis of scientific literature [1, 2, 5–8, 10–13] shows that efficiency of the project management system significantly increases if this system integrates risk management practices. Investment controlling (the application of methods of controlling in project management) is one of the best examples of such integration [5]. However, investment controlling is still an emerging field in project management and its theory and practical applications need further development. Our study shows that there is still

© Springer International Publishing AG 2016
O. Galinina et al. (Eds.): NEW2AN/ruSMART 2016, LNCS 9870, pp. 752–765, 2016.
DOI: 10.1007/978-3-319-46301-8_66

a lack of advanced risk management tools for such an important strategic field as building and maintenance of portfolio of projects [5, 6, 11].

In this paper, we develop a model of project risks rating (PRR) for the telecommunication company. This tool provides the managers with the qualitative assessment scores of project's risks. These scores will than utilized as one of the criteria to (1) range the projects by severity of exposures; (2) analyse the projects' fit into company's risk profile; and (3) choose the best portfolio of projects with the lowest risks given return.

2 Key Opportunities and Exposures of Global and Russian Telecommunication Industries

The global telecommunication industry has changed significantly in the past 10 years. These changes has been influenced by the rapid development of new technologies such as high-speed wireless Internet, mobile devices, big data, cloud technologies, etc. while the regulatory pressure has reduced thus stimulating new investments into the industry.

Our 2016–2020 outlook for the industry suggests that the volume of "traditional" voice services will continue to decline while high-speed communication, video streaming, cloud services for enterprises and Internet of Things will provide new growth opportunities for the industry's players. Rising competition and threats from newcomers will remain the main challenges for the industry [3, 5]. The other challenges include (1) viability of new technologies and adoption of new products by customers; (2) regulatory risks with regards of spectrum release framework, data privacy rules and regulators attitude to mergers and acquisitions; and (3) global macroeconomic uncertainties [3].

Capital intensity will persist in the industry in 2016–2020; the share of capital expenditures (capex) of revenue will sustain around 20 % [9]. The hostile environment described above requires telecoms to develop and maintain project management system aimed at (1) building an efficient project portfolio that will drive companies' competitive advantage but staying within the boundaries of the resource and risk appetite restrictions; (2) regularly tracking the execution of the projects and preventing the realisation of the risks or mitigating/curing their negative variances at the earliest possible stage. These tasks can be solved by the new edge project management system - investment controlling.

3 Definition and Functions of Investment Controlling

Controlling is a modern system of management which pivotal purpose is to create, develop and maintain a competitive advantage of the company. Its advantages over the other management systems are presented in [6].

Investment controlling is the application of methods of controlling to project management, the combination of processes, skills, tools and techniques, ensuring the achievement of project goals in the highly uncertain environment. [5]. It can be utilized both for management of a single project and of a portfolio of projects [1, 8]. The functions of investment controlling are listed in [5]. The implementation of investment controlling, despite of the costs, decreases the variances of actual time spent and costs vs. the strategic plans up to 50 % [5, 6]. This helps to achieve earlier payback of programs and projects (the components) in the portfolio as well as gaining the sustainable competitive advantage to the company.

Investment controlling can be split into two parts. The first is the directional part – project portfolio controlling (PPC) which is responsible for setting up the project portfolio. The second is the applied part – project controlling which is responsible for delivery of components in accordance to the strategic plan. The objectives, processes and tools of project controlling are described in details in [5, 6]; in this paper, we will discuss the project portfolio controlling and its risk assessment tools.

4   Reference Model of Project Portfolio Controlling

Project Portfolio Controlling (PPC) is a subsystem of investment controlling, which facilitates decision making around which portfolio's components should be developed, selected and executed to maximize the economic value of portfolio and, therefore, maintaining the company's return on investments in line with the goals set by strategic plan. The key objectives of PPC are: (1) increase the company's value and return on investment (ROI); (2) ensure strategic fit of components; (3) maintaining the company's strategic focus; (4) ensuring efficient allocation of resources among components; (5) keeping the desired risk profile of the portfolio; and (6) continuously monitoring of the portfolio perfomance and making adjustments in the portoflio. The reference model of project portfolio controlling is presented in Fig. 1.

Risk management is an integral part of this reference model. It is performed in the following sub-processes:

Sub-process A. Screening of new components. If component's integral risk score exceeds the company's risk appetite set in strategic plan for this type of components than the component is denied.

Sub-process B. Prioritization of components and evaluation of components' economic efficiency. Portfolio managers quantify the component's risks and use this assessment as an input to evaluate the sensitivity of economic efficiency of components (by using methods described in [1]). The component with the quantitative risk score above the company's tolerance level can be denied.

Sub-process C. Portfolio balancing and selection of the optimal portfolio. While constructing the set of alternative portfolios, the managers assess the risks of (i) insufficiency and/or conflict of resources to execute the portfolio. After the set of alternative portfolios is constricted, portfolio managers evaluate the risks scores of each

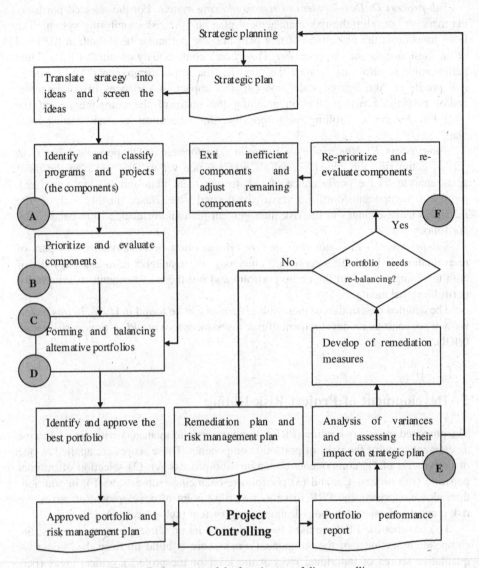

Fig. 1. Reference model of project portfolio controlling

alternative and reject those with the risk scores exceed the company's risk appetite. Portfolios' risks scores are also serve as one of the criteria to choose the optimal portfolio among alternatives.

Sub-process D. Development of risk monitoring system. For the selected portfolio, the managers develop the risk management plan and the risk monitoring system. The latter transforms the objectives of this plan into the particular tasks both at the level of the components and the portfolio. The system continuously monitors (i) the actual performance of each task against the plan; (ii) the actual components' and portfolio risk profile vs. the desired state; and (iii) the impact of variances in components' and/or portfolio's risk profile on attaining the goals of the components and the portfolio. Project controlling is responsible for execution of risk management plan.

Sub-process E. Monitoring of portfolio's performance. On periodical basis, as part of comparing the actual portfolio's performance versus strategic plan the managers analyse (i) the performance of risk management plan and (ii) the actual risk profiles of components/portfolio versus the desired state. Based on this analysis, the managers make changes in the risk management plan and/or decide to re-balance the portfolio.

Sub-process F. Re-evaluation and re-prioritization of components. As part of re-evaluation of portfolio's economic efficiency, the managers re-assess the risks of both the components and the entire portfolio and use these assessments as criteria for portfolio re-balancing.

The detailed description of these sub-processes can be found in [1, 8]. In this paper, we will concentrate on development of risk assessment tool of PPC – project risk rating (PRR).

5   Development of Project Risk Rating

The proposed project risk rating (PRR) supplies portfolio managers with the qualitative assessment score of the risks of portfolio components. These scores are applied as one of the criteria of (1) components screening (sub-process A); (2) selection of optimal portfolio (sub-process C); and (3) portfolio re-balancing (sub-process F). In addition, the ratios comprising the PRR serve as a starting point of development of pro-active risk control indicators for the risk monitoring system (sub-process D) [5].

To construct the PRR we used the scoring model principles. In these settings, the composite assessment of the component's risk score is build up from the "bricks" – qualitative scores of individual risks at the level of the project's critical areas (perspectives). For telecommunication companies, we consider the following perspectives: (1) macroeconomic environment; (2) market potential; (3) competition; (4) technology; (5) quality of project management; (6) supply chain and logistics; (7) force majeure events; (8) regulatory environment; (9) finance. The importance of each perspective for the component is set by portfolio managers and is reflected in its weight (w_i). The keep the model balanced we recommend (1) to assign 20 % to financial perspectives while the remaining 80 % should be equally distributed among the remaining eight perspectives.

In turn, risks at the level of each perspective are caused by various external and internal exposures, which can be estimated by the calculation of certain indicative variables (sub-factors). Importance of each sub-factor is reflected in its weight (σ_j) which is set by portfolio managers. PRR's formula is presented below.

$$PRR = \sum_{i=1}^{N} w_i * \sum_{j=1}^{M} \sigma_J * I_j \tag{1}$$

PRR – numerical score of PRR

I_j – numerical score of variable j

w_i – weight of the perspective i, $w_i \in [0,1]$, $\sum_{i=1}^{M} w_i = 1$

σ_j – weight of sub-factor j, $\sigma_i \in [0,1]$, $\sum_{j=1}^{M} \sigma_i = 1$

N – the number of perspectives

M – the number of sub-factors constituting the perspective

Based on literature review [1, 5, 6, 13] we developed a list of possible sub-factors for a generic telecom project. (Table 1).

Table 1. The list of sub-factors for PRR in telecommunication

Sub-factors	Method of calculation
Macroeconomics	
Average growth of real GDP (AGDP), %	$$AGDP = \sum_{t=0}^{T} \frac{GDP_t - GDP_{t-1}}{GDP_{t-1}} \tag{2}$$ GDP_t – projected growth in real GDP at year t, GDP_{t-1} – projected growth in real GDP at year t-1 T- project's lifespan, t$\in$[0,T]
Dynamics of growth of real GDP (DGDP),x	$$DGDP = \frac{ADGP_T}{ADGP_{-5}} \tag{3}$$ $ADGP_T$ – forecast of average growth of real GDP in the component's lifespan $ADGP_{-5}$ – average growth of real GDP in the five years prior to the launch of component
Cash flow elasticity to interest rate changes (CFE), x	$$CFE = \frac{\Delta CFO}{\Delta I} * \frac{I}{CFO} \tag{4}$$

Sub-factors	Method of calculation
	ΔCFO – change in aggregate component's cash flow from operations following the change in interest rates ΔI – change in interest rates I – aggregate interest expenses during component's lifespan CFO – aggregate component's cash flow from operations
Cash flow exposure to foreign currency risks (ECFO), %	$$ECFO = FCFO - FCFI - FDR \qquad (5)$$ ECFO – cash flow exposure to foreign currency risks, % FCFO - % of operating cash flow denominated in foreign currency FCFI - % of capex denominated in foreign currency FDR - % of debt repayment in foreign currency
	Market potential
Market capacity, %	$$MP = \frac{AP - CP}{AP} \times 100\% \qquad (6)$$ AP – maximum market capacity, units CP – current market capacity, units
Variability of demand (VD), %	$$VD = \frac{\sigma_d}{D_{av}} \times 100\% \qquad (7)$$ σ_d – standard deviation of demand in last 5 years, units D_{av} – average demand in the last 5 years, units
Breakeven point	Special calculation [6]
Average EBITDA margin (AEM), %	$$AEM = \frac{\sum_{t=T1}^{T} EBITDA_t}{\sum_{t=T1}^{T} R_t} \times 100\% \qquad (8)$$ $EBITDA_t$ – estimated EBITDA in year t T_1 – time of components' completion, years T – component's lifespan, years R_t –estimated revenue in year t, t$\in$[T_1,T]
	Competition
Expected change in competition (DC), x	$$DC = \frac{1 - \sum_{j=1}^{J1} m_{jt1}}{1 - \sum_{j=1}^{J2} m_{jt2}} \qquad (9)$$ T_1 – current state, years T_2 – expected state at the component's launch, years J1 – number of companies at the market at T_1

Sub-factors	Method of calculation
	$J2$ – number of companies at the market at T_2 m_j – market share of j-company at states T_1 and T_2, %
Intensity of competition (IC),x	$$IC = 1 - \left(N * \sqrt{\left(1/N \sum_{i=1}^{N} [m_i - (1/N)]^2 \right)} \right) \qquad (10)$$ N – number of companies to operate in the market m_i – market share of company I at project commercial stage
Comparison of time to the market with that of the most menacing competitor (TMC), x	$$TMC = \frac{T_c}{T_r} = \frac{T_{1c} + T_{2c} + T_{3c}}{T_{1r} + T_{2r} + T_{3r}} \qquad (11)$$ T_c – time to breakeven point of company's project T_r- time to breakeven point of projects of most menacing competitor T_{1c}. T_{1r} – duration of R&D stage of the company and competitor T_{2c}, T_{2r} – duration of stage "R&D – launch of sales" of company and competitor T_{3c}. T_{3r} – duration of stage "launch of sales – investments' payback" of company and competitor
	Technology
Risk of core technology in the project (TR), x	$$TR = max|f_i|_{i=1}^{F} \qquad (12)$$ f_i – expert assessment of risk of element i of core technology planned to be utilized in the component (core technology), using the scale in Table 3
Innovation success ratio (K_N), %	$$K_N = \frac{N_{ut}}{N_t} \times 100\% \qquad (13)$$ N_{ut} – number of unsuccessful/cancelled R&D projects in core technology by companies in the industry at the period t N_t – total number of R&D projects in core technology started by companies in the industry at the period t
Average length of R&D in core project technology (KT), x	$$K_T = \frac{\dfrac{\sum_{j=1}^{L_r} T_{rj}}{L_r}}{\dfrac{\sum_{i=1}^{L_c} T_{ci}}{L_c}} \qquad (14)$$ T_{ri} – length of R&D project j in core technology performed by the

Sub-factors	Method of calculation
	most menacing competitor
	L_r – the total number of R&D projects in core technology performed by the most menacing competitor
	T_{ci} – length of R&D project i in core technology performed by the company
	L_c – the total number of R&D projects in core technology performed by the company
	Quality of project management
Project schedule performance index (VT), %	$$VT = \frac{\sum_{m=1}^{L} \frac{ACT_m}{SCT_m}}{L} \times 100\% \qquad (15)$$ L – number of projects using the core technology in the last 5 years ACT_m – actual duration of m-project $m \in [1, L]$ SCT_m – budgeted duration of m-project
Personnel adequacy index (PA), x	$$PA = \frac{\sum_{z=1}^{Z} \frac{QP_z}{AP_z}}{Z} \times 100\% \qquad (16)$$ Z – total number of professions to complete the project z – index, denoting profession z$\in$[1,Z] QP_z – quantity of personnel at profession z in the company which has adequate skills to complete the project AP_z – total quantity of personnel of profession z in the company
Project cost performance index (CI), %	$$CI = \frac{\sum_{m=1}^{L} \frac{APC_m}{SPC_m}}{L} \times 100\% \qquad (17)$$ L – number of projects using the core technology in the last 5 years APC_m – actual cost of m-project m$\in$[1,L] SPC_m – budgeted cost of m-project
	Supply chain and logistics
Dependence on limited suppliers (NS), x	$$NS = \sum_{j=1}^{M} w_j * NS_j \qquad (18)$$ NS_j – number of suppliers delivering j-critical component for the component, j=1,M w_j- weight determining importance of j-component for the component, $w \in [0,1] \sum w = 1$

Sub-factors	Method of calculation
Suppliers, buyers and other partners' financial health (CR), x	CR_k - credit ratings of k-critical supplier assigned by international rating agencies (Moody's, S&P or Fitch) or determined internally
Probability of suppliers' error (PSE)	$$PSE = \sum_{l=1}^{L} w_l \frac{VE_l}{V_l} \qquad (19)$$ w_l – weight of l-supplier in the component's supply chain $w \in [0,1] \sum_1^L w_l = 1$ VE_l – volume of defected components shipped by l-supplier in the past 5 years V_l – total volume of components supplied by l-supplier L – total number of suppliers
Force Majeure Events	
Exposure to event risks (EER), x	$$EER = \sum_{i=1}^{N} w_i \, EE_i \qquad (20)$$ EE_i – assessment of severity of i-event risk (using scale in Table 3) w_i – weight, determining importance of i-event for the component; $w_i \in [0,1] \sum_1^N w_i = 1$ N – number of possible event risks
Exposure to risks related to data privacy and security (DPI), x	$$DPI = \frac{VC_p}{VC} \qquad (21)$$ VC_p – losses of peer companies in similar projects related to data privacy and cyber security issues in the last 5 years VC – losses of peer companies which related to data privacy and cyber security issues in the last 5 years
Institutional and regulatory environment	
Exposure to regulatory risks (ERR), x	$$ERR = \sum_{i=1}^{N} w_i \, ER_i \qquad (22)$$ ER_i – assessment of severity of i-regulation risk in the component: regulation of spectrum and operating licences, restrictions on acquisitions, etc. w_i – weight, determining importance of i-regulation for the component; $w_i \in [0,1] \sum_1^N w_i = 1$
Institutional framework and effectiveness at project's market (IFE),	$$IFE = \sum_{h=1}^{3} u_h \, FE_h \qquad (23)$$

Sub-factors	Method of calculation
%	FE_1 – World Bank Government Effectiveness Index
	FE_2 – World Bank Rule of Law Index
	FE_3 – World Bank Control of Corruption Index
	u_h – importance of each index for the project $u_h \epsilon [0,1], \sum_{h=1}^{3} u_h = 1$
Financial	
Average cash flow debt coverage (ACFDC), x	$$ACFDC = \left[\sum_{t=1}^{T} \frac{FFO_t}{D_t}\right] \bigg/ T \qquad (24)$$
	FFO_t- project cash flow from operations before changes in working capital in the period t
	D_t – average debt outstanding related to the component at period t
	T – component's lifespan
Minimal interest coverage (MIC), x	$$MIC = min_{t=1.T} \left(\frac{FFO_t}{I_t}\right) \qquad (25)$$
	I_t - interest payable in the period t
	T – component's lifespan
Average debt to capital ratio (ADC), x	$$ACFDC = \left[\sum_{t=1}^{T} \frac{D_t}{D_t + E_t}\right] \bigg/ T \qquad (26)$$
	D_t - average debt outstanding related to the component at period t
	Et – average equity in the component at period t
	T – component's lifespan
Funding requirement (FA), %	$$FA = \frac{Cash + CFAP + E + CD}{CFI + D + I} \times 100\% \qquad (27)$$
	Cash – cash available for component funding
	CFAP – cash flow available for component funding during investment stage
	E – equity
	CD – available committed debt facilities
	CFI – capital expenditures
	D – debt repayment during investment stage
	I – interest payment during investment stage

For constructing the PRR we used the following rating scale to denote the severity of the component's risks: very low (VL), low (L), medium-low (ML), medium-high (MH), high (H) and very high (VH).

To get this risk severity level in accordance to the scale, the portfolio managers, using their experience in the past projects, opinion of experts' panel and/or consultants (see description of methods of aligning expert opinions in [6, 11]) divide the entire set of possible values for each sub-factor into the intervals. Each interval matches the risk rating for the particular sub-factor (see example in Table 2).

Table 2. Example of splitting sub-factor's values into the intervals

Sub-factor	VL	L	ML	MH	H	VH
Minimal interest coverage ratio (ML)	≥10.5x	7.5x–10.5x	5.5x–7.5x	3.5x–5.5x	1.5x–3.5x	<1.5x

To determine the composite risk rating of the entire component, we convert each of sub-factor's risk rating into a numeric value based upon the scale below (Table 3).

Table 3. Mapping sub-factors' PRR to the risk model to numeric scores

VL	L	ML	MH	H	VH
1	3	6	9	12	15

The numerical score for each sub-factor is multiplied by its weight (σ_j) with the results then summed to produce the composite weighted factor score for the each perspective (FRR). Then, the numerical score for each perspective is multiplied by the weight of this perspective (w_i) with the outcome summed to produce the composite numerical PRR. The numerical FRRs and PRR then mapped back to a qualitative rating based on the following rule (Table 4).

Table 4. Mapping numerical scores of FRRs and PRR to qualitative scores

Qualitative FRRs and PRR	Aggregated perspective's or sub-factor's score
Very low	$X < 1.5$
Low	$1.5 \leq X < 4.5$
Medium-low	$4.5 \leq X < 7.5$
Medium-high	$7.5 \leq X < 10.5$
High	$10.5 \leq X < 13.5$
Very high	$X \geq 13.5$

The advantages of the PRR over the other existing risk assessment tools such as risk charts and risks maps are:

1. PRR allows project managers to analyse components' or portfolio's risks at the required level starting from individual risks at the level of each component's perspective up to the risks at the level of entire component
2. By using PRR, FRR and individual ratios from Table 1, project managers can (i) "track" the contribution of individual risks to the composite risks of project's perspectives and the entire project; and then (ii) allocate resources and management's time for risks' mitigation more efficiently by focusing on the most significant and menacing risks
3. The PRR's scorecard is well understood by portfolio managers as they have been already using these ratios in decision making while solving the other managerial tasks. The calculation of these ratios does not require the expensive IT applications. The source data to calculate these ratios is already contained in the company's management information systems
4. The ratios comprising PRR scorecard can be utilized as a base for developing of control indicators for the risks monitoring system. The latter provides the portfolio managers with the early warnings of upcoming threats and potential losses

6   Conclusion

In this paper, we developed a model of the project risks rating (PRR) for the companies operating in telecommunication industry. This tool is designed to supply portfolio managers with qualitative assessment of risks of the components (projects, programs) of project portfolio. The PRR's score will serve as one of the criteria (along with economic efficiency and/or others) of selecting the best projects/project portfolio among the alternatives.

In the first part of the paper, we analysed the key characteristics of telecommunication industry as well as the opportunities and challenges that telecoms will face in the next 3–5 years. We demonstrate that the companies, on the one hand, will continue to increase capital spending to capture new technological and market opportunities but, on the other hand, will operate in a hostile environment with elevated technological, regulatory and other risks. In this environment, conventional project portfolio management will no longer deliver the necessary return on investments.

In the second part of the paper, we presented the definition and function of investment controlling – the application of controlling to project portfolio management. We then developed the reference model of the project portfolio controlling, a subsystem of investment controlling responsible for the building of project portfolio from components (individual projects or programs), described its' main sub-processes and applications of risk management in each sub-process.

In the third part of paper, we developed the model of project risk rating. In this model, the composite assessment of component's risks is added up from qualitative scores of individual risks at the level of project's critical areas (perspectives) such as macroeconomic environment, market environment, competition, technological risks, etc.). We worked out the example of scorecard to estimate these individual risks and the rules of combining these estimates into the composite project's risk score. We also recommend that the PRR scorecard's ratios to be utilized as a base to develop control indicators for the risks monitoring system.

References

1. Anshin, V.M., Demkin, I.V., Nikonov, I.M., Tsar'kov, I.N.: Models of Portfolio Management in Uncertain Environment. MATI, Moscow (2007)
2. Chapman, C., Ward, S.: Project Risk Management Processes Techniques and Insights. Wiley, West Sussex (2003)
3. Ernst and Young. http://www.ey.com/Publication/vwLUAssets/ey-global-telecommuni cations-study-navigating-the-road-to-2020/$FILE/ey-global-telecommunications-study-navi gating-the-road-to-2020.pdf
4. Glukhov, V.V., Balashova, E.S.: Economics and Management in Infocommunnication: Tutorial. Piter, Saint Petersburg (2012)
5. Grishunin, S., Suloeva, S.: Project controlling in telecommunication industry. In: Balandin, S., Andreev, S., Koucheryavy, Y. (eds.) NEW2AN/ruSMART 2015. LNCS, vol. 9247, pp. 573–584. Springer, Heidelberg (2015)
6. Grishunin, S.V., Suloeva, S.B.: Strategic controlling and anti-crisis management. In: Strategy and Tactics of Anti-Crisis Management of a Firm. Specialnaya Literatura, Saint-Petersburg (1996)
7. Kendrick, T.: Identifying and Managing Project Risk. AMACON, New York (2015)
8. Kozlov, A.S.: Methodology of Management of Portfolios of Projects and Programs. Flinta, Moscow (2011)
9. Moody's Investors Service. https://moodys.com/researchdocumentcontentpage.aspx?docid= PBC_1006595
10. Nekrasova, T.P., Aksenova, E.E.: Economic Evaluation of Investments in Telecommuni- cation Industry. St. Petersburg State Polytechnic University, Saint Petersburg (2011)
11. Orlov, A.I.: The current state of risk controlling. J. Kuban State Agrarian Univ. 98(04), 933– 942 (2014)
12. Raz, T., Michael, E.: Use and benefits of tools for project risk management. Int. J. Project Manag. 19, 9–17 (2001)
13. Yescombe, E.R.: Principles of Project Finance. Academic Press, London (2002)

Innovation Venture Financing Projects in Information Technology

Alexander Bril$^{(\boxtimes)}$, Olga Kalinina$^{(\boxtimes)}$, and Olga Valebnikova$^{(\boxtimes)}$

Peter the Great St. Petersburg Polytechnic University, Saint-Petersburg, Russia
{b.a.r, olgakalinina}@bk.ru, olgavalebnikova@gmail.com

Abstract. The article deals with crucial issues of innovation venture financing of in the field of information technologies in Russian Federation. The classification of main types of business venture developed, depending on the financial resources attraction. Compiled and analyzed scheme of small innovative companies in terms of passing the stages of investment in the company. The method of "buy-back" cost calculation of the company financed by venture funds is presented. Special transmitter-receiver modules production startup at the enterprise, working in the field of IT-technologies is introduced as an example of economic indicators calculation and monitoring.

Keywords: Innovation · Venture business · Company valuation · Transmitter-receiver modules

1 Introduction

Large companies are abandoning the most risky, from their point of view, projects, and as known the higher the risk, the higher, if successful, returns. Therefore, these projects are very often taken by the engineer, who is the author of the technical part of the project. However, the success of his business requires funding, as well as organizational and commercial experience. The problem of "long money", i.e. loans for a period of 3–5 years or more, especially without a quality pledge is known. At least equal problem for risky innovation project initiator is finding of a qualified assistants and consultants on organizational and commercial part. A system of venture financing operates effectively in various countries to solve these problems.

2 Methodology

Venture financing is a long-term direct investments in small innovative enterprises with a high or relatively high risk in order to obtain a significant profit in the period from 3 to 7 years. This activity is supported by some large companies, and venture capital funds and innovative development networks are created [1–4]. The main objective of these organizations is finding new and effective ideas - innovations and the creation of conditions for their transformation into highly profitable businesses.

© Springer International Publishing AG 2016
O. Galinina et al. (Eds.): NEW2AN/ruSMART 2016, LNCS 9870, pp. 766–775, 2016.
DOI: 10.1007/978-3-319-46301-8_67

The main types of venture financing are presented in the Fig. 1.

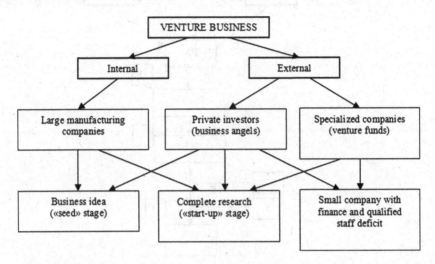

Fig. 1. The integrated circuit of venture business varieties, compiled by the authors

The development of domestic venture capital business is carried out mostly in large enterprises at the expense of the idea initiators resources. External venture financing is proceeded by raising funds from external sources: investment, insurance, charity and pension funds, State resources and private investors. Organization is carried out either through the creation of venture capital funds, or by individual activity of entrepreneurs, called business angels. At the same time, if not going into the industry details, the most widely used are three types of funding: applied research ("seed" stage according to the US-European classification of innovation development stages), the start-ups in the same classification and small innovative enterprises with lack of financial resources.

Venture capital investments for funding decisions require a feasibility study. It is necessary to take into account two main points. The first - is the availability of information for plan calculations, that is fundamentally different to those, for example, objects of funding as "seed" or existing small business. And the second - the funding decision rules of venture capital funds, industrial companies or business angels. If the business angel - a private investor decides whether to participate in innovative projects or not mostly based on personal experience and risk assessment, then venture capital funds have specially designed guidelines for the selection of investment projects. The methodology for assessing the effectiveness of future investments only for venture capital funds is presented in this paper.

Venture companies income formed by the "buy-back" system, i.e., the sale of joint venture company shares in innovative enterprise after its 3–7 years.

Diagram of the small innovative company organization with the release of the fund and repurchase stages is presented in Fig. 2.

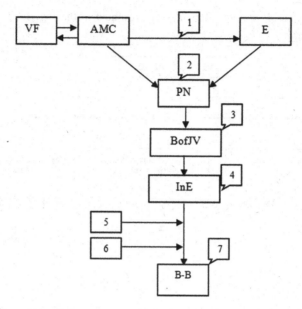

Fig. 2. The scheme of small innovative enterprise organization and operation, compiled by the authors

The scheme contains the following legend:

VF – Venture fund, AMC – asset management company, E – entrepreneur, PN – Primary Negotiations, BofJV – beginning of joint venture, InE- Innovative enterprise, B-B – buy-back procedure;

1 – primary selection of innovative ideas and entrepreneurs procedure;

2 – procedure for the first meeting and clarification of entrepreneur and his company background;

3 – due diligence, auditing, financial analysis and evaluation of actual and planned enterprise value, reconciliation of the data and preparation for new innovative businesses registration procedure;

4 – organizational issues, the initial funding in accordance with the business plan and start of joint venture;

5 – financing a new venture in accordance with the plan and the operational schedule, experts involvement in innovation management, joint organization of production and sales;

6 – attraction of credit funds for the development of the company;

7 – preparation and organization of joint venture share buy-back in innovation company.

Organizational and technological and economic issues are solved at each stage. From the financial and economic point of view, enlarged or preliminary business plans are considered at the first stage. The second stage specifies the initial information on the project and the applicant company, in the case of agreed estimates reception joint work program is developed.

The third step is to determine the value of the applicant company, the amount of shareholders equity, the cost of the joint venture at the beginning of financing and at the time of the agreed repayment date. This is the most difficult part of venture company and entrepreneur negotiations. A revised business plan for the development of joint venture should be developed and agreed in this process, which is a basis for further sources of financing, order and buy-back price. Innovation screenings occurs often at this stage due to inability to obtain consistent estimates. According to experts [1], it is this stage of venture project evaluation where most contention between investors and company founders appears, and at this stage breaks down up to 40 % transactions. A sign of reaching an agreement is considered to be a divergence of views on the value of the company not more than 20 %. The main objective of the feasibility study and analysis of the company's value lies in the achievement of such an agreement.

"Buy-back" system is realized through three main ways:

– Management buyout (sale of venture capital company shares to existing share-holders and management);
– External repurchase (sale of venture capital company shares to outside investors);
– The sale of venture capital company shares by initial public offering (IPO).

3 Main Body

The most difficult issue of venture business is the selection procedure for innovation financing. Effective organization of this process involves the forced screening and presentation to funding the best and most promising ideas in the field of IT.

Experience and statistics show that out of several thousand variants of innovative ideas and entrepreneurs primary selection for the procedure of the first meeting and data clarification remains about 700–800 options. After this stage another 500 options are eliminated. In the end, after the initial joint activities (due diligence - Procedure 3, Fig. 2) stays about 12 to 14 options of innovative companies to invest.

The literature offered and used in practice a large number of methods and approaches for determining the value of the company and venture investors attraction [6, 7]. The objective is to systematize these methods in relation to different stages of venture capital financing and, most importantly, in relation to the different level of innovation. Radical innovations associated with the new worldwide or Russian prod-ucts and processes that are more risky and, therefore, the results are more difficult to predict. For such innovations plan calculations of cost indicators should be led on the basis of simple aggregated methods. For simple innovations related to substantial and sometimes with minor modifications of known processes and products more detailed methods can be used.

Experience allows to define the basic requirements for the content and parameters of financial and economic calculations at all selection stages and joint work of IT-innovative enterprises (according to Fig. 2).

In the first stage - the procedure of primary selection of innovative ideas and entrepreneurs - the main criterion is the idea of the project. Nevertheless, financial and economic part of the business plan is considered almost always. It is advisable to limit calculation by simple indicators of commercial efficiency (simple payback period and rate of return), focusing on the plan calculations reliability on forecasted timing and volume of sales of new products. In these matters the owners of ideas tend to increased optimism. Business plan cost indicators are calculated based on the cost approach.

Risk assessment at this stage, should be carried out based 4–5 situations analysis, including optimistic and pessimistic variants.

The second stage - the first meeting and data clarification about the entrepreneur and his company. Economic calculations during this procedure specify the accuracy of cost parameters and preliminary risk assessment. Availability of prototypes, laboratory test results, special statistics are important for this stage.

In the third phase - the procedure of due diligence - plan economic calculations of venture financing project are carried out. This is a basic step to determine the actual cost of the innovative business before venture capital financing (pre-money) and justification of future innovative enterprise cost for repurchase price approval.

This step should be divided into two sub-step 3.1. and 3.2. The calculations in sub-step 3.1 aimed to determine the future value of the company subject to its independent development, without the use of venture capital financing.

Sub-step 3.2 is focused on the future value of innovation company calculation based on venture capital attraction, the resulting synergy effects and the effect of attracting specialists of the management company MC (specialists in innovation management, production organization and sales of new products). An important question here is dependence of the new innovative enterprise value (procedure 4 Fig. 2), not only from sum of capital, but also from the advice of experienced investors. The results of studies [1, 3, 4] shows that 40 % of the sample investors who were passive, failed more often than active investors; and the more active investor is in terms of the frequency of interaction with the owner, the more likely investment led to positive results.

Using two approaches to assess the future value of the company (with and without venture capital) allows to use the method of comparative effectiveness in establishing the agreed price of buy-back and making decision on venture capital financing in innovative enterprise (see. Further example of the calculation for the special electronic transmitters module equipment startup).

Sub-step 3.1 defines the value of pre-money based on the cost approach after audit and evaluation of existing businesses financial condition. Calculation of future value (compared to the cost of repurchase) is based on income approach. The initial amount is determined by the pre-money, discount rate is set based on the profitability of the existing assets of the enterprise or as a cumulative value based on the CAPM model. If there is a sales statistics, and IPO of similar enterprises (by size and degree of innovation) future added value will be determined on the basis of a comparative approach.

In sub-step 3.2 the value of post-money should be determined, where appreciation of the company by attracting venture capital must be taken into account, synergies and participation of the management company in the future business development. Appreciation coefficients system is included in management companies regulatory framework of most developed venture capital funds. Determination of these factors can be associated with the franchises cost statistics in similar activities.

On the basis of the agreed post-money value, the calculation of the company value at the time of the planned exit of innovative enterprise venture fund can be started. There are several approaches for the assessment. The simplest of these are venture approach and method of future value.

Venture approach is more simple in calculation, it does not require to determine future cash flow for calculation the company value. Venture capital average profitability statistics and planned rate of return of a particular venture capital fund is necessary to implement it. Then the value of the company's venture capital exit point C_{b-b} can be determined as follows:

$$C_{b-b} = post - money.(1+i)^n \tag{1}$$

where i – comparison rate, taken at the level of the planned fund rate of return based on average data (20–30 % per year);

n - planning period in years.

Comparison rate determination is a key calculation moment, requires special justification.

Buy-back cost is determined by the planned share of venture fund in a new enterprise shareholders' equity.

Future value method requires more detailed and time-consuming calculation. It is based on the construction of tables and graphs of planned cash flows and takes into account the credit, income and payments of innovative enterprises in joint working process with a venture capital company. Comparison rate here is defined as the weighted average cost of capital used for the project WACC. Risk assessment is carried out based on a sensitivity analysis to changes in the future value of the planned sales volumes, costs and comparison rate.

As a result of economic calculations in the third stage of venture capital financing quite a broad base of financial and economic indicators for harmonization and clarification is created. The relatively simple algorithms of presented methods and existing software products allows to focus on future value monitoring on ensuring the reliability of the source data.

Table 1 presents proposed methods of calculating the value indicators in a systematic way.

Table 1. Methods for calculating the value of the company at different stages

Stage	Methods for calculating the value of the company		
	«pre – money»	«post – money»	At the buy-back moment
3.1	Cost-based	equals «pre – money»	Profitable comparative
3.2	Cost-based	Regulatory appreciation	Venture future value

This payment system allows to create an information base for clarification and consistent assessment of actual and planned cost of the enterprise, the organizational issues solution and preparation for new innovative enterprises registration.

4 Example of Project Calculations

As an example of the proposed economic calculation method, consider a start-up for the production of special transceiver modules to existing small business.

The company has the equipment, personnel and documentation for small-scale production units. The value of assets, determined on the basis of cost method «pre - money» is 1.6 million rubles.

Analysis of the market and competitors indicated that the planned sales could reach over the next three years, from 3 to 4 thousand pcs. per year, with selling prices from 45 to 55 thousand rubles per unit.

The calculations of the enterprise capacity show that production of new units in the amount of 250 pieces can be mastered per year at a cost of 49 thousand rubles per unit. By the third year the production volume can be adjusted (without additional major investment) up to 320 pcs. at a cost of 48 thousand rubles per unit.

Venture capital fund, which is ready to invest in the project is not more than 75 % of its total cost is found.

Further calculations show that the acquisition of special equipment for the installation of elements and arrangement of additional workplaces in the amount of 4.5 mln. rubles can increase plant capacity to 1100 units by the third year of the project. Production cost of one unit at the same time can be reduced in the first year to 45 thousand rubles per unit and in a third year to 43.5 thousand rubles per unit.

There is a need to conduct economic calculations and risk assessment in order to determine a more effective solution to the current enterprise: organization of independent small-scale production or raising venture capital fund to increase capacity by 3.5 times up to 1100 units per year.

In the first version (small-scale production) in accordance with the procedure «post - money» equals «pre - money» 1.6 million rubles. On the basis of the income method future enterprise value F at the end of the third year is determined. Comparison rate is based on WACC model - weighted average cost of capital for the project - 20 %. The selling price of the module is stated at 50 thousand rubles per unit, first year sales volumes 250 units, second year - 300 units and in the third year - 320 units. In turn, production cost is 49.0 thousand rubles per unit, 48.5 thousand rubles per unit and 48.0 thousand rubles per unit. Depreciation rate is 25 % per year, using the linear method of depreciation; respectively, the residual value of assets at the end of the third year is 0.4 million rubles. F value is the following, million rubles:

$$F = 0.65 \cdot (1+0.20)^2 + 0.85 \cdot (1+0.20) + 1.04 + 0.4 = 3.396 \qquad (2)$$

To expand the value of the enterprise information at the end of the third year additional simple methods as "50 % of sales" or "fivefold profit on sales" could be used.

The volume of sales at the end of the third year of the project is $320 \times 50 = 16.0$ million rubles. Then the future enterprise value is estimated to be $16 \times 0.5 = 8$ million rubles. When calculating based on a "fivefold sales profit" method, future costs is 3.2 million rubles, which is largely consistent with the results of calculations by the income method.

In the **second variant of development** - creation of a joint venture with a venture capital fund - «post - money» will increase compared to «pre - money» by a factor of rise in price in accordance with the planned increase in return on sales by 6.3 % and amount to $(1.6 + 4.5) \times 1.063 = 6.48$ million rubles.

Next similarly, value of the company at the end of the third year is counted by the venture method based on calculation of future value. The output of the joint venture business is planned for the end of the third year.

Venture capital investing into the project in the amount of 4.5 million rubles, with an average yield of 30 % per year, expected value of its stake in the joint venture at the end of the third year should reach, million rubles:

$$F = 4.5 \cdot (1 + 0.30)^3 = 9.89 \tag{3}$$

If the share in the joint venture $4.5/(4.5 + 1.6) = 73.8$ %, then the value of the joint venture at the end of the third year will be $9.89/73.8 = 13.4$ million rubles. And value of the share attributable to the company that owns innovation, will be $13.4 - 9.89 = 3.51$ million rubles. This value is calculated on the basis of venture capital approach, it is very slightly higher than the main results of the first embodiment from 3.2 to 3.396 million rubles.

To calculate the value of the company at the end of the third year on the basis of a methodology for assessing the future cost it is necessary to know the selling price of modules and build a cash flow schedule in accordance with the same calculation as in the first embodiment.

In the venture version of the project it is planned to reduce production cost of modules from an average of 48.5 thousand rubles per unit to 43.5 thousand rubles per unit or by 10.3 %. In this connection more competitive selling price of the module is planned 48.5 thousand rubles per unit, sales volumes in the first year - 900 units, in the second - 1000 units and the third - 1100 units. Production cost is 45.0 thousand rubles per unit, 44.0 thousand rubles per unit and 43.5 thousand rubles per unit respectively each year. The depreciation rate is 25 % per year, using the linear method of depreciation respectively, the residual value of assets at the end of the third year is 1.62 million rubles. Comparison rate is adopted similarly to the first embodiment at a level of 20 %. F value is the following, million rubles:

$$F = 4.72 \cdot (1 + 0.20)^2 + 6.12 \cdot (1 + 0.20) + 7.22 + 1.62 = 22.98 \tag{4}$$

In this calculation the value of share attributable to the company-owner of innovation, will be $22.98 \times 0.262 = 6.02$ million rubles, which is much higher than the results of calculations of the first embodiment of the venture, and using venture method of calculation.

Enterprise cost information clarification at the end of the third year on the basis of the "50 % of the sales volume" or " fivefold sales profit" methods, provides additional data for analysis.

The volume of sales at the end of the third year of the project is $1100 \times 48.5 = 53.35$ million rubles. Then, future enterprise value is estimated at $53.35 \times 0.5 = 26.67$ million rubles. When calculating on a "fivefold sales profit" future value will be 26.0 million rubles, which is greatly exceeds the results of the first option, and the calculations on venture basis. This is due to the fact that the calculation of future value is taken into account synergies from the creation of joint venture fund company. A significant difference in the planned economic performance requires additional risk assessment and initial information adopted for calculation. The calculation results are shown in Table 2.

Table 2. The results of economic calculations for the two development options

Project indicators	The project of own module production	Venture project
The volume of sales in the third year of the project: units million rubles	320 16.0	1100 53.35
Production cost in the third year of project, million rubles	15.36	47.85
Comparison rate (cost of capital WACC), %	20	30
Enterprise value «pre – money», million rubles	1.6	1.6
Enterprise value «post – money», million rubles, including the share of the enterprise	1.6 1.6	6.48 1.70
Enterprise value at the end of the third year, million rubles Venture method, including the share of the enterprise Income method (future value), including the share of the enterprise «50 % of sales volume» method, including the share of the enterprise «fivefold sales profit» method, including the share of the enterprise	– – 3.396 – 8.0 – 3.20 –	13.40 3.51 22.98 6.02 26.67 6.99 26.0 6.812
Net present value of the project NPV, including the share of the enterprise	2.10 –	14.52 3.80

In addition to assessing the buy-back cost the effectiveness of the projects on the basis of determining the net present value NPV is also calculated in p. 7 Table 2. It confirms high efficiency of commercial joint venture with a venture capital fund.

It should be noted that two considered options for the development of special modules production can be supplemented by projects attracting "business angels", investors, bank lending or other methods of business organization. The presented method allows monitoring of planned situations with relatively simple compared to those recommended in the literature [5] economic indicators calculations.

5 Conclusion

The results of calculations for the two variants of special modules production and monitoring of economic indicators suggest the following conclusions for making a decision on the development of the enterprise:

1. It is necessary to carry out a refined risk assessment: the first variant is the lack of competitiveness, in the case of analogues and substitutes, due to the high cost and the planned sales price, in the second option - the possibility of organizing an effective buy-back (or external management buy-out) at the high cost of the enterprise.
2. The essential difference in the planned value of the company at the time (the end of the third year of the project), obtained on the basis of different calculation methods, requires more detailed study of information sources, particularly with regard to the comparison rate (cost of capital for the project) and the expected production cost of new products.
3. Monitoring of the project economic performance allows more soundly build organizational section of the business plan and calculate break-even in phases of the project.

References

1. Venture Accelerators: the Network of Innovation Development. Technical report of JSC "RBC" and the World Bank, Moscow, 276, p. (2013). http://www.rusventure.ru/ru/programm/analytics
2. Popov, A., Roosenboom, P.: Venture capital and new business creation international evidence. J. Bank. Finan. 37(12), 4695–4710 (2013)
3. Woolley, J.L., Bruno, A.V., Carlson, E.D.: Social venture business model archetypes: five vehicles for creating economic and social value. J. Manag. Glob. Sustain. 2, 7–30 (2013)
4. Killing, J.P.: Strategies for Joint Venture Success. Routledge Library Editions: International Business, Routledge (2013). vol. 22, 132 p
5. Rogova, E.M., Sapozhnikova, M.A.: Problems of value Russian innovative companies evaluation at early stages of development by venture investors. Econ. Sci. 185(6–1), 150–158 (2013). Problems of Value Russian Innovative Companies Evaluation at Early Stages of Development by Venture Investors
6. Nekrasova, T., Leventsov, V., Axionova, E.: Forecasting of investments into wireless telecommunication systems. In: Balandin, S., Andreev, S., Koucheryavy, Y. (eds.) NEW2AN/ruSMART 2014. LNCS, vol. 8638, pp. 519–525. Springer, Heidelberg (2014)
7. Nikolova, L.V., Kuporov, J.J., Rodionov, D.G.: Risk management of innovation projects in the context of globalization. Int. J. Econ. Finan. Issues 5(3S), 68–72 (2015)

Information Risk Analysis
for Logistics Systems

Elena Velichko[1(✉)], Constantine Korikov[1],
Anatoliy Korobeynikov[2,3], Aleksey Grishentsev[2],
and Mihail Fedosovsky[2]

[1] Peter the Great St. Petersburg Polytechnic University, St. Petersburg, Russia
velichko-spbstu@yandex.ru,
korikov.constantine@spbstu.ru
[2] Saint Petersburg National Research University of Information Technologies,
Mechanics and Optics, St. Petersburg, Russia
Korobeynikov_A_G@mail.ru, grishentcev@ya.ru
[3] St. Petersburg Institute of Terrestrial Magnetism,
Ionosphere and Radio Wave Propagation of the RAS, St. Petersburg, Russia

Abstract. The algorithm for calculation of the information risk is suggested. The algorithm takes into account the flows of all the components of the transportation system, the quality of their interaction, restrictions, and probability distribution. By using this algorithm the logistics company or intelligent transportation system gets the data on the assessment of the information risks, which helps the decision maker to decide whether it is reasonable to conclude the contract or not.

Keywords: Information risk analysis · Algorithm · Logistics information system

1 Introduction

The importance of management of information risks is undoubtful. Effective management of material and related (information, financial) flows aimed at achieving corporate objectives with optimal resource expenditures is a key factor for the success in today's business. Selection of the optimal relations between the risk and business activity level and the profitability and reliability based on a risk assessment is a significant part of the business decision-making process.

Information technologies have transformed many industries and now are transforming transportation systems. Almost all modern transportation systems use information technologies and specialized software products to solve the problem of rationalization of product flows and related financial and informational flows. An important aspect of this problem is management of risks in the supply chains [1, 2].

Risk management is the process of risk identifying and assessing and taking steps to reduce the risk to an acceptable level [3]. Organizations use the risk assessment, the first step in the risk management methodology, to determine the extent of the potential threat, vulnerabilities, and the risk associated with an information technology system. The output of this process helps to identify appropriate controls for reducing or

© Springer International Publishing AG 2016
O. Galinina et al. (Eds.): NEW2AN/ruSMART 2016, LNCS 9870, pp. 776–785, 2016.
DOI: 10.1007/978-3-319-46301-8_68

eliminating the risk in support decision making in business. In solving the problems of purchase, transportation, and distribution of goods (transport tasks) intelligent transportation system (ITS) take into account the risk factors, the effects of which should be considered in the process of information flow management [3, 4].

This paper considers an algorithm for the calculation of the information risks that arise in the transportation and distribution of material resources in logistics systems. The algorithm relies on the mathematical apparatus of Markov chains based on the Kolmogorov differential equations.

2   Algorithm for Assessment of Information Risks

Optimization of information flows is accompanied by various information risks. The Information Risk (IR) is the risk of loss (in the form of damage or lost profit) resulting from the use of information technologies by a company. In this context the IR is obtained by multiplying the maximum possible damage by the probability of passing through the entire route. IR is a function of time. In other words, IR can be presented in terms of the damage expressed in monetary units by which the logistics information system operates.

The transportation task involves the solution of the problem of delivery of goods to specified places at specified times. It is assumed that the routes to the points of delivery and the capacities of these routes are known. By analyzing the maximum loss Sloss in the case of nonperformance of the order and the imposition of penalties and also the acceptable risk approved in a particular logistics company, the decision maker (DM) must accept or reject the order for the transportation of goods.

The available models of solution of such tasks of the calculation of information risks typically use standard methods [3–7]. However, these models do not use the mathematical apparatus of Markov chains based on Kolmogorov differential equations [4] which allows one to increase the reliability of the assessment and which has demonstrated its efficiency. Such an approach to the solution of the problem of IR assessment is new.

Let us give a formalized statement of the problem in the general form. There are N points at different locations (district, city, several cities, etc.). We refer to one point as a point of departure, and the remaining N-1 points are called points of delivery. (If there are k points of departure, it is necessary to solve the problem k times).

The means of communication between all points are given. There can be both a one-way or two-way traffic between the points. In addition, we set the function of traffic capacity between the vertices (points) $f\_prop_{ij}(t)$ that depends on the time of the day. The function is expressed by dimensionless values ranging from 0 to 1.

We assume that a logistics company has received a commercial offer to enter into a contract for the delivery of goods from the point of departure to n ($n < N$) points of delivery. In addition, the shipping and delivery times are strictly defined. The decision maker who knows the admissible risk approved by the logistics company (for example,

30 monetary units) and takes into account the information from intelligent transportation system, the most important part of which is the IR calculated by ITS, accepts or rejects the proposal. After this the problem is considered to be solved.

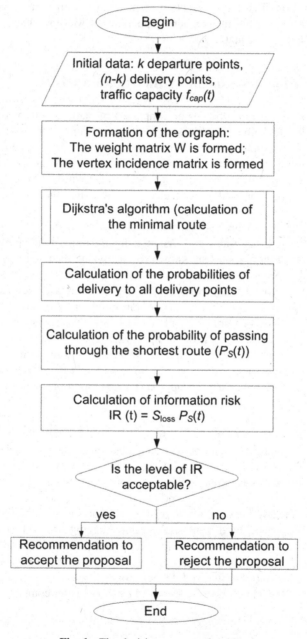

Fig. 1. The decision support algorithm

The decision support algorithm for assessment of information risks we suggest can be represented by the diagram shown in Fig. 1.

Let us consider basic steps of the algorithm.

1. Formation of the Digraph

A mathematical model in the form of a digraph (for instance, the or graph) shown in Fig. 2) is built. The choice of the digraph is based on the initial data on existing routes, and a possible type of traffic (one-way or two-way) is taken into account. In the digraph the two-way traffic is shown by bi-directional arrows and one-way traffic is shown by unidirectional arrows.

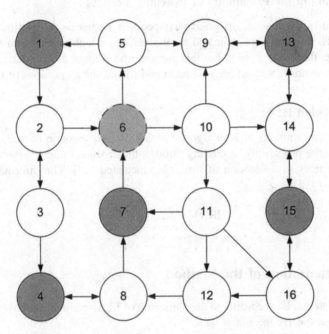

Fig. 2. Schematic of the vertices of departure and delivery 6 – vertices of departure. 1, 4, 7, 13 and 15 – vertices of delivery

The points of departure and destination are vertices G_1 and the roads are the arcs that connect the vertices. In the case of one-way traffic, different weights are assigned to the arcs connecting node i and node j and node j and node i. If there is no traffic between the points, the weight of the corresponding arc is considered to be infinite. In this case there is no bi-directional or unidirectional arrow at the digraph.

The weight matrix $\mathbf{W}$ is formed. Element w_{ij} is taken to be the time required to move cargo from node i to node j multiplied by the road capacity between the i-th and j-th vertices $f\_prop_{ij}(t)$ [8–10]. The time is equal to the distance between the i-th and

j-th vertices divided by the allowed velocity between these vertices. All the necessary data are available in the database of the ITS. By using the information on the connections between the vertices obtained from the ITS database, the vertex incidence matrix $A_{vert.inc.}$ is formed [11]. The digraph is formed on the basis of information on all n vertices of destination.

2. Calculation of the Shortest Route

At this stage, the Dijkstra's algorithm is performed n times, and the shortest route in the digraph is calculated. As a result, the shortest route passing through all the given vertices is formed.

3. Calculation of the Probability of Reaching Vertices

The probability of delivery of goods to specified points as a function of time is calculated. These points are indicated by the vertices which are finite at the portions that constitute the shortest route. For this purpose, the Kolmogorov systems of ordinary differential equations (SODE) are generated and solved for all portions of the resulting route.

4. Calculation of IR

By using all the solutions of Kolmogorov SODE and the theorem of multiplication of probabilities, the probability of passing through the shortest route ($P_S(t)$) including all the given vertices as a function of time t is calculated [16]. The information risk is calculated as [3, 12]

$$\mathrm{IR}(t) = S_{\mathrm{loss}} \cdot P_S(t). \tag{1}$$

3   Implementation of the Method

Let us calculate the IR according to the algorithm of IR assessment we have developed. First of all we specify the initial data.

We assume that there is a supplier (vertices 6) of goods. At 2 PM he got an offer to conclude a contract for the delivery of goods to five consumers which are at different locations (vertices 1, 4, 7, 13, and 15). We assume that the maximum loss S_{loss} is 80,000 rubles. The acceptable risk is 30,000 rubles.

The decision maker rejects or accepts the proposal on the basis of the assessment of IR with the help of ITS.

According to the algorithm given above, the calculation of IR is performed as follows.

Step 1. Figure 2 shows schematically the vertices of departure and delivery.

We specify, according to this figure, the vertex incidence matrix A [11]. Element $a_{i,}$ $_j{\in}A$ shows the number of arcs emerging from the i-th node and coming to the j-th node.

$$A = \begin{pmatrix}
0 & 1 & 0 & 0 & 0 & 0 & 0 & 0 & 0 & 0 & 0 & 0 & 0 & 0 & 0 & 0 \\
0 & 0 & 1 & 0 & 0 & 1 & 0 & 0 & 0 & 0 & 0 & 0 & 0 & 0 & 0 & 0 \\
0 & 1 & 0 & 1 & 0 & 0 & 0 & 0 & 0 & 0 & 0 & 0 & 0 & 0 & 0 & 0 \\
0 & 0 & 0 & 0 & 0 & 0 & 0 & 1 & 0 & 0 & 0 & 0 & 0 & 0 & 0 & 0 \\
1 & 0 & 0 & 0 & 0 & 0 & 0 & 0 & 1 & 0 & 0 & 0 & 0 & 0 & 0 & 0 \\
0 & 0 & 0 & 0 & 1 & 0 & 0 & 0 & 0 & 1 & 0 & 0 & 0 & 0 & 0 & 0 \\
0 & 0 & 0 & 0 & 0 & 1 & 0 & 0 & 0 & 0 & 0 & 0 & 0 & 0 & 0 & 0 \\
0 & 0 & 0 & 1 & 0 & 0 & 1 & 0 & 0 & 0 & 0 & 0 & 0 & 0 & 0 & 0 \\
0 & 0 & 0 & 0 & 0 & 0 & 0 & 0 & 0 & 0 & 0 & 0 & 1 & 0 & 0 & 0 \\
0 & 0 & 0 & 0 & 0 & 0 & 0 & 0 & 1 & 0 & 1 & 0 & 0 & 0 & 0 & 0 \\
0 & 0 & 0 & 0 & 0 & 0 & 1 & 0 & 0 & 0 & 0 & 1 & 0 & 0 & 0 & 1 \\
0 & 0 & 0 & 0 & 0 & 0 & 0 & 1 & 0 & 0 & 0 & 0 & 0 & 0 & 0 & 0 \\
0 & 0 & 0 & 0 & 0 & 0 & 0 & 0 & 1 & 0 & 0 & 0 & 0 & 1 & 0 & 0 \\
0 & 0 & 0 & 0 & 0 & 0 & 0 & 0 & 0 & 0 & 0 & 0 & 1 & 0 & 1 & 0 \\
0 & 0 & 0 & 0 & 0 & 0 & 0 & 0 & 0 & 0 & 0 & 0 & 0 & 1 & 0 & 1 \\
0 & 0 & 0 & 0 & 0 & 0 & 0 & 0 & 0 & 0 & 1 & 0 & 0 & 1 & 0 & 0
\end{pmatrix}$$

Let us calculate the matrix of weights W. To this end, we specify the distance between the vertices (in the form of a list where the first number in the square brackets shows the number of the node from which the arc of the digraph emerges and the second number shows the number of the node to which the arc of the digraph comes. After the square brackets there is the number indicating the weight of the arc, i.e., the distance between the vertices in km.

$\{[1,2],12\},\{[2,3],11\},\{[2,6],11\},\{[3,2],11\},\{[3,4],14\},\{[4,8],12\},\{[5,1],7\},\{[5,9],$ $11\},\{[6,5],14\},\{[6,10],11\},\{[7,6],8\},\{[8,4],12\},\{[8,7],10\},\{[9,13],8\},\{[10,9],11\},$ $\{[10,11],8\},\{[11,7],15\},\{[11,12],6\},\{[11,16],5\},$ $[12,8],12\},$ $\{[13,9],8\},$ $\{[13,14],11\},$ $\{[14,13],11\},\{[14,15],8\},$ $\{[15,14],8\},\{[15,16],7\},$ $\{[16,12],6\},$ $\{[16,15],7\}.$

For simplicity, we assume that the road capacity between the points $f\_prop_{ij}(t)$ is the same at all routes and is equal to 0.8. We also assume that the allowed speed in this area is the same everywhere and is equal to 50 km/h.

Then we calculate the matrix of weights W and construct the digraph shown in Fig. 3.

By applying the algorithm developed above, we calculate the shortest route including vertices 1, 4, 7, 13 and 15. This route is as follows

$[[6,5,1],[1,2,3,4],\ [4,8,7],\ [7,6,10,9,13],\ [13,14,15]].$

Step 2. According to the data obtained at the first step, we write the Kolmogorov SODE.

$$\begin{cases}
\dot{P}_1(t) = P_5(t) \\
\dot{P}_5(t) = -P_5(t) + P_6(t) \\
\dot{P}_6(t) = -P_6(t) \\
P_6(0) = 1,\ P_1(0) = P_5(0) = 0,\ P_1(t) + P_5(t) + P_6(t) = 1.
\end{cases}$$

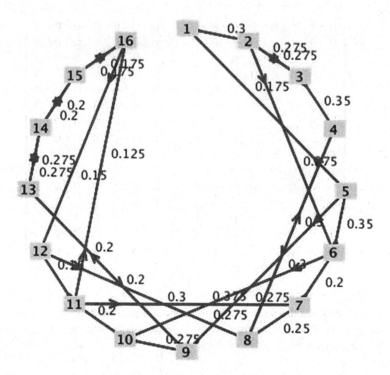

Fig. 3. Digraph built from the data of the example

$$\begin{cases} \dot{P}_1(t) = -P_1(t) \\ \dot{P}_2(t) = -P_2(t) + P_1(t) \\ \dot{P}_3(t) = -P_3(t) + P_2(t) \\ \dot{P}_4(t) = P_3(t) \\ P_1(0) = 1, \ P_2(0) = P_3(0) = P_4(0) = 0, \ P_1(t) + P_2(t) + P_3(t) + P_4(t) = 1. \end{cases}$$

$$\begin{cases} \dot{P}_4(t) = -P_4(t) \\ \dot{P}_7(t) = P_8(t) \\ \dot{P}_8(t) = -P_8(t) + P_4(t) \\ P_4(0) = 1, \ P_7(0) = P_8(0) = 0, \ P_4(t) + P_7(t) + P_8(t) = 1. \end{cases}$$

$$\begin{cases} \dot{P}_6(t) = -P_6(t) + P_7(t) \\ \dot{P}_7(t) = -P_7(t) \\ \dot{P}_9(t) = -P_9(t) + P_{10}(t) \\ \dot{P}_{10}(t) = -P_{10}(t) + P_6(t) \\ \dot{P}_{13}(t) = P_9(t) \\ P_7(0) = 1, \ P_6(0) = P_9(0) = P_{10}(0) = P_{13}(0) = 0, \ P_6(t) + P_7(t) + P_9(t) + P_{10}(t) + P_{13}(t) = 1. \end{cases}$$

$$\begin{cases} \dot{P}_{13}(t) = -P_{13}(t) \\ \dot{P}_{14}(t) = -P_{14}(t) + P_{13}(t) \\ \dot{P}_{15}(t) = P_{14}(t) \\ P_{13}(0) = 1, \, P_{14}(0) = P_{15}(0) = 0, \, P_{13}(t) + P_{14}(t) + P_{15}(t) = 1. \end{cases}$$

We solve these SODE. The result of the solution of SODE will be the probability of reaching a particular node vs time. In this example analytical solutions are possible.

Step 3. By using all the solutions of Kolmogorov SODE obtained at Step 2 and the multiplication theorem of probabilities, we calculate the probability of passing through the shortest route $(P_S(t))$ including given vertices as a function of time t (general probability). After this we calculate IR [3, 12]:

$$IR(t) = S_{loss} \times PS(t). \tag{2}$$

The results obtained at this Step are shown in Figs. 4 and 5 in a graphic form.

As one can see from the data presented in Fig. 5, the accessible risk equal to 30 monetary units is achieved if the terms of the contract on the delivery of the goods state that the delivery time may be greater than 9.1 h. Therefore, in this case the decision maker has the data which can reduce the risks at the conclusion of the contract.

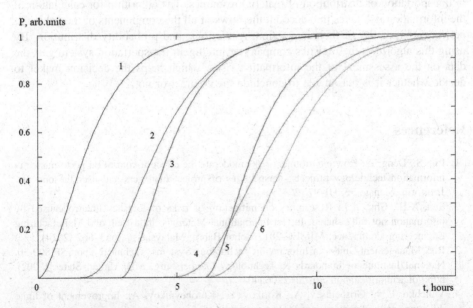

Fig. 4. Probabilities of reaching specified vertices and general probability as a function of time for passage through the shortest route: 1 – vertex 1, 2 – vertex 4, 3 – vertex 7, 4 – vertex 13, 5 – vertex 15, 6 – total probability

IR, 10^3 m.u.

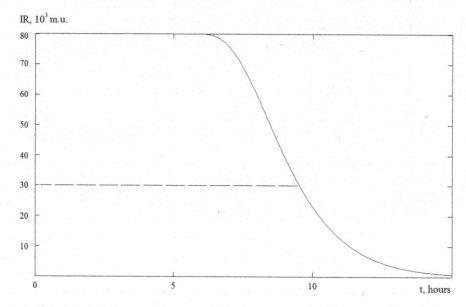

Fig. 5. Variation in information risk as a function of passage through the shortest route

4 Conclusion

The algorithm suggested in the paper allows one to assess the information risks arising in transportation or distribution of material resources. The algorithm for calculation of the information risk takes into account the flows of all the components of the vehicular network, the quality of their interaction, restrictions, and probability distribution. By using this algorithm the logistics company or intelligent transportation system gets the data on the assessment of the information risks, which helps the decision maker to decide whether it is reasonable to conclude the contract or not.

References

1. Liu, S., Deng, Z.: How environment risks moderate the effect of control on performance in information technology projects: perspectives of project managers and user liaisons. Int. J. Inform. Manag. **35**, 80–97 (2015)
2. Sun, X.H., Chu, X.J.: Researches on mitigation of risks of logistics finance caused by information not fully shared. In: 1st International Materials, Industrial, and Manufacturing Engineering Conference, MIMEC-2013, Johor Bahru, Malaysia, pp. 663–667 (2014)
3. Risk Management Guide for Information Technology Systems. Technical report SP 800-30. National Institute of Standards & Technology Gaithersburg, MD, United States (2002). http://dl.acm.org/citation.cfm?id=2206240
4. Velichko, E.N., Grishentsev, A., Korikov, K., Korobeynikov, A.: Improvement of finite difference method convergence for increasing the efficiency of modeling in communications. In: Balandin, S., Andreev, S., Koucheryavy, Y. (eds.) NEW2AN/ruSMART 2014. LNCS, vol. 8638, pp. 591–597. Springer, Heidelberg (2014)

5. Korobeynikov, A.G., Grishentcev, A.Y., Komarova, I.E., Ashevskii, D.Y., Aleksanin, S.A., Markina, G.L.: Mathematical model for calculating risk information for information and logistics system. Sci. Tech. J. Inform. Tech. Mech. Opt. **15**, 538–545 (2015)
6. Shih, K.-H., Cheng, C.-C., Wang, Y.-H.: Financial information fraud risk warning for manufacturing industry - using logistic regression and neural network. Rom. J. Econ. Forecast. **14**, 54–71 (2014)
7. Branger, N., Kraft, H., Meinerding, C.: Partial information about contagion risk, self-exciting processes and portfolio optimization. J. Econ. Dyn. Control **39**, 18–36 (2014)
8. Li, W., Kuang, H.: An approach to evaluating the logistics-financing service risk with hesitant fuzzy information. Int. J. Digit. Content Tech. Appl. **6**, 17–23 (2012)
9. Li, W.: Research on logistics-financing service risk evaluation with linguistic information. Int. J. Digit. Content Tech. Appl. **6**, 10–16 (2012)
10. Bali, O., Gümüş, S., Kaya, I.: A multi-period decision making procedure based on intuitionistic fuzzy sets for selection among third-party logistics providers. J. Multiple-Valued Logic Soft Comput. **24**, 547–569 (2015)
11. Valiente, G.: Algorithms on Trees and Graphs. Springer, Barselona (2002)
12. Ceci, C., Colaneri, K., Cretarola, A.: A benchmark approach to risk-minimization under partial information. Insur.: Math. Econ. **55**, 129–146 (2014)

Author Index

Printed in the United States
By Bookmasters